Lecture Notes in Computer Science 11941

More information about this series at http://www.springer.com/series/7412

Bhabesh Deka · Pradipta Maji ·
Sushmita Mitra · Dhruba Kumar Bhattacharyya ·
Prabin Kumar Bora · Sankar Kumar Pal (Eds.)

Pattern Recognition and Machine Intelligence

8th International Conference, PReMI 2019
Tezpur, India, December 17–20, 2019
Proceedings, Part I

Springer

Editors
Bhabesh Deka
Tezpur University
Tezpur, India

Pradipta Maji
Indian Statistical Institute
Kolkata, India

Sushmita Mitra
Indian Statistical Institute
Kolkata, India

Dhruba Kumar Bhattacharyya
Tezpur University
Tezpur, India

Prabin Kumar Bora
Indian Institute of Technology Guwahati
Guwahati, India

Sankar Kumar Pal
Indian Statistical Institute
Kolkata, India

ISSN 0302-9743 ISSN 1611-3349 (electronic)
Lecture Notes in Computer Science
ISBN 978-3-030-34868-7 ISBN 978-3-030-34869-4 (eBook)
https://doi.org/10.1007/978-3-030-34869-4

LNCS Sublibrary: SL6 – Image Processing, Computer Vision, Pattern Recognition, and Graphics

This Springer imprint is published by the registered company Springer Nature Switzerland AG
The registered company address is: Gewerbestrasse 11, 6330 Cham, Switzerland

Foreword

Welcome to PReMI 2019, the 8th International Conference on Pattern Recognition and Machine Intelligence. We were glad to have you among us at this prestigious event to share exciting results of your pattern recognition and machine intelligence research. This year we were fortunate to have the conference located in a historic city of Northeastern India–Tezpur. We hope that you enjoyed this idyllic place on the bank of the mighty Brahmaputra and that you took the opportunity to enjoy the beauty and uniqueness of this cultural hub of Assam. The Technical Program Committee worked very hard to put together an outstanding program of 90 full-length oral papers and 41 short oral-cum-poster presentations that aptly reflect the recent accomplishments in machine intelligence and pattern recognition research from across the globe. This represents a significant growth in the number of presentations compared to previous years. Although, we had planned to adopt the single-session format so that everyone could attend all oral presentations and learn from them, due to the large number of high-quality papers from multiple sub-themes, we decided to have parallel sessions.

The conference began with two parallel tutorial sessions followed by another, led by three eminent academic experts. The first session was on 'Compressed Sensing' by Dr. Ajit Rajwade from IIT Bombay. The second session was led by Prof. Chiranjib Bhattacharyya from IISc Bangalore on 'Machine Learning and its Application in Industrial Problem Solving.' The third session was delivered by Dr. M. Tanveer from IIT Indore on 'Machine Algorithms for the Diagnosis of Alzheimer's Disease.' The main conference began the next day led by two plenary talks, five invited talks, and two industry talks, followed by the oral presentations.

The first plenary talk was delivered by Prof. Witold Pedrycz of University of Alberta, Canada on 'Granular Artificial Intelligence' to highlight its applications in modeling environments and pattern recognition. Prof. Jayaram Udupa of University of Pennsylvania, Philadelphia, USA delivered the next plenary talk on 'Biomedical Imaging.' The technical sessions were organized in parallel oral presentations under ten major themes: Bioinformatics, Biomedical Signal Processing, Deep Learning, Soft Computing, Image and Video processing, Information Retrieval, Machine Learning and Pattern Recognition, Remote Sensing and Signal Processing, and Smart and Intelligent Sensors. Each of these sessions began with an invited academic or industry talk. There were five invited academic talks and two industry talks. Prof. Pushpak Bhattacharyya of IIT Patna delivered an invited talk on 'Imparting Sentiment and Politeness on Computers.' Prof. C. V. Jawahar of IIIT Hyderabad delivered his invited talk on 'Beyond Text Detection and Recognition: Emerging Opportunities in Scene Understanding.' Another invited talk on 'Cognitive Analysis using Physiological Sensing' was delivered by Prof. S. K. Saha of Jadavpur University. The fourth invited talk focused on 'The Rise of Hate Content in Social Media,' and was presented by Dr. Animesh Mukherjee of IIT Kharagpur. Another invited talk on wireless networks was delivered by Prof. Sudip Misra of IIT Kharagpur.

PReMI 2019 was also be an excellent platform for facilitating industry-academic collaboration. As in previous years, this year's event included industry talks to better expose academic researchers to real-life problems and recent application-oriented developments. The first industry talk was delivered by Dr. Praneeth Netrapalli from Microsoft Research India. His talk focused on 'How to Escape Saddle Point Efficiently.' The second industry talk was on 'Machine Learning and its Applications in Remote Sensing Data Classification,' was delivered by Dr. Anil Kumar from the Department of Photogrammetry and Remote Sensing, Indian Institute of Remote Sensing, Indian Space Research Organization (ISRO), Dehradun, India.

Another attractive and distinguishing feature of PReMI 2019 was the inclusion of a Doctoral Symposium in the name of late Prof. C. A. Murthy, which was held on the first day of the conference. This forum was useful for PhD students to showcase their research outcomes and seek advice and mentorship from prominent scientists and engineers. Although we received a good number of submissions for this symposium, we were only able to select 12 presentations.

Finally, we feel privileged to acknowledge and appreciate the expertise and hard work of the various committees, sub-committees, and individuals of PReMI 2019, who were deeply involved in making this prestigious event an outstanding success. It was truly a memorable experience for us to work with such a professional group. Our sincere and heartfelt thanks to all.

Please enjoy the proceedings!

December 2019

Sushmita Mitra
Dhruba Kumar Bhattacharyya
Prabin Kumar Bora

Preface

Recent technological advancements have played a prominent role in directing Information Technology research towards making ‘intelligent’ machines. Traditionally, intelligence has been commonly associated with humans as an intellectual characteristic, by virtue of which they demonstrate the ability to transcend trivial computations or decisions. Intelligent computations or decisions act as a driving force to deal with problems from a wide range of domains. Intelligent machines have the ability to acquire crucial knowledge from the environment, which enables them to learn and draw significant inferences based on evidences. Since these machines are knowledge-oriented, they possess the ability to generalize which makes them quite reliable. This volume covers all these aspects of machine intelligence, as an outcome of PReMI 2019, an international conference on Pattern Recognition and Machine Intelligence, held at Tezpur University, India, during December 17–20, 2019. It includes 90 full-length and 41 short papers from across the globe. It aims to provide a comprehensive and in-depth discussion of the contemporary research trends in the domain of pattern recognition and machine intelligence. The conference began with two plenary talks followed by five invited talks and oral presentations. The first plenary talk on ‘Granular Artificial Intelligence’ highlighted its applications in modelling environments and pattern recognition and was delivered by Prof. Witold Pedrycz of University of Alberta, Canada. The second plenary talk by Prof. Jayaram Udupa of University of Pennsylvania, Philadelphia, USA focused on ‘Biomedical Imaging.’

The technical sessions included parallel oral presentations under six major themes: Machine Learning and Deep Learning, Bioinformatics and Medical Imaging, Pattern Recognition and Remote Sensing, Intelligent Sensor and Information Retrieval, Signal, Image, and Video Processing, and Evolutionary and Soft Computing. The sessions included five academic talks and two industry talks. Prof. P. Bhattacharyya of IIT Patna delivered the first invited talk on ‘Imparting Sentiment and Politeness on Computers.’ Another invited talk on ‘Beyond Text Detection and Recognition: Emerging Opportunities in Scene Understanding’ was by Prof. C. V. Jawahar of IIIT Hyderabad. Prof. S. K. Saha of Jadavpur University focused his talk on ‘Cognitive Analysis using Physiological Sensing.’ The fourth invited talk, presented by Dr. Animesh Mukherjee of IIT Kharagpur, was on ‘The Rise of Hate Content in Social Media.’ Another invited talk on wireless networks was delivered by Prof. Sudip Misra of IIT Kharagpur. As in previous editions, PReMI 2019 intended to bridge the gap between academia and industry by adopting a collaborative approach in the relevant fields. As in previous years, PReMI 2019 included two industry talks to provide better exposure to academic researchers in real-life problems and recent application-oriented developments. Dr. Praneeth Netrapalli from Microsoft Research India deliberated on ‘How to Escape Saddle Point Efficiently.’ Another industry talk by Dr. Anil Kumar from the Department of Photogrammetry and Remote Sensing, Indian Institute of Remote Sensing,

Indian Space Research Organization (ISRO), Dehradun, India, focused on 'Machine Learning and its Applications in Remote Sensing Data Classification.'

Section I of this volume primarily deals with Machine Learning and Deep Learning and their applications in diverse domains. Kannadasan et al. propose an approach to predict performance indices in Computer Numerical Control (CNC) milling using regression trees. Subramanyam et al. introduce a machine learning based method to detect dyscalculia. Dammu and Surampudi explore the temporal dynamics of the brain using Variational Bayes Hidden Markov model with applications in autism. Dutta et al. present a robust dense or sparse crowd identification technique based on classifier fusion. Buckchash and Raman present a novel sampling algorithm for sustained Self-supervised pre-training for Temporal Order Verification. John et al. analyze the retraining conditions of a network after pruning layer-wise. Jain and Phophalia introduce M-ary Random Forest algorithm, where multiple features are used for splitting at a time instead of just one in the traditional Random Forest algorithm. Jain et al. propose a dynamic weighing scheme for Random Forest algorithm. Alam and Sobha present a method for instance ranking using data complexity measures for training set selection. Karthik and Katika introduce an identity independent face anti-spoofing technique based on Random scan patterns. Das et al. propose a method for automatic attribute profile construction for spectral-spatial classification of hyperspectral images. Rathi et al. propose an enhanced depression detection method from facial cues using univariate feature selection techniques. Shanmugam and Tamilselvan propose a game theoretic approach to design an efficient mechanism for identifying a trustworthy cloud service provider by classifying the providers based on how they cooperate to form a coalition. Prasad et al. propose an incremental k-means clustering method to improve the quality of clusters. This session also includes several significant contributions on Deep Learning. Challa et al. propose a multi-class deep all CNN architecture for detection of diabetic retinopathy using retinal fundus images. Maiti et al. aim to provide a solution to the problem faced in real-time vehicle detection in aerial images using skip-connected convolutional network. Repala and Dubey use Convolutional Neural Networks for unsupervised depth estimation. Basavaraju et al. introduce a Deep CNN based technique with minimum number of skip connections to derive a High Resolution image from a Low Resolution image. Mazumdar et al. propose a technique to detect image manipulations using Siamese Convolution Neural Networks. Mishra et al. present a deep learning based model comprising of causal convolutional layers for load forecasting. Trivedi et al. present a technique for facial expression recognition using multichannel CNN. Gupta et al. propose a data driven sensing approach for action recognition. Sharma et al. propose a method for gradually growing residual and self-attention based dense deep back propagation network for large scale super-resolution image. M J et al. propose a method to solve 3D object classification on point cloud data using 3D Grid Convolutional Neural Networks (GCNN). Singh et al. introduce a new stegananalysis approach to learn prominent features and avoid loss of stego signals using densely connected convolutional network. Rastogi and Gangnani outline a generalized semi-supervised learning framework for multi-category classification with generative adversarial networks.

Section II covers topics from the field of Bioinformatics and Medical Imaging. Mahapatra and Mukherjee introduce GRAphical Footprint for classifying species in large scale genomics. While Pant and Paul present an effective clustering method to produce enriched gene clusters using biological knowledge, Paul et al. report the impact of continuous evolution of Gene Ontology on similarity measures. Kakati et al. introduce DEGNet, a deep neural network to predict the up-regulating and down-regulating genes from Parkinson's and breast cancer RNA-seq datasets. Saha et al. perform a survival analysis study with the integration of RNA-seq and clinical data to identify breast cancer subtype specific genes. Barnwal et al. present a deep learning based Optical Coherence Tomography (OCT) image classifier for Intra Vitreal Anti-VEGF therapy. In another effort, Patowary and Bhattacharyya present an effective method for biomarker identification of Esophageal Squamous Cell Carcinoma using integrative analysis of Differentially Expressed genes. Jana et al. present a gene selection technique using a modified Particle Swarm Optimization approach. This session also includes several noteworthy contributions in the field of Biomedical Applications. Baruah et al. propose a model to simulate the effects of signal interference by mathematically modeling a pair of dendritic fibers using cable equation. While Dasgupta et al. cluster EEG signals of epileptic patients and normal individuals using feed forward neural networks, Dammu et al. report a new approach that uses brain dynamics in the classification of autistic and neurotypical subjects using rs-fMRI data. While Singh et al. use Stockwell transform to detect Dysrhythmia in ECG, Kumar et al. propose an energy efficient MECG reconstruction method. Chowdhury et al. investigate the changes in the brain network dynamics between alcoholic and non-alcoholic groups using electro-encephalographic signals. Das and Mahanta analyze segmentation techniques for cell identification from biopsy tissue samples of childhood medulloblastoma microscopic images based on conventional machine learning methods. Mahanta et al. introduce a method for automatic counting of platelets and white blood cells from blood smear images. Kumar and Maji present an effective method for automatic recognition of virus particles in Transmission Electron Microscopy (TEM) images. Roy et al. use Deep Convolutional Neural Network to detect Necrosis in mice liver tissue by classifying microscopic images after dividing them into small patches in the preprocessing phase. Deshpande and Bhatt employ Bayesian deep learning to register the medical images that are rendered corrupt by nonlinear geometric distortions. Das and Mahanta develop a novel method where the features of Adenocarcinoma and Squamous Cell Carcinoma of a histological image are taken from various statistical and mathematical models implemented on the coefficients of wavelet transform of an image. Banerjee et al. propose a new iterative method to remove the effects of motion artifacts from multiple non-simultaneous angiographic projection. Datta and Deka propose a Compressed Sensing based parallel MRI reconstruction method. Deka et al. introduce Single-image Super Resolution technique for diffusion-weighted and spectroscopic MR images. Konar et al. propose a Quantum-Inspired Bidirectional Self-Organizing Neural Network architecture for fully automatic segmentation of T1-weighted contrast enhanced MR images. Sasmal et al. present a two-class classification of colonoscopic polyps using multi directions and multi frequency texture analysis. Section III is focused on Pattern Recognition and Remote Sensing. Ahmed and Nath introduce a modified Conditional FP-tree, an efficient frequent pattern mining approach that uses both bottom-up and top-down approach to generate frequent patterns. Malla and Bhavani present a method to predict link weights for directed Weighted Signed Networks using features from network and it's dual. Dhar et al. present a content resemblance based Author Identification System using ensemble learning techniques. Baruah and Bharali present a comparative study of the airline networks on India with ANI based on some network parameters. Chittaragi and Koolagudi present a spectral feature-based dialect classification method from stop consonants. Saini et al. introduce neighborhood concepts to enhance the traditional Self-organizing maps based multi-label classification. This session also includes several interesting contributions in the field of Remote Sensing. Gadhiya et al. propose a multi frequency PolSAR image classification algorithm with stacked autoencoder based feature extraction and superpixel generation. Parikh et al. introduce an ensemble technique for land cover classification problems. Sarmah and Kalita present a supervised band selection method for hyperspectral images using information gain ratio and clustering. Baruah et al. present a non-sub-sampled shearlet transform based remote sensing image retrieval technique. Hire et al. propose a perception based navigation approach for autonomous ground vehicles using Convolutional Neural Networks (CNN). Shah et al. introduce a deep learning architecture using Capsule Network for automatic target recognition from Synthetic Aperture Radar images. Rohit and Mishra present an end-to-end trainable model of Generative Adversarial Networks used to hide audio data in images. Mankad et al. investigate feature reduction techniques for replay anti-spoofing in voice biometrics.

Session IV includes several interesting research outcomes in the field of Intelligent Sensor and Information Retrieval. Devi et al. propose a method where two layers of CNT-BioFET are fabricated for Creatinine detection. While Hazarika et al. present the modeling and analyzing of long-term drift observed in ISFET, Jena et al. present Significance-based Gate Level Pruning (SGLP) technique to design an approximate adder circuit for image processing application. Senapati and Sahu model mathematically a MOS-based patch electrode multilayered capacitive sensor. Nath et al. propose the design, implementation and working of a syringe based automated fluid infusion system integrated with micro-channel platforms for lab-on-a-chip application. Several significant works on Information Retrieval will also be discussed in this session. Chakrabarty et al. report a Joke Recommendation system, based primarily on Collaborative Filtering and Joke-Reader segmentation influenced by the similarity of the user's preference patterns. Vijai Kumar et al. propose TagEmbedSVD, a tag-based Cross-Domain Collaborative model aimed to enhance personalized recommendations in the cross-domain setting. Two of the many important aspects of recommendation systems are diversity and long tail item recommendation, which can be improved by an efficient method as presented by Agarwal et al. Pahal et al. introduce a context-aware reasoning framework that caters to the need and preferences of a group of users in a smart home environment by providing contextually relevant recommendations. Anil et al. The authors introduce a method to apply meta-path for network embedding in mining heterogeneous DBLP network. Kumar et al. introduce two techniques, one to automatically detect and filter null tweets in the Twitter data and another to identify sarcastic tweets using context within a tweet. Yumnam and Sharma propose a grammar-driven approach to parse Manipuri language using Earley's parsing algorithm. Gomathinayagam et al. introduce an information fusion-based approach for query expansion for news articles retrieval. Kumar and Singh propose a prioritized named entity driven LDA, a variant of topic modeling method LDA, to address the issue of overlapping topics by prioritizing named entities related to the topics. Yadav et al. perform sentiment analysis to extract sentiments from a piece of text using supervised and unsupervised approach. Kundu et al. propose a method for finding active experts for a new question in order to improve the effectiveness of a question routing process in community question answering services.

Section V covers topics from the fields of Signal, Image, and Video Processing. Saikia et al. describe a case study involving a framework to classify facies categories in a reservoir using seismic data by employing machine learning models. Sahoo and Dandapat introduce a technique to analyze the changes in speech source signals under physical exercise induced out-of-breadth condition. Mukherjee et al. present a long short-term memory-recurrent neural network for segregating musical chords from clips of short durations which can aid in automatic transcription. Kamble et al. propose a novel Teager energy based sub-band features for audio acoustic scene detection and classification. Pj et al. propose an audio replay attack detection technique using non-voiced audio segments and deep learning models like CNN to classify the audio as genuine or replay. Jyotishi and Dandapat use inverse filtering-based technique to present a novel feature to represent the amount of nasalization present in a vowel. Bhat and Shekar propose an approach for iris recognition by learning fragile bits on multi-patches using monogenic riesz signals. Baghel et al. present a shouted and normal speech classification detection mechanism using 1D CNN architecture. Qadir et al. aim to provide a quantitative analysis of electroencephalographic-based cognitive load while driving in Virtual Reality (VR) environment compared to a fixed non-VR environment. A good number of contributions on Image and Video Processing are also part of this session. Mukherjee et al. propose an unsupervised detection technique for mine regions with reclamation activity from satellite images. Meetei et al. introduces a text detection technique in natural scene and document images in Manipuri and Mizo. Mukherjee et al. propose a method to generate segmented surface of a mapping environment in modeling 3D objects from Lidar data. Koringa and Mitra propose a class similarity based orthogonal neighborhood preserving projection technique for recognition of images with face and hand-written numerals. Mondal et al. introduce a novel combinatorial algorithm for segmentation of articulated components of 3D digital objects using Curve Skeleton. Rajpal et al. propose EAI-Net, an effective and accurate iris segmentation network based on U-Net architecture. Memon et al. use ANN and XGBoost algorithms to classify data for land cover categorization in Mumbai region. Selvam and Mishra introduce a multi-scale attention aided multi-resolution feature extractor baseline network for human pose estimation. Tiwari et al. present a CNN-based method to detect splicing forgery in images using camera-specific features. Vishwakarma et al. propose a multi-focus image fusion technique using sparse representation and modified difference. Biswas and Barma focus on the quality assessment of agricultural product based on microscopic image, generated by Foldscope. Kagalkar et al. propose a learning-based pipeline with image clustering and image selection methods for 3D reconstruction of heritage sites. Sharma and Dey describe a method to analyze the texture quality of fingerprint images using Local Contrast Phase Descriptor. Adarsh et al. present a novel framework for detecting and filling missing regions in point cloud data using clustering techniques. Mullah et al. present a sparsity regularization based spatial-spectral super-resolution method for multispectral remote sensing image. Chetia et al. introduce a quantum image edge detection technique based on four directional sobel operator. Hatibaruah et al. propose a method for texture image retrieval using multiple low and high pass filters and decoded sparse local binary pattern. Devi and Borah propose a new feature extraction approach where both multi-layer and multi-model features are extracted from pre-trained CNNs for aerial scene classification. Bhowmik et al. present a novel high-order Vector of Locally Aggregated Descriptors (VLAD) with increased discriminative power for scalable image retrieval. Pal et al. perform Image-based analysis of patterns formed in drying drops of a colloidal solution. Choudhury and Sarma propose a two-stage framework for detection and segmentation of writing events in air-written Assamese characters. Purwar et al. present an innovative method to offer a promising deterrent against alcohol abuse using Augmented Reality (AR) as a tool. Roy and Bag introduce a detection mechanism for handwritten document forgery by analyzing writer's handwriting. Prabhu et al. use it as a feature extractor for face recognition techniques. Naosekpam et al. propose a novel scalable non-parametric scene parsing system based on super-pixels correspondence. Saurav et al. introduce an approach towards automatic autonomous vision-based powerline segmentation in aerial images using convolutional neural networks. Bhunia et al. present a new correlation filter based visual object tracking method to improve accuracy and robustness of trackers. Kumar et al. introduce an efficient way to detect objects in 360° videos for robust tracking. Singh and Sharma introduce an effective hybrid change detection algorithm in real-time applications using centre-symmetric local binary patterns. Jana et al. present a novel multi-tier fusion strategy for event classification in unconstrained videos using deep neural networks.

In Session VI, several significant contributions in the field of Evolutionary and Soft Computing are reported. Dhanalakshmy et al. analyze and empirically validate the impact of mutation scale factor parameter of the Differential Evolution algorithm. Pournami et al. discuss a scheme to modify the conventional PSO algorithm and use it to present an image registration algorithm. Lipare and Edla propose an approach where a shuffling strategy is applied to PSO algorithm for improving energy efficiency in WSN. Srivatsa et al. propose a GA based solution to solve the classic Sudoku problem. Indu et al. present a critical analysis of an existing prominent graph model of evolutionary algorithms. Singh and Bhukya present an evolutionary approach based on a steady-state GA for selection of multi-point relays in Mobile Ad-Hoc networks. Shaji et al. propose a new Aggregated Rank Removal Heuristic applied to adaptive large neighborhood search to solve Work-over Rig Scheduling Problem. Saharia and Sarmah report a method for optimal design of a DC-DC converter with the goal of minimizing overall losses. Bandagar et al. propose a MapReduce based distributed/parallel approach for standalone fuzzy-rough attribute reduction algorithm. Bar et al. attempt to find an optimal rough set reduct having least number of induced equivalence classes or granules with the help of A* Search algorithm.

We are very grateful to all the esteemed reviewers who were deeply involved in the review process and helped improve the quality of the research contributions. We are privileged and delighted to acknowledge the continuous support received from various committees, sub-committees, and Springer LNCS in preparing this volume of the prestigious PReMI 2019. Thanks are also due to Sushant Kumar, Muddashir, and

Kabita for taking on the major load of secretarial jobs. Finally, we also extend our heartfelt thanks to all those who helped host the PReMI 2019 reviewing process on the EasyChair.org site. We hope that PReMI 2019 was an academically productive conference and you will find the proceedings to be a valuable source of reference for your ongoing and future research.

We hope you will enjoy the proceedings!

December 2019

Pradipta Maji
Bhabesh Deka
Sushmita Mitra
Dhruba Kumar Bhattacharyya
Prabin Kumar Bora
Sankar Kumar Pal

Message from the Honorary General Chair

I am delighted to see that the eighth edition of the biennial International Conference on Pattern Recognition and Machine Intelligence, PReMI 2019, was held, for the first time, in the north-east region of our country at Tezpur University, Assam, India, during December 17–20, 2019. Assam is the most vibrant among the eight states in the north-east, rich in natural resources, and has a great touristic charm. PReMI 2017 was held in the year that marked the 125th birthday of late Prof. Prasanta Chandra Mahalanobis, the founder of our Indian Statistical Institute (ISI). PReMI 2019, on the other hand, was organized when our ISI was preparing to celebrate the birth centenary of another doyen in statistics, namely, Prof. C. R. Rao, a living legendary. Prof. Rao has always been an inspiration to us, and was associated in different capacities with PReMI.

Since its inception in 2005, PReMI has always drawn big responses globally in terms of paper submission. This year, PReMI 2019 was no exception. It has a nice blend of plenary and invited talks, and high-quality research papers, covering different facets of pattern recognition and machine intelligence with real-life applications. Both classical and modern computing paradigms are explored. Special emphasis has been given to contemporary research areas such as big data analytics, deep learning, AI, Internet of Things, and Smart and Intelligent Sensors through both regular and special sessions. Some pre-conference tutorials were also arranged for the beginners. All this made PReMI 2019 an ideal state-of-the-art platform for researchers and practitioners to exchange ideas and enrich their knowledge.

I thank all the participants, speakers, reviewers, different chairs, and members of various committees for making this event a grand success. My thanks are due to the sponsors for their support, and Springer for publishing the PReMI proceedings under the prestigious LNCS series. Last, but not the least, I sincerely acknowledge the support of Tezpur University in hosting the event. I believe, the participants had an academically fruitful and enjoyable stay in Tezpur.

December 2019 Sankar Kumar Pal

Organization

Conference Committee

Patrons

Vinod Kumar Jain	Tezpur University, India
Sanghamitra Bandyopadhyay	Indian Statistical Institute, Kolkata, India

Honorary General Chair

Sankar Kumar Pal	Indian Statistical Institute, Kolkata, India

General Co-chairs

Sushmita Mitra	Indian Statistical Institute, Kolkata, India
Dhruba K. Bhattacharyya	Tezpur University, India
Prabin K. Bora	Indian Institute of Technology Guwahati, India

Program Co-chairs

Bhabesh Deka	Tezpur University, India
Pradipta Maji	Indian Statistical Institute, Kolkata, India

Organizing Co-chairs

Partha Pratim Sahu	Tezpur University, India
Kuntal Ghosh	Indian Statistical Institute, Kolkata, India
Prithwijit Guha	Indian Institute of Technology Guwahati, India

Joint Organizing Co-chair

Vijay Kumar Nath	Tezpur University, India

Plenary Co-chairs

Ashish Ghosh	Indian Statistical Institute, Kolkata, India
Nityananda Sarma	Tezpur University, India

Industry Liaisons

Manabendra Bhuyan	Tezpur University, India
P. L. N. Raju	NESAC (ISRO), India
Darpa Saurav Jyethi	Indian Statistical Institute, NE Centre, Tezpur, India

International Liaisons

Tianrui Li	Southwest Jiaotong University, China

E. J. Ientilucci	RIT, Rochester, USA
Sergei O. Kuznetsov	HSE, Moscow, Russia

Tutorial Co-chairs

Malay Bhattacharyya	Indian Statistical Institute, Kolkata, India
Arijit Sur	Indian Institute of Technology Guwahati, India

Publication Co-chairs

Sarat Saharia	Tezpur University, India
Swati Choudhury	Indian Statistical Institute, Kolkata, India

Publicity Co-chairs

Manas Kamal Bhuyan	Indian Institute of Technology Guwahati, India
Utpal Sharma	Tezpur University, India
Sanjit Maitra	Indian Statistical Institute, NE Centre, Tezpur, India

Webpage Chair

Santanu Maity	Tezpur University, India

Advisory Committee

Anil K. Jain	Michigan State University, USA
B. L. Deekshatulu	Jawaharlal Nehru Technological University, India
Andrzej Skowron	University of Warsaw, Poland
Rama Chellappa	University of Maryland, USA
Witold Pedrycz	University of Alberta, Canada
David W. Aha	Naval Research Laboratory, USA
B. Yegnanarayana	IIIT Hyderbad, India
D. K. Saikia	Tezpur University, India
Jiming Liu	Hong Kong Baptist University, China
Ronald Yager	Iona College, USA
Henryk Rybinski	Warsaw University of Technology, Poland
Jayaram Udupa	University of Pennsylvania, USA
Malay K. Kundu	Indian Statistical Institute, Kolkata, India
Sukumar Nandi	IIT Guwahati, India
S. N. Biswas	Indian Statistical Institute, Kolkata, India

Technical Program Committee

A. R. Vasudevan	NIT Calicut, India
Aditya Bagchi	ISI, Kolkata, India
Ajay Agarwal	CSIR-CEERI, Pilani, India
Alok Kanti Deb	IIT Kharagpur, India
Alwyn Roshan Pais	NIT Surathkal, India

Amit Sethi	IIT Bombay, India
Amita Barik	NIT Durgapur, India
Amrita Chaturvedi	IIT(BHU), Varanasi, India
Ananda Shankar Chowdhury	Jadavpur University, Kolkata, India
Animesh Mukherjee	IIT Kharagpur, India
Anindya Halder	NEHU, Shillong, India
Anjana Kakoti Mahanta	Gauhati University, India
Anubha Gupta	IIIT Delhi, India
Anuj Sharma	Punjab University, India
Arindam Karmakar	Tezpur University, India
Arnab Bhattacharya	IIT Kanpur, India
Arun Kumar Pujari	Central University of Rajasthan, India
Aruna Tiwari	IIT Indore, India
Ashish Anand	IIT Guwahati, India
Asif Ekbal	IIT Patna, India
Asim Banerjee	DA-IICT, Gandhinagar, India
Asit Kumar Das	IIEST, Shibpur, India
Ayesha Choudhary	JNU, Delhi, India
B. Surendiran	NIT, Puducherry, India
Bhabatosh Chanda	ISI, Kolkata, India
Bhargab Bhattacharya	ISI, Kolkata, India
Bhogeswar Borah	Tezpur University, India
Birmohan Singh	SLIET, Punjab, India
Debanjan Das	IIIT Naya Raipur, India
Debasis Chaudhuri	DIC, DRDO, Panagarh, India
Debnath Pal	IISc, Bangalore, India
Deepak Mishra	IIST, Trivandarum, India
Devanur S. Guru	University of Mysore, India
Dinabandhu Bhandari	Heritage Institute of Technology, Kolkata, India
Dinesh Bhatia	NEHU, Shillong, India
Dipti Patra	NIT Rourkela, India
Dominik Slezak	University of Warsaw, Poland
Francesco Masulli	University of Genova, Italy
Goutam Chakraborty	Iwate Prefectural University, Japan
Hari Om	IIT (ISM), Dhanbad, India
Hrishikesh Venkataraman	IIIT Sri City, India
Indira Ghosh	JNU, Delhi, India
Jagadeesh Kakarla	IIIT-D&M, Kancheepuram, India
Jainendra Shukla	IIIT Delhi, India
Jamuna Kanta Sing	Jadavpur University, India
Jayanta Mukhopadhyay	IIT Kharagpur, India
Joydeep Chandra	IIT Patna, India
Kamal Sarkar	Jadavpur University, India
Kandarpa Kumar Sarma	Gauhati University, India
K. Manglem Singh	NIT Manipur, India
Krishna P. Miyapuram	IIT Gandhinagar, India

Lawrence Hall	University of South Florida, USA
Laxmidhar Behera	IIT Kanpur, India
Lipika Dey	TCS Innovation Lab, Delhi, India
Manish Shrivastava	IIIT Hyderabad, India
Mario Koeppen	Kyushu Institute of Technology, Japan
Minakshi Banerjee	RCC Institute of Information Technology, Kolkata, India
Mita Nasipuri	Jadavpur University, India
Mohammadi Zaki	DA-IICT, Gandhinagar, India
Mohit Dua	NIT Kurukshetra, India
Mohua Banerjee	IIT Kanpur, India
Muhammad Abdullah Adnan	BUET, Bangladesh
Muhammad Abulaish	South Asian University, New Delhi, India
Mulagala Sandhya	NIT Warangal, India
Nagesh Kolagani	IIIT Sri City, India
Navajit Saikia	AEC, Guwahati, India
Niloy Ganguly	IIT Kharagpur, India
P. N. Girija	University of Hyderabad, India
Pabitra Mitra	IIT Kharagpur, India
Pankaj B. Agarwal	CSIR-CEERI, Pilani, India
Pankaj Kumar Sa	NIT Rourkela, India
Partha Bhowmick	IIT Kharagpur, India
Partha Pratim Das	IIT Kharagpur, India
Patrick Siarry	University Paris-Est Créteil, France
Pinaki Mitra	IIT Guwahati, India
Pisipati Radha Krishna	NIT Warangal, India
Pradip Kumar Das	IIT Guwahati, India
Pradipta Kumar Nanda	ITER, SOA University, Bhubaneswar, India
Pranab Goswami	IIT Guwahati, India
Praveen Kumar	IIT Guwahati, India
Prerana Mukherjee	IIIT Sri City, India
Punam Saha	University of Iowa, USA
Rahul Katarya	DTU, Delhi, India
Rajarshi Pal	IDRBT, Hyderabad, India
Rajat Kumar De	ISI, Kolkata, India
Rajendra Prasath	IIIT Sri City, India
Rajib Kumar Jha	IIT Patna, India
Rama Rao Nidamanuri	IIST, Trivandrum, India
Rohit Sinha	IIT Guwahati, India
S. Jaya Nirmala	NIT Trichy, India
Samarendra Dandapat	IIT Guwahati, India
Sambhunath Biswas	ISI, Kolkata, India
Samit Bhattacharya	IIT Guwahati, India
Sanjoy Kumar Saha	Jadavpur University, India
Sarif Naik	Philips Research India, Bangalore, India
Saroj Kumar Meher	ISI, Bangalore, India

Satish Chand	JNU, Delhi, India
Shajulin Benedict	IIIT Kottayam, India
Shanmuganathan Raman	IIT Gandhinagar, India
Sharad Sinha	IIT Goa, India
Shiv Ram Dubey	IIIT Sri City, India
Shubhra Sankar Ray	ISI, Kolkata, India
Shyamanta M. Hazarika	IIT Guwahati, India
Sivaselvan Balasubramanian	IIIT-D&M, Kancheepuram, India
Snehashish Chakraverty	NIT Rourkela, India
Snehasis Mukherjee	IIIT Sri City, India
Soma Biswas	IISc, Bangalore, India
Somnath Dey	IIT Indore, India
Sourangshu Bhattacharya	IIT Kharagpur, India
Srilatha Chebrolu	NIT Andhra Pradesh, India
Srimanta Mandal	DA-IICT, Gandhinagar, India
Srinivas Padmanabhuni	IIT Tirupati, India
Sriparna Saha	IIT Patna, India
Subhash Chandra Yadav	Central University of Jharkhand, India
Sudip Paul	NEHU, Shillong, India
Sujit Das	NIT Warangal, India
Sukanta Das	IIEST, Shibpur, India
Sukhendu Das	IIT Madras, India
Sukumar Nandi	IIT Guwahati, India
Suman Mitra	DA-IICT, Gandhinagar, India
Susmita Ghosh	Jadavpur University, India
Suyash P. Awate	IIT Bombay, India
Swagatam Das	ISI, Kolkata, India
Swanirbhar Majumder	Tripura University, India
Swapan Kumar Parui	ISI, Kolkata, India
Swarnajyoti Patra	Tezpur University, India
Tanmoy Som	IIT(BHU), Varanasi, India
Tony Thomas	IIITM-K, Kerala, India
Ujjwal Bhattacharya	ISI, Kolkata, India
Utpal Garain	ISI, Kolkata, India
V. M. Manikandan	IIIT Kottayam, India
V. Susheela Devi	IISc, Bangalore, India
Varun Bajaj	IIIT-D&M, Jabalpur, India
Vijay Bhaskar Semwal	MANIT, Bhopal, India
Vinod Pankajakshan	IIT Roorkee, India
Viswanath Pulabaigari	IIIT Sri City, India

Additional Reviewers

Abhijit Dasgupta
Abhirup Banerjee
Abhishek Bal
Abhishek Sharma
Airy Sanjeev
Ajoy Mondal
Alexy Bhowmick
Amalesh Gope
Amaresh Sahoo
Anirban Lekharu
Ankita Mandal
Anwesha Law
Aparajita Khan
Apurba Sarkar
Arindam Biswas
Arpan K. Maiti
Ashish Phophalia
Ashish Sahani
Atul Negi
Avatharam Ganivada
Avinash Chouhan
B. S. Daya Sagar
Bappaditya Chakraborty
Barnam J. Saharia
Bhabesh Nath
Bikramjit Choudhury
Binayak Dutta
Chandra Das
Debadatta Pati
Debamita Kumar
Debasis Mazumdar
Debasish Das
Debasrita Chakraborty
Debjyoti Bhattacharjee
Debojit Boro
Deepak Gupta
Deepika Hazarika
Dibyajyoti Chutia
Dipankar Kundu
Dipen Deka
Ekta Shah
Gaurav Harit
Haradhan Chel
Helal U. Mullah
Ibotombi S. Sarangthem
Indrani Kar
Jaswanth N.
Jay Prakash
Jaya Sil
Jayanta K. Pal
Jaybrata Chakraborty
K. C. Nanaiah
K. Himabindu
Kannan Karthik
Kaushik Das Sharma
Kaushik Deva Sarma
Kishor Upla
Kishorjit Nongmeikapam
L. N. Sharma
Lipi B. Mahanta
M. Srinivas
M. V. Satish Kumar
Manish Sharma
Manuj Kumar Hazarika
Milind Padalkar
Minakshi Gogoi
Monalisa Pal
Nabajyoti Mazumdar
Nabajyoti Medhi
Nayan Moni Kakoty
Nayandeep Deka Baruah
Nazrul Hoque
Nilanjan Chattaraj
P. V. S. S. R. Chandra Mouli
Pankaj Barah
Pankaj Kumar
Parag Chaudhuri
Parag K. Guha Thakurta
Paragmoni Kalita
Partha Garai
Partho S. Mukherjee
Pranav Kumar Singh
Prasun Dutta
Pratima Panigrahi
Ragini (CUJ)
Rajiv Goswami
Rahul Roy
Rajesh Saha
Rajeswari Sridhar
Rakcinpha Hatibaruah
Ram Sarkar
Reshma Rastogi
Riku Chutia
Rishika Sen
Rosy Sarmah
Rupam Bhattacharyya
Rutu Parekh
Sai Charan Addanki
Saikat Kumar Jana
Sanasam Ranbir Singh
Sanghamitra Nath
Sanjit Maitra
Sankha Subhra Nag
Sathisha Basavaraju
Shaswati Roy
Shilpi Bose
Shobhanjana Kalita
Sibaji Gaj
Siddhartha S. Satapathy
Sipra Das Bit
Smriti Kumar Sinha
Soumen Bera
Subhadip Boral
Sudeb Das
Sudip Das
Sujoy Chatterjee
Sujoy M. Roy
Suman Mahapatra
Sumant Pushp
Sumit Datta
Supratim Gupta
Sushant Kumar
Sushant Kumar Behera
Sushil Kumar
Sushmita Paul
Swalpa Kumar Roy
Swarup Chattopadhyay
Swati Banerjee
Tandra Pal
Thoudam Doren Singh
Vivek K. Mehta
Yash Agrawal

Sponsoring Organizations

Endorsed by

IAPR International Association for Pattern Recognition

Technical Co-sponsors

Computing by Cognition Centre for Soft Computing Research, ISI

International Rough Set Society

Web Intelligence Consortium

Springer Springer International Publishing

Financial Sponsors

Diamond Sponsors

North Eastern Council
Government of India

ऑयल इंडिया लिमिटेड
Oil India Limited
Conquering Newer Horizons

North Eastern Council

Oil India Limited

Gold Sponsor

Indian Space Research Organization

Silver Sponsor

Council of Scientific & Industrial Research

Abstracts of Invited Talks

Granular Artificial Intelligence: A New Avenue of Artificial Intelligence for Modeling Environment and Pattern Recognition

Witold Pedrycz

Department of Electrical and Computer Engineering, University of Alberta, Edmonton, Canada
wpedrycz@ualberta.ca

Recent advancements in Artificial Intelligence fall under the umbrella of industrial facets of AI (Industrial AI, for short) and explainable AI (XAI). We advocate that in the realization of these two pursuits, information granules and Granular Computing play a significant role. First, it is shown that information granularity is of paramount relevance in building meaningful linkages between real-world data and symbols commonly encountered in AI processing. Second, we stress that a suitable level of abstraction (information granularity) becomes essential to support user-oriented framework of design and functioning AI artifacts. In both cases, central to all pursuits is a process of formation of information granules and their prudent characterization. We discuss a comprehensive approach to the development of information granules by means of the principle of justifiable granularity; here various construction scenarios are discussed. In the sequel, we look at the generative and discriminative aspects of information granules supporting their further usage in the formation of granular artifacts, especially pattern classifiers. A symbolic manifestation of information granules is put forward and analyzed from the perspective of semantically sound descriptors of data and relationships among data.

Imparting Sentiment and Politeness on Computers

Pushpak Bhattacharyya[1,2]

[1] Department of Computer Science and Engineering, Indian Institute of Technology Patna, India
[2] Department of Computer Science and Engineering, Indian Institute of Technology Bombay, India
pb@cse.iitb.ac.in

In this talk we will describe the attempts made at making machines "more human" by giving them sentiment and politeness abilities. We will give a perspective on automatic sentiment and emotion analysis, with a description of our work in this area, touching upon the challenging problems of sarcasm arising from numbers, and multitask and multimodal sentiment and emotion analysis. Subsequently we touch upon the interesting problem of "making computers polite" and our recent work on this. We will end with noting the places of rule based, classical ML based and deep learning-based approaches in NLP.

Beyond Text Detection and Recognition: Emerging Opportunities in Scene Understanding

C. V. Jawahar

Centre for Visual Information Technology, IIIT Hyderabad, India
jawahar@iiit.ac.in

Recent years have seen major advances in the performance of reading text in natural outdoor. Methods for text detection and recognition, are reporting very high quantitative performances on popular benchmarks. In this talk, we discuss a set of opportunities in scene understanding where text plays a critical role. Many associated challenges, ongoing research and emerging opportunities for research are discussed.

Cognitive Analysis Using Physiological Sensing

Sanjoy Kumar Saha

Department of Computer Science and Engineering, Jadavpur University, Kolkata, India
sks_ju@yahoo.co.in

Cognitive load is a measure of the processing done using working memory of the brain. The effectiveness of an activity is dependent on the amount of cognitive load experienced by an individual. Subjected to a task, assessing the cognitive load of an individual may be useful in evaluating the individual and/or task. Physiological sensing can help in cognitive analysis. EEG, GSR, PPG sensors, Eyegaze tracker can sense different aspects. Talk will mostly focus on EEG signal and eye gaze data and their processing. Their applicability in assessing the readability of text materials will also be highlighted.

The Rise of Hate Content in Social Media

Animesh Mukherjee

Department of Computer Science and Engineering,
Indian Institute of Technology, Kharagpur, India
animeshm@gmail.com

The recent online world has seen an upheaval in the fake news, misinformation, misbehavior and hate speech targeted toward communities, race and gender. This has resulted in severe consequences. Reports say, that the last US election was heavily influenced by the social media (https://www.bbc.com/news/technology-46590890). The EU referendum similarly was under social media influence (https://www.referendumanalysis.eu/eu-referendum-analysis-2016/section-7-socialmedia/impact-of-social-media-on-the-outcome-of-the-eu-referendum/). Facebook has been considered responsible for the spread of unprecedented volume of hate content resulting into Rohingya genocide. The Pittsburg Synagogue shooter was an active member of the extremist social media website GAB where he continuously posted anti-Semitic comments finally resulting into the shooting. Similar cases have been reported for the Tamil Muslim community in Sri Lanka, attacks on refugees in Germany and the Charleston church shooting incident. A concise report of the events and the damages caused thereby is present in this article from the Council on Foreign Relations (https://www.cfr.org/backgrounder/hatespeech-social-media-global-comparisons).

Since 2017, we have started to put in focused efforts to tackle this problem using computational techniques. Note that this is not a computer science problem per se; this is a much larger socio-political problem. As a first work we show how simple opinion conflicts among social media users can lead to abusive behavior (CSCW 2018, https://techxplore.com/news/2018-10-convolutional-neural-networkabuse-incivility.html). As a next step we investigate the GAB social network and show how hateful users are much more densely connected among each other compared to others; how the messages posted by hateful users spread far, wide and deep into the social network compared to the normal users (ACM WebSci 2019). Consequently, we propose a solution to this problem; as such suspending hateful accounts or deleting hate messages is not a very elegant solution since this curbs the freedom of speech. More speech to counter hate speech has been thought to be the best solution to fight this problem. In a recent work we characterize the properties of such counter speech and show how they vary across target communities (ICWSM 2019). Presently, we are also investigating how the hate patterns change if they are allowed to evolve in an unmoderated environment. To our surprise we observe that hatespeech is steadily increasing and new users joining are exposed to hate content at an increased and faster rate. Further the language of the whole community is driven to the language of the hate speakers. We believe that this is just the beginning. There are many challenges that need to be yet addressed. For instance, we plan to investigate how misinformation and hate content are related - do they influence each other? Can containing one of them automatically contain the other to some extent at least? In similar lines, how does hate content affect gender and cause gender/sex related discrimination/crime.

In this talk, we will try to present a summary of some of the above experiences that we have had in the last few years relating to our ventures into hate content analysis in social media.

How to Escape Saddle Points Efficiently?

Praneeth Netrapalli

Microsoft Research India
praneeth@microsoft.com

Non-convex optimization is ubiquitous in modern machine learning applications. Gradient descent based algorithms (such as gradient descent or Nesterov's accelerated gradient descent) are widely used in practice since they have been observed to perform well on these problems. From a theoretical standpoint however, this is quite surprising since these nonconvex problems are teeming with highly suboptimal saddle points, and it is well known that gradient descent (and variants) can get stuck in such saddle points. A key question that arises in resolving this apparent paradox is whether gradient descent based algorithms escape saddle points where the Hessian has negative eigenvalues and converge to points where Hessian is positive semidefinite (such points are called second-order stationary points) efficiently. We answer this question in the affirmative by showing the following: (1) Perturbed gradient descent converges to second-order stationary points in a number of iterations which depends only poly-logarithmically on dimension (i.e., it is almost "dimension-free"). The convergence rate of this procedure matches the wellknown convergence rate of gradient descent to first-order stationary points, up to log factors, and (2) A variant of Nesterov's accelerated gradient descent converges to second-order stationary points at a faster rate than perturbed gradient descent. The key technical components behind these results are (1) understanding geometry around saddle points and (2) designing a potential function for Nesterov's accelerated gradient descent for non-convex problems.

Machine Learning and Its Application in Remote Sensing Data Classification Applications

Anil Kumar

Indian Institute of Remote Sensing, Dehradun, India
anil@iirs.gov.in

From 2000 onwards sub-pixel or later called soft classification has been explored very extensively. Later learning based algorithm taken over when modified forms of learning algorithms were proposed. Artificial Neural networks (ANN) is a generic name for a large class of machine learning algorithms, most of them are trained with an algorithm called back propagation. In the late eighties, early to mid-nineties, dominating algorithm in neural nets was fully connected neural networks. These types of networks have a large number of parameters, and so do not scale well. But convolutional neural networks (CNN) are not considered to be fully connected neural nets. CNNs have convolution and pooling layers, whereas ANN have only fully connected layers, which is a key difference. Moreover, there are many other parameters which can make difference like number of layers, kernel size, learning rate etc. While applying Possibilistic c-Means (PCM) fuzzy based classifier homogeneity within class was less while observing learning based classifiers homogeneity was found more. Best class identification with respect to homogeneity within class was found in CNN soft output as shown in the figure. With this it gives a path to explore various deep leaning algorithms in various applications of earth observation data like; multi-sensor temporal data in crop/forest species identification, remote sensing time series data analysis. As learning based algorithms require large size of training data, but in remote sensing domain it's difficult to generate large training data sets. This issue also has been resolved in this research work.

Brief History of Topic Models

Chiranjib Bhattacharyya

Department of Computer Science and Automation, Indian Institute of Science, Bangalore, India
chiru@iisc.ac.in

The success of Machine Learning is often attributed to Supervised Learning models. Comparatively, progress on learning models without supervision is limited. However, Unsupervised Learning has the potential of unlocking a whole new class of applications. In this tutorial we will discuss Topic models, an important class of Unsupervised Learning Models which have proven to be extremely successful in practice. The tutorial will discuss a self-contained introduction to learning Latent variable models (LVMs) and will discuss Topic models as a special case of LVMs. Time permitting, recent results on deriving sample complexity of Topic models will be also covered.

Introduction to Compressed Sensing

Ajit Rajwade

Department of Computer Science and Engineering,
Indian Institute of Technology Bombay, India
ajitvr@cse.iitb.ac.in

Compressed sensing is a relatively new sensing paradigm that proposes acquisition of images directly in compressed format. This is different from the more conventional sensing methods where the entire 2D image array is first measured followed by JPEG/MPEG compression after acquisition. Compressed sensing basically aims to reduce *acquisition time*. It has shown great results in speeding up MRI (magnetic resonance imaging) acquisition where time is critical, in improving frame rates of videos, and in general in improving acquisition rates in a variety of imaging modalities.

Central to compressed sensing is the solution to a seemingly under-determined system of linear equations, i.e. a system of equations where the number of unknowns (n) is greater than the number of knowns (m). Hence at first glance, there will be infinitely many solutions. However the theory of compressed sensing states that if the vector of unknowns is sparse, and the system's sensing matrix obeys certain properties, then the system is provably well-posed and unique solutions can be guaranteed. Moreover, the theory also states that the solution can be computed efficiently, and is robust to measurement noise or slight deviations from sparsity.

In this talk, I will give an introduction to the above concepts. I will also introduce a few applications, and enumerate a few research challenges/directions.

Novel Support Vector Machine Algorithms for the Diagnosis of Alzheimer's Disease

M. Tanveer

Discipline of Mathematics, Indian Institute of Technology Indore, India
mtanveer@iiti.ac.in

Support vector machine (SVM) has provided excellent performance and has been widely used in real world classification problems due to many attractive features and promising empirical performance. Different from constructing two parallel hyperplanes in SVM, recently several non-parallel hyperplane classifiers have been proposed for classification and regression problems. In this talk, we will discuss novel non-parallel SVM algorithms and their applications to Alzheimer's disease. Numerical experiments clearly show that non-parallel SVM algorithms outperform traditional SVM algorithms. This talk concludes that non-parallel SVM variants could be the viable alternative for classification problems.

Contents – Part I

Pattern Recognition

Machine Learning

Deep Learning

Contents – Part II

Information Retrieval

Smart and Intelligent Sensors

Pattern Recognition

Identity Independent Face Anti-spoofing Based on Random Scan Patterns

Kannan Karthik(✉) and Balaji Rao Katika

Department of Electronics and Electrical Engineering,
Indian Institute of Technology Guwahati, Guwahati 781039, Assam, India
{k.karthik,k.balaji}@iitg.ac.in

Abstract. Conventional face anti-spoofing paradigms tend to operate on plain facial profiles and learn either the natural face space alone (one-class training problem) or both the natural face space as well as the spoof sample space (2-class training problem). However, this rigidity with respect to spatially constrained measurements, makes the base feature or statistic vulnerable to noise related to pose and camera perspective/orientational and scale changes. Noting that the sharpness profile computed on a natural face is largely independent of the pose and perspective change, it is imperative that the measurements be extracted in an identity independent setting by ignoring the pose/perspective variation. To facilitate this, we have deployed a 2-dimensional random walk for capturing lower order pixel correlation statistics from natural faces, with virtually no perceptual interference. The proposed identity independent frame has surpassed the state of the art with reference to a 3D mask dataset (image oriented, isolated frame setting), with an EER of 2.25% without auto-population and an EER of 0.45% with auto-population.

Keywords: Anti-spoofing · 3*D* mask · 2D random walk · Scan-patterns · Auto-population · Outlier detection · Prosthetic

1 Introduction

Since most facial recognition systems operate based on notion of perceptual similarity of images of human faces, a majority of them, cannot tell the difference between a spoofed version of a face, versus, a naturally captured facial image. It is therefore important to have a counter-spoofing layer that sits above the facial recognition system, which attacks the environment linked to the image acquisition process and attempts to anticipate any form of spoofing. Much of the literature, related to counter-spoofing, has been directed to a specific form of facial spoofing called planar spoofing [11], wherein the impersonator (X) tends to present planar images or printed photos of the target subject (Y), (who is being impersonated). This a type of geometrically constrained spoofing, in which, the identity of the targeted individual (Y), is embedded in the form of a planar intensity variation induced either by double printing or by capturing an image

B. Deka et al. (Eds.): PReMI 2019, LNCS 11941, pp. 3–12, 2019.
https://doi.org/10.1007/978-3-030-34869-4_1

of a natural photograph. This form of spoofing-model, albeit trivial, is enticing from a research perspective inviting solutions on multiple fronts, some related to sharpness reduction [6], some related to base-image quality degradation [4], some linked to model-specific, geometrically induced distortions [5,7] etc.

With the co-existence of a diverse entertainment industry, a rapid advancement in cosmetic technology and the inclusive growth and evolution of touch-up artists, facial spoofing frames have evolved considerably. Every individual tends to possess a distinct facial surface-contour, which, can be captured either overtly or surreptitiously. This surface contour, can be used to synthesize a prosthetic of that individual's face. This prosthetic can be designed either using paper-craft [2] or rigid plastic or some semi-elastic form of material. However, most prosthetics are customized according to the target individual who is being impersonated (viz. Y) and are usually de-linked from the identity of the individual who is the impersonator (X), to ensure that his/her identity is not revealed during the counter-spoofing or authentication process. This opens up the problem to the counter-spoofing community thanks to the following conjecture:

Claim-1: Given an impersonator X and a target subject Y, the prosthetic is designed to mimic the surface contour of Y and has very little to do with the surface contour of X. This is to ensure identity masking from the point of the view of the attacker (X). The only way this can be achieved, is by ensuring that this physical facial re-mapping or (physical face-morphing), is of a many-to-one type. Thus a single prosthetic designed to impersonate Y, can fit multiple individuals of the X-type. In other words, one mask is designed to fit many. This makes the prosthetic, presented as a synthetic surface contour of X, an over-smoothed approximation of X's facial profile with some depth information. One therefore anticipates a reduction in facial image sharpness as far as image of the prosthetic of X is concerned. This sharpness variation can be captured by performing a gradient based analysis.

In this paper, we focus on literature connected with prosthetic based facial spoofing. A 3D-mask dataset was developed using paper craft models in Erdogmus et al. [2], wherein the prosthetics were customized to target different subjects. Some examples of this are shown in Fig. 1. The natural faces of the subjects are shown in Fig. 1(b)(row-2), while their corresponding spoofed versions with prosthetics are shown in Fig. 1(a)(row-1). It is obvious that paper craft model has been cleverly designed to mimic the surface contours including the ocular and nasal profiles of each targeted subject. In Erdogmus et al. [2], the base feature used for recognition was the Local Binary Pattern (LBP) along with its variants. The 3D-Mask dataset was analyzed both as a sequence of static images, and also as a video sequence, by extending the LBP analysis to include both time and space differentials. A 2-class SVM was finally constructed by learning the prosthetic as well as the genuine face spaces, coupled with the decision boundary/surface. Spoof-detection was done by extracting the same features from a typical query test-face and checking its position with respect to the two reference clusters. In a video-based setting associated with the 3D mask database,

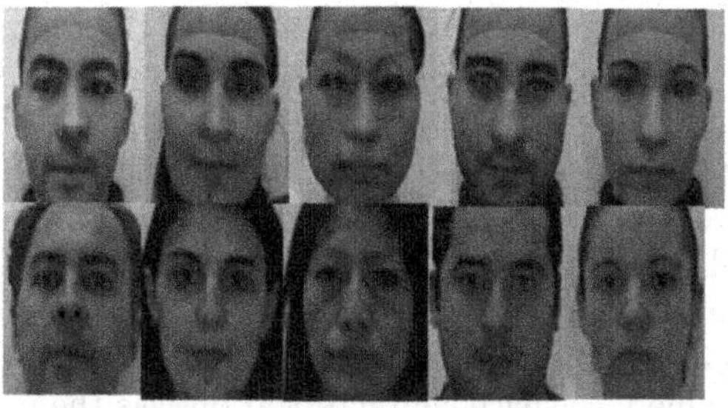

Fig. 1. (a) Examples of 3D mask faces for different subjects, taken from 3D mask database [2] (b) Examples of their corresponding real genuine samples.

more options exist, since it is possible to deploy space-time micro-feature analysis to search for liveliness in the facial profile, consistencies and naturalness in expression changes etc. Optical flow methods were used in Feng et al. [3] to detect differences in dynamism with respect to texture between an imposter and a genuine subject. A deep-learning network for attacking multi-biometric spoofing including facial spoofing was developed by Menotti et al. [9], but again the learning was two sided and assumed availability of samples related to the spoofing process. Wen et al. [10], proposed a mixed bag of features ranging from intensity and gradient all the way up to those which captured color and texture, with the objective of covering the complete gamut of statistics, which would help segregate the genuine face class from all forms of spoofing. However, once again, this arrangement demanded availability of spoof training samples, necessary for constructing a 2-class SVM. This existing frame had several issues:

- Very often the nature, texture and structure of the customized prosthetic may not be known. This implies that spoof-class training samples may not be available. Hence, it is important to shift and restrict the training process to the genuine face sample set, where the acquisition procedure, naturally captured facial profile coupled with the local statistics remains predictable.
- Since LBP features are highly localized in space and are registered, pose deviations and scale changes because of facial migrations with respect to the camera will lead to a contortion of measurements. This will interfere with the counter-spoofing procedure. We term this form of interference as perceptual interference, which arise when the measurements are registered in space.

The first problem related to absence of a spoof model, can be addressed through an inlier space characterization procedure by learning the space spanned by genuine natural facial images from different subjects, for different poses and mild illumination variations. This inlier space characterization was done through a query feature ranking procedure, in relation to the genuine face feature set, to detect outliers in [7]. Genuine face space characterization coupled with anomaly detection in a much more general setting by constructing a one-class SVM was done in Arashloo et al. [1].

While this arrangement was designed to care of the first problem related to the absence of a proper spoofing model, they were applied to planar spoofing alone. In both these papers [1,7], the measurements were registered in space either by gridding the image or by computing statistics in specific spatial zones, whose locations were largely static. They thus proved to be ineffective, when confronted with 3D-spoof models, wherein the prosthetics attempted to mimic the depth profile in the imposter's face.

Attacking this 3D-spoofing problem, with a single sided training procedure, involving only genuine face space characterization was the main challenge. This led to the proposed architecture which was placed on an identity independent setting. The rest of the paper is organized as follows: In Sect. 2 we propose a new paradigm based on identity independent feature auto-population through random scan patterns. Section 3 validates the choice of randomly scanned feature and builds a one-class SVM to characterize the space of natural faces. Experimental results and comparison with the state of the art are in Sect. 4.

2 Proposed Paradigm and Architecture

The anti-spoofing problem is a typical frame wherein the nature of impersonation remains unknown in practice. By treating this problem as a form of planar image or printed photo spoofing, the problem becomes analytically tractable, mainly because of physical constraints. To make the analysis model independent, without compromising on the robustness of the detection process, it is important to change the paradigm or the manner in which the measurements are gathered.

Claim 2: We claim that most anti-spoofing systems work best in an identity independent setting, wherein the measurements or features extracted are taken in such a way that perceptual relevance is given the least importance. However, the residual correlation or some other statistics, which may be derived from this dissolved identity, carry necessary information regarding the environment or channel in which the information has been captured to perform anti-spoofing. This identity dissolution, in our case, is performed using a constrained shuffle of pixels in the spatial domain using a 2-dimensional random walk. This 2-D random walk has been inspired by Space Filling Curves [8], which was originally devised for retaining the compressibility of video signals after encryption.

Claim 3: By auto-populating the each facial image profile with these scans, it is possible to construct several variations of the same profile, which essentially carry the same pixel-correlation information, with minimal content interference. Thus a single facial profile is transformed into an ensemble of scans produced using independent 2-D random walk patterns. This ensemble carries significant information regarding the sharpness profile of the facial image and will have sufficient information to segregate 3D mask profiles from natural facial images.

2.1 Random Scan Algorithm

Between a complete shuffle and a raster scan, one can find a judicious trade-off between feature transparency and conservation versus the size of the auto-populated set based on the type of scan. With a perfect shuffle of pixel complete pixel correlation structure is lost, while in the case of a raster scan this correlation structure is preserved. However, in the case of the raster scan, the peeling is done in a regularized fashion and hence only one instance of the facial profile can be made available, imparting a significant rigidity to the generation of statistics. A trade in terms of conserving the correlation profile in pixels, while at the same time permitting multiple shuffle trajectories is via the randomized correlated scan.

Figure 2 shows two typical variations in the scan patterns, executed over the same image F_i. Eventually when the scan is completed, a 1-D vector $\bar{X}_i$ is generated as a function of the randomizer/key KEY_i:

$$\bar{X}_i = CSCAN(F_i, KEY_i) \tag{1}$$

where, $CSCAN(.,.)$, represents the randomized correlated scan algorithm based on a 2-dimensional random walk directed by a key sequence, KEY_i. While perceptual identity is lost in the unregistered feature vector $\bar{X}_i$ structure and format of the data captured is conserved. Gradient and sharpness features can now be computed on the top of this randomly scanned intensity feature. The key sequence carries information pertaining to the direction/trajectory of the random walk. In case, there is an abrupt termination of the walk, the key sequence also stores information related to the pixel jump. In a nutshell, the key sequence is a sequence of location pointers forming a linked list. To reduce scanning complexity, F_i is a down sampled version of the parent facial image. Let $\bar{X}_i = [x_{i,1}, x_{i,2}, ..., x_{i,n}]$. It is to be noted that not all the correlated scans are contiguous in terms of random walk.

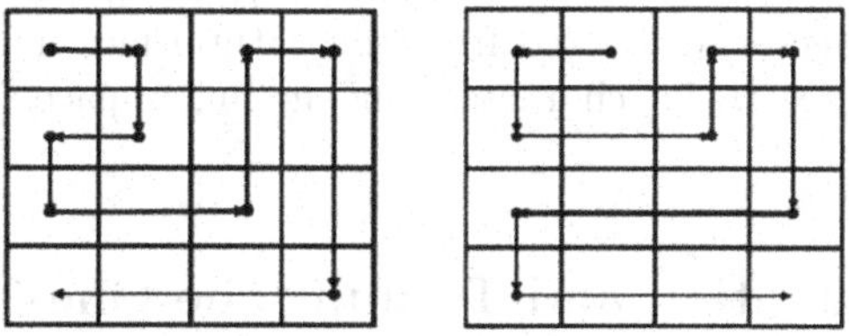

Fig. 2. Random but correlated scans for same facial image F_i (two different walks executed on the same facial image).

When the pointer in the 2-D random walk either walks into a corner or bumps into its own tail, it may encounter an abrupt termination. At this point, the pointer must hop to a new free cell within the same grid and resume the random walk. This process continues till all the pixels within the image grid are traversed. If $N \times N$ is the size of the image, the length of the scanned vector $\bar{X}_i$ is $n = N^2$ and the path length is $n - 1$ units. The walk is rectangular in nature and diagonal transitions are not allowed. Because of this, the primary scanned vector $\bar{X}_i$ must be median filtered using a 1×3 ($w = 3$) window, to iron out singularities. The scanned vector becomes smoother with a larger window size w at the expense of a loss of detail and an un-necessary alteration of natural pixel correlation statistics. It is important that the median filter does interfere with the accuracy of the natural image statistics. Hence, the optimal choice for w is three. The pre-processed statistic is given by,

$$\bar{X}_{MED,i} = MEDIAN[\bar{X}_i, w] \tag{2}$$

with median filter window size, $w = 3$.

2.2 Final Differential Statistic

Based on earlier conjectures and observations, it is clear that the prosthetic arrangement is likely to have a smoother surface contour as compared to the natural face (partly owing to CLAIM-1 in Sect. 1). This is based on the one-mask fits many assumption, the mask designed to dissolve the identity of the imposter (X), while emulating the identity of the target (Y), who is being impersonated. Hence, a simple differential feature which captures the first or second order pixel derivative, will be sufficient to discriminate between a natural face as compared to one which has a prosthetic. The natural face is expected to have a greater roughness (culminating in a greater and more heterogeneous sharpness profile) as compared to that of the prosthetic. Let $\bar{D}_{X,i}$ be the differential statistic computed on the median filtered 1D sequence. If $\bar{D}_{X,i} = [d_{i,1}, d_{i,2}, ..., d_{i,n}]^T$ and $\bar{X}_{MED,i} = [x_{MED,i,1}, x_{MED,i,2}, ..., x_{MED,i,n}]$,

$$d_{i,r} = x_{MED,i,r} - x_{MED,i,(r-1)} \tag{3}$$

for $r \in \{1, 2, ..., n\}$ and with initial conditions, $x_{MED,i,(0)} = 0$. The vector, $\bar{D}_{X,i}$ is the final feature vector, extracted from the natural face image class alone, is fed to a one-class SVM [1] for characterizing the inlier space [7] (or the natural face space).

3 Feature Validation and Training the One-Class SVM

Feature validation is done by splitting the 3DMAD dataset [2] (composition given in Table 1), into natural faces and prosthetic based images. The base feature used for this comparison is the norm of the final differential vector, $\bar{D}_{X,i}$, which is given by,

$$E_i = ||\bar{D}_{X,i}||_2 = \sqrt{d_{i,1}^2 + d_{i,2}^2 + ... + d_{i,n}^2} \tag{4}$$

The conditional distributions, $f_{E/NATURAL}(e)$ and $f_{E/SPOOF}(e)$ are computed on the same scale in Fig. 3, for the 3D-MASK dataset (these are essentially

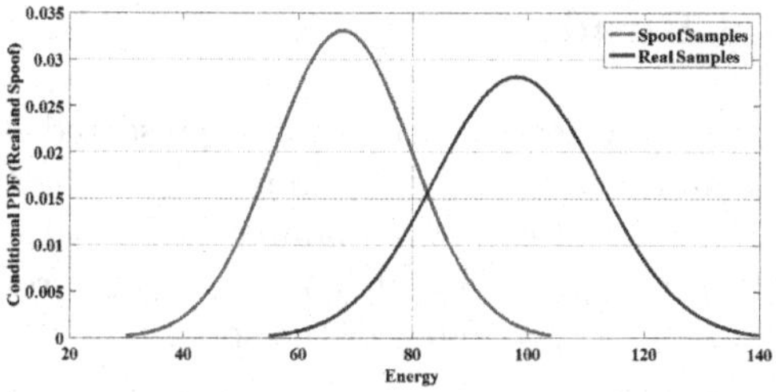

Fig. 3. Conditional distributions of the differential energy feature for natural and spoof samples, computed from the 3DMAD dataset. (Color figure online)

conditional histograms which have been interpolated to impart smoothness to the functions). In Fig. 3, the conditional distribution shown in blue corresponds to the genuine face space energy profile, while that shown in red corresponds to the energy profile generated from the prosthetic samples. As expected, the differential statistics produced from the natural face space have a larger mean and larger variance (because of the increased roughness and intensity diversity), while that of the prosthetic shows a smaller mean and variance (owing to over-smoothing stemming from the one-mask fits all claim). While the conditional distributions demonstrate the feature separability and ability of the random scans to conserve the lower order correlation statistics present in the image, the impact of the of the random scan in obscuring the identity of the individual subjects is demonstrated in Fig. 4. Notice that the scanned versions presented for simplicity as a 2-D shuffled version in Fig. 4(b, d), have no resemblance to their corresponding un-scanned counterparts (Fig. 4(a, c)). Thus, the processing and feature extraction is done in truly an identity independent setting.

Table 1. Description and composition of 3D mask database [2]

3D mask [2]	No of subjects	No of poses/subject
Faces with 3D masks	17	50
Natural faces	17	50

The set of natural faces from the 3DMAD database is split into a one-class training set for characterizing the inlier space and a test set which comprises of both natural faces as well as spoofed faces using the prosthetic. Given final differential base feature vectors $\bar{D}_{X,i}$, $i \in \{1, 2, ..., N_{GEN,T}\}$, where, $N_{GEN,T}$ is the number of training images from the genuine and natural face space. A one-class SVM [2], is constructed by building a hyper-sphere around the genuine multi-dimensional base differential features vector set corresponding to genuine face images, with an α-trim outlier fraction set to 10%.

4 Experimental Results

The 3DMAD database [2], whose composition is presented in Table 1, is split into three sets: (i) Genuine face set for training, $x\%$ of the total genuine face space; (ii) Genuine face set for testing, remaining $(100 - x)\%$ of the remaining genuine face space; (iii) Spoof samples ONLY for testing, from the paper-craft based prosthetic arrangement, $y = 100\%$ of the spoof set.

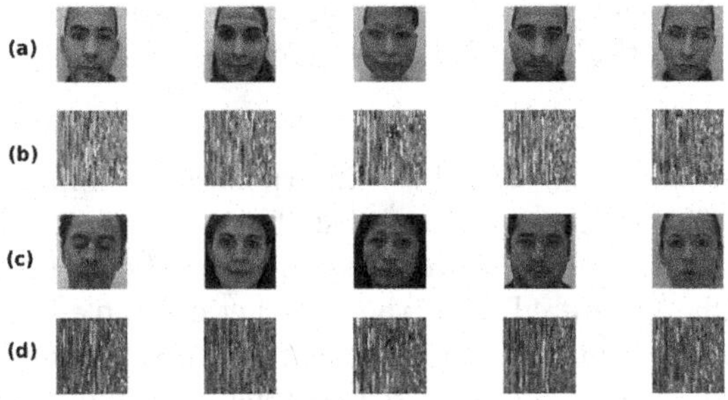

Fig. 4. (a) Samples of 3D mask faces (b) Random walk features extracted for mask faces (c) Samples of real genuine face (d) corresponding random walk features.

Table 2. Error rates for different trim factors α and database splits with 3D mask set. Note that best results are obtained for $\alpha = 10\%$.

Inlier ($x(\%)$)/Outlier samples ($y(\%)$)	EER@ $\alpha = 5\%$	EER@ $\alpha = 10\%$	EER@ $\alpha = 15\%$	EER@ $\alpha = 20\%$
$x = 30\%, (100 - x) = 70\%, y = 100\%$	5.9291	5.0115	8.4520	10.4653
$x = 40\%, (100 - x) = 60\%, y = 100\%$	5.8956	3.4048	5.2192	8.0694
$x = 50\%, (100 - x) = 50\%, y = 100\%$	3.2860	2.2594	4.2459	5.8806
$x = 60\%, (100 - x) = 40\%, y = 100\%$	2.5087	1.3078	3.3742	4.1446
$x = 70\%, (100 - x) = 30\%, y = 100\%$	1.2161	0.1852	3.0428	3.0760

Selection of the trim factor in the one-class SVM α, is a careful tradeoff between extent of generalization of the natural face space versus weeding out spoof samples which are likely to be close in structure with respect to the natural space. With limited training samples, the need for generalization calls for an expansion of the hyper-sphere (or a reduction of α), while the urge to weed out almost all spoof samples as outliers, demands a compaction or a contraction of the hyper-sphere (or an increase in α). Either way there will be mis-classifications either in the form of false positives or in the form of false negatives. Somewhere in between there is compromise and this optimal trimming factor was found to be $\alpha = 10\%$, as the outlier fraction. This is visible in Table 2, wherein best results are obtained for a trim factor of $\alpha = 10\%$. For a specific inlier (genuine space) training fraction $x\%$ (viz. along a specific row in Table 2), the Equal error rate (EER) decreases and then increases when α is varied from 5% to 20%, with the minima hovering around $\alpha = 10\%$. Note that in this table no auto-population is done using the random scans. The EER therefore is slightly on the higher side for $x = 50\%$, 50% natural face samples for training, wherein the minimum EER (corresponding to $\alpha = 10\%$) was found to be 2.25%.

4.1 Auto-Population Results and Comparisons

It is natural to deploy the proposed random scan tool to derive statistically equivalent but identity independent representations of the same natural facial profile. Thus, every natural face training image is converted into an ensemble of scans which carry equivalent statistical information pertaining to the lower order pixel correlation profile. Since the walk is randomized each realization of the parent image is distinct and provides a unique perspective. Results are therefore expected to improve considerably with this form of auto-population. The effective number of training samples is magnified by a significant amount, viz. by a scale factor N_{SCAN}. Impact of different ensemble sizes (or scale factor N_{SCAN}) is shown in Table 3. Note that $N_{SCAN} = 1$, corresponds to results without auto-population and the EER numbers are expected to drop from left to right along a specific row. Saturation is expected beyond a certain point as the additional scans carry no new information for characterizing the inlier space. For $x = 50\%$, 50% face training, the lowest EER is obtained for $N_{SCAN} = 20$, highlighted in bold in Table 3, with a percentage of $EER = 0.43\%$, which is way below the number obtained in the same row corresponding to $N_{SCAN} = 1$, which is, $EER = 2.25\%$.

A fair comparison is possible only when the state of the art algorithms are compared on an image analysis front (with or without implicit auto-population) but applied to the 3DMAD database. It is unfair to compare video processing algorithms which attempt to detect liveliness in faces by examining wrinkle and crease line dynamics to track consistency in emotional transitions of subjects. The only paper that fits this constraint is the original work by Erdogmus et al. [2]. Both the random scan versions of the proposed algorithm with and without auto-population out-perform the state of the art. This validates the identity independent paradigm (Table 4).

Table 3. Performance with optimal trim factor, $\alpha = 10\%$ and auto-population using the proposed random scan algorithm. EER results saturate beyond a certain point.

Inlier ($x(\%)$)/ Outlier samples ($y(\%)$)	EER@ $\alpha = 10\%$				
	$N_{Scan} = 1$	$N_{Scan} = 5$	$N_{Scan} = 10$	$N_{Scan} = 20$	$N_{Scan} = 30$
$x = 30\%, (1-x) = 70\%, y = 100\%$	5.0115	2.0163	2.1085	2.1059	2.1426
$x = 40\%, (1-x) = 60\%, y = 100\%$	3.4048	1.2387	0.4527	0.4560	0.3486
$x = 50\%, (1-x) = 50\%, y = 100\%$	2.2594	**0.9489**	**0.6657**	**0.4310**	**0.4510**
$x = 60\%, (1-x) = 40\%, y = 100\%$	1.3078	0.5489	0.2626	0.2441	0.2605
$x = 70\%, (1-x) = 30\%, y = 100\%$	0.1852	0.1131	0.0871	0.0776	0.0731

Table 4. Comparison with the state of the art, which has used the 3DMAD dataset as a sequence of images. Training fraction, 50% from the natural face space.

Algorithm	Classifier	EER %
Ergodomus et al. [2]	SVM	4.92
Proposed random scan (NO auto-population $N_{SCAN} = 1$)	SVM	**2.25**
Proposed random scan (with auto-population $N_{SCAN} = 20$)	SVM	**0.4310**

5 Conclusions

This paper proposes an identity independent paradigm for facial anti-spoofing based by deploying 2-D random walks, to preserve the lower order pixel correlation in images, while dissolving the identity of subjects. With the suppression of perceptual interference, stemming from this form of constrained shuffle of pixels, results have improved significantly, in relation to the state of the art techniques. The EER rates with and without auto-population for this identity independent frame have been found to be 2.25% and 0.45% respectively.

References

1. Arashloo, S.R., Kittler, J., Christmas, W.: An anomaly detection approach to face spoofing detection: a new formulation and evaluation protocol. IEEE Access **5**, 13868–13882 (2017)
2. Erdogmus, N., Marcel, S.: Spoofing face recognition with 3D masks. IEEE Trans. Inf. Forensics Secur. **9**(7), 1084–1097 (2014)
3. Feng, L., et al.: Integration of image quality and motion cues for face anti-spoofing: a neural network approach. J. Vis. Commun. Image Represent. **38**, 451–460 (2016). http://www.sciencedirect.com/science/article/pii/S1047320316300244
4. Galbally, J., Marcel, S.: Face anti-spoofing based on general image quality assessment. In: 2014 22nd International Conference on Pattern Recognition (ICPR), pp. 1173–1178. IEEE (2014)
5. Garcia, D.C., de Queiroz, R.L.: Face-spoofing 2D-detection based on moiré-pattern analysis. IEEE Trans. Inf. Forensics Secur. **10**(4), 778–786 (2015)
6. Karthik, K., Katika, B.R.: Face anti-spoofing based on sharpness profiles. In: 2017 IEEE International Conference on Industrial and Information Systems (ICIIS), pp. 1–6. IEEE (2017)
7. Karthik, K., Katika, B.R.: Image quality assessment based outlier detection for face anti-spoofing. In: 2017 International Conference on Communication Systems, Computing and IT Applications (CSCITA), pp. 72–77. IEEE (2017)
8. Matias, Y., Shamir, A.: A video scrambling technique based on space filling curves (Extended Abstract). In: Pomerance, C. (ed.) CRYPTO 1987. LNCS, vol. 293, pp. 398–417. Springer, Heidelberg (1988). https://doi.org/10.1007/3-540-48184-2_35
9. Menotti, D., et al.: Deep representations for iris, face, and fingerprint spoofing detection. IEEE Trans. Inf. Forensics Secur. **10**(4), 864–879 (2015)
10. Wen, D., Han, H., Jain, A.K.: Face spoof detection with image distortion analysis. IEEE Trans. Inf. Forensics Secur. **10**(4), 746–761 (2015)
11. Zhang, Z., Yan, J., Liu, S., Lei, Z., Yi, D., Li, S.Z.: A face antispoofing database with diverse attacks. In: IEEE International Conference on Biometrics (ICB), pp. 26–31 (2012)

Automatic Attribute Profiles for Spectral-Spatial Classification of Hyperspectral Images

Arundhati Das, Kaushal Bhardwaj, and Swarnajyoti Patra(✉)

Department of CSE, Tezpur University, Tezpur 784028, India
{arundha,kauscsp,swpatra}@tezu.ernet.in

Abstract. Attribute profiles integrate spectral and spatial information present in an image. The construction of attribute profiles is based on attribute filtering which requires proper threshold values. In the literature, only a few approaches are available to automatically detect the threshold values. Among them, the recently presented state-of-the-art method overcomes the prior limitations but is computationally demanding. In this paper, we present a simple and computationally efficient method to detect the threshold values automatically. The proposed method obtains the candidate threshold values directly from the tree representation of the image and selects suitable threshold values in two stages. In the first stage, we separate the larger attribute values to preserve important components and then the lower attribute values are clustered in the second stage to detect the final threshold values. Using these threshold values attribute profiles are constructed for two real hyperspectral data sets considering three different attributes. The experimental results demonstrate that the proposed method is effective and faster than the state-of-the-art method.

Keywords: Hyperspectral image · Attribute profiles · Threshold selection · Spectral-spatial classification

1 Introduction

Hyperspectral images (HSIs) are acquired in hundreds of contiguous channels with a fine spectral resolution that allow accurate class-discrimination among the surface objects. This classification accuracy is further improved by integrating the spectral and spatial information of the HSI. An effective way of integrating spectral and spatial information is by constructing attribute profiles (APs). The characterization of spectral-spatial information with APs has gained tremendous popularity over the last decade [3,6]. APs are concatenation of original image and the filtered images obtained by applying attribute filters (AFs). AFs process a given image based on its connected components while preserving the geometry

The RPS-NER research grant from AICTE, New Delhi supports this work in part.

B. Deka et al. (Eds.): PReMI 2019, LNCS 11941, pp. 13–21, 2019.
https://doi.org/10.1007/978-3-030-34869-4_2

of the objects. It can merge the objects to their background based on comparison of values for any characteristic (attribute) that can be computed for a connected component with a predefined threshold value [10]. Several attributes related to the shape, scale or gray-level of the image are suggested in the literature [6,10]. A sequence of threshold values helps in multiscale characterization of the image providing a set of filtering results which on concatenation along with the original image create an attribute profile (AP) [3]. For hyperspectral images an extended AP (EAP) is generated by concatenating the APs constructed for each of its component images in the reduced dimension [4,6].

One major issue in the construction of APs is the selection of proper threshold values employed during attribute filtering. To address this issue, in the literature few methods attempted to construct large profile and select the informative filtered images from it [1,9]. In these approaches we need to create the large profile considering manual sampling of large number of threshold values. With a goal for creating low dimensional APs having sufficient spectral-spatial information, a small number of techniques are present in the literature that detect threshold values automatically [2,5,7,8]. In [7] an interesting approach is presented that is sensitive to the preliminary clustering or classification for detection of candidate threshold values. Approaches in [5] and [8] are concentrated on single attributes namely area and standard deviation respectively. Another interesting approach is presented by Cavallaro *et al.* in [2] that overcomes the prior limitations and automatically detects threshold values by exploiting regression to approximate the curve generated after computing a measure for all the possible threshold values. This method is computationally demanding.

In this paper, we propose a simple and computationally efficient method to detect the threshold values automatically. The proposed method represents the given image using a tree structure. This tree structure represents the nested connected components of the image which can be exploited for obtaining attribute values as candidate threshold values. Then, on the unique candidate threshold values two-stage clustering operation is performed. In the first stage to preserve important connected components the larger attribute values are separated from the lower attribute values that mostly represent noise. In the second stage the lower attribute values are clustered into required number of groups whose representatives are selected as final threshold values to construct APs. Such constructed APs are compact in size and represent sufficient spectral-spatial information. Experiments are conducted on two real hyperspectral images that confirm the effectiveness and efficiency of the proposed method over the state-of-the-art method.

2 Basics of Attribute Profiles

The attribute profiles are constructed based on the connected operators called attribute filters (AFs). AFs process a gray-scale image by filtering the connected components (CCs) from the image. For this, the gray-scale image is represented with the help of a tree structure. Among the several tree representations available in literature, max tree (min tree) is considered in this paper [10]. Next, the nodes of the tree are filtered based on a criterion formulated considering an attribute and a threshold value. Finally, the filtered image is obtained from the filtered tree. As introduced in [10], AFs when uses max-tree representation for filtering bright objects is termed as attribute thinning; whereas, its dual attribute thickening uses min-tree representation to filter dark objects. An AP consists of the original gray-scale image I and the attribute thinning γ_{λ_i} and thickening ϕ_{λ_i} results obtained considering a sequence of threshold values λ_i. It is defined as:

$$AP(I) = \{\phi_{\lambda_t}(I), \phi_{\lambda_{t-1}}(I), ..., \phi_{\lambda_1}(I), I, \gamma_{\lambda_1}(I), \gamma_{\lambda_2}(I), ..., \gamma_{\lambda_t}(I)\} \quad (1)$$

For spectral-spatial analysis of an HSI H, an EAP is created by concatenating APs constructed on first ℓ component images (C_j) that are most informative. An EAP can be formulated as:

$$EAP(H) = \{AP(C_1), AP(C_2), ..., AP(C_\ell)\} \quad (2)$$

3 Proposed Method

In this paper, we propose a simple and computationally efficient method for automatic selection of the suitable threshold values to construct attribute profiles by overcoming the prior limitations of the state-of-the-art. The proposed architecture is shown in Fig. 1. Initially, first ℓ informative components of HSI are extracted using PCA. For each component, a max-tree as well as a min-tree is created. In a max-tree (min-tree) the nodes at different level depict the nested CCs of the gray-level image with the whole image at the root and the smallest CCs at the leaves. For the constructed trees, separate sets of threshold values are automatically selected by the proposed method which are used to construct an AP. APs constructed for each component image are concatenated to form an EAP. The constructed EAP has rich spectral-spatial content which is fed to SVM classification.

In order to select the threshold values for each PC, the tree representation of the given PC is processed to retrieve the attribute values of each node. These

---

Algorithm 1. Proposed threshold selection for attribute filtering of an image

---

Input: Tree T, Attribute A, Number of thresholds K
Output: Set of K thresholds.

1: Obtain the attribute value for each node of the tree T and store in a vector V.
2: Retain only the unique values of V in sorted order.
3: Cluster the vector V into two groups: $G1$ (low to medium range values) and $G2$ (higher range values).
4: Cluster $G1$ into K groups.
5: Chose representatives from each of the K groups and return as thresholds.

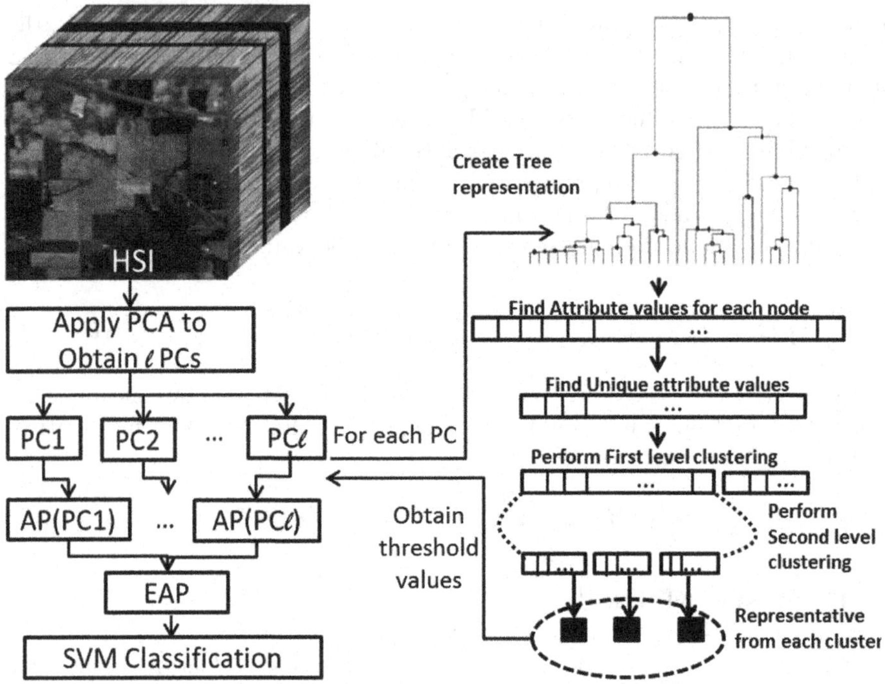

Fig. 1. Proposed architecture of automatic attribute profiles for HSI. Figure shows for each PC of HSI suitable threshold values are automatically obtained by exploiting the tree representation of component image and clustering technique.

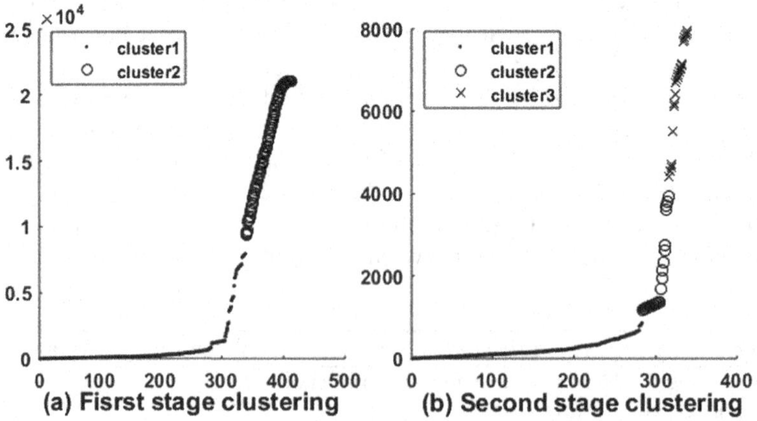

Fig. 2. The results of two-stage clustering for first PC of the Indian Pines data set considering area attribute. (a) All unique attribute values are clustered into two groups; (b) The first group of attribute values is clustered again into three groups to detect three thresholds (one from each cluster).

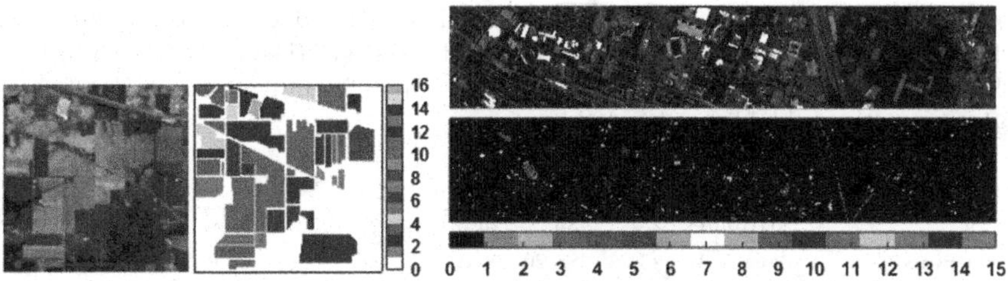

(a) Indian Pines (b) University of Houston

Fig. 3. Color composite and available reference ground truth for the HSIs (a) Indian Pines and (b) University of Houston.

attribute values are stored in a vector from which only unique attribute values are retained for further processing. Among the unique attribute values the smaller ones come from the CCs which are mostly noise whereas the larger ones depict important CCs that need to be preserved. In the first stage of the proposed method, we separate the larger attribute values from the smaller ones by applying a clustering algorithm to secure that the larger CCs do not get filtered. Figure 2(a) shows two groups of attribute values obtained for the max-tree corresponding to first PC of a real HSI with low to medium range of attribute values in one group and higher attribute values in the other group. In the next stage, only the group having lower attribute values is considered for further processing. This group is re-clustered into K groups where K is the number of required thresholds. An example with three clusters is shown in Fig. 2(b). The representatives of each cluster are chosen as final threshold values. Algorithm 1 summarizes the proposed method. For choosing representatives from each cluster the centroid of each cluster is a suitable option. However, the first cluster in the second stage may have extremely small values because of which it may not incorporate much spatial information, so the maximum attribute value from the first cluster and the centroids of the rest of the clusters are considered as threshold values for constructing APs.

4 Experimental Results

The effectiveness of the proposed method is assessed by the experimental results obtained on two different HSI data sets. The first HSI[1] is collected by the sensor AVIRIS. The imagery is from an agricultural land situated at Indian Pines, USA. Its size is 145 × 145 pixels with 20 m spatial resolution. The total number of spectral bands available for use after preprocessing is 200. The spectral coverage is ranging from 400–2500 nm. The second HSI data set[2] is collected by the sensor CASI. The imagery includes the campus of University of Houston spreading over

[1] Accessible at http://engineering.purdue.edu/~biehl/MultiSpec.
[2] Accessible at http://hyperspectral.ee.uh.edu.

Table 1. Average overall accuracy ($\overline{OA}$), related standard deviation (std) and kappa coefficient (kappa) for profiles constructed by the state-of-the-art and the proposed method considering three different attributes and two different hyperspectral data sets. The best results are highlighted in boldface.

Data set			EAP_{CC}		EAP_P		EAP_G		EAP_{prop}	
	Attributes		25	35	25	35	25	35	25	35
Indian Pines	Area	$\overline{OA}$	83.26	88.20	96.07	96.09	95.88	96.05	**96.60**	**96.91**
		kappa	0.8086	0.8652	0.9551	0.9554	0.9530	0.9550	**0.9612**	**0.9648**
		std	0.4167	0.3118	0.3345	**0.1752**	**0.2162**	0.2458	0.3031	0.2659
	DBB	$\overline{OA}$	82.20	87.19	93.25	96.24	89.20	95.92	**95.30**	**96.62**
		kappa	0.7964	0.8536	0.9230	0.9571	0.8767	0.9534	**0.9464**	**0.9614**
		std	0.2918	0.4421	**0.2125**	0.2322	0.3815	**0.1848**	0.2706	0.2979
	SD	$\overline{OA}$	86.28	93.15	**94.96**	94.55	94.10	94.35	94.06	**95.87**
		kappa	0.8433	0.9218	**0.9425**	0.9378	0.9326	0.9355	0.9322	**0.9529**
		std	0.4155	0.2944	**0.3020**	0.3705	0.3200	0.2003	0.3095	**0.1616**
University of Houston	Area	$\overline{OA}$	88.73	91.54	96.92	**97.57**	95.35	96.83	**97.08**	97.28
		kappa	0.8782	0.9086	0.9667	**0.9737**	0.9498	0.9657	**0.9684**	0.9706
		std	0.3396	0.1882	**0.1874**	**0.1863**	0.2703	0.2830	0.1881	0.1909
	DBB	$\overline{OA}$	88.56	90.93	93.68	97.36	91.48	96.46	**95.44**	**97.58**
		kappa	0.8763	0.9019	0.9316	0.9714	0.9079	0.9617	**0.9506**	**0.9739**
		std	**0.1815**	0.2598	0.3395	**0.1580**	0.2106	0.2470	0.2375	0.1949
	SD	$\overline{OA}$	91.29	94.89	95.06	95.85	93.26	95.53	**96.02**	**96.11**
		kappa	0.9058	0.9447	0.9465	0.9551	0.9270	0.9517	**0.9570**	**0.9579**
		std	0.2117	0.3947	0.2271	0.2672	**0.1581**	**0.2047**	0.1806	0.2297

some of its neighboring urban area in Texas, USA. The imagery has 144 spectral bands between the range 380–1050 nm. Each image is of 349 × 1905 pixels and 2.5 m spatial resolution. Figure 3 shows false color image alongside the related map showing the available reference samples. The proposed method is tested using the attributes *area*, *diagonal of bounding box* (DBB) and *standard deviation* (SD). The corresponding EAPs are constructed on the HSIs after reducing their dimension using principal component analysis (PCA) and considering the first 5 PCs those correspond to most of the cumulative variance in the original HSI data. The constructed profiles obtained by the proposed method are compared to those obtained by the recent state-of-the-art [2]. The different measures used in [2] (*number of connected components*, *pixel count* and *sum of gray-values*) generate different profiles referred as EAP_{CC}, EAP_P and EAP_G respectively.

For classification we have employed a one-against-all support vector machine (SVM) classifier with radial basis function (RBF) kernel. The SVM parameters $\{\sigma, C\}$ are obtained by applying grid search with five-fold cross-validation. The SVM is trained with 30% randomly selected labeled data for each class whereas the rest 70% labeled data are used for testing. The experimental results are reported after running ten times to nullify random effects of the results and taking average of the quality indices namely overall accuracy ($\overline{OA}$), the average kappa coefficient (kappa) and the standard deviation (std). The experiments are

Table 2. Computational time in seconds required to construct EAP of 35 in size by proposed and state-of-the-art method.

Attribute	Indian Pines				University of Houston			
	t_{CC}	t_P	t_G	t_{prop}	t_{CC}	t_P	t_G	t_{prop}
Area	874	869	854	17	8068	8258	8253	466
DBB	384	377	393	17	7525	6181	7632	463
SD	30859	31349	31442	18	1097565	1007517	1087365	564

carried out using 64-bit Matlab (R2015a) running on a workstation with CPU Intel(R) Xeon(R) 3.60 GHz and 16 GB RAM.

Table 1 demonstrates the classification results of the EAP_{prop} (created using the proposed method) and the EAP_{CC}, EAP_P and EAP_G considering two different HSI data sets. The profiles for both the proposed and state-of-the-art are constructed considering the first two (leading to profile size 25) and the first three (leading to profile size 35) automatically detected thresholds. One can see from the table that the accuracies delivered by the EAP_{prop} are mostly better than the EAP_{CC}, EAP_P and EAP_G in both the data sets. This signifies the potentiality of the proposed method in automatically generating compact as well as informative attribute profile for HSI classification. One can see that the first two thresholds are able to incorporate most of the spatial information because of which the next thresholds could incorporate only little additional information. This is visible from the difference of accuracies obtained considering two and three thresholds.

The results on computational time again show the advantage of the proposed technique over the state-of-the-art method. Table 2 demonstrates the computational time needed for construction of attribute profile of size 35 using both the recent state-of-the-art method and the proposed method. Time required for constructing EAP_{CC}, EAP_P, EAP_G and EAP_{prop} is indicated by t_{CC}, t_P, t_G and t_{prop} respectively. From the table, one can see that for all the attributes in both the data sets the proposed method is much faster than the state-of-the-art method. This confirms that the proposed method is more efficient than the state-of-the-art method. Note that, more time is needed by the state-of-the-art method for creating GCF after computing a measure of interest for all possible threshold values and employing regression to approximate the GCF curves. The time is also proportional to the number of candidate thresholds. This is visible from the time required in case of attribute SD in Table 2. In contrast, the proposed technique employs only a simple two-stage clustering to detect the threshold values. This makes the proposed technique faster.

Parameter Sensitivity: The proposed method employs two stage clustering for which different combinations of clustering algorithms are tested. Although, all the combinations have nominal difference, the best results as reported in Table 1 are obtained when hierarchical agglomerative clustering with single linkage is used at first stage and K-means is used at second Stage. For choosing representatives from each cluster, best results are obtained when we consider maximum attribute value in the first cluster and centroid in the rest.

5 Conclusion

This paper introduces a simple and computationally efficient method for automatically selecting threshold parameter values to construct attribute profiles. The proposed method automatically obtains the candidate threshold values exploiting the tree representation of an image and selects suitable threshold values in two stages. In the first stage, the significant connected components that possess higher attribute values are preserved by separating them from the lower values which mostly represent noise. In the second stage, the final threshold values are obtained by clustering the lower attribute values and considering their representatives. The APs constructed using these detected thresholds are compact in size and possess sufficient spectral-spatial information. Experiments are carried out on two real HSI data sets. The experimental results show that the proposed method has mostly better accuracy and is much faster than the state-of-the-art method.

References

1. Bhardwaj, K., Patra, S.: An unsupervised technique for optimal feature selection in attribute profiles for spectral-spatial classification of hyperspectral images. ISPRS J. Photogramm. Remote Sens. **138**, 139–150 (2018)
2. Cavallaro, G., Falco, N., Dalla Mura, M., Benediktsson, J.A.: Automatic attribute profiles. IEEE Trans. Image Process. **26**(4), 1859–1872 (2017)
3. Dalla Mura, M., Benediktsson, J.A., Waske, B., Bruzzone, L.: Morphological attribute profiles for the analysis of very high resolution images. IEEE Trans. Geosci. Remote Sens. **48**(10), 3747–3762 (2010)
4. Das, A., Bhardwaj, K., Patra, S.: Morphological complexity profile for the analysis of hyperspectral images. In: 2018 4th International Conference on Recent Advances in Information Technology (RAIT), pp. 1–6. IEEE (2018)
5. Ghamisi, P., Benediktsson, J.A., Sveinsson, J.R.: Automatic spectral-spatial classification framework based on attribute profiles and supervised feature extraction. IEEE Trans. Geosci. Remote Sens. **52**(9), 5771–5782 (2014)
6. Ghamisi, P., Dalla Mura, M., Benediktsson, J.A.: A survey on spectral-spatial classification techniques based on attribute profiles. IEEE Trans. Geosci. Remote Sens. **53**(5), 2335–2353 (2015)
7. Mahmood, Z., Thoonen, G., Scheunders, P.: Automatic threshold selection for morphological attribute profiles. In: 2012 IEEE International Geoscience and Remote Sensing Symposium (IGARSS), pp. 4946–4949. IEEE (2012)
8. Marpu, P.R., Pedergnana, M., Dalla Mura, M., Benediktsson, J.A., Bruzzone, L.: Automatic generation of standard deviation attribute profiles for spectral-spatial classification of remote sensing data. IEEE Geosci. Remote Sens. Lett. **10**(2), 293–297 (2013)

9. Pedergnana, M., Marpu, P.R., Dalla Mura, M., Benediktsson, J.A., Bruzzone, L.: A novel technique for optimal feature selection in attribute profiles based on genetic algorithms. IEEE Trans. Geosci. Remote Sens. **51**(6), 3514–3528 (2013)
10. Salembier, P., Oliveras, A., Garrido, L.: Antiextensive connected operators for image and sequence processing. IEEE Trans. Image Process. **7**(4), 555–570 (1998)

Enhanced Depression Detection from Facial Cues Using Univariate Feature Selection Techniques

Swati Rathi[1(✉)], Baljeet Kaur[1,2], and R. K. Agrawal[1]

[1] School of Computer and Systems Sciences, JNU, Delhi, India
swatirathi362@gmail.com, baljeetkaur26@hotmail.com, rkajnu@gmail.com
[2] Hansraj College, University of Delhi, Delhi, India

Abstract. Timely detection of depression and the accurate assessment of its severity are the two major challenges that face the medical community. To assist the clinicians, various objective measures are being explored by researchers. In literature, features extracted from the images or videos, are found relevant for detection of depression. Various feature extraction methods are suggested in literature. However, the high dimensionality of the features so obtained provide an overfitted learning model. This is handled in this work with the help of three popular univariate filter feature selection methods, which identify the reduced size of relevant subset of features. The combinations of univariate techniques with well-known classification and regression techniques are investigated. The performance of classification and regression techniques improved with the use of feature selection methods. Moreover, the proposed model has outperformed most of the video-based existing methods for identifying depression and determining its level of severity.

Keywords: Classification · Depression · Motion History Image · Regression · Univariate feature selection · Visual features

1 Introduction

Depression is a psychological disorder attributed to the presence of low mood and disinclination towards routine activities for a period generally longer than two weeks. It negatively impacts a persons well-being and is known to increase the risk of suicidal tendencies [1]. Of the 300 million people affected by depression globally, 57 million people (18%) belong to India [18]. Depression may be self-assessed using Patient Health Questionnaire (PHQ) [13] or may be determined through clinical interviews, which are based on Hamilton Depression Rating Scale [17]. These methods suffer from subjectivity and bias as they depend upon the honesty and willingness of the patient during interaction, and the clinicians ability to interpret the subjects response [25]. According to WHO, more than half of those affected by depression are misdiagnosed, thereby, increasing the false

B. Deka et al. (Eds.): PReMI 2019, LNCS 11941, pp. 22–29, 2019.
https://doi.org/10.1007/978-3-030-34869-4_3

positives and false negatives [1]. Hence, it is desirable to build a decision support system to aid the clinicians in making an accurate depression assessment, based on objective behavioral markers such as, the speech properties, facial expressions and body gestures which are less likely to be suppressed by the patient [10].

Research community has studied the role of non-verbal behavior like facial emotions, speech and semantic information for depression diagnosis and have established their correlation with depression [4,5,15]. Some of the suggested methods for depression detection are unimodal [19,29], while others are multimodal, that combine two or more modalities [5,15]. Though the multimodal systems perform better, they entail high time and space complexity. Moreover, acquisition of data from multiple modalities incurs high cost. Hence, it is desirable to build a unimodal depression detection system that is cost effective, simple and efficient, in terms of acquisition as well as building the model.

Many patients suffering from depression are either unable to articulate their feelings or hesitate in discussing them with the clinician. In such situations, facial expressions used for determining the human emotions [24] can be helpful. Mehrabian et al. [14] stated that in our day to day interaction, facial expressions are responsible for 55% of the total information exchange, while language is responsible for only 7% of the daily interaction. Hence, in our work, we propose to strengthen the unimodal depression detection system based on only the facial cues obtained through video recordings.

In literature, various features from the face have been extracted and studied for their correlation with depression [21]. Recently, Pampouchidou et al. [20], proposed a novel method: Landmark Motion History Image (LMHI) to represent the movement of facial landmarks across the video frames. They concluded that Histogram of Oriented Gradient (HOG) features extracted from the LMHI: FaceHOGs are the most relevant features for depression classification, with an F1 score of 0.5 and 0.9, for detecting the depressed and the non-depressed category respectively. The experiment had been performed on the Distress Analysis Interview Corpus - Wizard of Oz (DAIC-WOZ) dataset [23]. The high dimensionality of the FaceHOG feature set may be the reason for the low F1 depressed score. This can be handled effectively using feature selection techniques.

To the best of our knowledge, application of feature selection has not been explored much for depression detection using videos. In our work, we investigate three popular univariate filter feature selection techniques: Fisher Discriminant Ratio (FDR) [8], Mutual Information (MI) [7] and Pearson Correlation (PC) [22], to find the relevant set of FaceHOG features [20]. It is well-known that the learning algorithm plays a key role in the development of the decision model to achieve high performance. Since filter feature selection determines the relevant features independent of the learning algorithm, it becomes important to investigate which combination of univariate feature selection method and learning algorithm provides the maximum performance. Hence, we explored four well-known classifiers: Decision Tree (DT), Linear Discriminant Analysis (LDA), k Nearest Neighbor (KNN) and Support Vector Machine (SVM) [8]. In literature, many of the depression detection models are built as regression problem. Hence, we have also investigated four well-known regression techniques for the determination of depression severity level: Decision Tree (DTR), Linear (LR), Partial Least Square (PLSR) and Support Vector (SVR) [8]. Through exhaustive experiments, we determine the most suitable feature selection technique, the best performing classification/regression technique and the best combination of the univariate feature selection method and the classification/regression technique. Section 2 summarizes the related work and Sect. 3 presents the experiments and results. Finally, Sect. 4 concludes our work and gives future directions.

2 Related Work

Facial cues have been widely studied in correlation with the mental state of the person [9]. Cohn et al. [4] extracted features using facial action units and Active Appearance Model for prediction of depression. Meng et al. [15] captured facial dynamics from videos in a Motion History Histogram (MHH) image and computed Local Binary Pattern (LBP) and Edge Oriented Histogram (EOH) features for the depression prediction. Cummins et al. [5] extracted Space-Time Interest Points and Pyramid of HOG features from the videos to estimate depression. Jan et al. [12] extracted LBP, EOH and Local Phase Quantization (LPQ) features for each video frame and captured their change across frames using 1-D MHH. Dimension of the resulting 1-D MHH features was reduced using Principal Component Analysis. In [6], facial and head movements were computed from the facial landmarks of a video and Min-Redundancy Max-Relevance method was used to select the relevant features. Nasir et al. [16] computed polynomial parameterization of the visual features and reduced their dimensionality using Mutual Information Maximization, for depression classification. Yang et al. [30] proposed Histogram of Displacement range method to compute features from the facial landmarks. Combination of Deep Convolutional Neural Network (DCNN) and deep neural network model was used for depression detection. Pampouchidou et al. [19] compared the performance of LMHI with Motion History Image and Gabor-inhibited LMHI. LBP, HOG and LPQ features were extracted from these images for depression prediction. Zhu et al. [31] extracted appearance features from static frames of video and motion features from the optical flow images using DCNN for depression detection. Jazaery et al. [2] used 3D CNN and Recurrent Neural Network to learn the spatio-temporal features from videos for depression detection.

Fig. 1. Landmark Motion History Image of a healthy person with Id 310

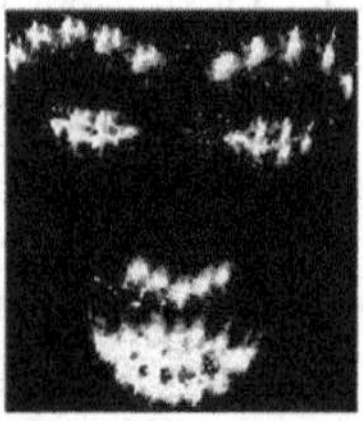

Fig. 2. Landmark Motion History Image of a depressed person with Id 321

3 Combination of Univariate Feature Selection and the Learning Algorithms

Pampouchidou et al. [20] extracted many visual feature sets viz. FaceLBP, FaceHOG, head motion, blinking rate etc. using the 2D facial landmarks given for each video frame. Using the method [20], we construct the LMHI of size 252×248 pixels, that represents the motion of 2D facial landmarks. Figures 1 and 2 are two example LMHI. To extract the HOG features from the image, it is divided into cells of size 32×32 pixels each and, gradient is computed for each pixel of the cell. A 9 bin histogram is then created to represent the contribution of all the pixels in a cell. To account for illumination variance, histogram normalization is done on blocks of size 2×2 cells. Cell histograms from all the blocks are concatenated to form the FaceHOG feature set of dimension 1296. Pampouchidou et al. concluded that FaceHOG features are the most relevant for classifying depression with an F1 Score of 0.5 and 0.9 for the depressed and non-depressed category respectively. However, dimensionality of the FaceHOG features is high in relation to the number of samples in the DAIC-WOZ dataset. This may cause overfitting of the decision model [3] and can possibly be the reason for a low F1 score for identifying the depressed individuals. To circumvent this problem, it is imperative to reduce the dimension of FaceHOG feature set. Using only a relevant subset of the features, helps to enhance the performance of the decision system. Several filter and wrapper methods are suggested in literature for feature selection [11]. Filter feature selection methods are much simpler than the wrapper approaches and help build a cost-effective system in terms of time and space. To our best knowledge, the univariate filter feature selection techniques have not been explored much for video-based depression detection. In this work, we determine a relevant subset of the FaceHOG features by using the three univariate filter feature selection techniques: Fisher Discriminant Ratio (FDR) [8], Mutual Information (MI) [7] and Pearson Correlation (PC) [22]. After obtaining the subset of relevant features, we apply four well-known classification techniques: DT, LDA, KNN and SVM, and four regression techniques: DTR, LR, PLSR and SVR, to determine the best combination of feature selection and learning algorithm, for effective depression identification (classification) and its severity estimation (regression) respectively.

4 Experiments and Results

All the experiments have been performed on the DAIC-WOZ dataset [23]. The data given had been partitioned into the training, development and test sets. Classification labels (depressed or non-depressed) and PHQ-8 scores (for regression) were given for all the sets except the test set, hence we train our model on the training set (107 samples) and test its efficacy on the development set (35 samples). Each univariate filter feature selection method, computes the relevance of each FaceHOG feature w.r.t. the response variable and ranks them in the descending order of their relevance score. The decision model is learned in the order of the ranked features incrementally and the minimum number of features that give the best performance are finally selected (#).

Table 1. Classification comparison

	Metric	WFS[a]	FDR	MI	PC
DT	F1 dep	0.40	**0.54**	0.44	0.44
	F1 ndep	0.67	**0.79**	0.72	0.72
	#	1296	125	**11**	39
LDA	F1 dep	0.18	**0.70**	0.62	0.64
	F1 ndep	0.62	**0.85**	0.73	0.76
	#	1296	**116**	209	495
KNN	F1 dep	0.23	**0.61**	0.40	0.54
	F1 ndep	0.73	**0.81**	0.76	0.79
	#	1296	**23**	90	685
SVM	F1 dep	0.60	**0.61**	0.50	0.58
	F1 ndep	**0.80**	0.77	0.73	0.78
	#	1296	**4**	43	5

[a] Without Feature Selection

Table 2. Regression comparison

	Metric	WFS[a]	FDR	MI	PC
DTR	MAE	6.58	**4.97**	5.12	5.05
	RMSE	7.99	**6.38**	6.61	6.8
	#	1296	216	**9**	33
LR	MAE	11.5	4.93	5.45	**4.64**
	RMSE	14.6	6.06	6.64	**5.98**
	#	1296	10	**3**	18
PLSR	MAE	5.53	**5.09**	5.19	5.15
	RMSE	6.84	6.41	6.76	**6.31**
	#	1296	24	61	**1**
SVR	MAE	5.45	**5.11**	5.2	5.15
	RMSE	6.73	**6.5**	6.62	6.53
	#	1296	792	364	**154**

[a] Without Feature Selection

Performances of the three feature selection techniques (FDR, MI, PC), in conjunction with the four classification techniques (DT, LDA, KNN, SVM) and the four regression techniques (DTR, LR, PLSR, SVR) are compared in Tables 1 and 2 respectively. The classification performance is shown in terms of F1 depressed score (F1 dep), F1 non-depressed score (F1 ndep) and the regression performance is shown in terms of Mean Absolute Error (MAE), Root Mean Square Error (RMSE). On application of the three univariate methods, the minimum number of features (#) for which maximum performance is achieved (F1 dep for classification and MAE for regression), is recorded. With each learning method, feature selection technique that gives the best performance is highlighted in bold. Following are the observations based on Table 1:

- Application of feature selection improves the performance of the model, except in the case of MI and PC with SVM.
- For each classification technique, FDR outperforms MI and PC in terms of F1 dep, F1 ndep and the number of selected features (except for DT).
- For each feature selection technique, LDA outperforms all other classifiers.
- FDR followed by LDA gives the best F1 Score (depressed and non-depressed). Both FDR and LDA are based on the Fisher criterion. However, FDR is unable to perform feature combination like LDA and LDA is unable to discard the irrelevant features like FDR. They complement each other and the combination gives better performance.

Following observations are based on Table 2:

- Use of feature selection improves the performance of all regression methods.
- FDR outperforms MI and PC in combination with all the regression techniques except with LR.

Table 3. Comparison of the proposed model with the state-of-the-art

Classification F1 depressed (F1 Non-Depressed)	Regression MAE (RMSE)
0.70 (0.85) Proposed	4.60 (5.90) [26]
0.63 (0.89) [16]	**4.64 (5.98) Proposed**
0.62 (0.77) [26]	5.33 (6.45) [28]
0.53 (mean) [28]	5.88 (7.13) [23]
0.50 (0.90) [20]	6.47 (7.86) [16]
0.50 (0.90) [27]	

- PC based feature selection followed by LR gives the best performance in terms MAE and RMSE. PC selects those features which have high degree of linear correlation with the response variable, and LR models the linear relationship between features and the response variable. Hence, the combination of the two provides better performance.

Table 3 compares the best results of the proposed combination technique with the existing methods for depression detection based on visual cues. The proposed combination of FDR and LDA outperforms all the classification models suggested in literature. For, depression severity estimation, the combination of PC and LR outperforms most of the existing regression models.

5 Conclusion

Features captured from video data are relevant for depression detection. They are a strong contender as a unimodal technique, that is capable of supporting clinicians for monitoring patients and correctly assessing the severity of their problem. Due to the high dimensionality of the extracted features from videos, the complexity of the decision models built for depression detection is high. Also, it provides an overfitted learning model. To circumvent this, we have employed univariate filter feature selection methods to reduce the dimensionality of features required to build the depression detection systems. Four well known classifiers and four regression methods have been successfully explored in combination with the feature selection techniques, and the role of feature selection has been emphasized. The relevant features obtained using FDR are transformed by the LDA classifier making the combination of FDR and LDA most appropriate for video-based depression classification. To diagnose the depression severity, PC in combination with the LR has been found to be the most suitable as both PC and LR are based on the linear correlation between the features and the response variable. The proposed combinations for classification and regression for video-based depression detection, outperform most of the existing results.

Future work will focus on reducing the dimension of other visual features obtained from the video data, and identify those features which are relevant for the task of depression detection. We will also explore advanced feature selection techniques which will not only eliminate irrelevant features with respect to the response variable, but also remove the redundant/correlated features.

References

1. Depression. https://www.who.int/news-room/fact-sheets/detail/depression. Accessed 29 Apr 2019
2. Al Jazaery, M., Guo, G.: Video-based depression level analysis by encoding deep spatiotemporal features. IEEE Trans. Affect. Comput. (2018). https://doi.org/10.1109/TAFFC.2018.2870884
3. Bellman, R.: Curse of Dimensionality. Adaptive Control Processes: A Guided Tour. Princeton University Press, Princeton (1961)
4. Cohn, J.F., et al.: Detecting depression from facial actions and vocal prosody. In: 2009 3rd International Conference on Affective Computing and Intelligent Interaction and Workshops, pp. 1–7. IEEE (2009)
5. Cummins, N., Joshi, J., Dhall, A., Sethu, V., Goecke, R., Epps, J.: Diagnosis of depression by behavioural signals: a multimodal approach. In: Proceedings of the 3rd ACM International Workshop on Audio/Visual Emotion Challenge, pp. 11–20. ACM (2013)
6. Dibeklioğlu, H., Hammal, Z., Yang, Y., Cohn, J.F.: Multimodal detection of depression in clinical interviews. In: Proceedings of the 2015 ACM on International Conference on Multimodal Interaction, pp. 307–310. ACM (2015)
7. Ding, C., Peng, H.: Minimum redundancy feature selection from microarray gene expression data. J. Bioinf. Comput. Biol. **3**(02), 185–205 (2005)
8. Duda, R.O., Hart, P.E., Stork, D.G.: Pattern Classification. Wiley, Hoboken (2012)
9. Ekman, R.: What the Face Reveals: Basic and Applied Studies of Spontaneous Expression Using the Facial Action Coding System (FACS). Oxford University Press, Oxford (1997)
10. Ellgring, H.: Non-Verbal Communication in Depression. Cambridge University Press, Cambridge (2007)
11. Guyon, I., Elisseeff, A.: An introduction to variable and feature selection. J. Mach. Learn. Res. **3**(Mar), 1157–1182 (2003)
12. Jan, A., Meng, H., Gaus, Y.F.A., Zhang, F., Turabzadeh, S.: Automatic depression scale prediction using facial expression dynamics and regression. In: Proceedings of the 4th International Workshop on Audio/Visual Emotion Challenge, pp. 73–80. ACM (2014)
13. Kroenke, K., Spitzer, R.L., Williams, J.B.: The PHQ-9: validity of a brief depression severity measure. J. Gen. Intern. Med. **16**(9), 606–613 (2001)
14. Mehrabian, A., Russell, J.A.: An Approach to Environmental Psychology. The MIT Press, Cambridge (1974)
15. Meng, H., Huang, D., Wang, H., Yang, H., Ai-Shuraifi, M., Wang, Y.: Depression recognition based on dynamic facial and vocal expression features using partial least square regression. In: Proceedings of the 3rd ACM International Workshop on Audio/Visual Emotion Challenge, pp. 21–30. ACM (2013)
16. Nasir, M., Jati, A., Shivakumar, P.G., Nallan Chakravarthula, S., Georgiou, P.: Multimodal and multiresolution depression detection from speech and facial landmark features. In: Proceedings of the 6th International Workshop on Audio/Visual Emotion Challenge, pp. 43–50. ACM (2016)

17. Nutt, D.: The Hamilton depression scale- accelerator or break on antidepressant drug discovery? J. Neurol. Neurosurg. Psychiatry **85**, 119–120 (2014). https://doi.org/10.1136/jnnp-2013-306984
18. Organization, W.H., et al.: Depression and other common mental disorders: global health estimates. Technical report, World Health Organization (2017)
19. Pampouchidou, A., et al.: Quantitative comparison of motion history image variants for video-based depression assessment. EURASIP J. Image Video Process. **2017**(1), 64 (2017)
20. Pampouchidou, A., et al.: Depression assessment by fusing high and low level features from audio, video, and text. In: Proceedings of the 6th International Workshop on Audio/Visual Emotion Challenge, pp. 27–34. ACM (2016)
21. Pampouchidou, A., et al.: Automatic assessment of depression based on visual cues: a systematic review. IEEE Trans. Affect. Comput. (2017). https://doi.org/10.1109/TAFFC.2017.2724035
22. Pearson, K.: Notes on the history of correlation. Biometrika **13**(1), 25–45 (1920)
23. Ringeval, F., et al.: AVEC 2017: real-life depression, and affect recognition workshop and challenge. In: Proceedings of the 7th Annual Workshop on Audio/Visual Emotion Challenge, pp. 3–9. ACM (2017)
24. Sariyanidi, E., Gunes, H., Cavallaro, A.: Automatic analysis of facial affect: a survey of registration, representation, and recognition. IEEE Trans. Pattern Anal. Mach. Intell. **37**(6), 1113–1133 (2014)
25. Schumann, I., Schneider, A., Kantert, C., Löwe, B., Linde, K.: Physicians attitudes, diagnostic process and barriers regarding depression diagnosis in primary care: a systematic review of qualitative studies. Fam. Pract. **29**(3), 255–263 (2011)
26. Sun, B., et al.: A random forest regression method with selected-text feature for depression assessment. In: Proceedings of the 7th Annual Workshop on Audio/Visual Emotion Challenge, pp. 61–68. ACM (2017)
27. Valstar, M., et al.: AVEC 2016: depression, mood, and emotion recognition workshop and challenge. In: Proceedings of the 6th International Workshop on Audio/visual Emotion Challenge, pp. 3–10. ACM (2016)
28. Williamson, J.R., et al.: Detecting depression using vocal, facial and semantic communication cues. In: Proceedings of the 6th International Workshop on Audio/Visual Emotion Challenge, pp. 11–18. ACM (2016)
29. Williamson, J.R., Quatieri, T.F., Helfer, B.S., Horwitz, R., Yu, B., Mehta, D.D.: Vocal biomarkers of depression based on motor incoordination. In: Proceedings of the 3rd ACM International Workshop on Audio/Visual Emotion Challenge, pp. 41–48. ACM (2013)
30. Yang, L., Jiang, D., Xia, X., Pei, E., Oveneke, M.C., Sahli, H.: Multimodal measurement of depression using deep learning models. In: Proceedings of the 7th Annual Workshop on Audio/Visual Emotion Challenge, pp. 53–59. ACM (2017)
31. Zhu, Y., Shang, Y., Shao, Z., Guo, G.: Automated depression diagnosis based on deep networks to encode facial appearance and dynamics. IEEE Trans. Affect. Comput. **9**(4), 578–584 (2017)

Trustworthy Cloud Federation Through Cooperative Game Using QoS Assessment

Shanmugam Udhayakumar[1(✉)] and Tamilselvan Latha[2(✉)]

[1] Department of Computer Science and Engineering, Saveetha School of Engineering, Saveetha Institute of Medical and Technical Sciences, Chennai, India
mailtoudhay@gmail.com

[2] Department of Information Technology, B.S. Abdur Rahman Crescent Institute of Science and Technology, Chennai, India
latha.tamil@crescent.education

Abstract. Growing demand for cloud applications has imposed a great challenge for users in finding a suitable cloud service providers. Trustworthy cloud computing has been a promising solution for effectively judging a secure provider by monitoring the behavior of the environment. However, designing an efficient mechanism for identifying a trustworthy cloud service provider has been a problem in this heterogeneous cloud. In this paper, we address the problem through the game theoretic approach by classifying the providers, based on how they cooperate to form a coalition. Trustable characteristics like sharing of information, the formation of network community, and other quality of service parameters ensure a highly trustworthy service. These functions form the element of cooperative game theory for sharing the payoffs in a manner that a federation through coalition gets benefited. The model proves that there exists an equilibrium, where the property of core, exits and hence, any mistrusted members wishing to break the coalition are nullified.

Keywords: Trusted computing · Cooperative game theory · Cloud federation · Behavioral trust

1 Introduction

Information technology has gained a tremendous advantage in doing business through cloud computing. The service delivery model of the cloud has evolved from a simple application delivery model to complex nature where a multinational organization can be easily set up with minimal time and effort. Though cloud is being adopted extensively by many startups and small and medium enterprises, still numerous issues are encompassing it. Specifically, trust, privacy, and security are the major issues that needs to be addressed immediately for effective adoption. Privacy relates to the protection of personal data within the preview of the owner and within a designated boundary. Security is all about protecting the data from unauthorized access and destruction or modification of the data. But trust relates to how good the ecosystem

B. Deka et al. (Eds.): PReMI 2019, LNCS 11941, pp. 30–37, 2019.
https://doi.org/10.1007/978-3-030-34869-4_4

behaves as expected, and how trust can ensure the confidence among users through integrity assurance. Hence privacy and security objective could be achieved only if the environment is trustworthy. The trusted nature of the environment could be achieved through direct and indirect trust [1]. Direct trust relates to the experience of the agent who is assessing the parameters of a particular entity based on regular observations and indirect trust is all about how others can influence the information being passed.

Our proposed model takes into account direct observation of behavior for the formation of cloud federation. Federation of cloud can be achieved through the cooperation of like-minded service providers to get into a mutual agreement for the purpose of cost sharing or to provide the resources. This cost-sharing method is derived though the game-theoretic approach. Game theory is a study of mathematical models of conflict and cooperation between rational decision-makers [2]. It models the strategic scenario between the players and analyses its behavior for future coordination. Generally classified into cooperative and non-cooperative, where the later approach utilizes the competition between individual players and the former i.e., cooperative game theoretic model ensures that every player has the right to form a pre-play communication, to make any mutual agreement [3]. The agreements can be of either to improve their strategies or share the cost-benefit. Hence in an environment like cloud, competitions outcome can destroy the providers for gain. Therefore for a trustworthy cloud service, the cooperative game would be the ideal choice to adopt for mutual gain.

2 Literature Survey

Trustworthy multi-cloud communities can maximize the benefits and minimizes the misbehavior and collusion attack amongst cloud players [4]. Existence of Nash Equilibrium and revenue sharing mechanism through game theory has been considered for edge based cloud computing system [5]. A game theory based trust measurement model is proposed for social networking to solve free-riding problem through punishment mechanism [6]. In another work [7], based on multimedia application delivery model in the cloud, it aims at minimizing the penalties due to violation of service quality by any untrusted providers. The federation dynamically provides VM instances to users with Quality of Services (QoS) guarantee and satisfies fairness and stability property. An extensive survey towards trusted cloud computing focusing on security, reliability, dependability, and many more parameters related to improving the trust is carried out [8]. A trust model for assessing the cloud service providers based on their trust category and then assuring it through attestation is proposed [9, 10]. Thus it can be said that computing in cloud to identify the service providers requires in depth knowledge on game theory and its associated properties. Thereby, trustworthy cloud service providers can easily be combined to form a coalition and bind them into a joint agreement so that they do not move to form another coalition.

3 Cooperative Game Theory Model

By formulating a coalition game and define a proper payoff characteristic function for that coalition, can mitigate the insider attacks namely collusion attacks. Preventing collusion attack by forming a coalition of trusted partners in the cloud will increase the trustworthiness of players. If we can able to cooperate then there are more advantages then being alone.

3.1 Motivation for Game Theoretic Solution

(i) The trustworthy nature of cloud environment can be modeled as a coalitional game with transferable payoff (N, υ), where, N is a finite set of players, indexed, by i, and υ: $2^N \longmapsto R$. Here υ can be said as a trustworthy characteristics function that is associated with every non empty subset S of N a real number $\upsilon(S)$. It means that the function υ is the quantified trusted level each coalition S can achieve.

(ii) The objective of modeling this trusted grand coalition is

a. To define the trustworthy characteristics of every possible coalition.
b. To prove that this game will give a stable coalition outcome, where no other coalition will obtain a better trusted outcome for its members.
c. Identify malicious service providers, who could not possibly join the coalition is said to be under high suspicion.
d. Also, the coalition can be dissolved at any other later time and can regroup to form a new coalition based on the policy and imputation strategy.

3.2 Trustworthy Characteristics

Trust is an uncertain principle, where the states of any cloud service providers (CSP) are not fully identified and the information is mostly imperfect. In this scenario, the information about a provider can be assessed based on how much he has cooperated with another, during his past experience. The objective is to identify any provider's capability to interact. To avoid detection, a malicious machine does not coordinate with other peers. Trusted machines are one, who interact and coordinate for some process then can jointly agree on certain terms and conditions, which is an important principle for trust. A CSP is said to have more trust if he has **more networked peers** with him (Fig. 1).

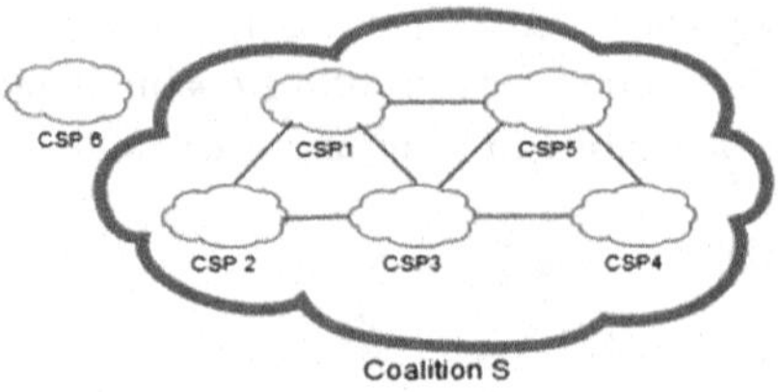

Fig. 1. Networked CSP to form Coalition S

According to game theory, every player communicates and shares the information that is observed during the previous interactions. This helps them to make an assured decision, whether or not to cooperate with other members or nodes. The third parameter to assess the trusted nature of the service provider is the maximum quality service completion nature of the provider. It implies that for every service that is being delivered before the coalition, its completion objective is assessed. The completion parameters are Timely Response, Downloaded Size, Successful Service Initialization, Successful Service Completion, Log File Stored, User's Service Satisfaction Report, Security and Privacy Objective in SLA Agreement. Any coalition game must be modeled with correct payoff value, which in our case is the trustworthy characteristics function $\upsilon(S)$. Thus our trustworthy characteristics function has three components.

Case 1: Maximum Networked Members
For N players in a network, all the possible coalition S is 2^N, i.e., $S \in 2^N$, and the number of nodes in it is |S|. Then the total possible networked members would be |S| − 1, who can at the maximum interact or get networked with a particular service provider. At any time t, the trustworthy characteristics function for the networked members are

$$N_t(S) = |S| - 1 \tag{1}$$

Case 2: Maximum Interaction to Share the Information through Cooperation
For identifying the maximum cooperation by a provider with that of his peer group, we need to identify the probability of interaction taken place. This, in turn, is the probability of every other provider giving an admission policy. Suppose for any provider *i*, it will have a log table which will contain the history of interactions made HI(*i*). In every interaction that *i* make will have the details of *j* and amount of cooperation that it has made, given in-terms of cooperation probability $P_{i,j}$. As already discussed, as the size of the coalition is high, then every player would be more tolerant and robust. So, we assume that the size is maximum. Then we can define the trustworthy characteristics function as.

$$C_t(S) = max_{j \in S}\left\{\sum i \in I \,.\, p_{ij} \middle| I = \left\{i \in S,\, i \neq j,\, p_{ij} \neq 0\right\}\right\} \tag{2}$$

The Eq. 2, defines that *i* is a player who is currently under the coalition I which in turn is a subset of S. The cooperation is between *i* and *j*, whose interaction during the past history is noted and hence the cooperation probability is assumed to be non-zero. However, a case of the initial state, when the players are new for the coalition, the trustworthy characteristic function is not determined.

Case 3: Maximum Services that have been delivered correctly
Identifying the maximum service quality completion by a provider, we need to assess various QoS parameters. Even though there are numerous QoS, many of them require verification at that instance of time. This verification if it is done by a peer provider through a standardized API, then the intention of the coalition would succeed. The purpose of the coalition S is that, to jointly agree on utility share by all the providers.

The utility can be maximized only if, every player in the coalition have the authority to assess the performance of every other player. Through this way, the players can improve, coordinate with peers and provide the best service for its customers.

$$Q_t(S) = max_{i,j \in S} \begin{cases} \sum_{k=1}^{n} \left(We \,.\, Q_{ij}^{pk}(t-\delta t) + (1-We) \,.\, Q_{ij}^{dk}(t) \right) & if\ i\ checks\ j \\ Q_{ij}^{k}(t-\delta t) & else \end{cases} \quad (3)$$

W_e, is the weight associated with the previous evaluation of the quality metrics.

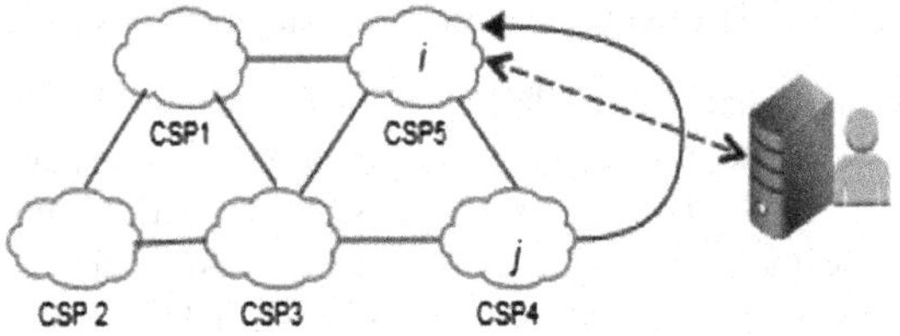

Fig. 2. CSP j assess the QoS of i for interaction with the consumer

Q_{ij}^{pk} is the previous observation of the quality score for various quality metrics k, which is evaluated at time $t - \delta t$, by j for i. If player i does not check j then the normal observation Q_{ij}^{k} of either i or j is considered, which is extremely low or neglected.

Q_{ij}^{dk} is the direct observation of the quality score for various quality metrics k, evaluated at time t, by *j* for *i*, as shown in Fig. 2. The weight associated with this component is $(1 - W_e)$. Assuming that every verification of *j*, towards *i*, leads to a positive result of a negative value. Then to evaluate it, we simply perform a normalization process of the successful positive trust value using.

$$Q_{ij}^{dk}(t) = \frac{No.\ of\ Success}{No.\ of\ Success + No.\ of\ Failure}$$

For example, for a total of 5 direct observations made at time t, if the success is 3 then 3/3 + 2 = 0.6 is the trust generated by *j* towards *i*, for a specific quality metrics say, the number of successful initialization (k_s). To evaluate,Q_{ij}, we may require to monitor the metrics associated with each provider and should be properly weighted. This quality measure $Q_{ij}(t)$ is an evaluation of quality by *j* towards *i*, estimated based on the observation made at time t. and represented as a real number in the range [0, 1], where 1 indicates complete trust and 0 indicates distrust. In case the assessment is carried by any other player other than the peer j, then the Weightage factor W_e need not be considered, just the previous observation be taken. ***Considerations:*** The size of the coalition S should be more than 1, because if S = 1 then, it means that there are no peer coalition. Moreover, Eqs. 1 and 2, will become invalid, having the trustworthy characteristics $\upsilon(S) = 0$.

If |S| = 1, then υ(S) = 0, Hence, we need a linear combination of the three metrics,

$$\upsilon(S) = \begin{cases} 0 & |S| = 1 \\ \alpha Nt(S) + \beta Ct(S) + \delta Qt(S), & |S| \geq 2 \end{cases} \tag{4}$$

Where, α, β, γ are weightage factors and $\alpha + \beta + \gamma = 1$, the weights are calculated based on relative preference over others using the mathematical algorithm of Analytical Hierarchy Processing (AHP).

4 Proof of Game Theory

Let us prove that our trustworthy characteristics function υ(S), satisfies these solution concepts for making our model a truly game theoretic solution.

Theorem 1: Individual Rationality: A player in the coalition should receive more than what he will gain by not entering into the coalition.

In order to investigate the individual rationality, we need to identify the individual payoff for a particular player *i*, before joining the coalition. Also, the share of payoff received by *i*, after joining the coalition.

Definition: For a coalition with no nodes to join the coalition, the value of |S| = 1, and hence the characteristics function for networking becomes $N_t(S) = 0$. Therefore, for an individual player, to get the maximum benefit should network with all other members of that coalition. Thus the networking function for an individual payer is

$$N_t^S(i) = \max(|S|) - 1 \tag{5}$$

We have, the cooperation parameter, where any node admitted into the coalition gets the maximum probability of cooperation, therefore

$$C_t^S(i) = max_{j \in S} \cdot p_{ij} \tag{6}$$

Similarly, for any successful transaction requires the delivery of quality of services

$$Q_t^S(i) = max_{i,j \in S}\, Q_{ij} \tag{7}$$

The imputation *x* is then the linear combination of all the individual players payoff, thus its trustworthy share is defined as

$$x_t^S(\mathrm{i}) = \frac{1}{|S|}\left(\alpha N_t^S(i) + \beta C_t^S(\mathrm{i}) + \gamma Q_t^S(\mathrm{i})\right) \tag{8}$$

The above equation proves that, the overall imputation, if divided by the total number of players in the coalition will certainly be greater than zero. Hence the payoff share for any player who is in the coalition is more than that of not being in the coalition.

Theorem 2: Core: A Set of feasible allocation or a stable state, which cannot be improved upon by another coalition.

Even though a player gets benefited through joining a coalition, rather than staying alone, the player would jump from one coalition to another.

Definition: For any transferable utility cooperative game (N, υ), where N denotes the set of players and υ is the characteristics function (in our case it is the trust function). An imputation y is dominant over the existing imputation or payoff *x*, where y is the payoff for the coalition C, where, $C \in 2^N$, such that each player in *C* prefers *y*, because $x_i \leq y_i$ for all $i \in C$. This implies that any player *i* who is getting benefited in the C, will stand to threaten C to leave the coalition. This is the player who is the most untrusted, or distrustful member. To overcome such an opportunity, there should exist equilibrium, called the *core* which is non-empty, where a set of imputations are not being dominated by any other better coalition.

Proof: The sum of the payoff of all the members in the coalition must be larger than the value of that coalition, then this equilibrium point is the core of the game.

$$\sum x_i^S(\mathrm{i}) \geq \upsilon_t(S), \text{ for all } S \in 2^N \tag{9}$$

Through this equation, we can state that the trustworthy characteristics of the cloud providers have a core and, the possibility of any provider leaving the coalition will fail, and also his chances of getting into another coalition are also very difficult.

5 Implementation and Results

A simple Image Processing Service is built to provide such a kind of cooperation by making it mandatory to select at least 3 services out of 8. Since the implementation is setup on a real-time cloud environment developed using OpenStack, the number of services being initialized is depended on the number of VM on the test system, in our case it is a Quad core CPU with 8 GB RAM. The result for the trust policy where it evaluates the three attributes to identify the cooperative nature is presented in Table 1.

Table 1. Evaluation of cooperation trust value for TIPS

Cloud service providers	Maximum networked services	Maximum interactions cooperation	Maximum services delivered correctly	Cooperation trust value
Pixlr	5/8	6/8	0.58	0.42
Fotor	4/8	5/8	0.70	0.42
PicMonke	4/8	6/8	0.7	0.44
Prisma	7/8	6/8	0.8	0.54
BeFun	3/8	4/8	0.72	0.38
OneDrive	3/8	4/8	0.58	0.34
AWS	7/8	7/8	0.76	0.54
G-Drive	6/8	7/8	0.67	0.49

Thus, all CSPs gain equally by coordinating with peer members, also the cooperation trust value dynamically changes for every interaction made. The experimental results prove that all have a balanced score with imputation policy.

6 Conclusion

Cloud has advanced to such an extent that there are numerous service providers for a single application; hence clients need to select a provider who can deliver the service as promised. Only a trusted service provider is capable and certain that his service would deliver what has been promised. To ensure the trustworthy nature of CSP, it is necessary to assess and coordinate the providers to form a coalition so that they are bound to certain conditions. Thus our work has proposed and modeled a theoretical approach for cloud federation through cooperative game theory. The payoff share and properties of core in the game, evidently describes that the players are well off if they are within the coalition agreeing to the imputation policy of the coalition. Else, they are sidelined to work it out alone, which can lead to malicious intent to subvert the coalition. In future, the model can be simulated for an application service delivery by CSP and compared for their share. Further, in a real-time open stack implementation, the model can be placed as an initial criterion for behavior assessment, where every provider must form a coalition to be rated with a trust value.

References

1. Shaikh, R., Sasikumar, M.: Trust model for measuring security strength of cloud computing service. Procedia Comput. Sci. **45**, 380–389 (2015)
2. Myerson, R.: Game Theory: Analysis of Conflict. Harvard University Press, Cambridge (1991)
3. Nisan, N., Roughgarden, T., Tardos, E., Vazirani, V.V.: Algorithmic Game Theory, pp. 385–410. Cambridge University Press, Cambridge (2007)
4. Omar, W., Jamal, B., Hadi, O., Azzam, M.: Towards trustworthy multi-cloud services communities: a trust-based hedonic coalitional game. IEEE Trans. Serv. Comput. **11**, 184–201 (2018)
5. Cao, Z., Zhang, H., Liu, B., Sheng, B.: A game theoretic framework for revenue sharing in edge-cloud computing system. In: IEEE IPCCC (2018)
6. Wang, Y., Cai, Z., Yin, G., Gao, Y., Tong, X., Han, Q.: A game theory-based trust measurement model for social networks. Comput. Soc. Netw. **6** (2016)
7. Hassan, M.M., Alelaiwi, A., Alamri, A.: A dynamic and efficient coalition formation game in cloud federation for multimedia applications. In: International Conference on Grid & Cloud Computing and Applications, pp. 71–77 (2015)
8. Chiregi, M., Navimipour, J.: Cloud computing and trust evaluation: a systematic literature review of the state-of-the-art mechanisms. J. Electr. Syst. Inf. Technol. **5**, 608–622 (2018)
9. Shanmugam, U., Tamilselvan, L.: Trusted computing model with attestation to assure security for software services in a cloud environment. Int. J. Intell. Eng. Syst. **10**, 144–153 (2017)
10. Shanmugam, U., Chandrasekaran, S., Tamilselvan, L., Fareez, A.: An adaptive trust model for software services in hybrid cloud environment. In: WSEAS International Conference on Computers. Recent Researches in Computer Science, pp. 497–503 (2011)

Incremental k-Means Method

Rabinder Kumar Prasad[1], Rosy Sarmah[2(✉)], and Subrata Chakraborty[3(✉)]

[1] Department of CSE, Dibrugarh University, Dibrugarh 786004, Assam, India
rkp@dibru.ac.in

[2] Department of CSE, Tezpur University, Tezpur 784028, Assam, India
rosy8@tezu.ernet.in

[3] Department of Statistics, Dibrugarh University, Dibrugarh 786004, Assam, India
subrata_stats@dibru.ac.in

Abstract. In the last few decades, k-means has evolved as one of the most prominent data analysis method used by the researchers. However, proper selection of k number of centroids is essential for acquiring a good quality of clusters which is difficult to ascertain when the value of k is high. To overcome the initialization problem of k-means method, we propose an incremental k-means clustering method that improves the quality of the clusters in terms of reducing the Sum of Squared Error (SSE_{total}). Comprehensive experimentation in comparison to traditional k-means and its newer versions is performed to evaluate the performance of the proposed method on synthetically generated datasets and some real-world datasets. Our experiments shows that the proposed method gives a much better result when compared to its counterparts.

Keywords: k-means · Sum of squared error · Improving results

1 Introduction

The prime objective of the k-means method is to divide the datasets D into k number of clusters by minimizing the Sum of Squared Error [8]. It is a more effective method if the initial k centroids are selected from respective k clusters. The selection of initial k centroids is another challenging task of k-means that produces different clusters for every set of different initial centroids. Thus, the initial random k centroids influence the quality of clustering. To overcome this problem, we proposed an incremental k-means clustering method (incrementalKMN) that attempts to improve the quality of the result in terms of minimizing the SSE_{total} error. The proposed method starts with a single seed and in each iteration; it selects one seed from the maximum $SSE_{partial}$ of a cluster (explained later in Sect. 3) and repeats the steps until it reaches k number of centroids.

2 Related Work

Among the partition-based clustering methods, k-means is the pioneering and most popular method that finds its use in almost every field of science and

B. Deka et al. (Eds.): PReMI 2019, LNCS 11941, pp. 38–46, 2019.
https://doi.org/10.1007/978-3-030-34869-4_5

engineering. As the k-means method is too sensitive to initial centroids, many procedures were introduced by the researchers to initialize centroids for k-means [4]. A Partitioning Around Medoids (PAM) method was introduced in [7], which starts with a random selection of k number of data objects. In each step, a swap operation is made between a selected data object and a non-selected data object, if the cost of swapping is an enhancement of the quality of the clustering. A Maximin method introduced in [6] that selects the first centroid arbitrarily, and the ith center, i.e. ($i \in \{2, 3, ..., k\}$) is selected from the point that has the greatest minimum-distance to the previously chosen centers. In [3], the clustering process starts with randomly dividing the given data set into J subsets. These subsets are grouped by the k-means method producing J sets of intermediate centers with k points each. These center sets are joined into a superset and clustered by k-means into J stages, each stage initialized with a different center set. The members of the center set that gives the minimum total sum of squared error (SSE_{total}) are then taken as the final centers. In [2], k-means++ is introduced, that selects the first centroid arbitrarily and the i-th center i.e. ($i \in 2, 3, ..., k$) is selected from $x \in X$ with a probability of $\frac{D(X)^2}{\sum_{x \in X} D(X)^2}$, where $D(X)$ indicates the least-distance from a point x to the earlier nominated centroids. To resolve the initialization problem of k-means method, a MinMax k-means method is introduced in [12], that changes the objective function, as an alternative of using the sum of squared error. It starts with randomly selected k number of centroids and tries to minimize the maximum intra-cluster variance using a potential objective function Eps_{max} instead of the sum of the intra-cluster variances. Besides, each cluster allots a weight, such that clusters with larger intra-cluster variance are allotted higher weights, and a weighted version of the sum of the intra-cluster variances are clustered automatically. It is less affected by initialization, and it produces a high-quality solution, even with faulty initial centroids. A variance-based method is introduced in [1], that initially sorts the points on the attribute based on the highest variance and then divides them into k number of groups along with the same dimension. The centers are then selected to be the points that correlate with the medians of these groups. Note that this method contempt all attributes, but one with the highest variance probably useful only for data sets in which the changeability is more in a single dimension. In [11], initially, a kd-tree of the data objects is created to perform density estimation, and then a modified maximin method is used to select k centroids from densely inhabited leaf buckets. Here, the computational cost is dominated by kd-tree construction. An efficient k -means clustering filtering algorithm using density based initial cluster centers (RDBI) is introduced in [10], it improves the performance of k-means filtering method by locating the centroid value at a dense area of the data set and also avoid outlier as the centroid. The dense area is identifying by representing the data objects in a kd-tree. A new approach IKM-+ introduced in [9], iteratively improves the quality of the result produced by k-means by eliminating one cluster and separating another and again re-clustering. It uses some heuristic method for faster processing. From, our brief selected survey, we conclude that proper initialization of k number of centroids is a challenging task, especially when the number of clusters is high. In this paper, an incremental k-means (incrementalKMN) clustering method is proposed, which gives k desired number of clusters even when the number of clusters is high. In the next section, we discuss the preliminaries of incrementalKMN method.

3 Preliminaries

Our incremental k-means (incrementalKMN) is based on the k-means [8] method, which is the most popular method. k-means uses a total sum of squared error (SSE_{total}) as an objective function, which tries to minimize the SSE_{total} of each cluster. The SSE_{total} is computed as:

$$SSE_{total} = \sum_{i=1}^{k} \sum_{x \in C_i} (c_i - x)^2 \tag{1}$$

In other way, for $i \in \{1, 2, ...k\}$ and k is the number of clusters then $SSE_{total} = SSE_1 + SSE_2 +, ...SSE_k$, where SSE_i is considered as partial SSE. In this paper, partial SSE will be reffered to as $SSE_{partial}$ of cluster c_i, and can be computed using the following equation:

$$SSE_{partial_{c_i}} = \sum_{x \in C_i} (c_i - x)^2 \tag{2}$$

where, c_i is a mean of the ith cluster.

To minimize the SSE_{total}, and producing k-number of centroids, the k-means method follows the following steps:

Step 1: Select k numbers of centroids arbitrarily as initial centroids.
Step 2: Assign each data objects to its nearest centroid.
Step 3: Recompute the centroids C.
Step 4: Repeat step 2 and 3 until the centroids no longer change.

Even though the k-means method is more effective and widely used method and the performance in terms of acquired SSE_{total} is strongly dependent on the selection of the initial centroids. Due to arbitrarily initial centroids selection, it can not produces desired clusters. Therefore, selection of initial centroids is very important for maintaining the quality of clustering.

4 Our Proposed IncrementalKMN Method

Here, we introduced an incremental k-means (incrementalKMN) clustering method which produces k number of clusters effectively. The central concept of the proposed method is, in each iteration, it increments the seed values by one until it reaches k number of seeds. Initially, the first seed is selected as a mean data object of a given dataset D. In each iteration, the ith center i.e.

$(i \in 2, 3, ..., k)$ is selected from maximum $SSE_{partial}$ of a cluster c_i. The ith center is a maximum distance from the data object and the centroid of maximum $SSE_{partial}$ cluster. Similarly, repeat the iteration until it reaches k number of seeds. Finally, we formed k desired number of clusters and also minimized the SSE_{total} error.

4.1 The IncrementalKMN Method

We introduce an incremental method for selecting centers for the k-means algorithm. Let $SSE_{partial}$ and SSE_{total} denotes the partial sum of squared error of a cluster and summation of the partial sum of squared error of all the clusters in a given dataset are computed in each iteration. Then we describe the following algorithm, which we call incrementalKMN method:

Algorithm 1: incrementalKMN(D,k) Clustering Method

Input: Datasets $D \leftarrow \{d_1, d_2, \ldots\ldots\ldots\ldots d_n\}$ and the value of k.
Output: k number of desired clusters.

$i \leftarrow 1$
$C \leftarrow \{\Phi\}$
$c_i \leftarrow Find\_Mean(D)$ `//` c_i `is the mean data object of` D.
$C \leftarrow C \cup \{c_i\}$
$Assign\_Cluster\_id(D, C, i)$ `//` `Data objects in` D `are assigned Cluster_id starting from 1 to` i `based on the centroids in` C. `At any point of time` C `consists of` i `number of centroids.`
while $i < k$ **do**
  $i \leftarrow i + 1$
  $c_p \leftarrow Max\_SSE_{partial}(C)$ `//` c_p `is the centroid of the cluster with maximum` $SSE_{partial}$ `value.`
  $c_i \leftarrow Max\_Dist\_from\_Centroid(D, c_p)$ `//` c_i `is the object at maximum distance from` c_p.
  $C \leftarrow C \cup \{c_i\}$ `// Update the Centroid List C.`
  $Assign\_Cluster\_id(D, C, i)$
  $Compute\_Centroid(D, C)$ `// Centroids are recomputed based on the cluster_ids assigned to objects.`
Compute SSE_{total} `// Compute` SSE_{total} `of desired clusters`
End

In Fig. 1(a), represents the DATA1 self-generated dataset with seven number of distinct clusters. Now, apply the incrementalKMN method on it. Initially, the first seed, i.e., initial centroid $C = \{c_1\}$ is assigned as a mean of the given dataset, which shows in Fig. 1(a). In each iteration, compute the SSE_{total} of the clusters and also assigned the $SSE_{partial}$ to each centroid. Now we compute the maximum $SSE_{partial}$ of centroid c_1, in this case, the SSE_{total} and $SSE_{partial}$ both are equal and considered as a maximum $SSE_{partial}$. Then compute the farthest distance between data objects and its centroid c_1 of a maximum $SSE_{partial}$

of a cluster, select the farthest distance data object i.e., second centroid c_2 and update the centroid list C i.e. $C = \{c_1, c_2\}$, which shows in Fig. 1(b) and update the centroid. In Fig. 1(c), c_1 shows the maximum $SSE_{partial}$ compared with c_2 now select the data object from cluster c_1 whose distance is maximum with respect to centroid c_1 and update the centroid C i.e. $C = \{c_1, c_2, c_3\}$ and recompute the centroid values. Similarly, in Fig. 1(d)–(h), continue the process until we have selected k number of centroids. Finally, we obtain k-number of clusters and also minimize the error in terms of SSE_{total}.

5 Experimental Analysis

In this section, we evaluate the performance of incrementalKMN k-means method with k-means, k-means++ [2], PAM [7], IKM-+ [9] partition-based clustering methods. We have used two different types of datasets. The first type of datasets are two-dimensional synthetic datasets [5], which shows the precision of methods graphically, where the result of the method can be identified as the location of centroids. The second type of datasets is used from online UCI machine learning repository. Results are averaged over 40 runs of each of the competing algorithms on synthetically generated datasets and the average SSE_{total} of each method presented in Table 1, where k determines the number of clusters. Table, 1 shows that PAM, IKM-+ and the incrementalKMN methods radically reduces the SSE_{total}. However, the incrementalKMN method provides better results in terms of SSE_{total} as compared with PAM and IKM-+ method, and also it increases the accuracy of clusters.

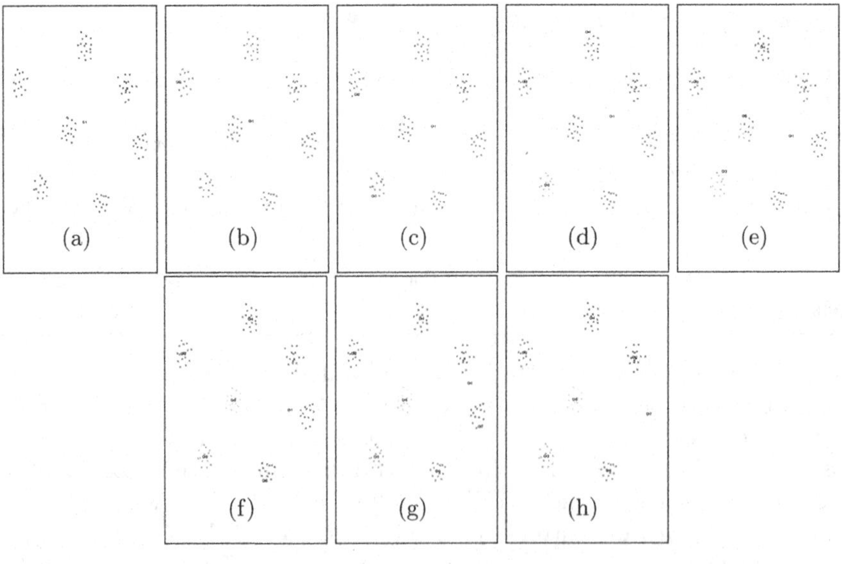

Fig. 1. The process of IncrementalKMN clustering method on DATA1 dataset

Here, we have visualized the results of some datasets, and we select the high quality SSE_{total} of each dataset for k-means, k-means++, PAM, IKM-+ and worst solution of SSE_{total} of the incrementalKMN method. For S-series datasets S-1, the quality of results of the proposed method is better then the results of k-means and k-means++ method, however, PAM and IKM-+ method gives the same results, which can be seen in Fig. 2(a)–(e). Similarly, for A-series A-3 dataset, the quality of results of the proposed method is better than the k-means and k-means++ method, which is reported in Fig. 3. However, comparing with SSE_{toatal}, the proposed method is better than the PAM and IKM-+ method, which is reported in Table 1. Due to the restriction of paper size, we could not report the result for all the datasets. In Fig. 4, the results of incrementalKMN method for S-2, S-3, S-4, A = 1, A-2, Brich-1, Brich-2, D31 and R15 datasets is reported. We observe, k-means++ method is more effective as compared with the k-means method, especially when the number of clusters is low. The clustering quality of k-means and k-means++ method degrades when the number of clusters is high. From Table 1, we observed that the proposed method is more effective, especially when the number of clusters is high and also useful for reducing the SSE_{total}. Therefore, we can say that incrementalKMN method is better than the competing methods.

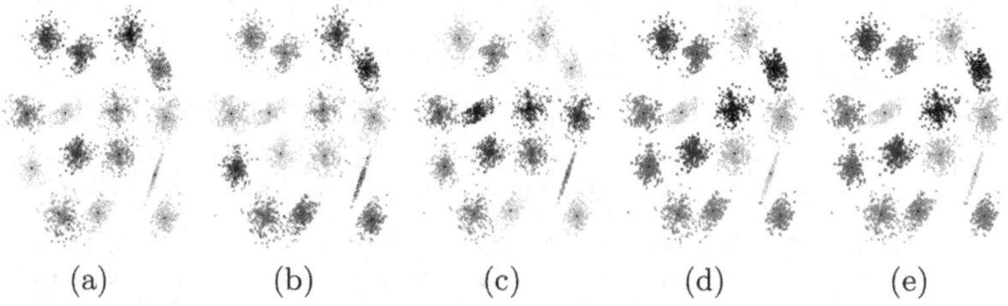

Fig. 2. For S-Set S1 dataset, the high quality result of (a) k-means, (b)k-means++(c)PAM (d)IKM-+ and (c)Worst result of incrementalKMN

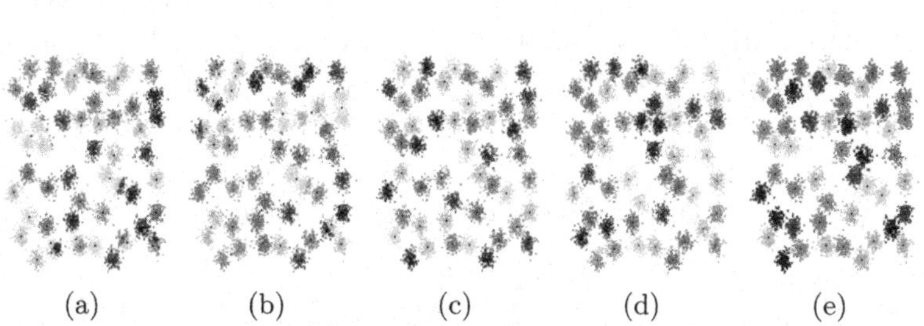

Fig. 3. For A-Set A3 dataset, the high quality result of (a) k-means, (b)k-means++(c)PAM (d)IKM-+ and (c)Worst result of incrementalKMN

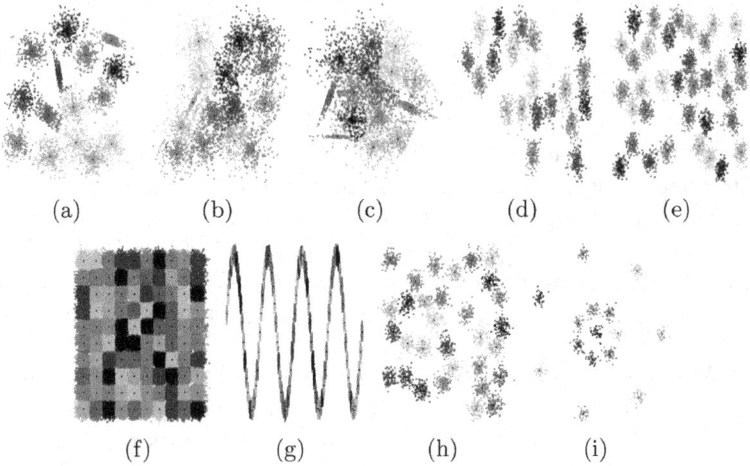

Fig. 4. Result of incrementalKMN method for S-2(a), S-3(b), S-4(c), A-1(d), A-2(e)Brich-1(f), Brich-2(g), D31(h) and R15(i) datasets

Table 1. Results on synthetic dataset

Serial No.	Data set	Average of SSE_{total}				
		k-means	k-means++	PAM	IKM-+	IncrementalKMN
1	Data1(k = 7)	2.59E+3	2.01E+3	2.04E+3	2.04E+3	**2.01E+3**
2	S-Set(S1)(k = 15)	1.88E+13	1.36E+13	8.96E+12	8.92E+12	**8.91E+12**
3	S-Set(S2)(k = 15)	2.13E+13	1.87E+13	1.33E+13	1.33E+13	**1.32E+13**
4	S-Set(S3)(k = 15)	1.96E+13	1.88E+13	1.70E+13	1.69E+13	**1.68E+13**
5	S-Set(S4)(k = 15)	1.72E+13	1.56E+13	1.58E+13	1.57E+13	**1.56E+13**
6	A-Set(A1)(k = 20)	2.11E+10	1.98E+10	1.22E+10	1.22E+10	**1.21E+10**
7	A-Set(A2)(k = 35)	3.48E+10	3.06E+10	2.03E+10	2.03E+10	**2.01E+10**
8	A-Set(A3)(k = 50)	5.23E+10	4.55E+10	2.91E+10	2.90E+10	**2.88E+10**
9	Brich1(k = 100)	1.25E+14	9.35E+13	—	9.28E+13	**9.25E+13**
10	Brich2(k = 100)	7.69E+13	4.89E+12	—	4.75E+11	**4.55E+11**
11	D31(k = 31)	5.59E+3	4.97E+3	3.43E+3	3.41E+3	**3.39E+3**
12	D15(k = 15)	4.45E+2	3.95E+2	1.12E+2	1.12E+2	**1.08E+2**

Table 2. Results on real-world dataset

Serial No.	Data set	Average of SSE_{total}				
		k-means	k-means++	PAM	IKM-+	IncrementalKMN
1	Iris(k = 3)	9.93E+01	8.48E+01	8.46E+01	7.89E+01	**7.86E+01**
2	Seed(k = 3)	5.87E+2	7.07E+02	6.05E+02	6.94E+02	**5.84E+02**
3	Letter Recognition(k = 26)	6.49E+05	6.23E+05	6.60E+05	6.16E+05	**6.15E+05**
4	Musk(k = 2)	6.10E+09	6.06E+09	6.02E+09	5.92E+09	**5.90E+09**

Table 3. Adjusted Rand Index on synthetic and real dataset

Serial No.	Data set	Adjusted Rand Index				
		k-means	k-means++	PAM	IKM-+	IncrementalKMN
1	D31	0.9820231	0.9851990	0.9971239	–	**0.9973272**
2	R15	0.9867001	0.9462493	**0.9991208**	–	**0.9991208**
3	Iris	0.8797315	0.8797315	0.8797315	–	**0.8922595**
4	Seed	0.8743677	0.8743677	0.8713602	–	**0.8753602**
5	Letter Recognition	0.93116598	0.92783476	0.92841268	–	**0.93767423**
6	Musk	0.50310242	0.50310242	0.50295906	–	**0.6185838**

The result of quality of clusters based on SSE_{total} of k-means, k-means++, PAM, IKM-+ and proposed method for average of 40 repeatations on each of the real-world datasets is shown in Table 2, where k determines the number of clusters. In the proposed method, the average SSE_{total} is slightly less than the IKM-+ [9] method and much lesser than for the other competing methods. Therefore, we can say that incrementalKMN gives better clusters with respect to SSE_{total}.

We have also compared our algorithm and its counterparts using the external validity index, *Adjusted Rand Index*. Table 3, shows that the adjusted Rand Index on synthetic datasets (R15 and D31) and real datasets is high as compared with its competitors.

6 Conclusion

In this paper, we proposed an incremental k-means method, which is partition-based clustering method for handling initialization problem of k-means. In each iteration, it increments the centroid list by adding one data object from given dataset based on the maximum distance between data objects and centroid of the cluster with $SSE_{partial}$. In terms of average SSE_{total} of synthetically generated datasets and real datasets, the incremental method is less as compared to its competitors. Similarly, in terms of *Adjusted Rand Index*, the incremental method is high as compared to its competitors. So, the incremental method could give a guaranteed solution even if, the number of clusters is high. For many datasets, its poor result is better than its competing algorithm. The time complexity of this method is linear order.

In real-world datasets, need further work different distance similarity measures on high dimension.

References

1. Al-Daoud, M.B.: A new algorithm for cluster initialization. In: WEC 2005: The Second World Enformatika Conference (2005)

2. Arthur, D., Vassilvitskii, S.: k-means++: the advantages of careful seeding. In: Proceedings of the Eighteenth Annual ACM-SIAM Symposium on Discrete Algorithms, pp. 1027–1035. Society for Industrial and Applied Mathematics (2007)
3. Bradley, P.S., Fayyad, U.M.: Refining initial points for k-means clustering. In: ICML, vol. 98, pp. 91–99. Citeseer (1998)
4. Celebi, M.E., Kingravi, H.A., Vela, P.A.: A comparative study of efficient initialization methods for the k-means clustering algorithm. Expert Syst. Appl. **40**(1), 200–210 (2013)
5. Fränti, P., Sieranoja, S.: K-means properties on six clustering benchmark datasets (2018). http://cs.uef.fi/sipu/datasets/
6. Gonzalez, T.F.: Clustering to minimize the maximum intercluster distance. Theoret. Comput. Sci. **38**, 293–306 (1985)
7. Hadi, A.S., Kaufman, L., Rousseeuw, P.J.: Finding groups in data: an introduction to cluster analysis. Technometrics **34**(1), 111 (1992)
8. Han, J., Pei, J., Kamber, M.: Data Mining: Concepts and Techniques. Elsevier, Amsterdam (2011)
9. Ismkhan, H.: Ik-means-+: an iterative clustering algorithm based on an enhanced version of the k-means. Pattern Recogn. **79**, 402–413 (2018)
10. Kumar, K.M., Reddy, A.R.M.: An efficient k-means clustering filtering algorithm using density based initial cluster centers. Inf. Sci. **418**, 286–301 (2017)
11. Redmond, S.J., Heneghan, C.: A method for initialising the K-means clustering algorithm using kd-trees. Pattern Recogn. Lett. **28**(8), 965–973 (2007)
12. Tzortzis, G., Likas, A.: The MinMax k-means clustering algorithm. Pattern Recogn. **47**(7), 2505–2516 (2014)

Modified FP-Growth: An Efficient Frequent Pattern Mining Approach from FP-Tree

Shafiul Alom Ahmed(✉) and Bhabesh Nath

Tezpur University, Napaam, Tezpur 784028, Assam, India
tezu.shafiul@gmail.com, bnath@tezu.ernet.in

Abstract. Prefix-tree based FP-growth algorithm is a two step process: *construction of frequent pattern tree (FP-tree)* and then *generates the frequent patterns* from the tree. After constructing the FP-tree, if we merely use the conditional FP-trees (CFP-tree) to generate the patterns of frequent items, we may encounter the problem of recursive CFP-tree construction and a huge number of redundant itemset generation. Which also leads to huge search space and massive memory requirement. In this paper, we have proposed a new data structure layout called Modified Conditional FP-tree (MCFP-tree). Moreover, we have proposed a new pattern growth algorithm called *Modified FP-Growth* (MFP-Growth), which uses both top-down and bottom-up approaches to efficiently generate the frequent patterns without recursively constructing the MCFP-tree. During mining phase only one MCFP-tree is maintained in main memory at any instance and immediately deleted or discarded from the memory after performing the mining. From the experimental analysis, it is noticed that the proposed MFP-Growth algorithm requires less memory to construct the MCFP-tree as compared to conditional FP-tree. Moreover, the execution of the MFP-Growth method is found significantly faster than the traditional FP-Growth as it does not generate redundant patterns.

Keywords: Association Rule (AR) · FP-growth · Frequent Pattern (FP) · FP-tree · Pattern Mining (PM) · Data Mining (DM) · Frequent Itemset (FI)

1 Introduction

Frequent pattern (FP or FI) mining is considered as an fundamental problem of DM. It has been extensively exercised in some major DM operations, such as ARM, sequential patterns, classification, max and closed FP and clustering and has applications in many areas such as market-basket analysis, bioinformatics and web mining etc. The problem of mining FIs was first discussed by Agrawal *et al.* in 1993 (Apriori Algorithm) [1]. But the major problem of Apriori algorithm is that it generates huge number of candidate itemsets and also uses multiple

B. Deka et al. (Eds.): PReMI 2019, LNCS 11941, pp. 47–55, 2019.
https://doi.org/10.1007/978-3-030-34869-4_6

database scan. Later on the researchers have discovered many association rule mining techniques with candidate generation approach but most of the algorithms suffers from same problems. In 2000 Han *et al.* [6] proposed an prefix path tree based approach called FP-Growth. This method performs the rule mining in two steps. First it constructs a compressed tree data structure called **FP_tree** using only two database scans. Secondly, it recursively constructs conditional frequent pattern tree (CFP-tree), a special kind of projected data structure to generate the frequent patterns for each individual frequent item of the database. Though the FP-Growth algorithm has been considered as one of the best and fastest frequent pattern generation algorithms; still it has a major disadvantage. For each individual frequent item of FP-tree, FP-growth recursively constructs the conditional FP-trees. Which also leads to huge search space and massive memory requirement.

In this work, an efficient CFP-tree data structure called Modified Conditional FP-tree (MCFP-tree) has been introduced. Unlike conditional FP-tree, the MCFP-tree is constructed in the reverse order of the header table to improve the mining process. Moreover, a new tree-traversal algorithm have been proposed to perform the mining process faster. It is not required to recursively construct the MCFP-tree for a single frequent item of the header table hence improves the performance. We have compared our technique with FP-growth and the experimental result shows that the modified FP-growth outperforms FP-growth.

2 Related Work

FP-tree is a special prefix-tree data structure, used by **FP-Growth** [6] algorithm to efficiently store the dataset information. Though, the performance of FP-Growth is noteworthy with respect to other existing pattern mining approaches, but the algorithm has some crucial drawbacks also. FP-Growth takes a huge amount of time to recursively construct the conditional FP-trees, particularly when the dimensionality of the dataset is high and the minimum support threshold value is low. If the *min-supp* value decreases then performance of FP-Growth also demotes and at some point of time with very low minimum support, it becomes almost similar to Apriori. Therefore, in case of low support threshold, the recursive mining of the CFP-trees, decreases the pattern mining performance unexpectedly. Therefore in 1997 an effective method was proposed by Park et al. [11] called *nonordfp*. It uses an unordered array of pairs (*child node name, child node index*) to map the children, so that the traversing becomes easier, just reading an array sequentially. But, this method works better only if the number of entries for the mapped arrays are very small. Otherwise, since the array needs a sequential memory space it becomes infeasible for huge or dynamic datasets. Many variants of FP-Growth algorithm have been found in literature such as CFP-Growth [12], Improved FP-Growth [8], CT-PRO [13], COFI-tree [3], Inverted Matrix [2], FP-Growth* [4], H-mine [10], Opportunistic Projection [9] and many significant work can be found in [5,7] also. In this work, an efficient method called Modified FP-Growth has been proposed to enhance the frequent pattern mining.

3 Proposed Method: Modified FP-growth

The Modified FP-growth method for mining FPs can be described in three phases. The working principle of MFP-Growth is illustrated with the help of a small dataset D [Table 1] and suppose the support threshold be 20%.

- ***Phase 1. FP-tree Construction:*** The proposed MFP-growth utilizes the conventional FP-tree construction algorithm to construct the FP-tree. Initially, the dataset D is scanned once to fetch the support count of individual item. The frequent items are then added to the header table with respect to their frequency descending order excluding the infrequent items. Then MFP-Growth constructs FP-tree for dataset D as depicted in Fig. 1.

Table 1. Transactional dataset (D)

TID	ITEMS
1	{a, b}
2	{b, c, d}
3	{a, c, d, e}
4	{a, d, e}
5	{a, b, c}
6	{a, b, c, d}
7	{a}
8	{a, b, c}
9	{a, b, d}
10	{b, c, e}

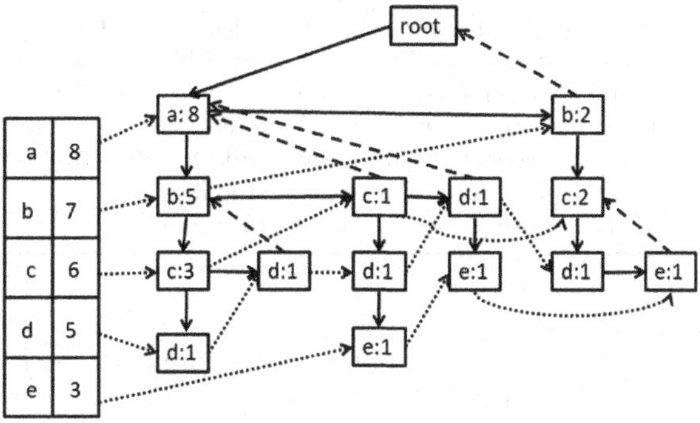

Fig. 1. FP-tree for D.

- ***Phase 2. Construction of MCFP-tree:*** Then the proposed method constructs an effective conditional FP-tree, called MCFP-tree to minimize the mining complexity. Like conditional FP-tree nodes, the MCFP-tree nodes also consists of: ItemLabel, NodeCount represents the number of transactions shared the path, ParentNodeLink, ChildNodeLink and SiblingNodeLink, additionally it contains two other informations namely RealativeItemCount and index. If the support of a node x in a path P_i is P_i(x), then all the node's support along the prefix path from node x to the root(excluding) is considered as P_i(x). The proposed method inserts the prefix path frequent items in the opposite order of conditional FP-tree. Therefore, the Modified CFP-Tree structure consists of a root node labelled with the frequent item for which the CFP-Tree is constructed and it can be described as the reverse CFP-Tree. The complete procedure of MCFP-tree construction illustrated in Algorithm 1.

Algorithm 1. Procedure: Modified_CFP_Tree-Construction(FP-Tree, X)

input: Minimum support count (*minsup*), *FP-Tree* and the *item (X)* .
output: MCFP-tree of the item X .

1: Derive all the prefix paths P_i for item X (excluding) from the FP-tree and find the frequency counts of each individual items along the prefix paths of X.
2: Discard the infrequent items and create a HeaderTable containing the frequent items.
3: Define the root of the MCFP-tree: root(X) and set the same item node link.
4: **for** each prefix path P_i of item X **do**
5:     tempRoot = root;
6:     **for** each item I_k in P_i **do**
7:         **if** I_k is frequent and present in the jth position of the header table **then**
8:             ***Call*** tempRoot = insert-MCFP-tree (I_k, P_i(x), tempRoot,j);
9:         **end if**
10:     **end for**
11: **end for**
12: **Procedure: insert-MCFP-tree**(I, *Count*, tempRoot, *index*)
13: ***if*** tempRoot has a child_node with label I ***then*** increment the frequency count of the child_node and HeaderTable[index].Count by adding the frequency count *Count* and set tempRoot = child_node;
14: ***else*** create a new child_node of tempRoot with label I, RelativeItemCount as 0 and set the frequency count of the child_node as *Count* and set HeaderTable[index].Count = *Count* and tempRoot = child_node ;
15: ***return***(tempRoot); //Return to step 9.

- Let us consider, say we are to mine all the frequent patterns for the item 'e' from the above FP-tree (Fig. 1).
  * First, the algorithm derives the set of all the prefix paths P_i(e) one by one in a bottom-up approach and derives the frequency count of each individual items along the prefix paths. Then the frequent items are inserted in to the header table and the infrequent items are excluded.
  * The algorithm constructs the MCFP-tree with the root node label 'e' and sets the support of the root(e) as the total support count of item 'e' from the header table of FP-tree. The corresponding MCFP-tree for item 'e' is shown in Fig. 2a. The corresponding conditional FP-tree for the item 'e' is shown in Fig. 2b.

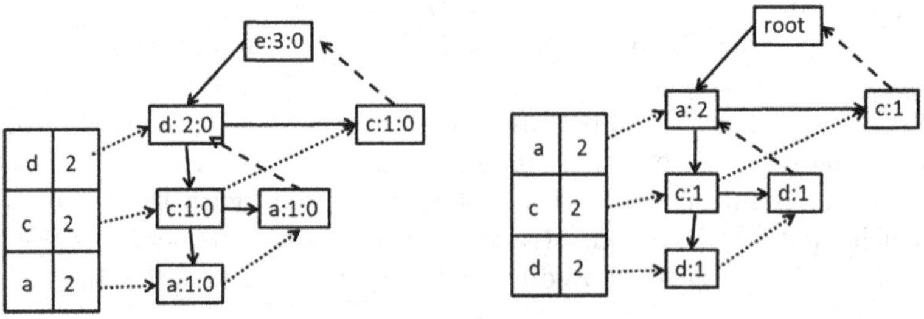

(a) (b)

Fig. 2. MCFP-tree and conditional FP-tree for the item 'e'

Algorithm 2: Modified_FP-Growth

input: Minimum support count (*minsup*),
Modified_CFP-Tree.
output: Set of frequent patterns.

1: **if** *MCFP − tree* contains a single path Z **then**
2: **for** each combination β of the nodes in Z **do**
3: generate pattern $\beta \cup \gamma$ with support = minimum support count of the item in β.
4: **end for**
5: Define a List[] data structure to store frequent item and its support count.
6: List[0]=HeaderTable[0] //stores root node item and its support count.
7: **else**
8: **for** i = 2 to n; where n is the size of HeaderTable **do**
9: *temp =* HeaderTable[i]→SameItemNodeLink;
10: **while** temp!=NULL **do**
11: **for** each node P in the prefix path of temp (excluding root) **do**
12: Set P →realCount += temp→Count.
13: **end for**
14: Set Set temp→Count = 0.
15: temp = temp→SameItemLink;
16: **end while**
17: **for** j = 1 to i-1 **do**
18: temp = HeadetTable[j]→SameItemNodeLink;
19: List[1]=HeaderTable[j]
20: **while** temp!=NULL **do**
21: **for** each node P in the prefix path of temp **do**
22: Set HeaderTable[P →index]→Count += P →realCount
23: P →realCount = 0.
24: **end for**
25: temp = temp→SameItemLink;
26: **end while**
27: l=1;
28: **for** k=1 to i-1 **do**
29: **if** HeaderTable[k]→Count ≥ *minsup* **then**
30: List[l] = HeaderTable[k];
31: **end if**
32: HeaderTable[k]→.Count=0;
33: **end for**
34: **for** each combination α of the nodes in List[] **do**
35: generate pattern $\alpha \cup \gamma$ with support = minimum support count of the item in α. //$\alpha = \{List[0], List[1]\}$
36: **end for**
37: **end for**
38: **end for**
39: **end if**

- *Phase 3. Mining the MCFP-tree:* After constructing the MCFP-Tree, the next phase is the extraction of frequent patterns from the MCFP-tree. In this section, we have introduced an enhanced FP-growth method called Modified_FP-Growth. FP-Growth algorithm uses bottom-up scanning to recursively reconstruct the conditional FP-trees to generate the FPs for a single FI of the header table. On the other hand Modified_FP-Growth constructs a single for individual item of the header table of FP-tree and employs a bottom-up tree-traversing method to efficiently generate the FIs from the MCFP-tree without recursively constructing the MCFP-trees. The algorithm proposed for mining the FPs from the MCFP-Tree is illustrated in the Algorithm 2.
The procedure for mining the FIs for item 'e' from the MCFP-tree (Fig. 2a) by the proposed method is depicted bellow.

For node 'd', relCount('e') = 2 ≥ *minsupp*. Therefore, pattern {'e', 'd':2} is generated from Fig. 3. For item 'c', relCount('d') = 1≤ *minsupp* and relCount('e') = 2 ≥ *minsupp* as shown in Fig. 4 and it generates pattern {'e', 'c':2}. Similarly, for item 'a', item 'c' is infrequent as shown in Fig. 5. Therefore, it generates {'e', 'a':2} and {'e', 'd', 'a':2}. After mining all the frequent patterns from a MCFP-tree, the tree is deleted from the main memory to construct the MCFP-tree for other items of the header table.

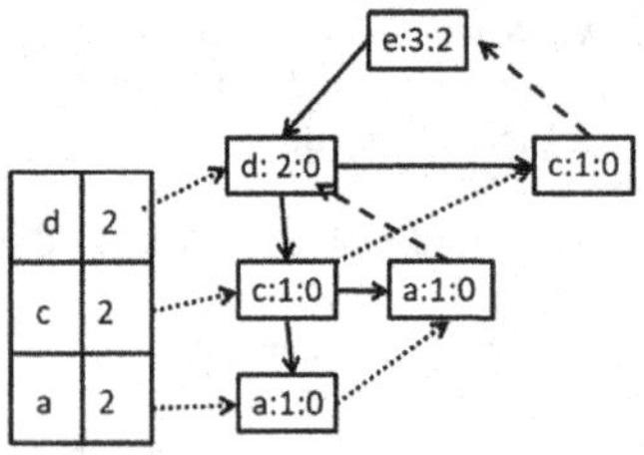

Fig. 3. For item ‘d’

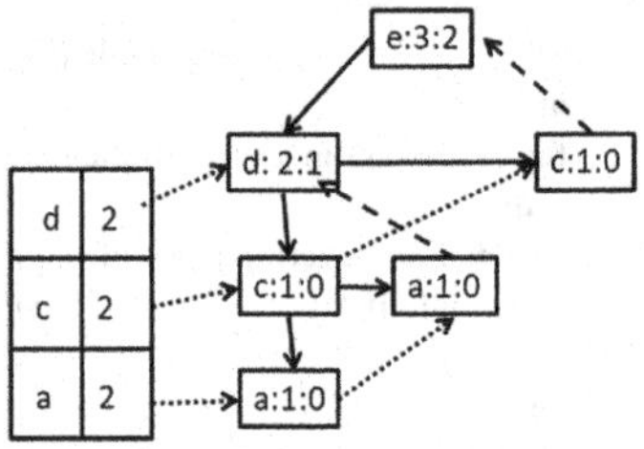

Fig. 4. For item ‘c’

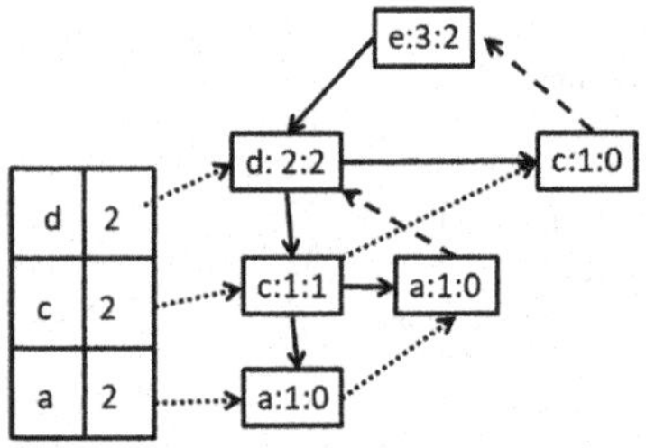

Fig. 5. For item ‘a’

4 Performance Analysis

The performance of MCFP-Growth method is evaluated in terms of total time taken to execute the algorithm and memory requirement with respect to the FP-Growth technique. Experiments were performed on a machine with 3.2 GHz Intel i7 processor and 8 GB memory and 64-bit Linux operating system. The algorithms are implemented in C language and executed without running any background process. To analyse the performance of MCFP-Growth, we have used both real and synthetic datasets. To justify the effectiveness of MCFP-Growth algorithm, we have used dense datasets as well as sparse dataset also. The datasets mentioned in Table 2 are collected from UCI and FIMI repository.

Table 2. Datasets used

Dataset	Category	Average length	Number of transaction	Number of item	Type
T40I10D100k	Synthetic	40	100,000	1000	Sparse
Connect-4	Real	43	67,557	129	Dense

To compare the performance of both the approaches in terms of execution time, three set of experiments have been performed for each support threshold value. Then the execution time for each support threshold is considered as the average execution time of the three experiments for both the approaches. The execution time reported, is the total of MCFP-tree construction and time taken by the Modified_FP-Growth algorithm to mine the MCFP-tree. A comparison of total number of pattern generated and execution time for different datasets of MCFP-Growth and the FP-Growth have been illustrated in Table 3.

Table 3. Execution time

Dataset	Support (in %)	FP-Growth	MCFP-Growth	No. of patterns
		Mining time (secs)	Mining time (secs)	
T40I10D100K	.01	59.33	39.64	65236
	.02	27.90	24.30	2293
	.03	19.74	17.05	793
	.04	14.29	11.04	440
	.05	9.09	7.44	316
Connect-4	.5	44.80	41.26	88316229
	.6	37.12	29.24	21250671
	.7	32.51	26.05	4129839
	.8	26.58	20.04	533975
	.9	17.29	12.54	27127

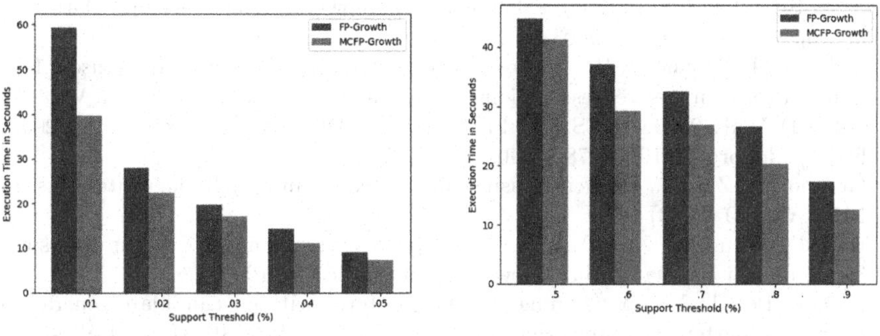

(a) Dataset : T40I10D100K (b) Dataset : Connect-4

Fig. 6. Execution time for different thresholds

As depicted in Fig. 6, FP-Growth algorithm invests more time as compared to the proposed method. For each item of the header table of FP-tree, the proposed method constructs a single MCFP-tree. But on the contrary, for each conditional FP-tree, FP-Growth algorithm recursively constructs conditional FP-trees for each frequent item of the header table. Therefore, recursive construction of trees lead to more number of tree node construction. If we consider the Dataset T40I10D100K, for .01% support threshold the header table size is 755. For the 755 header items the proposed method creates total 8736968 number of nodes to construct the 755 MCFP-tree, but for the same support threshold FP-Growth creates 8830100 number of tree nodes. Therefore, for each item of the header table, FP-Growth requires on average 123 more nodes as compared to the proposed method.

5 Conclusion

The major advantage of incorporating the relCount variable to the node structure of the proposed MCFP-tree is that unlike FP-Growth, it is not required to recursively construct the conditional FP-trees and hence reduces the tree construction time. We have also proposed and implemented the MCFP-Growth method for efficiently mining all the FIs in large datasets. The performance analysis shows that it efficiently computes the complete set of FIs and outperforms FP-Growth with respect to execution time and memory requirement.

References

1. Agrawal, R., Imielinski, T., Swami, A.: Tmining association rules between sets of items in large databases. In: ACM SIGMOD International Conference on Management of Data, vol. 22, pp. 207–216 (1993)
2. El-Hajj, M., Zaïane, O.R.: Inverted matrix: efficient discovery of frequent items in large datasets in the context of interactive mining. In: Proceedings of the Ninth ACM SIGKDD International Conference on Knowledge Discovery and Data Mining, pp. 109–118. ACM (2003)
3. El-Hajj, M., Zaïane, O.R.: Non-recursive generation of frequent K-itemsets from frequent pattern tree representations. In: Kambayashi, Y., Mohania, M., Wöß, W. (eds.) DaWaK 2003. LNCS, vol. 2737, pp. 371–380. Springer, Heidelberg (2003). https://doi.org/10.1007/978-3-540-45228-7_37
4. Grahne, G., Zhu, J.: Efficiently using prefix-trees in mining frequent itemsets. In: FIMI, vol. 90 (2003)
5. Han, J., Cheng, H., Xin, D., Yan, X.: Frequent pattern mining: current status and future directions. Data Min. Knowl. Disc. **15**(1), 55–86 (2007)
6. Han, J., Pei, J., Yin, Y.: Mining frequent patterns without candidate generation: a frequent-pattern tree approach. In: Proceedings of ACMSIGMOD, Dallas, TX, pp. 1–12 (2000)
7. Han, J., Pei, J., Yin, Y., Mao, R.: Mining frequent patterns without candidate generation: a frequent-pattern tree approach. Data Min. Knowl. Disc. **8**(1), 53–87 (2004)
8. Lin, K.C., Liao, I.E., Chen, Z.S.: An improved frequent pattern growth method for mining association rules. Expert Syst. Appl. **38**(2011), 5154–5161 (2011)
9. Liu, J., Pan, Y., Wang, K., Han, J.: Mining frequent item sets by opportunistic projection. In: Proceedings of the Eighth ACM SIGKDD International Conference on Knowledge Discovery and Data Mining, pp. 229–238. ACM (2002)
10. Pei, J., Han, J., Lu, H., Nishio, S., Tang, S., Yang, D.: H-mine: hyper-structure mining of frequent patterns in large databases. In: Proceedings 2001 IEEE International Conference on Data Mining, pp. 441–448. IEEE (2001)

11. Racz, B.: Nonordfp: an FP-growth variation without rebuilding the FP-tree. In: Proceedings of IEEE ICDM Workshop on Frequent Itemset Mining Implementations (2004)
12. Schlegel, B., Gemulla, R., Lehner, W.: Memory-efficient frequent-itemset mining. In: Proceedings of the 14th International Conference on Extending Database Technology, pp. 461–472. ACM (2011)
13. Sucahyo, Y.G., Gopalan, R.P.: CT-PRO: a bottom-up non recursive frequent itemset mining algorithm using compressed FP-tree data structure. In: FIMI, vol. 4, pp. 212–223 (2004)

Link Weight Prediction for Directed WSN Using Features from Network and Its Dual

Ritwik Malla(✉) and S. Durga Bhavani

School of Computer and Information Sciences, University of Hyderabad, Hyderabad, India
ritwikmalla@gmail.com, sdbcs@uohyd.ernet.in

Abstract. Link prediction problem in social networks is a very popular problem that has been addressed as an unsupervised as well as supervised classification problem. Recently a related problem called link weight prediction problem has been proposed. Link weight prediction on Weighted Signed Networks (WSNs) holds great significance as these are semantically meaningful networks. We consider two groups of features from the literature - edge-to-vertex dual graph features and fairness-goodness scores in order to propose a supervised framework for weight prediction that uses fewer features than those used in the literature. Experimentation has been done using three different feature sets and on three real world weighted signed networks. Rigorous assessment of performance using (i) Leave-one-out cross validation and (ii) N% edge removal methods has been carried out. We show that the performance of Gradient Boosted Decision Tree (GBDT) regression model is superior to the results presented in the literature. Further the model is able to achieve superior weight prediction scores with significantly lower number of features.

Keywords: Link prediction · Line graph · Centrality measures · Fairness and goodness · Supervised regression

1 Introduction

Liben Nowell et al. formulated the problem of link prediction [9] which attempts to predict links that are either missing or that are highly likely to appear in future. In the literature, several unweighted similarity measures for link prediction have been proposed - Adamic Adar (AA), Jaccard Coefficient (JC), PropFlow (PF), Katz (KZ), etc. [14]. Weighted versions of these measures have also been proposed and utilized for link prediction in weighted networks [7,11,13].

Only recently [3], the problem of prediction of link along with weight has been addressed using these weighted similarity measures. In the literature [16] it has been argued that separately designed algorithms for link and weight prediction

B. Deka et al. (Eds.): PReMI 2019, LNCS 11941, pp. 56–64, 2019.
https://doi.org/10.1007/978-3-030-34869-4_7

should yield better results in comparison to a single algorithm designed for both the link and weight prediction.

Link weight prediction attempts to predict the weight of links with missing weight information or links predicted to appear in the future. It has potential applications in recommendation systems, sentiment analysis, network evolution, etc. Importance of weight of link cannot be overstated. There is an earlier work on prediction of sign of the links [8], which is a simpler problem compared to the weight prediction.

The networks may be modeled as graphs of different types such as weighted, unweighted, multi-graph, directed, undirected, etc. In this work, we restrict our attention to weighted signed networks (WSNs) which are weighted graphs where the edge labels contain positive or negative values. WSNs can represent the real world asymmetric nature of human relationships, where person A may like person B but person B may not like person A.

In machine learning, link prediction can be formulated as a binary classification problem [4] where the links are either present or absent and weight prediction can be formulated as a regression problem that yields continuous weight values. In this paper we address the problem of weight prediction of edges in different WSNs.

2 Problem Statement

A WSN is modeled as a directed weighted graph $G(V, E, W)$ where V is the set of vertices, E is the set of edges connecting vertex pairs and $W(e)$ is the set of edge weights, where $e \in E$ and weight takes $-ve$ or $+ve$ sign with $W(e) \in [-1, 1]$. The problem of weight prediction attempts to predict the weights of the edges $e \in E$, where edges e have their weight information missing.

3 Related Literature

J. Zhao et al. propose that weight between any two nodes that do not have a direct edge can be calculated by multiplying the weights of the paths linking them [15].

Fu et al. [3] treat link weight prediction problem as a supervised regression problem where weight is predicted for an undirected network. Fu et al. use a edge-to-vertex dual representation of the graph, called line graph, in which edges of the original graph are transformed into vertices in line graph so that vertex centrality indices can be used to predict link weights. Further the authors argue that weight information may not be captured fully using only similarity indices, hence a combination of similarity indices and node centrality indices have been used for weight prediction.

Kumar et al. [6] are the first authors to predict weights on weighted signed networks (WSNs). Kumar et al. used 11 special features including triadic balance, bias and deserve, along with their proposed new features called goodness and fairness measures. They conduct supervised regression in order to predict the weights on the links.

4 Proposed Approach

We basically take the approach of Fu et al. who take two groups of features, one from the original graph and the other from the edge-to-vertex dual (line graph) in order to perform link weight prediction. We extend this approach to directed weighted signed networks. We choose path based node centrality indices that are relevant for directed graphs such as Page Rank and Betweenness Centrality as features on the line graph. We find that the unsupervised measures proposed by Kumar et al. such as Fairness and Goodness measures [6] are good candidates for graph features.

We use Fairness-Goodness (FG) measures of Kumar et al. as the original graph features along with node centrality features on the line graph and propose a supervised regression approach.

The work done by Kumar et al. has been considered as the baseline method and we investigate if the performance can be improved as well as if the number of features can be reduced in the supervised regression approach to weight prediction.

A flowchart depicting our approach is shown in Fig. 1.

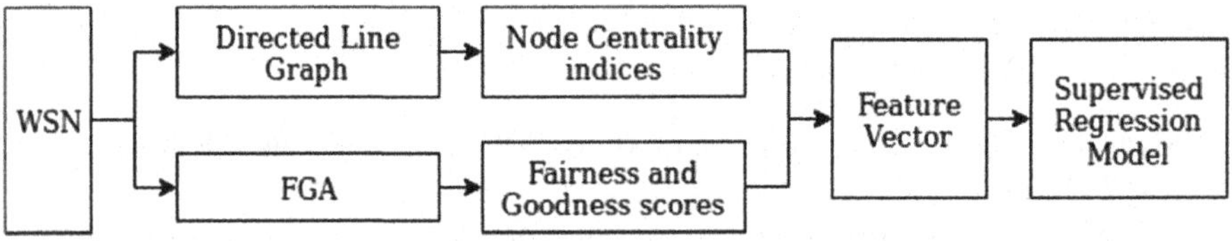

Fig. 1. Proposed approach

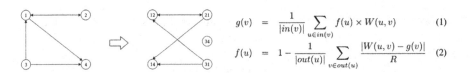

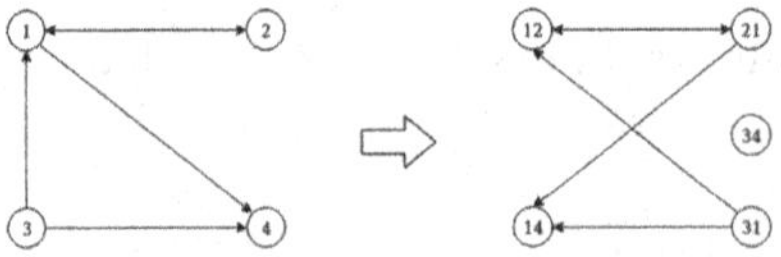

Fig. 2. Line graph transformation

$$g(v) = \frac{1}{|in(v)|} \sum_{u \in in(v)} f(u) \times W(u, v) \quad (1)$$

$$f(u) = 1 - \frac{1}{|out(u)|} \sum_{v \in out(u)} \frac{|W(u, v) - g(v)|}{R} \quad (2)$$

4.1 Directed Line Graph

The approach has to be carefully tuned for directed graphs. We use the notion of directed line graph in the following way:

If the directed edges $(u, v), (v, w) \in E$ then an edge is added between the vertices (uv, vw) in the line graph. An example for line graph construction that depicts the conversion of original directed graph to directed line graph is given (See Fig. 2).

4.2 Fairness Goodness Measures

The central recursive equations of fairness goodness algorithm(FGA) of Kumar et al. [6] are given in Eqs. (1) and (2) respectively.

Goodness of a vertex v, $g(v)$, is the mean weighted incident rating over the links $(u, v) \in E$ of $G(V, E)$. $g(v)$ represents the trust that the in-neighbours of v place on vertex v. Therefore, $g(v)$ is high if more number of fair nodes rate vertex v. $g(v)$ is considered to be an estimate of $w(u, v)$ and fairness of vertex u, $f(u)$, is high if it has more number of good out-neighbors.

The fairness and goodness scores are mutually recursive and the FGA algorithm stops when in consecutive iterations for all vertices, the difference between the consecutive scores of $f(u)$ and $g(v)$ is less than an error threshold. Fairness scores lie in [0, 1]. Goodness scores and edge weights lie in $[-1, 1]$. The largest difference between a weight and goodness score is the range of difference, R = 2.

4.3 Feature Extraction

Feature extraction is done by calculating *fairness and goodness* (FG) features on the original graph and *node centrality* features on the line graph (Line). We used LPMade [10] tool to convert the line graph from an edge list to the network file format and compute the standard path based node centrality indices like node betweenness centrality and pagerank [14]. Fairness and goodness measures have been computed using the code made available by the authors [2].

Three different feature sets have been considered for the weight prediction experimentation.

1. **Feature set I (Line + FG)** This experiment builds a supervised learning model for directed link (u, v) using two FG measures - $f(u) * g(v)$, $g(v)$ and two Line measures - page rank and node betweenness centrality measures.
2. **Feature set II (Line)** This experiment is done using only the two Line measures, namely page rank and node betweenness centrality on the line graph.
3. **Feature set III (FG)** This experiment uses only the two FG measures $f(u) * g(v)$ and $g(v)$.

5 Experiments and Results

Table 1 gives a concise description of the three real world weighted signed network (WSN) datasets used by Kumar et al. These data sets have been chosen in order to compare the performance with the work of Kumar et al. [6]. The original datasets are available on SNAP and the normalized datasets after data preprocessing are made available by the authors at [2]. Bitcoin-Alpha (BTC Alpha) and Bitcoin-OTC (OTC Net) are the two directed signed bitcoin exchange networks that have been considered in this paper where each vertex is a Bitcoin user and each edge is the trust rating that they give to each other. Wikipedia

Requests for Adminship (Wiki RfA) network is a directed signed network where each vertex is a Wikipedia user and each edge (A, B) represents the weight corresponding to the vote that user A gave to user B.

5.1 Datasets

Table 1. Weighted signed networks (WSNs)

Network	Vertices	Edges
BTC Alpha	3783	24,186
OTC Net	5881	35,592
Wiki RfA	9654	104,554

5.2 Graph Preprocessing

The transformation of original graph to the corresponding line graph should satisfy the criteria that the number of vertices of line graph is the same as the number of edges in the original graph. Since the original graph is a directed graph, the transformation can lead to isolated vertices in the line graph. Hence the isolated vertices in the line graph and the corresponding directed edges in the original graph are removed from further consideration.

The impact of isolated vertices removal from line graph is shown in Table 2.

Table 2. Undirected and directed line graphs on isolated vertices removal

	BTC Alpha			OTC Net			Wiki RfA		
	Vertices	Edges	File size	Vertices	Edges	File size	Vertices	Edges	File size
Undirected line graph	24184	2518684	28.2 MB	35591	4693528	54.1 MB	104554	11116567	131.5 MB
Directed line graph	24177	1256332	14.1 MB	35586	2301858	26.5 MB	99307	3564619	41.9 MB

5.3 Evaluation

Two standard metrics being used to calculate the performance of link weight prediction are *Root Mean Square Error (RMSE)* and *Pearson Correlation Coefficient (PCC)*. *RMSE* is the error calculated between the actual and predicted weight and the range of the value lies in between 0 to 2, as edge weights are between −1 and 1. Therefore, the value 0 is the best and 2 is the worst. PCC is a measure of the linear correlation between two variables X and Y. The range for *PCC* lies between −1 to 1 and 0 denotes randomness.

5.4 Regression Model

Three regression models namely, *Linear Regression* and *Elastic Net* and *Gradient Boosted Decision Tree (GBDT)* have been used. The first two models are considered in order to compare the results of our approach with those of Kumar et al. And among the various experiments we carried out with other standard models available in WEKA [1], *GBDT* is chosen for its superior performance in comparison with the rest of the models.

Gradient boosting is an ensemble of typically weak prediction models like decision trees and further that uses boosting. The implementations have been done in WEKA. For squared error loss function, *GBDT* has been implemented using *Additive Regression*, *Bootstrap Aggregating (Bagging)* and *Reduced Error Pruning Tree (REPTree)*. For the GBDT experimentation, we consider default values of $batchSize$ = 100, $numIterations$ = 10, $shrinkage$ = 1.0, $seed$ = 1 and $maxDepth = 6$. We found that depth restricted tree gives better prediction scores in comparison to that of depth unrestricted tree.

The robustness of the models is tested thoroughly following the approach of Kumar et al. using the methods given below:

1. **Leave-one-out cross validation.** This is an extreme form of cross validation where number of folds equals the number of instances. The number of models being built for training/testing is equal to the number of instances in the dataset and hence involves a large amount of experimentation.
2. **N% edge removal.** The network data is split into train and test sets where test dataset is of size N% edges, where N is taken from 10 to 90 with a step size of 10. The model is trained with remaining $(100 - N)$ edges and the calculation of the prediction scores on test set is conducted. The process is repeated for each split 10 times with a random seed value from 1 to 10. And the mean of the scores is computed and tabulated.

5.5 Results and Analysis

The results obtained by the supervised regression model ($AdditiveRegression + Bagging + REPTree$) are reported here. Also, we conducted experiments by replacing *REPTree* with *Random Tree* to insert randomness in the learning process and found that both models show equal performance. Therefore, we only consider the former model.

Leave-One-Out Cross Validation. The results given in Tables 3, 4 and 5 show that *GBDT* regression model achieves an improvement in PCC of 4%, 9% and 5% in case of BTC Alpha, OTC Net and Wiki RfA respectively in comparison to all the models considered by Kumar et al. Note that, the *GBDT* model built using feature set I is composed of only 4 features whereas the model of Kumar et al. is built using 13 features. It is interesting to note that just by taking the two FG features of Kumar et al. in a supervised framework, the performance of *GBDT* model outperforms the best results of Kumar et al.

Table 3. Leave-one-out cross validation results for weight prediction on BTC Alpha

	Number of features	Linear regression		Elastic net		GBDT	
		RMSE	PCC	RMSE	PCC	RMSE	PCC
Feature set I	4	0.24	0.59	0.24	0.59	**0.22**	**0.66**
Feature set II	2	0.29	0.09	0.29	0.09	0.28	0.25
Feature set III	2	0.24	0.58	0.24	0.58	0.23	0.63
Kumar et al. supervised	13	**0.22**	0.62	0.24	0.60	–	–

Table 4. Leave-one-out cross validation results for weight prediction on OTC Net

	Number of features	Linear regression		Elastic net		GBDT	
		RMSE	PCC	RMSE	PCC	RMSE	PCC
Feature set I	4	0.27	0.66	0.27	0.65	**0.24**	**0.75**
Feature set II	2	0.35	0.15	0.35	0.15	0.32	0.45
Feature set III	2	0.27	0.65	0.27	0.65	0.25	0.72
Kumar et al. supervised	13	0.26	0.66	0.27	0.65	–	–

Now, comparing the performance of different feature sets among the models chosen by Kumar et al., we observe the following. *Linear Regression* and *Elastic Net* models with feature set I (4 features) closely approximate the results of Kumar et al. (13 features) for *PCC* and *RMSE*. It is to be noted that in Tables 4 and 5 the results obtained using only FG features (Feature set III) are almost equal to that of Kumar et al.

Clearly, Feature set I which combines Line and graph features outperforms all the other combinations. It is interesting to note that though line graph features by themselves perform poorly and FG measures perform very well, their combination gives slightly improved results. For example in case of OTC Net in Table 4, ($Line + FG$) gives a PCC value of 0.75 whereas Kumar et al. obtain a PCC of 0.65 using 13 features. In fact, among the models tested *GBDT* gives the lowest RMSE of 0.24 as compared to 0.26 obtained by Kumar et al. A similar trend can be seen for the other two datasets in Tables 3 and 5.

N% Edge Removal. In Fig. 3, the performance using N% edge removal of four models: *Linear Regression*, *Elastic Net* and two *GBDT* models are plotted. The two GBDT models are built using Line features, one generated using undirected line graph *GBDT_U* and the other using directed line graph *GBDT_D*. In Wiki RfA, *GBDT_D* model achieves better scores compared to that of *GBDT_U* model. In BTC Alpha and OTC Net, *GBDT_U* model achieves comparable scores to that of *GBDT_D* model. And in all the networks, *GBDT* models give lower *RMSE* and higher *PCC* in comparison to the *Linear Regression* and *Elastic Net* models. Therefore, we do not include BTC Alpha and OTC Net plots. And *GBDT_D* is preferable since undirected line graph is not scalable as can be seen in Table 2.

Table 5. Leave-one-out cross validation results for weight prediction on Wiki RfA

	Number of features	Linear regression		Elastic net		GBDT	
		RMSE	PCC	RMSE	PCC	RMSE	PCC
Feature set I	4	0.23	0.49	0.23	0.50	**0.22**	**0.55**
Feature set II	2	0.26	0.04	0.26	0.04	0.25	0.24
Feature set III	2	0.23	0.49	0.23	0.50	**0.22**	0.54
Kumar et al. supervised	13	**0.22**	0.50	0.23	0.47	–	–

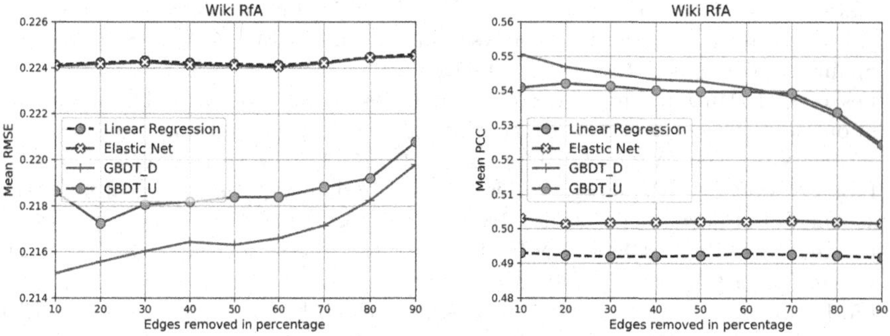

Fig. 3. N% edge removal performance evaluation in terms of RMSE and PCC

6 Conclusion

We have demonstrated that *GBDT* models outperform *Linear Regression* and *Elastic Net.* Our (*Line* + *FG*) approach which takes only four features (Feature Set I) and applies *GBDT* model performs better than that of Kumar et al. which considers 13 features. In case of WSNs, even though *GBDT_D* and *GBDT_U* give comparable results, *GBDT_D* approach has significant computational advantage over *GBDT_U*.

This work is done on a static snapshot of the network. The results need to be validated further using larger networks. We would like to apply these ideas to predict ratings on a recommendation network. And since temporal information in the network is utilized to enhance the performance of link prediction [5,12], a similar approach for weight prediction may be a useful direction to pursue.

References

1. Weka project webpage. https://www.cs.waikato.ac.nz/ml/weka/. Accessed 30 May 2019
2. WSN code and data. https://cs.stanford.edu/~srijan/wsn/. Accessed 30 May 2019
3. Fu, C., et al.: Link weight prediction using supervised learning methods and its application to yelp layered network. IEEE Trans. Knowl. Data Eng. **30**(8), 1507–1518 (2018)

4. Hasan, M.A., Chaoji, V., Salem, S., Zaki, M.: Link prediction using supervised learning. In: Proceedings of SDM 06 workshop on Link Analysis, Counterterrorism and Security, Bethesda, Maryland, USA (2006)
5. Jaya Lakshmi, T., Durga Bhavani, S.: Temporal probabilistic measure for link prediction in collaborative networks. Appl. Intell. **47**(1), 83–95 (2017)
6. Kumar, S., Spezzano, F., Subrahmanian, V.S., Faloutsos, C.: Edge weight prediction in weighted signed networks. In: 2016 IEEE 16th International Conference on Data Mining (ICDM), pp. 221–230, Barcelona, Spain, December 2016
7. Lakshmi, T.J., Bhavani, S.D.: Link prediction measures in various types of information networks: a review. In: IEEE/ACM 2018 International Conference on Advances in Social Networks Analysis and Mining, ASONAM 2018, Barcelona, Spain, 28–31 August 2018, pp. 1160–1167 (2018)
8. Leskovec, J., Huttenlocher, D., Kleinberg, J.: Predicting positive and negative links in online social networks. In: Proceedings of the 19th International Conference on World Wide Web, WWW 2010, pp. 641–650. ACM, New York (2010)
9. Liben-Nowell, D., Kleinberg, J.: The link-prediction problem for social networks. J. Am. Soc. Inform. Sci. Technol. **58**(7), 1019–1031 (2007)
10. Lichtenwalter, R.N., Chawla, N.V.: LPmade: link prediction made easy. J. Mach. Learn. Res. **12**, 2489–2492 (2011)
11. Lü, L., Zhou, T.: Link prediction in weighted networks: the role of weak ties. EPL (Europhys. Lett.) **89**(1), 18001 (2010)
12. Munasinghe, L., Ichise, R.: Time aware index for link prediction in social networks. In: Cuzzocrea, A., Dayal, U. (eds.) DaWaK 2011. LNCS, vol. 6862, pp. 342–353. Springer, Heidelberg (2011). https://doi.org/10.1007/978-3-642-23544-3_26
13. Murata, T., Moriyasu, S.: Link prediction of social networks based on weighted proximity measures. In: IEEE/WIC/ACM International Conference on Web Intelligence (WI 2007), pp. 85–88. CA, USA, November 2007
14. Newman, M.: Networks : An Introduction, 1st edn. Oxford University Press, Oxford (2010)
15. Zhao, J., et al.: Prediction of links and weights in networks by reliable routes. Sci. Rep. **5**, 12261 (2015)
16. Zhu, B., Xia, Y., Zhang, X.J.: Weight prediction in complex networks based on neighbor set. Sci. Rep. **6**, 38080 (2016)

An Ensemble Learning Based Author Identification System

Ankita Dhar[1(✉)], Himadri Mukherjee[1], Sk. Md. Obaidullah[2], and Kaushik Roy[1]

[1] Department of Computer Science, West Bengal State University, Kolkata, India
ankita.ankie@gmail.com, himadrim027@gmail.com, kaushik.mrg@gmail.com
[2] Department of Computer Science and Engineering, Aliah University, Kolkata, India
sk.obaidullah@gmail.com

Abstract. Author identification is an emerging domain in the area of Natural Language Processing (NLP) that allows us to identify the respective author of a particular piece of text. Every author had some unique characteristics of writing that involves their signature style of applying specific terms, making their piece of art distinct and noticeable and also there exists an extended story behind the linguistic and stylistic analysis in the identification of authors. In this paper we aim to produce a content resemblance based author identification system (AIS) using ensemble learning model for identification of author for a given piece of text. The experiment was tested on approximately 6,000 passages from 26 authors obtained from Bangla literature. Experimental results reported that the proposed technique performed better compare to the state-of-the art methods.

Keywords: Author identification · Content resemblance · Rotation forest · Ensemble learning

1 Introduction

Authors basically possess a unique writing style that is quite noticeable in their articles. The writing style is not always relevant with the topic of the article and can be identified by the readers. The task of identifying the author from a set of different authors based on a piece of text that may vary from short to big articles is author identification. Author identification has various applications associated with it such as intelligence-detaining the messages connecting the terrorists; civil law-copyright and estate disagreement; criminal law-identification of writers of provoking and ransom letters; cyber security-finding out the authors behind the virus source code; detection of plagiarism and identifying ghost writer. Earlier, author identification problem generally relied on huge texts whereas the recent works shows the focus is on short texts because of the blooming trends of social media that uses short text messages [14]. Identification of authors can be classified in two means: closed category and open category [11]. The closed category

B. Deka et al. (Eds.): PReMI 2019, LNCS 11941, pp. 65–73, 2019.
https://doi.org/10.1007/978-3-030-34869-4_8

means represents the identifying author in a group while the open category problem leaves the author unspecified in the series. Most of the works followed closed category procedure that are said to be easier compared to open category means.

While dealing with text documents in Bangla, a lot of challenges need to be faced because of the unavailability of standard resources and tools. Bangla being the 6th most popular language in whole world [6], demands the development of the technologies that help the users in predicting and choosing the respective author for a piece of text from a set of literature articles being considered based on various lexical, syntactic and analytical features. Also, Bangla based system can assist the users who are not well accomplished in western languages to utilize the information technology efficiently. These facts motivated us to devote our time and knowledge in this area of interest. Thus, development of an automatic author identification system in Bangla will have a great social impact especially in south Asia region. In this study, author identification has been considered as a supervised learning procedure, especially a task of multiple domain categorization problems. However, unlike traditional text representation schemes, we introduce a content based resemblance method for extraction of features and using rotation forest-an ensemble learning algorithm, the proposed model outperformed the existing models by obtaining an average accuracy of 98.75%. Also, we have encountered various articles from different authors having similar styles of writing their piece of art which made our task more difficult as well as challenging.

The rest of the paper is structured as follows: Sect. 2 provides a brief literature survey followed by Sect. 3 illustrating the proposed methodology; in Sect. 4, the experimental results have been analysed and comparative study with the existing works have been provided; and Sect. 5 concludes the paper showing some future works in this area.

2 Related Works

A general overview on author identification problem from the literature study have been observed that author identification has not widely been explored in Bangla and it is relatively an innovative field. We came across few works, for instance, Das and Mitra [4] have also worked with 36 text documents from 3 authors: Rabindranath Tagore, Sukanta Bhattacharya and Bankim Chandra Chattopadhyay. Uni-gram and bi-gram features were used and reported an accuracy of 90% for uni-gram and 100% for bi-gram feature. However, the database was too small to produce such genuine conclusions. Further, the authors being considered had different writing styles which made their study quite easy. Chakraborty and Choudhury [3] proposed 3 graph based methods by considering the association of series of character, use of certain expressions and formation of the sentences in a text document for the generation of individual graphs for an author. For each method, the graphs for different authors were clustered for developing a weighted graph and a simple graph traversal algorithm was used for author identification task. The experiment was tested on the articles of 6 authors and reported a maximum accuracy of 94.98% which they claimed to be 9.89%

higher than the standard method even for short texts. Phani et al. [10] worked with 3,000 passages, 1000 passages each from 3 Bengali authors: Sarat Chandra Chattopadhyay, Rabindranath Tagore and Bankim Chandra Chattopadhyay based on character n-grams, feature selection, ranking and analysis, and learning curve to evaluate the connection between the train and test accuracy. The experimental results show their proposed model obtained maximum accuracy of 98% based on character bi-gram feature. Rakshit et al. [12] delved into the articles for automatic extraction of linguistic and stylistic knowledge which are essential for the interpretation and differentiation of poems. They have worked with semantic based features for categorization of Bangla poems into 4 classes: devotional, love, nature and nationalism using SVM classifier and obtained an accuracy of 56.8% but they concluded that for poem categorization, word features are not sufficient and thus used semantic based features together with stylistic features and obtained an accuracy of 92.3%. Anisuzzaman and Salam [2] tested a hybrid approach by fusing n-gram and Naïve Bayes and achieved accuracy of 95%.

However, a number of similar works have been carried out for English, Marathi, Arabic and others. For instance, Madigan et al. [8] introduced scattered Bayesian logistic regression along with bag of words and bi-gram POS based features for identification of 27,342 articles from 114 authors. Alharthi et al. [1] proposed CNN based approach using different feature vectors for identifying the documents of Litrec dataset and obtained an average accuracy around 60%. Digamberrao and Prasad [5] developed two methods based on lexical and stylistic feature extraction schemes for author identification of 15 philosophical documents in Marathi from 5 authors and obtained 80% accuracy based on optimization with J48 algorithm. Otoom et al. [9] worked on a database comprises of 12 features and 456 articles from 7 authors. They combined the features with machine learning algorithms and achieved an accuracy of 82% for identification of the respective authors.

3 Proposed Methodology

The literature articles were first tokenized and then stopwords were removed followed by content resemblance measure and thereafter rotation forest algorithm was used to identify the author of the given piece of texts. The overview of the proposed system is diagrammatically demonstrated in Fig. 1.

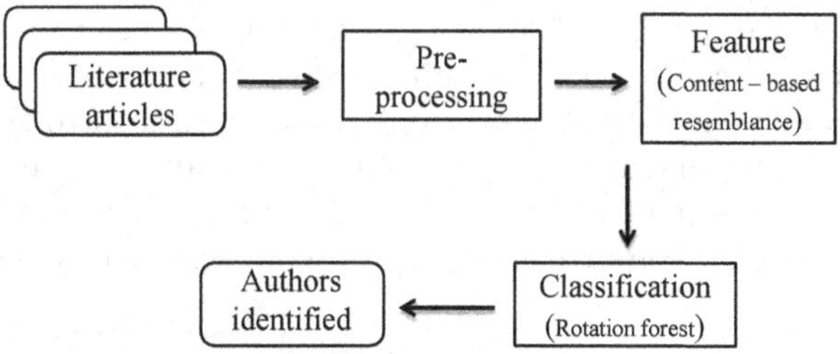

Fig. 1. The overview of the proposed system.

3.1 Dataset

Data is the most important element of an experiment. It is very essential for the database to incorporate real world properties and disparity in order to develop robust systems. From the literature study it has been found that there is no standard author identification database is available in Bangla consisting of sufficient amount of text documents from adequate amount of authors and thus we had to build our own database. The present work includes approximately 6,000 passages written by 26 authors such as Rabindranath Tagore (1), Kazi Nazrul Islam (2), Sarat Chandra Chattopadhyay (3), Bankim Chandra Chattopadhyay (4), Jibananda Das (5), Sunil Gangopadhyay (6), Humayun Ahmed (7), Vivekanada (8), Sukanta Bhattacharya (9), Sukumar Ray (10) and others in Bangla literature from www.ebanglalibrary.com. The database consists of male and female authors of various ages with some similarity in the writing styles and also has an ample amount of works such as novels, poems, essays, songs, dramas, short and big stories and others thoroughly been digitized to be used for author identification tasks.

3.2 Pre-processing

Digital text documents are generated using a linear sequence of characters, words and phrases. Prior to the further processing or analysis on the data, the raw texts were segregated into linguistic element called 'token'. The process of segregation of sentences into such tokens is termed as 'tokenization'. To extract the tokens, they were split depending on 'space' delimiter in our experiment. After the task of segregation of sentences, the total count of tokens becomes 6,79,343. Since each and every token do not contribute enough useful information that can be prove to be useful in further classification task, thus there is no need to retain those elements in the feature set and hence stopwords were removed from the set of tokens to clean and filter the dataset. Since the selection of stopwords is problem specific, therefore in the present work, 355 Bangla stopwords have been used provided in [15]. The total count of tokens results in 5,57,214 after removal of stopwords.

3.3 Feature

In the experiment a feature extraction technique has been proposed based on the resemblance of the contents between two text documents and named it as 'content resemblance measure' feature. The feature has been proposed due to the fact that in some cases it has been observed that the contents of the articles written by two different authors have similarities which lead to the misclassification. Thus we focus on extracting those relevant features that can be used for reducing the change of these ambiguities in the contents. The features based on content resemblance measure $Con\_Sim(x, y)$ are determined by the Eq. 1.

$$Con\_Sim(x, y) = \frac{\sum_{p \in D(x|y), q \in D(y|x)} S\_Cos(p, q)}{|D(x|y)| * |D(y|x)|} \tag{1}$$

where $D(x|y)$ denotes the set of contents for document x but not for document y, $D(y|x)$ denotes the set of contents for document y but not for document x and $S\_Cos(p,q)$ is the similarity between the contents of two text documents calculated based on soft cosine concept. A soft cosine measures the similarities between a pair of features. The standard cosine similarity metric treated the features as individual feature, whereas the soft cosine considers the similarity of features between two vectors, that helps to establish the concept of soft cosine as well as the similarity between two features.

$$S\_Cos(p,q) = \frac{\sum_{m,n}^{L} sim_{mn} a_m b_n}{\sqrt{\sum_{m,n}^{L} sim_{mn} a_m a_n}\sqrt{\sum_{m,n}^{L} sim_{mn} b_m b_n}} \tag{2}$$

where sim_{mn} is the similarity between feature m and n and L is the total data.

3.4 Ensemble Learning Model

Rotation forest classifier is a tree-based ensemble learning model with some major differences to random forest classifier and implements transformation on the subsets of attributes before the construction of each tree. The two major differences of rotation forest lies with the transformation of the attributes into sets of principle components; and secondly it utilizes the algorithm of decision tree (J48) by default. However, random forest and bagging provides encouraging outcomes while dealing with large number of ensembles whereas rotation forest is particularly developed to deal with a small number of ensembles. In rotation forest algorithm, the number of trees needed in the forest is determined by the user. It was proved to be useful compare to other ensembles such as AdaBoost, bagging and random forest on a number of standard databases [13].

4 Results and Discussion

4.1 Our Results

An experimental set up was made in order to compare rotation forest algorithm with three other ensemble models mentioned above in Sect. 3.4 on the dataset consisting of around 6,000 passages with 5,57,214 tokens after pre-processing. In all ensemble models, J48 was used as the default classifier, an extended version of C4.5, except in the random forest algorithm that generates the tree in such a way that random choice of a feature can be made at each node. As PCA is being used as default parameter for projection filter, hence the discrete features were changed into numeric features for rotation forest classifier. The decision tree classification algorithm implements an error-based pruning model. The confidence value required while pruning the tree can be given by the user which is by default set as 0.25 and in most of the cases provides better outcomes, thus, it was left unchanged in our present work. We have evaluated the experiment using 5-fold cross validations, keeping J48 classifier fixed along with changing projection filter and the results are provided in Table 1.

Table 1. Accuracy obtained using J48 classifier along with changing projection filter.

Classifier	Projection filter	No. of leaves	Size of tree	Accuracy (%)
J48	**Principal component analysis**	41	81	**96.75**
	Random projection	59	117	93.50
	StringtoWord Vector	53	105	76.75
	Standardize	53	105	75.50
	Cartesian product	53	105	76.75

Since maximum accuracy of 96.75% was obtained using PCA as projection filter, therefore, we have further studied the experiment keeping PCA constant and changing the classification algorithm to best first tree (BFTree). Based on the way of pruning, the results are provided in Table 2.

Table 2. Accuracy obtained using BFTree classifier and PCA projection filter.

Classifier	Projection filter	Pruning	No. of leaves	Size of tree	Accuracy (%)
Best	Principal	**Post-pruned**	53	27	**98.75**
first	Component	Un-pruned	55	28	97.75
tree	Analysis	Pre-pruned	53	27	98.00

The maximum accuracy of 98.75% is obtained using BFTree classifier for post-pruned scenario along with PCA as projection filter and the number of leaves and size of tree being 53 and 27. The average percentage of correct identification for all the authors is shown in Fig. 2 where it can be observed that there is a fall of accuracy for identification of Rabindranath Tagore, Sarat Chandra Chattopadhyay and Bankim Chandra Chattopadhyay due to the reason of their presence during the golden era of Bengali Renaissance and were having similar style of writing. Another confusion has been observed between the writings of Syed Shamsul Haq and Shamsur Rahman and between Syed Shamsul Haq and Samares Mazumder.

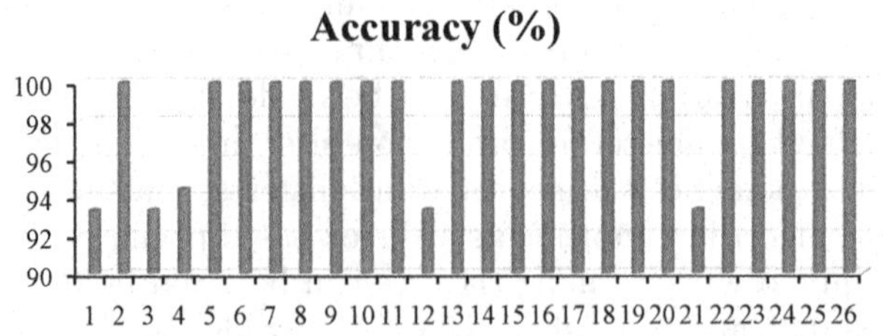

Fig. 2. The percentages of correct identification of all the authors.

We have compared the performance of the system with bagging, AdaBoost and random forest ensembles as well using WEKA [7] and the obtained accuracies together with precision, recall, F1 measure and Kappa statistics (statistic for measuring the score of homogeneity in the ratings provided by the judges and considers the similarity arises by chance) given for all the classifiers are presented in Table 3. We have considered BFTree classifier for bagging and AdaBoost ensembles as we have obtained maximum accuracy of 98.75% using BFTree for rotation forest. It can be observed from the Table that the performance of rotation forest is better among all the ensemble models being considered for the present experiment.

Table 3. Comparison among various ensemble learning models.

Ensembles	Precision	Recall	F1 measure	Kappa statistics	Accuracy (in %)
Rotation forest	**0.988**	**0.988**	**0.988**	**0.987**	**98.75**
AdaBoost	0.933	0.930	0.930	0.927	93.00
Bagging	0.886	0.880	0.880	0.875	88.00
Random forest	0.699	0.700	0.690	0.688	70.00

4.2 Comparison with Existing Methods

The working methodology have been compared with some of the recent works performed in Bangla. Our proposed approach outperformed all other existing methods for Bangla in terms of recognition accuracy. Also the dataset being developed for the present experiment is larger compare to all other works. The accuracies obtained for the existing methods along with our proposed model is demonstrated in Table 4.

Table 4. Comparison of the working methodology with other existing approaches.

Reference	Used feature	Accuracy (in %)
Chakraborty and Choudhury [3]	Graph based	94.98
Phani et al. [10]	Character n-gram	98.00
Proposed work	**Content resemblance based**	**98.75**

5 Conclusion

An ensemble learning based author identification system has been proposed in this paper which is capable of capturing the expressive behaviours of 26 authors and efficiently assigns the text pieces belonging to the respective authors using a content resemblance measure feature. The proposed system performed better compare to the existing approaches developed by others for the task. We have also harvested quite a large database in Bengali consists of around 6,000 passages which can be made publicly available on request. In future, we would like to explore some linguistically inspired features that can help in order to develop an automatic author identification system in Bangla. Considering the literature survey, it can be said that this study appears to be the first well grounded experiment for identification of authors in Bangla, especially in terms of number of authors as well as number of texts. Also in future, we would like to study the proposed methodology to other applications of NLP such as analysis of sentiments from blogs and tweets, detection of plagiarism in Bangla and others.

Acknowledgement. One of the authors thank DST for INSPIRE fellowship. We thank Dr. Amitabha Biswas, Asst. Prof. of Bengali Dept, WBSU and Ankita Shaw and Shilpa Roy, students of Comp. Sc. Dept, WBSU for their help while developing the dataset. We also thank Debajyoti Bose, Dept. of Electrical, Power & Energy, University of Petroleum and Energy Studies for his help during the preparation of this manuscript.

References

1. Alharthi, H., Inkpen, D., Szpakowicz, S.: Authorship identification for literary book recommendations. In: Proceedings of the International Conference on Computational Linguistics, pp. 390–400 (2018)
2. Anisuzzaman, D.M., Salam, A.: Authorship attribution for Bengali language using the fusion of N-gram and Naïve bayes algorithms. Int. J. Inf. Technol. Comput. Sci. **10**, 11–21 (2018)
3. Chakraborty, T., Choudhury, P.: Authorship identification in Bengali language: a graph based approach. In: Proceedings of the International Conference on Advances in Social Networks Analysis and Mining, pp. 443–446 (2016)
4. Das, S., Mitra, P.: Author identification in Bengali literary works. In: Kuznetsov, S.O., Mandal, D.P., Kundu, M.K., Pal, S.K. (eds.) PReMI 2011. LNCS, vol. 6744, pp. 220–226. Springer, Heidelberg (2011). https://doi.org/10.1007/978-3-642-21786-9_37
5. Digamberrao, K.S., Prasad, R.S.: Author identification using sequential minimal optimization with rule-based decision tree on Indian literature in Marathi. Procedia Comput. Sci. **132**, 1086–1101 (2018)
6. Ethnologue: (2019). https://www.ethnologue.com/language/ben
7. Hall, M., Frank, E., Holmes, G., Pfahringer, B., Reutemann, P., Witten, I.H.: The WEKA data mining software: an update. ACM SIGKDD Explor. Newslett. **11**(1), 10–18 (2009)
8. Madigan, D., Genkin, A., Lewis, D.D., Argamon, S., Fradkin, D., Ye, L.: Author identification on the large scale. In: Proceedings of the Meeting of the Classification Society of North America (2005)
9. Otoom, A.F., Abdullah, E.E., Jaafer, S., Hamdallh, A., Amer, D.: Towards author identification of Arabic text articles. In: Proceedings of the International Conference on Information and Communication Systems, pp. 1–4 (2014)
10. Phani, S., Lahiri, S., Biswas, A.: Authorship attribution in Bengali language. In: Proceedings of the 12th International Conference on Natural Language Processing, pp. 100–105 (2015)

11. Potha, N., Stamatatos, E.: A profile-based method for authorship verification. In: Likas, A., Blekas, K., Kalles, D. (eds.) SETN 2014. LNCS (LNAI), vol. 8445, pp. 313–326. Springer, Cham (2014). https://doi.org/10.1007/978-3-319-07064-3_25
12. Rakshit, G., Ghosh, A., Bhattacharyya, P., Haffari, G.: Automated analysis of Bangla poetry for classification and poet identification. In: Proceedings of the International Conference on Natural Language Processing, pp. 247–253 (2015)
13. Rodriguez, J.J., Kuncheva, L.I., Alonso, C.J.: Rotation forest: a new classifier ensemble method. IEEE Trans. Pattern Anal. Mach. Intell. **28**(10), 1619–1630 (2006)
14. Sapkal, K., Shrawankar, U.: Transliteration of secured SMS to Indian regional language. Procedia Comput. Sci. **78**, 748–755 (2016)
15. Stopwords: (2019). https://www.isical.ac.in/~fire/data/stopwords_list_ben.txt

Comparison of the Airline Networks of India with ANI Based on Network Parameters

Dimpee Baruah(✉) and Ankur Bharali

Department of Mathematics,
Dibrugarh University, Dibrugarh 786004, Assam, India
dimpeebaruah2005@gmail.com, a.bharali@dibru.ac.in

Abstract. Transportation systems play an important role in maintaining the balance of modern civilization. Among them, one of the essential transport systems is air transportation system. So the study and proper maintenance of the air transportation network, either domestic or global, prove to be very much essential. The present paper provides an analysis of three major airline networks of India viz. Indigo, Jet Airways, and Air India from the perspectives of structure and robustness. The study is done by adopting complex network approach. A comparison of these three major airlines with Airport Network of India (ANI) based on some network parameters is also analysed.

Keywords: Airline network · Robustness · Complex network · Network parameters · ANI

1 Introduction

In present time, one of the essential transport systems is the air transportation system. And, it is one of the inseparable elements of our present societies for their high level of mobility. So the study of air transportation network plays a vital role for proper maintenance of the system. Many studies have been done to analyze various countries air transportation network such as US [10], China [11], India [2], Brazil [5] etc. The studies have been done to analyze the infrastructure, connectivity, flow of traffic etc. of the network. As the dependency on these networks is increased, so the study of the robustness of the network becomes essential. Robustness is the ability of a network to continue to perform properly when it is subject to failures or attacks [7]. Robustness is one of the most anticipated properties of any transportation network. The level of vulnerability to which a network can expose through random and most central nodes or/and link failures, can be evaluated by studying the robustness of the network during different node or edge failures [1]. The robustness analysis of an airline network shows how vulnerable and fragile the network is when the network facing unintentional errors (random attack) and intentional attacks [12].

In this paper a comparison of three major airline networks of India viz. Indigo, Jet airways, Air India, with ANI based on network parameters is analysed. Firstly, the paper provides an analysis of robustness of the airlines. The daily networks of four different European airlines, during the busiest week, were analyzed by Han et al. [8]. And they have found all airline networks have scale-free and small-world properties

B. Deka et al. (Eds.): PReMI 2019, LNCS 11941, pp. 74–81, 2019.
https://doi.org/10.1007/978-3-030-34869-4_9

[8]. Also Reggiani et al. in 2009 [13] and in 2010 [14] investigated the Lufthansa network (LH) as both the worldwide and European networks; and they also found the networks were scale-free in both the cases. Again, multiple airlines in different alliances and parts of the whole world were studied by Lordan (2014) [12]. In 2015, Wijdeveld studied the robustness of 17 European airline networks for both error or random failure and targeted attacks [16].

The remaining parts of this paper is structured as follows: next section includes some parameters used in complex network analysis followed by an investigation on Airport Network of India (ANI) and three airline networks and then calculation of the values of various metrics of these networks is done. In the next section, the robustness analysis of the three airline networks have been reported, followed by their comparison with ANI. Conclusions are presented in the last section.

2 Some Parameters Used in Complex Network Analysis

Now, some definitions of some network parameters, which are frequently used in complex network analysis, are discussed (Table 1).

Table 1. Some network parameters

Measures	Formula	Variables	Description
Diameter	$d = max_{i,j \in V} L_{ij}$	L_{ij} = the shortest path (SP) between the nodes i and j	It is the longest SP between any two nodes in the network
Average Shortest Path Length (ASPL) [4]	$L = \frac{1}{N(N-1)} \sum_{i,j=1,\, i \neq j}^{N} L_{ij}$	N = the total number of vertices in the network	It is the average of all the shortest paths
Network Clustering Coefficient (NCC) [5]	$C = \frac{3 \times n_t}{n_{ct}}$	n_t = the number of triangles in the network n_{ct} = the number of connected triples of vertices	It measures the cliquishness of a node
Betweenness [6]	$\eta_u = \sum_{s \neq u \in V} \sum_{t \neq u \in V} \frac{\sigma_{st}(u)}{\sigma_{st}}$	σ_{st} = number of SP from vertex s to t $\sigma_{st}(u)$ = number of SP from s to t that pass through u	The Betweenness of a node is the number of SP passing through the node
Reachability [9]	$R_i = \frac{n_i}{N-1}$	n_i = number of nodes reachable from the node i	It is the probability of connectivity between any two nodes
Network Criticality [15]	$\tau = 2NTrace(L^{+})$	N = number of nodes L^{+} = Moore-Penrose inverse of Laplacian matrix L of the network.	Smaller the value of τ higher is the robustness of the network
Graph Density or Network Density [5]	$D = \frac{2M}{N(N-1)}$	M = number of edges N = number of nodes	It is a ratio of the number of edges to the number of edges in a complete graph of same size

3 An Investigation on ANI and Three Airlines Networks

In ANI and airline networks, nodes are the domestic airports and there exists a link between any two nodes (airports) if there is at least one direct flight connecting them per week. The number of such flight per week between the airports (nodes) is taken as the weight of the link in the network. Some dummy links are also added to the networks to make the networks symmetric. For the analysis of airline networks of India, three major airlines, viz. Indigo, Jet Airways and Air India are considered. All the air movement data for the networks are consider for the year 2016, obtained from Airports Authority of India (www.aai.aero). The values of the network parameters of ANI [3] and the three airline networks under consideration are given in Table 2 (Figs. 1 and 2).

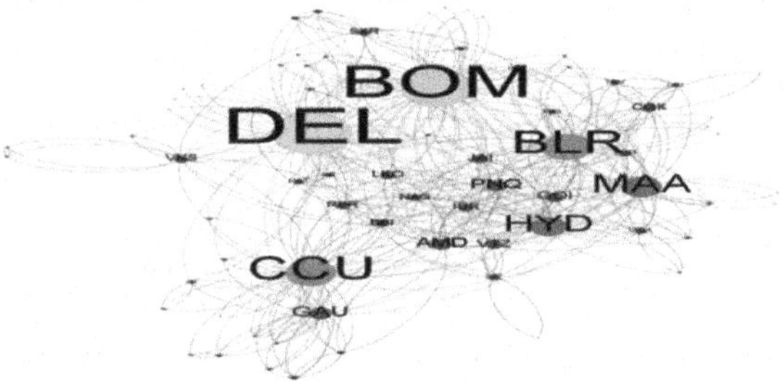

Fig. 1. Structure of ANI

Table 2. Network parameters of ANI and three airlines

Network parameters	ANI	Indigo	Jet Airways	Air India
Number of nodes (Airports)	79	37	46	53
Directed links (edges)	496	278	162	192
Connected components	1	1	1	1
Diameter	4	3	5	4
Average degree	6.279	7.514	3.522	3.623
APL	2.262	1.872	2.413	2.327
NCC	0.605	0.761	0.533	0.526
Network density	0.08	0.209	0.078	0.07
Network criticality	1.1334×10^3	730.3659	2.8582×10^3	3.5302×10^3

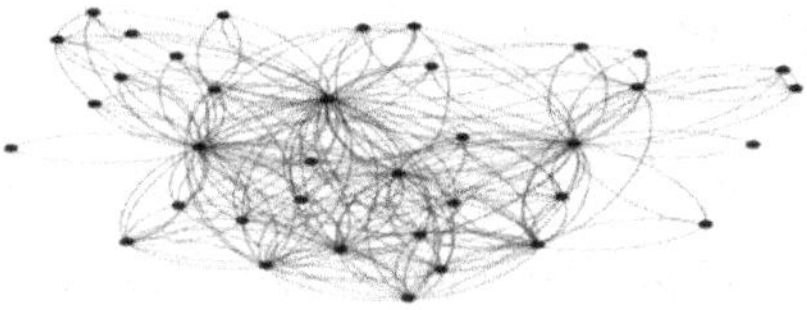

Indigo

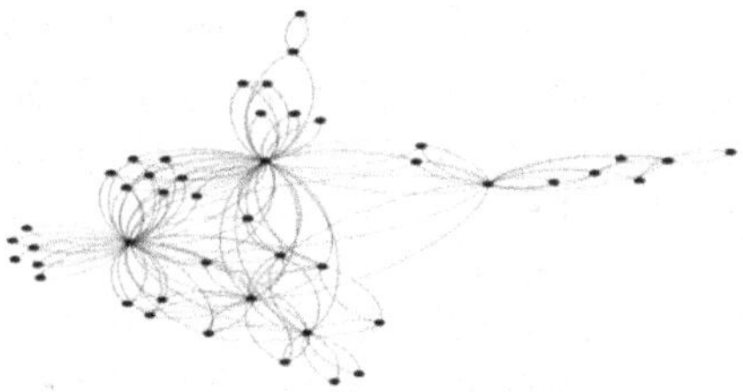

Jet Airways

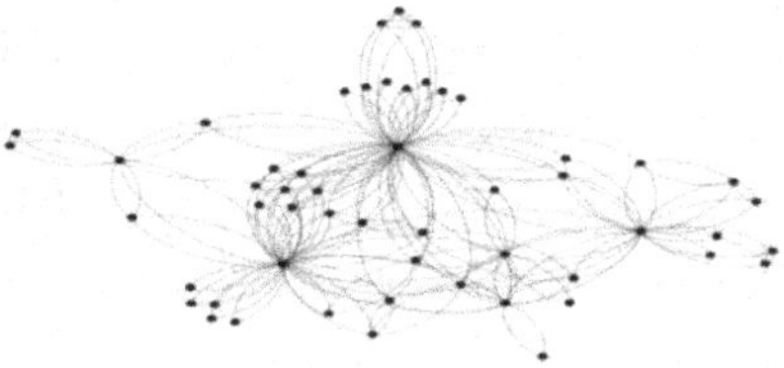

Air India

Fig. 2. Structure of the three airlines

4 Robustness Analysis of the Airline Networks

Robustness analysis of the networks can evaluate the effect of targeted attack (e.g. attack by terrorists on an airport) and random attack or failures (e.g. inclement weather) of the networks. In the analysis for targeted attack, we remove five highest degree nodes (airports) from each airline networks and observe the change in the average path length, clustering coefficient, network criticality, Reachability and to study the effect of random failures, we remove any five randomly selected airports from each airline networks and calculate the network parameters. Table 3 shows the change of network parameters (in percentage) of the three airline networks after removal of key node based on degree. The negative sign indicates decrement of the parameter whereas positive sign indicates the increment of the values. We also calculate these measures after removal of the nodes sequentially, based on relatively high degree or high betweenness i.e., in order to evaluate the robustness of the airlines networks, each time a node with high degree or betweenness is isolated (removed), the centrality measures are calculated again for all the remaining connected nodes (airports). Then from the remaining nodes, the airport with highest centrality is selected for removal in the next step and so on. The process will continue until removal of five airports. Figures 3 and 4 show the change of APL and NCC after consecutive removal of five nodes (airports) based on degree and betweenness and consecutive removal of five random nodes (airports) respectively.

Table 3. Change of network parameters after removal of a key node

Airline network	Airport removed	Number of nodes	Number of edges	$\Delta L\%$	$\Delta C\%$	$\Delta\tau\%$	$\Delta R\%$
Indigo	Delhi	36	216	7.37	−14.98	22.80	0
	Mumbai	35	224	0.80	−6.31	2.68	−2.78
	Kolkata	35	232	3.26	−0.39	2.6	−2.78
	Bangalore	36	238	1.34	0.39	−3.29	0
	Hyderabad	36	248	0.91	−0.66	−6.45	0
Jet Airways	Mumbai	39	110	3.07	−37.71	0.88	−23.18
	Delhi	42	114	0.04	−46.34	0.65	−26.67
	Bangalore	45	138	1.70	−16.89	7.96	0
	Kolkata	44	142	−8.00	−16.89	7.78	−21.92
	Chennai	43	144	−1.41	1.69	−5.7	−6.62
Air India	Delhi	45	126	6.06	−32.13	2.86	−33.14
	Mumbai	47	142	2.19	−26.81	5.43	−16.70
	Kolkata	50	162	0.86	−20.72	4.34	−12.7
	Chennai	51	176	0.17	0.57	−4.08	−1.92
	Hyderabad	51	176	0.17	−3.04	−2.77	−1.92

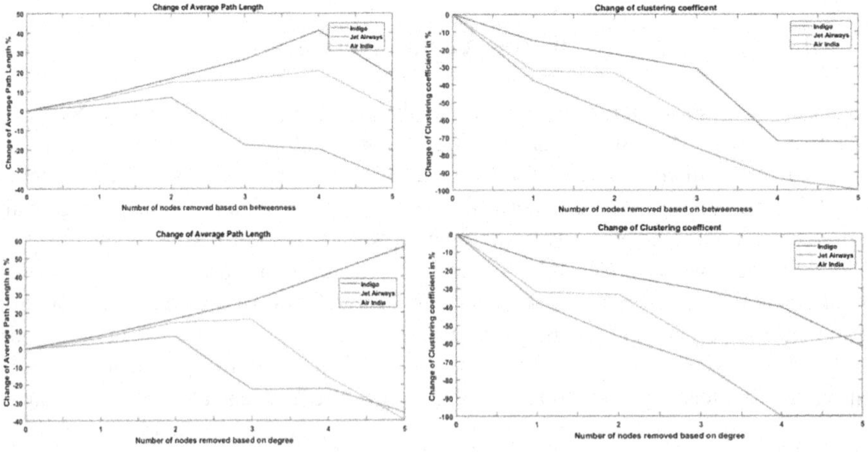

Fig. 3. Change of APL and NCC after consecutive removal of five nodes (airports) based on betweenness and degree.

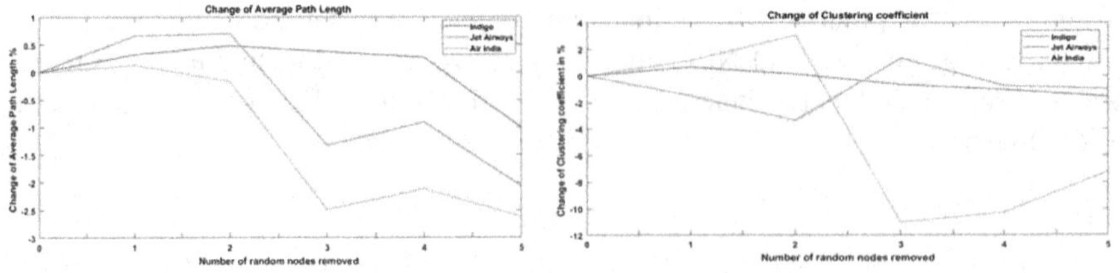

Fig. 4. Change of the APL and NCC after consecutive removal of random nodes (airports)

5 Comparison of Airline Network with ANI

In this section, a comparative study is carried out between the three airline networks and ANI. From structural perspective, we calculate the Correlation coefficient between the network parameters of the three airline networks and ANI. From the Table 4, we observe that all the three airline networks are positively correlated but Indigo network is most strongly correlated with ANI. And from robustness perspective, a comparison is made based on network parameters of the three airline networks and ANI by evaluating the value of R^2 (Coefficient of determination) for two cases. First one is evaluated for the values of the parameters after removal of one node and second one is for the values of the parameters after removal of five nodes. In case of removal of one node, Air India seems to have marginally better consensus with ANI than the other airline networks (see Fig. 5). However in case of multiple removals of nodes, we observe that Indigo network has sufficiently higher value of R^2 than Jet Airways and Air India (see Fig. 6).

Table 4. Correlation between network parameters of ANI and three airline networks (Structural perspective)

	ANI	Indigo	Jet Airways	Air India
ANI	1	0.9984	0.9391	0.9296
Indigo		1	0.9491	0.9484
Jet Airways			1	0.9998
Air India				1

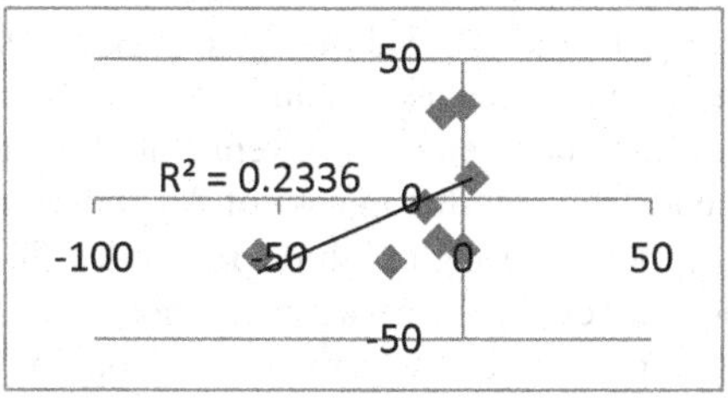

Indigo

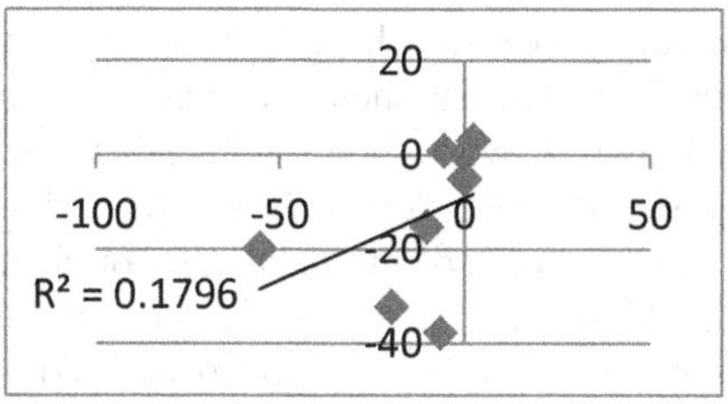

Jet Airways

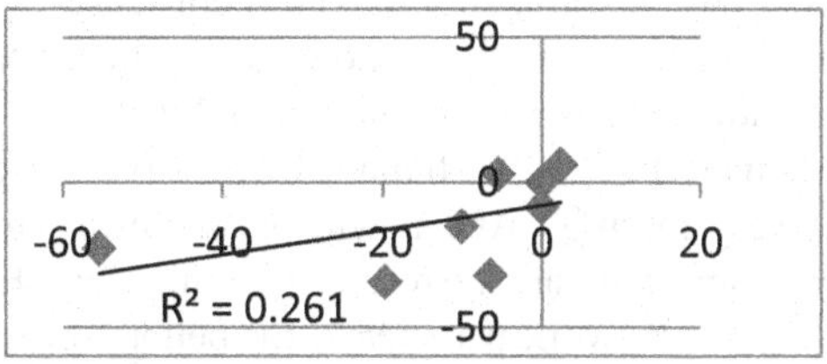

Air India

Fig. 5. Comparison based on network parameter after removal of one node from the networks

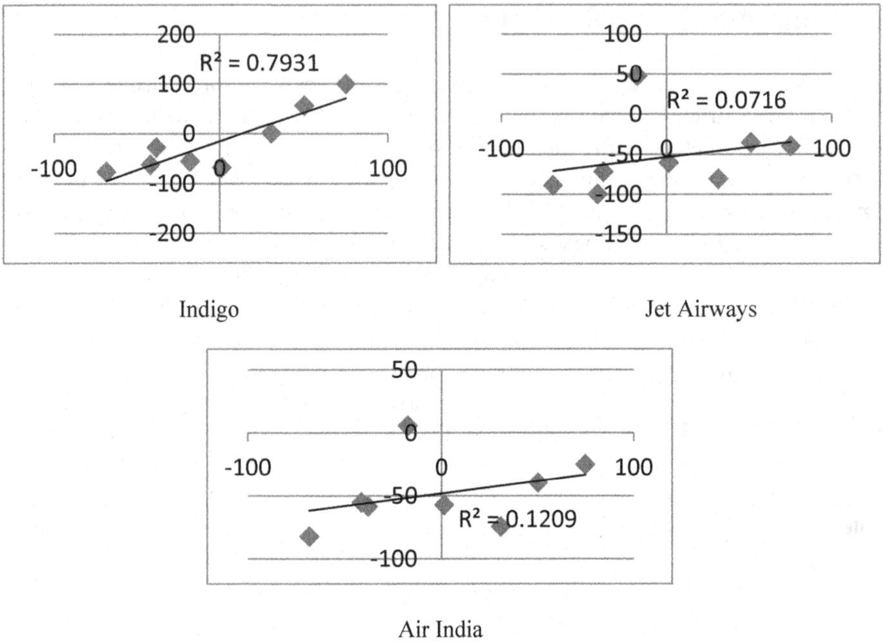

Fig. 6. Comparison based on network parameters after removal of five nodes from the networks

6 Conclusion

In this study, an analysis of the three major airline networks is carried out, which gives a fair idea about the whole ANI. It is expected that after removal of high degree nodes, average path lengths should increase and clustering coefficients should be decrease. But only Indigo network shows an increment in average path length after removal of nodes (airports) based on degree. It may be attributed to the fact that in case of Air India and Jet Airways most of the paths are not always routed through high degree nodes (like Delhi, Mumbai), rather these airlines provide direct flights between the not so high degree nodes (airports). Also from the comparison of network parameters of the three airline networks with ANI, we observe that from structural perspective Indigo network is strongly correlated with ANI. From robustness perspective, although for removal of only one node, Air India has shown a slightly better consensus with ANI, in case of multiple removals of nodes Indigo has again shown a way better correlation with ANI than rest of the airlines under consideration. Though in this study we consider the three most popular Airlines namely Indigo, Air India and Jet Airways, Jet Airways stops all its operations in India very recently. And it will be interesting to perform a similar study without Jet Airways and compare the results with this work. For further study, we can also consider some more network parameters for better correlation analysis.

References

1. Albert, R., Jeong, H., Barabási, A.-L.: Error and attack tolerance of complex networks. Nature **406**, 378–382 (2000)
2. Bagler, G.: Analysis of the airport network of India as a complex weighted network. Phys. A **387**(12), 2972–2980 (2008)
3. Bharali, A., Baruah, D.: On structure and robustness of airport network of India. Res. Rev.: Discret. Math. Struct. **5**(2), 26–33 (2018)
4. Barabási, A.L., Albert, R.: Emergence of scaling in random networks. Science **286**, 509–512 (1999)
5. Couto, G.S., Couto Da Silva, A.P., Ruiz, L.B., Benevenuto, F.: Structural properties of the Brazilian air transportation network. Ann. Brazil Acad. Sci. **87**(3), 1653–1674 (2015)
6. Freeman, L.: Centrality in social networks conceptual classification. Soc. Netw. **1**, 215–239 (1978)
7. Gribble, S.: Robustness in complex systems. In: Proceedings of the 8th Workshop on Hot Topics in Operation Systems (Hot OS-VIII) (2001)
8. Han, D.D., Qian, J.H., Liu, J.G.: Network topology and correlation features affiliated with European airline companies. Phys. A: Stat. Mech. Appl. **388**(1), 71–81 (2009)
9. Hossain, M., Alam, S., Rees, T., Abbass, H.: Australian airport network robustness analysis: a complex network approach. In: Proceedings of Australasian Transport Research Forum, pp. 1–21 (2013)
10. Chi, L.-P., et al.: Structural properties of US flight network. Chin. Phys. Lett. **20**(8), 1393–1396 (2003)
11. Li, W., Cai, X.: Statistical analysis of airport network of China. Phys. Rev. E **69**(4), 1–6 (2004)
12. Lordan, O., Sallan, J.M., Simo, P., Gonzalez-Prieto, D.: Robustness of airline alliance route networks. Commun. Nonlinear Sci. Numer. Simul. **22**(1–3), 587–595 (2015)
13. Reggiani, A., Signoretti, S., Nijkamp, P., Cento, A.: Network measures in civil air transport: a case study of Lufthansa. In: Naimzada, A.K., Stefani, S., Torriero, A. (eds.) Networks, Topology and Dynamics. LNE, pp. 257–282. Springer, Heidelberg (2009). https://doi.org/10.1007/978-3-540-68409-1_14
14. Reggiani, A., Nijkamp, P., Cento, A.: Connectivity and concentration in airline networks: a complexity analysis of Lufthansa network. Eur. J. Inf. Syst. **19**(4), 449–461 (2010)
15. Tizghadam, A., Leon-Garcia, A.: On robust traffic engineering in core networks. In: Global Telecommunications Conference, IEEE GLOBECOM, pp. 1–6 (2008)
16. Wijdeveld, S.J.: Airline and alliance networks topology and robustness from a complex network approach. Master thesis (2015). http://resolver.tudelft.nl/uuid:7bfc759c-e2bc-422e-ae69-3277d31c307f. Accessed 31 Aug 2019

Spectral Feature Based Kannada Dialect Classification from Stop Consonants

Nagaratna B. Chittaragi[1,2](✉), Pradyoth Hegde[1], Siva Krishna P. Mothukuri[1], and Shashidhar G. Koolagudi[1]

[1] Department of Computer Science and Engineering, National Institute of Technology Karnataka, Surathkal, Karnataka, India
nbchittaragi@gmail.com, pradyothhegde@gmail.com, msivakrish@gmail.com, koolagudi@nitk.edu.in

[2] Department of Information Science and Engineering, Siddaganga Institute of Technology, Tumkur, Karnataka, India

Abstract. This study focuses on the investigation of the significance of stop consonants in view of the classification of Kannada dialects. Majority of the studies proposed have shown the existence of evidential differences in the pronunciation of vowels across dialects. However, consonant based studies on dialect processing are found to be comparatively lesser. In this work, eight stop consonants are used for characterization of five Kannada dialects. Acoustic characteristics such as cepstral coefficients, formant frequencies, spectral flux, and rolloff features are explored from spectral analysis of stops. The consonant dataset is derived from standard Kannada dialect dataset consisting of 2417 consonants obtained from 16 native speakers from each dialect. Support vector machine (SVM) and decision tree-based extreme gradient boosting (XGB) ensemble classification methods are employed for automatic recognition of Kannada dialects. The research findings show that the stops existing for shorter duration also convey dialectal linguistic cues. Combination of spectral properties has contributed to the identification of distinct dialect-specific information across Kannada dialects.

Keywords: Kannada dialect classification · Stop consonants · Spectral features · SVM · XGB

1 Introduction

Automatic identification of dialects may be considered as a prominent research area in the speech community due to its implications to automatic speech recognition (ASR) systems. The primary task of the dialect identification is to recognize the speaker's regional variety of the language spoken. Due to overlaps in vocabulary and phonetic similarities among the dialects, dialect identification problem is considered to be more challenging than language identification [1].

The present paper explores the identification of five Kannada dialects from the stop consonant utterances. The Kannada language is an official language

B. Deka et al. (Eds.): PReMI 2019, LNCS 11941, pp. 82–90, 2019.
https://doi.org/10.1007/978-3-030-34869-4_10

spoken in Karnataka state. Very few dialect related studies are found in literature w.r.t. Kannada language. Stop consonants represent one of the broad categories of phones. The production of a stop involves a complete closure of the oral cavity followed by the release in the form of noise burst. The stop consonants are differentiated from each other in terms of the manner of articulation (whether voiced and aspirated) and the place of articulation. Because of their short duration, the classification of stops is a challenging problem.

Literature gives comparatively fewer works available for consonant-based dialect processing.

A study is proposed with the investigation of the existence of dialectal information in the bursts during the production of the stop consonants of two Greek dialects, namely, standard Greek and Cypriot [15]. A study has analyzed the pre-nasalization differences during the production of the voiced stop consonants among two dialects of Greek. The acoustic cues of voice onset time (VOT) and F0 are captured to represent the laryngeal contrast of Korean stops between Chonnam Korean and Seoul Korean dialects. These acoustic features have effectively modelled the cross-dialectal variations among the stops [7]. Cretan and Thessalonikan dialects of Modern Greek are considered and measured amplitude and duration features. This model has suggested the more complex and interactive influences of dialects, gender, and stress in the realization of pre-nasality in the voiced stops [10]. Few systems can be found for addressing dialect processing of Kannada language. An author has considered only two dialects of Kannada earlier and used only pronunciation variations among vowels. They used acoustic features for their analysis [13].

However, very few systems have been found for analysis of Kannada consonants from the dialect perspectives. A study has included the analysis of dialectal variations among four primitive dialects for the task of speech recognition. However, this has not included any analysis w.r.t. stop consonants [9]. Research findings with consonant-based Kannada dialect processing are found to be limited because of the following reasons. 1. Availability of Standard dialect dataset for Kannada language. 2. Accurate segmentation of stops is a difficult task due to the duration of the stops. 3. Plosives comprise the stops and affricates and are considered particularly challenging to recognize because of their highly dynamic characteristics differs with vowels. And this has motivated to consider this problem for dialectal analysis.

In this study, eight un-aspirated unvoiced and voiced consonants namely, /p/, /b/, /T/, /t/, /D/, /d/, /k/, and /g/ are used for classification of Kannada dialects. These eight consonants are commonly known as plosive sounds, as they are produced due to the constriction occurred at different regions in the mouth. Among these, /k/ & /g/ are velar, /p/ & /b/ are labial, /t/ & /d/ are dental, and /T/ & /D/ are retroflex consonants. Plosives of Kannada have typically observed with silent period in the closure phase, as closure duration of /k/, /t/, /T/ and /p/ are longer when compared to /g/, /d/, /D/ and /b/. Also, voiced plosives show the voicing bar at the lower frequencies. A stronger vertical spike is observed showing release burst for /k/, /t/ and /p/.

Table 1. Kannada dialect dataset

	CENK	CSTK	HYDK	MUBK	STHK
No. of speakers	30	34	37	26	29
Gender (Male+Female)	18+12	19+15	25+12	12+14	16+13
Duration (in min.)	112	132	120	130	128

Occurrences of both palatal stops in natural communication are comparatively rare in Kannada language. Hence, in this work, four types of stops produced at the velar, dental, retroflex, and labial consonants are analyzed individually to see their significances across five dialects. Spectral characteristics representing features such as the first three formant frequencies, Mel frequency cepstral coefficients (MFCCs), spectral flux, centroid, rolloff, and entropy are extracted from stops. Dialect classification systems are developed with SVM and XGB ensemble techniques.

The present paper is organized as follows. Consonant dataset details are provided in Sect. 2. Details of extracted spectral features and employed classification models are discussed in Sect. 3. Information regarding the experiments carried out, results and analysis are given in Sect. 4. Conclusions of the current work are presented in Sect. 5 along with future directions.

2 Consonant Based Dialect Dataset

The consonant dataset used in this work is derived from spontaneous Kannada dialect dataset [5]. It consists of five distinct Kannada dialects spoken across Karnataka, namely: Central Karnataka Region (CENK), Coastal Karnataka Region (CSTK), Hyderabad Karnataka Region (HYDK), Mumbai Karnataka Region (MUBK), and Southern Karnataka Region (STHK). Text-independent spontaneous speech is recorded from native dialect speakers. The age group of speakers lies between 21–72 years. Majority of the speakers are from rural areas, and they are not moved to any other places for a long time. An interview style is followed to make speakers talk continuously in a reasonably quiet environment. Detailed information regarding Kannada dialect dataset is presented in Table 1.

It is observed that the duration of plosive sounds is concise; hence, segmentation and extraction of significant dialect-specific features from them is a tedious task. Hence, in this work, all consonants collected are in terms of monosyllables (CV units), and plosive sounds are combined with vowels /a/, /u/, /i/ or /o/. Since co-articulation between them is comparatively less [11]. The /CV/ tokens are identified in spoken utterances manually using Praat tool. The onset of stop burst and offset of the vowels are identified through simultaneous inspections of both waveform and the spectrogram [2]. /CV/ syllables are segmented by detecting burst onset, /CV/ transition and complete vowel utterances. Majority of the /CV/ units are chosen from the word-initial position. A plosive dataset considered in this study consists of total 2417 stops extracted from 16 (9 Female + 7

Male) speakers from each dialect. The total number of consonant clips available for each dialect are as follows, CENK-455, CSTK-478, HYDK-484, MUBK-501, STHK-499.

3 Spectral Feature Extraction and Classification Models

In this paper, acoustic characteristics those significantly differentiates stops consonants of five dialects of Kannada are extracted. Features such as formants, standard MFCCs, spectral flux, centroid, rolloff, and entropy are explored to represent spectral behavior. These features try to capture vocal tract variations among different consonants across Kannada dialects.

3.1 Spectral Feature Extraction

Formants: Due to the oral cavity is closed during the constriction at the specific place at vocal tract, formants may not be available. However, closure is followed by the release of noise burst, due to which front cavity is excited by a sudden reduction in downstream. This, in turn, leads in shifting of formants either upwards or downward along with amplitude. This is depending on the place of constriction of the consonant and the following vowel. Three formant frequencies play a significant role in identifying different vowels. Similarly, vowels, along with stops, can also encapsulate variations occur during the pronunciation variations [14]. LPC based McCandless formant tracking algorithm is employed to extract the three formant frequencies from plosives with a 10 ms overlapped 20 ms frame [12]. LPC is a widely used method for formant extraction due to its compact and accurate computation. Figure 1 presents the utterance of /k/ consonant manually segmented from the word "Kannada" from five male speakers of five Kannada dialects.

MFCCs: Spectral changes between the five Kannada dialects are captured using MFCC features. These features have proven to capture the vocal tract variations of speech signal successfully. The coefficients extracted resembles the human auditory system. 13 MFCCs are extracted from a speech signal using block processing approach from a 20 ms frame with a shift of 10 ms.

Spectral Flux: Timbre is the speaker-specific feature of the sound unit that helps to compare the similarity of two speech utterances. The spectral flux usually corresponds to a perceptual roughness of sound. In this work, flux feature is computed and used to measure the spectral changes existing between two successive frames. It is computed by extracting the power spectra of one frame against the same of the previous one [8].

$$Fl_{(i,i-1)} = \sum_{k=1}^{Wf_L} (EN_i(k)) - (EN_{i-1}(k))^2 \tag{1}$$

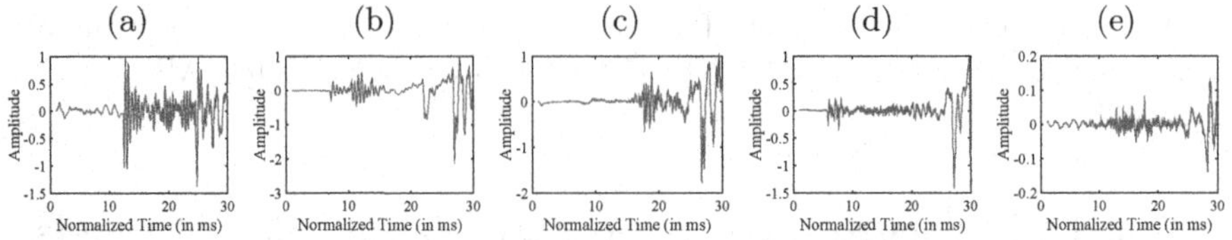

Fig. 1. Utterance of /k/ manually segmented from word "Kannada" five male speakers of five Kannada dialects, (a) CENK, (b) CSTK, (c) HYDK, (d) MUBK, (e) STHK

where $EN_i(k) = \frac{X_i(k)}{\sum_{l=1}^{Wf_L} X_i(l)}$, here $EN_i(k)$ is the kth normalized DFT coefficient at the ith frame, Wf_L is the frame size.

Spectral Entropy: Spectral entropy of a signal measures the distribution of spectral power. Spectral entropy is used to capture the abrupt changes within the energy levels of consonants. While computing spectral entropy of a frame, corresponding spectrum is divided into L sub-bands (bins). The energy E_f of the fth sub-band, for f = 0, . . ., L-1 is calculated using Eq. (2). Then, energies of all bins are normalized by dividing with the spectral energy of the whole frame, i.e., $ef = \frac{E_f}{\sum_{f=0}^{L-1} E_f}$, the entropy of each normalized energy value is calculated using the Eq. (3)

$$E(i) = \frac{1}{Wf_L} \sum_{k=1}^{w_L} |x_i(k)|^2 \tag{2}$$

$$H = -\sum_{f=0}^{L-1} ef.\log_2(ef) \tag{3}$$

In this work, the value of L is set to 10 indicating that each frame is divided into 10 bins.

Spectral Roll-Off: Spectral rolloff feature is treated as a spectral shape descriptor of an audio signal and it is used to differentiate voiced and unvoiced sounds. This feature is defined as the frequency below which a certain percentage (generally 90%–95%) of the magnitude distribution of the spectrum is concentrated.

3.2 Classification Models

In this work, standard SVM and multiple classifiers based XGB algorithms are used for the implementation of dialect recognition systems from stops. The SVM classification method is employed to capture dialect-specific cues. It is designed with the one-versus-rest approach to handle the 5-class pattern classification problem. Radial basis function (RBF) kernel function is used for separating hyperplane with the maximal margin in a high dimensional feature space [3]. Apart from this, nowadays, ensemble algorithms are gaining popularity. These

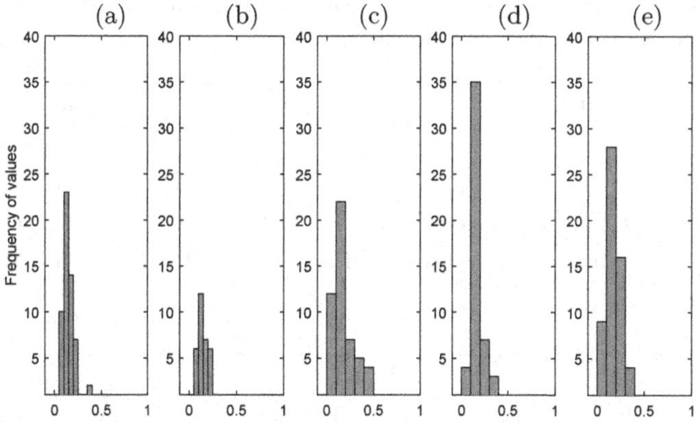

Fig. 2. Spectral rolloff variations among five Kannada dialects, (a)-CENK, (b)-CSTK, (c)-HYDK, (d)-MUBK, (e)-STHK

are powerful prediction and classification techniques in enhancing the performance with a combination of multiple classifiers over single classification methods. However, use of these methods for speech processing in specific for dialect identification is rarely found [6]. Hence, XGB method is used for classifying five dialects. Among ensembles, the most commonly used gradient boosting involves three significant steps: The first step is the selection of a suitable loss function; here, multi-class *logloss* is used as the problem addressed is a classification problem. Next step is choosing a base classifier, and decision trees are used in this paper, where trees are constructed using the greedy approach. Few parameters such as best split, number of leaf nodes, maximum levels are fine-tuned to produce a better performance. In the third step, trees are added one at a time; a gradient descent procedure is used for minimization of loss during the addition of trees. The XG boost library is used for implementation to handle five dialect classes [4].

4 Experiments and Results

This section provides details of the spectral analysis of stops through features explored from stops from five dialects along with complete details of experiments and results obtained.

From Fig. 1, it can be noticed that there are variations in the pronunciation of consonant /k/ across five dialects. Varying length and energy of the burst regions can be seen among five dialects. Spectral rolloff is generally treated as a spectral shape descriptor of an audio signal and it is usually used for discrimination of voiced and unvoiced sounds. In this study, the histograms are plotted for the spectral rolloff feature and are presented in Fig. 2. The rolloff parameters are extracted from /k/ stops spoken from five dialect speakers. From these histograms, it has been observed that the spectral rolloff value distributions

Table 2. Average dialect recognition performance from **consonants** level utterances using SVM and XGB methods (Accuracies in %)

Sl. no.	Features extracted	Kannada dialects	
		SVM	XGB
1	MFCCs	73.33	78.00
2	Spectral flux, rolloff, centroid, F1 and F2	59.37	68.76
3	Spectral flux, rolloff, centroid, F1 and F2 + MFCCs	77.65	**78.33**

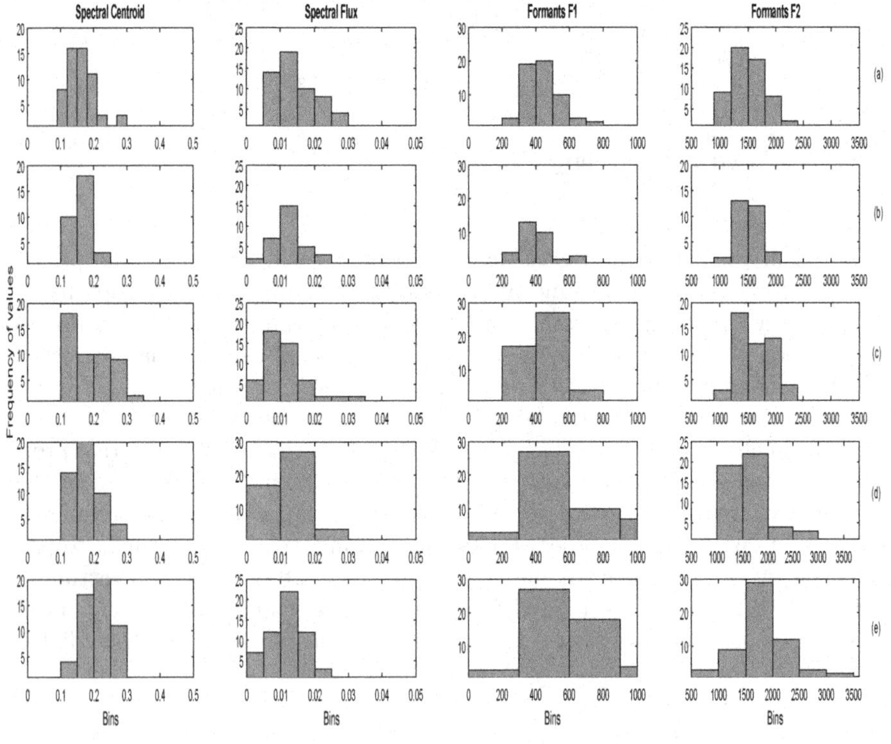

Fig. 3. Histograms of four spectral features, drawn for utterance of /k/ from five dialect speakers. (a) CENK, (b) CSTK, (c). HYDK, (d) MUBK, (e) STHK

are comparatively lower with CSTK and CENK dialects over three remaining dialects. It is also observed that the dialect MUBK and STHK dialect is with higher values of the spectral rolloff sequence. Besides, the variation is more intense for these two dialects. Whereas, CSTK and CENK dialects are noticed with lower values of spectral rolloff.

In order to show the differences across dialects with consonant /k/, centroid, flux, and two formants (F1 and F2) features are considered. The same above mentioned histogram is used. Figure 3 is drawn to show the spectral variations across dialects. Formants F1 and F2 are seen to be the distinguishing features among dialects with /k/ stops. However, the spectral centroid is seen with lesser variations. Even spectral flux feature considered is also contributing to several differences among dialects. Based on the analysis of the features as mentioned above in the characterization of dialects, several experiments are conducted by using these features for classification of dialects. In this work, single and ensemble SVM classification methods are employed for the development of Kannada dialect classification systems. Average dialect classification performance obtained from consonant utterances are presented in Table 2.

From the results obtained from the Table 2, it is observed that MFCCs are seen to be powerful features in classification of dialects from very shorter utterances such as stop consonants. Even spectral features are also captured the dialectal cues from stops. The combination of features has demonstrated an accuracy of 78.33%, which is slightly higher than MFCC features alone. However, from these analyses, it is noticed that dialect specific evidence are present even at consonant level utterances. Spectral attributes can effectively model the dialectal variations of stop consonants.

5 Conclusions

The present paper has proposed a Kannada dialect identification system from the stop consonants. For this purpose, spectral features are extracted from the shorter stop consonants to capture dialect-specific features. Stop consonant dataset is used in this is created from Kannada dialect dataset. SVM and XGB classification models are employed for automatic recognition of dialects. Combination of spectral features has demonstrated the better classification performance over MFCC features alone. A highest of 78.33% of dialect recognition performance is achieved with the use of stop consonants. However, in this paper, consonants considered are associated with vowels sounds. In the future, consonants alone can be used for classification of dialects. Dialect-specific features can be explored from the burst regions of the consonants as they consist of more relevant dialectal information. Apart from spectral analysis, excitation source features can also enhance dialect recognition performance.

Acknowledgment. This work is supported by DST-GOI (Department of Science and Technology, Government of India) sponsored project entitled *Characterization and Identification of Dialects in the Kannada Language.*

References

1. Biadsy, F.: Automatic dialect and accent recognition and its application to speech recognition. Ph.D. thesis, Columbia University (2011)
2. Boersma, P.: Praat, a system for doing phonetics by computer. Glot Int. **5**(10), 341–345 (2002)
3. Chang, C.C., Lin, C.J.: LIBSVM: a library for support vector machines. ACM Trans. Intell. Syst. Technol. (TIST) **2**(3), 27 (2011)

4. Chen, T., Guestrin, C.: XGBoost: a scalable tree boosting system. In: Twenty second International Conference on Knowledge Discovery and Data Mining, pp. 785–794. ACM (2016)
5. Chittaragi, N.B., Limaye, A., Chandana, N.T., Annappa, B., Koolagudi, S.G.: Automatic text-independent kannada dialect identification system. In: Satapathy, S.C., Bhateja, V., Somanah, R., Yang, X.-S., Senkerik, R. (eds.) Information Systems Design and Intelligent Applications. AISC, vol. 863, pp. 79–87. Springer, Singapore (2019). https://doi.org/10.1007/978-981-13-3338-5_8
6. Chittaragi, N.B., Prakash, A., Koolagudi, S.G.: Dialect identification using spectral and prosodic features on single and ensemble classifiers. Arab. J. Sci. Eng. **43**(8), 4289–4302 (2018)
7. Choi, H.: Acoustic cues for the Korean stop contrast-Dialectal variation. Citeseer (2013)
8. Giannakopoulos, T., Pikrakis, A.: Introduction to Audio Analysis: A MATLAB Approach. Academic Press, Cambridge (2014)
9. Hemakumar, G., Punithavalli, M., Thippeswamy, K.: Speech recognition system for different Kannada dialects. Int. J. Sci. Res. Comput. Sci. Eng. Inform. Technol. **2**(5), 180–188 (2017)
10. Jong Kong, E., Syrika, A., Edwards, J.R.: Voiced stop prenasalization in two dialects of greek. J. Acoust. Soc. Am. **132**(5), 3439–3452 (2012)
11. Kalaiah, M.K., Bhat, J.S.: Effect of vowel context on the recognition of initial consonants in kannada. J. Audiol. Otol. **21**(3), 146 (2017)
12. McCandless, S.: An algorithm for automatic formant extraction using linear prediction spectra. IEEE Trans. Acoust. Speech Signal Process. **22**(2), 135–141 (1974)
13. Nagesha, K.S., Kumar, G.H.: Acoustic-phonetic analysis of Kannada accents. Tata Institute of Fundamental Research, Mumbai (2010)
14. Reetz, H., Jongman, A.: Phonetics Transcription, Production, Aoustics and Perception. Wiley Blackwell, Hoboken (2009)
15. Themistocleous, C.: The bursts of stops can convey dialectal information. J. Acoust. Soc. Am. **140**(4), EL334–EL339 (2016)

Incorporation of Neighborhood Concept in Enhancing SOM Based Multi-label Classification

Naveen Saini(✉), Sriparna Saha, and Pushpak Bhattacharyya

Indian Institute of Technology Patna, Bihar, India
{naveen.pcs16,sriparna,pb}@iitp.ac.in

Abstract. The self-organizing map (SOM), which is a type of neural network, helps in the exploratory phase of data mining by projecting the input data into a lower-dimensional map consisting of a grid of neurons. In recent years, SOM has also been applied for classification of data points. The prominent utility of SOM based classification is evident from the use of no labeled data during training. In this paper, a self-organizing map based algorithm is proposed to solve the multi-label classification problem, named as ML-SOM. SOM follows an unsupervised training process to learn the topological structure of the training points. At testing-phase, a testing instance can be mapped to a specific neuron in the network and it's label can be determined using the training instances mapped to that specific neuron and nearby neurons. Thus in this paper, we have considered the neighborhood information of SOM to determine the label vector of testing instances. Experiments were performed on five multi-labeled datasets and performance of the proposed system is compared with various state-of-the-art methods showing competitive performance. Results are also validated using statistical significance t-test.

Keywords: Self-organizing Map · Multi-label classification · Unsupervised methods · Neighborhood function · Gaussian function

1 Introduction

In general, classifying an object or instance, x_i, in machine learning refers to assigning a single label out of a set of disjoint labels, L. This type of task is called as single label classification problem. But, in real-life, we can encounter with some classification problems where each of the data instances, x_i, may belong to more than one class and the task is to predict multiple labels of that instance and thus it can be referred to as multi-label classification (MLC) [5]. Now-a-days, MLC has become a hot research area and its applications can be found in various real-life domains like bioinformatics [7], image classification, etc. In document classification problem (DCP), one document may belong to more than one category, for example, biology and computer science. Therefore, DCP can also be considered as a MLC problem.

B. Deka et al. (Eds.): PReMI 2019, LNCS 11941, pp. 91–99, 2019.
https://doi.org/10.1007/978-3-030-34869-4_11

In the literature, several techniques are developed to deal with multi-label classification problems. As discussed in ref. [2], two types of existing approaches are there: algorithm dependent and algorithm independent. In algorithm independent approach, traditional classifiers are used which transform the multi-label classification problem into a set of single-label classification problems. While, this is not the case in algorithm dependent approach for multi-label task. In algorithm independent approach, each classifier is associated to some class and used to solve binary classification problem. This developed method is called as Binary-Relevance (BR) [3]. But, BR method has some drawbacks: it considers the classes independent of each other, which is not always true. Some of the examples of multi-label classifiers belonging to BR category are: Support Vector Machine-BR, J48 Decision tree-BR and k-Nearest Neighbor-BR. Later, some improvements over BR method were proposed [1]. Another version of algorithm-independent approach is LP (Label-Powerset) transformation method. All the classes allocated to a particular instance are combined into a unique and new class by considering correlations among-st classes. But this method increases the number of classes. An example of algorithm dependent approach is ML-kNN [8] in which for each instance, k-nearest classes are determined. Then principle of maximum posteriori is used to find the classes of a new instance.

Most of the proposed algorithms for multi-label classification are supervised in nature which require some labeled data for training the models. Therefore, there is a need to develop an unsupervised/semi-supervised framework to deal with multi-label classification which can achieve comparable results or can outperform the supervised methods. In the current paper, unsupervised neural network, called as self-organizing map [4], is used for proposing a multi-label classification framework which does not require any labeled data at the time of training. Self-organizing Map (SOM) is a neural network model consisting of two layers: input and output. Output layer is a grid of neurons arranged in low-dimensional manner. The principle of SOM states that the input patterns which are similar to each other in the input space appear next to each other in the neuron space. This is due to cooperation and adaptation process of SOM. Thus, it can be used for classification purpose. Usually, low dimensional space consists of 2−d grid of neurons. Let $T = \{x_1, x_2....x_H\}$ be a set of H training samples in n-dimensional space, then each neuron (or map unit) u ∈ D (number of neurons) has: (a) a predefined position in the output space: $z^u = (z_1^u, z_2^u)$; (b) a weight vector $w^u = [w_1^u, w_2^u, \ldots, w_n^u]$. It is important to note that dimension of weight vector of a neuron should be equal to vector dimension of input vector to perform mapping.

Label information of the testing instances is used while checking the performance of the system. However, there are many previous works on supervised SOM that make use of labeled data during training. We can use the supervised SOM to increase the performance of multi-label classification task, but, generating labeled data is a time consuming and cost sensitive process. Therefore, in this paper, we adopted the unsupervised SOM.

Recently, [2] have proposed a SOM based method for multi-label classification and used the traditional SOM training algorithm. But, it suffers from following drawbacks: (a) during testing an instance, it considers only the winning (mapping) neuron to decide its label vector. Label vector can be obtained by first averaging the label vectors of training instances mapped to that winning neuron and then some threshold value can be utilized to decide the class labels. The neighborhood information captured by SOM, one of its key-characteristics, was not utilized at the time of generating the label vector of any test-instance. In the current study we incorporated the use of neighbor-hood information during the testing phase of SOM based multi-label classification framework. In our proposed algorithm, we have also varied the number of neighboring neurons. (b) Authors have not given any information regarding the parameters like SOM training parameters, threshold value, used in their algorithm, therefore, sensitivity analysis is performed in our framework to determine the best values of the parameters used.

Experiments were performed on five multi-labeled datasets and results are compared with various existing supervised and unsupervised methods for MLC. Results illustrate that our system is superior to previously existing SOM based multi-label classifier and some other existing methods. Rest of the sections are organized as below: Sect. 2 discusses the proposed framework. Section 3 and Sect. 4 describe the experimental setup and discussion of results, respectively. Finally, Sect. 5 concludes the paper.

2 Proposed Methodology for Multi-label Classification Using Self Organizing Map

In this section, we first present the self-organizing map and its procedure of mapping instances to neurons. Then we will discuss the proposed classification algorithm (ML-SOM) for multi-label instances.

1. Representation of Label Vector: In multi-labeled data, an instance x_i may belong to more than one class. Therefore, label vector (LV_i) of instance x_i will be represented by a binary vector and the size of the binary vector will be equal to the number of classes. The kth position of the label vector will be 1 if that particular instance belongs to class 'k', otherwise, it will be 0. For example, if there are 10 classes and an instance belongs to first, third, fifth, seventh and ninth classes, then binary vector will be represented as $[1, 0, 1, 0, 1, 0, 1, 0, 1, 0]$.

2. SOM Training Algorithm: Training of SOM starts after initializing the weight vectors of neurons denoted as $w = \{w_1, w_2.....w_D\}$, where D is the number of neurons. These weight vectors are chosen randomly from the available training samples. In this paper, we have used the sequential learning algorithm for training of SOM. However, batch algorithm can also be used for the same task. Basic steps of the learning approach of SOM based multi-label classification can be found in the paper [2] and the brief overview of the algorithm is presented below. Let $Maxiter$ be the maximum number of SOM training iterations, η_0 and σ_0 be the initial learning rate and neighborhood radius, respectively, which decrease continuously as the number of iterations increases. For each data sample presented to the network, first its winning neuron is determined using the shortest Euclidean distance criterion. Then, neighboring neurons around the winning neuron are determined using the position vectors of the neurons. The best choice of selecting neighborhood is Gaussian function which is represented as $h_{uj} = \exp(-\frac{d_{u,j}^2}{2\sigma^2})$, where, u is a winning neuron index and j is the neighboring neuron index, d_{uj} is the Euclidean distance between neuron u and neuron j using position vectors, σ is the neighborhood radius and calculated as $\sigma = \sigma_0 * (1 - \frac{t}{Maxiter})$, where, t is the current iteration number. Finally, weights of the winning and neighboring neurons are updated as $w^u = w^u + \eta \times h_{u',j} \times (x - w^u)$, where, $\eta = \eta_0 * (1 - \frac{t}{Maxiter})$. It was done so that they come close to the input samples presented to the network and form a cluster of similar neurons around the winning neuron. Note that for a particular neuron, initially all the remaining neurons are neighbors (represented by σ) and this will keep on decreasing as training iteration continues.

*3. **Multi-label Data Classification Procedure:*** When a test instance x_i is presented to the trained SOM network, then firstly mapping is performed *i.e.*, its closest neuron 'b' is determined. Now, the label vector of the test instance x_i can be determined using the following steps:

(a) Perform averaging of label vectors of the training instances mapped to the closest neuron 'b' and nearby (adjacent neighbors) neurons 'N'. The obtained vector is called prototype vector (PV_i) and the kth value of the prototype vector indicates the probability of test instance belonging to class 'k'. The kth value is represented as $PV_{i,k} = \frac{|H_{w,k}|+|H_{N,k}|}{|H_w|+|H_N|}$, where, $|H_{w,k}|$ and $|H_{N,k}|$ are the set of training instances mapped to closest and nearby neurons, respectively, belonging to class 'k', $|H_w|$ and $|H_N|$ are the total number of training instances mapped to closest and nearby neurons, respectively.

(b) If the probability of a class in the obtained prototype vector is greater than or equal to some threshold, then probability value will be replaced by 1, otherwise it will be replaced by 0. The obtained binary vector will be the label vector of the testing instance.

Extracting Neighboring Neurons Around Winning Neuron: To decide the neighbors around the wining neuron, Euclidean distances between winning neuron and other neurons are calculated using position vectors. For example: suppose there are 9 neurons (indices $= 0, 1, 2, \ldots, 8$) arranged in 3×3 grid having position vectors $\{(0, 0), (0, 1), (0, 2), (1, 0), (1, 1), (1, 2), (2, 0), (2, 1), (2, 2)\}$. If index $= 3$ is the winning neuron, then indices of neighboring neurons will be $\{0, 1, 4, 6, 7\}$ as they are adjacent to the winning neuron. In general, if other neurons have distances less than or equal to 1.414, then those neurons will be adjacent neighboring neurons.

3 Experimental Setup

For our experimentation, we have chosen various multi-labeled data sets having varied number of labels. These data sets are publicly[1] available and related to different domains like audio, biology, images. Brief description of the datasets used in our experiment can be found in the paper [2]. These datasets are divided into 70% training data and 30% testing data. For the purpose of comparison, seven supervised algorithms and one recently proposed unsupervised algorithm are used. Supervised algorithms include: J48 Decision tree, Support Vector Machine (SVM), k-Nearest Neighbors (kNN), Multi-label k-Nearest Neighbor (MLkNN) [8], Back-propagation Multi-label Learning (BPMLL) [6] which is

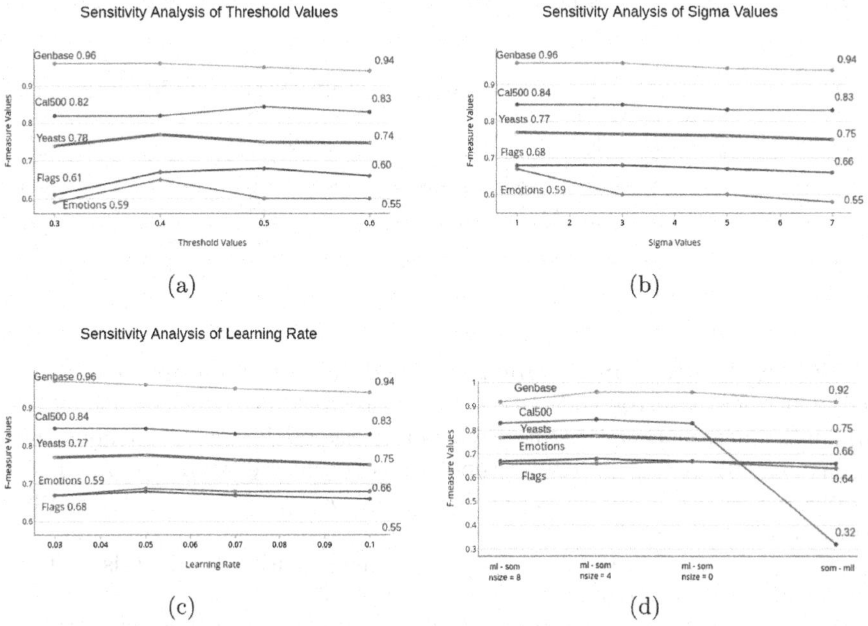

Fig. 1. Sensitivity analysis on (a) threshold value; (b) σ_0; (c) η_0 for different datasets; (d) F-measure values obtained using ML-SOM method by varying the neighborhood sizes in comparison with SOM-MLL

Table 1. Precision values obtained by different multi-label classification techniques

Dataset	ML-SOM	SOM-MLL	SVM-BR	J48-BR	KNN-BR	SVM-LP	J48-LP	KNN-LP	BPMLL	MLkNN
cal500	0.84 ± 0.03	0.60 ± 0.02	0.62 ± 0.07	0.45 ± 0.08	0.35 ± 0.04	0.34 ± 0.05	0.34 ± 0.04	0.35 ± 0.04	0.35 ± 0.04	0.60 ± 0.06
emotions	0.68 ± 0.06	0.63 ± 0.07	0.68 ± 0.10	0.59 ± 0.13	0.63 ± 0.11	0.68 ± 0.16	0.58 ± 0.15	0.63 ± 0.11	0.64 ± 0.12	0.70 ± 0.16
flags	0.75 ± 0.05	0.68 ± 0.05	0.72 ± 0.06	0.69 ± 0.14	0.68 ± 0.16	0.69 ± 0.12	0.66 ± 0.14	0.68 ± 0.16	0.69 ± 0.08	0.72 ± 0.10
genbase	0.96 ± 0.03	0.93 ± 0.03	0.99 ± 0.02	0.99 ± 0.03	0.99 ± 0.02	0.99 ± 0.02	0.99 ± 0.03	0.99 ± 0.02	0.04 ± 0.04	0.98 ± 0.05
yeast	0.80 ± 0.03	0.71 ± 0.01	0.72 ± 0.06	0.60 ± 0.06	0.60 ± 0.07	0.66 ± 0.05	0.54 ± 0.06	0.60 ± 0.07	0.62 ± 0.05	0.72 ± 0.04
Average	0.80	0.71	0.75	0.66	0.65	0.67	0.62	0.65	0.47	0.74

[1] http://mulan.sourceforge.net/datasets-mlc.html.

a Neural network based dependent method. Unsupervised algorithm includes SOM-MLL [2] which is based on self-organizing map. For evaluation, three well known measures namely, Precision, Recall, and, F1-measure [7] are utilized in our approach. Their descriptions and mathematical formulations can be found in the paper [7].

Table 2. Recall values obtained by different multi-label classification techniques

Dataset	ML-SOM	SOM-MLL	SVM-BR	J48-BR	KNN-BR	SVM-LP	J48-LP	KNN-LP	BPMLL	MLkNN
cal500	0.86 ± 0.03	0.23 ± 0.01	0.23 ± 0.04	0.29 ± 0.07	0.35 ± 0.06	0.35 ± 0.06	0.34 ± 0.05	0.35 ± 0.06	0.72 ± 0.05	0.22 ± 0.05
emotions	0.67 ± 0.03	0.60 ± 0.05	0.66 ± 0.11	0.57 ± 0.10	0.63 ± 0.08	0.71 ± 0.09	0.58 ± 0.17	0.63 ± 0.08	0.73 ± 0.11	0.63 ± 0.18
flags	0.67 ± 0.04	0.65 ± 0.06	0.76 ± 0.16	0.74 ± 0.12	0.65 ± 0.14	0.68 ± 0.10	0.66 ± 0.15	0.65 ± 0.18	0.76 ± 0.12	0.76 ± 0.17
genbase	0.96 ± 0.03	0.92 ± 0.03	0.99 ± 0.02	0.99 ± 0.02	0.99 ± 0.02	0.99 ± 0.03	0.98 ± 0.04	0.99 ± 0.02	0.66 ± 0.03	0.95 ± 0.05
yeast	0.77 ± 0.03	0.54 ± 0.01	0.58 ± 0.03	0.58 ± 0.07	0.60 ± 0.06	0.62 ± 0.04	0.54 ± 0.07	0.60 ± 0.06	0.69 ± 0.05	0.59 ± 0.07
Average	0.79	0.59	0.64	0.63	0.64	0.67	0.62	0.64	0.71	0.63

Table 3. F-measure values obtained by different multi-label classification techniques

Dataset	ML-SOM	SOM-MLL	SVM-BR	J48-BR	KNN-BR	SVM-LP	J48-LP	KNN-LP	BPMLL	MLkNN
cal500	0.84 ± 0.04	0.32 ± 0.01	0.34 ± 0.07	0.34 ± 0.07	0.34 ± 0.05	0.34 ± 0.05	0.33 ± 0.05	0.34 ± 0.05	0.45 ± 0.03	0.32 ± 0.06
emotions	0.65 ± 0.03	0.60 ± 0.06	0.60 ± 0.11	0.55 ± 0.08	0.60 ± 0.06	0.67 ± 0.12	0.55 ± 0.14	0.60 ± 0.08	0.66 ± 0.10	0.63 ± 0.16
flags	0.66 ± 0.04	0.64 ± 0.05	0.73 ± 0.11	0.70 ± 0.13	0.65 ± 0.16	0.67 ± 0.09	0.66 ± 0.15	0.65 ± 0.15	0.70 ± 0.10	0.73 ± 0.11
genbase	0.96 ± 0.03	0.92 ± 0.04	0.99 ± 0.02	0.99 ± 0.03	0.99 ± 0.02	0.99 ± 0.03	0.99 ± 0.04	0.99 ± 0.02	0.06 ± 0.06	0.96 ± 0.05
yeast	0.77 ± 0.03	0.59 ± 0.01	0.61 ± 0.03	0.56 ± 0.06	0.57 ± 0.07	0.62 ± 0.04	0.51 ± 0.06	0.57 ± 0.07	0.63 ± 0.06	0.62 ± 0.05
Average	0.78	0.61	0.65	0.63	0.63	0.66	0.61	0.63	0.50	0.65

Sensitivity Analysis on Parameters Used: In SOM, there are 5 parameters used, which are: grid topology (hexagonal or rectangular), initial learning rate (η_0), initial neighborhood radius (σ_0), neighborhood function and number of neurons. In our work, rectangular topology is considered, while the number of neurons is kept as $5 \times 5 = 25$. During generation of label vector for the test instance, some *threshold* value is used. For each data set, to select the best values of η_0, σ_0 and threshold, we have performed sensitivity analysis on these parameters. For this analysis, we have varied the values of one parameter while keeping others as fixed. For example, to determine the best value of η_0, we have executed the experiment with varied value of η_0, while keeping σ_0 and *threshold* values fixed. The value of η_0 at which we got the best value of F-measure is considered as the best value of η_0. Now to determine the best value of σ_0, we have fixed η_0 (equals to the best value obtained) and *threshold* values. The value of σ_0 at which we attained the best value of F-measure is considered as the best value of σ_0. Similar experiments are executed to determine the best value of *threshold*. Thus, the F-measure values obtained by varying parameters, *threshold*, σ_0 and η_0, are shown in Fig. 1(a), (b) and (c), respectively. Following are values obtained for different datasets: (a) flags: threshold $= 0.4$, $\sigma_0 = 1$ and $\eta_0 = 0.04$; (b) emotions: threshold $= 0.4$, $\sigma_0 = 1$ and $\eta_0 = 0.05$; (c) cal500: threshold $= 0.5$, $\sigma_0 = 1$ and $\eta_0 = 0.04$; (d) yeast: threshold $= 0.4$, $\sigma_0 = 1$ and $\eta_0 = 0.05$; (e) genbase: threshold $= 0.3$, $\sigma_0 = 1$ and $\eta_0 = 0.04$.

In paper [2], grid topology was taken as *hexagonal*, while in our approach, it is taken as *rectangular* to see the performance improvement, but, the number of neurons are kept fixed in the grid *i.e.,* 25. The results reported in this paper by ML-SOM methods are the average values over 10 runs and framework is implemented on a Intel Core i7 CPU 3.60 GHz with 4 GB of RAM on Ubuntu.

4 Discussion of Results

Results obtained by our proposed methods (ML-SOM) for Precision, Recall and F-measure on five datasets are shown in Tables 1, 2 and 3, respectively. In these tables, algorithm-independent approaches are represented by Binary-relevance (BR) or Label-Powerset (LP). We have performed the experiment by varying the size of neighboring neurons (NS) around the winning neuron, i.e., number of adjacent neurons to winning neuron as: 8, 4, and 0, using the best values of the parameters obtained after sensitivity analysis for different datasets. Here, NS = 8 means, neighboring neurons will be at distance less than or equal to 1.414, NS = 4 means neurons lying within distance of 1, NS = 0 means, no neighboring neuron is considered. In the tables, results shown are corresponding to NS = 4 (as we achieve good results using this size, see Fig. 1(d)). It is surprising to see that in SOM-MLL method, label vector of a testing instance is calculated using label vectors of training instances mapped to winning neuron of the testing instance or we can say they have taken NS = 0. This proves that incorporation of neighborhood information in determining the label vector of testing instance helps in getting better results.

Considering the Precision value in Table 1, ML-SOM method gives best results for *cal500*, *flags* and *yeast* data sets in comparison to supervised and upsupervised methods. While for remaining datasets, MLkNN performs the best. Regarding recall value shown in Table 2, for *cal500* and *yeast* data sets, our method performs the best. While for *emotions*, *flags* and *genbase* data sets, BPMLL, SVM-BR and SVM-LP perform better in comparison to other methods. After observing the F-measure table (see Table 3), we can conclude that our system performs best for *cal500* and *yeast* data sets. But, if we consider the average precision, recall and F-measure values of our proposed method, ML-SOM, then those are better than all remaining methods' average precision, recall and F-measure values which proves the overall effectiveness of the proposed method.

Figure 1(d) shows the F-measure value obtained by our proposed approach vs. different neighborhood sizes (NS) in comparison with SOM-MLL which considers NS = 0. This figure clearly indicates that when NS = 4 is considered, the best value of F-measure is obtained by our ML-SOM method and this is better than F-measure value of SOM-MLL method.

In general, experimental results show that our unsupervised method achieve competitive (or in some cases, best) performance in comparison to supervised algorithms. Our proposed methods obtained best results in comparison to SOM-MLL method, which is an unsupervised method for multi-label classification. To check the superiority of our proposed ML-SOM method, statistical hypothesis test[2] is conducted at the 5% significance level. It checks whether the improvements obtained by the proposed approach are happened by chance or those are statistically significant. This t-test provides p-value. Smallest p-value indicates that the proposed approach is better than others. This test is conducted using the F-measure values obtained by the proposed method reported in Table 3 over other unsupervised methods, namely, SOM-MLL. The p-values obtained for *cal500, emotions, flags, genbase*, and *yeast* data sets, are 0.00001, 0.000522, 0.065792, 0.001394 and 0.00001, respectively, out of which the p-values 0.00001, 0.000522 and 0.00001 for *cal500, emotions, yeast* data sets evidently support our results. The higher p-values for remaining datasets are due to competitive F-measure by our method over SOM-MLL as can be seen in Table 3.

5 Conclusion

In the current paper, a self-organizing map based algorithm is proposed to solve the multi-label classification problem (ML-SOM). The principle of SOM is utilized in our framework which states that similar instances will map to nearby neurons in the grid. During training of SOM, no label information was used. It is used only for checking the performance of the system. For classification of a test instance, first it is mapped to closest neuron and then neighboring neurons are detected around the closest neuron using position vectors. Finally, its label vector is determined using the closest and neighboring neurons.

The proposed method was tested on five multi-labeled datasets related to different domains and results are compared with various supervised and unsupervised methods. Obtained experimental results proved that the incorporation of neighboring neurons in finding the label vector of a test instance enhances the system performance. Our system suffers from the problem of fixed number of neurons in the neuron grid. This should be determined adaptively/dynamically as per the training data. In future, we would like to apply this approach for solving different real-life problems of NLP domain.

References

1. Alvares-Cherman, E., Metz, J., Monard, M.C.: Incorporating label dependency into the binary relevance framework for multi-label classification. Expert Syst. Appl. **39**(2), 1647–1655 (2012)
2. Colombini, G.G., de Abreu, I.B.M., Cerri, R.: A self-organizing map-based method for multi-label classification. In: 2017 International Joint Conference on Neural Networks (IJCNN), pp. 4291–4298. IEEE (2017)
3. Read, J., Pfahringer, B., Holmes, G., Frank, E.: Classifier chains for multi-label classification. Mach. Learn. **85**(3), 333 (2011)
4. Simon, H.: Neural networks and learning machines: a comprehensive foundation (2008)

[2] https://www.socscistatistics.com/tests/studentttest/default.aspx.

5. Tsoumakas, G., Katakis, I.: Multi-label classification: an overview. Int. J. Data Warehous. Min. (IJDWM) **3**(3), 1–13 (2007)
6. Tsoumakas, G., Spyromitros-Xioufis, E., Vilcek, J., Vlahavas, I.: MULAN: a Java library for multi-label learning. J. Mach. Learn. Res. **12**(Jul), 2411–2414 (2011)
7. Venkatesan, R., Er, M.J.: Multi-label classification method based on extreme learning machines. In: 2014 13th International Conference on Control Automation Robotics and Vision (ICARCV), pp. 619–624. IEEE (2014)
8. Zhang, M.L., Zhou, Z.H.: A k-nearest neighbor based algorithm for multi-label classification. GrC **5**, 718–721 (2005)

Machine Learning

Prediction of Performance Indexes in CNC Milling Using Regression Trees

Kannadasan Kalidasan, Damodar Reddy Edla(✉), and Annushree Bablani

Department of Computer Science and Engineering,
National Institute of Technology Goa, Farmagudi, India
kannadasankk@gmail.com, {dr.reddy,annubablani}@nitgoa.ac.in

Abstract. Machine Learning (ML) is a major application of artificial intelligence which has its importance in all fields of engineering. ML models learn automatically from the dataset and makes intelligent decisions and predictions. Computer Numerical Control (CNC) plays a vital role in manufacturing parts. Each parts manufactured need desired performance index values depend on its usage. Surface roughness, geometric tolerances are major performance index values. The deviations of the performance index values arises because of controllable and uncontrollable parameters. To adjust the parameters, there is a need to find relation between controlled parameters and their performance index values. Thus, we are motivated to design a Machine Learning model for the problem. In this work, we have proposed a regression tree based model which predicts the performance index values by taking the CNC machining parameters as the input. The regression tree built can be useful for the manufacturers for achieving the desired performance index values.

Keywords: Regression tree · Prediction · CNC milling · Surface roughness · Geometric tolerances

1 Introduction

The modern machining industries aim to produce high quality product which has desired dimensional accuracy, surface roughness in a cost and time effective manner. CNC machining plays a vital role in modern industry for making products on scales from small individual parts like screw, mold to large, heavy-duty operations such as military and automobile products [4]. CNC milling is one of the major machining operations which makes complex shapes and finishing of the machined products. Surface roughness and the milling accuracy are two major output parameters regarded as the index for evaluating performance of milled product. ML models are capable of learning the non linear relationships between input and output variables. Hence, ML Models can be used in the CNC milling process to model the relationship between input and output parameters [1].

B. Deka et al. (Eds.): PReMI 2019, LNCS 11941, pp. 103–110, 2019.
https://doi.org/10.1007/978-3-030-34869-4_12

Several studies have been done in the literature which investigate the effect of milling parameters on performance index values. The effect of milling parameters on the surface roughness is modelled using Gaussian Process Regression (GPR) and the surface roughness value is predicted using the milling parameters [11]. A statistical multiple regression technique has been proposed for the prediction of surface roughness using aluminium 6061 alloy for end milling process [7]. A statistical model to analyze the effect of machining parameters and geometrical parameters on the vibration amplitude is proposed in [10]. A second order mathematical model have been proposed and the results were verified using ANOVA. An ANN based prediction model for predicting surface roughness of the CNC face milled product is proposed in [1], where authors have used Taguchi method for designing the experiments and have considered various factors such as depth of cut, cutting speed, engagement and wear of cutting tool, etc. The surface roughness of the face milled product is predicted using feed forward ANN. Although ANNs are capable of learning non-linear relationships, the process of training and learning happens in a black-box manner. Hence, for improved interpretability and understandability of the model, in this paper we have used regression trees for prediction of the performance index values.

Most of the papers studied about the effect of input parameters on the surface roughness. To the best of our knowledge, no studies have been done which investigate the effect of input parameters on geometric tolerances such as parallelism and perpendicularity.

2 Methodology

2.1 Regression Trees

Regression trees [2,6] are one of the machine learning approaches which helps in construction of prediction models using data. The prediction models are developed by partitioning the data and then fitting a small model on that partitioned data. The procedure of partitioning is recursively followed thereby giving various set of small models. The procedure is graphically represented as tree hence called decision tree. The decision tree consists of nodes as features, branch as rules and leaf as the output. Decision Tree have clear representation of the information which makes it one of the most useful machine learning methods for linear problems. Classification and Regression Tree [2] is one of the algorithm for decision tree which uses Gini Index to determine the point of partition of the data [3]. The linear regression model can be represented as:

$$y = w_0 + \sum_{j=1}^{k} w_j x_j$$

where y is a output, $x_1, x_2, \ldots, x_k$ are independent variables, $w_1, w_2, \ldots, w_k$ are regression coefficients. The squared error is used as the cost function for the regression tree.

$$Error = \sum_{j=1}^{k} (\hat{y} - y)^2$$

where $\hat{y}$ and y represents the predicted output and actual output respectively. CART algorithm derives rules from the information and it does not require any pre-specified values. For the prediction of values, it takes "if and then" rules predicates. Each branch represents a conditional statement, each leaf node is derived by joining "if" statements from root node to leaf node.

In this paper, we have used regression tree for the prediction of performance indexes, as the tree representation created using CART algorithm provides more understandability about the system. Also, the interpretability of the model is high with respect to the tree with which rules can be derived. Thus the manufacturer can decide values of the input parameters accordingly.

2.2 Milling Process

Milling is a type of CNC machining in which the rotating cutter move towards the workpiece and removes the material. The milling is accomplished by varying the parameters such as spindle speed, direction, pressure etc. Milling has various types which includes pocket milling, slot milling, drilling etc. Slot milling is considered for the experiments for building the predicition model. The input parameters considered for the model are spindle speed, feed rate, depth of cut. In the milling process, slots of dimensions 33 mm × 6 mm × 3 mm were made with the combinations of the input parameters. The combination of values is chosen by considering four levels for each input parameter. With the levels for each input value, 96 combinations are chosen. The milling is accomplished in aluminium alloy (Al 7075) by varying the three input parameters such as feed rate (FR) (mm/rev), spindle speed (SS) (rpm), depth of cut (DC) (mm) in MAX MILLPLUS milling machine. The steps followed in the process are listed below.

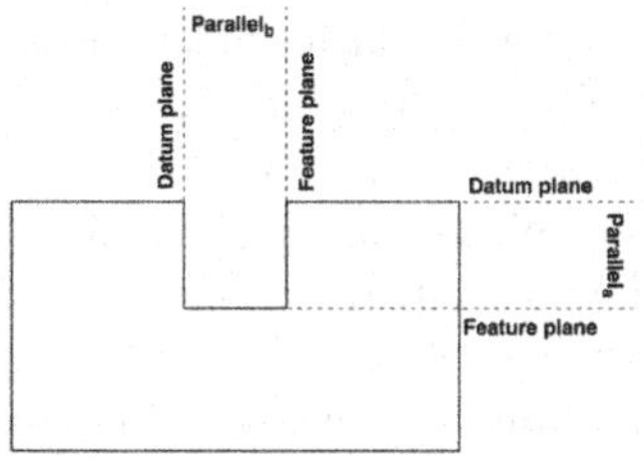

(a) Geometric Tolerances - Parallel

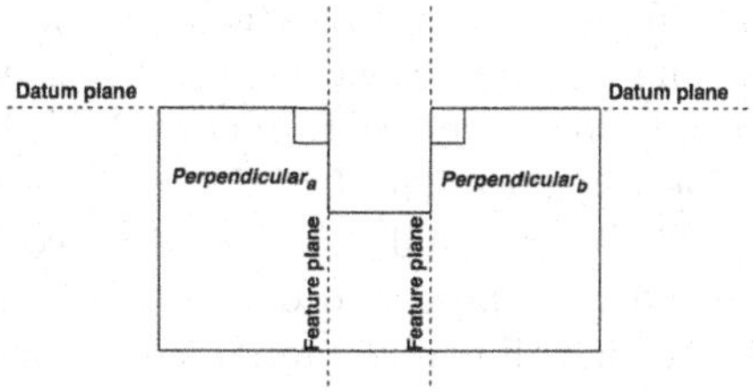

(b) Geometric Tolerances - Perpendicularity

Fig. 1. Graphical representation of geometric tolerances

1. The aluminium alloy workpiece is face milled for the smoother and flat surface such that the measurement errors and surface errors can be minimized.
2. The dimension of the slots and the parameters are given in G-code files to the CNC. It should be noted that each slot has different values of parameters but the same dimension.
3. The surface roughness value of each slot is measured using surface roughness tester.
4. Geometric tolerances of the milled product is measured using CMM as follows: The parallelism is measured considering the surface of the object as datum plane and the milled surface as feature plane. The value measured is referred as '$parallel_a$'. The parallelism of the two sides of the milled surfaces considering one of the surface as datum plane and other as feature plane, the value is measured, which is referred as '$parallel_b$'. The graphical representation of the two parameters $parallelism_a$ and $parallelism_b$ is as shown in the Fig. 1.
5. The perpendicularity is measured considering the surface of the object being datum plane and two sides of the milled slot being feature planes, two values are measured which are referred a '$perpendicular_a$' and '$perpendicular_b$' respectively. The graphical representation of the two parameters $perpendicular_a$ and $perpendicular_b$ is as shown in the Fig. 1.

3 Experimental Results

In this section, the specifications of the machine used for the experiments, analysis of data and the results obtained are discussed. The experimental data has been collected by performing experiments using 3 - axis vertical axis machine MTAB MAXMILLPLUS as shown in the Fig. 2(a). SINUMERIK 808 controller is used in the machine which is designed by SIEMENS [9]. After milling process, the average surface roughness (R_a) data is collected using surface roughness tester belongs to Mitutuyo SJ-210 series as shown in Fig. 2(c). The surface roughness tester can measure upto 17.5 mm with three different measuring speeds such as 0.25, 0.5, 0.75 mm/s which are widely used in the literature [5].

The geometric tolerances such as parallelism and perpendicularity of the slots are collected using SPECTRA CMM designed by ACCURATE which is shown in the Fig. 2(b). The CMM can be operated in both motorized or CNC mode [8]. The major milling parameters considered for the milling process are feed rate, spindle speed and depth of cut. Aluminium alloy (Al 7075) is used as the object for milling. The minimum values considered for FR, SS and DC are 0.08 mm/rev, 2000 rpm and 0.1mm respectively. The maximum values considered for FR, SS and DC are 0.2 mm/rev, 6000 rpm and 0.8 mm respectively. The values considered are found to be the optimized range for the milling process in aluminium alloy [1,4,7,10].

The regression tree model built is validated using 10-fold cross validation method. The mean square error and the root mean square error of the model is calculated. Figure 3(a), (b), (c), (d), (e) shows the regression trees created for the

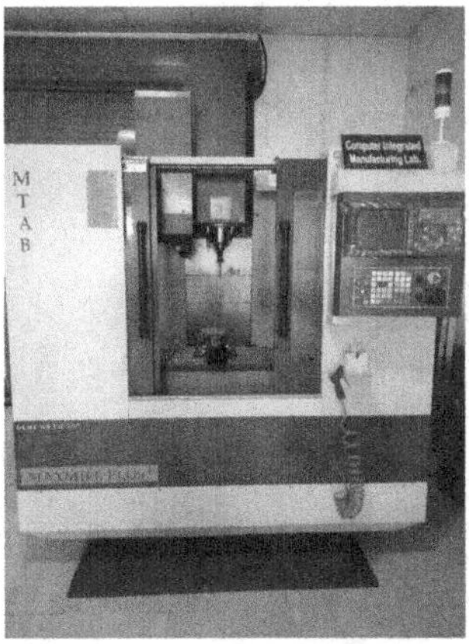

(a) 3 - axis CNC vertical milling machine

(b) Co-ordinate Measuring Machine (CMM)

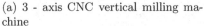

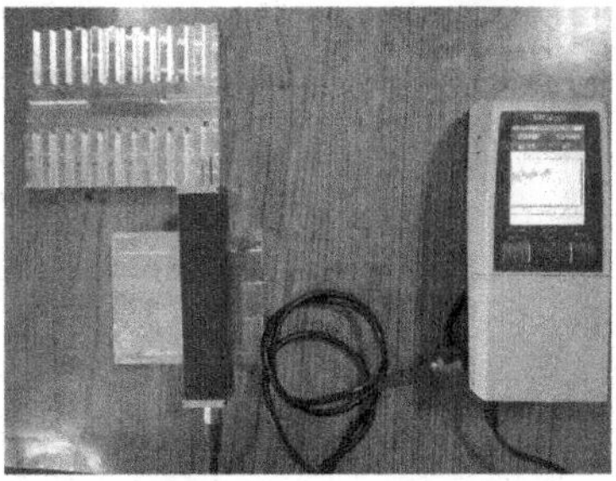

(c) Surface Roughness Tester

Fig. 2. Machines used for the experiments

prediction of the average surface roughness, $parallel_a$, $parallel_b$, $perpendicular_a$, $perpendicular_b$ respectively. From the Fig. 3(a), it can be seen that the first split of the data, the information gain of DC is higher compared to other parameters. The impact of depth of cut is higher in surface roughness value compared to other parameters. Also, the surface roughness value is higher when depth of cut is more than 0.22 mm. For the geometric tolerance - parallelism, feed rate has higher impact other than any parameters which can be observed from the Fig. 3(b), (c).

Figures 4, 5, 6, 7 and 8 represents the actual value and predicted value of the all the output parameters average surface roughness, $parallel_a$, $parallel_b$, $perpendicular_a$, $perpendicular_b$ respectively. It can be observed from the figures that the decision tree model predicted the output parameter values with lesser error. The actual value line almost overlaps estimate line which symbolizes decision tree model is trained properly to learn the relationship present in the data. The validation root mean square error obtained for all performance indexes are 0.5107 mm, 0.3659 mm, 0.0551 mm, 0.6334 mm, 1.6157 mm respectively.

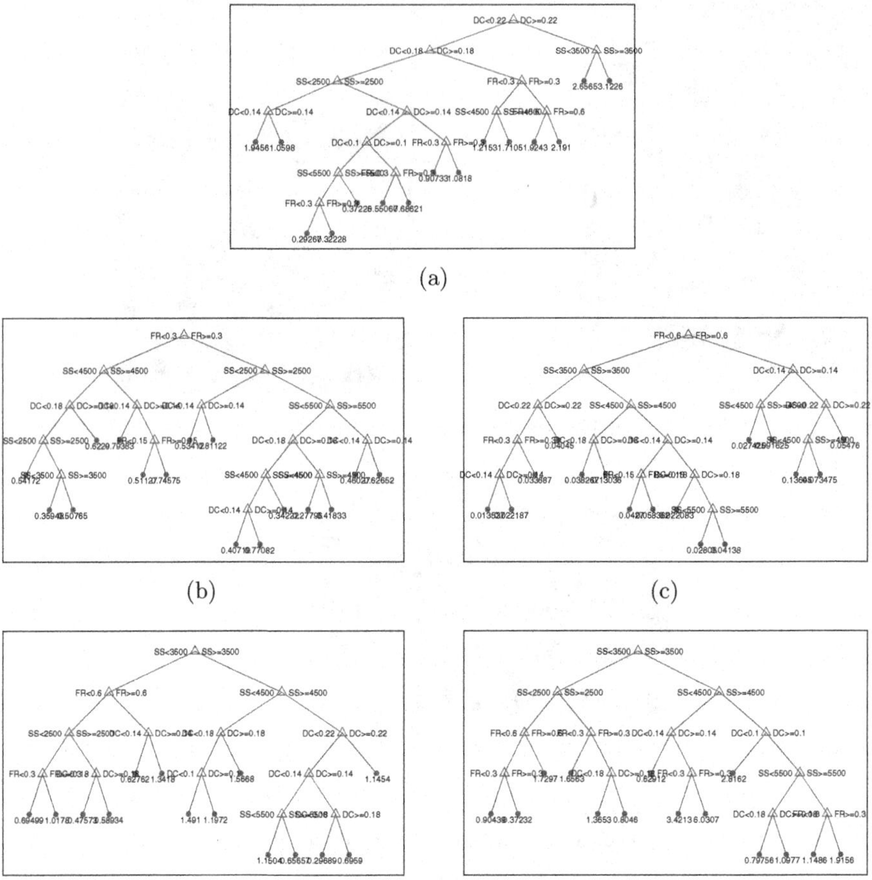

(a)

(b)

(c)

(d)

(e)

Fig. 3. Regression trees for all the output variables

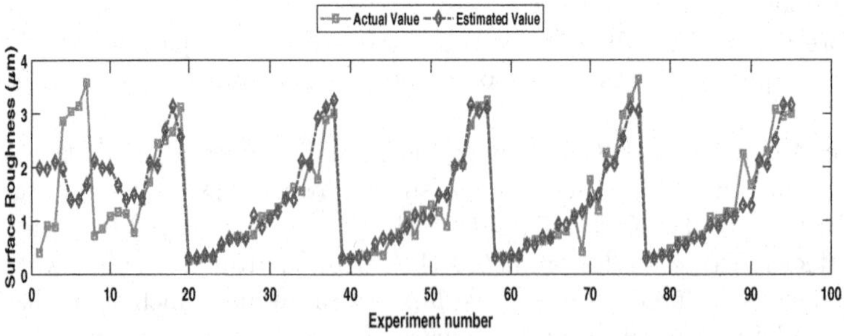

Fig. 4. Actual vs estimated value - surface roughness

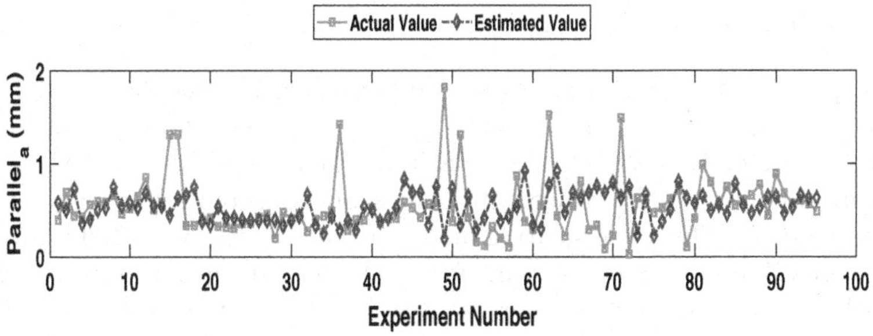

Fig. 5. Actual vs estimated value - $parallel_a$

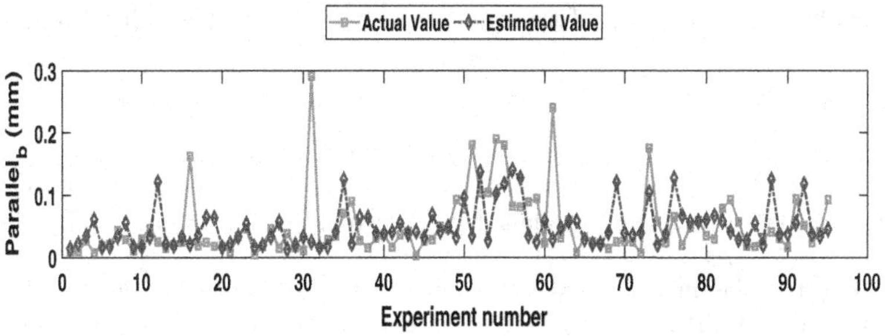

Fig. 6. Actual vs estimated value - $parallel_b$

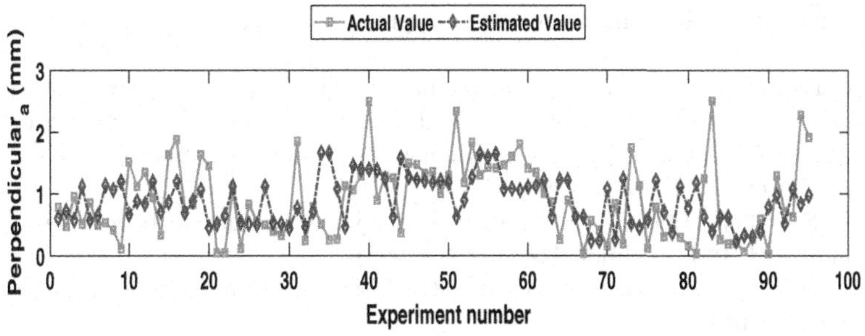

Fig. 7. Actual vs estimated value - $perpendicular_a$

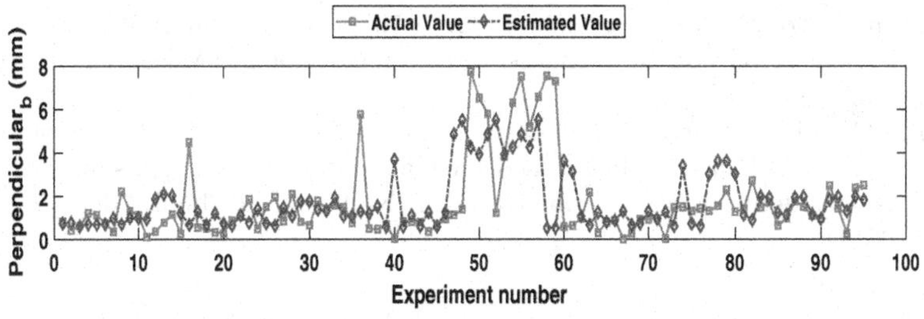

Fig. 8. Actual vs estimated value - $perpendicular_b$

4 Conclusion

In this paper, we proposed a prediction model using regression tree to predict performance index values such as average surface roughness and geometric tolerances of slot milled product. The experiments were conducted in 3 - axis vertical milling machine. The performance index values are collected using surface roughness tester and co-ordinate measuring machine. The dataset collected is analysed and the regression tree model is built for prediction. Results shows that model achieves lesser root mean square error. Also, to the best of our knowledge no studies have built a model which predicts the geometric tolerances of the milled product. Our proposed model will be useful in manufacturing the parts with desired performance index values, where manufacturer can choose the parameters using the model instead of trial and error which is not an efficient way.

References

1. Benardos, P., Vosniakos, G.C.: Prediction of surface roughness in CNC face milling using neural networks and taguchi's design of experiments. Robot. Comput.-Integr. Manuf. **18**(5–6), 343–354 (2002)
2. Breiman, L.: Classification and Regression Trees. Routledge, Abingdon (2017)
3. Chen, J., Lin, Y., Leu, Y.: Predictive model based on decision tree combined multiple regressions. In: 2017 13th International Conference on Natural Computation, Fuzzy Systems and Knowledge Discovery (ICNC-FSKD), pp. 1855–1858, July 2017
4. Chiu, H.W., Lee, C.H.: Prediction of machining accuracy and surface quality for CNC machine tools using data driven approach. Adv. Eng. Softw. **114**, 246–257 (2017)
5. Corporation, M.: Portable surface roughness tester surftest SJ-210 series (2018). https://www.mitutoyo.co.jp/eng/support/service/catalog/03/E4388_178.pdf
6. Loh, W.Y.: Classification and regression trees. Wiley Interdisc. Rev.: Data Min. Knowl. Discov. **1**(1), 14–23 (2011)
7. Lou, M.S., Chen, J.C., Li, C.M.: Surface roughness prediction technique for CNC end-milling. J. Ind. Technol. **15**(1), 1–6 (1998)
8. ASSP Ltd.: CMM spectra (2018). http://www.accurategauging.com/product-item/spectra-lite
9. MTAB Engineers Pvt. Ltd.: CNC vertical machining center (2018) https://mtabindia.com/brochure/maxmill.pdf
10. Subramanian, M., Sakthivel, M., Sooryaprakash, K., Sudhakaran, R.: Optimization of end mill tool geometry parameters for Al7075-T6 machining operations based on vibration amplitude by response surface methodology. Measurement **46**(10), 4005–4022 (2013)
11. Zhang, G., Li, J., Chen, Y., Huang, Y., Shao, X., Li, M.: Prediction of surface roughness in end face milling based on Gaussian process regression and cause analysis considering tool vibration. Int. J. Adv. Manuf. Technol. **75**(9–12), 1357–1370 (2014)

Dyscalculia Detection Using Machine Learning

Alka Subramanyam[1], Sonakshi Jyrwa[1], Juhi M. Bansinghani[2], Sarthak J. Dadhakar[2], Trena V. Dhingra[2], Umesh R. Ramchandani[2](✉), and Sharmila Sengupta[2]

[1] Department of Psychiatry, B.Y.L. Nair Ch. Hospital, Mumbai, India
alka.subramanyam@gmail.com, jyrwasonakshi@gmail.com
[2] Department of Computer Engineering, Vivekanand Education Society's Institute of Technology, Mumbai, India
{2015juhi.bansinghani,2015sarthak.dadhakar,2015trena.dhingra, 2015umesh.ramchandani,sharmila.sengupta}@ves.ac.in

Abstract. A great amount of research is going on in the detection of learning disabilities, but the detection of Dyscalculia remains a tedious and time-consuming task even today. Various tests are conducted to detect if the patient has Dyscalculia and each test has to be evaluated manually as the scores alone are not sufficient to determine it. In some cases, Curriculum-Based Tests [CBTs] or Wide Range Achievement Tests [WRAT] or both need to be conducted after analysis of the results of the Woodcock-Johnson Tests. As a collaborative project between the Department of Psychiatry B.Y.L. Nair Ch. Hospital and Department of Computer Engineering, Vivekanand Education Society's Institute of Technology a system is developed to help improve the detection of Dyscalculia. The Woodcock-Johnson Tests of Achievements are conducted by the doctors and the results of these tests determine the learning disability.

Keywords: Decision tree · Dyscalculia · Learning disability · Machine learning · Random forest · Woodcock-Johnson tests of achievements

1 Introduction

Specific learning disorder (SLD) is a neurodevelopmental disorder characterized by difficulties in learning and using academic skills such as reading, writing and calculations despite the adequate socio-cultural opportunity, intact vision and hearing, normal intelligence and conventional schooling [5,15,17]. Dyscalculia is a type of SLD with specific impairment in Mathematics. It is an alternate term used to refer to a pattern of difficulties characterized by difficulty in numerical processing, learning and memorization of mathematical facts, mathematical reasoning and fluency [7]. About 5–15% of school-going children in India have SLD. There is a dearth of studies in India on SLD and Dyscalculia specifically.

The inability to process information can interfere with learning primitive skills such as reading, writing and/or mathematics. Along with those skills it

B. Deka et al. (Eds.): PReMI 2019, LNCS 11941, pp. 111–120, 2019.
https://doi.org/10.1007/978-3-030-34869-4_13

can also interfere with much more complex skills such as time planning, attention, abstract reasoning, long or short term memory and organization. Learning disabilities are categorized into three types:

a. Dyslexia: It is a type of learning disability that hinders a person's ability to read. A dyslexic person faces problems while reading.
b. Dysgraphia: It refers to the difficulty with writing. A dysgraphic person faces problems while thinking, writing their thoughts down, spelling, grammar and memory.
c. Dyscalculia: It refers to the difficulty with calculations and mathematics. A person with Dyscalculia faces problems with numeric calculations and math reasoning. The three disabilities are very different yet very closely correlated and thus, it can become difficult to separate one from the other.

There are various tests to detect these learning disabilities, one of them is the Woodcock-Johnson Tests of Achievements. It is quite effective in detecting Dyslexia and Dysgraphia, but in a few cases, even the results of this test cause skepticism. In such cases, one has to turn to Curriculum-Based Tests [CBTs] or Wide Range Achievement Tests or both to identify Dyscalculia. Our goal is to use machine learning algorithms to accurately detect Dyscalculia.

2 Drawbacks of Existing System

The Woodcock-Johnson Tests of Achievement (WJ ACH) is a test designed to quantify the academic performance of not only children but also adults from age 2 to 95 and grades K.0 through 18.0. This test has 22 subtests for measuring five areas of academic achievement: reading, oral language, written language, math and knowledge. The standard battery comprises of seven subtests and the extended battery has 14. Additional subtests can provide supplemental scores. This study uses a math battery. It is ideal to examine progress in reading, writing and mathematics achievement areas [1].

Quantitative reasoning, computation skills, mental computation and math fluency are required to make mathematical calculations from addition to trigonometry. The test-taker is given a series of basic mathematical problems which include multiplication, division, decimals, fractions, basic algebra questions and so on. The Math and Calculations [Test 5], Applied Problems [Test 10], Quantitative Concepts [Test 18A] and Number Series [Test 18B] are not timed. The Math Fluency [Test 6] test is timed. The scores are evaluated on the basis of score received, grade and age. The invigilator assesses the types of mistakes made. If mistakes like $94 - 37 = 67$ are made, the child may have difficulty in understanding rudimentary mathematical concepts, like carrying over and borrowing. But some children have a tendency to get the problem wrong even when they grasp the mathematical concepts they are working with. For example, after doing a couple of subtraction problems, a child may solve the third problem as a subtraction problem too even though it's a division problem. This can point to attention issues [9].

The Wechsler IQ Test [WISC III and IV] has been administered to find the intellect of children. The test has a verbal and performance section. The score is calculated and noted down. If the score is beyond 130, it is considered excellent. A high score would be one that lies between 120 and 129. A score between 110 and 119 is considered moderate and average if it is less than 90. A score lower than 70 denotes borderline mental functionality and score below 69 denotes mental retardation [2].

Wide Range Achievement Test [WRAT] is an additional screening test that is used to determine if there is a need for a more inclusive achievement test. These tests refer to skills meant to be learned by individuals through direct instruction or intervention. It quantifies skills like spelling, reading and arithmetic. It is a timed test. It has two parts. The first part requires solving problems presented verbally related to reading number symbols, counting and solving arithmetic problems. The second consists of 40 arithmetic problems with a time frame of 15 min [3].

For the Diagnosis and Assessment of Dyscalculia, we have the Psychoeducational assessments and Curriculum-based Tests. Practical experience tells us that there are varying profiles of arithmetic skills. The test commonly used is the Woodcock-Johnson Psychoeducational battery designed to assess the basic arithmetic skills. The importance is not only paid to correct answers but also to the processing speed, solving strategies and qualitative assessment of performance. The above-mentioned procedure is followed at B.Y.L. Nair Ch. Hospital for assessment of dyscalculia and it may differ at different centers. The processing tests have not been standardized to the Indian children population and thereby results cannot be entirely relied upon and at times inconclusive despite our detailed clinical evaluation it indicates an impairment in mathematical skills. Curriculum-based Tests have to be invariably used to help identify children with Dyscalculia and supplement our findings. Being curriculum-based, it only determines if the child meets the learning objective of his/her grade level and does not fully reveal all the actual deficits. Also, the tests used are not able to delineate mathematical difficulties due to attentional/phonological deficits from the developmental dyscalculics and may lead to the overdiagnosis of Dyscalculia. The whole process of assessment hence is time consuming, tedious and vexing for the child.

Because of the inherent issues in the diagnostic tools available and the complex nature of Dyscalculia itself, it was necessary to develop a tool that would assist us in correctly assessing and diagnosing Dyscalculia in Indian children.

3 Literature Survey

The process of maturation of processing of numbers in our brain is a neuroplastic maturation process. It develops into a mature systematized complex neural network during the development from childhood to adolescence [11,18]. After studying the functional images of the brain we can say that multiple areas of the brain are involved in numerical-arithmetic skills acquisition and operation.

The maturity of these Domain-specific functions depends on the development of other areas such as attention and working memory (mental maths, multi-digit arithmetic), language, sensorimotor (finger counting) and visuospatial skills [11]. There may be a primary genetic vulnerability to impaired development of numerical functions, linguistics, visuospatial skills and executive functions or the maturation process being affected by epigenetic mediated environmental influences [8].

This justifies the correlation of Attention Deficit Hyperactivity Disorder [ADHD], Dyslexia with Dyscalculia. 20–60% of those with SLD have other learning difficulties/ disability such as ADHD and dyslexia [12,14,18]. Shavlev et al. [16], demonstrated Attention Deficit Disorder [ADD] in 32% of dyscalculics studied. Also, children with ADD noted to make mathematical errors secondary to impulsiveness and inattention. One empirical data noted the treatment of ADD with stimulant improved the calculating ability without any effect on rudimentary numerical skills [14].

Also, one study noted 52% variance in calculating ability was accounted by reading skills [10]. Therefore, deficient phonological skills in pre-school children were linked with unsatisfactory performance on calculation related questions in primary school [6]. Thus, a disorder of linguistic development is a risk factor for poor calculating ability [13].

4 Methodology

The Woodcock-Johnson Tests' results are used as the input and training data for the machine learning algorithms. The input data contains the results of the Math and Calculations [Test 5], Math Fluency [Test 6], Applied Problems [Test 10], Quantitative Concepts[Test 18A] and Number Series [Test 18B] of Woodcock-Johnson and also the results of Wide Range Achievement Test [WRAT]. The test result of Woodcock-Johnson Tests uses aggregated result and it cannot precisely detect whether the patient is having dyscalculia or not. For our model focuses on each question rather than the aggregated score in a particular section. This allows looking at the trend of each question for a particular grade. The inputs for a particular question is considered 1 if the question is attempted and answered correctly, considered 0 if the question is not attempted and considered −1 if the question is attempted but incorrect. Using the same format for all tests, the input data set has been collected for 650 patients. Random Forest Classification Algorithm has been used to train the model. Two distinct models are created. One model uses the above-mentioned tests from Woodcock-Johnson along with the WRAT [549 cases] and the other model uses only Woodcock-Johnson test [650 cases]. The reason for creating two different models is that WRAT is not conducted for all the patients. While using the system, doctors will input the results of the test and the outcome will be predicted.

The emerging trends in Machine Learning and Data Science can be used in the health sector to predict the outcomes depending on the various results of the test. We are using the dataset of already diagnosed patients and the Random

Forest algorithm to analyze and find out if the patient has Dyscalculia. Our system has been trained using the data collected from tests. The importance of all attributes has been determined. Figure 1 shows the flow diagram of the complete process of dyscalculia detection.

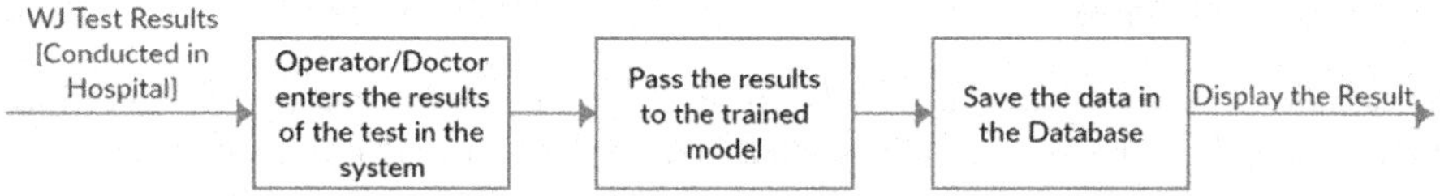

Fig. 1. Flow diagram.

The dataset consists of various factors that are responsible for the Dyscalculia of the child. Patients having dyscalculia face it difficult to solve certain questions of the WJ Test. Such questions are considered as an important factor for the distinction between a normal person and the person suffering from dyscalculia. Initially, the data of the prior patients were collected from the hospital. The data was then organized in CSV and later in the database and then we used it to train the model. The entire dataset has been split for training and testing. The following attributes are included in the dataset as shown in the table below Table 1.

Table 1. Dataset description

Sr no.	Feature	Description
1	Age	Age
2	Grade	The class in which the student was when diagnosed.
3	IQ	The IQ of the patient,
4	Test 5	The results of test 5 - Math and Calculation
5	Test 6	Math Fluency
6	Test 10	Applied Problems
7	Test 18A	Quantitative Concepts
8	Test 18B	Number series
9	WRAT-2	Wide Range Achievement test (if any)

5 Algorithm

Classification is a supervised learning approach in machine learning. A classification model attempts to draw some conclusions from the observed values that it learns from the input data which is given to it and then it uses the learning from the input data to classify new observations. Classification problems are used to classify examples into a given set of categories [4]. Our system performs classification using supervised learning model to determine if the patient has dyscalculia.

5.1 Decision Trees

Regression and classification models are use cases of the decision tree. Decision trees work well with categorical as well as numerical data. The data set is broken down into subsets later the decision tree is incrementally developed. The tree consists of one root node, intermediate nodes and leaf nodes. A decision node has two or more branches. The leaf node represents the classification of the node. The intermediate nodes are the child and parent nodes of the nodes above and below it respectively. The topmost node is the root node and the best predictor. We have generated a decision tree using the Random Forest Algorithm considering all the features in the dataset. Figures 4 and 5 show the decision trees for the prediction of dyscalculia.

5.2 Random Forest

Random forests are an assembly of random decision forests, like decision trees they work well with tasks like classification and regression. The algorithm is used to construct multiple decision trees during the training time and it outputs the class that is the mode of the classes (classification) or mean prediction (regression) of individual trees. Random decision forests are used to correct the habit of decision trees of overfitting to their training data. For each decision tree, the importance of each node can be calculated using Gini importance (gi), but the assumption is that it is a binary tree, W rnode (right node) and W lnode (left node) is calculated using the Eq. 1.

$$gi_j = w_{lnode(j)}C_{lnode(j)} - w_{rnode(j)}C_{rnode(j)} \tag{1}$$

The importance of feature is the decrease of node impurity divided by the probability of reaching towards the node. The probability of reaching towards the node can be calculated by the count of samples that reach the node, divided by the total count of samples. The importance of the feature can be determined by how higher is the value. Feature Importance (fx) for node y is calculated using the Eq. 2.

$$fx_y = \frac{\sum_{y:nodeysplitsonfeaturex} gx_y}{\sum_{z \in allnodes} gx_z} \tag{2}$$

Our system has made use of the above equations to calculate the feature importance which is shown in Fig. 3.

6 Results

The model has been trained and tested. It has an accuracy of 99.87% when trained without the results of the WRAT tests and 99.94% when trained with it. It highlights the importance of individual questions on the result of the test i.e. if the person has Dyscalculia or not. The result of the test is the confidence percentage of the model that the child has Dyscalculia.

6.1 Determination of Efficiency

The primary goal of the project is to find a set of questions from the current tests which will help to find out if the child has Dyscalculia. Figure 3 shows the importance of different features.

6.2 Determination of Accuracy

The accuracy of the model can be found out by splitting the dataset as training and testing data. Testing data can be the same as training data. Labels are not considered in the training data. We have split training and testing data in a 70:30 ratio. Patients of the same grade and similar IQ (intelligence quotient) are used in testing to compare the results. The accuracy in the context of whether it is actually detecting Dyscalculia can be determined when these sets of questions alone can help to predict Dyscalculia.

Furthermore, the model is tested on fresh data for new patients and it could successfully detect the Dyscalculia. Figure 3 shows the decision tree generated by the Random Forest Classification Algorithm. This tree includes WJ Johnson Test results along with WRAT results as input. Figure 4 shows the decision tree generated by the Random Forest Classification Algorithm. This tree includes only WJ Johnson Test results as input.

The graph in Fig. 5 shows the analysis of Question 27 of Test 5 and grade 9. Total patients having dyscalculia and having attempted question right, wrong or not attempted is specified in the graph. Question 27 of Test 5 was selected as it has high importance (Fig. 2).

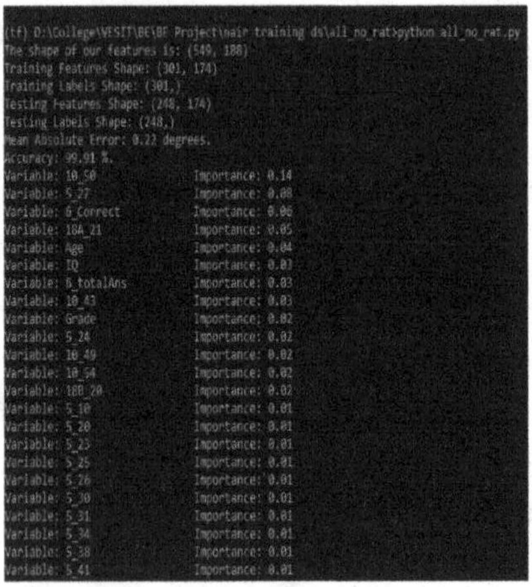

(a) First

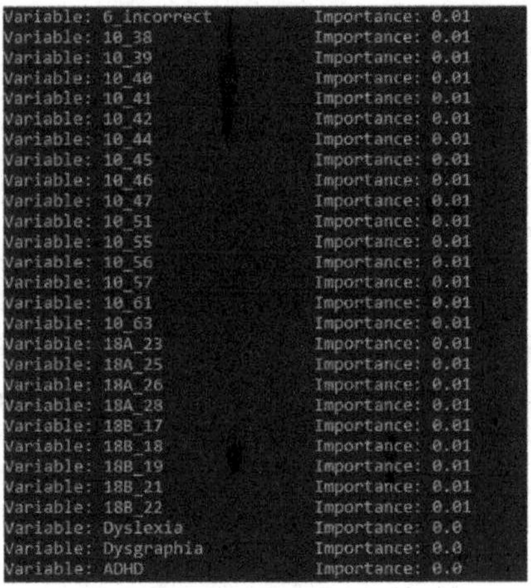

(b) Second

Fig. 2. Caption

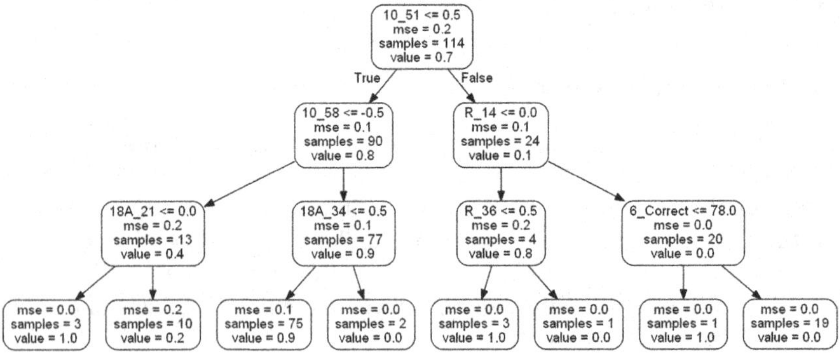

Fig. 3. Decision tree with WRAT as input.

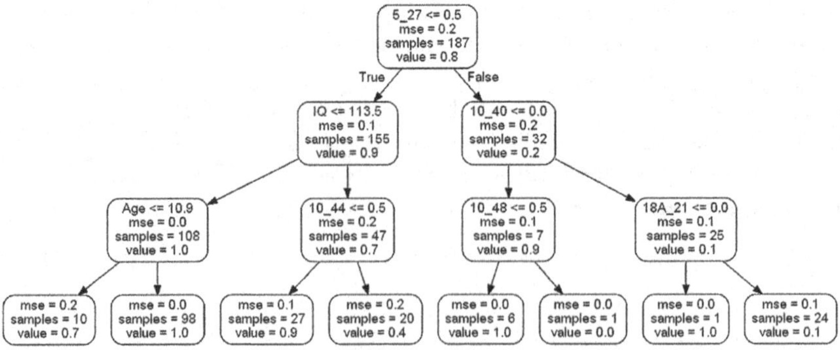

Fig. 4. Decision tree without WRAT as input.

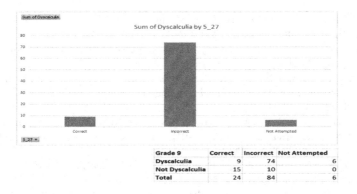

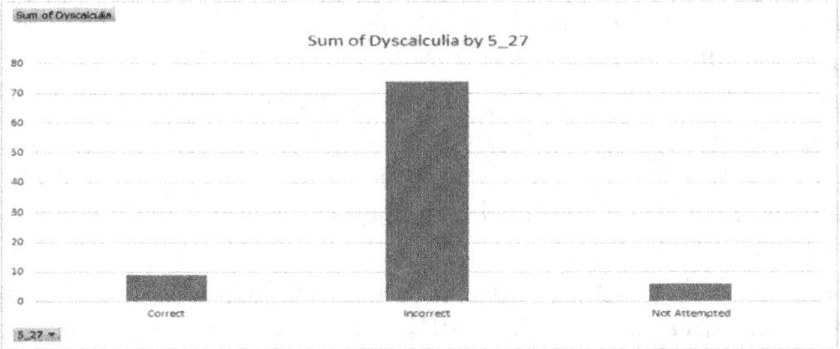

Grade 9	Correct	Incorrect	Not Attempted
Dyscalculia	9	74	6
Not Dyscalculia	15	10	0
Total	24	84	6

Fig. 5. Analysis of Grade 9 and Test 5 Question 27.

7 Conclusion

This tool will help in correctly assessing and diagnosing Dyscalculia among children in India. It will not only reduce the time spent on detecting Dyscalculia, but it will also ensure that the results are more efficient. The current process of assessment is time consuming, tedious and vexing for the child and therefore, a machine learning approach would save time for the medical experts and patients alike, thereby offering speedy diagnosis and earlier intervention in Dyscalculia.

References

1. https://pdfs.semanticscholar.org/2a09/6dd797994de82198f4dc982d92aa831169bd.pdf
2. Wechsler intelligence scale. https://geniustests.com/about-wechsler/wechsler-intelligence-scale
3. Wide range achievement test. http://www.minddisorders.com/Py-Z/Wide-Range-Achievement-Test.html
4. Neurol (2001)
5. Desmonet, J.F., Taylor, M.J., Chaix, Y.: Development dyslexia. Lancet **363**, 1451–1460 (2004). https://www.ncbi.nlm.nih.gov/pubmed/15121410
6. Jordan, J. A., Wylie, J., Mulhern, G.: Phonological awareness and mathematical difficulty: a longitudinal perspective (2010)
7. Karande, S., Bhosrekar, K., Kulkarni, M., Thakker, A.: Health-related quality of life of children with newly diagnosed specific learning disability. J. Trop. Pediatr. **55**, 160–169 (2009)
8. Kaufmann, L., Von Aster, M.: The diagnosis and management of dyscalculia. Dtsch. Arztebl. Int. **109**(45), 767–778 (2012). https://www.ncbi.nlm.nih.gov/pmc/articles/PMC3514770/
9. Kelly, K.: Types of tests for dyscalculia. https://www.understood.org/en/school-learning/evaluations/types-of-tests/test-for-Dyscalculia
10. Korhonen, J.G., Linnanmaki, K., Aunio, P.: Language and mathematical performance: a comparison of lower secondary school students with different levels of mathematical skills (2011)

11. Kucian, K., Kaufmann, L.: A developmental model of number representations. Behav. Brain Sci. **32**, 340–341 (2009)
12. Landerl, K., Moll, K.: Comorbidity of learning disorders: prevalence and familial transmission. J. Child Psychol. Psychiatry **5**, 287–294 (2010)
13. Manor, O., Shalev, R.S., Joseph, A., Gross-Tsur, V.: Arithmetic skills in kindergarten children with developmental language disorders. Eur. J. Paediatr. Neurol. **5**, 71–77 (2001)
14. Rubinstein, O., Bedard, A.C., Tannock, R.: Methylphenidate improves general but not core numerical abilities in ADHD children with co-morbid Dyscalculia or mathematical difficulties (2008)
15. Karande, S.: Current challenges in managing specific learning disability in Indian children. J. Postgrad. Med. **54**, 75 (2008)
16. Shavlev, R.S., Auerbach, J., Gross-Tsur, V.: Developmental Dyscalculia behavioural and attentional aspects: a research note (1995)
17. Shaywitz, S.: Dyslexia. N. Engl. J. Med. **338**, 307–312 (1998)
18. Von Aster, M.G., Shalev, R.: Number development and developmental Dyscalculia. Dev. Med. Child Neurol. **49**, 868–873 (2007)

Temporal Dynamics of the Brain Using Variational Bayes Hidden Markov Models: Application in Autism

Preetam Srikar Dammu(✉) and Raju Surampudi Bapi

School of Computer and Information Sciences, University of Hyderabad, Hyderabad, India
preetam.srikar@gmail.com, raju.bapi@iiit.ac.in

Abstract. Investigating the functional connectivity (FC) patterns of the brain using resting-state functional magnetic resonance imaging (rs-fMRI) has been instrumental in revealing the effects of neurological disorders. Several studies have established that brain connectivity is dynamic in nature, and that brain diseases have an impact on both FC and its temporal properties. Various computational techniques have been proposed in the literature for modeling brain dynamics, yet most of these approaches have limitations that hinder the process of building accurate models. In this work, we explore a promising approach using Hidden Markov Models with Variational Bayesian Inference (VB-HMM) proposed by Ryali et al. (PLoS computational biology 12 (12), e1005138). A comprehensive study has been conducted quantifying useful statistical properties of the time-varying brain states and their underlying network configurations, providing insights on the influence of Autism on the functioning of the brain. This work focuses on the triple network model which consists of three major intrinsic connectivity networks (ICNs) that are known to play important roles in higher-order cognition. Autistic individuals demonstrated higher persistence in brain states possessing inter-network interactions in comparison to neurotypical subjects.

Keywords: Autism · Dynamic FC · Variational Bayes HMM

1 Introduction

Research has indicated that the brain's Functional Connectivity (FC) does not remain stationary over time, and that it has multiple states distinguished by unique connectivity patterns [1,3,7,18,19]. Efforts are being made to characterize how neurological ailments affect brain connectivity and its dynamic properties [10].

Several studies have shown that Autism Spectrum Disorder (ASD) alters brain connectivity [8,9]. Accuracy has been improved by taking the dynamic property of FC into account in identifying subjects with ASD [12]. This suggests that brain dynamics possess noteworthy information about the disorder. In this work, our three primary goals are:

B. Deka et al. (Eds.): PReMI 2019, LNCS 11941, pp. 121–130, 2019.
https://doi.org/10.1007/978-3-030-34869-4_14

1. To model brain dynamics in autistic children.
2. To examine if there is any deviation in brain dynamics of autistic subjects in comparison to typically developing ones of similar demographics.
3. To investigate how autism affects the temporal properties of the brain if any anomalies caused by autism are discovered.

The triple network model consists of three Intrinsic Connectivity Networks (ICNs), namely Default Mode Network (DMN), Salience Network (SN) and Central Executive Network (CEN) [15]. It has been proven to be of primary importance in understanding higher-order cognitive function and dysfunction [11]. Aberrations in the engagement and disengagement of these three core networks have been shown to play a significant role in neurological disorders such as schizophrenia, depression, anxiety, dementia and autism [5,6,11,16]. Supported by previous research claims indicating that neurological disorders have an impact on the functioning of the triple network model, we focus our work on SN, CEN, and DMN.

Despite the existence of multiple approaches to model brain dynamics, majority of them have not focused on computing useful statistical measurements of the temporal properties of these time-varying states. Moreover, many of the approaches rely on the usage of sliding windows, which might lead to erroneous assumptions of the number of brain states. A recent framework leveraging the power of Hidden Markov Models (HMMs) with Variational Bayesian Inference has been proposed by Ryali et al. [15] that addresses the aforementioned setbacks by accurately estimating several useful parameters. It also eliminates the need for arbitrary assumptions on the number of brain states. In their work, they have used the VB-HMM framework to show how interactions between DMN, SN and CEN mature with age by conducting a study on adult and child cohorts [15].

As a natural extension, we thought of characterizing how autism influences the brain dynamics in children aged between 6.5 and 14. It has been found that children diagnosed with ASD exhibit a deviation in brain connectivity in comparison to typically developing ones of similar demographics. In this paper, several insightful properties of the brain states have been quantified and discussed. Such detailed analysis on the effects of autism is believed to be contributory to better comprehending of the disorder.

2 Methods

2.1 Variational Bayes Hidden Markov Models (VB-HMM)

VB-HMM is a framework that employs Bayesian Inference approximation to handle large scale problems that would otherwise be intractable using traditional ML-HMMs [13], and was proposed by Ryali et al. [15]. Notations and equations used in this section to describe VB-HMM are adapted from their paper [15].

Notations: Matrices are represented using uppercase, while scalars and vectors are represented using lowercase. Let T be the number of time samples, and S be the number of subjects. y_t^s is an M dimensional time sample for subject s at time t, where M is number of ROIs (Region of Interests), then $Y = \left\{\{y_t^s\}_{t=1}^T\right\}_{s=1}^S$ represents the observed voxel time series data. Let $Z = \left\{\{z_t^s\}_{t=1}^T\right\}_{s=1}^S$ be the underlying hidden discrete states, where z_t^s is the state label for subject s at time t.

Z is a first order Markov chain with transition matrix A and initial distribution π. The probability of the observation y_t given its state is assumed to be a multivariate Gaussian distribution with mean μ_k and covariance Σ_k.

Let the unknown parameters of the HMM model be $\Phi = \{\pi, A, \Theta\}$, where $\Theta = \{\mu_k, \Sigma_k\}_{k=1}^K$. Then, $p(Y, Z, \Phi)$ will be the joint probability distribution of the observations, hidden states, and the required HMM parameters. The objective is to model the HMM parameters Φ accurately. The traditional maximum likelihood approach may result in inaccurate characterization of the parameters since it requires that the number of hidden states to be specified a priori. With the help of Bayesian inference, high precision approximations can be attained.

Let us consider $p(Z, \Phi|Y)$ to be the true posterior and $q(Z, \Phi|Y)$ to be any arbitrary probability distribution. The Kullback-Leibler (KL) divergence is used to measure of how different q is from the true posterior distribution:

$$KL(q|p) = -\int dZ d\Phi q(Z, \Phi|Y) \log \frac{p(Z, \Phi|Y)}{q(Z, \Phi|Y)} \tag{1}$$

The Kullback-Leibler measure is greater than or equal to zero at all times and becomes zero only when both of the distributions in comparison are identical. Now, the log of marginal distribution of observations Y can be expressed as

$$\log P(Y) = F(q) + KL(q||p) \tag{2}$$

where $F(q)$ is the *negative free energy*, which is computed as

$$F(q) = \int dZ d\Phi q(Z, \Phi|Y) \log \frac{p(Y, Z, \Phi)}{q(Z, \Phi|Y)} \tag{3}$$

A strict lower bound is defined by $F(q)$ on $\log P(Y)$, and the objective of Bayesian approximation is to model q such that the lower bound $F(q)$ is maximized. The posterior distribution is estimated using Expectation-Maximization (EM) algorithm similar to the Baum-Welch algorithm. Further details on the workings of VB-HMM can be found in Ryali's paper [15].

Viterbi decoding algorithm is used for obtaining optimal hidden state sequence. For each distinct state, mean lifetime and occupancy rate are calculated from the optimal state sequence. The transition probabilities are estimated by the VB-HMM algorithm during the maximization step.

2.2 Flow of VB-HMM Analysis

The steps used in conducting the study are listed below in a sequential manner:

1. Build an HMM model for the observed fMRI time series data using the VB-HMM algorithm ($\Phi = \{\pi, A, \Theta\}$ is estimated).

2. Use Viterbi decoding for estimating the underlying hidden state sequence with help of the parameters learned.
3. Compute mean lifetimes and occupancy rates of the dynamic brain states.
4. Compute FC matrices for each state from the estimated covariance matrices.
5. Apply Louvain Community Detection algorithm on the FC matrices to detect their underlying community structures.
6. Combine states with similar underlying community structure and compute temporal properties of the resultant states.

Finally, we analyze the impacts of ASD by comparing the brain dynamics of autistic and typically developing cohorts.

3 Data Description

Preprocessed Data. Preprocessed fMRI time series data has been obtained from the "ABIDE preprocessed" project [4]. Data preprocessed with *Configurable Pipeline for the Analysis of Connectomes (CPAC)* using the *Automated Anatomical Labeling (AAL)* parcellation has been used in conducting the study. A detailed report on the preprocessing steps can be obtained from the ABIDE website (*http://preprocessed-connectomes-project.org/abide/*).

For every subject, 176 frames were acquired with a repetition time (TR) of 2s and the scan duration used in this study is approximately six minutes (352s).

Regions of Interest (ROIs). While each of the three ICNs consist of several regions, few representative regions have been selected based on their importance in higher-order cognition. The regions used to represent SN are Right Insula (Insula_R) and Right Midcingulate Area (Cingulum_Mid_R). For CEN, the regions used are Right Middle Frontal Gyrus (Frontal_Mid_R) and Right Inferior Parietal Lobule (Parietal_Inf_R). And for DMN, Right and Left Precuneus (Precuneus_L and Precuneus_R), and Right middle frontal gyrus-orbital part (Frontal_Med_Orb_R) have been selected. These regions have been identified as crucial for representing their respective networks by a previously published study [17].

Description of Cohorts. To study the impacts of ASD in children, 20 typically developing individuals and 20 individuals clinically classified as autistic were selected. These two cohorts were matched in age, gender and IQ demographics. ADOS (Autism Diagnostic Observation Schedule) total scores for ASD subjects ranged from 8 to 19.

Cohort	Subjects	Age (Mean ± Std, Range)	FIQ Score (Mean ± Std, Range)
NYU ASD	20, Male	10.41 ± 1.68, 6.5–14	108.15 ± 8.74, 95–130
NYU TD	20, Male	9.94 ± 1.91, 6.5–14	111.09 ± 7.19, 95–130

4 Experiments and Results

To address the sensitivity of HMMs to initialization, the experiment is run 100 times with random initializations and the model with maximum lower bound $F(q)$ is selected. The FC matrices are computed from estimated covariance matrices of each state and the interdependence of every pair of ROIs is represented by Pearson Correlation values. To investigate the properties of underlying connectivity patterns of each brain state, the Louvain Community Detection algorithm [2] present in the Brain Connectivity Toolbox [14] has been employed and

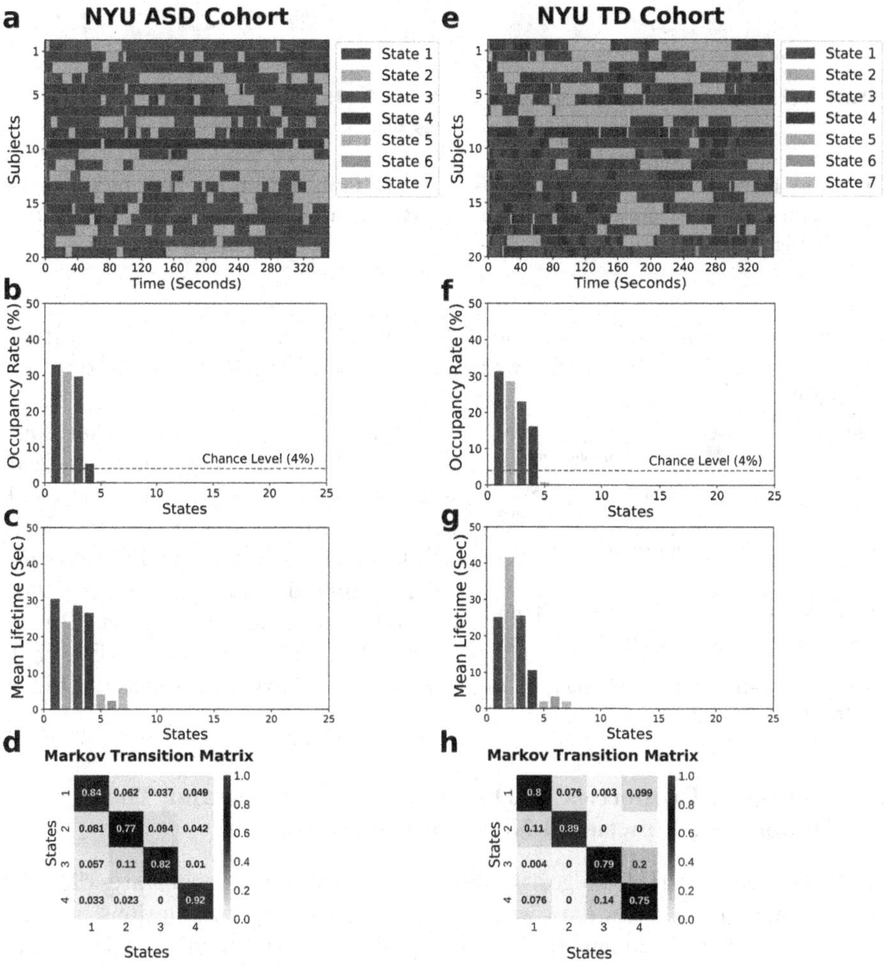

Fig. 1. Temporal properties of dynamic brain states in **ASD & TD Cohorts:** (a, e) Time evolution of dynamic states in each subject. (b, f) Occupancy Rates (c, g) Mean Lifetimes (d, h) Transition Probabilities of states with occupancies above chance level

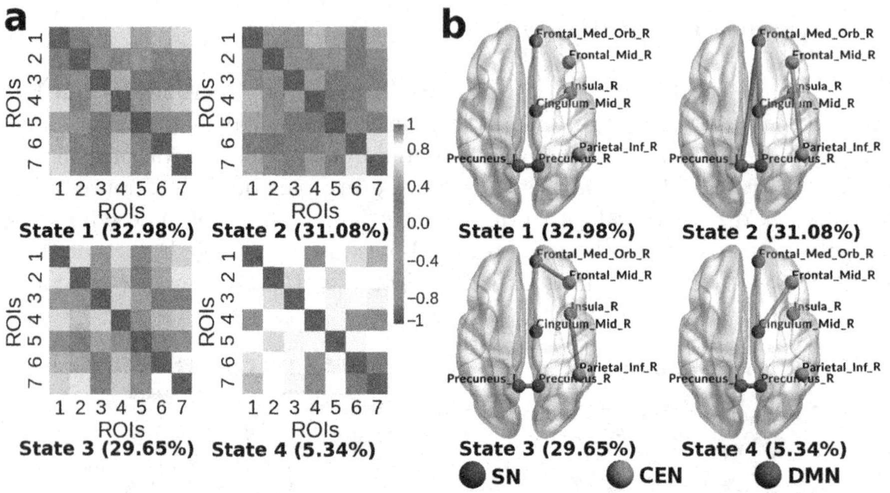

Fig. 2. (a) Dynamic FCs and (b) DFNs of **ASD Cohort**. States with occupancies above chance level are shown here. In (a), for ROI names of corresponding ROI numbers, see Fig. 3. In (b), edges of the same color belong to the same community.

this revealed interactions among the ICNs, and also between the regions within them. The BrainNet Viewer tool was used for visualizing the communities discovered [20].

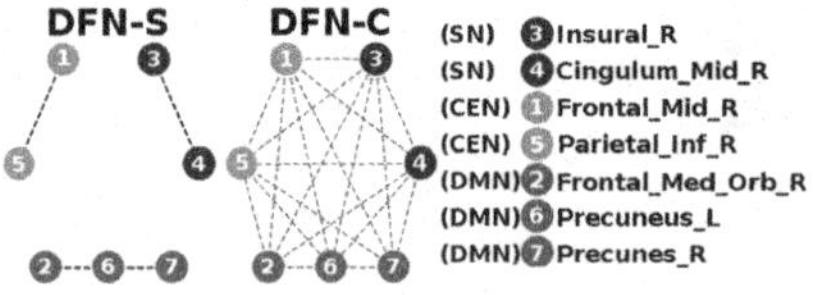

Fig. 3. Segregated DFN (DFN-S) and Cross-Connected DFN (DFN-C)

Brain states possessing similar connectivity patterns have been merged into two distinct dynamic functional network configurations. States in which SN, CEN and DMN formed fully segregated communities are grouped together and referred to as ***segregated DFNs (DFN-S)*** and the remaining DFNs that possess cross-network interactions are referred to as ***cross-connected DFNs (DFN-C)*** (see Fig. 3).

4.1 Temporal Properties of Dynamic Brain States and Their Underlying Functional Network Configurations

For allowing the VB-HMM algorithm to discover number of brain states directly from the fMRI time-series data without making arbitrary assumptions, we initialize the number of hidden states to a large upper-bound number. As a result, all of the excess states would be assigned zero occupancy leaving us with only relevant states. In this work, we set an upper-bound of 25 states (leading to an initial chance level of 4% for each state) and only seven non-zero occupancy states were found in both of the cohorts.

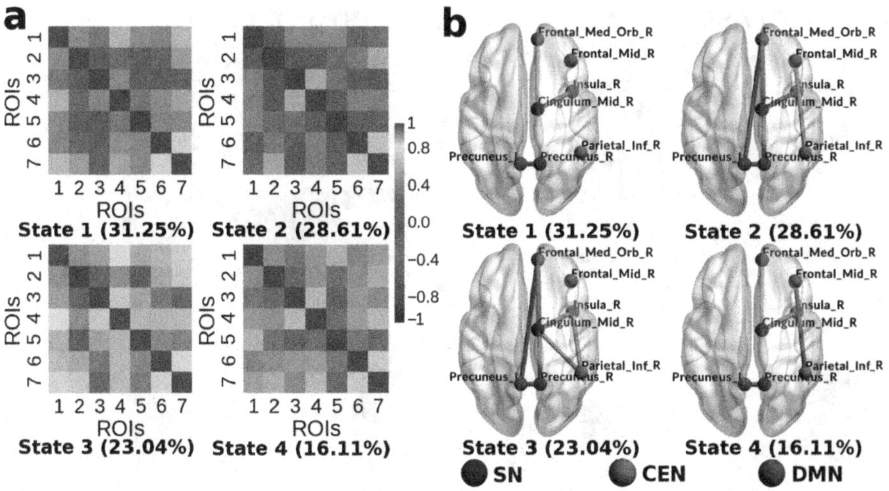

Fig. 4. (a) Dynamic FCs and (b) DFNs of **TD Cohort**. States with occupancies above chance level are shown here. In (a), for ROI names of corresponding ROI numbers, see Fig. 3. In (b), edges of the same color belong to the same community.

Dynamic Brain States. In both ASD and TD cohorts, four states with an occupancy rate above the initial chance level were found while remaining states had negligible occupancy. We can observe that the top two states in both of the cohorts have similar occupancy rates (close to 30%). The third and fourth states of each cohort had dissimilar occupancy rates, suggesting a difference in patterns of brain connectivity between ASD and TD cohorts. The mean lifetime of the top four states in ASD Cohort ranged from 24 to 30 s, while in TD Cohort it ranged from 10 to 42 s indicating a varying level of persistence in brain states. High values along the principal diagonal of the markov transition matrices indicate the stability of brain states (see Fig. 1).

Dynamic Network Configurations. Deeper insights about the brain dynamics are revealed by studying the underlying functional connectivity configurations. It is interesting to note that the first two states in both ASD and TD cohorts have the same underlying connectivity pattern (see Figs. 2 and 4), suggesting that these states are universally present across all of the subjects unaffected by the disease. Moreover, these states formed communities within the ICNs and had no interactions across them. DFN-S configuration (corresponding to states 1, 2 and 5 in ASD cohort and 1, 2 and 4 in TD cohort) in ASD cohort had significantly less occupancy in comparison to TD cohort. It is also observed that DFN-S has higher mean lifetime in healthy individuals, demonstrating higher persistence and stability of DFN-S in typically developing subjects (see Fig. 5).

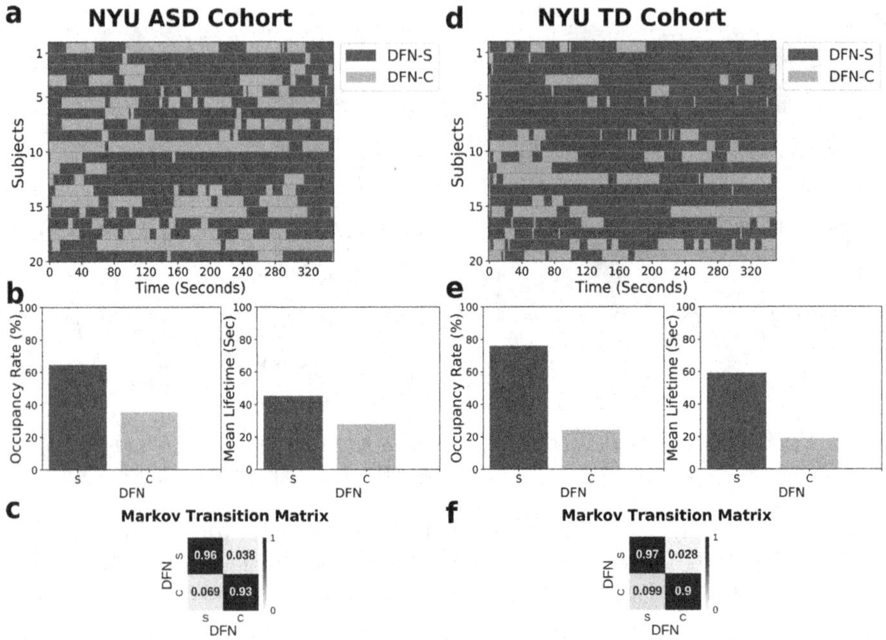

Fig. 5. Temporal properties of DFN-S & DFN-C in **ASD & TD Cohorts:** (a, d) Time evolution of the two configurations in each subject. (b, e) Occupancy Rates & Mean Lifetimes (c, f) Transition Probabilities

5 Discussion

Recent studies have proven that SN, CEN and DMN formed segregated communities only intermittently [15]. It has also been shown that children's resting brain activity spent less time in segregated functional network configurations compared to adults [15].

In addition to these earlier observations, our findings now indicate that children with ASD have lesser occupancy and shorter mean lifetime of DFN-S, exhibiting a clear deviation in brain dynamics as a result of autism. Narrowing the focus to a set of regions with well-documented roles in higher-order cognition enabled us to highlight the existence of impairment in the working memory and cognitive control. All three ICNs of the triple network model have been investigated extensively for their involvement in behavioral attributes, information processing and cognitive functioning. SN plays a crucial role in identifying and responding to salient external stimuli and internal events, whereas the CEN is responsible for high-level cognitive functions such as planning, decision making, and the control of attention and working memory [11]. The Default Mode Network forms an integrated system for self-related cognitive activity, including autobiographical, self-monitoring and social functions. Dynamic interactions among these ICNs govern attention transition and access to domain-general and domain-specific cognitive resources [11]. Impaired functioning of these networks can be held responsible for difficulties in social interaction, communication, behavior and sensory sensitivities, all of which are the most common traits of autism.

Investigating the dynamic changes in brain connectivity due to the presence of ASD at a subject level is beneficial in understanding the nature of the breakdown in function of the subject being studied. Autism refers to a wide spectrum of mental and behavioral conditions, often many of these being unique to the subject. The subject level time evolution of the brain states and their underlying configurations generated in this study can be used to identify atypical characteristics in the functioning of these networks in the corresponding subject.

6 Conclusion

In summary, we have modeled the effects of autism on the temporal dynamics of the brain. Patterns and trends that seemed unusual when compared to typically developing subjects have been identified and the underlying network configurations associated with these aberrations have been discussed.

Acknowledgement. We would like to thank Sriniwas Govinda Surampudi and Joyneel Misra of CVIT, IIIT-Hyderabad for providing valuable references and also for their suggestions. The VB-HMM *scripts* made available publicly by Ryali et al. were used in conducting the experiments in this work [15].

References

1. Allen, E.A., Damaraju, E., Plis, S.M., Erhardt, E.B., Eichele, T., Calhoun, V.D.: Tracking whole-brain connectivity dynamics in the resting state. Cereb. Cortex **24**(3), 663–676 (2014)
2. Blondel, V.D., Guillaume, J.L., Lambiotte, R., Lefebvre, E.: Fast unfolding of communities in large networks. J. Stat. Mech: Theory Exp. **2008**(10), P10008 (2008)
3. Chang, C., Glover, G.H.: Time-frequency dynamics of resting-state brain connectivity measured with fMRI. Neuroimage **50**(1), 81–98 (2010)
4. Craddock, C., et al.: The neuro bureau preprocessing initiative: open sharing of preprocessed neuroimaging data and derivatives. Neuroinformatics (41) (2013)
5. Fox, M.D., Raichle, M.E.: Spontaneous fluctuations in brain activity observed with functional magnetic resonance imaging. Nat. Rev. Neurosci. **8**, 700–711 (2007). https://doi.org/10.1038/nrn2201
6. Greicius, M.D., Krasnow, B., Reiss, A.L., Menon, V.: Functional connectivity in the resting brain: a network analysis of the default mode hypothesis. Proc. Nat. Acad. Sci. **100**(1), 253–258 (2003)
7. Hutchison, R.M., et al.: Dynamic functional connectivity: promise, issues, and interpretations. Neuroimage **80**, 360–378 (2013)
8. Just, M.A., Keller, T.A., Kana, R.K.: A theory of autism based on frontal-posterior underconnectivity. In: Development and Brain Systems in Autism, pp. 35–63 (2013)

9. Koshino, H., Carpenter, P.A., Minshew, N.J., Cherkassky, V.L., Keller, T.A., Just, M.A.: Functional connectivity in an fmri working memory task in high-functioning autism. Neuroimage **24**(3), 810–821 (2005)
10. Ma, S., Calhoun, V.D., Phlypo, R., Adalı, T.: Dynamic changes of spatial functional network connectivity in healthy individuals and schizophrenia patients using independent vector analysis. NeuroImage **90**, 196–206 (2014)
11. Menon, V.: Large-scale brain networks and psychopathology: a unifying triple network model. Trends Cogn. Sci. **15**(10), 483–506 (2011)
12. Price, T., Wee, C.-Y., Gao, W., Shen, D.: Multiple-network classification of childhood autism using functional connectivity dynamics. In: Golland, P., Hata, N., Barillot, C., Hornegger, J., Howe, R. (eds.) MICCAI 2014. LNCS, vol. 8675, pp. 177–184. Springer, Cham (2014). https://doi.org/10.1007/978-3-319-10443-0_23
13. Rabiner, L.R.: A tutorial on hidden Markov models and selected applications in speech recognition. Proc. IEEE **77**(2), 257–286 (1989)
14. Rubinov, M., Sporns, O.: Complex network measures of brain connectivity: uses and interpretations. Neuroimage **52**(3), 1059–1069 (2010)
15. Ryali, S., et al.: Temporal dynamics and developmental maturation of salience, default and central-executive network interactions revealed by variational bayes hidden Markov modeling. PLoS Comput. Biol. **12**(12), e1005138 (2016)
16. Seeley, W.W., et al.: Dissociable intrinsic connectivity networks for salience processing and executive control. J. Neurosci. **27**(9), 2349–2356 (2007)
17. Smith, S.M., et al.: Correspondence of the brain's functional architecture during activation and rest. Proc. Nat. Acad. Sci. **106**(31), 13040–13045 (2009)
18. Surampudi, S.G., Misra, J., Deco, G., Bapi, R.S., Sharma, A., Roy, D.: Resting state dynamics meets anatomical structure: temporal multiple kernel learning (tMKL) model. NeuroImage **184**, 609–620 (2019)
19. Surampudi, S.G., Naik, S., Surampudi, R.B., Jirsa, V.K., Sharma, A., Roy, D.: Multiple kernel learning model for relating structural and functional connectivity in the brain. Sci. Rep. **8**(1), 3265 (2018)
20. Xia, M., Wang, J., He, Y.: BrainNet viewer: a network visualization tool for human brain connectomics. PLoS One **8**(7), e68910 (2013)

Robust Identification of Dense or Sparse Crowd Based on Classifier Fusion

Saikat Dutta[1(✉)], Soumya Kanti Naskar[1], Sanjoy Kumar Saha[1], and Bhabatosh Chanda[2]

[1] Jadavpur University, Kolkata, India
saikat.dutta779@gmail.com, rijunaskar@gmail.com, sks_ju@yahoo.co.in
[2] Indian Statistical Institute, Kolkata, India
chanda@isical.ac.in

Abstract. For a video surveillance system, crowd behavior analysis and crowd managing are important tasks. Along with the event in which crowd participates, its volume and density are also important in managing the crowd. Hence, characterizing the crowd as dense or sparse is an essential component of a crowd handling system. In this context, most of the existing methods try to estimate the headcount. Unlike those, the proposed method exploits the domain-knowledge based low-level features to classify the crowd image as dense or sparse. We present three simple systems working with three different feature sets. These are all free from the burden of background estimation. Experiments are carried on a dataset formed by taking the images from UCF-CC50 and SanghaiTech. Performance of all three feature sets are satisfactory, and Corner-Point based methodology provides the best result.

Keywords: Crowd density · Crowd classification · Dense or sparse crowd

1 Introduction

Crowd management has become an important task and video surveillance systems can be of great help in this context. In daily life people may gather at various public places like railway station and market place, and also for different activities or events like sports and cultural. To ensure safety and proper management, crowd behavior analysis is crucial. The behavioral anomaly of the crowd depends not only on the nature of the participating group but also on the crowd volume and *density*. Hence, estimating these parameters through video surveillance system is an important step towards crowd behavior analysis and management. In this paper, we present three novel methods for classifying the crowd image as dense or sparse using domain knowledge based low level features. Finally, classifiers are fused to develop a robust system.

The paper is organized as follows. This brief introduction is followed by a review of past work presented in Sect. 2. Proposed methodology is elaborated in Sect. 3. Section 4 presents the experimental results and discussion. Concluding remarks are sited in Sect. 5.

B. Deka et al. (Eds.): PReMI 2019, LNCS 11941, pp. 131–139, 2019.
https://doi.org/10.1007/978-3-030-34869-4_15

2 Past Work

A large variety of methods exists in the literature. Some works are based on still images and some are on videos. Some works focus only on dense crowd images. One of the main approaches towards crowd density estimation is to count the population. This approach [1,2] can be sub-grouped as *human detection based* and *motion based*. In human detection based approach [3], the challenge lies in designing the human detector. and subsequent counting is straight forward. In motion based approach, the number of components with independent motion is taken as the count [4,5].

Marana et al. [6] used texture features in the form of Gray-level Dependence Matrices (GLDM) and applied Self Organizing Map (SOM) to classify crowd images to different density categories ranging from very low to very high. Li et al. [7] applied head-detector on the segmented foreground to obtain the count. Cheriyadat et al. [4] worked on image sequence with moving crowd, where low-level feature points are tracked, and regions with coherent motion are detected as objects for counting. SIFT features are also used for crowd detection in [8]. Corner points based methods are widely used to count the number of moving people [5,9]. Subburaman et al. [3] used gradient orientation features at interest points and Adaboost classifier. Jiang [10] proposed an improvisation on the regression based crowd counting mechanism. Idrees et al. [11] proposed a hybrid approach for highly dense crowd image, where head detector and interest point based count were combined with Fourier analysis. Hafeezallah et al. [12] introduced the curvelet frame change detection which enhances the statistical features for counting the individuals in the crowd.

In recent times convolutional neural network (CNN) is being used for crowd density estimation [13,14]. The network is trained with known crowd patches and then adapt it for target scenario. It is well known that obtaining a meaningful result from deep learning based method requires a huge training set whose distribution should be good representative of the population from which test (target) data would be drawn. Such a training set may not always be available. Second, it is observed that though a considerable variety of methods exists, there is not a single method that can handle all sorts of crowds. Moreover, some methods can handles image(s) of dense crowd only. Thus, characterizing a crowd as dense or sparse at the onset is essential in choosing an optimal strategy. In this work, we attempt to develop a robust system that can classify crowd image(s) into dense or sparse based on a small training set.

3 Proposed Methodology

In this work, we try to determine whether a crowd seen in an image is dense or sparse. Here, crowd image is conceived as texture image, and dense crowd image appears to be fine (micro) texture, while sparse crowd mimics coarse (macro) texture. Thus, sparse/dense crowd classification degenerates to fine/coarse texture classification. This motivates us to look for a variety of texture descriptors suitable for the task. Here we consider three different texture descriptors. First two try to rely on fractal dimension; whereas, the last one is based on count of interest (corner) points over. Feature extraction processes are detailed as follows.

Example of sparse crowd Example of dense crowd

Fig. 1. Sample images from the dataset

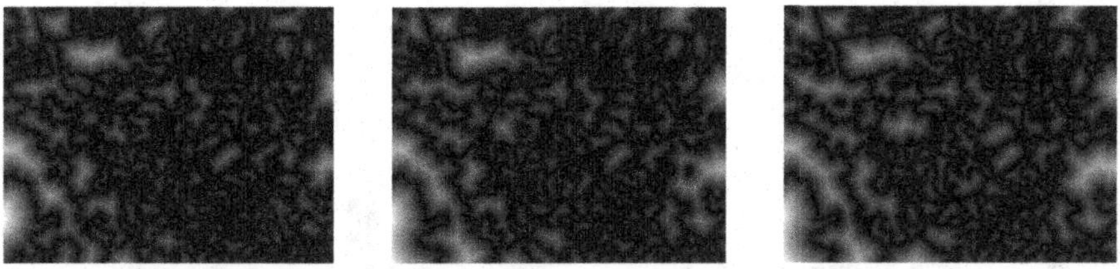

Dist. trans. with Thr=30 Dist. trans. with Thr=35 Dist. trans. with Thr=40

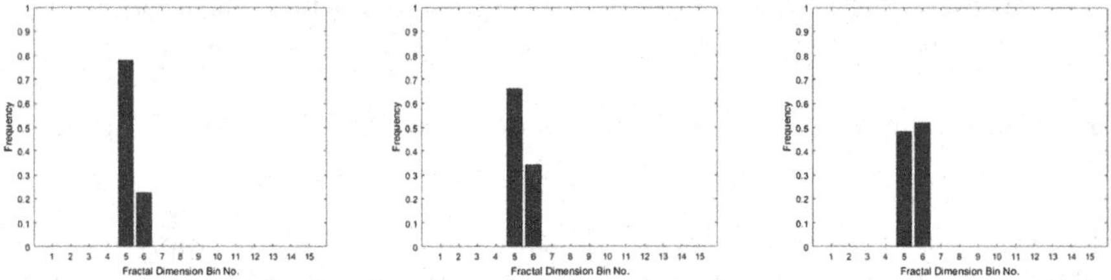

Histo. with Thr=30 Histo. with Thr=35 Histo. with Thr=40

Fig. 2. Distance transform based descriptor for example sparse crowd in Fig. 1

Descriptor Based on Distance Transform and Fractal Dimension: First, color image is converted into gray-scale image and segmented using morphological watershed algorithm [15,16], where gray-scale value of a pixel represents altitude at that location. The watershed line surrounds each region depicting a uniform surface feature. For Dense crowd images, a large number of small segments are obtained; while for sparse crowd, segments are large and small in number. Watershed algorithm produces a binary image with distinct regions with watershed line in-between. It may noted that one could have used any other segmentation scheme that generates closed contour.

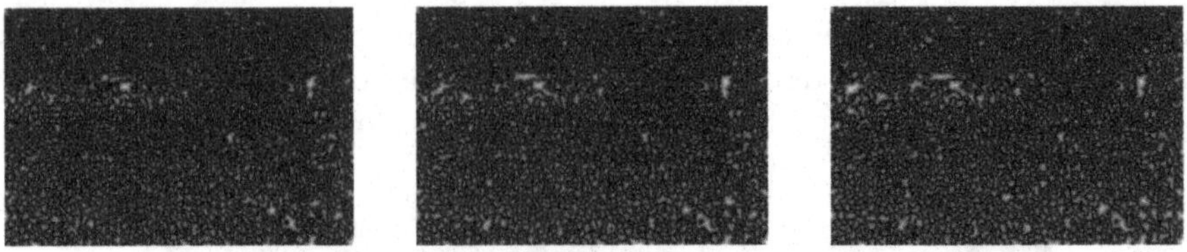

Dist. trans. with Thr=30 Dist. trans. with Thr=35 Dist. trans. with Thr=40

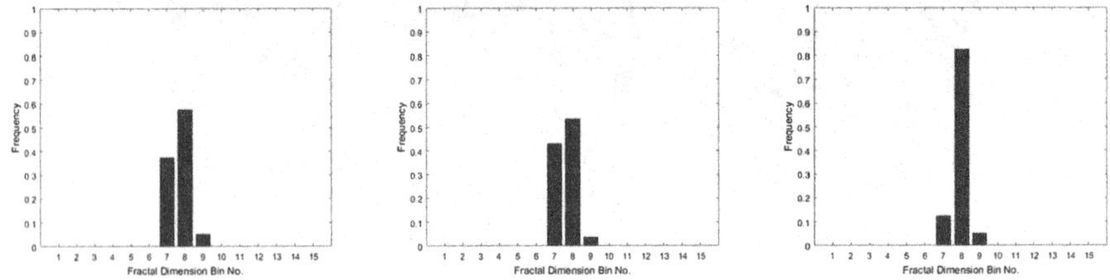

Histo. with Thr=30 Histo. with Thr=35 Histo. with Thr=40

Fig. 3. Distance transform based descriptor for example dense crowd in Fig. 1

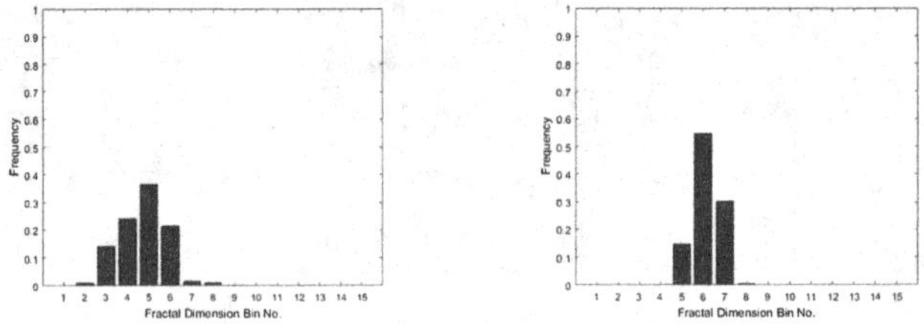

Histogram of Fractal Dimension for Sparse Crowd in Figure 1 Histogram of Fractal Dimension for Dense Crowd in Figure 1

Fig. 4. Fractal dimension descriptor for example sparse and dense crowds in Fig. 1

To extract texture feature from the said binary image, we apply distance transform [17]. The result of the transform is a two-dimensional matrix (say, T) of the same size as the image and a matrix element denotes the distance of the corresponding pixel from nearest watershed line. Hence, it reveals a kind (fine or coarse) of texture. Finally, texture feature is extracted from distance matrix in terms of fractal dimension. Note that, fractal dimension has already been used for texture segmentation [18,19]. It indicates roughness and self-similarity in the image. For a dense image, more self-similarity is expected compared to a sparse one. T is divided into $K \times K$ patches with a stride of K/p. Fractal dimension is computed over each patch. A normalized histogram of these fractal dimensions is taken as feature vector. Here, we empirically decide $K = 100$ and $p = 2$.

Watershed algorithm has a parameter that controls the segmentation process, and its selection is data-dependent and is a non-trivial task. Impact of different threshold values on segmentation will vary depending on the crowd density and the variation pattern can be an indicator of density. In our work, we take three

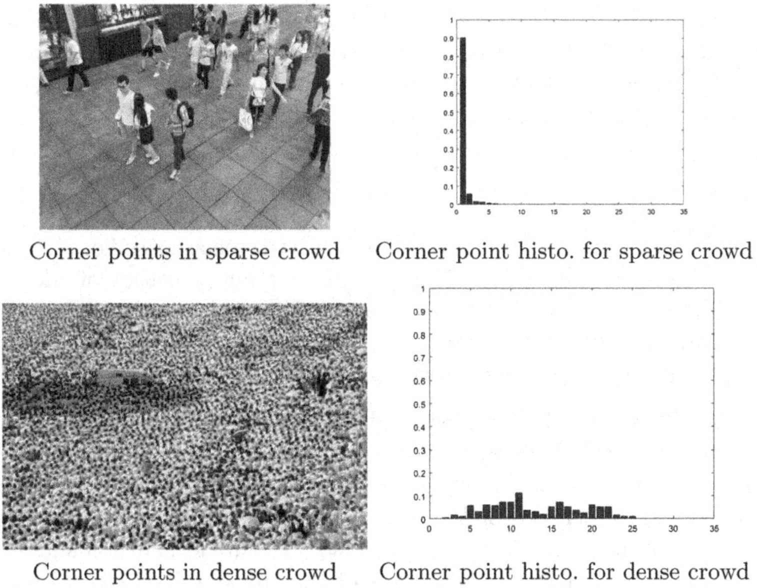
Corner points in sparse crowd
Corner point histo. for sparse crowd
Corner points in dense crowd
Corner point histo. for dense crowd

Fig. 5. Corner point based descriptor for example sparse and dense crowd in Fig. 1

threshold values: 40, 35 and 30 which are chosen empirically and applied to all the images in the dataset. Corresponding histograms are concatenated to form image texture descriptor. Figure 1 show sample sparse and dense images. Corresponding distance transform matrices and histograms are shown in Figs. 2 and 3. It is evident that the fractal dimension distribution is different the two types of crowd.

Descriptor Based on Fractal Dimension: Above algorithm is intuitively very promising, but it gets affected by certain issues. For example, we expect large and less number of segments in the binarized sparse crowd image. But the assumption fails in case of textured background. In order to get rid of it we drop the segmentation step. Fractal dimension is computed over each patch of gray-level image and these values are summarized into a normalized histogram of fractal dimension. The histograms of fractal dimension for sample sparse and dense crowd are shown in Fig. 4.

Descriptor Based on Corner Point: Fractal dimension based feature is global in nature and bears impact of background texture. To reduce such influence and to incorporate local character we focus on corner point based descriptor. Number of such points in a small patch of a dense crowd image is usually higher than that of sparse crowd image.

We extract corner points using Harris-Stephens algorithm [20]. Sensitivity factor is taken as 0.05. Then image is divided into patches as before. For each patch, corner points are counted. Histogram of normalized count is taken as the descriptor. The histograms of example crowd images are shown in Fig. 5.

Usually for a dense crowd, the non-zero histogram bins spread over large counts, whereas sparse crowd they are usually restricted to lower range of counts with strong peak. To reduce the effect of noise, an edge preserving smoothing [21] can be applied as pre-processing.

3.1 Classification

For all the three descriptors, we have used Decision tree as classifier [22]. During training, data is split at each decision node based on maximization of information gain at child nodes. During test, a simple condition is tested on feature at each node and corresponding branch is taken. This process goes on recursively and eventually a leaf node is reached based on which we predict the class-label.

Fusing the Classifier: It is understood from the description of features that some of them are supplementary and some are redundant too. Second, the classifier must be robust. That means, standard deviation of various test run must be as low as possible. So it may worth exploring the fusion of the classifiers based on these features. We have tried both kind of fusion: feature level fusion and decision level fusion. In the former case, three sets of features obtained based on (i) distance transform and fractal dimension, (ii) fractal dimension, and (iii) count of corner points are concatenated together to form a single feature vector, which is then fed to the classifier. In the latter case, output or decision obtained from each of the classifiers using three different feature sets as stated above are combined through an artificial neural network with three input nodes, two output nodes and a hidden layer. Results of fused classifiers are also reported.

4 Experimental Results and Discussion

We have performed the experiments on a machine with Intel®Core™i5-5200U CPU and 4 GB RAM. All the codes are written in MATLAB®.

Although there are many public datasets for crowd counting and tracking, dataset for crowd density based classification is not readily available, at least, to the best of our knowledge. Hence, we have created a dataset by collecting images from UCF-CC50 dataset [23] and SanghaiTech dataset [24]. The images are selected manually in a manner such that the pictures mostly contain the region of interest, *i.e.*, spaces where crowd is actually present. Multiple raters were employed to categorize these clearly as *dense* or *sparse*. Based on the raters opinion ground-truth is associated with each image as label. Final label is assigned to each image based on majority voting. The dataset thus prepared contains 64 dense and 64 sparse crowd images to avoid imbalance in dataset of either type.

To run the experiment with the given dataset, we have randomly partitioned the dataset of each category into two halves, trained the model, i.e., decision tree classifier with one half and test on the other half. This is done 50 times and an average score of accuracy is reported in Table 1 as a quantitative measure of performance of the proposed system.

For comparison among the descriptors, experiment is done for each descriptor separately and average classification accuracy is shown in the first three rows of Table 1. Table 1 reveals that accuracy due to corner point based descriptor (96.18%) is significantly higher than that of the fractal dimension based descriptors (80.06% and 88.59%). Second, lower standard deviation of the former indicates that this descriptor develops more robust descriptors compared to the other two. We tried to work with other widely used classifiers like neural network and SVM. But, the performance was poor and that can be attributed to limited dataset. For the same reason also we could not explore deep learning approach.

Table 1. Classification accuracy for different descriptors

	Sparce accuracy	Dense accuracy	Overall accuracy
Distance transform descriptor	82.56% ± 10.18%	77.56% ± 8.15%	80.06% ± 4.65%
Fractal dimension descriptor	88.06% ± 8.64%	89.12% ± 6.06%	88.59% ± 4.34%
Corner points descriptor	96.43% ± 2.60%	95.93% ± 3.64%	96.18% ± 1.89%
Feature level fusion	95.00% ± 3.68%	93.81% ± 4.21%	94.41% ± 2.21%
Decision level fusion	95.5% ± 4.77%	91.93% ± 7.66%	93.72% ± 4.31%
MCNN [24]	94.09% ± 2.42%	97.18% ± 2.18%	91.0% ± 4.78%

As suggested earlier, we have explored both feature level and decision level fusion of classifier.

Results are shown in 4th and 5th rows of Table 1. It is revealed that though in both cases robustness is improved, it cannot exceed the performance of corner point based descriptor. These indicates that fractal dimension based features are complementary to corner based descriptors and do not add any value while they are fused. Second, performance of decision level fusion and feature level fusion are same in terms of statistical significance.

We have compared the performance with Multi-column CNN (MCNN) used in [24]. The pretrained network is used to prepare the density map for the images of our dataset and that is used as input to neural network with one hidden layer. Results in Table 1 shows that accuracy of MCNN is less than corner point based descriptor and fused classifiers (both feature level and decision level).

5 Conclusion

In this work we have presented a simple method to classify a crowd image as dense or sparse. Proposed method exploits three different descriptors based om domain knowledge. It is found that among those features, interest point based feature performs best because it includes local information. Most important part is that neither of the features require interest region segmentation nor background subtraction. It is also seen that classifier fusion leads to more robustness or less variation in performance. But as these methods rely on texture information, a texture-heavy sparse crowd image may be wrongly classified as dense one. This issue may be addressed in future. Moreover, dataset can be further enhanced to include more variety and also to utilize deep learning. However, the work shows proposed feature based methodology has good potential in classifying the crowd as dense or sparse.

References

1. Ali, S., Nishino, K., Manocha, D., Shah, M.: Modeling, simulation and visual analysis of crowds: a multidisciplinary perspective. In: Ali, S., Nishino, K., Manocha, D., Shah, M. (eds.) Modeling, Simulation and Visual Analysis of Crowds. TISVC, vol. 11, pp. 1–19. Springer, New York (2013). https://doi.org/10.1007/978-1-4614-8483-7_1
2. Hashemzadeh, M., Pan, G., Yao, M.: Counting moving people in crowds using motion statistics of feature-points. Multimed. Tools Appl. **72**(1), 453–487 (2014)
3. Subburaman, V.B., Descamps, A., Carincotte, C.: Counting people in the crowd using a generic head detector. In: 2012 IEEE Ninth International Conference on Advanced Video and Signal-Based Surveillance (AVSS), pp. 470–475. IEEE (2012)
4. Cheriyadat, A. M., Bhaduri, B.L., Radke, R.J.: Detecting multiple moving objects in crowded environments with coherent motion regions. In: Computer Vision and Pattern Recognition Workshops (2008)
5. Albiol, A., Silla, M.J., Albiol, A., Mossi, J.M.: Video analysis using corner motion statistics. In: IEEE International Workshop on Performance Evaluation of Tracking and Surveillance, pp. 31–38 (2009)
6. Marana, A.N., Velastin, S.A., Costa, L.D.F., Lotufo, R.: Automatic estimation of crowd density using texture. Saf. Sci. **28**(3), 165–175 (1998)
7. Li, M., Zhang, Z., Huang, K., Tan, T.: Estimating the number of people in crowded scenes by mid based foreground segmentation and head-shoulder detection. In: International Conference on In Pattern Recognition (ICPR) (2008)
8. Arandjelovic, O.: Crowd detection from still images. In BMVC 2008: Proceedings of the British Machine Vision Association Conference, pp. 1–10. BMVA Press (2008)
9. Dittrich, F., Koerich, A., Oliveira, L.: People counting in crowded scenes using multiple cameras. In: 2012 19th International Conference on Systems, Signals and Image Processing (IWSSIP), pp. 138–141. IEEE (2012)
10. Mei, J.: An improved method of crowd counting based on regression (2013)
11. Idrees, H., Saleemi, I., Seibert, C., Shah, M.: Multi-source multi-scale counting in extremely dense crowd images. In: IEEE Conference on Computer Vision and Pattern Recognition (CVPR) (2013)
12. Hafeezallah, A., Abu-Bakar, S.: Crowd counting using statistical features based on curvelet frame change detection. Multimed. Tools Appl. **76**(14), 15777–15799 (2017)
13. Zhang, C., Li, H., Wang, X., Yang, X.: Cross-scene crowd counting via deep convolutional neural networks. In: 2015 IEEE Conference on Computer Vision and Pattern Recognition (CVPR), pp. 833–841. IEEE (2015)
14. Han, K., Wan, W., Yao, H., Hou, L.: Image crowd counting using convolutional neural network and Markov random field. arXiv preprint arXiv:1706.03686 (2017)
15. Couprie, M., Bertrand, G.: Topological gray-scale watershed transformation. In: Vision Geometry Vi, vol. 3168, pp. 136–147. International Society for Optics and Photonics (1997)

16. Bertrand, G.: On topological watersheds. J. Math. Imaging Vis. **22**(2–3), 217–230 (2005)
17. Rosenfeld, A., Pfaltz, J.L.: Sequential operations in digital picture processing. J. ACM (JACM) **13**(4), 471–494 (1966)
18. Keller, J.M., Chen, S., Crownover, R.M.: Texture description and segmentation through fractal geometry. Comput. Vis. Graph. Image Processing **45**(2), 150–166 (1989)
19. Chaudhuri, B.B., Sarkar, N.: Texture segmentation using fractal dimension. IEEE Trans. Pattern Anal. Mach. Intell. **17**(1), 72–77 (1995)
20. Harris, C., Stephens, M.: A combined corner and edge detector. In: Alvey Vision Conference, vol. 15, no. 50. Citeseer, pp. 10–5244 (1988)
21. Huang, T., Yang, G., Tang, G.: A fast two-dimensional median filtering algorithm. IEEE Trans. Acoust. Speech Signal Process. **27**(1), 13–18 (1979)
22. Breiman, L.: Classification and Regression Trees. Routledge, London (2017)
23. Idrees, H., Saleemi, I., Seibert, C., Shah, M.: Multi-source multi-scale counting in extremely dense crowd images. In: Proceedings of the IEEE Conference on Computer Vision and Pattern Recognition, pp. 2547–2554 (2013)
24. Zhang, Y., Zhou, D., Chen, S., Gao, S., Ma, Y.: Single-image crowd counting via multi-column convolutional neural network. In: Proceedings of the IEEE Conference on Computer Vision and Pattern Recognition, pp. 589–597 (2016)

Sustained Self-Supervised Pretraining for Temporal Order Verification

Himanshu Buckchash(✉) and Balasubramanian Raman

Machine Vision Lab, Department of Computer Science and Engineering, Indian Institute of Technology Roorkee, Roorkee, India
hbuckchash@cs.iitr.ac.in, balarfma@iitr.ac.in

Abstract. Self-Supervised Pretraining (SSP) has been shown to boost performance for video related tasks such as action recognition and pose estimation. It captures important spatiotemporal constraints which act as an implicit regularizer. This work seeks to leverage upon temporal derivatives and a novel sampling algorithm for sustained (long term) SSP. Main limitations of our baseline approach are – its inadequacy to capture sustained temporal information, weaker sampling algorithm, and the need for parameter tuning. This work analyzes the Temporal Order Verification (TOV) problem in detail, by incorporating multiple temporal derivatives for temporal information amplification and using a novel sampling algorithm that does not need manual parameter adjustment. The key idea is that image-only tuples contain less information and become virtually indiscriminating in case of cyclic events, this can be attenuated by fusing temporal derivatives with the image-only tuples. We explore a few simple yet powerful variants for TOV. One variant uses Motion History Images (MHI), others use optical flow. The proposed TOV algorithm has been compared with previous works along with validation on challenging benchmarks – HMDB51 and UCF101.

Keywords: Self-Supervised pretraining · Temporal order verification · Action recognition

1 Introduction

SSP leverages the colossal amount of unlabeled data to provide an initial weight configuration which noticeably improves the performance of a model during its supervised fine-tuning. Applications of SSP can be found in different domains such as Action Recognition (AR), Natural Language Processing (NLP), and so forth. SSP ensures that the weights are not domain-specific, they readily generalize on closely related domains as well. For example: SSP for action recognition does well for pose estimation [13]; similarly, a model pretrained for question answering does well for commonsense reasoning, textual entailment, semantic similarity [16], use of pretrained word embeddings for multiple NLP tasks [2,14,15].

B. Deka et al. (Eds.): PReMI 2019, LNCS 11941, pp. 140–149, 2019.
https://doi.org/10.1007/978-3-030-34869-4_16

SSP for action recognition can be posed as a Temporal Order Verification (TOV) task [13]. TOV requires an unsupervised algorithm to generate a tuple of frames such that few of them are in valid temporal order (positive tuple) and few are out of order (negative tuple). These tuples are used to train a deep learning model that uses a binary log loss that helps to learn the pose information while trying to assert whether the order of the tuple is valid or not. As an example, Fig. 1 shows a positive tuple sampled for cartwheel action. In this figure, swapping the second and third frame will result in a negative tuple.

Fig. 1. A tuple sampled (at frame number – 5, 16, 23) for the cartwheel action.

In a recent study by researchers at OpenAI, it is shown that SSP boosts the performance of supervised tasks, and the learning is transferable to multiple related domains [16]. Similarly, it is shown by Wang *et al.* that SSP boosts performance for supervised action recognition and pose estimation [13,21]. These works provide convincing results for pairing supervised learning with SSP. This work builds upon the TOV work done by Misra *et al.* [13], and explores the challenges of SSP of deep models in the context of action recognition.

Psychologically, it is proved that the spatiotemporal signals provide significant information for answering questions based on the temporal ordering of spatial data [3,17]. Information can be sampled from spatiotemporal data, retaining the temporal order, and can be utilized for reasoning about the pose or trajectory of some object. This idea has been utilized by previous researchers [7,9,13,18,21]. The main emphasis is on sampling data in order of temporal constraints and using a discriminative model to learn the distribution of spatiotemporal information by auto-generating the positive and negative labels for data. A recurrent theme is to use the frames sampled at appropriate time-steps and using them for training a neural network, which can then infer about the sequence or ranking of the frames [13,21]. This approach can be further leveraged for solving action recognition or pose estimation problem [5,8,10,12,18,22,23].

The concept of tuple order verification has been recently applied for learning the context in videos and images [6,13,20,21]. Doersch *et al.* used the context of images for learning parts and object categories. They model the SSP problem as teaching a classifier about the relative placement of object patches in an image. This pretrained representation is then used to discover several categories of objects without any supervision [4]. However, their work cannot be directly applied to videos. Misra *et al.* have adapted their idea from images to videos [13]. They model action recognition task as – learning to order the temporal information. They sample frames from high motion instances in a video and then train a triplet CNN with shared weights in a Siamese fashion to learn the order of the sampled frames (Fig. 1).

The main contribution of [13] is that unlike [21] they do not consider insertion of random samples for constructing positive and negative tuples, instead they sample frames from high motion window. This appears logically correct. However, they use image-only tuples, coupled with their sampling algorithm which does not capture *valleys* in the flow (as discussed under Sect. 2), which leaves room for degeneracy in performance. This work targets these two issues. First, we present temporal derivative fusion with image-only tuples for tighter temporal constraints, second, we provide a novel sampling algorithm that takes into account both the *valleys* (regions with low optical flow magnitude) and the *peaks* (regions with high optical flow magnitude). The sampling algorithm does not require manual parameter adjustment.

After performing an in-depth study on temporal derivatives and tuple sampling, the main contributions of this work are (1) Algorithm for fusion of multiple temporal derivatives with image-only tuples for persistent temporal dependencies. (2) A novel sampling algorithm that considers both *peaks* and *valleys* in optical flow. We found that considering both – peaks and valleys – improves results for cyclic events such as dribbling, clapping *etc.* (3) Proposed approach is made independent of manual parameter tuning, this increases the generalizability of our work. In addition to these, proposed work has been empirically and qualitatively validated on the two challenging action recognition datasets – HMDB51 [11], UCF101 [19].

Fig. 2. Example of tuples in case of sampling with and without valleys.

2 Analysis of Tuple Order Verification

It was observed by [13] that if the temporal windows are very far apart then there is a high probability of repetition of the same pose, especially in case of cyclic events. For cyclic actions such as clap, dribble, pullup, situp *etc.*, there

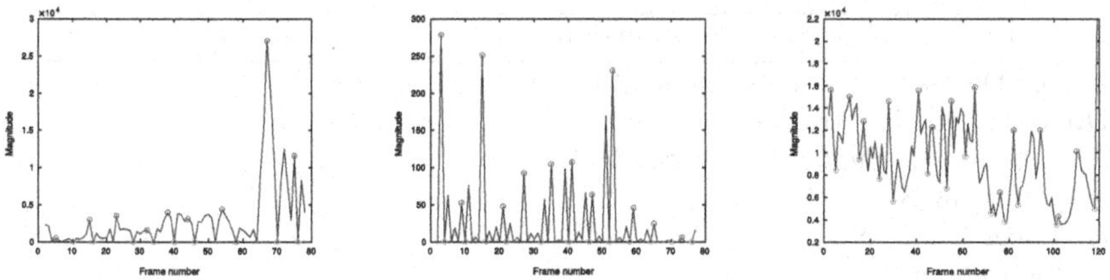

(a) Cartwheel (b) Clap (c) Dribble

Fig. 3. Optical flow signatures calculated with Frobenius norm for different actions.

exists a pause and movement cycle, where the action starts and finishes and then repeats itself. Unlike [13], instead of sampling from high motion windows (peaks), we found that it is also useful to consider the low motion windows (valleys). These motion windows are observed by taking Frobenius norm of optical flow. Frobenius norm for cartwheel action can be seen in Fig. 3, the corresponding frames are shown in Fig. 1. It can be seen in Fig. 2 first row, sampling frames from a dribbling action clip results similar information in consecutive frames, however, considering valleys during sampling helps alleviate the problem. Some events do not give enough time to capture valleys, such as chewing. Discrimination of the training tuples for these events is very confusing even for humans.

2.1 Proposed Method

Misra and Wang [13,21] observed that the triplet network performs as a better constraint on the latent representation of data by avoiding convergence of two points (in latent space) on to a single point. It was also noticed that taking up more than three frames does not provide any performance boost. Hence, a triplet Siamese model has been considered in this work (Fig. 4).

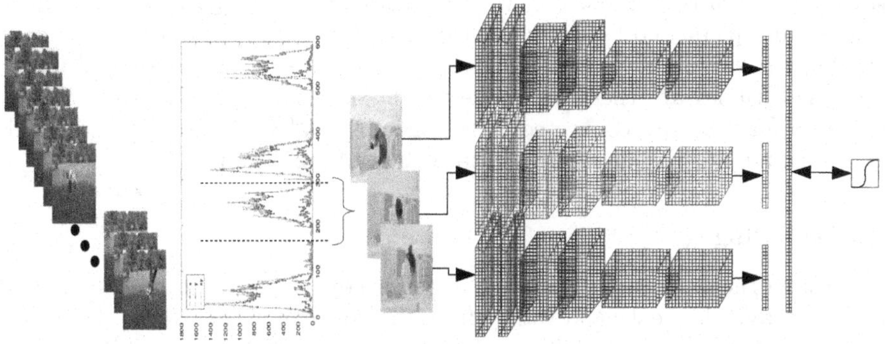

Fig. 4. Model of the proposed SSP technique showing how the Siamese-triplet network is trained on an action clip.

Algorithm 1. Self-supervised pretraining

Input: Dataset $\mathcal{D}$, Siamese model $\mathcal{M}$
Output: Set of pretrained weights $\mathcal{W}$

```
P, N ← SampleTuple(V) | ∀ video V ∈ D;
Split P, N as S_train, S_val, S_test sets ;                    // 70:10:20
// Train using S_train
foreach epoch do
    foreach mini_batch do
        Y ← M(batch_size, TempoDeriv(a, b, c)) ;   // TempoDeriv() appends
          the specific temporal derivative of tuple−(a, b, c)
        L ← −(1/N) Σ_{i=1}^{N} y_i · log(p(y_i)) + (1 − y_i) · log(1 − p(y_i));
        Backprop(M, Mean(L));
    end
end
```

Temporal derivatives, such as Motion History Images (MHI) or Optical Flow (OF), alone provide much discriminatory information, this was especially observed on UCF101 by [5]. The entire SSP algorithm by [13] depends upon the choice of three frames (a tuple). Each frame can be seen as a window of information; it follows that we have three windows. Choosing the right frame at each of the three windows is crucial for temporal inference. A single frame resembles a very narrow sized temporal window. This fact coupled with the cyclic nature of events and arbitrary motion spikes, makes this small window very vulnerable, and a slight miss in sampling can make room for an invalid training tuple. The key idea in this work is to make the information persist (a little longer) at each of these three temporal windows. A larger temporal window serves as a tighter temporal constraint by capturing more information at each sampling step. Following this idea, we have fused the temporal derivatives with static pose information for constructing a sustained SSP algorithm.

Unlike [13] the proposed SSP approach samples the sets of positive and negative tuples from both peaks and valleys as described by Algorithm 2. We first sample quintuplets, and for every quintuplet (a, b, c, d, e), — (b, c, d), (d, c, b) are considered positive and (b, a, d), (b, e, d), (d, a, b), (d, e, b) are considered negative. These tuples are then fed to the weight-shared Siamese network, whose outputs from the fully connected layers of individual CNNs are concatenated prior to training with binary cross-entropy loss (Algorithm 1).

2.2 Sampling

The positive/negative tuple sampling is performed according to the Algorithm 2. First we find the peaks and valleys from the flow magnitude. Next, the peaks are clustered, and the ones falling in the lowest cluster are removed. Only one peak is kept in a radius of size Minimum Peak Distance (MPD). This results in a sequence of local peaks and valleys. Subsequently, sum of squared distance (SSD) is used for pruning consecutive similar frames, using a threshold σ, which

Algorithm 2. Tuple sampling

```
Input: Video 𝒱
Output: Set of tuples, positive 𝒫 and negative 𝒩
ℱ ← f_1, f_2, ..., f_n frames from 𝒱 with stride of 2;
ℱ_of ← OpticalFlow(ℱ);
ℱ_peak ← Peaks(ℱ_of) with MPD ← 4;
ℱ_valley ← Valleys(ℱ_of);
Cluster ℱ_peak with k-means, drop lowest magnitude cluster;
i ← 1 ;                                          // Index of first peak
k ← 2 ;                                          // Index of second peak
while k ≤ End(ℱ_peak) do
    𝒮 ← j | ∀ j ∈ (ℱ_valley) ∧ ℱ_peak(i) ≺ ℱ_valley(j) ≺ ℱ_peak(k);
    Append CentralPeak(S) to ℱ_valley-new;
    i ← k;
    k ← k + 1;
end
ℱ_pv ← ℱ_peak + ℱ_valley-new ;                   // Combine
a ← 1 ;                                          // Index of first frame in ℱ_pv
b ← 2 ;                                          // Index of second frame in ℱ_pv
while b ≤ End(ℱ_pv) do
    For consecutive frame pair (a, b) ∈ ℱ_pv;
    if S(a, b) < σ then
        b ← b + 1 ;      // Sum of squared distance less than threshold σ
    else
        a ← b;   b ← b + 1;
    end
end
foreach quintuplet − (a, b, c, d, e) of consecutive frames a, b, c, d, e from ℱ_pv do
    𝒫 ← (b, c, d), (d, c, b);
    𝒩 ← (b, a, d), (b, e, d), (d, a, b), (d, e, b);
end
```

is determined empirically. Following this, peaks and valleys are combined and quintuplets are formed. For every quintuplet (a, b, c, d, e), — (b, c, d), (d, c, b) are considered positive and (b, a, d), (b, e, d), (d, a, b), (d, e, b) are considered negative.

3 Experiments

All experiments have been performed on split 1 of UCF101 [19] and HMDB51 [11]. [13] is used as a baseline for our experiments. Due to the high computational requirement of Siamese networks, a non-bulky CaffeNet like architecture is considered for the proof of concept. It is trained from scratch for all experiments with variable learning rate, dropout, momentum, and a batch size of 64. 650K tuples were sampled for UCF101, 350K tuples were sampled for HMDB51. We experimented with three variants of temporal derivatives. tRGB uses only RGB tuples, tBW uses only gray-scale image tuples, tOF-BW early-fuses gray-scale

Table 1. Results of tuple prediction on HMDB51 and UCF101 datasets.

Dataset	tRGB	tBW	tOF-BW	tOF-RGB	tMHI-RGB	[13]	[21]
UCF101	69.1	67.3	71.2	**74.8**	28.6	67.3	61.2
HMDB51	35.6	33.6	35.4	**38.2**	15.9	34.1	30.8

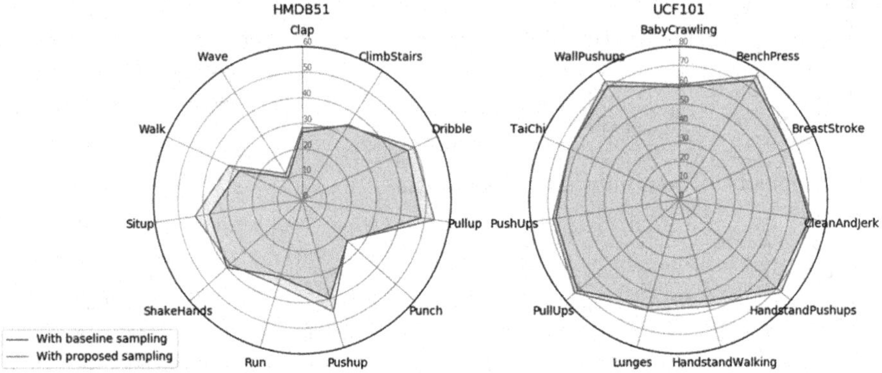

Fig. 5. Comparison of the proposed sampling algorithm with the baseline sampling on two challenging datasets – HMDB51 & UCF101.

frames with the flow derivatives, tOF-RGB early-fuses RGB frames with the flow, tMHI-RGB uses Motion History Images (MHI) with short temporal window [1]. All of these variants are tested with the proposed sampling algorithm.

3.1 Results Analysis

The reversed order of the tuples is also deemed as valid since it preserves temporal constraints. It was also observed that forward and backward tuples are naturally discovered during the sampling of cyclic actions. Semantic level fusion of flow captures details complementary to the spatial data. Tuple verification by Misra *et al.* captures pose, however this variant of tuple verification forces the model to learn motion transformation (Fig. 7). It can be inferred from the Table 1 that fusion of temporal derivatives (tOF-RGB, tOF-BW), significantly boosts the overall performance of SSP. When combined with spatial data, the flow acts as a saliency by preserving the motion information about the parts (Fig. 7). The best results are obtained by tOF-RGB model which outperforms

Table 2. Comparison of sampling methods for TOV.

Dataset	[13]	tRGB
UCF101	67.3	**69.1**
HMDB51	34.1	**35.6**

Fig. 6. Results of retrieved nearest neighbors against four query images. Top row results by tOF-RGB model, bottom row for a model initialized with random weights.

[13] by over 7% and [21] by over 13%, as reported in the Table 1. tMHI-RGB performs miserably because it tends to clutter the image with a lot of information which reduces meaningful pose information. tBW and tRGB have small performance difference; it suggests that the model learns pose and is invariant to color. tOF-BW model not only learns pose information but also learns the flow transformation parameters. Table 2 shows the difference between the baseline sampling algorithm [13] and the proposed sampling for RGB tuples. For further clarity on the performance of the tuple sampling algorithm, class-wise comparison of cyclic actions is reported in Fig. 5. It can be seen that for both of the datasets, the proposed sampling algorithm performs better than the baseline sampling.

Figure 6 shows the frames retrieved for a nearest-neighbor query on the tOF-RGB model, in comparison to the baseline model which is initialized with random weights. We see that the tOF-RGB model captures the pose information better than the model with random weights. This establishes the significance of SSP.

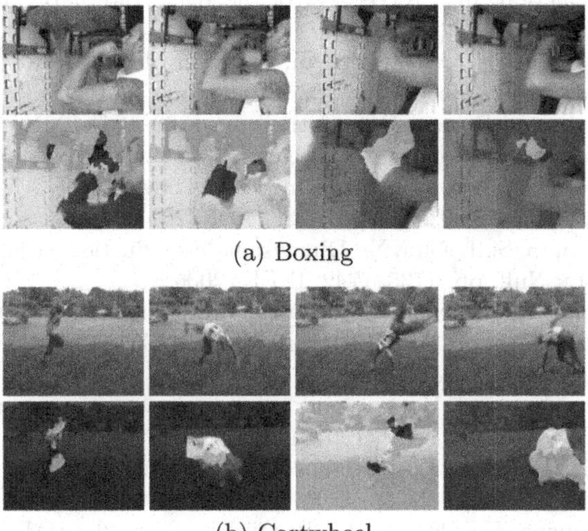

(a) Boxing

(b) Cartwheel

Fig. 7. Triplets having fusion of flow and spatial data for creation of positive tuples.

4 Conclusion

The whole purpose of having TOV is to be able to order or rank the data. The ability to rank or order the temporal data is in proportion to the ability to discriminate. To achieve this, each element of tuple-to-be-ordered should store information which is as discriminatory as possible. It was noticed that the spatial information could be augmented with temporal derivatives for each input to the Siamese-triplet. To attain this, multiple approaches were explored in this work. tOF-RGB achieved best results using proposed sampling along with flow and RGB information fusion. It could be concluded that the temporal derivatives provide a better representation for estimation of the pose. In the future, focus can be on the elongation of the temporal windows. To achieve this, 3D convolutions can be explored. The main takeaways of this work are: (1) Temporal derivatives are a strong prior for ordering (2) The combination of flow and spatial information is better than each considered individually, as we see that it acts as a salient pair (3) For sampling, both peaks and valleys should be considered for capturing the cyclic actions.

References

1. Ahad, M.A.R.: Motion History Images for Action Recognition and Understanding. Springer (2012)
2. Chen, D., Manning, C.: A fast and accurate dependency parser using neural networks. In: Proceedings of the 2014 Conference on Empirical Methods in Natural Language Processing (EMNLP), pp. 740–750 (2014)
3. Cleeremans, A., McClelland, J.L.: Learning the structure of event sequences. J. Exp. Psychol. Gen. **120**(3), 235 (1991)
4. Doersch, C., Gupta, A., Efros, A.A.: Unsupervised visual representation learning by context prediction. In: Proceedings of the IEEE International Conference on Computer Vision, pp. 1422–1430 (2015)
5. Feichtenhofer, C., Pinz, A., Zisserman, A.: Convolutional two-stream network fusion for video action recognition. In: Proceedings of the IEEE Conference on Computer Vision and Pattern Recognition, pp. 1933–1941 (2016)
6. Goroshin, R., Bruna, J., Tompson, J., Eigen, D., LeCun, Y.: Unsupervised learning of spatiotemporally coherent metrics. In: Proceedings of the IEEE International Conference on Computer Vision, pp. 4086–4093 (2015)
7. Hadsell, R., Chopra, S., LeCun, Y.: Dimensionality reduction by learning an invariant mapping. In: Null, pp. 1735–1742. IEEE (2006)
8. Jain, M., Jegou, H., Bouthemy, P.: Better exploiting motion for better action recognition. In: Proceedings of the IEEE Conference on Computer Vision and Pattern Recognition, pp. 2555–2562 (2013)
9. Jayaraman, D., Grauman, K.: Learning image representations equivariant to ego-motion. In: Proceedings of ICCV (2015)
10. Kingma, D.P., Welling, M.: Auto-encoding variational bayes. arXiv preprint arXiv:1312.6114 (2013)
11. Kuehne, H., Jhuang, H., Garrote, E., Poggio, T., Serre, T.: HMDB: a large video database for human motion recognition. In: 2011 IEEE International Conference on Computer Vision (ICCV), pp. 2556–2563. IEEE (2011)

12. Le, Q.V.: Building high-level features using large scale unsupervised learning. In: 2013 IEEE International Conference on Acoustics, Speech and Signal Processing (ICASSP), pp. 8595–8598. IEEE (2013)
13. Misra, I., Zitnick, C.L., Hebert, M.: Shuffle and learn: unsupervised learning using temporal order verification. In: Leibe, B., Matas, J., Sebe, N., Welling, M. (eds.) ECCV 2016. LNCS, vol. 9905, pp. 527–544. Springer, Cham (2016). https://doi.org/10.1007/978-3-319-46448-0_32
14. Pennington, J., Socher, R., Manning, C.: Glove: global vectors for word representation. In: Proceedings of the 2014 Conference on Empirical Methods in Natural Language Processing (EMNLP), pp. 1532–1543 (2014)
15. Qi, Y., Sachan, D.S., Felix, M., Padmanabhan, S.J., Neubig, G.: When and why are pre-trained word embeddings useful for neural machine translation? arXiv preprint arXiv:1804.06323 (2018)
16. Radford, A., Narasimhan, K., Salimans, T., Sutskever, I.: Improving language understanding by generative pre-training. https://s3-us-west-2.amazonaws.com/openai-assets/research-covers/language-unsupervised/language_understanding_paper.pdf (2018)
17. Reber, A.S.: Implicit learning and tacit knowledge. J. Exp. Psychol. Gen. **118**(3), 219 (1989)
18. Simonyan, K., Zisserman, A.: Two-stream convolutional networks for action recognition in videos. In: Advances in Neural Information Processing Systems, pp. 568–576 (2014)
19. Soomro, K., Zamir, A.R., Shah, M.: Ucf101: a dataset of 101 human actions classes from videos in the wild. arXiv preprint arXiv:1212.0402 (2012)
20. Srivastava, N., Mansimov, E., Salakhudinov, R.: Unsupervised learning of video representations using LSTMS. In: International Conference on Machine Learning, pp. 843–852 (2015)
21. Wang, X., Gupta, A.: Unsupervised learning of visual representations using videos. In: Proceedings of the IEEE International Conference on Computer Vision, pp. 2794–2802 (2015)
22. Wang, X., Gupta, A.: Generative image modeling using style and structure adversarial networks. In: Leibe, B., Matas, J., Sebe, N., Welling, M. (eds.) ECCV 2016. LNCS, vol. 9908, pp. 318–335. Springer, Cham (2016). https://doi.org/10.1007/978-3-319-46493-0_20
23. Yang, W., Gao, Y., Cao, L., Yang, M., Shi, Y.: mPadal: a joint local-and-global multi-view feature selection method for activity recognition. Appl. Intell. **41**(3), 776–790 (2014)

Retraining Conditions: How Much to Retrain a Network After Pruning?

Soumya Sara John(✉), Deepak Mishra, and J. Sheeba Rani

Indian Institute of Space Science and Technology,
Thiruvananthapuram 695547, India
soumyasara100@gmail.com,
{deepak.mishra,sheeba}@iist.ac.in

Abstract. Restoring the desired performance of a pruned model requires a fine-tuning step, which lets the network relearn using the training data, except that the parameters are initialised to the pruned parameters. This relearning procedure is a key component in deciding the time taken in building a hardware-friendly architecture. This paper analyses the fine-tuning or retraining step after pruning the network layer-wise and derives lower bounds for the number of epochs the network will take based on the amount of pruning done. Analyses on the propagation of errors through the layers while pruning layer-wise is also performed and a new parameter named 'Net Deviation' is proposed which can be used to estimate how good a pruning algorithm is. This parameter could be an alternative to 'test accuracy' that is normally used. Net Deviation can be calculated while pruning, using the same data that was used in the pruning procedure. Similar to the test accuracy degradation for different amounts of pruning, the net deviation curves help compare the pruning methods. As an example, a comparison between Random pruning, Weight magnitude based pruning and Clustered pruning is performed on LeNet-300-100 and LeNet-5 architectures using Net Deviation. Results indicate clustered pruning to be a better option than random approach, for higher compression.

Keywords: Pruning · Model compression · Deep neural networks · Retraining network

1 Introduction

Neural Networks have been widely used to solve different real world problems in different arenas because of its remarkable function approximation capability. Recently, Deep Neural Networks have attained much awaited attention across various discipline. Despite of all the advances in the field of Deep Learning, including the arrival of better and improved optimization algorithms and GPUs, several questions are still puzzling. One of them is the optimum architecture size. This includes, how one should decide the number of nodes and number of layers for a network to solve a particular problem using the data set. Both the deep and

B. Deka et al. (Eds.): PReMI 2019, LNCS 11941, pp. 150–160, 2019.
https://doi.org/10.1007/978-3-030-34869-4_17

shallow networks have their own share of advantages and disadvantages. This makes it difficult for the network designer to create the optimum architecture. Shallow networks exhibit better generalization performance and learn faster, but have a higher tendency to overfit [5]. Deeper networks like LeNet-300-100 or LeNet-5 [12], form complex decision boundaries but avoids overfitting.

Model Compression techniques are inspired by the fault tolerance property to network damage conditions, seen in larger networks. Pruning is one such model compression technique. Introducing damage to the network purposefully will compromise its performance in terms of accuracy. However, a procedure of retraining can be used to regain the original performance. In general, the percentage reduction in accuracy is proportional to the amount of damage made. When the damage to the network is rigorous, the network requires more retraining to regain the desired accuracy on the particular data set. [18] conducted experiments comparing the accuracy of large, but pruned models (large-sparse) with their smaller, but dense (small-dense) counterparts and gave results stating that the large-sparse models outperforms the small-dense models with $10\times$ reduction in the number of non zero parameters with minimal reduction in accuracy.

Major Contributions. The contributions of this paper are summarized as follows: (1) We derive theoretical bounds for the number of epochs a pruned network will require to reach the original performance, relative to the number of epochs the original unpruned network had taken to reach the same performance. (2) We derive a relation bounding the error that will be present at the output based on layer-wise error propagation due to the pruning done in different layers. (3) A new parameter 'Net Deviation' is proposed, that could serve as a measure to select the appropriate pruning method for a particular network and data, by comparing the net deviation curves for these methods for different percentage of pruning. This parameter could be an alternative to 'test accuracy' that is normally used. Net Deviation is calculated while pruning, using the same data that was used for pruning. The detailed proofs of the stated theorems are given in the Appendix.

2 Related Works

Research in the area of architecture selection has led to different pruning approaches. Recently, obtaining a desired network architecture received significant attention from various researchers [10,14,17] and [11]. Network pruning was found to be a viable and popular alternative to optimize an architecture. This research can be dichotomised into two categories, with and without retraining.

2.1 Without Retraining

The weights which contribute maximum to the output must not be disturbed, if no retraining is required. In order to achieve this, [7] have used the idea of Core-sets that could be found through SVD or Structured Sparse PCA or an activation-based procedure. Even when this method could provide a high compression, the matrix decomposition complexity involved could be higher for larger networks.

2.2 With Retraining

Most of the research works focus on methods that have a retraining or fine-tuning step after pruning. Such methods can be classified again as given below:

As an Optimization Procedure. In this category, retraining is defined as a procedure over the trained model to find the best sparse approximation of the trained model, that doesn't reduce the overall accuracy of the network. [4] does the same in two steps- one to learn the weights that can approximate a sparser matrix (L step) and another to compress the sparser matrix again (C step). [1] does model compression by considering model compression as a convex optimization algorithm. Minor fine tuning is also done at the end of this retraining procedure.

Without Any Optimization Procedure. A pruning scheme without any optimization procedure, delves into two things: either to keep the prominent nodes or to remove redundant nodes using some relevant criteria. Prominent nodes are defined as the nodes that contribute the most to the output layer nodes. These nodes could be defined based on the weight connections or gradients, as seen in [3,8,9,13] and [6]. Redundant nodes can be removed by clustering nodes that give similar output as done in [14]. Data-free methods also exist, like [15] that does not use the data to calculate the pruning measure.

3 Analysis of Retraining Step of a Sparse Neural Network

3.1 Preliminaries

Consider a Multi-Layer Perceptron with 'L' layers, with $n^{(l)}$ nodes in layer l. Corresponding weights and biases are denoted as $\mathbf{W}^{(1)}, \mathbf{W}^{(2)}, ..., \mathbf{W}^{(L-1)}$ and $\mathbf{B}^{(2)}, \mathbf{B}^{(3)}, ..., \mathbf{B}^{(L)}$ respectively, where $\mathbf{W}^{(l)} \in \mathbb{R}^{n^{(l)} \times n^{(l+1)}}$ and $\mathbf{B}^{(l)} \in \mathbb{R}^{n^{(l)} \times 1}$. The loss function is denoted as $L(\mathbf{W})$. This could be mean-squared error or cross-entropy loss, defined on both the weights and biases, using the labels and the predicted outputs.

3.2 Error Propagation in Sparse Neural Network

Pruning process can be made parallel if the same is done layer wise. For the same cause, the layer wise error bound, with respect to the overall allowed error needs to be known. This section hence looks into the individual contributions of the change in the parameter matrices in each layer to the final output error. The output of the neural network is given as

$$\mathbf{Y}^{(L)} = f(\mathbf{W}^{(L-1)^T}\mathbf{Y}^{(L-1)} + \mathbf{B}^{(L)}) \tag{1}$$

The deviation introduced in the output error $\delta\mathbf{Y}^{(L)}$, due to pruning the parameter matrices $\mathbf{W}^{(L-1)}$ by $\delta\mathbf{W}^{(L-1)}$, can be bounded as

$$||\delta\mathbf{Y}^{(L)}|| \leq ||\frac{\partial\mathbf{Y}^{(L)}}{\partial\mathbf{W}^{(L-1)}}\delta\mathbf{W}^{(L-1)}|| + ||\frac{\partial\mathbf{Y}^{(L)}}{\partial\mathbf{Y}^{(L-1)}}\delta\mathbf{Y}^{(L-1)}|| \tag{2}$$

Theorem 1. *Assuming that the input layer is left untouched, the output error introduced by pruning the trained network $N(\{\boldsymbol{W}_l\}_{l=1}^{L}, \boldsymbol{X})$ will always be upper bounded by the following relation,*

$$||\delta\boldsymbol{Y}^{(L)}|| \leq \sum_{l=2}^{L}\left[\prod_{\substack{i=l+1 \\ (l\neq L)}}^{L}||\frac{\partial\boldsymbol{Y}^{(i)}}{\partial\boldsymbol{Y}^{(i-1)}}||\right]||\frac{\partial\boldsymbol{Y}^{(l)}}{\partial\boldsymbol{W}^{(l-1)}}||||\delta\boldsymbol{W}^{(l-1)}|| \tag{3}$$

The above relation essentially explains the accumulation of the error in each layer to produce the error in the final layer i.e., if ϵ is the total allowed error in the final layer, then it can be bounded by the sum of individual layer errors, ϵ_l ($l = 2, 3, ..., L$) as shown below:

$$\epsilon \leq \epsilon_2 + \epsilon_3 + ... + \epsilon_L \tag{4}$$

The above equation sets apart error bounds on different layers and will be of much help in optimisation-based pruning techniques.

$\epsilon_L = ||\frac{\partial\mathbf{Y}_L}{\partial\mathbf{W}_L}\delta\mathbf{W}_L||$

and $\epsilon_l = \left[\prod_{i=l+1}^{L}||\frac{\partial\mathbf{Y}^{(i)}}{\partial\mathbf{Y}^{(i-1)}}||\right]||\frac{\partial\mathbf{Y}^{(l)}}{\partial\mathbf{W}^{(l-1)}}||||\delta\mathbf{W}^{(l-1)}||$, for $l \neq L$

The assumption of $\epsilon_l = 0$ results in a simple relation given in Eq. (5), which can help in explaining two design practices used in classification networks:

$$\delta\mathbf{W}^{(l-1)^T}\mathbf{Y}^{(l-1)} = \mathbf{0} \tag{5}$$

1. *An optimised structure of Multi-Layer Perceptrons used for classification will have $n^{(l)} \geq n^{(l+1)}$, where $n^{(l)}$ denotes the number of nodes in layer l and $l = 2, 3, ..., L$.*
   Since $\mathbf{W}^{(l)} \in \mathbb{R}^{n^{(l)} x n^{(l+1)}}$ and $\mathbf{Y}^{(l)} \in \mathbb{R}^{n^{(l)} \times 1}$, for Eq. (5) to have a solution, $n^{(l)} \geq n^{(l+1)}$. Thus the minimum number of nodes the hidden layers can have equally to help the network train well from the data, is the number of nodes in the output layer.

2. *Data dependent approaches results in better compression models.*
Each neural network is unique because of its architecture and the data it was trained on. Any pruning approach must not change the behaviour of the network with respect to the application it was destined to perform. Consider $\mathbf{Y}^{(l)} \in \mathbb{R}^{n^{(l)} \times B}$, where B is the batch size and $\mathbf{W}^{(l)} \in \mathbb{R}^{n^{(l)} \times n^{(l+1)}}$. For Eq. (5) to be satisfied, the column space of $\mathbf{Y}^{(l)}$ must lie in the null space of $\delta\mathbf{W}^{(l)}$ and vice-versa. This implies that the entries for the appropriate $\delta\mathbf{W}^{(l)}$ can be obtained when the pruning measure is coined based on the features obtained from that layer.

3.3 Net Deviation(D)

Different pruning algorithms are currently present for model compression. A measure, similar to test accuracy, for comparing different pruning approaches based on the compression ratios, is the normalized difference between the obtained error difference and the bound, which is defined as Net Deviation, given in (6). An example explaining the use of D, is given in Sect. 4.1.

$$D = ||\delta\mathbf{Y}^{(L)}|| - \sum_{l=2}^{L}\left[\prod_{\substack{i=l+1 \\ (l \neq L)}}^{L} ||\frac{\partial\mathbf{Y}^{(i)}}{\partial\mathbf{Y}^{(i-1)}}||\right] ||\frac{\partial\mathbf{Y}^{(l)}}{\partial\mathbf{W}^{(l-1)}}|| ||\delta\mathbf{W}^{(l-1)}|| \tag{6}$$

3.4 Theoretical Bounds on the Number of Epochs for Retraining a Sparse Neural Network

Assume that the loss function is continuously differentiable and strictly convex. The losses decide the number of epochs the network takes to reach convergence and hence, the number of epochs to reach convergence can be viewed to be directly proportional to the loss. If the total number of parameters in the network is M, the parameters can be made p-sparse in $\frac{M!}{p!(M-p)!}$ number of ways. Hence, the bounds provided below must be understood in the average sense. Adding a regularisation term still keeps the loss function strongly convex, if the initial loss function is strongly convex. This makes Theorem 2 and all the accompanying relations valid even for loss functions with regularisation terms.

Theorem 2. *Given a trained network $N(\{\boldsymbol{W}_l\}_{l=1}^{L}, \boldsymbol{X})$, trained from initial weights $\boldsymbol{W}_{initial}$ using $t_{initial}$ epochs. For fine-tuning the sparse network $N_{sparse}(\{\boldsymbol{W}_l\}_{l=1}^{L}, \boldsymbol{X})$, there exists a positive integer γ that lower bounds the number of epochs (t_{sparse}) to attain the original performance as,*

$$t_{sparse} \geq \frac{\gamma\mu_1}{\mu_2}\left[\frac{||\nabla L(\boldsymbol{W}_{initial})||^2}{||\nabla L(\boldsymbol{W}_{sparse})||^2}\right] t_{initial} \tag{7}$$

When there are different hidden layers, the gradients would follow the chain rule and the following equation can be incorporated in (7) to obtain the bound.

$$||\nabla L(\mathbf{W})||^2 = ||\nabla L(\mathbf{W}^{(1)})||^2 + ||\nabla L(\mathbf{W}^{(2)})||^2 + ... + ||\nabla L(\mathbf{W}^{(L-1)})||^2 \quad (8)$$

The definitions for μ_1 and μ_2 vary for connection and node pruning and are given below. The equations are written for pruning the trained model with final parameter matrix $\mathbf{W}^*$. $\mathbf{W}^{'}$ could be either the initial or sparse matrix.

Connection Pruning: Taking advantage of the fact that connection pruning results in a sparse parameter matrix of the same size as that of the unpruned network, μ can be defined as:

$$\mu^{'} \leq \frac{||\nabla L(\mathbf{W}^{'}) - \nabla L(\mathbf{W}^*)||}{||\mathbf{W}^{'} - \mathbf{W}^*||} \quad (9)$$

Node and Filter Pruning: Node and filter pruning reduces the rank of the parameter matrix and hence Eq. (9) cannot be used. PL inequality is used instead.

$$\mu^{'} \leq \frac{||\nabla L(\mathbf{W}^{'})||^2}{||L(\mathbf{W}^{'}) - L(\mathbf{W}*)||} \quad (10)$$

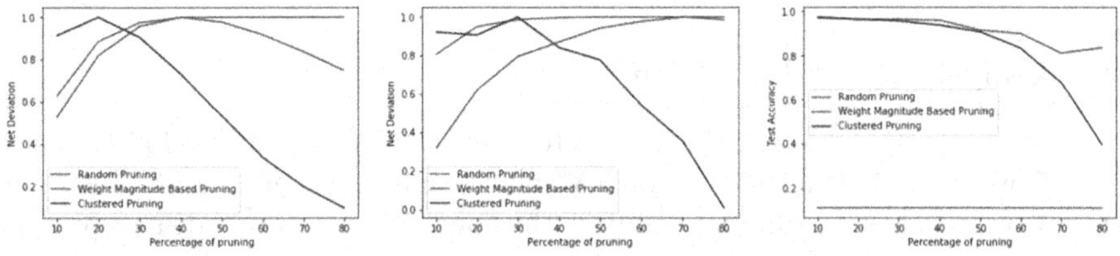

(a) Pruning on LeNet-300-100 (b) Pruning on LeNet-5 CNN (c) Pruning on LeNet-300-100

Fig. 1. Comparison of the three pruning methods for different percentage of pruning using Net Deviation ((a) and (b)) and Test Accuracy in (c). (a) and (c) are results on LeNet-300-100 and (b) are of LeNet-5

4 Experimental Results and Discussion

Simulations validating the theorems stated were performed on two popular networks: LeNet-300-100 and LeNet-5 [12], both trained on MNIST digit data set and had test accuracies of 97.77% and 97.65% respectively.

Table 1. The number of epochs taken for fine-tuning LeNet-300-100 pruned at different pruning and sparsity ratios.

Node pruning			Connection pruning		
Pruning ratio	Theoretical lower bound	Average epochs taken	Sparsity ratio	Theoretical lower bound	Average epochs taken
0.1	0.23	1.7	0.1	0.24	2.2
0.5	2.6	2.6	0.6	1.48	2.4
0.9	19.71	50	0.9	2.26	2.8

Table 2. The number of epochs taken for fine-tuning LeNet-5 pruned at different pruning and sparsity ratios.

Node pruning			Connection pruning		
Pruning ratio	Theoretical lower bound	Average epochs taken	Sparsity ratio	Theoretical lower bound	Average epochs taken
0.1	0.52	2	0.1	1.26	2.3
0.5	5.049	13	0.6	2.783	2.8
0.9	18.623	25.6	0.9	1.62	2.9

4.1 Analysis of Net Deviation

To explain the application of the parameter 'Net Deviation', LeNet-300-100 and LeNet-5 were pruned using Random, Weight magnitude based and Clustered Pruning approaches. In random pruning, the connections were made sparse randomly, while in clustered pruning, the features of each layer were clustered to the required pruning level. One out of each node or filter in the cluster is kept. The second method chose the nodes that had higher weight magnitude connections. The results for different percentage of pruning in an average sense, are given in Fig. 1, which explains that, for lower level of pruning, D is lower for random pruning approach. But for higher pruning or higher model compression, random pruning is not a good pruning method to look into. Net deviation is calculated using the same batch of data that was used for pruning. A similar comparison has been done on LeNet-300-100 using test accuracy as the parameter and the results are shown in Fig. 1(c). It could be seen that when choosing the appropriate method for pruning, for a particular data set and network, test accuracy does not give much information with respect to the amount of compression or the percentage of pruning. Because of similar inferences, random pruning could be applied by a user, who wants smaller compression, with lower computational complexity as random pruning is computationally less expensive than clustered pruning.

4.2 Theoretical Bounds on the Number of Epochs for Retraining a Sparse Neural Network

Both the networks were pruned randomly with the same seed in two ways: Connection Pruning and Node Pruning. Tables 1 and 2 show the results obtained at different pruning ratios for LeNet-300-100 and LeNet-5 respectively. For LeNet-300-100 pruned at pruning ratios 0.1, 0.5 and 0.9 and for sparsity ratios of 0.1, 0.6 and 0.9, the value of γ was obtained as 0.01, 1 and 100 and 1e-4, 0.2 and 0.5, respectively. Similarly for LeNet-5, γ was found to be 1e-5, 2e-4 and 5e-5 for pruning ratios 0.1, 0.5 and 0.9 and 3e-5, 1e-6 and 1e-5 for sparsity ratios 0.1, 0.6 and 0.9 respectively. The results validate the bound provided in Theorem 2.

5 Conclusions

This paper has theoretically derived and experimentally validated the amount of retraining that would be required after pruning, in terms of the relative number of epochs. Also, the propagation of errors through the layers, due to pruning different layers is analysed and a bound to the amount of error that the layers contribute was derived. The parameter 'Net Deviation' can be used to study different pruning approaches and hence can be used as a criteria for ranking different pruning approaches. If not completely avoided, reducing the number of epochs linked to retraining the network will reduce the computational complexity involved with training a Neural Network. An empirical formula to calculate the γ parameter that bounds the number of epochs required for retraining, is considered as a future work.

A Appendix

A.1 Proof of Theorem 1

From [16], for a function with more than one variable $q(x, y)$, when the uncertainties in x and y are independent and random, the uncertainty in q can be written as

$$\delta q^2 = \left[\frac{\partial q}{\partial x}\delta x\right]^2 + \left[\frac{\partial q}{\partial y}\delta y\right]^2 \tag{11}$$

Applying triangular inequality, the following equation is valid.

$$\delta q \leq |\frac{\partial q}{\partial x}|\delta x + |\frac{\partial q}{\partial y}|\delta y \tag{12}$$

In the concept of pruning of neural networks, the weight changes and output changes in earlier layers are independent to each other. The output of the neural network is

$$\mathbf{Y}^{(L)} = f(\mathbf{W}^{(L-1)^T}\mathbf{Y}^{(L-1)} + \mathbf{B}^{(L)}) \tag{13}$$

The function $f(.)$ can be sigmoid or relu or softmax function as per the layer considered. Usually, the softmax function is used in the output layer. Assuming no change in the bias of the layers,

$$||\delta \mathbf{Y}^{(L)}||^2 = ||\frac{\partial \mathbf{Y}^{(L)}}{\partial \mathbf{W}^{(L-1)}}\delta \mathbf{W}^{(L-1)}||^2 + ||\frac{\partial \mathbf{Y}^{(L)}}{\partial \mathbf{Y}^{(L-1)}}\delta \mathbf{Y}^{(L-1)}||^2 \tag{14}$$

and

$$||\delta \mathbf{Y}^{(L)}|| \leq ||\frac{\partial \mathbf{Y}^{(L)}}{\partial \mathbf{W}^{(L-1)}}\delta \mathbf{W}^{(L-1)}|| + ||\frac{\partial \mathbf{Y}^{(L)}}{\partial \mathbf{Y}^{(L-1)}}\delta \mathbf{Y}^{(L-1)}|| \tag{15}$$

Accumulating the effect of weight changes in all the layers and assuming no change in the input layer, we get

$$\begin{aligned} ||\delta \mathbf{Y}^{(L)}|| \leq ||\frac{\partial \mathbf{Y}^{(L)}}{\partial \mathbf{W}^{(L-1)}}\delta \mathbf{W}^{(L-1)}|| + ||\frac{\partial \mathbf{Y}^{(L)}}{\partial \mathbf{Y}^{(L-1)}}\frac{\partial \mathbf{Y}^{(L-1)}}{\partial \mathbf{W}^{(L-2)}}\delta \mathbf{W}^{(L-2)}|| + ... \\ + \left[\prod_{i=3}^{L}||\frac{\partial \mathbf{Y}^{(i)}}{\partial \mathbf{Y}^{(i-1)}}||\right]||\frac{\partial \mathbf{Y}^{2}}{\partial \mathbf{W}^{1}}\delta \mathbf{W}^{1}|| \end{aligned}$$

This is given in Theorem 1.

$$||\delta \mathbf{Y}^{(L)}|| \leq \sum_{l=2}^{L}\left[\prod_{\substack{i=l+1 \\ (l\neq L)}}^{L}||\frac{\partial \mathbf{Y}^{(i)}}{\partial \mathbf{Y}^{(i-1)}}||\right]||\frac{\partial \mathbf{Y}^{(l)}}{\partial \mathbf{W}^{(l-1)}}||||\delta \mathbf{W}^{(l-1)}|| \tag{16}$$

Hence the proof. Suppose the output error is bounded by ϵ, which corresponds to the LHS in (16) given above. Expanding the RHS,

$$\begin{aligned} \epsilon_1 &= 0 \\ \epsilon_2 &= \left[\prod_{i=3}^{L}||\frac{\partial \mathbf{Y}^{(i)}}{\partial \mathbf{Y}^{(i-1)}}||\right]||\frac{\partial \mathbf{Y}^{2}}{\partial \mathbf{W}^{1}}\delta \mathbf{W}^{1}|| \\ \epsilon_L &= ||\frac{\partial \mathbf{Y}^{(L)}}{\partial \mathbf{W}^{(L-1)}}\delta \mathbf{W}^{(L-1)}|| \end{aligned}$$

Hence, the result given in (4) is obtained, which could also be seen in [2].

$$\epsilon \leq \epsilon_1 + \epsilon_2 + ... + \epsilon_L \tag{17}$$

An alternative expression can be found in a similar pattern of derivation starting with (14)

$$\epsilon^2 = \epsilon_1^2 + \epsilon_2^2 + ... + \epsilon_L^2 \tag{18}$$

The proof of (5) is quite obvious. Assuming that the error in each layer is approximately zero, $\epsilon_L = 0$ and that the function f(.) is sigmoid,

$$\begin{aligned} ||\frac{\partial \mathbf{Y}^{(L)}}{\partial \mathbf{W}^{(L-1)}}\delta \mathbf{W}^{(L-1)}|| &= 0 \\ ||\mathbf{Y}^{(L)} \odot (\mathbf{1} - \mathbf{Y}^{(L)})||||\delta \mathbf{W}^{(L-1)^T}\mathbf{Y}^{(L-1)}|| &= 0 \\ \delta \mathbf{W}^{(L-1)^T}\mathbf{Y}^{(L-1)} &= \mathbf{0} \end{aligned}$$

$\odot$ refers to the Hadamard product of matrices. Similar results can be obtained for other layers as well.

A.2 Proof of Theorem 2

Assume that the loss function is continuously differentiable and strictly convex. By the strong convexity assumption, the Polyak-Lojasiewicz (PL) inequality can be implied, which is stated as follows:

For a continuously differentiable and strongly convex function f, over parameter x, with minimum point x^*,

$$\frac{1}{2}||\nabla f(x)||^2 \geq \mu(f(x) - f(x^*)), \forall x \tag{19}$$

Using the above relation on the loss function $L(W)$,

1. Initial training of the network, till convergence

$$\frac{1}{2}||\nabla L(W_{initial})||^2 \geq \mu_1(L(W_{initial}) - L(W^*)) \tag{20}$$

2. Fine-tuning the sparse network, to reach back to the original performance

$$\frac{1}{2}||\nabla L(W_{sparse})||^2 \geq \mu_2(L(W_{sparse}) - L(W^*)) \tag{21}$$

The number of epochs to reach convergence is directly proportional to the difference in losses and can be written as:

$$t_{initial} \propto ||L(W_{initial} - L(W^*)|| \tag{22}$$

$$t_{sparse} \propto ||L(W_{sparse} - L(W^*)|| \tag{23}$$

Combining (20–23) given above, and with a γ to accommodate the proportionality in (22) and (23),

$$t_{sparse} \geq \frac{\gamma\mu_1}{\mu_2}\left[\frac{||\nabla L(W_{initial})||^2}{||\nabla L(W_{sparse})||^2}\right]t_{initial} \tag{24}$$

References

1. Aghasi, A., Abdi, A., Nguyen, N., Romberg, J.: Net-trim: convex pruning of deep neural networks with performance guarantee. In: Advances in Neural Information Processing Systems, pp. 3177–3186 (2017)
2. Aghasi, A., Abdi, A., Romberg, J.: Fast convex pruning of deep neural networks. arXiv preprint arXiv:1806.06457 (2018)
3. Augasta, M.G., Kathirvalavakumar, T.: A novel pruning algorithm for optimizing feedforward neural network of classification problems. Neural Process. Lett. **34**(3), 241 (2011)

4. Carreira-Perpinán, M.A., Idelbayev, Y.: Learning-compression algorithms for neural net pruning. In: Proceedings of the IEEE Conference on Computer Vision and Pattern Recognition, pp. 8532–8541 (2018)
5. Castellano, G., Fanelli, A.M., Pelillo, M.: An iterative pruning algorithm for feedforward neural networks. IEEE Trans. Neural Netw. **8**(3), 519–531 (1997)
6. Dong, X., Chen, S., Pan, S.: Learning to prune deep neural networks via layer-wise optimal brain surgeon. In: Advances in Neural Information Processing Systems, pp. 4857–4867 (2017)
7. Dubey, A., Chatterjee, M., Ahuja, N.: Coreset-based neural network compression. In: Proceedings of the European Conference on Computer Vision (ECCV), pp. 454–470 (2018)
8. Engelbrecht, A.P.: A new pruning heuristic based on variance analysis of sensitivity information. IEEE Trans. Neural Netw. **12**(6), 1386–1399 (2001)
9. Hagiwara, M.: Removal of hidden units and weights for back propagation networks. In: Proceedings of 1993 International Conference on Neural Networks (IJCNN-1993), Nagoya, Japan, vol. 1, pp. 351–354. IEEE (1993)
10. Han, S., Mao, H., Dally, W.J.: Deep compression: compressing deep neural networks with pruning, trained quantization and huffman coding. arXiv preprint arXiv:1510.00149 (2015)
11. Huynh, L.N., Lee, Y., Balan, R.K.: D-pruner: filter-based pruning method for deep convolutional neural network. In: Proceedings of the 2nd International Workshop on Embedded and Mobile Deep Learning, pp. 7–12. ACM (2018)
12. LeCun, Y., Bottou, L., Bengio, Y., Haffner, P., et al.: Gradient-based learning applied to document recognition. Proc. IEEE **86**(11), 2278–2324 (1998)
13. LeCun, Y., Denker, J.S., Solla, S.A.: Optimal brain damage. In: Advances in Neural Information Processing Systems, pp. 598–605 (1990)
14. Li, L., Xu, Y., Zhu, J.: Filter level pruning based on similar feature extraction for convolutional neural networks. IEICE Trans. Inf. Syst. **101**(4), 1203–1206 (2018)
15. Srinivas, S., Babu, R.V.: Data-free parameter pruning for deep neural networks. arXiv preprint arXiv:1507.06149 (2015)
16. Taylor, J.R.: Error Analysis. University Science Books, Sausalito (1997)
17. Tung, F., Mori, G.: Clip-q: deep network compression learning by in-parallel pruning-quantization. In: Proceedings of the IEEE Conference on Computer Vision and Pattern Recognition, pp. 7873–7882 (2018)
18. Zhu, M., Gupta, S.: To prune, or not to prune: exploring the efficacy of pruning for model compression. arXiv preprint arXiv:1710.01878 (2017)

M-ary Random Forest

Vikas Jain and Ashish Phophalia(✉)

Indian Institute of Information Technology, Vadodara, Gandhinagar, India
{201671001,ashish_p}@iiitvadodara.ac.in

Abstract. Random forest (RF) is a supervised, ensemble of decision trees method. Each decision tree recursively partitions the feature space into two disjoint sub-regions using axis parallel splits until each sub-region becomes homogeneous with respect to a particular class or reach to a stoppage criterion. The conventional RF uses one feature at a time for splitting. Therefore, it does not consider the feature inter-dependency. Keeping this aim in mind, the current paper introduces an approach to perform multi-features splitting. This partition the feature space into M-regions using axis parallel splits. Therefore, the forest created using this is named as M-ary Random Forest (MaRF). The suitability of the proposed method is tested over the various heterogeneous UCI datasets. Experimental results show that the proposed MaRF is performing better for both classification and regression. The proposed MaRF method has also been tested over Hyperspectral imaging (HSI) for classification and it has shown satisfactory improvement with respect to other state-of-the-art methods.

Keywords: Classification · Ensemble method · Hyperspectral imaging · Random forest

1 Introduction

Random forest (RF) is an ensemble-based, supervised machine learning algorithm [5][1]. It consists of numerous randomized decision trees as an atomic unit used for classification and regression problems. RF can be implemented and executed as the parallel threads, hence it is fast and easy to implement. It has been used for various domains, like medical imaging, pattern recognition and classification [7] etc.

A decision tree in RF is built during the training phase using bootstrap sampling. The performance of the decision tree depends on several important parameters like splitting criteria, feature selection, number of trees, and the number of instances on the leaf node. However, the best choice of these parameters is not answered precisely [10,14]. This motivated various methods to come up with a heuristic approach in building the decision tree and hence RF. For example, to reduce the computation and to improve the accuracy, Geurts et al.

[1] Referred to as conventional random forest throughout the text.

B. Deka et al. (Eds.): PReMI 2019, LNCS 11941, pp. 161–169, 2019.
https://doi.org/10.1007/978-3-030-34869-4_18

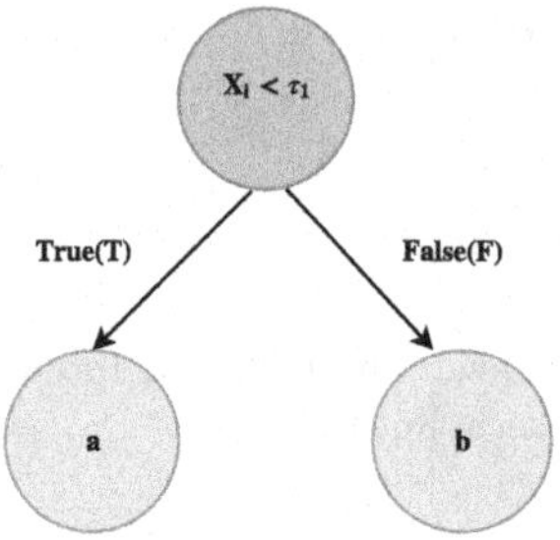

Fig. 1. An example of binary decision tree.

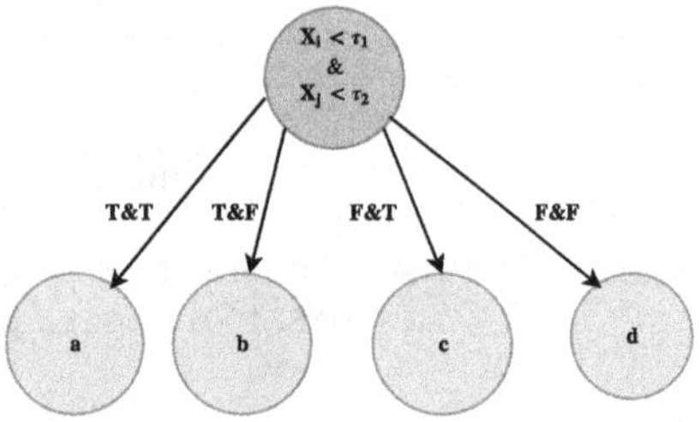

Fig. 2. An example to 4-ary decision tree.

[9] introduced randomness for both attribute selection and for splitting point choice so that it can reduce the variance as compared to weaker randomization approach. Paul et al. have proposed a method to reduce the unimportant features and to limit the number of trees [15]. In addition, there has been some work done on proving the consistency of RF and leveraging dependency on the data by several researchers [3,4,8,17]. Denil et. al. [8] used Poisson distribution in feature selection for growing a tree, whereas Wang et al. [17], has proposed a Bernoulli Random Forest (BRF) framework incorporating Bernoulli distribution for the feature and splitting point selection. In recent years, the success of deep neural network has inspired other learners to benefit from deep, layered architecture. Therefore, Zhou et al. [20] proposed the Deep RF whose performance is robust to the hyper-parameters settings.

Murthy et al. [13] have proposed an oblique decision tree. It splits the feature space using a hyperplane defined by a linear combination of the feature variables. There may exist much domain where one or two oblique hyperplanes will give the best classification. In such situations, the axis parallel split has to approximate the correct model with a staircase structure. But, the computational cost of the oblique decision tree is exponential, which makes it an NP-hard problem [13]. Wickramarachchi et al. [18] proposed a new way to induce oblique decision tree using the eigen vectors of the estimated covariance matrices of the respective classes. In both methods, parameters tuning is more time consuming, therefore it takes a longer time to find the best fit hyperplane. However, it captures the linear relationship between the features. We have proposed an M-ary random forest (MaRF) approach. It uses 'N' independent features at a time to partition the feature space into 2^N regions. The proposed approach is tractable and takes less time in comparison to the oblique decision tree.

The remaining paper has been arranged in the following manner: Sect. 2, presents the proposed MaRF approach. Section 3, discuss the implementation details, and performance analysis over UCI and HSI dataset. It has been concluded in Sect. 4.

2 Proposed Approach

In conventional RF [5], each decision tree is designed as a binary tree using axis parallel splits. It partitions the feature space into two subspace, with feature

Algorithm 1. M_ary_Decision_Tree()

```
Input: Training Subsample
Output: M_ary_Decison_Tree
while (Stoppage_Criteria == False) do
    if (X_i < τ1 & X_j < τ2) then
        goto subtree "a";
    else
        if ((X_i < τ1 & X_j >= τ2)) then
            goto subtree "b";
        else
            if ((X_i >= τ1 & X_j < τ2)) then
                goto subtree "c";
            else
                ((X_i >= τ1 & X_j >= τ2)) goto subtree "d";
            end
        end
    end
end
```

X_i, and a threshold value τ_1. The selection of feature is done on the basis of the optimum value of splitting criterion. If $X_i < \tau_1$, go to subtree "*a*" otherwise goto subtree "*b*", as shown in Fig. 1. At any internal node, there are at most two subtrees. However, the binary decision tree is unable to capture feature dependency. Therefore, M-ary RF is proposed. It uses "N" independent features at a time, to divide the feature space into a maximum of 2^N sub-space. It computes the splitting criterion value for all possible combination of "N" features to decide features for splitting. For example, consider $N = 2$, it will divide the feature space, into $2^2 = 4$ sub-space, refer Fig. 2. Let X_i and X_j are two selected features with their threshold values as τ_1 and τ_2 at an internal node of M-ary decision tree, which are selected as features for splitting. If both $X_i < \tau_1$ and $X_j < \tau_2$ are true (T) then divide the data into subtree "*a*". If first is true and second is false (F) then divide into subtree "*b*". If first is F and second is T then divide into subtree "*c*" and if both are F then divide into subtree "*d*". Refer Algorithm 1 for designing the M-ary decision tree. In MaRF, decision trees are constructed as M-ary decision tree, later the prediction is done on the basis of majority voting.

3 Experiments and Results

The proposed method has been rigorously tested over twenty-one datasets for both classification and regression tasks. It has also been tested over real-life application using Hyperspectral imaging (HSI) dataset.

3.1 Datasets

We have UCI datasets [2] for evaluation of the proposed method. These datasets are varying in terms of number of classes, number of features, and number of

Table 1. Classification accuracy (in %) comparison between state-of-the-art methods and proposed MaRF with average over 10 iterations (High value is preferred)

SN	Dataset	Dimension	# Classes	RF	Biau08 [4]	Biau12 [3]	Denil [8]	BRF [17]	Oblique	MaRF
1	Transfusion	748*5	2	72.2	68.92	70.27	72.97	77.7	72.8	**73.8**
2	Spambase	4601*57	2	91.1	60.59	60.59	**94.4**	94.1	92.7	**94.4**
3	CVR	435*16	2	88.8	51.86	61.4	94.42	**95.58**	92.3	95.1
4	Madelon	2600*500	2	60.6	49.27	50.31	54.81	**69.23**	65.7	63.83
5	Wine	178*13	3	96.9	40.59	41.18	96.47	**97.65**	85.1	97.2
6	CMC	1473*9	3	50.29	42.72	42.65	53.6	**54.63**	47.1	51.6
7	Verbetral	310*6	3	80.9	48.39	48.39	82.26	**82.58**	53.7	81.9
8	Connect-4	67557*42	3	64.58	64.52	65.47	66.19	76.19	67.4	**81.8**
9	Vehicle	946*18	4	72.6	27.98	23.1	68.81	71.67	72.1	**74.3**
10	Zoo	101*17	7	85.3	50	41	80	85	70	**89.6**
11	Abalone	4177*8	29	27.2	16.05	16.52	26.23	26.44	25.5	26.6

Table 2. MSE comparison between state-of-the-art methods and proposed MaRF with average over 10 iterations (Least value is preferred)

SN	Dataset	Dimension	RF[5]	Biau08 [4]	Biau12 [3]	Denil [8]	BRF [17]	MaRF
1	Slump	103*10	42.91	62.30	62.31	60.19	55.67	**35.5**
2	Servo	167*4	2.83	3.19	2.39	2.26	2.07	**0.71**
3	Automobile	205*26	1.65	1.53	1.51	1.46	1.41	**0.9**
4	Yacht	308*7	54.73	229.86	225.97	150.58	128.85	**26.18**
5	Housing	506*14	32.17	85.50	82.97	81.62	77.81	**24.41**
6	Student	649*33	49.75	9.83	9.81	9.38	8.93	**4.87**
7	Concrete	1030*9	132.32	279.13	279.70	278.64	275.56	**72.46**
8	Wine quality	4898*12	0.61	0.81	0.81	0.67	**0.51**	0.59
9	Airfoil	1503*6	29.9	66.66	47.73	43.47	38.57	**22.07**
10	Energy_y1	768*8	2.41	64.11	40.71	24.53	19.85	**1.59**

instances and hence, are heterogeneous in nature. The ratio of the number of instances to the number of features in these benchmark datasets has high variations. The detailed description of the datasets are given in Tables 1 and 2 for classification and regression respectively.

3.2 Parameters

There are four main parameters associated with decision tree, namely: (1) the number of trees n_{tree}, (2) the number of minimum instances at leaf node n_{min}, (3) the train-test ratio in which dataset is divided into training set and test set, (4) the maximum tree depth T_{depth}. In our experiment, the value of n_{tree} is kept as 45, which has been decided empirically. The n_{min} is kept as 5 and the train-test ratio is kept as 0.7. The T_{depth} is varying for each datasets. Each of the experiment has been repeated over 10 iterations and the average mean value is computed.

3.3 Performance Analysis

The results generated with the proposed MaRF are compared with conventional RF [5], and the recent state-of-the-art methods for the classification and regres-

Table 3. Comparison of T_{depth} between conventional RF and MaRF

SN	Dataset	Dimension	# Classes	RF	MaRF
1	Wine	178*13	3	12	**7**
2	Verbetral	310*6	3	19	**10**
3	Vehicle	946*18	4	26	**18**
4	Zoo	101*17	7	25	**19**

sion datasets. The highest learning performance among these comparisons is marked in boldface for each dataset. For classification performance analysis, the experiment has been conducted with the eleven well known UCI datasets. The proposed method is showing improvement for nine datasets out of eleven in comparison to Biau08 [4], Biau12 [3], and Denil [8] state-of-the-art methods. To investigate the effect of choosing the N-independent features for partitioning over the binary splits, it has been compared to conventional RF. One can observe from the Table 1, the proposed MaRF is showing improvement for all the datasets except for Abalone as compare to conventional RF. We have also computed the depth of tree T_{depth}, for few datasets and compared with conventional RF, as shown in Table 3. One can observe that using MaRF approach, the decision tree has grown to lesser depth as compared to conventional RF. Therefore, the testing time in MaRF would be less as compared to conventional RF.

In regression, it can be observed from Table 2 that MaRF achieves the significant reduction in MSE on all of the datasets as compare to all state-of-the-art methods. In particular, one can observe that for Yacht, Student and Concrete dataset, the proposed method has reduced the MSE by more than 40%. From Table 2, one can observe the proposed MaRF method is showing improvement for all categorical and numerical datasets.

3.4 Computation Cost Analysis

In conventional RF, consider the case of extremely randomized tree [9] such that $k = \sqrt{q}$ features are selected at each node, here 'q' is the number of features. Suppose, there are 'p' number of instances then, the time complexity to construct a tree would be $\mathcal{O}(k \cdot p)$ [12]. In case of an oblique decision tree, the possible number of distinct hyperplane would be $\binom{p}{k}$ and each feature value could be selected in 2^k possibilities. Therefore, the computation cost would be $\mathcal{O}(2^k \cdot \binom{p}{k})$ [13]. Therefore, it is an NP-hard problem. In case of MaRF, suppose $N = 2$ features are used at each node for splitting. Hence, it would search in $\binom{k}{2}$ ways to choose two best features. In general, the computation cost would be $\mathcal{O}(\binom{k}{N} \cdot p)$. The MaRF approach does not require any parameters tuning for multi-feature splitting as required in the oblique decision tree.

3.5 Real Life Application: Hyperspectral Image Classification

In this section, we evaluate the performance of MaRF on two publicly available benchmarks HSI datasets, named Indian Pines and Pavia University [1]. Indian Pines dataset is captured with 224 spectral bands in the wavelength range from 0.4 to 2.5μm. It has 145 $\times$ 145 pixels and spatial resolution is 20 m/pixel. The 200 band are selected for classification after 24 noisy bands being removed. The reference map has 10366 pixels that belong to 16 classes [1]. Pavia University data has 115 spectral bands in the wavelength range from 0.43 to 0.86μm. It has 610 $\times$ 340 pixels with high spatial resolution of 1.3 m/pixel. The 12 noisy bands are removed, and the remaining 103 bands are selected for classification. The referenced map have 42776 pixels that belong to 9 classes [1].

All the parameters are kept the same, except the train-test ratio. Since due to the limited availability of the labeled training samples in HSI make the classification task more challenging [11]. Therefore, only a limited set of instances are chosen for the training part. Hence, 15 instances from each class of Indian Pines, and 50 instances per class for Pavia University are chosen for training and rest used for testing. The performance of the algorithms is measured using overall accuracy (OA), average accuracy (AA), and kappa coefficient (κ) [16]. The OA calculate the ratio of the number of correctly classified samples to the total number of test samples. The AA is the average percentage of correctly classified samples for each class. The value of the κ coefficient is used for the consistency check. High value is preferred for all the three measures [16].

The proposed MaRF method is compared with the SVM-3DG [6] and CRF [19]. Cao et al. [6] proposed an approach to use spatial information as prior for extracting spatial features before classification as well as for labeling the class in post-processing. The one can observe from the Table 4 that for Indian Pines image, proposed MaRF method has improved the overall accuracy of predicting the correct class labels by more than 4% value. Also, it is showing the improvement in consistency with a rise in κ coefficient by 5% value. However, the proposed approach is generating poor class-wise efficiency for the small class. This is due to the class imbalance problem. For the Pavia University image, the proposed MaRF method is showing the improvement for all (OA, AA, and κ coefficient) the measuring parameters. If we compare with the conventional RF, then also the propose MaRF method is showing significant improvement for class-wise accuracy and for all other measured parameters. The results are encouraging in the sense like even, the SVM-3DG [6] uses extracted features for classification, and CRF [19] uses feature selection still the proposed MaRF is outperforming which is directly applied over the pixels without extracting any features. One can also observe the visual impact of the corresponding classification map using Fig. 3.

Table 4. Class Specific Accuracy, OA, AA and kappa coefficient (κ) (in %) obtained by SVM-3DG [6], CRF [19], RF [5] and MaRF for Indian Pines and Pavia University Dataset (# Iterations = 10)

Class	Indian Pines							Pavia University						
	Total	Train	Test	SVM-3DG	CRF	RF	MaRF	Total	Train	Test	SVM-3DG	CRF	RF	MaRF
1	46	15	31	**96.77**	70.75	1.93	15.2	6631	50	6616	97.39	86.68	96.67	**98.1**
2	1428	15	1413	58.46	79.72	96.67	**97.4**	18649	50	18634	97.27	96.2	99.03	**99.7**
3	830	15	815	**93.37**	69.91	61.16	72.5	2099	50	2084	89.41	**93.4**	68.21	89.4
4	237	15	222	96.40	**98.84**	12.92	25.5	3064	50	3049	97.25	**99.63**	70.27	97.5
5	483	15	468	86.11	**91.42**	81.92	86.5	1345	50	1330	**99.61**	86.93	94.2	99.1
6	730	15	715	95.80	94.69	**97.16**	95.7	5029	50	5014	**98.41**	98.09	43.21	97.7
7	28	15	13	**100**	83.59	0	0	1330	50	1315	98.20	**99.82**	55.65	87.8
8	478	15	463	**100**	53.18	98.42	97.6	3682	50	3667	84.00	86.63	93.48	**97.1**
9	20	15	5	**100**	90.00	0	0	947	50	932	**99.89**	72.79	99.73	98.5
10	972	15	957	68.86	81.87	**86.9**	85.5	-	-	-	-	-	-	-
11	2455	15	2440	78.57	86.98	95.3	**95.4**	-	-	-	-	-	-	-
12	593	15	578	**96.89**	47.37	44.77	86.3	-	-	-	-	-	-	-
13	205	15	190	94.21	90.14	**95.47**	90	-	-	-	-	-	-	-
14	1265	15	1250	77.84	96.55	**98.12**	96.9	-			-	-	-	-
15	386	15	371	**95.42**	63.32	22.64	52.5	-	-		-	-	-	-
16	93	15	78	**98.72**	89.04	49.35	78.2	-	-	-	-	-	-	-
OA	-	-	-	81.12	82.20	83.68	**88.44**	-	-	-	96.06	82.23	86.6	**97.9**
AA	-	-	-	**89.84**	80.48	58.93	67.2	-	-	-	95.71	91.13	80.1	**96.1**
κ	-	-	-	78.64	79.55	81	**86**	-	-	-	94.78	76.61	84.1	**97.3**
Total	**10249**	**240**	**10009**	-	-	-	-	**42776**	**450**	**42641**	-	-	-	-

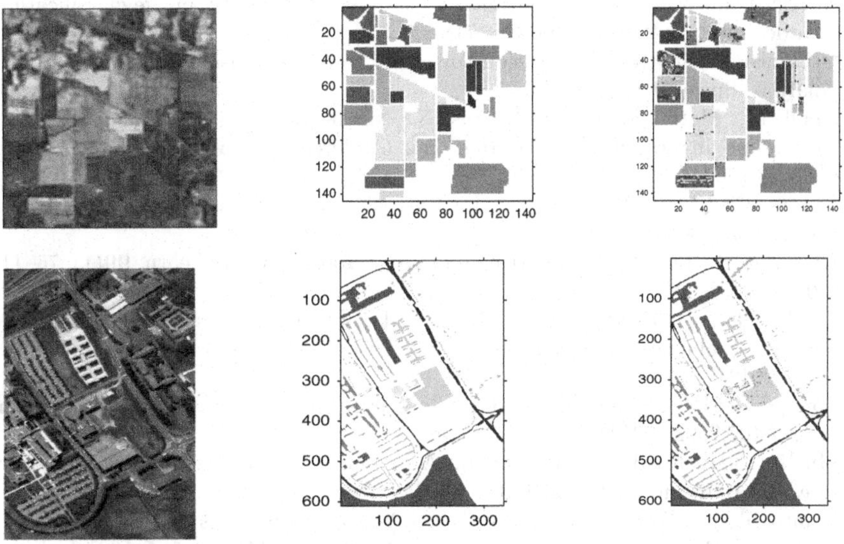

Fig. 3. Classification map of Indian Pines (Row 1) and Pavia University (Row 2) images, showing the original map, ground truth and MaRF for the test samples from left to right respectively

4 Conclusion

In this paper, we have proposed an M-ary random forest (MaRF) based on multi-feature splitting. The proposed MaRF approach is tested over several well known heterogeneous datasets from the UCI repository. It has shown promising results for both classification and regression kind of tasks as compared to conventional RF and other state-of-the-art methods. In MaRF, decision trees grow up to lesser depth as compare to the conventional RF, which results in reducing the testing time as well. The MaRF has also been tested over Hyperspectral datasets. The results showed significant improvement in terms of AA, OA, and kappa coefficient as compared to state-of-the-art methods. Overall, the experimental results show that the proposed method can provide another direction to the world for further exploration.

References

1. Indian pines and pavia university dataset. http://lesun.weebly.com/hyperspectral-data-set.html. Accessed 15 Jan 2019
2. UCI repository. https://archive.ics.uci.edu/ml/index.php. Accessed 15 Nov 2018
3. Biau, G.: Analysis of a random forests model. J. Mach. Learn. Res. **13**(Apr), 1063–1095 (2012)
4. Biau, G., Devroye, L., Lugosi, G.: Consistency of random forests and other averaging classifiers. J. Mach. Learn. Res. **9**(Sep), 2015–2033 (2008)
5. Breiman, L.: Random forests. Mach. Learn. **45**(1), 5–32 (2001)
6. Cao, X., Xu, L., Meng, D., Zhao, Q., Xu, Z.: Integration of 3-dimensional discrete wavelet transform and markov random field for hyperspectral image classification. Neurocomputing **226**, 90–100 (2017)
7. Criminisi, A., Shotton, J.: Decision Forests for Computer Vision and Medical Image Analysis. Springer (2013)
8. Denil, M., Matheson, D., De Freitas, N.: Narrowing the gap: random forests in theory and in practice. In: International Conference on Machine Learning, pp. 665–673 (2014)
9. Geurts, P., Ernst, D., Wehenkel, L.: Extremely randomized trees. Mach. Learn. **63**(1), 3–42 (2006)
10. Ishwaran, H.: The effect of splitting on random forests. Mach. Learn. **99**(1), 75–118 (2015)
11. Ji, R., Gao, Y., Hong, R., Liu, Q., Tao, D., Li, X.: Spectral-spatial constraint hyperspectral image classification. IEEE Trans. Geosci. Remote Sens. **52**(3), 1811–1824 (2014)
12. Louppe, G.: Understanding random forests: from theory to practice. arXiv preprint arXiv:1407.7502 (2014)
13. Murthy, S.K., Kasif, S., Salzberg, S.: A system for induction of oblique decision trees. J. Artif. Intell. Res. **2**, 1–32 (1994)
14. Oshiro, T.M., Perez, P.S., Baranauskas, J.A.: How many trees in a random forest? In: Perner, P. (ed.) MLDM 2012. LNCS (LNAI), vol. 7376, pp. 154–168. Springer, Heidelberg (2012). https://doi.org/10.1007/978-3-642-31537-4_13
15. Paul, A., Mukherjee, D.P., Das, P., Gangopadhyay, A., Chintha, A.R., Kundu, S.: Improved random forest for classification. IEEE Trans. Image Process. **27**(8), 4012–4024 (2018)

16. Wang, L., Zhao, C.: Hyperspectral Image Processing. Springer (2016)
17. Wang, Y., Xia, S.T., Tang, Q., Wu, J., Zhu, X.: A novel consistent random forest framework: Bernoulli random forests. IEEE Trans. Neural Netw. Learn. Syst. **29**(8), 3510–3523 (2018)
18. Wickramarachchi, D., Robertson, B., Reale, M., Price, C., Brown, J.: HHCART: an oblique decision tree. Comput. Stat. Data Anal. **96**, 12–23 (2016)
19. Zhang, Y., Cao, G., Li, X., Wang, B.: Cascaded random forest for hyperspectral image classification. IEEE J. Sel. Top. Appl. Earth Obs. Remote Sens. **11**(4), 1082–1094 (2018)
20. Zhou, Z.H., Feng, J.: Deep forest: towards an alternative to deep neural networks. arXiv preprint arXiv:1702.08835 (2017)

Exponentially Weighted Random Forest

Vikas Jain, Jaya Sharma, Kriti Singhal, and Ashish Phophalia(✉)

Indian Institute of Information Technology, Vadodara, Gandhinagar, India
{201671001,201551021,201551024,ashish_p}@iiitvadodara.ac.in

Abstract. Random forest (RF) is a supervised, non-parametric, ensemble-based machine learning method used for classification and regression task. It is easy in terms of implementation and scalable, hence attracting many researchers. Being an ensemble-based method, it considers equal weights/votes to all atomic units i.e. decision trees. However, this may not be true always for varying test cases. Hence, the correlation between decision tree and data samples are explored in the recent past to take care of such issues. In this paper, a dynamic weighing scheme is proposed between test samples and decision tree in RF. The correlation is defined in terms of similarity between the test case and the decision tree using exponential distribution. Hence, the proposed method named as Exponentially Weighted Random Forest (EWRF). The performance of the proposed method is rigorously tested over benchmark datasets from the UCI repository for both classification and regression tasks.

Keywords: Classification · Ensemble method · Random forest

1 Introduction

Random forest (RF) is an ensemble-based, supervised machine learning algorithm proposed by Leo Brieman [6][1]. It consists of numerous randomized decision trees to solve classification and regression problems. In RF, decision trees are constructed independently. Therefore, RF can be implemented and executed as parallel threads, hence it is fast and easy to implement. It has been used for various domains like brain tumor segmentation, Alzheimer detection, face recognition, human pose detection, object detection etc [7].

A decision tree in RF is built during the training phase using the bagging concept. A decision tree has several important parameters like predefined splitting criteria, tree depth and the number of elements on the leaf node. However, the best choice of these parameters is not answered precisely yet [7,10]. This motivated various methods to come up with the heuristic approach in building the decision tree and hence RF. The method proposed by Paul et al. [15] converges with reduced and important features, and derived the bound for the

[1] Referred to as conventional random forest throughout the text.

J. Sharma and K. Singhal—Equal contribution.

B. Deka et al. (Eds.): PReMI 2019, LNCS 11941, pp. 170–178, 2019.
https://doi.org/10.1007/978-3-030-34869-4_19

number of trees. In addition, there has been some work done on proving the consistency of RF and leveraging dependency on the data by several researchers [4,5,9,16]. Denil et. al. [9] used Poisson distribution in feature selection for growing a tree, whereas Wang et al. [16], has proposed a Bernoulli Random Forest (BRF) framework incorporating Bernoulli distribution for the feature and splitting point selection.

The conventional RF assigns equal weights to the votes casted by each individual tree [6]. Hence, the prediction is made based on the majority voting. However, in the real-life scenario, a dataset may have a huge number of features, but the percentage of truly informative features may be less. Therefore, the contribution of such decision trees, which are populated by less informative attributes may be less. Hence, all the trees in a forest are not equally contributing to the better classification [8]. Therefore, instead of assigning a fixed weight to the decision tree, the dynamic weight should be assigned. Paul et al. [13] have proposed a method to compute the weights during the training phase and assigns a fixed weight to each decision tree. The mechanism proposed by Winham et al. [17] and Liu et al. [12], both computes the weight either based on the performance of tree computed using OOB samples or using a feature weighing scheme. Akash et al. [2] compute the confidence as weight in RF using the entropy or Gini score calculated during the tree construction. However, these methods do not talk about the relationship of these weights with test samples. Therefore, a dynamic weighing scheme is proposed in this paper. It computes the similarity between test cases and the decision tree using exponential distribution. Therefore, the forest formed is named as Exponentially Weighted Random Forest (EWRF).

The remainder of this paper is organized as follows: Sect. 2, describes RF as a classifier and regression and problem associated with conventional RF. Section 3, presents the proposed EWRF approach. Section 4, discuss the implementation details and performance. It has been concluded in Sect. 5.

2 Random Forest

Random forest built upon decision trees as an atomic units. Each decision tree either behaves as a classifier for classification or as a regressor to predict the output for regression task. Given a dataset $\mathbb{D} = \{(X_1, C_1), (X_2, C_2), ..., (X_M, C_M)\}$ with M number of instances such that $X_i \in \mathbb{R}^N$ with N number of attributes. Let the dataset is having class labels as $C_i \in \{Y_1, Y_2, ..., Y_C\}$. Initially, dataset $\mathbb{D}$, is partitioned into training set $\mathbb{D}_1$, having M' number of instances ($M' < M$), and testing set $\mathbb{D}_2$, having remaining instances. Decision trees are constructed using training samples along with bootstrap sampling (random sampling along with replacement) as described in [6].

2.1 Random Forest as Classifier

Random forest assigns the class value based on the proportion of the individual class values present at the leaf node.

The class distribution for the jth class at the terminal node h, in the decision tree t, for the test case X, can be represented as:

$$p_{j,h}^{t} = \frac{1}{n_h} \sum_{i \in h} \mathbb{I}(Y_i = j) \tag{1}$$

here: n_h is total number of instances in the terminal node h. $\mathbb{I}(\cdot)$ is an Indicator function.

Based on maximum class distribution, the class value j, is assigned by the decision tree t, for the test case X, by the following equation:

$$\hat{Y}_j^t = \max_{1 \leq j \leq C} \{p_{j,h}^t\} \tag{2}$$

To assign the final class value based on majority voting in conventional RF, first count the predicted class by each decision tree for the test case X, using the following equation:

$$C(Y_i = j) = \sum_{t=1}^{n_{tree}} \mathbb{1} \cdot \mathbb{J}(\hat{Y}_j^t) \tag{3}$$

here, $\mathbb{J}(\cdot)$ is an indicator function. Finally, based on majority voting, RF assigns the final class value using Eq. (4).

$$\hat{Y} = \max_{1 \leq j \leq C} \{C(Y_i = j)\} \tag{4}$$

2.2 Random Forest as Regressor

In regressor task, decision trees have to predict the outcome. In the regression dataset, the outcome value associated with each instance is a single real value i.e. $\mathbf{Y_i} \in \mathbf{R}$. In order to construct RF as a regressor, Mean Squared Error (MSE) is used as the splitting criterion. Once all the decision trees are constructed, the test instance is passed to each decision tree. Based on the decision tree node values, test instance follows either left or right subtree and reaches to the leaf node. The predicted value is the mean value of instances present at the leaf node. The predicted value for a test case $\mathbf{X}$, at a terminal node h, by the decision tree t, is the mean value of instances present within the leaf node. It can be calculated as:

$$\hat{Y}_h^t = \frac{1}{n_h} \sum_{y_i \in h} Y_i \tag{5}$$

Finally the predicted value by the RF is the average of values predicted by each trees. Hence, the overall prediction made by forest can be computed as:

$$\hat{Y} = \frac{1}{n_{tree}} \sum_{t=1}^{n_{tree}} \mathbb{1} \cdot \hat{Y}_h^t \tag{6}$$

2.3 Problem with Conventional Random Forest

Random forest classifier to be effective, each decision tree must have reasonably good classification performance and trees must be diverse and weakly correlated [14]. The diversity is obtained by randomly choosing training instances and attributes for each tree. However, a decision tree can not always contribute effectively to each and every test instance. Considering a dataset with a high ratio of less informative attributes, the performance of RF gets significantly affected. This is due to the equal contribution of decision trees while performing majority voting. In such cases, performance can be increased by reducing the contribution of decision trees whose nodes are populated by non-informative attributes and assigning a dynamic weight to the decision trees [3,11].

3 Proposed Method

The proposed EWRF consists of two steps. In the first step, decision trees are constructed as described in conventional RF [6]. In the second step, the exponential weight score is calculated as described in following subsections.

3.1 Exponential Weight Score Calculation

During the testing phase, test samples are passed to each and every decision tree in the forest. Let F_i is the feature value for splitting at an internal node of a decision tree t. A test sample $X = \{a_1, a_2, ..., a_j, ..., a_N\}$, is passed to a decision tree. It is guided either to the left ($a_j^X \leq \tau$) or right ($a_j^X > \tau$) subtree, based on threshold τ, and move down until it reaches to the leaf node of decision tree t. The sum of the squared distance between corresponding attribute values in the test sample X, and participating nodes F_i, in the path of the decision tree t, is calculated as follows:

$$d = \sum ||F_i - a_j^X||_2; \forall F_i \in t; a_j \in X$$

Thus, we have $\{d_1, d_2, ..., d_{n_{tree}}\}$ distances computed for each test sample, with respect to all decision trees. The smaller the value of d for the decision tree, the more will be the similarity between tree and test case till that node, and hence the corresponding will be high weight value. This has been shown in Fig. 1. In the proposed EWRF, the weight associated with each decision tree directly proportional to the similarity between the test instance and decision tree. Hence, the weight associated with a decision tree is computed using an exponential distribution measure to maintain such a relationship. In this way, the weight of each decision tree for incoming test cases may vary. The exponential tree weight score is calculated as follows:

$$W_{\mathbf{X}}^t = \frac{1}{Z} \exp\left\{ -\frac{\sum ||F_i - a_j^X||_2}{\alpha} \right\} \tag{7}$$

Algorithm 1. Prediction(X)

Input: n_{tree} = # trees, and test case X_i
Output: Predicted class / output value
for *tree* t, **to** n_{tree} **do**
- Calculate the sum of difference of distance d, between the attribute value of test case and corresponding attributes values of participating nodes of decision tree in the path followed by test case ;
- Calculate the Exponential weight score for each tree t, using Equation (7) and store it into a list ;
- For classification, store the class value with maximum proportion, refer Equation (2) OR
For regression, store the predicted value as mean of instances present at leaf node, refer Equation (6) ;

end
- Normalize the weight score calculated for each tree t, and assign this value to the concerned tree ;
- Multiply the Exponential weight score to each decision tree and perform majority voting, refer equation (8) and (9);
- Return:
The class value, for Classification
OR
The predicted output value, for Regression ;

where Z is the normalizing term, which is the sum of weights of all dsecision trees. The α value is one of the hyper-parameter to control the weight score. For classification, the Eq. (3) is turned out to be as:

$$C(Y = j) = \sum_{t=1}^{n_{tree}} (W_{\mathbf{X}}^t) \cdot J(\hat{Y}_j^t) \tag{8}$$

For regression, the Eq. (6) is turned out to be as:

$$\hat{Y} = \frac{1}{n_{tree}} \sum_{t=1}^{n_{tree}} (W_{\mathbf{X}}^t) \cdot \hat{Y}_h^t \tag{9}$$

At last, weighted voting is performed using Eqs. (8) and (9) for predicting output in classification and regression tasks respectively, shown in Fig. 2. The pseudo code for predicting the class or regression value is provided in Algorithm 1.

4 Experimental Results

This section is comprised of datasets, implementation details, and performance analysis of EWRF compared to conventional RF, and state-of-the-art methods.

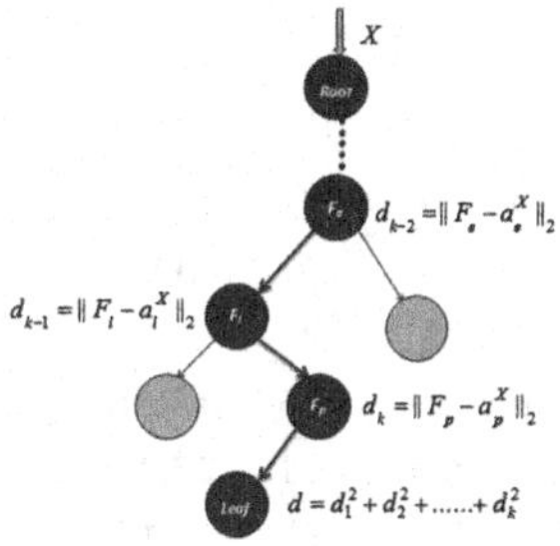

Fig. 1. An example to show the calculation of distance during testing. In this example an test instance X follows the path marked as bold blue lines (F_e, F_l and F_p) to reach up to leaf node. The distance is calculated at the corresponding node in the path followed by the test case. At the root node, all distances are sum up to get the final distance between test case and decision tree. (Color figure online)

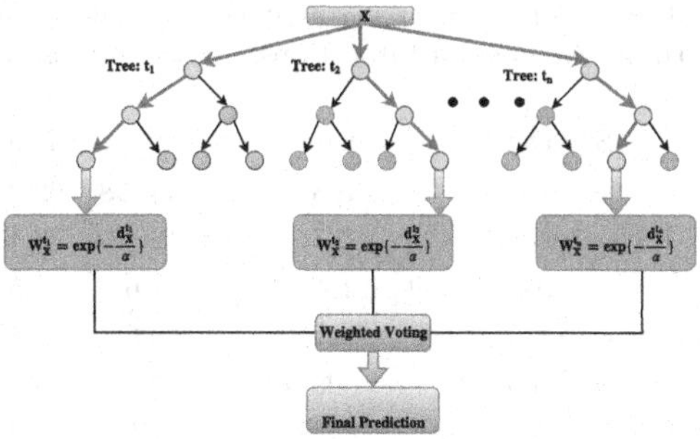

Fig. 2. The proposed EWRF method to show how exponentially weighted score is calculated by different decision trees, for the given test instance. Further, weighted voting is performed for final prediction

4.1 Datasets and Implementation Details

The experiments have been conducted over the benchmark datasets, which are publicly available over the UCI repository [1]. These datasets are from a variety of domains and have different combinations of numerical attribute values. These datasets vary in terms of the number of classes, features, and instances for rigorous testing of the proposed method.

There are five main parameters for conducting the experiments: (1) the number of trees n_{tree}, (2) the number of minimum instances at leaf node n_{min}, (3) the sample ratio in which dataset is divided into training set and test set, (4) the maximum tree depth T_{depth}, and (5) value of α for computation of exponential weighing score. The value of n_{tree} is decided empirically. The experiments have been done over Vehicle, Wine, and Abalone datasets with n_{tree} in the range of 10 to 100 with a step size of 10. We have observed that beyond $n_{tree} = 50$, the accuracy saturates, so it is kept as 50 in all experiments. The n_{min} is kept as 5 and the sample ratio for dividing the datasets into training and testing is kept as 0.5. These values are taken from the state-of-the-art methods for a fair comparison. Experiments have been done with different values of T_{depth} and the results are quoted with the depth, where accuracy is better among different trials. The value of α is chosen as 0.45 for classification and 0.75 for regression. It is also decided by experimenting with different values of $\alpha = \{0.15, 0.45, 0.75, 1.0\}$. Each of the experiment is repeated 10 times with the random selection of training and testing subsets.

Table 1. MSE comparison between state-of-the-art methods and proposed EWRF with average over 10 iterations (least value is the best)

SN	Dataset	Dimension	RF	Biau08	Biau12	Denil	BRF	EWRF
1	Slump	103*10	**41.58**	62.30	62.31	60.19	55.67	45.63
2	Servo	167*4	2.58	3.19	2.39	2.26	2.07	**0.91**
3	Automobile	205*26	1.62	1.53	1.51	1.46	1.41	**1.18**
4	Yacht	308*7	50.6	229.86	225.97	150.58	128.85	**35.88**
5	Housing	506*14	27.7	85.50	82.97	81.62	77.81	**27.1**
6	Student	649*33	41.9	9.83	9.81	9.38	8.93	**4.1**
7	Concrete	1030*9	130.5	279.13	279.70	278.64	275.56	**125.54**
8	Wine quality	4898*12	0.57	0.81	0.81	0.67	**0.51**	0.57
9	Airfoil	1503*6	28.9	66.66	47.73	43.47	38.57	**26.2**
10	Energy_y1	768*8	**2.33**	64.11	40.71	24.53	19.85	5.1

Table 2. Classification accuracy comparison between state-of-the-art methods and proposed EWRF with average over 10 iterations (high value is the best)

SN	Dataset	Dimension	# Classes	RF	Biau08	Biau12	Denil	BRF	EWRF
1	Transfusion	748*5	2	72.2	68.92	70.27	72.97	**77.7**	72.9
2	Spambase	4601*57	2	91.1	60.59	60.59	**94.4**	94.1	90.7
3	CVR	435*16	2	88.8	51.86	61.4	94.4	**95.6**	94.4
4	Madelon	2600*500	2	60.6	49.27	50.31	54.81	**69.2**	58.8
5	Wine	178*13	3	96.9	40.59	41.18	96.47	97.7	**98.3**
6	CMC	1473*9	3	50.29	42.72	42.65	53.6	54.6	**55.2**
7	Verbetral	310*6	3	80.9	48.39	48.39	82.26	82.3	**82.9**
8	Connect-4	67557*42	3	64.58	64.52	65.47	66.19	76.19	**77.1**
9	Vehicle	946*18	4	72.6	27.98	23.1	68.81	71.67	**73.5**
10	Zoo	101*17	7	85.3	50	41	80	85	**87.2**
11	Abalone	4177*8	29	**27.1**	16.05	16.52	26.23	26.44	**27.1**

4.2 Performance Analysis

The results generated with the proposed EWRF are compared to the conventional RF [6], and the state-of-the-art methods, i.e. four variants of random forest Biau08 [5], Biau12 [4], Denil [9] and BRF [16] for the regression and classification datasets. The highest learning performance among these comparisons is marked in boldface for each dataset.

In regression, it can be observed from Table 1 that EWRF achieves the significant reduction in MSE on seven datasets out of ten datasets. In particular, one can observe that for the Concrete dataset, Biau08 [5], Biau12 [4], Denil [9], and BRF [16] have almost same MSE value. However, there is more than 50% reduction in MSE for Yacht, Concrete, and Housing datasets. The proposed method has also shown improvement for datasets having large number of classes like

Student, and Automobile. From Table 1, it is clear that the proposed method has shown much improvement over the compared state-of-the-art methods.

For classification, the comparison between the existing state-of-the-art methods and proposed EWRF is shown in Table 2. It can be seen that EWRF is showing improvement as compare to Biau08 [5], Biau12 [4] and Denil [9] for all the classification data except for Spambase. In comparison with BRF [16], the proposed method is showing improvement for seven datasets out of eleven datasets. In comparison to conventional RF, the proposed EWRF is showing improvement for nine datasets out of eleven datasets.

5 Conclusion

The conventional Random Forest (RF) assigns equal weights to the votes cast by each individual tree. Also, the approaches proposed in the past assigns weights to every decision tree during the training phase only. In this paper, we have explored the dynamic relationship between test samples and decision trees, based on which aggregation/weighted voting is performed. Thus, weights derived in EWRF are dynamic in nature. The proposed method is tested over various heterogeneous datasets and compared to state-of-the-art competitors. The proposed method has shown improvement for both regression and classification tasks.

References

1. UCI repository. https://archive.ics.uci.edu/ml/index.php. Accessed 15 Nov 2018
2. Akash, P.S., Kadir, M.E., Ali, A.A., Tawhid, M.N.A., Shoyaib, M.: Introducing confidence as a weight in random forest. In: 2019 International Conference on Robotics, Electrical and Signal Processing Techniques (ICREST), pp. 611–616. IEEE (2019)
3. Amaratunga, D., Cabrera, J., Lee, Y.S.: Enriched random forests. Bioinformatics **24**(18), 2010–2014 (2008)
4. Biau, G.: Analysis of a random forests model. J. Mach. Learn. Res. **13**(Apr), 1063–1095 (2012)
5. Biau, G., Devroye, L., Lugosi, G.: Consistency of random forests and other averaging classifiers. J. Mach. Learn. Res. **9**(Sep), 2015–2033 (2008)
6. Breiman, L.: Random forests. Mach. Learn. **45**(1), 5–32 (2001)
7. Criminisi, A., Shotton, J.: Decision Forests for Computer Vision and Medical Image Analysis. Springer (2013)
8. Deng, H., Runger, G.: Feature selection via regularized trees. In: The 2012 International Joint Conference on Neural Networks (IJCNN), pp. 1–8. IEEE (2012)
9. Denil, M., Matheson, D., De Freitas, N.: Narrowing the gap: random forests in theory and in practice. In: International Conference on Machine Learning, pp. 665–673 (2014)
10. Ishwaran, H.: The effect of splitting on random forests. Mach. Learn. **99**(1), 75–118 (2015)
11. Kulkarni, V.Y., Sinha, P.K., Petare, M.C.: Weighted hybrid decision tree model for random forest classifier. J. Inst. Eng. (India): Ser. B **97**(2), 209–217 (2016)

12. Liu, Y., Zhao, H.: Variable importance-weighted random forests. Quant. Biol. **5**(4), 338–351 (2017)
13. Paul, A., Mukherjee, D.P.: Enhanced random forest for mitosis detection. In: Proceedings of the 2014 Indian Conference on Computer Vision Graphics and Image Processing, p. 85. ACM (2014)
14. Paul, A., Mukherjee, D.P.: Reinforced random forest. In: Proceedings of the Tenth Indian Conference on Computer Vision, Graphics and Image Processing, p. 1. ACM (2016)
15. Paul, A., Mukherjee, D.P., Das, P., Gangopadhyay, A., Chintha, A.R., Kundu, S.: Improved random forest for classification. IEEE Trans. Image Process. **27**(8), 4012–4024 (2018)
16. Wang, Y., Xia, S.T., Tang, Q., Wu, J., Zhu, X.: A novel consistent random forest framework: Bernoulli random forests. IEEE Trans. Neural Netw. Learn. Syst. **29**(8), 3510–3523 (2018)
17. Winham, S.J., Freimuth, R.R., Biernacka, J.M.: A weighted random forests approach to improve predictive performance. Stat. Anal. Data Min.: ASA Data Sci. J. **6**(6), 496–505 (2013)

Instance Ranking Using Data Complexity Measures for Training Set Selection

Junaid Alam and T. Sobha Rani(✉)

School of Computer and Information Sciences, University of Hyderabad, Hyderabad, India
mdjunaidalamqureshi@gmail.com, tsrcs@uohyd.ernet.in

Abstract. A classifier's performance is dependent on the training set provided for the training. Hence training set selection holds an important place in the classification task. This training set selection plays an important role in improving the performance of the classifier and reducing the time taken for training. This can be done using various methods like algorithms, data-handling techniques, cost-sensitive methods, ensembles and so on. In this work, one of the data complexity measures, Maximum Fisher's discriminant ratio (F1), has been used to determine the good training instances. This measure discriminates any two classes using a specific feature by comparing the class means and variances. This measure in particular provides the overlap between the classes. In the first phase, F1 of the whole data set is calculated. After that, F1 using leave-one-out method is computed to rank each of the instances. Finally, the instances that lower the F1 value are all removed as a batch from the data set. According to F1, a small value represents a strong overlap between the classes. Therefore if those instances that cause more overlap are removed then overlap will reduce further. Empirically demonstrated in this work, the efficacy of the proposed reduction algorithm (DRF1) using 4 different classifiers (Random Forest, Decision Tree-C5.0, SVM and kNN) and 6 data sets (Pima, Musk, Sonar, Winequality(R and W) and Wisconsin). The results confirm that the DRF1 leads to a promising improvement in kappa statistics and classification accuracy with the training set selection using data complexity measure. Approximately 18–50% reduction is achieved. There is a huge reduction of training time also.

Keywords: Maximum Fisher's discriminant ratio · Classification · Batch removal · Kappa statistics · Instance ranking

1 Introduction

Classification accuracy heavily depends on the training data that is provided for the training. There are two ways in which this can be done to obtain a higher accuracy. One is to select features that are most relevant and non-redundant. The other one is to select a set of compact and small training sets. Behaviour

B. Deka et al. (Eds.): PReMI 2019, LNCS 11941, pp. 179–188, 2019.
https://doi.org/10.1007/978-3-030-34869-4_20

of the machine learning model is entirely dependent on the training set chosen from all possible sets. This training set selection can play an important role in improving the accuracy of the machine learning technique and reduces time it takes for the training.

A compact and small training set may improve the accuracy of the classifier. This process requires the removal of irrelevant and redundant instances from the original set. This is an NP hard problem. Hence several methods are proposed in the literature.

Training set selection can be done using different ways such as algorithms, data-handling techniques, cost-sensitive methods, ensembles and so on. In this paper, a method has been proposed to identify the set of good instances, or in other words remove the bad instances/samples from the training set. This is a filter approach.

The paper is organized in the following way. In Sect. 2, literature survey provides the current sate-of-the-art for the selection of the training sets. In Sect. 3, a pre-processing reduction algorithm (DRF1) to remove the bad instances from the training set is proposed. In Sect. 4, experimental setup such as tools, data sets, classifiers, results and so on are provided. In Sect. 5, the results are thoroughly analyzed. In Sect. 6, conclusions on the proposed reduction algorithm (DRF1) are given.

2 Literature Survey

- Nearest neighbor editing rules All of these algorithms use the nearest neighbor rule, but each algorithm has some unique criteria to reduce the instances [19]. They are Condensed Nearest Neighbor rule (CNN), Reduced Nearest Neighbor rule (RNN), Edited Nearest Neighbor rule (ENN), Selective Nearest Neighbor rule (SNN), Variable Similarity Metric (VSM) and All k-NN.
- Instance based learning methods These are instance based algorithms [1]. They are Instance Based-2 (IB2), IB3, IB4, IB5, Shrink (Subtractive) Algorithm, Model Class Selection (MCS), Typical Instance Based Learning (TIBL), Random Mutation Hill Climbing (RMHC) and Encoding Length (EL).
- Ordered removal algorithms Decremental Reduction Optimization Procedure (DROP1) is identical to RNN, DROP2 uses more information and ordering the removal, DROP3 filters noise, DROP4 filters noise more carefully, DROP5 smoothens the decision boundary, Decremental Encoding Length (DEL) is identical to DROP3 [19].
- Silhouette coefficient measure In order to select instances that are more helpful in classification Silhouette index is used [6]. Here the instances are ranked based on silhouette index. It was found that instances of high relevance are located at the root of the class, and instances of less relevance will be in the peripheral places close to the other classes. It eliminates both instances with less or high relevance value, while retaining mid-ranked instances.

– Instance reduction methods It is used to detect and remove redundant instance that can reduce the number of tuples in a data set [1]. In numerosity reduction methods, the data are replaced or approximated by the alternative [8]. They are Rough Set Attribute Reduction (RSAR), Fuzzy Rough Feature Selection (FRFS) and so on.

2.1 Data Complexity Measures

The predictability or performance of the classifier depends strongly on the training data complexity [5]. Ho and Basu [15], define some data complexity measures for the binary class data sets. Molinada et al. [21] expansion, some problems for two or more classes in Ho and Basu definition of measurement are discussed. They analyzed the effect of these generalized measures using two prototype selection algorithms and comment that Fisher's discriminant ratio for prototype selection is most effective. These measures can be divided into 3 broad categories. Measures of separation of classes [15], Measures to find the overlap between the classes, measures based on geometry, topology and multiplication density [21]. Of these measures, in this work, an overlap measure called as "Ratio of maximum Fisher's discriminant (F1)" is used to rank the instances.

2.2 Factors Influencing Classification

Some of the factors which affect the classifiers are [2–4]: concentrated on the amount of overlap in different dimensions from different classes, included in the shape of decision boundary from different classes, included in the size of the classes, the proximity between classes.

3 Proposed Reduction Algorithm (DRF1)

Proposed reduction algorithm (DRF1) uses ratio of maximum Fisher's discriminant(F1) to rank the instances while using leave-one-out method. All the instances which are meeting the criteria, where the F1 value is lowered by the inclusion of the instances are removed as a batch.

3.1 Ratio of Maximum Fisher's Discriminant (F1)

Ratio of Fisher's discriminant was presented by Ho and Basu [15] and is used to calculate to show how two classes differ along a specific feature.

$$f_{\mathrm{i}} = \frac{(\mu_1 - \mu_2)^2}{\sigma_1^2 + \sigma_2^2}$$

where μ_1 and μ_2 are means and σ_1 and σ_2 are variances for class1 and class2 respectively. For 1-D, it is $\mathrm{f_i}$, if considering multi-dimensional data of dimensions N, then it is $\max(\mathrm{f_i})$, which is F1, where i is 1, 2, 3, ..., N. High value of F1 represents weak overlap between the classes. Range of F1 is $[0 \text{ to } +\infty]$.

3.2 Leave-one-out Algorithm

Baseline of the algorithm is to remove bad instances from the data set D. If D2 is a bad instance, this algorithm removes noisy instances from the data sets, since a noisy instance is usually associated with D2. These noisy instances are mostly from a different class, and such data is more likely to be classified correctly without D2. On the other hand, removing some instances near the discriminating line may make the classifier to classify incorrectly. Because most of their neighbors can be mis-classified. Thus it is for the algorithm to keep the non-noisy boundary instances. On the border cases, there is usually a collection of boundary instances. In each case, the closest neighboring majority is rightly classified.

3.3 Batch Removal Algorithm

Batch removal is also given by Wilson and Martinez [19]. Here, each and every instance is checked to see whether it meets the removal criteria before removing any of them. Then all those instances which satisfy the criteria are discarded. Depending on the leave-one-out method, removal of all instances from the data set is carried out in one go.

Data Reduction using F1 (DRF1) algorithm

Input ⇐ D (Data set, which is N dimensional)
Output ⇐ D1 (A batch of good set of instances), D2 (A batch of bad set of instances)

---

for each feature n **in N for D**

$$F_{\mathrm{n}} = \frac{(\mu_1 - \mu_2)^2}{\sigma_1^2 + \sigma_2^2}$$

F1=max(F_{n})
for each instance i **in D**
  Find F1 and store in $F1_{\mathrm{i}}$, using Leave-one-out method
D1 ← ϕ, D2 ← ϕ
for each instance i **in D**
  if F1>$F1_{\mathrm{i}}$ **then**
    D2 ← D2 ∪ i
  else
    D1 ← D1 ∪ i
return D1 and D2

---

4 Experimental Setup

Computer system is used for all experimentation with Intel(R) Core(TM) i5-2400 processor CPU @3.10GHz, Architecture x86_64 with 4GB RAM, Operating System running Ubuntu 18.04.2. Configuration has no role in the experimentation.

Weka [20] is used for classification, and R package [9], [10–14, 16, 17] are used for carrying out the F1 measure and the proposed data removal algorithm. 4

classifiers as Random Forest(RF), Decision Tree-C5.0(C5.0), SVM and kNN are used to validate the results. 25 runs are conducted for each data set and average results are presented here. Data is split into training set and test set in the ratio of 3:1 in each run. Table 1 shows the data sets taken from UCI repository [7] for experimentation.

Table 1. Data sets used in the experimentation.

Data Set	# of Features	# of Instances	Classes
Pima	9	768	Diabetic (268) Non-diabetic (500)
Musk	167	6598	Musk (1017) Non-musk (5581)
Sonar	61	208	M (111) R (97)
Wisconsin	10	342	Benign (223) Malignamt (119)
Winequality (R)	12	1599	Good (855) Bad (744)
Winequality (W)	12	4898	Good (3258) Bad (1640)

4.1 Validation Measures

In order to measure the performance of the reduction algorithm, two measures are chosen. One is accuracy and the other is kappa statistics.

Accuracy(P_A) of the classifier is calculated using confusion matrix. Confusion matrix is created by $T1$, $T2$, $F1$ and $F2$, where $T1$ is True Class1, $T2$ is True Class2, $F1$ is False Class1 and $F2$ is False Class2.

$$P_A = \frac{T1 + T2}{T1 + T2 + F1 + F2}$$

Kappa Statistics (P_K) of the classifiers is calculated by accuracy and random agreement(P_R) [18]. Random agreement calculated by probability of Yes(P_Y) and No(P_N), where

$$P_Y = \frac{(T1 + F1) * (T1 + F2)}{(T1 + T2 + F1 + F2)^2}, P_N = \frac{(T2 + F2) * (F1 + F2)}{(T1 + T2 + F1 + F2)^2}$$

$$P_R = P_Y + P_N, P_K = \frac{P_A - P_R}{1 - P_R}$$

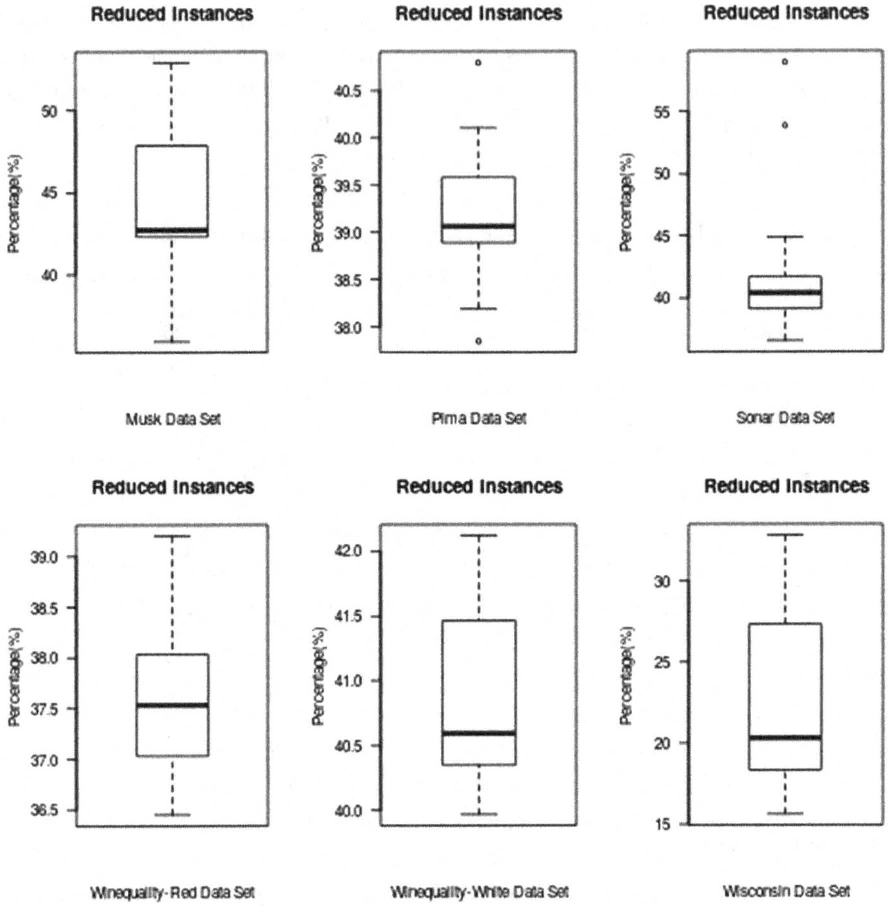

Fig. 1. Reduction rate achieved using the proposed DRF1 algorithm.

4.2 Results

Figures 2 and 3 represent all 4 combinations such as Train_Before (Training set before reduction), Train_After (Training set after reduction), Test_Before (Test set before reduction) and Test_After (Test set after reduction). Figure 1 shows the data reduction rates achieved using the DRF1.

After analyzing P_{A} and P_{K} in Figs. 2 and 3, Musk and Wisconsin gave similar improvements for all 4 classifiers in the training set. Pima, Sonar, Winequality-(R and W) also gave similar improvements, but only for 3 classifiers like RF, C5.0 and SVM in the training set. Pima improves the kNN classifier but this type of improvement not for other data sets.

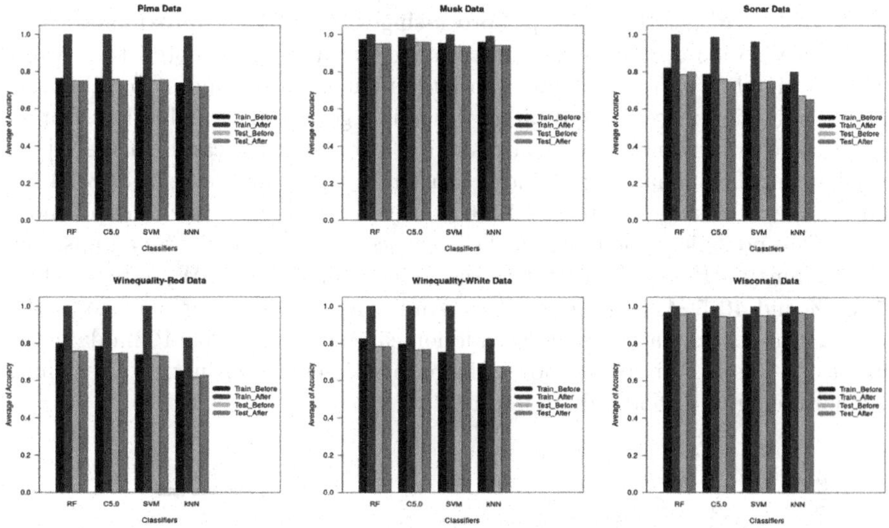

Fig. 2. Classification accuracy(P_{A}) for 6 different data sets using the classifiers RF, C5.0, SVM, kNN.

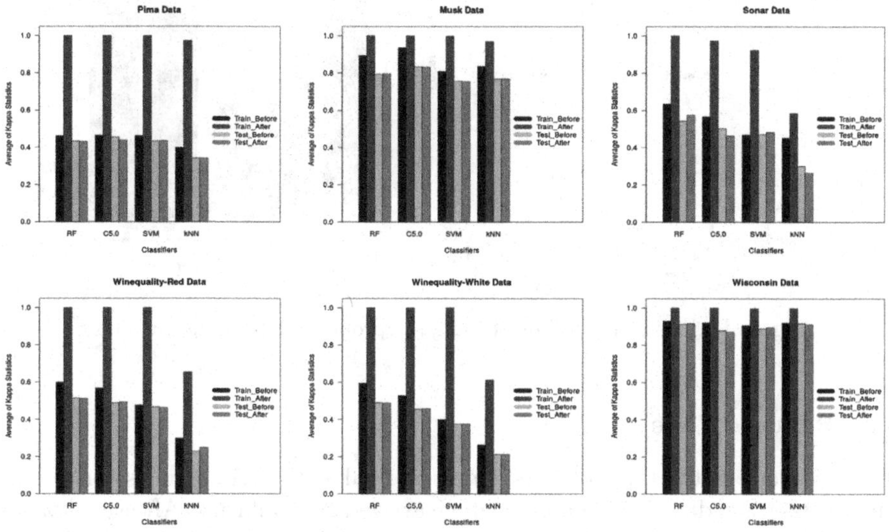

Fig. 3. Kappa Statistics(P_{K}) for 6 different data sets using the classifiers RF, C5.0, SVM, kNN.

5 Analysis

5.1 Reduction Rate

After analyzing Fig. 1, it is clear that, for Musk it provides a reduction of 41–48%, which achieves good compaction rate in the training set, and the mean reduction rate is **42%**. It means that after pre-processing using DRF1, approximately 58% of instances forms a core set as the training set. It is to be noted that Musk has large number of features and also large number of instances. For Pima, it offers a reduction rate of **39–39.5%**, which is a good rate of reduction. But as the variance is very low for this data set, a stable reduction can be achieved for this data set. For Sonar, it provides variation in the reduction rate of 37–42%, and it's mean reduction rate is **39–40%**. But in some runs reduction of more than 50% is possible, depending on the sequence of the instances. So an unstable reduction may achieved for this data set. Winequality-(R and W) is reduced by **37–38%** and **40.5–41.5%** respectively. And in all runs there appears to be the same % of reduction, and a stable reduction can be achieved like Pima data set. On the other hand, Wisconsin shows a large variation in the reduction rate, which may not then give a good reduction rate.

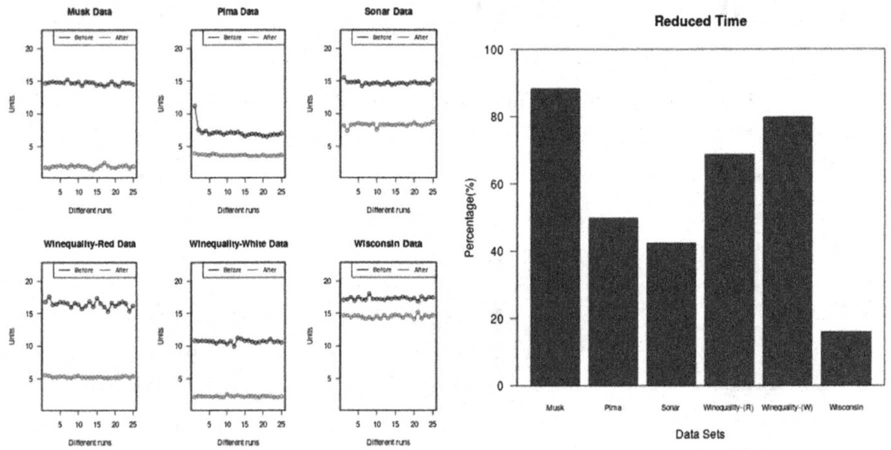

Fig. 4. Reduced time achieved using the proposed DRF1 algorithm.

5.2 Reduced Time

In Fig. 4, there are two types of plots. One is original reduced time (units) with each run, another is reduced time in percentage as average of all runs. After analyzing Fig. 4, Musk, Winequality (R) and Winequality (W) have shown a reduction of training times by 88%, 69% and 80%, and the training times after reduction are **12%, 31% and 20%** respectively. This is a huge reduction as compared to the original training time. It is to be noted that Musk and Winequality have large number of features and Musk has also large number of instances. Pima requires less than **50%** time for training after using DRF1. Sonar and Wisconsin require **58% and 84%** respectively now, which is quite less since both data sets have less data points. If large data sets are used, using DRF1 the time will be reduced considerably.

6 Conclusions

This work proposes a reduction algorithm (DRF1) for selecting the best training set after ranking the instances. After using the DRF1, data sets show promising improvements in time taken for the training, the kappa statistics and the accuracy of the classification for the training sets. Also, the test sets give equal or better results with these reduced sets. This reduction seems to be removing confusing instances near the discrimination line that is border line instances. More experiments need to be done to remove the redundant instances. All the experiments were conducted for 2-class problems. So, this can be extended to multi-class problems. Other data complexity measures can be explored for training set selection.

References

1. Aha, D.W., Kibler, D., Albert, M.K.: Instance-based learning algorithms. Mach. Learn. **6**(1), 37–66 (1991). https://doi.org/10.1007/BF00153759
2. Basu, M., Ho, T.K.: The learning behavior of single neuron classifiers on linearly separable or nonseparable input. In: IJCNN 1999, International Joint Conference on Neural Networks. Proceedings (Cat. No. 99CH36339), vol. 2, pp. 1259–1264, July 1999. https://doi.org/10.1109/IJCNN.1999.831142
3. Basu, M., Ho, T.: Data Complexity in Pattern Recognition, January 2006. https://doi.org/10.1007/978-1-84628-172-3
4. Baumgartner, R., Somorjai, R.: Data complexity assesment in undersampled classification of high dimensional biomedical data. Pattern Recogn. Lett. **27**, 1383–1389 (2006). https://doi.org/10.1016/j.patrec.2006.01.006
5. Cano, J.R.: Analysis of data complexity measures for classification. Expert Syst. Appl. **40**(12), 4820–4831 (2013). https://doi.org/10.1016/j.eswa.2013.02.025
6. Dey, D., Solorio, T., Montes y Gómez, M., Escalante, H.J.: Instance selection in text classification using the silhouette coefficient measure. In: Batyrshin, I., Sidorov, G. (eds.) MICAI 2011. LNCS (LNAI), vol. 7094, pp. 357–369. Springer, Heidelberg (2011). https://doi.org/10.1007/978-3-642-25324-9_31
7. Dua, D., Graff, C.: UCI machine learning repository (2017). http://archive.ics.uci.edu/ml
8. El-hasnony, I.M., Bakry, H.M.E., Saleh, A.A.: Article: comparative study among data reduction techniques over classification accuracy. Int. J. Comput. Appl. **122**(2), 9–15 (2015)
9. Hornik, K., Buchta, C., Zeileis, A.: Open-source machine learning: R meets Weka. Comput. Stat. **24**(2), 225–232 (2009). https://doi.org/10.1007/s00180-008-0119-7
10. Hothorn, T., Zeileis, A.: partykit: a modular toolkit for recursive partytioning in R. J. Mach. Learn. Res. **16**, 3905–3909 (2015). http://jmlr.org/papers/v16/hothorn15a.html
11. Kuhn, M.: caret: Classification and Regression Training (2018). https://CRAN.R-project.org/package=caret, r package version 6.0-80
12. Kuhn, M., Quinlan, R.: C50: C5.0 Decision Trees and Rule-Based Models (2018). https://CRAN.R-project.org/package=C50, r package version 0.1.2
13. Liaw, A., Wiener, M.: Classification and Regression by random forest. R News **2**(3), 18–22 (2002). , http://CRAN.R-project.org/doc/Rnews/

14. Meyer, D., Dimitriadou, E., Hornik, K., Weingessel, A., Leisch, F.: Misc Functions of the Department of Statistics, Probability Theory Group (Formerly: E1071), TU Wien (2018). https://CRAN.R-project.org/package=e1071, r package version 1.7-0
15. Ho, T.K., Basu, M.: Complexity measures of supervised classification problems. IEEE Trans. Pattern Anal. Mach. Intell. **24**(3), 289–300 (2002). https://doi.org/10.1109/34.990132
16. Urbanek, S.: rJava: Low-Level R to Java Interface (2018). https://CRAN.R-project.org/package=rJava, r package version 0.9-10
17. Weihs, C., Ligges, U., Luebke, K., Raabe, N.: klaR analyzing german business cycles. In: Baier, D., Decker, R., Schmidt-Thieme, L. (eds.) Data Analysis and Decision Support, pp. 335–343. Springer, Berlin (2005). https://doi.org/10.1007/3-540-28397-8_36
18. Wikipedia contributors: Cohen's kappa – Wikipedia, the free encyclopedia (2019). https://en.wikipedia.org/w/index.php?title=Cohen%27s_kappa&oldid=895198761. Accessed 31 May 2019
19. Wilson, D.R., Martinez, T.R.: Reduction techniques for instance-based learningalgorithms. Mach. Learn. **38**(3), 257–286 (2000). https://doi.org/10.1023/A:1007626913721
20. Witten, I.H., Frank, E.: Data Mining: Practical Machine Learning Tools and Techniques, 2nd edn. Morgan Kaufmann, San Francisco (2005)
21. Li, Y., Dong, M., Kothari, R.: Classifiability-based omnivariate decision trees. IEEE Trans. Neural Netw. **16**(6), 1547–1560 (2005). https://doi.org/10.1109/TNN.2005.852864

Deep Learning

A Multi-class Deep All-CNN for Detection of Diabetic Retinopathy Using Retinal Fundus Images

Uday Kiran Challa(✉), Pavankumar Yellamraju, and Jignesh S. Bhatt

Indian Institute of Information Technology Vadodara, Gandhinagar, India
{201551081,201552028,jignesh.bhatt}@iiitvadodara.ac.in

Abstract. Diabetic retinopathy (DR) is a diabetes complication that affects retina due to diabetes mellitus. Funduscopy capture the phenomena, however, it needs to be augmented by effective algorithms that enable detection of severity levels of DR from fundus images. Researchers have mostly proposed conventional convolutional neural networks (CNNs) for binary classification with a very few attempts on multi-class detection of DR. In this paper, unlike other approaches, we propose a deep All-CNN architecture for the detection of DR and its five levels. We first correct the fundus images against the sensor parameters by using Gaussian filters and perform blending in order to highlight foreground features. This is followed by removal of retinal boundaries that further helps the detection process. The proposed pre-processing helps in visualization of intrinsic features of the images as well as builds trust in predictions using the model. Thus pre-processed images are fed into the proposed All-CNN architecture. This has 10 convolutional layers and a softmax layer for the final classification. It includes three convolutional layers with kernel size 3×3 at strides of 2 which are designed to work as pooling layers; while two convolutional layers with kernel size 1×1 at strides of 1 are constructed to act as fully connected layers. We apply the proposed methodology on the publicly available Kaggle dataset that has the five-class labeled information. Our model is trained on 30000 retinal fundus images while tested on 3000 images. The proposed architecture is able to achieve 86.64% accuracy, loss of 0.46, and average F1 score of 0.6318 for the five classes. Unlike other architectures, our approach: (1) is All-CNN architecture, (2) provides multi-class information of DR, and (3) has outperformed existing approaches in terms of accuracy, loss and F1 score.

Keywords: All-CNN · Deep learning · Diabetic retinopathy · Multi-class classification · Retinal fundus images

1 Introduction

Worldwide diabetic patients are about 280 millions which are estimated to be doubled by 2025 [1]. It is predicted that more than 30% of the diabetic patients

B. Deka et al. (Eds.): PReMI 2019, LNCS 11941, pp. 191–199, 2019.
https://doi.org/10.1007/978-3-030-34869-4_21

will be within the Asia Pacific region by the end of 2025 [2]. In particular, according to the World Health Organization (WHO), 31.7 million people were affected by diabetes mellitus (DM) in India in the year 2000 and this figure is estimated to rise upto 79.4 million by 2030, largest across the world [3].

One of the serious drawbacks of diabetes is an eye disease called diabetic retinopathy (DR) that damages retina leading to the blindness. It is estimated that globally around 15000 to 39000 people loose their sights because of the DR. In the world, around 14.6% aged 40 years and above suffered with the DR after a 5-year duration of diabetes [4]. Note that the DR affects upto 80% of those who had diabetes for 20 years or more [5]. Hence this urge for prevention, early detection, and diagnosis of DR as well as its severity scales.

In this paper, we focus on developing algorithmic approach for multi-class classification of the DR from retinal fundus images. Note that the researchers have mostly worked on developing algorithms for DR or No-DR [6–8]. However, according to the DR severity scale, it is divided into five classes: No DR, mild non proliferative DR (NPDR), moderate NPDR, severe NPDR, and proliferative DR (PDR). In particular, features like microaneurysms, hemorrhages and exudates are observed in retina during the NPDR stage. In PDR stage, abnormal blood vessels are formed at the back of eye called neovascularization that bursts and bleeds, and finally leads to the blur vision. Hence it is important to detect the stages of DR.

In the recent years, deep convolutional neural network (CNN) has shown excellent performance in information processing for medical data [9–11]. One among the CNNs is All-CNN, where pooling and fully connected layers are replaced by the convolutional layers. Experimental results in [12] has shown that inspite of its simplicity, All-CNN performs better than the conventional CNN with a minimum error rate for CIFAR-10 dataset. In this paper, we propose a deep learning based All-CNN architecture for the multi-class classification of diabetic retinopathy on publicly available Kaggle dataset [13].

2 Related Work

Recently classification of diabetic retinopathy becomes an active area of research. In [14], images are augmented by rotation and translation for class balance, and pre-processed using histogram normalization. AlexNet, VGG16 and GoogLeNet CNN classifiers are trained on these images; and the GoogLeNet has achieved 95% recall for binary classification and 75% recall for 3-class classification.

In [15–18] 5-class classification of DR is performed on Kaggle dataset, however, they use conventional CNNs and/or transfer learning based approaches. In [15] 30000 coloured retinal fundus images are used for training. The images are augmented, normalized and denoised as a pre-processing step. Generated features are fed to award winning CNN architecture for 5-class classification and claimed accuracy upto 85% when validated on 3000 images. In [16], colour normalized and augmented 80000 retinal fundus images are trained on the model and the model has been validated on 5000 images with accuracy of 75%. Retinal fundus images are pre-processed using non-local means denoising (NLMD) for denoising and augmented for the class balance in [17]. For 5-class classification 22.7%, 40%, 41% are achieved by the Baseline model, AlexNet and GoogLeNet respectively. A pre-trained Inception-v3 network is tested for 5-class classification of DR which achieved an accuracy of 48.2% using transfer learning in [18].

3 Data Description and Proposed Pre-processing

In this paper, we have used publicly available Kaggle dataset provided by EyePACS organization [13]. This dataset consists of 35126 retinal fundus images of the left and right eyes. In this experiment, we have randomly chosen 20000 images for training and 3000 images for the test. Average size of each image is 3000×2000 pixels. The statistics of the DR levels among the images available in Kaggle dataset is listed in Table 1.

3.1 Proposed Pre-processing

This is a crucial step in classification as it greatly helps in the final decision. We propose a three-step pre-processing on the given dataset as follows:

Data Augmentation: Referring to Table 1, one can see that the given dataset has class imbalance, i.e., 73.50% of images are having the class 'No DR' while remaining class images are 26.50%. Though this is the natural statistics of the data, however, it often introduces bias while training a model. To avoid this, we augment the data (label 1, 2, 3, and 4) by flipping (vertically and horizontally) and rotating images at 45°, 90°, 135°, and 180° in order to match with data label 0. This turns the total 30000 training images with equal class proportion as shown in Table 2.

Table 1. Statistics of DR levels in Kaggle dataset [13].

Class	Label	#Images	Percentage
No DR	0	25810	73.50%
Mild NPDR	1	2443	6.90%
Moderate NPDR	2	5292	15.10%
Severe NPDR	3	873	2.50%
PDR	4	708	2.00%

Table 2. Data Augmentation.

Label	#Images	Percentage
0	6600	22.00%
1	6300	21.00%
2	6300	21.00%
3	5300	17.66%
4	5500	18.33%

Corrections Against Sensor Parameters: Typically the retinal fundus images are sensitive to the camera settings including sensor resolution, and also affected by different degrees of noise due to the lightening effects. To mitigate the lightening effects, local average colour of an image is subtracted so that it gets mapped to 50% gray level. This is achieved by first passing the image through a Gaussian filter and then the image is blended. We perform this step as,

$$I = \alpha I_1 + \beta I_2 + \gamma, \tag{1}$$

where I is the filtered image, I_1 is the given image that should be overlyed on image I_2 which is a Gaussian blurred image, α and β are transparency levels, and γ is a scalar. In order to create this intermediate image, the images are first scaled to have same radius (300 pixels), then Eq. (1) is modified as $I = 4I_1 - 4I_2 + 128$. By choosing $\alpha = 4$ and $\beta = -4$ we ensure removal of background while extracting foreground features. A scalar $\gamma = 128$ is added to the weighted sum of two images in order to map local average colour to 50% gray.

Removal of Boundaries: It is a well-known fact that generally boundaries do not help in extracting features from images. Thus we remove the boundaries of the retinal images by clipping it to 95%. Resultant size of images is 224 × 224 pixels.

Now we present results of the proposed pre-processing steps in some challenging cases in Fig. 1. The left panel of Fig. 1(a) shows a dark image with a poor visual quality which is converted to a better contrast image after the pre-processing as shown in the right panel. In Fig. 1(b) an image (left panel) is shown which is overexposed to light and after the pre-processing (right panel) it is converted to 50% gray level. In Fig. 1(c) noisy image (left panel) with different colour effects is denoised (right panel). Finally Fig. 1(d) shows an image in which foreground features in the given image (left panel) are highlighted in the processed image (right panel) after reducing background effects.

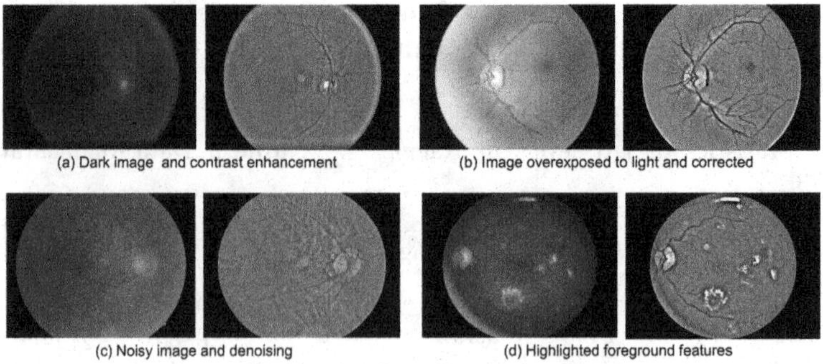

(a) Dark image and contrast enhancement
(b) Image overexposed to light and corrected
(c) Noisy image and denoising
(d) Highlighted foreground features

Fig. 1. Results for proposed pre-processing on fundus images in some challenging cases (a)–(d) : (left panel) given image and (right panel) corresponding pre-processed image.

4 Proposed All-CNN for 5-Class DR Detection

In this section, we present our proposed deep All-CNN architecture for the 5-class classification of diabetic retinopathy using retinal fundus images. The All-CNN architecture is inspired from [12].

4.1 Proposed All-CNN

Figure 2 shows the proposed All-CNN architecture. It has 10 convolutional layers and a softmax layer for the 5-class classification. Note that in the proposed model, pooling and fully connected layers are modified by convolutional layers. Increased convolutional layers are perceived to enable the network to learn deeper structures. Hence, it unifies the entire model as convolution operation leading to a simpler architecture. Rectified linear units (ReLu) activation is used at each convolutional layer for incorporating non-linearity in the model. See that the coloured (RGB) pre-processed images (discussed in Sect. 3) of size 224×224 are given as the input.

In the first convolution layer 32 filters (or kernels) each of size 7×7 are used. Since features in retina are wide spreaded, we use large kernel size in order to extract the local dependencies. Following this, second convolution layer with 32 filters each of size 5×5 are used. Convolution layer 3 is acting as a pooling layer. We use kernel size 3×3 at strides of 2 (performs overlapping convolutions) for dimensionality reduction. Mostly features like exudates and microaneurysms are present nearby. So overlapping convolutions capture such features. Further, fourth convolution layer with 64 number of 3×3 filters is used. This is followed by the fifth convolution layer with 64 number of 3×3 filters at strides of 2, which acts as another pooling layer. Sixth convolution layer has 128 number of 3×3 filters. This is followed by seventh convolution layer with 128 number of 3×3 filters at strides of 2, which again acts as a pooling layer. In eighth convolution layer we use 128 number of 3×3 filters. Finally, ninth and tenth convolution

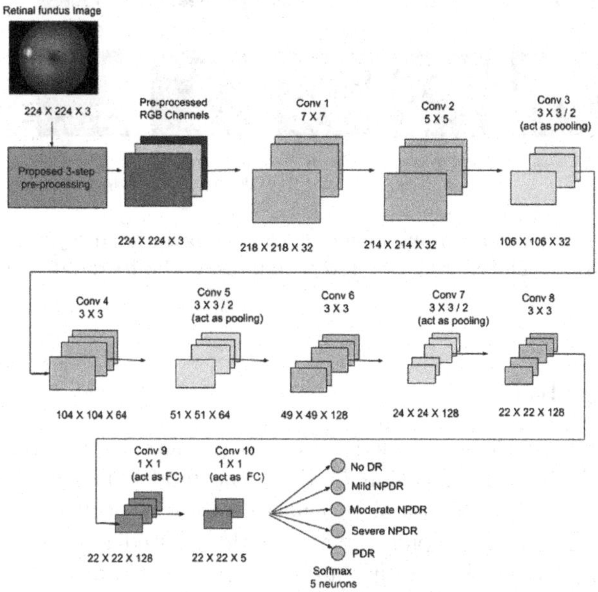

Fig. 2. Proposed deep All-CNN for 5-class detection of diabetic retinopathy.

layers are incorporated to work as the fully connected layers. Ninth convolution layer with 128 number of 1×1 filters at strides of 1 and tenth convolution layer with 5 number of 1×1 filters at strides of 1 are used. Finally, output from these layers is given to the softmax layer that provides five class predictions corresponding to the DR scale (Fig. 2). Batch normalization is added after each convolutional layer to increase the stability of network. Hence each layer now independently learn features from other layers and thus learn deeper features. Dropout with probability 0.5 is added after the batch normalization at each convolutional layer so that overfitting is avoided and the network is regularized.

Figure 3 shows saliency maps generated at each convolutional layer of the proposed All-CNN (Fig. 2). As shown in Fig. 3, saliency maps allow the medical diagnosis through a detailed visual inspection. They are generated by manipulating gradients of the data that are formed in forward and backward passes. Gradient of the output category is computed with respect to the input image. In Fig. 3 one can see that the gradients are highlighted showing the presence of intrinsic features like microaneurysms, soft exudates and hard exudates which are responsible for the DR levels.

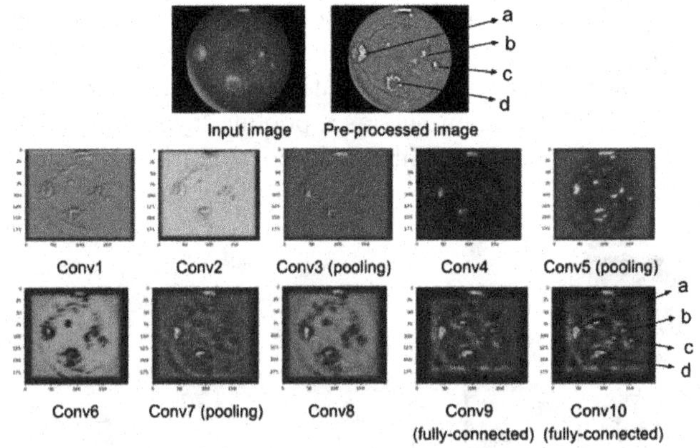

Fig. 3. Saliency maps of an input image at each convolutional layer of the proposed All-CNN (Fig. 3): a - Optic Disc, b - Hard exudates, c - Soft exudates and d - Microaneueysms.

4.2 Training/Test

A supervised training is performed using the proposed architectures shown in Fig. 2 for the 5-class classification of DR. The model uses the cross entropy as loss in the objective function. Stochastic gradient descent optimization is used to minimize the discrepancy between the desired and estimated output of the network. The weights are updated using the backpropogation algorithm. Once the training is done, test images are applied as input for the multi-class classification.

5 Results

In this section, we present the results obtained by proposed architecture on Kaggle dataset. Experiments are conducted on Nvidia Tesla k80 machine with standard memory of 14GB RAM. We make use of 30000 images for training and 3000 images for the test from Kaggle dataset. We run our proposed All-CNN architecture for 200 epochs with a batch size 32 and achieved the accuracy of 86.64% and loss of 0.46. We first calculate confusion matrix and F1 score on the test data. We calculate Recall, Precision, F1 score (Table 3: left panel) using the values in confusion matrix (Table 3: right panel). One can observe that we achieve average F1 score of 0.6318 over the 5-class classification on the test data.

Classification results of the proposed methods and the comparative performance with state-of-art are listed in Table 4. Our proposed architecture performs better detection when compared to other algorithms. It should be noted that algorithms in [15, 16] for 5-class classification use large number of convolutional, Maxpooling and fully connected layers when compared to our proposed All-CNN architecture. Improved pre-processing by the proposed 3-steps and better classification by the proposed All-CNN classifier enable better results.

Table 3. Left panel: Test Scores; and Right panel: Confusion matrix of proposed All-CNN.

Class label	Recall	Precision	F1 score
0	0.964	0.910	0.936
1	0.480	0.451	0.473
2	0.525	0.750	0.612
3	0.570	0.511	0.537
4	0.567	0.638	0.601

Predicted labels (columns); *True labels* (rows)

	0	**1**	**2**	**3**	**4**
0	2053	36	27	14	0
1	72	96	29	0	3
2	123	78	252	18	9
3	6	4	16	57	17
4	4	0	12	23	51

Table 4. Comparative performance for diabetic retinopathy classification.

Algorithm	Classification	Architecture	Dataset (samples)	Accuracy (%)
Ghosh et al. [15]	2-class; 5-class	Award winning CNN architecture	Kaggle (30000)	95%; claimed upto 85%
Pratt et al. [16]	5-class	Conventional CNN	Kaggle (80000)	75%
Marco Alban and Tanner Gilligan (2016) [17]	2-class, 3-class, 5-class	GoogLeNet	Kaggle	77%, 58%, 45%
Lam et al. [14]	2-class; 3-class	GoogLeNet	Kaggle (35000), Messidor-1 (1200)	95%; 75%
Masood et al. [18]	5-class	Inception-v3	Kaggle (3108)	48.2%
Proposed	**5-class**	**All-CNN (proposed)**	**Kaggle (33000)**	**86.64%**

6 Conclusion

We present a novel deep All-CNN architecture for detection of diabetic retinopathy at the five severity scales. The proposed pre-processing extracts intrinsic features leading to insights for different stages in the disease. Our approach outperformed existing approaches in terms of accuracy, loss, and F1 score. Hence, we have developed a simpler and effective architecture that is useful for practical applications in medical imaging.

References

1. Tajunisah, I., et al.: Awareness of eye complications and prevalence of retinopathy in the first visit to eye clinic among type 2 diabetic patients. Int. J. Ophthalmol. **4**(5), 519–524 (2011)
2. Cockram, C.S.: The epidemiology of diabetes mellitus in the Asia-Pacific region. Hong Kong Med. J. **6**(1), 43–52 (2000)
3. Wild, S., et al.: Global prevalence of diabetes: estimates for the year 2000 and projections for 2030. Diab. Care **27**(5), 1047–1053 (2004)
4. The Second National Health and Morbidity Survey: Diabetes mellitus among adults aged 30 years and above, p. 2008. Institute of Public Health (IPH), Ministry of Health, Malaysia (2007)
5. Kertes, P.J., Johnson, T.M. (eds.): Evidence Based Eye Care. John Wiley and Sons, Lippincott Williams & Wilkins (2007)
6. Gulshan, V, et al.: Development and validation of a deep learning algorithm for detection of diabetic retinopathy in retinal fundus photographs. Jama, **316**(22), 2402–2410 (2016)
7. Sohini, R., et al.: DREAM: diabetic retinopathy analysis using machine learning. IEEE J. Biomed. Health Inf. **18**(5), 1717–1728 (2013)
8. Lin, Z., et al.: A framework for identifying diabetic retinopathy based on anti-noise detection and attention-based fusion. In: Frangi, A.F., Schnabel, J.A., Davatzikos, C., Alberola-López, C., Fichtinger, G. (eds.) MICCAI 2018. LNCS, vol. 11071, pp. 74–82. Springer, Cham (2018). https://doi.org/10.1007/978-3-030-00934-2_9
9. Vogl, W.-D., et al.: Spatio-temporal signatures to predict retinal disease recurrence. In: Ourselin, S., Alexander, D.C., Westin, C.-F., Cardoso, M.J. (eds.) IPMI 2015. LNCS, vol. 9123, pp. 152–163. Springer, Cham (2015). https://doi.org/10.1007/978-3-319-19992-4_12
10. Cao, C., et al.: Deep learning and its applications in biomedicine. Genomics, Proteomics Bioinf. **16**(1), 17–32 (2018)
11. Litjens, G., et al.: A survey on deep learning in medical image analysis. Med. Image Anal. **42**, 60–88 (2017)
12. Springenberg, J., et al.: Striving for simplicity: the all convolutional net. arXiv preprint arXiv:1412.6806 (2014)
13. https://www.kaggle.com/c/diabetic-retinopathy-detection/data . Accessed Jan 2017
14. Lam, C., et al.: Automated detection of diabetic retinopathy using deep learning. In: 2018 AMIA Summits on Translational Science Proceedings, pp. 147–155 (2017)
15. Ghosh, R., et al.: Automatic detection and classification of diabetic retinopathy stages using CNN. In: 2017 4th International Conference on Signal Processing and Integrated Networks (SPIN), pp. 550–554 (2017)

16. Pratt, H., et al.: Convolutional neural networks for diabetic retinopathy. In: International Conference On Medical Imaging Understanding and Analysis (MIUA) 2016, vol. 90, pp. 200–205, December 2016
17. Marco Alban Stanford. Automated detection of diabetic retinopathy using fluorescein angiography photographs, technical report, stanford university (2016)
18. Masood, S., et al.: Identification of diabetic retinopathy in eye images using transfer learning. In: 2017 International Conference on Computing, Communication and Automation (ICCCA), pp. 1183–1187 (2017)

Real-Time Vehicle Detection in Aerial Images Using Skip-Connected Convolution Network with Region Proposal Networks

Somsukla Maiti[1(✉)], Prashant Gidde[1], Sumeet Saurav[1], Sanjay Singh[1], Dhiraj[1], and Santanu Chaudhury[2]

[1] CSIR-CEERI, Pilani, Rajasthan, India
[2] IIT Jodhpur, Jheepasani, India
{somsukla,sanjay}@ceeri.res.in

Abstract. Detection of objects in aerial images has gained significant attention in recent years, due to its extensive needs in civilian and military reconnaissance and surveillance applications. With the advent of Unmanned Aerial Vehicles (UAV), the scope of performing such surveillance task has increased. The small size of the objects in aerial images makes it very difficult to detect them. Two-stage Region based Convolutional Neural Network framework for object detection has been proved quite effective. The main problem with these frameworks is the low speed as compared to the one class object detectors due to the computation complexity in generating the region proposals. Region-based methods suffer from poor localization of the objects that leads to a significant number of false positives. This paper aims to provide a solution to the problem faced in real-time vehicle detection in aerial images and videos. The proposed approach used hyper maps generated by skip connected Convolutional network. The hyper feature maps are then passed through region proposal network to generate object like proposals accurately. The issue of detecting objects similar to background is addressed by modifying the loss function of the proposal network. The performance of the proposed network has been evaluated on the publicly available VEDAI dataset.

Keywords: Vehicle detection · Hyper maps · Skip connected · Region Proposal Network (Rpn) · Aerial images · Aerial videos

1 Introduction

There is a growing need of aerial surveillance for civil and military purposes in today's world. This helps in maintaining the decorum of an area while simultaneously ensuring the safety and security of the place. Aerial Surveillance using drones/UAVs have been proved to be very useful in this context. Visual Inspection of areas is the most common practice used in surveillance. Use of an UAV

B. Deka et al. (Eds.): PReMI 2019, LNCS 11941, pp. 200–208, 2019.
https://doi.org/10.1007/978-3-030-34869-4_22

to monitor and stream videos to the base station works as a better alternative as compared to the regular visual surveillance through CCTV system. Videos taken from a higher altitude provides a better visibility of the area. It also adds up to the advantage of tracking high speed objects with ease without increasing its own pace.

Real-time monitoring of unwanted and suspicious activities happening in an area is an important concern. This involves Detection and Recognition of objects followed by subsequent tracking of identified suspicious objects. Detection of objects, such as, vehicles, animals, is a challenging task in aerial images due to the small size of objects with respect to the complete frame. This often creates a misclassification scenario for the detector. The misclassification mainly happens due to the fact that there are several small surrounding structures and ambient background objects that look very similar to the objects of interest from the high altitude. Another major issue that makes the task of a detector more difficult is the use of low resolution cameras with UAVs. This is mainly due the limitation of the payload handling capacity of the drone.

Object detection in aerial images is performed by finding the most salient features of the objects and use them for further processing. Many Saliency based approaches have been developed which are effective to find small objects with salient spectral features but did not perform well in real-time aerial videos. The requirement of a real-time object detection has pushed the research towards using deep learning approaches. It has paced up the performance of object detection and recognition in real-time. Many Convolutional Neural Network (CNN) based architectures have been proposed in recent works that can detect objects with high accuracy. Along with high accuracy, the detectors has also boosted the speed of detection. The convolutional framework of region proposal networks helped in detecting very small objects even when the object is partially occluded or looks similar to the background. The use of high resolution camera payloads with the UAVs has substantially improved the quality of aerial image datasets and consequently has reduced the chances of misclassification.

2 Related Works

Vehicle detection in aerial images has been an active area of research since last three decades. Most of the object detection algorithms use sliding window approach to generate candidate regions. The candidate regions which are similar to the object properties are considered as the detected objects. But these approaches are very time-consuming and computationally heavy as these methods use several different sized windows and slides over the entire image. Thus these techniques are not well suited for real-time object detection in videos. Region proposals provide computationally less complex solution for object detection. Several region proposal methods have been developed and the performance has improved a lot. Many saliency based approaches [1] have been developed that produce good results in images where number of objects is not very high and the foreground objects are significantly different from the background. But these methods fail to provide good results in real-time object detection problem.

To overcome the problem of dedicated feature extraction in images containing multiple objects, several deep learning based approaches have been developed in the last decade. Several CNN based frameworks [3] have been proposed that showed good results for object detection. Zhu et al. have developed a CNN architecture [17] based on the AlexNet framework with Selective Search with a simple modification using empirically set threshold range. A CNN based GoogleNet architecture has been adopted in [12] to detect objects in UCMerced dataset and classify the objects based on a threshold based decision process. A CNN based salient object detection has been proposed in [6], which uses nonlinear regression for refinement of saliency map generated. The CNN based object detection architectures can be broadly classified into two categories: one-stage and two-stage detectors. The one-stage detectors, namely, YOLO (you only look once), SSD (single shot multi-box detector) etc., provide very fast detection. But these detectors fail to detect small size objects. The two-stage detectors, on the other hand, are very accurate but detect objects at a subsequently low speed due to high computational complexity. CNN based object proposal classification has been proposed in [11] which uses region proposal network (RPN) and performs very accurate object detection. To incorporate location invariance, position-sensitive score maps have been computed for every proposal [2]. Due to good accuracy, several researchers have adopted the concept of RPN in their work [5,13] for object detection in aerial images in DLR-3K and VEDAI [9] datasets. The accuracy of two-stage object detection has been improved further by adopting the faster RCNN like framework and modifying it by using hierarchical feature maps [4,15]. A coupled R-CNN based vehicle proposal network has been proposed in [14] that reported good accuracy in aerial images in DLR-3K dataset. The above mentioned two-stage classifiers perform well but are noticeably slower as compared to the one-stage classifiers. YOLO [10] and SSD [8] provides very fast detection by using only one CNN architecture for both classification and localization of the objects. But these methods fail to detect objects in aerial images as size of objects are very small as compared to the size of proposals. To provide a solution to this problem, focal loss [7] has been introduced by Lin et al. which involves a scaling factor that puts more weightage to hard classifiable objects as compared to easily classifiable objects.

The current approaches for vehicle detection are mainly focused on aerial images and thus speed factor is not taken much in consideration in these detectors. Even the methods, that perform fast detection of vehicles, are unable to produce very accurate results. This sums up to the unavailability of a suitable vehicle detection framework for real-time aerial videos. In the following section, a two-stage CNN based framework have been proposed that addresses both speed and accuracy issue.

3 Proposed Framework

The proposed object detection framework is a two-stage model that uses a skip connected convolutional network followed by a region proposal network. It generates features from images using convolution layers. The features derived are then used to generate multiple object-like regions with scores by applying a region proposal network. The proposed framework as shown in Fig. 1. It uses first 5 convolutional layers of the ZF-Net architecture [16]. The features from the shallow layer provides low level information about the images. The deeper layers compute more fine details about the image. The features from the shallow layers and the deep layers are merged into a single feature map to define new hyper feature map. This incorporates low level details and deep level highly semantic representation of data together.

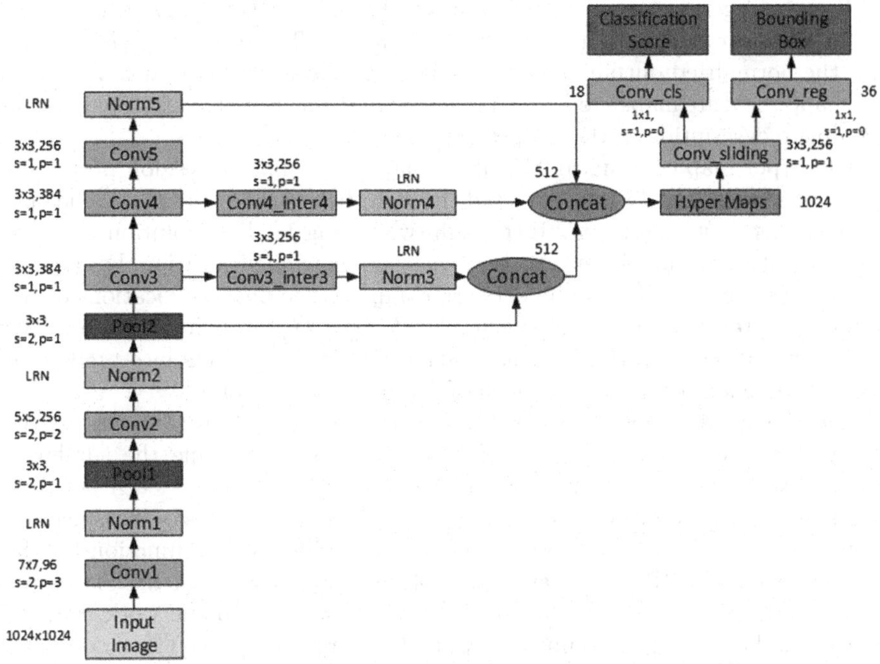

Fig. 1. Proposed convolutional hyper maps based framework

The output of each convolution layer is passed through Rectified Linear Unit (ReLU) activation function to induce non-linearity in the data. The ReLU output is normalized using Local Response Normalization (LRN). The output of corresponding units of the same convolution layers from different kernels are used to normalize the value of every unit in the image. The normalization function is defined by Eq. 1, where $a_{i,j}^k$ is the output of the k-th kernel at image location (i, j), $b_{i,j}^k$ is the output of LRN at location (i, j), N is the total number of convolution kernels in a layer and n is the size of the normalization window. The output of the normalization layer in *conv*1 and *conv*2 is passed through Max pooling layer is performed that helps in adapting the detector to adapt to the small rotational changes of the objects.

$$b_{i,j}^{k} = a_{i,j}^{k} / (\kappa + \alpha \sum_{l=max(0,k-n/2)}^{min(N-1,k+n/2)} (a_{i,j}^{l})^2)^{\beta} \tag{1}$$

The shallow and deep layers are first scaled to an intermediate size. The ReLU normalized output of the third convolution layer is passed through an intermediate convolutional layer with 256 kernels. The ReLU output of the *conv3inter3* layer is normalized using LRN and concatenated with *Pool2* output, to define the shallow level features of images. Similarly, the ReLU output of the fourth convolution layer is passed through an intermediate convolutional layer with 256 kernels to generate *conv4inter4* output. The output is then passed through ReLU nonlinear unit and then further normalized. These outputs are merged with the normalized output of *conv5* to generate deep feature maps. The deep and shallow feature maps are merged together to define Hyper Maps that provide a complete description of the images.

The hyper map contains 1024 feature maps. To define the region proposals, sliding window of size 3×3 is traversed through the entire feature map. The output of sliding operation is passed through two sibling 1×1 convolution layers to consider features from all the kernels at each sliding window location. For each of the sliding window location, one network computes the possible locations of the proposals in terms of a vector, $(x_0, y_0, width, height)$, through regression. The other convolution network performs classification of the objects into predefined classes and generates a score for each of the predicted region.

The predicted regions are compared with the ground truth boxes in the images. Intersection over Union (IoU) metric is used to define the similarity of the predicted regions to the ground truth. The total loss at each epoch is computed using a composite loss function that includes the loss of classification as well as regression network. We have employed different loss function for the two components in RPN. We have used Softmax classifier for calculation of loss in score generation of each predicted region of the classification network. Smooth L1 loss has been used to compute loss in the regression layer. The smooth L1 and cross entropy loss functions have been described in Eqs. 3 and 3 respectively. The overall loss function is defined in Eq. 2 where L_{cls} represents classification loss and L_{reg} represents regression loss. The parameter α in Eq. 2 represents a trade-off factor between classification and regression loss.

$$L_{total} = L_{cls} + \alpha L_{reg} \tag{2}$$

$$L_{reg} = \begin{cases} 0.5x^2 & , for |x| < 1 \\ |x| - 0.5 & , otherwise \end{cases}$$

$$L_{cls} = - \sum_{i} y_i log(p_i) \tag{3}$$

The size of final hyper maps are large enough as compared to the deeper convolutional layers. The size constraint reduces the computational complexity of the algorithm and thus helps in improving the speed of detection. But the increased size of feature map adds up to the problem of detection of more number of region proposals. Most of the proposals detected are prone to be a part of the background and thus scope of detecting false positives increases. To handle the number of false positives and solve the problem, the concept of Focal Loss has been used as mentioned in [7]. The loss function has been described in Eq. 4. The parameter γ decides on the scaling factor to keep or neglect the hard classified examples.

$$FocalLoss(p_i) = -(1 - p_i)^{\gamma} log(p_i) \tag{4}$$

Thus in the Proposed Framework V2, the cross entropy loss function has been replaced by the focal loss function. We have presented the experimental results for the two frameworks in the following section. The Proposed Framework V1 uses cross entropy loss and Proposed Framework V2 uses focal loss.

4 Experimental Results

The proposed architecture is trained using publicly available VEDAI dataset. The dataset consists of various backgrounds such as agrarian, rural and urban areas. The dataset is available in two different image sizes 512×512 and 1024×1024 with Annotation. For our experiment, we selected the VEDAI 1024 dataset that contains 1268 images of size 1024×1024 in .png format. We have used 935 images for training the proposed model while the remaining images have been utilized for testing the model. We have trained our model by using pre-trained weights of a ZF-Net model obtained from training the ZF-Net on the ImageNet dataset. The pre-trained ZF-Net model consists of lot of good lower level features which are important in feature extraction for RPN.

For the proposed framework, values of the hyperparameters in the LRN are taken as $\alpha = 0.00005$, $\beta = 0.75$ and $\kappa = 0$. The value of the parameter n is taken as 3. The hyperparameters for focal loss computation is taken as $\gamma = 2$ and $\alpha = 0.25$. This optimal pair of values provide a correct balance between the hard classified and well classified classes. We have further used the trained models to generate inference on aerial videos. The videos were captured at 25 fps and the background in the videos were completely different from the train image scenarios. The proposed framework has been compared with Faster RCNN architecture. The mean average precision (mAP) and the speed of the Faster RCNN, the Proposed Framework V1 with cross entropy loss and the Proposed Framework V2 with focal loss has been enlisted in Table 1.

The output of the Proposed Framework V1 on the VEDAI aerial images is shown Fig. 2. The Framework V2 improves both the speed and accuracy of detection. The Proposed Framework V2 can thus be used as a suitable model for vehicle detection in aerial images and aerial videos.

Table 1. mAP and detection speed of different object detection frameworks

Method	mAP	Speed (fps)
SSD-Inception V2	0.460	7
Faster RCNN	0.454	5
Proposed Framework1	0.644	14
Proposed Framework2	0.659	14

Fig. 2. Object detection results on test images using proposed framework V1

5 Conclusion

In this paper, we have proposed a two-stage hierarchical region-based CNN framework for detection of vehicles in aerial images and videos. The designed hyper maps based framework produces very accurate and fast vehicle detection result. Scalability of the object size in the videos has also been addressed using the proposed framework due to the use of the shallow and deep layer features. The Proposed Framework V1 produces very accurate results, but the speed of operation is slow as compared to Proposed Framework V2. The inclusion of focal loss in the Proposed Framework V2 helps in reducing the number of proposals per frame by solving the class imbalance problem and thus improves the speed of operation. Thus Proposed Framework V2 generates superior result in terms of accuracy as well as speed and thus can be used as a suitable vehicle detector in aerial videos. However, the proposed framework sometimes does not detect all the vehicles in subsequent frames in aerial videos. The mAP can be further improved to address the problem and improve the accuracy of detection.

References

1. Borji, A., Cheng, M.-M., Jiang, H., Li, J.: Salient object detection: a benchmark. IEEE Trans. Image Process. **24**(12), 5706–5722 (2015)
2. Jifeng Dai, Yi Li, Kaiming He, and Jian Sun. R-FCN: object detection via region-based fully convolutional networks. In: Advances in Neural Information Processing Systems, pp. 379–387 (2016)
3. De Oliveira, D.C., Wehrmeister, M.A.: Towards real-time people recognition on aerial imagery using convolutional neural networks. In: 2016 IEEE 19th International Symposium on Real-Time Distributed Computing (ISORC), pp. 27–34. IEEE (2016)
4. Deng, Z., Sun, H., Zhou, S., Zhao, J., Zou, H.: Toward fast and accurate vehicle detection in aerial images using coupled region-based convolutional neural networks. IEEE J. Sel. Top. Appl. Earth Obs. Remote. Sens. **10**(8), 3652–3664 (2017)
5. Lee, J., Wang, J., Crandall, D., Šabanović, S., Fox, G.: Real-time, cloud-based object detection for unmanned aerial vehicles. In: 2017 First IEEE International Conference on Robotic Computing (IRC), pp. 36–43. IEEE (2017)
6. Li, X., et al.: DeepSaliency: multi-task deep neural network model for salient object detection. IEEE Trans. Image Process. **25**(8), 3919–3930 (2016)
7. Lin, T.-Y., Goyal, P., Girshick, R., He, K., Dollár, P.: Focal loss for dense object detection. In: Proceedings of the IEEE International Conference on Computer Vision, pp. 2980–2988 (2017)
8. Liu, W., et al.: SSD: single shot MultiBox detector. In: Leibe, B., Matas, J., Sebe, N., Welling, M. (eds.) ECCV 2016. LNCS, vol. 9905, pp. 21–37. Springer, Cham (2016). https://doi.org/10.1007/978-3-319-46448-0_2
9. Razakarivony, S., Jurie, F.: Vehicle detection in aerial imagery: a small target detection benchmark. J. Vis. Commun. Image Represent. **34**, 187–203 (2016)
10. Redmon, J., Divvala, S., Girshick, R., Farhadi, A.: You only look once: unified, real-time object detection. In: Proceedings of the IEEE Conference on Computer Vision and Pattern Recognition, pp. 779–788 (2016)
11. Ren, S., He, K., Girshick, R., Sun, J.: Faster R-CNN: towards real-time object detection with region proposal networks. In: Advances in Neural Information Processing Systems, pp. 91–99 (2015)
12. Ševo, I., Avramović, A.: Convolutional neural network based automatic object detection on aerial images. IEEE Geosci. Remote. Sens. Lett. **13**(5), 740–744 (2016)
13. Sommer, L.W., Schuchert, T., Beyerer, J.: Fast deep vehicle detection in aerial images. In: 2017 IEEE Winter Conference on Applications of Computer Vision (WACV), pp. 311–319. IEEE (2017)
14. Tang, T., Zhou, S., Deng, Z., Lei, L., Zou, H.: Fast multidirectional vehicle detection on aerial images using region based convolutional neural networks. In: 2017 IEEE International Geoscience and Remote Sensing Symposium (IGARSS), pp. 1844–1847. IEEE (2017)
15. Tang, T., Zhou, S., Deng, Z., Zou, H., Lei, L.: Vehicle detection in aerial images based on region convolutional neural networks and hard negative example mining. Sensors **17**(2), 336 (2017)

16. Zeiler, M.D., Fergus, R.: Visualizing and understanding convolutional networks. In: Fleet, D., Pajdla, T., Schiele, B., Tuytelaars, T. (eds.) ECCV 2014. LNCS, vol. 8689, pp. 818–833. Springer, Cham (2014). https://doi.org/10.1007/978-3-319-10590-1_53
17. Zhu, H., Chen, X., Dai, W., Fu, K., Ye, Q., Jiao, J.: Orientation robust object detection in aerial images using deep convolutional neural network. In: 2015 IEEE International Conference on Image Processing (ICIP), pp. 3735–3739. IEEE (2015)

Dual CNN Models for Unsupervised Monocular Depth Estimation

Vamshi Krishna Repala and Shiv Ram Dubey(✉)

Computer Vision Group, Indian Institute of Information Technology, Sri City, Chittoor, Andhra Pradesh, India
{vamshi.r14,srdubey}@iiits.in

Abstract. The unsupervised depth estimation is the recent trend by utilizing the binocular stereo images to get rid of depth map ground truth. In unsupervised depth computation, the disparity images are generated by training the CNN with an image reconstruction loss. In this paper, a dual CNN based model is presented for unsupervised depth estimation with 6 losses (DNM6) with individual CNN for each view to generate the corresponding disparity map. The proposed dual CNN model is also extended with 12 losses (DNM12) by utilizing the cross disparities. The presented DNM6 and DNM12 models are experimented over KITTI driving and Cityscapes urban database and compared with the recent state-of-the-art result of unsupervised depth estimation.

Keywords: Dual CNN · Depth estimation · Unsupervised · Deep learning

1 Introduction

The image based depth estimation of scene is a very active research area in the field of computer vision. The depth map from images can be estimated in various ways like structure from motion [14], multi-view stereo [19], monocular methods [17], single-image methods [18], etc. The deep learning and convolutional neural networks (CNNs) based methods perform outstanding in most of the problems of computer vision such as image classification [10], facial micro-expression recognition [15], face anti-spoofing [13], hyper-spectral image classification [16], image-to-image transformation [9], colon cancer nuclei classification [1], etc. Inspired from the success of deep learning, several researchers also tried to utilize the CNN for the depth prediction, specially in monocular imaging conditions. These approaches are classified mainly in three categories namely learning-based stereo [21,23], supervised single view depth estimation [3,11], and unsupervised depth estimation [4,6]. The stereo image pairs and ground truth disparity data are needed in order to train the learning-based stereo models. In real scenario, creating such data is very difficult. Moreover, these methods generally create the artificial data which can not represent the real challenges appearing in natural images and depth maps. The supervised single view depth estimation methods

B. Deka et al. (Eds.): PReMI 2019, LNCS 11941, pp. 209–217, 2019.
https://doi.org/10.1007/978-3-030-34869-4_23

also use ground truth depth to train the model. The main hurdle in supervised approaches is availability and creation of ground truth depth maps which is always not available in real applications.

The unsupervised depth estimation methods do not need any ground truth depth maps. Basically, they utilize the underlying theory of epipolar constraints [7]. Recently, Garg et al. used auto-encoder deep CNN to predict the inverse depth map (i.e. disparity) from left image [4]. They computed a warp image (i.e. reconstructed left image) from disparity map and right image. Finally, the error between original and reconstructed left image is used as the loss to train the whole setup in unsupervised manner. This approach is further improved by Godard et al. by incorporating the left-right consistency [6]. In left-right consistency, basically two depth maps (i.e. left and right) are generated using auto-encoder only from the left input image. The left input image is used with generated right depth map and the right image is used with generated left depth map to reconstruct the right and left images respectively. Zhou et al. [22] utilized the concepts of unsupervised image depth estimation proposed in [3] and [6] to tackle the monocular depth and camera motion estimation in unstructured video sequences in unsupervised learning framework. In one of the recent work, the 3D loss such as photometric quality of frame reconstructions is combined with 2D loss such as pixel-wise or gradient-based loss for learning the depth and ego-motion from monocular video in unsupervised manner [12].

While the unsupervised based methods have gained the attention in recent times, there is still need of discovering better suited unsupervised networks and loss functions. Through this paper, we propose a dual CNN based model for unsupervised monocular image depth estimation by utilizing the 6 losses (DNM6). We also extend the dual CNN model with 12 losses and generate DNM12 architecture to improve the quality of depth maps. The appearance matching loss, disparity smoothness loss and left-right consistency loss are used in this paper. The rest of the paper is structured by presenting the proposed dual CNN models DNM6 and DNM12 in Sect. 2, the experimental results and analysis in Sect. 3, and the concluding remarks in Sect. 4.

2 Proposed Methodology

2.1 Dual Network Model with 6 Losses (DNM6)

The proposed idea of dual network model (DNM) using CNN is illustrated in Fig. 1. This model is based on the 6 losses, thus referred as the DNM6 model. The DNM6 model has two CNN one for each left and right images of stereo pair. During training, the left image I^l and right image I^r are considered as the inputs to the left CNN named as CNN-L and right CNN named as CNN-R respectively. The $I_{i,j}$ refers to the $(i,j)^{th}$ co-ordinate of image I. It is assumed that both I^l and I^r images are captured in similar settings. Both CNN's are based on the auto-encoder algorithm and combined these two networks named as dual network. The CNN architecture (in both CNNs) is taken from the Godard et al. [6]. The CNN-L predicts the left disparity map d^l, whereas the CNN-R

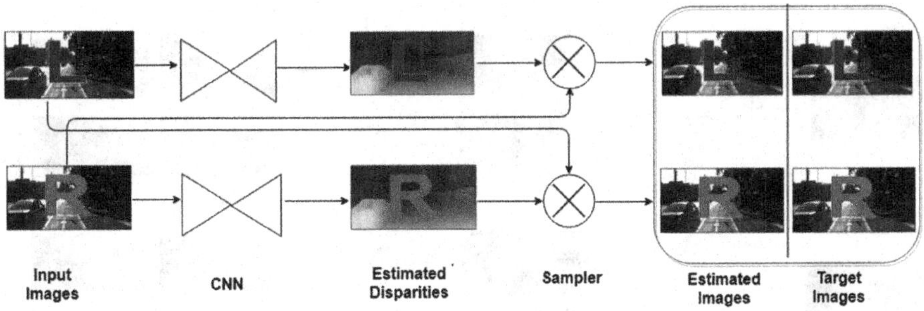

Fig. 1. Pictorial representation of proposed dual network model with 6 losses (DNM6)

predicts the right disparity map d^r. The $d_{i,j}$ refers to disparity value at $(i,j)^{th}$ co-ordinate of disparity map d. In order to reconstruct the left and right image from left and right disparity maps (d^l and d^r), the bilinear sampling from the Spatial Transform Networks [8] is used in this paper. The similar approach is also followed in [6] for reconstruction from disparity map. The left image is reconstructed from the left disparity map d^l and input right image I^r, whereas the right image is reconstructed from the right disparity map d^r and input left image I^l as shown in the Fig. 1. The reconstructed left and right images are referred as $\hat{I}^l$ and $\hat{I}^r$ respectively throughout the paper. We also used the loss functions (C) such as appearance matching loss (C_{ap}), disparity smoothness loss (C_{ds}) and left-right consistency loss (C_{lr}) similar to [6] but in dual network framework. The loss functions are defined below.

Appearance Matching loss: To enforce the appearance of estimated images must be similar to the input image, a combination of L1 norm and Structural Similarity Index Metric (SSIM) [20] loss term is used for both left and right images, defined as [6],

$$C_{ap}^{\beta} = \frac{1}{N}\sum_{i,j} \alpha \frac{1 - SSIM(I_{ij}^{\beta}, \hat{I}_{ij}^{\beta})}{2} + (1-\alpha) \parallel I_{ij}^{\beta} - \hat{I}_{ij}^{\beta} \parallel \tag{1}$$

where $\beta \in \{l, r\}$, C_{ap}^{l} refers appearance matching loss between estimated left image and input left image and C_{ap}^{r} refers appearance matching loss between estimated right image and input right image and α represents the weight between SSIM and L1 norm.

Disparity Smoothness Loss: The image gradient based disparity smoothness loss is computed from both disparity maps to ensure the estimated disparity map should be smooth. Similar to [6], the disparity smoothness loss is given as,

$$C_{ds}^{\beta} = \frac{1}{N}\sum_{i,j} |\partial_x d_{ij}^{\beta}| e^{-||\partial_x I_{ij}^{\beta}||} + |\partial_y d_{ij}^{\beta}| e^{-||\partial_y I_{ij}^{\beta}||} \tag{2}$$

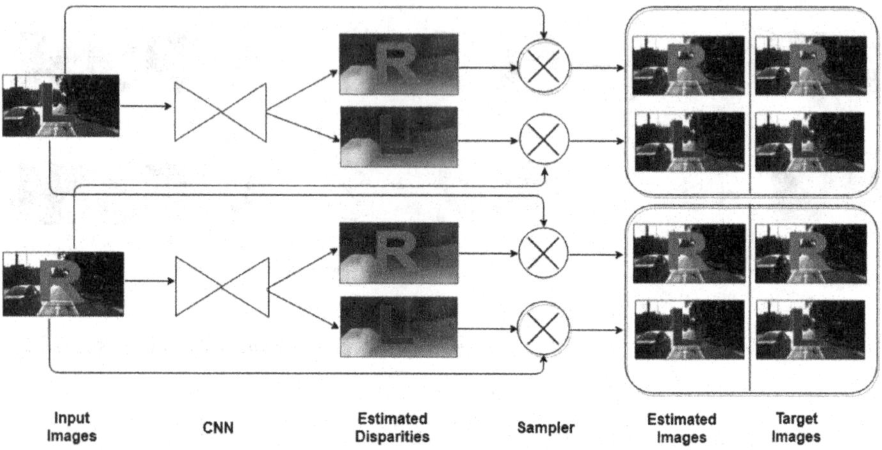

Fig. 2. Pictorial representation of our dual network model with 12 losses (DNM12)

where $\beta \in \{l, r\}$, C^l_{ds} refers the disparity smoothness loss of left disparity map d^l estimated by CNN-L, C^r_{ds} refers the disparity smoothness loss of right disparity map d^r estimated by CNN-R and ∂ is the partial derivative.

Left Right Consistency Loss: To maintain the estimated left disparity map d^l and right disparity map d^r to be consistent, the L1 term penalties on estimated disparities similar to [6] are computed between d^l and d^r as follows,

$$C_{lr} = \frac{1}{N}\sum_{i,j} |d^l_{ij} - d^r_{ij+d^l_{ij}}| \quad \text{and} \quad C_{rl} = \frac{1}{N}\sum_{i,j} |d^r_{ij} - d^l_{ij+d^r_{ij}}| \tag{3}$$

where C_{lr} and C_{rl} refer the left to right and right to left consistency losses respectively.

Similar to Godard et al. [6], four output scales s in both left and right CNNs are used in this paper in order to make the loss functions more robust. The combined cost function C_s at scale s including all above losses i.e. appearance matching losses C^l_{ap} and C^r_{ap}, disparity smoothness losses C^l_{ds} and C^r_{ds} and left-right consistency losses C_{lr} and C_{rl} is given as $C_s = \alpha_{ap}(C^l_{ap} + C^r_{ap}) + \alpha_{ds}(C^l_{ds} + C^r_{ds}) + \alpha_{lr}(C_{lr} + C_{rl})$. The final Cost/Loss function for proposed DNM6 model is computed as $C = \sum_{s=1}^{4} C_s$ at different output scales from s = 1 to 4 similar to [6]. At testing time, a single left image, I^l is needed as the input to the left CNN (i.e., CNN-L) and it predicts the disparity map d^l from the trained network. Note that, the right CNN with input I^r can also be used to predict the disparity map d^r. Once disparity map d (i.e. d^l or d^r) is computed, it can be converted into depth map (D) as $D = \frac{f \times B}{d}$, where f represents the focal length and B is the baseline between stereo cameras.

2.2 Dual Network Model with 12 Losses (DNM12)

In our previous DNM6 model, disparity maps are estimated from each network individually, whereas in this DNM12 model, the left-right cross disparity mapping is also proposed as depicted in Fig. 2. The left and right CNN networks of DNM6 are extended to generate two output disparities (i.e. left and right) from each CNN. Similar to Godard et al. [6], it generates both left and right disparity maps from a single image. During training, the left image I^l and right image I^r of stereo pair are provided as inputs to the left CNN (CNN-L) and right CNN (CNN-R) respectively. In DNM12 architecture, both the CNN's predict the left and right disparities independently as illustrated in Fig. 2. Here, we consider d^{l_l} and d^{l_r} as the left and right disparity maps respectively estimated by the left CNN-L and similarly d^{r_l} and d^{r_r} as the left and right disparity maps respectively estimated by the right CNN-R. As shown in the Fig. 2, four bilinear samplers are used for reconstructing the two output left images $\hat{I}^{l_l}$ and $\hat{I}^{r_l}$ corresponding to left input image and two output right images $\hat{I}^{l_r}$ and $\hat{I}^{r_r}$ corresponding to right input image. The $\hat{I}^{l_l}$ uses d^{l_l} and I^r, $\hat{I}^{l_r}$ uses d^{l_r} and I^l, $\hat{I}^{r_l}$ uses d^{r_l} and I^r, and $\hat{I}^{r_r}$ uses d^{r_r} and I^l. In DNM12, four appearance matching losses, four disparity smoothness losses and four left-right consistency losses are considered.

The *Four Appearance Matching Losses* are defined as follows,

$$C_{ap}^{\beta_\gamma} = \frac{1}{N} \sum_{i,j} \alpha \frac{1 - SSIM(I_{ij}^{\gamma}, \hat{I}_{ij}^{\beta_\gamma})}{2} + (1-\alpha) \parallel I_{ij}^{\gamma} - \hat{I}_{ij}^{\beta_\gamma} \parallel \tag{4}$$

where $\beta \in \{l, r\}$, $\gamma \in \{l, r\}$, $C_{ap}^{l_l}$ and $C_{ap}^{l_r}$ are the appearance matching losses for left CNN-L and $C_{ap}^{r_l}$, $C_{ap}^{r_r}$ are the appearance matching losses for right CNN-R. The total appearance matching loss is given by $C_{ap} = (C_{ap}^{l_l} + C_{ap}^{l_r} + C_{ap}^{r_l} + C_{ap}^{r_r})$.

The *Four Disparity Smoothness Losses* are computed as follows,

$$C_{ds}^{\beta_\gamma} = \frac{1}{N} \sum_{i,j} |\partial_x d_{ij}^{\beta_\gamma}| e^{-\|\partial_x I_{ij}^{\beta}\|} + |\partial_y d_{ij}^{\beta_\gamma}| e^{-\|\partial_y I_{ij}^{\beta}\|} \tag{5}$$

where $\beta \in \{l, r\}$, $\gamma \in \{l, r\}$, $C_{ds}^{l_l}$, $C_{ds}^{l_r}$ are the disparity smoothness losses for left CNN-L and $C_{ds}^{r_l}$, $C_{ds}^{r_r}$ are the disparity smoothness losses for right CNN-R. The total disparity smoothness loss is computed as $C_{ds} = (C_{ds}^{l_l} + C_{ds}^{l_r} + C_{ds}^{r_l} + C_{ds}^{r_r})$.

The *Four Left-Right Consistency Losses* are calculated as follows,

$$C_{lr}^{l} = \frac{1}{N} \sum_{i,j} |d_{ij}^{l_l} - d_{ij+d_{ij}^{l_l}}^{l_r}|, \quad \text{and} \quad C_{rl}^{l} = \frac{1}{N} \sum_{i,j} |d_{ij}^{l_r} - d_{ij+d_{ij}^{l_r}}^{l_l}| \tag{6}$$

$$C_{lr}^{r} = \frac{1}{N} \sum_{i,j} |d_{ij}^{r_l} - d_{ij+d_{ij}^{r_l}}^{r_r}|, \quad \text{and} \quad C_{rl}^{r} = \frac{1}{N} \sum_{i,j} |d_{ij}^{r_r} - d_{ij+d_{ij}^{r_r}}^{r_l}| \tag{7}$$

where C_{lr}^{l}, C_{rl}^{l} are the left-right and right-left consistency losses for left CNN-L and C_{lr}^{r}, C_{rl}^{r} are the left-right and right-left consistency losses for right CNN-R. The total left-right consistency loss is calculated as $C_{lr} = (C_{lr}^{l} + C_{rl}^{l} + C_{lr}^{r} + C_{rl}^{r})$.

Similar to DNM6, the total Loss function in DNM12 is also defined as $C = \sum_{s=1}^{4} C_s$ at different output scales from s = 1 to 4, where C_s at a particular scale is computed by weighted sum of all losses as $C_s = \alpha_{ap} \times C_{ap} + \alpha_{ds} \times C_{ds} + \alpha_{lr} \times C_{lr}$. The same procedure as provided in previous DNM6 model is followed in DNM12 also for testing, a single image is taken as input to either CNN-L or CNN-R and it predicts the disparity map from the trained network which is converted into depth map.

Table 1. Experimental results by using proposed dual CNN based DNM6 and DNM12 models for unsupervised depth estimation over KITTI benchmark database. The training is done over KITTI training images and the evaluation is done over KITTI test images. In this table, pp denotes the post-processing. The best results without post-processing are highlighted in bold face.

	Lower is better					Higher is better		
Method	Abs Rel	Sq Rel	RMSE	RMSE log	d1-all	a1	a2	a3
Godard et al. [6] No LR	0.123	1.417	6.315	0.220	30.318	0.841	0.937	0.973
Godard et al. [6]	0.124	1.388	6.125	0.217	**30.272**	0.841	0.936	0.975
DNM6 Model	0.1223	1.4004	6.162	0.214	31.050	**0.848**	**0.941**	**0.976**
DNM12 Model	**0.1221**	**1.3058**	**6.069**	**0.213**	31.455	0.841	0.939	**0.976**
DNM6 Model PP	0.1157	1.2037	5.830	0.203	30.004	0.852	0.945	0.979
DNM12 Model PP	0.1157	1.1404	5.772	0.203	30.342	0.848	0.944	0.979

3 Experimental Results and Analysis

We have used the standard datasets such as KITTI and Cityscapes for the experiments. The KITTI database [5] consists of stereo pairs from different scenes. Similar to Godard's work [6], 29,000 stereo pairs are used for training and 200 high-quality images are used as the test cases along with its depth maps. The Cityscapes database [2] contains the stereo pairs captured for autonomous driving. Similar to Godard's work [6], we have used the 22,973 stereo pairs for training after cropping each image such that the 80% of the height is preserved and the car hoods are removed. Similar to [6], we have used the same 200 KITTI stereo images for testing over Cityscapes database.

The CNN architectures in our network are same as in Godard et al. [6]. The proposed DNM6 and DNM12 models are implemented in TensorFlow which contains 62 million trainable parameters. We have used following parameters, $\alpha = 0.85$, $\alpha_{ap} = 1$, $\alpha_{ds} = 0.1$, $\alpha_{lr} = 1.0$ and learning rate $\lambda = 10^{-4}$ for first 30 epochs and 0.5 x 10^{-4} for next 10 epochs and 0.25 x 10^{-4} for the last 10 epochs. The data augmentation is done on fly, similar to [6]. During test time, a post-processing is performed to reduce the effect of stereo dis-occlusions similar to [6].

In both DNM6 and DNM12 methods, the estimated disparity map $d(x)$ is further converted into depth map as $D(x) = \frac{fB}{d(x)}$, where f is the focal length and B is the baseline. The evaluation of both models are done with the estimated depth maps $D(x)$ and provided ground truth depth maps $G(x)$. The evaluation metrics are same as in [6] such as Absolute Relative difference (**Abs Rel**), Squared Relative difference (**Sq Rel**), Root Mean Square Error (**RMSE**), **RMSE log**, and **d1-all**. The lower values of these metrics represent the better performance. We also measured the *Accuracy metrics* (i.e., **a1**, **a2**, and **a3** similar to [6]) for which higher is better.

The results are reported in Table 1 over KITTI database and compared with very recent state-of-the-art unsupervised method proposed by Godard et al. [6] with and without left-right (LR) consistency. Note that the lower values of **Abs Rel**, **Sq Rel**, **RMSE**, **RMSE log**, and **d1-all** and the higher values of accuracies **a1**, **a2**, and **a3** represent the better performance. The performance of proposed DNM6 and DNM12 methods are also tested with a pre-procesing (**PP**) step to reduce the effect of stereo dis-occlusions [6]. The best results without PP are highlighted in bold face in Table 1. It can be easily observed that the proposed dual CNN based models i.e. both DNM6 and DNM12 perform better than Godard et al. [6] with and without left-right consistency. The **Abs Rel**, **Sq Rel**, **RMSE**, **RMSE log**, and **d1-all** values are generally lower and accuracies **a1**, **a2**, and **a3** are higher for the proposed DNM6 and DNM12 methods. It is also noticed that DNM12 completely outperforms the Godard et al. [6] in all terms except **d1-all**. The performance of DNM6 model is improved in terms of the **Abs Rel**, **RMSE**, **a1**, **a2**, and **a3** as compared to the Godard model. The DNM12 model exhibits the better performance as compared to the DNM6 model in all terms except accuracies. As for as accuracies are concerned, the DNM6 model is superior as compared to DNM12 model because generating right disparity from left image and left disparity from right image is not suited for pixel level thresholding. This is also seen that the performance of proposed models improved significantly with post-processing step over KITTI database.

Table 2. Experimental results by using proposed dual CNN based DNM6 and DNM12 models for unsupervised depth estimation over Cityscapes benchmark database. The training is done over Cityscapes training images and the evaluation is done over KITTI test images. In this table, pp denotes the post-processing. The best results without post-processing are highlighted in bold face.

	Lower is better					Higher is better		
Method	Abs Rel	Sq Rel	RMSE	RMSE log	d1-all	**a1**	**a2**	**a3**
Godard et al. [6]	0.699	10.060	14.445	0.542	94.757	0.053	0.326	0.862
DNM6 Model	0.2704	3.7637	9.186	0.326	64.215	0.649	0.864	0.941
DNM12 Model	**0.2661**	**3.6491**	**8.915**	**0.316**	**61.163**	**0.669**	**0.875**	**0.946**
DNM6 Model PP	0.2474	2.9781	8.406	0.300	63.780	0.663	0.881	0.954
DNM12 Model PP	0.2396	2.8945	8.178	0.289	58.733	0.687	0.889	0.959

The results comparison of proposed models with Godard et al. [6] over Cityscapes database is illustrated in Table 2. In this Table, the training is performed over Cityscapes database, whereas the test images are same as in KITTI database. It is noticed from this experiment that the proposed models are superior than Godard et al. [6] over Cityscapes database in all terms. Moreover, the DNM12 model performs better than DNM6 model. As for as both databases are concerned, the results of proposed models over KITTI database is better than the Cityscapes database. The possible reason can be the difference between the camera calibration between training and testing databases. The similar observations are also made by Godard et al. [6]. The post-processing step enhances the performance of proposed DNM6 and DNM12 models over Cityscapes database.

4 Conclusion

In this paper, the dual CNN based models DNM6 and DNM12 are presented for unsupervised monocular depth estimation. The dual network models used two different CNNs (CNN-L and CNN-R) for left and right images of training stereo pairs respectively. In DNM6 and DNM12, total 6 and 12 losses are used, respectively. The results are computed over benchmark KITTI and Cityscapes databases and compared with the recent left-right consistency based method. It is observed that the DNM12 outperforms the existing method left-right consistency method. It is also observed that the DNM12 model improves the performance over DNM6 model in most of the cases. The post-processing step further boosts the performance of proposed models.

Acknowledgement. This research is supported by Science and Engineering Research Board (SERB), Govt. of India through Project Sanction Number ECR/2017/000082.

References

1. Basha, S.S., Ghosh, S., Babu, K.K., Dubey, S.R., Pulabaigari, V., Mukherjee, S.: RCCNet: an efficient convolutional neural network for histological routine colon cancer nuclei classification. In: IEEE ICARCV, pp. 1222–1227. IEEE (2018)
2. Cordts, M., et al.: The cityscapes dataset for semantic urban scene understanding. In: IEEE CVPR, pp. 3213–3223 (2016)
3. Eigen, D., Puhrsch, C., Fergus, R.: Depth map prediction from a single image using a multi-scale deep network. In: NIPS, pp. 2366–2374 (2014)
4. Garg, R., BG, V.K., Carneiro, G., Reid, I.: Unsupervised CNN for single view depth estimation: geometry to the rescue. In: ECCV, pp. 740–756 (2016)
5. Geiger, A., Lenz, P., Urtasun, R.: Are we ready for autonomous driving? The KITTI vision benchmark suite. In: IEEE CVPR, pp. 3354–3361 (2012)
6. Godard, C., Mac Aodha, O., Brostow, G.J.: Unsupervised monocular depth estimation with left-right consistency. In: IEEE CVPR (2017)
7. Hartley, R., Zisserman, A.: Multiple View Geometry in Computer Vision. Cambridge University Press, Cambridge (2003)
8. Jaderberg, M., Simonyan, K., Zisserman, A., et al.: Spatial transformer networks. In: NIPS, pp. 2017–2025 (2015)

9. Kancharagunta, K.B., Dubey, S.R.: CSGAN: cyclic-synthesized generative adversarial networks for image-to-image transformation. arXiv preprint arXiv:1901.03554 (2019)
10. Krizhevsky, A., Sutskever, I., Hinton, G.E.: ImageNet classification with deep convolutional neural networks. In: NIPS, pp. 1097–1105 (2012)
11. Liu, F., Shen, C., Lin, G., Reid, I.: Learning depth from single monocular images using deep convolutional neural fields. IEEE TPAMI **38**(10), 2024–2039 (2016)
12. Mahjourian, R., Wicke, M., Angelova, A.: Unsupervised learning of depth and ego-motion from monocular video using 3D geometric constraints. arXiv:1802.05522 (2018)
13. Nagpal, C., Dubey, S.R.: A performance evaluation of convolutional neural networks for face anti spoofing. arXiv preprint arXiv:1805.04176 (2018)
14. Nistér, D.: Preemptive ransac for live structure and motion estimation. Mach. Vis. Appl. **16**(5), 321–329 (2005)
15. Reddy, S.P.T., Karri, S.T., Dubey, S.R., Mukherjee, S.: Spontaneous facial micro-expression recognition using 3D spatiotemporal convolutional neural networks. arXiv preprint arXiv:1904.01390 (2019)
16. Roy, S.K., Krishna, G., Dubey, S.R., Chaudhuri, B.B.: HybridSN: exploring 3D–2D CNN feature hierarchy for hyperspectral image classification. arXiv preprint arXiv:1902.06701 (2019)
17. Saxena, A., Chung, S.H., Ng, A.Y.: Learning depth from single monocular images. In: NIPS, pp. 1161–1168 (2006)
18. Saxena, A., Chung, S.H., Ng, A.Y.: 3-D depth reconstruction from a single still image. IJCV **76**(1), 53–69 (2008)
19. Seitz, S.M., Curless, B., Diebel, J., Scharstein, D., Szeliski, R.: A comparison and evaluation of multi-view stereo reconstruction algorithms. IEEE CVPR **1**, 519–528 (2006)
20. Wang, Z., Bovik, A.C., Sheikh, H.R., Simoncelli, E.P.: Image quality assessment: from error visibility to structural similarity. IEEE TIP **13**(4), 600–612 (2004)
21. Zbontar, J., LeCun, Y.: Stereo matching by training a convolutional neural network to compare image patches. J. Mach. Learn. Res. **17**(1–32), 2 (2016)
22. Zhou, T., Brown, M., Snavely, N., Lowe, D.G.: Unsupervised learning of depth and ego-motion from video. In: IEEE CVPR (2017)
23. Zhou, T., Krahenbuhl, P., Aubry, M., Huang, Q., Efros, A.A.: Learning dense correspondence via 3D-guided cycle consistency. In: IEEE CVPR, pp. 117–126 (2016)

Deep CNN for Single Image Super Resolution Using Skip Connections

Sathisha Basavaraju(✉), Smriti Bahuguna, Sibaji Gaj, and Arijit Sur

Indian Institute of Technology Guwahati, Guwahati, India
b.sathisha@iitg.ac.in

Abstract. With the current progress in deep learning domain, quite a few deep learning based models have been developed to address the challenges of Single Image Super Resolution (SISR) task. Recent trend in SISR models is towards increasing the depth of the model to achieve a high accuracy through multiple contexts learned at various depth of the network. Skip connections between different layers of the deep network help in utilising multi-context features and also help to address the issue of vanishing gradients but at the cost of increased computational complexity. In this paper, a deep CNN with a minimum number of skip connections is devised to derive a High-Resolution (HR) image from a Low-Resolution (LR) image. A detailed experimental analysis is carried out on skip connection parameters for a given depth to optimise network parameters. The proposed model gives better results compared to existing methods for multiple scales on benchmark data-sets.

Keywords: Residual learning · Skip connections · Single image super resolution · Deep learning

1 Introduction

SISR is a well-known task that targets to predict HR image from a noisy and blurred LR image. Various methods based on interpolation [20], reconstruction [19], and learning algorithms [3,5,6,11,14,18] have been proposed to generate super resolution from single image. Higher resolution image is useful for a vast span of applications including diagnosis of medical imaging, security and surveillance, satellite imaging and forensics, etc. Image resolution task can be classified into various categories like temporal resolution, spectral resolution, spatial resolution etc [3,19]. In this paper, the focus is given on the spatial resolution of an image which refers to the number of independent pixels per unit area or the pixel density of the image. SISR is an complex inverse operation as there exists a numerous number of HR image solutions for a given low resolution image. Further, objects within images have various angles of appearance, scales, and aspect-ratios. Hence, multi-context features extracted from deep networks would provide better clues for predicting HR information. In spite of this,

B. Deka et al. (Eds.): PReMI 2019, LNCS 11941, pp. 218–225, 2019.
https://doi.org/10.1007/978-3-030-34869-4_24

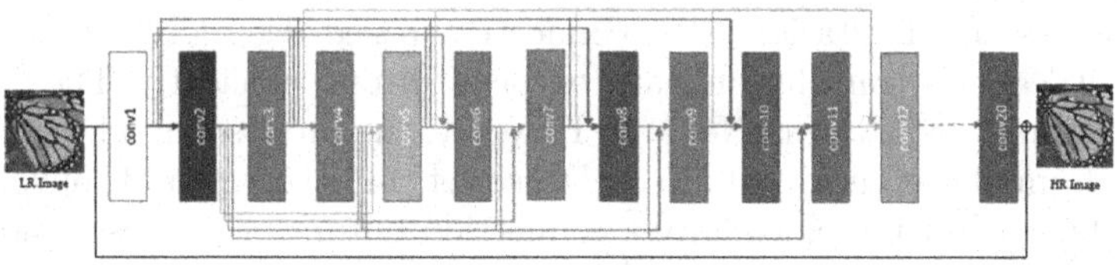

Fig. 1. Basic architecture of proposed model. The span s is set to 5, each convolutional layer has 64 kernels (growth-rate K), with kernel size 3×3 and depth of the network is set to 20.

multi-context features are not considered for reconstruction in recent deep learning techniques including VDSR [6], LapSRN [8], and EDSR [10]. To solve these challenges, a new SISR model is devised. The proposed model ensures the complete use of multi-context features by enabling every layer to access features from previous layers through multiple skip connections.

Remaining part of the paper is arranged as follows. Recent works on SISR are presented in Sect. 2. In Sect. 3, the proposed SISR model is explained. Section 4 details about the experimental set-up and corresponding results. The paper is concluded in Sect. 5.

2 Related Work

In recent literature related to single image super resolution, Dong et al. [3] attempted first to develop a deep learning model to map interpolated LR images to their corresponding HR images. Inspired by this baseline model, residual learning based deep networks are introduced in VDSR [6] and IRCNN [17] by piling a many convolutional layers. For the first time, a very deep network with recursive learning is proposed for parameter sharing in DRCN [7]. Recursive blocks and memory blocks are introduced for deeper networks in DRRN [12] and Memnet [13] respectively. Dense blocks proposed in DenseNet [4] have direct links between its any pair of layers. Using these local links, every layer access features from all the previous layers which belong to the same block. Inspired by the improved performance of the densely connected network of [4] in object recognition task, SRDenseNet [14] achieved better performance in super-resolving scale-4 images. The above mentioned deep learning based SISR techniques have attained significant success over conventional techniques. However, most of these techniques not considered the multi-context information learned at various depth of deep network. In other words, all of these methods do not allow direct access to features learned from all its previous layers but to immediate predecessor layer's features alone. Though SRDenseNet [14] taken care of multi-context information, it is limited to local blocks. In other words, the network is divided into multiple blocks. Each layer within the block is provided with direct links to all its predecessor layers. However, the model does not have direct connections between intermediate layers of any two dense blocks.

To address this limitation, skip connection based deep CNN is devised to utilise multi-context convolution features for SISR task efficiently. The proposed model is inspired by SRDenseNet [14]. However, the proposed model differs significantly from SRDenseNet [14]. In SRDenseNet [14], every pair of convolutional layers within each dense block have a connection but in the proposed model few connections are considered. Also, SRDenseNet [14] limit the skip connections to within dense blocks. This constraint limits each layer to utilise the features only from the block it belongs to. The proposed architecture does not divide the network into such discrete blocks, instead it uses overlapping blocks that provides continuous inclusion and smooth transition of features across different convolutional layers. The detailed description of the proposed method is given in the following section.

3 Proposed Model

Deep CNN based SISR models aim to predict each pixel's high frequency details by means of neighbouring pixels' details i.e pixel's context. The pixel's context can be represented as $d \times k \times k$, where depth of the network is represented as d, and the height and width of the kernel is represented as k [6]. Most of the existing models allow every convolutional layers to use the single context provided by its previous layer to reconstruct the HR information. DenseNet proposed in [4] allows direct links among every pair of layers within the same dense block. This enables every layer to utilise multi-context features learned from its previous layers within the same dense block. However, it does not have direct connections between intermediate layers of any two dense blocks. Hence, multiple-context is limited to layers within dense block. In order to utilise multi-context information to perform SISR task, a novel architecture is proposed as shown in Fig. 1. Skip connections from previous layers are taken in alternating fashion to avoid model complexity. However, it is still being able to include features within a long range in the deep network. Any layer in the network can have the maximum number of alternative skip connections from its previous layers is defined as span. Figure 1 shows the architecture with the span of 5 skip connections. A group of all the layers present within the span of a particular layer can be treated as its block. Since the architecture is not divided into discrete blocks, a layer may present in more than one block. Therefore, the proposed architecture can be treated as a set of overlapping blocks. Feature maps for a given block are concatenated and fed as input to the following convolutional layer. If the span is kept as s, the output of ith convolutional layer using a ReLu activation function is expressed as shown in Eq. 1.

$$X_i = max(0, w_i * [X_{i-(2s-1)}, X_{i-(2s-3)}, ..., X_{i-1}] + b_i) \tag{1}$$

where $[X_{i-(2s-1)}, X_{i-(2s-3)}, ..., X_{i-1}]$ represents the concatenation of feature maps from alternating previous layers. Every layer in the proposed model, except for the last one, produces K feature maps, which is termed as the growth rate of the model. The growth rate helps in the flow of information between consecutive layers in a regularised manner. The variation in the results for different values of K is explained in Sect. 4. Further, for a span of s skip connections, a convolutional layer has a maximum of $s * K$ input channels to learn from. With this architecture, the proposed model uses multi-context features to construct an HR image for a LR image.

3.1 Loss Function and Evaluation Metric

Most of the image reconstruction models employed the mean-squared error as the loss function due to its convex and differentiable nature. Hence, the same is applied to the proposed model. The equation Eq. 2 shows the loss function for the proposed model.

$$l_2 = \sum_j ||HR\_Pred_j - HR\_Org_j||_2^2 \tag{2}$$

where HR_Pred_j and HR_Org_j represents the predicted and ground truth high resolution images of the jth sample. Most of the existing image reconstruction models use Peak Signal-to-Noise Ratio (PSNR) and Structural Similarity Index (SSIM) as assessment metrics. Therefore, the accuracy of the proposed model is also assessed by means of PSNR and SSIM.

4 Experiments and Results

4.1 Datasets

The proposed model is learned on two datasets: (1) dataset containing 91 images proposed in [15], (2) 200 images from the training set of Berkeley Segmentation Dataset [1]. The learned model is tested on standard datasets including Set5 [2], Set14 [16], B100 [1] and Urban100 [5]. Since the size of the image dataset is small (291), it is augmented using the techniques followed in [6] to avoid over-fitting problem. In the first level, the training dataset is increased in three ways: (1) Scaling: down-scale an image with scales 0.6, 0.7, 0.8 and 0.9 times of the original image. (2) Rotation: rotate an image by 90°, 180°, and 270°. (3) Flipping: flip an image horizontally and vertically. These operations increased the dataset to 5820 images in total. In the second level, input images are generated for three scales, i.e., ×2, ×3 and ×4 to train the proposed model to generate multiple scales. For this, an image is taken from the augmented dataset, down-scaled it by a factor of 2, 3 & 4 separately and then interpolated it back to the size of the original image using bicubic interpolation to obtain the interpolated LR images. In the third level, non-overlapping patches of size 41×41 are extracted from the LR images with the corresponding patches from the HR images.

4.2 Training

During the training process, the batch size is set to 64. Adam optimize is used to reduce the network loss with 0.0001 initial learning rate. The momentum parameter is fixed to 0.9 with the decay parameter as 10^{-4}. Learning rate is decayed by a factor of ten when the loss does not reduce for ten training steps and is finished when the learning rate goes below 10^{-6}. All the models were trained for nearly 80 epochs which took around two weeks per model on NVIDIA Tesla K40 GPU.

4.3 Study of Skip Connections

To understand the parameters related to skip connections, several combinations of parameters are tested. The models with continuous skip connections from the previous layers are practically not feasible to train efficiently and also lead to increase in parameters with complex architecture. To address this issue, some of the connections are removed while still adopting the idea of features propagation through a combination of several past contexts in a deep network. Every alternative connection is skipped from the previous layers and limited the number of connections to 5 ($s = 5$). The central idea behind this configuration is that lesser number of alternative skip connections can be able to fuse features from the lowest levels while predicting the residual image at the last level. Table 1 shows results for different values of span, s, for scale 2 of Set14 dataset. As the span is increased to 6 and 7, the results get started to saturate. Further, the span size is reduced to 4 and 3 and found that the results become almost the same although it has taken a more time for them to converge. From Table 1 it is evident that for a network of 20 layers, a span of at least 5 previous connections may be able to include all possible multi-context features from initial layers' low-level features to last layers' high-level features. Experiments are carried out with different growth rates and span size to verify the behaviour of the model for feature size. Figure 2 shows a comparison between two growth rates, 32 and 64. From Fig. 2, it is clear that for less value of s, increasing the growth rate helps in converging the training faster. But for larger values of s, this change in growth rate does not affect the convergence rate or the PSNR values any better. Similarly, the effect of depth of the network is also verified by varying layer numbers from 20 to 32, keeping the span as 5 alternating skip connections. To keep the width of the network proportional to its depth, three different growth rates, 16, 32 and 64 are considered. The results are presented in Table 2, where

Table 1. PSNR value (dB) for scale 2 of Set14 on 20-layer model for different values of span s

Growth rate, K	Span = 3	Span = 4	Span = 5	Span = 6	Span = 7
32	33.007	32.944	33.011	32.990	33.066
64	33.090	33.122	33.152	33.153	33.179

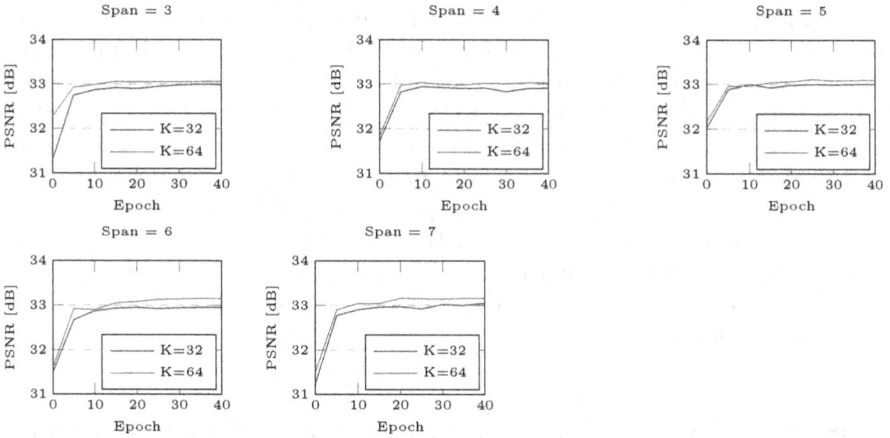

Fig. 2. Comparison of growth rate on Set14 dataset for scale 2

Table 2. PSNR value (dB) for scale 2 of Set14 using a span of 5 with different growth rates and model depths.

Growth rate, K	16	32	64
Proposed model with 20-Layers	32.916	33.011	33.152
Proposed model with 32-Layers	32.923	33.137	33.124

Table 3. Average PSNR/SSIM values of different SISR models for benchmark datasets for scale factors ×2, ×3 and ×4. Red text indicates the best and Blue indicates the second best

Dataset	Scale	Bicubic	SRCNN	VDSR	DRCN	LapSRN	SRDenseNet	Ours
Set5	×2	33.66/0.9299	36.66/0.9542	37.53/0.9587	37.63/0.9588	37.52/0.9591	–	37.678/0.9595
	×3	30.39/0.8682	32.75/0.9090	33.66/0.9213	33.82/0.9226	33.82/0.9227	–	33.798/0.9221
	×4	28.42/0.8104	30.48/0.8628	31.35/0.8838	31.53/0.8854	31.54/0.8855	32.02/0.8935	31.418/0.8832
Set14	×2	30.24/0.8688	32.42/0.9067	33.03/0.9124	33.04/0.9118	33.08/0.9130	–	33.179/0.9138
	×3	27.55/0.7742	29.28/0.8215	29.77/0.8314	29.76/0.8311	29.79/0.8320	–	29.857/0.8331
	×4	26.00/0.7027	27.49/0.7513	28.01/0.7674	28.02/0.7673	28.19/0.7720	28.50/0.7782	28.100/0.7690
B100	×2	29.56/0.8431	31.36/0.8879	31.90/0.8960	31.85/0.8942	31.80/0.8950	–	31.964/0.8976
	×3	27.21/0.7385	28.41/0.7863	28.82/0.7976	28.80/0.7963	28.82/0.7973	–	28.857/0.7997
	×4	25.96/0.6675	26.90/0.7101	27.29/0.7251	27.23/0.7233	27.32/0.7280	27.53/0.7337	27.305/0.7276
Urban100	×2	26.88/0.8403	29.50/0.8946	30.76/0.9140	30.75/0.9133	30.41/0.9101	–	30.857/0.9151
	×3	24.46/0.7349	26.24/0.7989	27.14/0.7989	27.15/0.8276	27.07/0.8272	–	27.204/0.8311
	×4	23.14/0.6577	24.52/0.7221	25.18/0.7524	25.14/0.7510	25.21/0.7553	26.05/0.7819	25.213/0.7565

it is visible that the proposed model with 32 layers performed similarly to the proposed model with 20 layers. Hence an increase in depth of the network may not always end with better results.

4.4 Performance Evaluation

For comparing results with other state-of-the-art models in SISR, proposed model with the following settings are considered; depth(D) = 20, span(s) =

7 and growth rate(K) = 64. The qualitative and quantitative analysis of the results from the proposed model compared with existing models is presented. The proposed model is compared with bicubic, and most recent state-of-the-art deep learning based models such as SRCNN [3], VDSR [6], DRCN [7], LapSRN [9] and SRDenseNet [14] on several datasets for scales 2, 3 and 4. The results indicate that the fusion of features from multiple contexts provides additional information in recreating HR image. The proposed model can be able to achieve satisfactory results and performs better than previous models in a few cases as shown in Table 3. Visual comparison with other models on scale 4 is shown in Fig. 3. From Fig. 3, it is visible that visually also our model performing better than previous models.

Fig. 3. Image "ppt3" from Set14 for scale 4

5 Conclusion

In this paper, deep CNN based SISR model is proposed to predict an high resolution image for a given low resolution image. The proposed model enables each of its convolutional layer to use multiple context information learned at various depths of the network using minimal skip connections. Various experiments are carried out to carefully imbibe multi-context features without making the model parameter-heavy. The experimental results showed that the proposed model has performed better than existing state-of-the-art methods in producing results for the benchmark datasets.

References

1. Arbelaez, P., Maire, M., Fowlkes, C., Malik, J.: Contour detection and hierarchical image segmentation. IEEE Trans. PAMI **33**(5), 898–916 (2011)
2. Bevilacqua, M., Roumy, A., Guillemot, C., Alberi-Morel, M.L.: Low-complexity single-image super-resolution based on nonnegative neighbor embedding (2012)

3. Dong, C., Loy, C.C., He, K., Tang, X.: Learning a deep convolutional network for image super-resolution. In: Fleet, D., Pajdla, T., Schiele, B., Tuytelaars, T. (eds.) ECCV 2014. LNCS, vol. 8692, pp. 184–199. Springer, Cham (2014). https://doi.org/10.1007/978-3-319-10593-2_13
4. Huang, G., Liu, Z., Van Der Maaten, L., Weinberger, K.Q.: Densely connected convolutional networks. In: 2017 IEEE Conference on CVPR, pp. 2261–2269. IEEE (2017)
5. Huang, J.B., Singh, A., Ahuja, N.: Single image super-resolution from transformed self-exemplars. In: Proceedings of the IEEE Conference on CVPR, pp. 5197–5206 (2015)
6. Kim, J., Kwon Lee, J., Mu Lee, K.: Accurate image super-resolution using very deep convolutional networks. In: Proceedings of the IEEE Conference on CVPR, pp. 1646–1654 (2016)
7. Kim, J., Kwon Lee, J., Mu Lee, K.: Deeply-recursive convolutional network for image super-resolution. In: Proceedings of the IEEE Conference on CVPR, pp. 1637–1645 (2016)
8. Lai, W.S., Huang, J.B., Ahuja, N., Yang, M.H.: Deep Laplacian pyramid networks for fast and accurate super-resolution. In: IEEE Conference on CVPR, vol. 2, p. 5 (2017)
9. Lai, W.S., Huang, J.B., Ahuja, N., Yang, M.H.: Fast and accurate image super-resolution with deep Laplacian pyramid networks. IIEEE Trans. PAMI (2018)
10. Lim, B., Son, S., Kim, H., et al.: Enhanced deep residual networks for single image super-resolution. In: The IEEE Conference on CVPR Workshops, vol. 1, p. 4 (2017)
11. Schulter, S., Leistner, C., Bischof, H.: Fast and accurate image upscaling with super-resolution forests. In: Proceedings of the IEEE Conference on CVPR, pp. 3791–3799 (2015)
12. Tai, Y., Yang, J., Liu, X.: Image super-resolution via deep recursive residual network. In: 2017 IEEE Conference on CVPR, pp. 2790–2798. IEEE (2017)
13. Tai, Y., Yang, J., Liu, X., Xu, C.: MemNet: a persistent memory network for image restoration. In: Proceedings of the IEEE Conference on CVPR, pp. 4539–4547 (2017)
14. Tong, T., Li, G., Liu, X., Gao, Q.: Image super-resolution using dense skip connections. In: ICCV, pp. 4809–4817. IEEE (2017)
15. Yang, J., Wright, J., Huang, T.S., Ma, Y.: Image super-resolution via sparse representation. IEEE Trans. Image Process. **19**(11), 2861–2873 (2010)
16. Zeyde, R., Elad, M., Protter, M.: On single image scale-up using sparse-representations. In: Boissonnat, J.-D., et al. (eds.) Curves and Surfaces 2010. LNCS, vol. 6920, pp. 711–730. Springer, Heidelberg (2012). https://doi.org/10.1007/978-3-642-27413-8_47
17. Zhang, K., Zuo, W., Gu, S., Zhang, L.: Learning deep CNN denoiser prior for image restoration. In: IEEE Conference on CVPR, vol. 2 (2017)
18. Zhang, K., Zuo, W., Zhang, L.: Learning a single convolutional super-resolution network for multiple degradations. In: IEEE Conference on CVPR, vol. 6 (2018)
19. Zhang, K., Gao, X., Tao, D., et al.: Single image super-resolution with non-local means and steering kernel regression. Image **11**, 12 (2012)
20. Zhang, L., Wu, X.: An edge-guided image interpolation algorithm via directional filtering and data fusion. IEEE Trans. Image Process. **15**(8), 2226–2238 (2006)

Detection of Image Manipulations Using Siamese Convolutional Neural Networks

Aniruddha Mazumdar(✉), Jaya Singh, Yosha Singh Tomar, and P. K. Bora

Department of Electronics and Electrical Engineering, Indian Institute of Technology Guwahati, Assam 781039, India
{m.aniruddha,prabin}@iitg.ac.in

Abstract. The processing history of an image can reveal the application of different types of image editing/manipulation operations applied to images and also can expose forgeries. This paper proposes a novel deep learning-based manipulation detection method using a siamese neural network. The advantage of the proposed method is that it can even detect manipulations not present in the training stage. The network is first trained to differentiate between different types of image editing operations. Once the network learns feature that can discriminate different image editing operations present in the training stage, the unknown manipulations are detected using the *one-shot classification* strategy. We show that the network can also check whether an image is downloaded from a social media platform or not. The experimental results validate the efficacy of the proposed method.

Keywords: Image forensics · CNN · Manipulation detection · Siamese network

1 Introduction

While creating an image forgery, the forger generally applies different types of image editing operations either to make the forged image look realistic or to remove the trace of the manipulations. An image uploaded to a social media platform is also modified using different post-processing and compression techniques. Therefore, it is an important problem to detect the presence of different image editing operations applied on images.

A number of methods are available in the literature for detecting different types of image manipulation operations: JPEG compression [2,13], median filtering [10], resampling [14], contrast enhancement [16] etc. Caldelli *et al.* [3] proposed a method to check whether an image is downloaded from social media platforms, e.g. Facebook, Instagram, or not. The method is based on the idea that when an image is uploaded to a social networking site, it undergoes JPEG compression for reducing its size. Therefore, the authors proposed to extract

B. Deka et al. (Eds.): PReMI 2019, LNCS 11941, pp. 226–233, 2019.
https://doi.org/10.1007/978-3-030-34869-4_25

features based on discrete cosine transform coefficients to detect the source of an image.

Although these methods are good at detecting different types of manipulations, methods from each of the categories work only under their respective assumptions about the traces of manipulations left by the forgery process. For example, the methods designed for detecting re-sampling operations cannot detect JPEG compression related traces. To handle this limitation, a few general purpose manipulation detection methods have been proposed, which aim at detecting many different manipulations in a single framework. The first general purpose forensics method was proposed by Qiu *et al.* [15], where different steganalysis features were used to detect different types of image processing operations. The method is based on the observation that different image editing operations destroy the natural statistics of the image pixels present in an authentic image in the same way steganography methods do while manipulating the pixels for embedding a message. Fan *et al.* [6] proposed another general-purpose forensics method for detecting different types of image editing operations. The authors proposed to create a Gaussian mixture model (GMM) of image patches corresponding to each editing operation. Then, the average log-likelihood of patches under the different GMMs corresponding to different classes are compared to decide the class of the patches.

Bayar and Stamm [1] proposed a deep learning-based general purpose forensics method for detecting different types of image manipulating operations. The image manipulation features are automatically learned from the training data by employing a convolutional neural network (CNN). The authors proposed a new convolutional layer, which suppresses the image content and enhances features important for detecting different editing operations. Although this method performs well, different manipulation operations have to be known exactly before training the network. In real forensics scenarios, there can be any image editing operations applied to an image. In this case, all the general purpose manipulation detection methods will fail as these methods assume that the operations are known before designing the algorithms.

This paper proposes a novel deep learning-based forensics method for detecting image manipulations. Instead of learning features to classify image patches to different manipulation classes as in [1], the proposed method learns the features which can whether they come from the same or different manipulation operations. For this, the proposed method employs a deep siamese CNN, which has twin CNNs accepting two image patches as the input and learns to classifies the patch pair as either identically processed (IP) or differently processed (DP). Once, the network learns to differentiate between different manipulations present in the, it is used to detect unknown manipulations using the *one-shot classification* strategy. Also we show the ability of the network in detecting the social media source of images.

2 Proposed Method

The universal forensics method proposed by Bayar and Stamm [1] shows that CNNs can automatically learn features important for detecting different image editing operations. However, there are some limitations of the method. For training the CNN, all the image editing operations should be known *a priori*. But, in a real forensics scenario, we may need to check whether an image is being processed by a particular type of editing operation which may not be present in the training stage. In this case, the exisitng methods [15], [6], [1] will fail. Therefore, there is a need to develop forensics method which can not only detect manipulations present in the training stage but also can generalize to unknown manipulations.

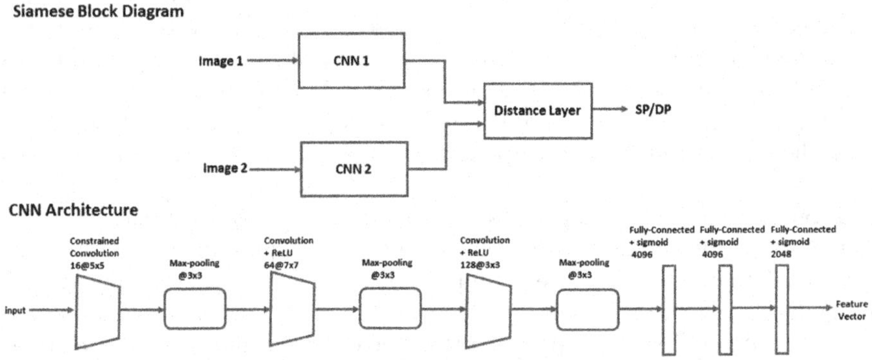

Fig. 1. The CNN architecture used in the proposed siamese network.

To overcome these limitations, we propose a distance metric learning method via a siamese neural network [11], which learns to check whether two image patches have undergone the same operation or two different operations. Once the network learns to discriminate between the manipulations present in the training stage, the network can be employed to check the presence of a manipulation not present in the training stage using the one-shot classification strategy. The reasons for the pair-wise classification image patches through distance metric learning technique is as follows: the distance metric learning technique enables the siamese network to learn more generic image manipulation related features than a simple CNN-based method [11]. This helps the network generalize to manipulations not present in the training stage.

Figure 1 shows the block diagram of the proposed framework. It has twin neural networks CNN1 and CNN2 sharing the same set of weights. It accepts two input images, which are independently processed by CNN1 and CNN2 and then a distance layer [11] computes a distance metric between the outputs of the twin networks. Because of the sharing of weights, CNN1 and CNN2 map two similar input images to very close points in the feature space. The proposed siamese CNN automatically learns the features that can check whether a pair of images has been similarly or differently manipulated.

2.1 Network Architecture

CNN. The first convolutional layer in each of CNN1 and CNN2 is a constrained convolutional layer [1]. The filters in the constrained convolutional layer are forced to learn a set of prediction error filters, which suppress image contents and produce prediction error. Each of CNN1 and CNN2 has the architecture shown in Fig. 1. It contains 3 convolutional layers, 2 max-pooling layers and 3 fully-connected layers. The first convolutional layer is the constrained convolutional layer [1] with 16 prediction error filters of size 5×5 and stride 1. This layer is followed by an unconstrained convolutional layer with 64 filters of size 7×7 with stride 2. The ReLU nonlinearity is applied element-wise to the output of this layer followed by the max-pooling layer with a kernel size 3×3 and stride 2. The output of this layer is fed to another unconstrained convolutional layer with 128 filters of size 3×3 and stride 1. The ReLU nonlinearity is applied element-wise to the output of this layer. It is followed by a max-pooling layer with a kernel size 3×3 and stride 2. This layer is followed by three fully-connected layers with 4096, 4096 and 2048 neurons respectively. The sigmoid non-linearity is used in each of these layers. The neurons in the fully-connected layers are dropped out [8] with a probability of 0.5 at each iteration of the training process. The output of the final fully-connected layer represents the features learned by the CNN.

Distance Layer. Given a pair of image patches $\mathbf{x}_1$ and $\mathbf{x}_2$ as input, CNN1 and CNN2 compute the feature vectors $\mathbf{f}_1$ and $\mathbf{f}_2$ respectively. A distance layer computes a distance metric between them, which is then fed to a single sigmoidal output neuron. This neuron computes the prediction of the input image patch pair as $p = \sigma(\sum_j \alpha_j \,|f_1(j) - f_2(j)|)$, where σ is the sigmoid non-linearity function and α_j is a learnable parameter representing the importance of each component of the feature vectors in the classification of the patch-pair.

2.2 Learning

The proposed siamese network is a binary classifier with label $y(\mathbf{x}_1, \mathbf{x}_2) = 1$ when both input image patches $\mathbf{x}_1$ and $\mathbf{x}_2$ come from the same manipulation class, and $y(\mathbf{x}_1, \mathbf{x}_2) = 0$ when $\mathbf{x}_1$ and $\mathbf{x}_2$ come from two different manipulation classes. The network is trained by minimising the average cross-entropy loss function C over a batch of pairs given by [12]

$$C = \frac{1}{M} \sum_{i=1}^{M} y(\mathbf{x}_1^i, \mathbf{x}_2^i) \log p(\mathbf{x}_1^i, \mathbf{x}_2^i) + (1 - y(\mathbf{x}_1^i, \mathbf{x}_2^i)) \log(1 - p(\mathbf{x}_1^i, \mathbf{x}_2^i)) \quad (1)$$

where M is the number of images in each batch. The parameters of the network are learnt in the training phase by minimising C using the stochastic gradient descent (SGD)-based backpropagation technique.

Table 1. Different manipulations considered in this paper

Manipulation	Detail
Gaussian blurring	Kernel size = 5×5 and standard deviation (σ) = 1.1
Median filtering	Kernel size = 5×5
Resampling	Scaling factor = 1.5 and bilinear interpolation
Noise addition	AWGN with standard deviation (σ) = 2
Gamma correction	Parameter (γ) = 1.5
JPEG compression	Quality Factor (QF) = 70

2.3 Manipulation Detection

Once the network learns to differentiate between different image editing operations, it is used to detect different manipulations either present or not present in the training stage using one-shot classification technique. In one-shot classification technique, there is at least one reference image from each class and the class of the test image is determined by computing pair-wise prediction of test image paired with the reference image from each of the classes. Let $\mathbf{I}$ be a test image and $\mathbf{I}_c, c = 1, 2, ..., C$ be the reference images, then the class of the test image, c^*, is computing as

$$c^* = \text{argmax}_c p_c \tag{2}$$

where, p_c is the pair-wise prediction for the pair corresponding to the reference image $\mathbf{I}_c$.

3 Experiments and Results

3.1 Dataset and Setup

A dataset was created using the unprocessed raw images taken from the Dresden Image Database [7]. The database contains more than 14, 000 images with resolutions of about 2000×3000 captured by 73 different digital cameras. We have saved 1566 images in the JPEG format with 100% quality factor (QF) and converted them into grayscale images by considering only the green channel of the images. We cropped image patches of size 150×150 from these images, resulting in 114, 000 unaltered image patches. Six different versions of these unaltered patches are created by editing them with the operations listed in Table 1.

The proposed system was implemented using the Keras [4] deep learning library on a Tesla K20c GPU with 5 GB of RAM. The Nadam optimiser [5] was used with the parameters set as: $learningrate(\eta) = 0.002$, $momentum(\mu) = 0.002$ and $decay = 0.005$ and $regularizationterm(\lambda) = 0.0001$. We have used the learning rate decay technique to converge to the minimum of C by reducing the fluctuations [17]. The training batch size was set to 16 images. We have used the batch normalisation technique [9] as it helps in achieving faster convergence and higher generalisation accuracy.

3.2 Results

In the first experiment, we have trained the network using the image patches coming from four different classes: original, Gaussian blurring, median filtering, and resampling. We randomly selected 40,000 patches from each class to create the training set. We sample 500,000 IP pairs of patches randomly where both image patches of a pair come from the same class (i.e. both patches come either from unaltered class or from the same manipulation class). Similarly, we sample 500,000 DP pairs randomly, where the two images of a pair come from two different classes. The validation set contains 10,000 IP pairs and 10,000 DP pairs and the test set contains 50,000 IP pairs and 50,000 DP pairs. The network training was stopped at 70,000 iterations as it started converging and saved the final parameters of the model for the future use. On this test set, the model achieves an accuracy of 99.26%. This experiment shows ability of the proposed siamese network in discriminate the different types of image editing operations.

In this experiment, we classify each of the four manipulation types individually, i.e. original, Gaussian blurring, median filtering and resampling using the one-shot classification technique described in Sect. 2.3. For this, we have created a test set by randomly sampling 10,000 images from each of the four manipulations. For comparison, we have implemented Bayar and Stamm's method and tested its performance on these test images. It should be noted that the size of image patches used in this experiment is 150×150. Table 2 shows the performance of the proposed siamese network along with the CNN-based method [1].

Table 2. Classification accuracies on different manipulation classes

Manipulation	Proposed method	Bayar and Stamm [1]
Original	**99.35**	99.01
Gaussian blurring	**99.51**	99.22
Median filtering	**99.64**	99.34
Resampling	**99.26**	98.88

Table 3. Classification accuracies on manipulations not present in the training stage

Manipulation	Accuracy (%)
AWGN ($\sigma = 2$)	95.09
Gamma correction ($\gamma = 1.5$)	92.17
JPEG compression ($QF = 70$)	93.6

The next experiment is carried out to see the ability of the proposed method in detecting manipulations not present at the training phase. The unknown

Table 4. Classification accuracies on images coming from different social media platforms

Platform	Accuracy (%)
Facebook	77.45
Instagram	77.85
Flickr	69.95

manipulations are detected using the one-shot classification technique. For this, we assume that we have at least one image from each of the manipulations that we want to detect. The test set contains 10,000 images coming from the following three manipulation: corrupting the images with additive white Gaussian noise (AWGN) and applying gamma correction, and JPEG compression with quality factor (QF) 70. As can be seen in Table 3, the network achieved classification accuracies of 95.09% and 92.17% and 93.6% on AWGN, gamma correction and JPEG classes respectively. From these results, it is evident that the network can generalize to images come from manipulation classes not used in the training stage. This is a huge advantage of the proposed method over the state-of-the-art method [15], [6], [1] as they can not be applied in this case.

An experiment was carried out to test the performance of the proposed method in checking if an image is downloaded from any of the following social media platforms: Facebook, Instagram, Flickr or not. We collected a set of 100 images downloaded from each platform. The images are divided into patches of size 150×150 to create a test set, and each patch is considered as an individual test image. The test set contains 3648 images in Facebook class, 3000 images in Instagram class, 4214 classes in Flickr class and 4000 images which are not downloaded from any platform. The images from each category are detected in the one-shot classification strategy. The detection accuracy are shown in Table 4. These results show the ability of the network in detecting the social media source of an image, without training specifically for the task. The above detection accuracies may be further improved by training the network with JPEG compressed images with different QF.

4 Conclusions and Future Work

In this paper, a novel image forensics method was proposed which can detect different types of image manipulations carried out on images. The proposed method employs a siamese CNN-based metric learning technique which takes a pair of image patches as input and learns to check whether they are identically or differently processed. Once, the network is trained the known/unknown manipulations are detected using the one-shot classification strategy. The experimental results show the ability of the method in detecting known/unknown manipulations. Also, the ability of the proposed method in detecting social media source of images are also established experimentally. The future work will involve further exploration of the universal nature of the proposed method.

References

1. Bayar, B., Stamm, M.C.: Constrained convolutional neural networks: a new approach towards general purpose image manipulation detection. IEEE Trans. Inf. Forensics Secur. **13**(11), 2691–2706 (2018)
2. Bianchi, T., Piva, A.: Image forgery localization via block-grained analysis of jpeg artifacts. IEEE Trans. Inf. Forensics Secur. **7**(3), 1003–1017 (2012)
3. Caldelli, R., Becarelli, R., Amerini, I.: Image origin classification based on social network provenance. IEEE Trans. Inf. Forensics Secur. **12**(6), 1299–1308 (2017)
4. Chollet, F., et al.: Keras (2015)
5. Dozat, T.: Incorporating Nesterov momentum into Adam (2016)
6. Fan, W., Wang, K., Cayre, F.: General-purpose image forensics using patch likelihood under image statistical models. In: 2015 IEEE International Workshop on Information Forensics and Security (WIFS), pp. 1–6. IEEE (2015)
7. Gloe, T., Böhme, R.: The 'Dresden image database' for benchmarking digital image forensics. In: Proceedings of the 2010 ACM Symposium on Applied Computing, pp. 1584–1590. ACM (2010)
8. Hinton, G.E., Srivastava, N., Krizhevsky, A., Sutskever, I., Salakhutdinov, R.R.: Improving neural networks by preventing co-adaptation of feature detectors. arXiv preprint arXiv:1207.0580 (2012)
9. Ioffe, S., Szegedy, C.: Batch normalization: accelerating deep network training by reducing internal covariate shift. arXiv preprint arXiv:1502.03167 (2015)
10. Kang, X., Stamm, M.C., Peng, A., Liu, K.R.: Robust median filtering forensics using an autoregressive model. IEEE Trans. Inf. Forensics Secur. **8**(9), 1456–1468 (2013)
11. Koch, G., Zemel, R., Salakhutdinov, R.: Siamese neural networks for one-shot image recognition. In: ICML Deep Learning Workshop, vol. 2 (2015)
12. LeCun, Y., Huang, F.J.: Loss functions for discriminative training of energy-based models. In: AIStats, vol. 6, p. 34 (2005)
13. Liu, Q.: An approach to detecting JPEG down-recompression and seam carving forgery under recompression anti-forensics. Pattern Recognit. **65**, 35–46 (2017)
14. Popescu, A.C., Farid, H.: Exposing digital forgeries by detecting traces of resampling. IEEE Trans. Signal Process. **53**(2), 758–767 (2005)
15. Qiu, X., Li, H., Luo, W., Huang, J.: A universal image forensic strategy based on steganalytic model. In: Proceedings of the 2nd ACM workshop on Information hiding and multimedia security, pp. 165–170. ACM (2014)
16. Stamm, M., Liu, K.R.: Blind forensics of contrast enhancement in digital images. In: 2008 15th IEEE International Conference on Image Processing, pp. 3112–3115. IEEE (2008)
17. Welling, M., Teh, Y.W.: Bayesian learning via stochastic gradient Langevin dynamics. In: Proceedings of the 28th International Conference on Machine Learning (ICML-11), pp. 681–688 (2011)

DaNSe: A Dilated Causal Convolutional Network Based Model for Load Forecasting

Kakuli Mishra[1(✉)], Srinka Basu[2], and Ujjwal Maulik[1]

[1] Department of Computer Science and Engineering, Jadavpur University, Jadavpur, Kolkata 700032, India
kakulimishra.94@gmail.com, umaulik.cse@jdvu.ac.in

[2] Department of Engineering and Technological Studies, Kalyani University, Kalyani 741235, India
srinka.basu@gmail.com

Abstract. Integration of renewable sources into energy grids has reduced carbon emission, but their intermittent nature is of major concern to the utilities. In order to provide an uninterrupted energy supply, a prior idea about the total possible electricity consumption of the consumers is a necessity. In this paper, we have introduced a deep learning based load forecasting model designed using dilated causal convolutional layers. The model can efficiently capture trends and multi-seasonality from historic load data. Proposed model gives encouraging results when tested on synthetic and real life time series datasets.

Keywords: Load forecasting · Smart grids · Deep learning · Dilated causal convolutional network · Renewable energy resource

1 Introduction

Load forecasting problem, commonly occurring in the context of smart grid systems deals with prediction of future energy demands of consumers based on their previous load consumption. There has been an extensive research in load forecasting problems, however prediction of future loads with high accuracy remains an open problem till date.

With the advancement of deep learning models, the complex patterns in sequential input data like in time series, can now be better identified over conventional machine learning models. Recently, deep neural network (DNN) based models have proved to be useful in load forecasting problems. In [14], authors use a DNN based model for load forecasting that was trained in two different ways - using a pre-trained restricted Boltzmann machine (RBM) and using the rectified linear unit (ReLu) without pre-training. To better capture the temporal dependencies from historical load data, some state-of-the-art deep learning models have been proposed - recurrent neural networks (RNN) [5,15,16], long short

B. Deka et al. (Eds.): PReMI 2019, LNCS 11941, pp. 234–241, 2019.
https://doi.org/10.1007/978-3-030-34869-4_26

term memory (LSTM) [9,10,13] and convolutional neural networks (CNN) [2,3]. In [10] LSTM based predictive models have been used for individual house level forecasts and aggregate level forecasts. Authors show that as the level of aggregation decreases, the accuracy drops but do not state a reason for the same. However later in [9], authors say that accuracy at individual house level can be improved if appliance readings of the house are included in training data. Rahman et al. in [13] has shown that in addition to short term dependencies, LSTM models can capture long term dependencies by obtaining long-term hour ahead forecasts in case of energy buildings. However the RNN and LSTM based models have long training time and suffers from overfitting due to vanishing gradient problem. The advantage of CNN over other two popular DNN based models is that CNN can be trained efficiently on a smaller training dataset without compromising the performance and the over fitting issue. Though the sliding filters in CNN helps to identify the patterns from historic load data for future load prediction, to access a broader range of history to capture the trend and seasonality, Borovykh et al. in [3] used dilated convolutional neural network (DCNN) for the first time in load forecasting problem. Inspired by Wavenet architecture [11], the DCNN based model in [3] called Augmented Wavenet, has a deep stack of dilated convolution layers which comprehends from a wide range of historic data when forecasting the future values.

In this paper we study the problem of load forecasting at the building level using deep CNNs where the forecasted values can follow the trend and seasonality present in historic data. Motivated by the Augmented Wavenet model [3], which can learn from the broader historic data, the proposed model - *D*il*a*ted Co*n*volutional Den*se* Network (DaNSe), is designed using multiple dilated causal convolutional layers with residuals and parameterized skip connections and multiple fully connected layers in the output. Dilation operation extensively captures the historic data for output predictions. The residuals and parameterized skip connections in each layer of the proposed model, speeds up convergence and train the deeper layers without over fitting. Reportedly, the only work that is similar to the proposed model is the SeriesNet architecture [12] that is an enhancement of the Augmented Wavenet model and use parameterized skip connections from each dilated convolutional layer to output layer. As compared to the SeriesNet architecture [12], the proposed model can better capture the non-linear trend and seasonality in the time series data resulting in improved accuracy. Experiments on synthetic and real life time series datasets show the improvement of proposed model over the existing SeriesNet model. This paper has been arranged as follows. Section 2 explains the state of the art techniques followed by architectural details of proposed model - DaNSe. Section 3 reports the experiment and results followed by conclusions and future work in Sect. 4.

2 Methodology

The proposed deep learning model *D*il*a*ted Co*n*volutional Den*se* Network (DaNSe), is designed using stacked dilated causal layers with residual connection and SeLU activation followed by fully-connected layers with ReLu activation. The key components of the DaNSe model is discussed below.

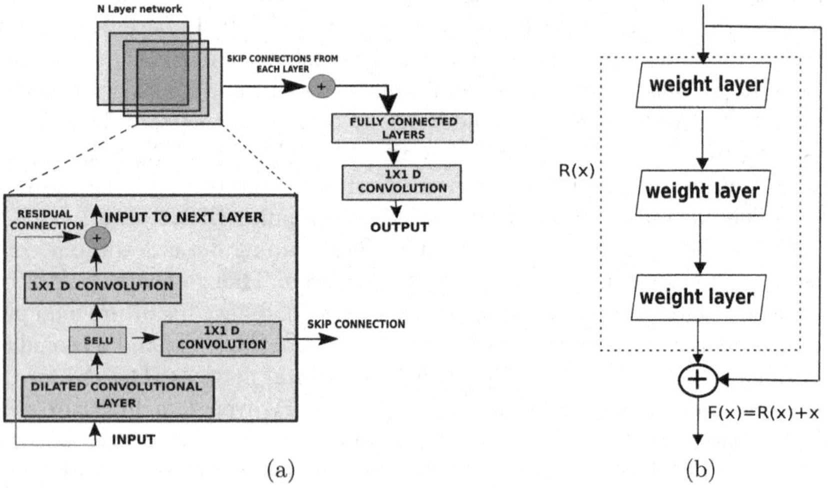

Fig. 1. (a) DaNSe architecture (b) A resnet block

- **Dilated Convolutional Neural Networks (DCNN):** The dilated convolution operator has been referred in the past as "convolution with dilated filter". Dilated filter is an up-sampling of convolution filter by injecting predefined gaps between the filter weights. The term causal with dilated networks intends to maintain the ordering in time series data [11]. Dilated convolution of two functions $f()$ and $g()$ in one-dimensional space, is represented as:

$$(f * g)(t) = \sum_{t=-\infty}^{\infty} f(t)g(t - lx) \tag{1}$$

  where the multiplier l is said to be the l-dilated convolution.
- **Residual Connections:** As the number of layers are increased in deep models, a degradation in accuracy signifies that the shallower counterpart of network is learning well but not the deeper counterpart. In order to construct the deeper counterpart of a shallower network, an idea of skip connections or residual connections between the layers has been proposed in [7]. If $F(x)$ is the underlying mapping of the model for input x, the stacked non-linear layers is used to fit another mapping: $R(x) = F(x) - x$, which is easier to optimize. Hence the original mapping is: $F(x) = R(x) + x$. Figure 1b shows a residual block.
- **Activations:** The activation functions used in the model are:

- Rectified linear unit (ReLu): A linear activation that will output the input, when it is positive else the output is zero. If x is the input $relu(x) = max(0, x)$.
- Scaled Exponential Linear Unit (SeLU): SeLU pushes the neuron activations towards zero mean and unit variance [8], integrating a self normalizing property. Activation function SeLU is represented as:

$$\mathrm{selu(x)} = \gamma \begin{cases} \alpha(\mathrm{e}^x - 1) \text{ for x} \leq 0 \\ x \text{ for x} \geq 0 \end{cases} \tag{2}$$

  where α and γ are the fixed parameters derived from inputs with mean 0 and standard deviation 1.

- ***D*il*a*ted Co*n*volutional Den*se*Network (DaNSe):** Fig. 1a shows the model architecture. Model has stacked dilated causal layers with residual connection from input to output, the sum of which is input to next dilated causal layer. The output of SeLU activation of each layer is parameterized by 1 × 1 convolution, which is then summed up and fed to fully-connected layers followed by 1 × 1 convolution for obtaining the final output. Instead to passing the sum directly into ReLu activation as in SeriesNet [12], fully connected layers helps reducing the sparsity of ReLu activation and gives accurate predictions. 80% dropout has been used in the last two layers of the model to reduce the over fitting problem.
  Model weights are learned by minimizing mean absolute error (MAE) with L2 regularization that penalize large weights to avoid over fitting. Model uses adaptive momentum (Adam) optimization technique where the weight updates are given as:

$$\Delta w_t = -\eta\left(v_t/\sqrt{s_t + \epsilon}\right)g_t \tag{3}$$

  where Δw_t is the gradient for weight w_t, η is the learning rate, v_t is exponential average of gradients along w_t, s_t is exponential average of squares of gradients along w_t, g_t is the gradient along w_t at time t and ϵ being a constant.

3 Experiment and Results

3.1 Experimental Setup

The data sets used in the experiment and the metrics used for measuring the performances are discussed below. The reported metric values are averaged over 10 runs.

- **Data Description:** Model has been tested on three different datasets - CIF 2016 competition dataset [6], CER-IRISH SME dataset [1] and a Turkish electricity data [4]. CIF-2016 benchmark dataset comprises of 72 real and synthetic time series with monthly periods varying length between 23 to 108 months. Commission of Energy Regulation (CER) IRISH comprises of half hourly power readings of 311 SMEs of Ireland during the period of 2009-10. Turkish electricity load data has daily load (in MW) for a period of nine years from 2000 to 2008 [4] with dual seasonality- a weekly and an yearly.

- **Error Metrics:** The error metrics used in the comparative study are SMAPE and distribution of APE as discussed below.
  1. Symmetric mean absolute percentage error (SMAPE): SMAPE is represented as below. SMAPE values ranges from 0 to 1. Lower valued SMAPE indicates better matching between forecasts and actuals.

$$SMAPE = \frac{1}{N}\sum_{i=1}^{N}\left(\frac{1}{n}\sum_{t=1}^{n}\frac{\mid F_t - A_t \mid}{(\mid F_t \mid + \mid A_t \mid)/2}\right) \tag{4}$$

     where, A_t and F_t are actual and forecast values respectively. n represents the length of time series and N is the number of individual time series values being considered.
  2. Absolute percentage error (APE): In order to quantify the error distribution in individual time series data we introduced APE, described below.

$$APE = \frac{1}{n}\sum_{t=1}^{n}\frac{\mid F_t - A_t \mid}{(\mid F_t \mid + \mid A_t \mid)/2} \tag{5}$$

     where, A_t and F_t are the actual and forecast values respectively. n represents the length of time series data.
- **Hyper-parameter selection and model training:** Initial weights of the proposed model are chosen randomly from a truncated normal distribution with zero mean and 0.05 standard deviation. L2 regularization factor of 0.001 is used. Proposed model has seven dilated causal layers with varying filter sizes. Initial three layers with filter width of 2 is intended to capture short duration trend or seasonal patterns in time series data, however to capture the longer duration trend, seasonal patterns and cyclic periodicities, filter width 4, seemed adequate. We used three fully connected layers, each with 32 hidden units and ReLu activations. The proposed model has been trained for 3000 epochs.
- **Data Preprocessing:** CIF-2016 data required no pre-processing, minimum pre-processing has been carried out on other two datasets. SME dataset had missing values which we replaced by moving averages. For our study, we converted the half-hourly power readings of SMEs into hourly data. Train and test sets for the CIF-2016 data has been predefined. For SMEs, we kept the last 24 h as test set and for the Turkish electricity data, we kept last one year as the test set.

3.2 Results and Analysis

As shown in the Table 1, proposed model DaNSe significantly outperforms SeriesNet for SME and Turkish electricity data while the performance is comparable for

CIF-2016 dataset. In case of SME and Turkish electricity data, DaNSe efficiently learns multiple seasonalities in the data giving better results. The time required for achieving this performance gain in DaNSe is same as that of SeriesNet architecture. DaNSe has also shown improved performance over other classical single layered CNN and LSTM models. Reason for improved performance in DaNSe is because the fully connected layers helps reducing the sparsity of ReLu activation giving an improved performance.

Table 1. Comparison of DaNSe and SeriesNet architecture for different datasets

Training data	Average SMAPE			
	DaNSe	SeriesNet	CNN	LSTM
SME	0.47	0.53	1.38	1.03
CIF-2016	0.24	0.22	0.72	1.96
Turkish electricity data	0.13	0.46	2	1.99

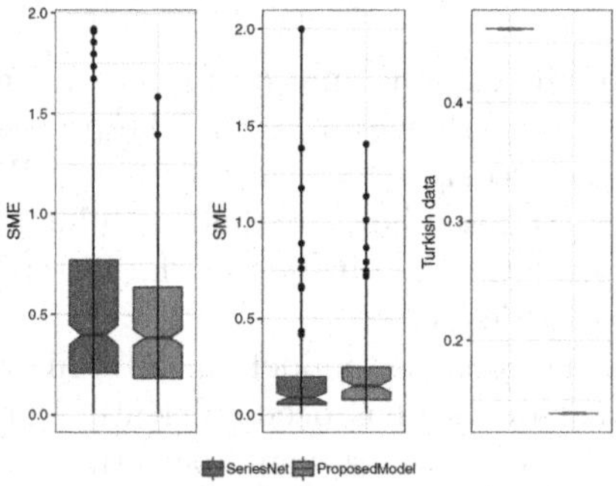

Fig. 2. Error distribution plot for SMEs, CIF-2016 and Turkish electricity data

We further analyze the error distribution for all the three datasets in terms of APE error, shown in Fig. 2. Outliers noticed in case of CIF-2016 data for both the models, is due to the time series that have less 6 months training data. In case of SMEs, we found high error rate for SeriesNet when there exists cyclic periodicity in the data. Turkish electricity has a single nine year dataset, DaNSe shows significant improvement as that of SeriesNet.

Further we analyze the performance of a few randomly selected time series from CIF-2016 and SME data. In Fig. 3, the actual versus predicted consumption is shown for 4 randomly chosen time series of CIF-2016 dataset. As shown in the Fig. 3, time series 13 and 19 has linear increasing trend while time series 6 and 30 has non-linear damping trend pattern. DaNSe architecture significantly outperformed in all cases particularly for non-linear damping pattern.

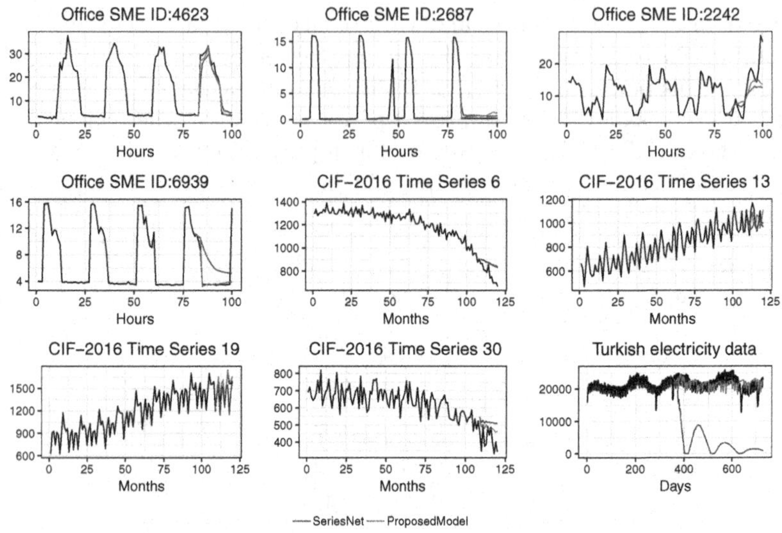

Fig. 3. Plot for comparison of DaNSe and SeriesNet in case of SME, CIF-2016 and Turkish electricity datasets.

The actual versus predicted consumption for 4 randomly chosen meter IDs of SME dataset has also been shown in Fig. 3. The proposed model is successful in capturing daily and weekly seasonality for meter ID 4623 and 6939 and 2687. Though for meter ID 2242, both methods exhibit degraded performance due to cyclic behavior in data while in rest of the meter IDs, proposed model significantly outperforms SeriesNet.

Figure 3 shows that the proposed model has 33% higher accuracy than the SeriesNet. The weekly seasonality has been efficiently learned by the proposed model as compared to that of yearly seasonal pattern.

4 Conclusions and Future Work

In this work, we propose a dilated causal convolutional network model with fully-connected layers for load forecasting. The proposed model named as DaNSe architecture achieves 33% higher accuracy over the existing SeriesNet architecture [12]. Plugging the fully connected layers with a combination of SeLU and ReLU non-linear activation has shown significant improvement as compared to that in SeriesNet architecture. The lower layers of the model, having small filter width learns the patterns with small periodicity in the data while the higher layers, with larger filter widths, learn the patterns with larger periodicity. The proposed model gives more accurate results in case of short as well as long seasonal patterns as compared to SeriesNet.

As future improvement to DaNSe model, we aim to explore its performance in case of varying forecast horizons, apply the distributed learning approaches and automate the hyper-parameter tuning. We also aim to explore the possibilities of incorporating gated memory layers capable to learn the cyclic periodicity present in the time series data.

References

1. Electricity smart metering technology trials findings report (2011). https://www.ucd.ie/t4cms/Electricity%20Smart%20Metering%20Technology%20Trials%20Findings%20Report.pdf
2. Amarasinghe, K., Marino, D.L., Manic, M.: Deep neural networks for energy load forecasting. In: 2017 IEEE 26th International Symposium on Industrial Electronics (ISIE), pp. 1483–1488. IEEE (2017)
3. Borovykh, A., Bohte, S., Oosterlee, C.W.: Conditional time series forecasting with convolutional neural networks. arXiv preprint arXiv:1703.04691 (2017)
4. De Livera, A.M., Hyndman, R.J., Snyder, R.D.: Forecasting time series with complex seasonal patterns using exponential smoothing. J. Am. Stat. Assoc. **106**(496), 1513–1527 (2011)
5. Fan, C., Wang, J., Gang, W., Li, S.: Assessment of deep recurrent neural network-based strategies for short-term building energy predictions. Appl. Energy **236**, 700–710 (2019)
6. Computational Intelligence in Forecasting: CIF 2016 (2016). http://irafm.osu.cz/cif/main.php
7. He, K., Zhang, X., Ren, S., Sun, J.: Deep residual learning for image recognition. In: Proceedings of the IEEE Conference on Computer Vision and Pattern Recognition, pp. 770–778 (2016)
8. Klambauer, G., Unterthiner, T., Mayr, A., Hochreiter, S.: Self-normalizing neural networks. In: Advances in Neural Information Processing Systems, pp. 971–980 (2017)
9. Kong, W., Dong, Z.Y., Hill, D.J., Luo, F., Xu, Y.: Short-term residential load forecasting based on resident behaviour learning. IEEE Trans. Power Syst. **33**(1), 1087–1088 (2018)
10. Kong, W., Dong, Z.Y., Jia, Y., Hill, D.J., Xu, Y., Zhang, Y.: Short-term residential load forecasting based on LSTM recurrent neural network. IEEE Trans. Smart Grid **10**, 841–851 (2017)
11. Oord, A.V.D., et al.: WaveNet: a generative model for raw audio. arXiv preprint arXiv:1609.03499 (2016)
12. Papadopoulos, K.: Seriesnet: a dilated causal convolutional neural network for forecasting. https://github.com/kristpapadopoulos/seriesnet (2018)
13. Rahman, A., Srikumar, V., Smith, A.D.: Predicting electricity consumption for commercial and residential buildings using deep recurrent neural networks. Appl. Energy **212**, 372–385 (2018)
14. Ryu, S., Noh, J., Kim, H.: Deep neural network based demand side short term load forecasting. Energies **10**(1), 3 (2016)
15. Shi, H., Xu, M., Li, R.: Deep learning for household load forecasting-a novel pooling deep RNN. IEEE Trans. Smart Grid **9**(5), 5271–5280 (2018)
16. Zhang, B., Wu, J.L., Chang, P.C.: A multiple time series-based recurrent neural network for short-term load forecasting. Soft. Comput. **22**(12), 4099–4112 (2018)

Multichannel CNN for Facial Expression Recognition

Prapti Trivedi(✉), Purva Mhasakar, Sujata, and Suman K. Mitra

Dhirubhai Ambani Institute of Information and Communication Technology, Gandhinagar, India
{201601020,201601082,201521003,suman_mitra}@daiict.ac.in

Abstract. In the past years there have been several attempts on the task of facial expression recognition. We have developed a new method based on the understanding of CNN and various image processing techniques. A multi-channel CNN architecture is proposed, which helps in performing improved facial expression recognition on frontal face images. For better feature extraction, fine tuning of images has been done by different preprocessing methods, namely Sobel edge detection, median filtering and Gaussian smoothing. Thereafter, the preprocessed images, have been fed in a novel manner in the proposed multi-channel CNN model. The model is evaluated on three challenging benchmark datasets - JAFFE, CK+ and Oulu-CASIA. The performance is comparable with various state-of-the-art approaches for facial expression recognition, which is evident from the results obtained.

Keywords: Facial expression recognition · Deep learning · convolutional neural network · Multi-channel architecture

1 Introduction

Expression is the most important mode of nonverbal communication between people. Human face is very expressive; we can express expressions such as happiness, anger, sadness, surprise, fear, disgust without the need of communicating through words. Analysis of facial expressions has been an attractive topic in the domain of computer vision. This is because it has a wide spectrum of potential applications such as behavioural analysis, psychology, human-computer interaction and so on. The process can be divided into three main tasks, i.e. face-detection, feature extraction and classification. For many years, the analysis has been done by using hand-crafted low-level descriptors, which are appearance based or geometric based [8]. These methods face challenges such as partial obstruction of facial regions, illumination variations, and head deflection. There have been advances in deep learning domain which solve these issues [8]. Deep learning methods such as CNN and RNN have been used for classification, feature extraction and classification. Researchers have been working towards improving the results of these algorithms. Yu et al. [21] fused the CNNs

B. Deka et al. (Eds.): PReMI 2019, LNCS 11941, pp. 242–249, 2019.
https://doi.org/10.1007/978-3-030-34869-4_27

by learning the set of weights of the network response. Zhao et al. advocated the use deep belief networks (DBNs) to automatically learn the features and the recognition of features was done using multilayer perceptron (MLP) [23]. Using two different types of CNN, Jung et al. [6] extracted the temporal appearance features and temporal geometry features from the image sequences. Song et al. [18] have used 3D-CNNs for 3D object detection task. To improve the performance of existing models, we have devised a new methodology to achieve improved results. In this paper, we employ CNN understanding to develop a novel architecture which leads to improvement in the accuracy of classification. Preprocessed images are fed into two channels and normal images are fed through the third channel. After each CNN channel has extracted its features, we concatenate the fully connected layers of these three CNN channels. By doing so, we are able to combine all the features extracted by different CNN pipelines on the given images and train them at the same time. This provides a significant boost to the performance since we have richer information about the images. We have applied the algorithm on widely used benchmark FER datasets such as Jaffe [12], CK+ [11] and Oulu-CASIA database [22].

2 Proposed Method

The proposed methodology has been divided into two subsections. The first section explains prepocessing techniques, while the second part elaborates the feature extraction and classification methods. Figure 1, provides an in depth understanding of the proposed approach, which will be discussed in the following subsections.

2.1 Preprocessing

Preprocessing plays an important role, since most of the datasets available have a lot of background data, which decreases the efficiency of recognising expressions. Moreover, focusing only on the face, and removing the unrelated information from the images leads to a better training process for the classifier. Hence we crop the frontal faces from the images, and enhance their quality of content by applying beneficial filter combinations.

We begin by cropping the images. For face detection, Haar-Cascade detection [3], is used. In this method, cascade of classifiers are applied on every image for a particular window size, and a window in which all the features get passed, is concluded to have a face region. After the images are cropped, image processing filters are applied, to remove noise and reduce the unnecessary information from the images. These are as follows:

A. Median filtering: In median filtering [13] for each pixel, first the neighbourhood pixels are located, the values of the pixels are sorted in ascending order and then the value of the pixel under consideration, is replaced by the computed median. This process is repeated for each pixel of the image.

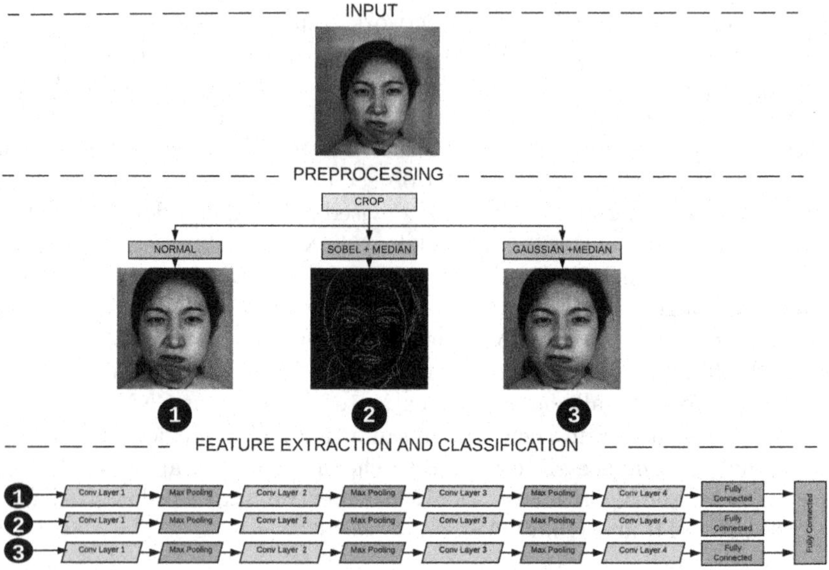

Fig. 1. The figure shows the design of the model of our proposed approach. The input images are prepocessed and fed into the multi-channel CNN model, as shown.

B. Gaussian smoothing filter: Gaussian smoothing filters [4] are also used in our proposed model for noise reduction. Unlike median filtering, Gaussian smoothing filtering is a linear method. For 2D images it has the isotropic (circularly symmetric) form:

$$G(\mathrm{x,y}) = \frac{1}{2\pi\sigma^2} e^{-\frac{x^2+y^2}{2\sigma^2}}$$

where σ is the degree of smoothing. This method is a weighted averaging of each pixels neighbourhood, and gives more weight to the central pixels than the farther neighbourhood pixels. On applying a Gaussian smoothing filter, we get an image with gentler smoothing and preserved edges. The standard deviation for Gaussian smoothing is set to 0.025, which is determined by method of experimentation.

C. Sobel Filter: Edges are regions in an image where there are sharp change in the brightness of the pixel. Sobel filters [7] are used to detect the nature of this change in intensity, and then conclude whether the region is an edge or not. Using this edge detection method, we convert the image to binary image, using 0.5 as the threshold value. The threshold value obtained by experimentation is chosen such that all the edges that are not stronger than threshold value are ignored. For each pixel, gradient approximations are calculated by convolving horizontal and vertical masks with the image and then the final value is the combined form of both G_x and G_y, according to the following formula:

$$|G| = \sqrt{G_x^2 + G_y^2} \quad A = \tan^{-1}(\mathrm{G}_y/G_x)$$

Here G_x and G_y are the horizontal and vertical derivatives respectively and the orientation of the gradient i.e. A is calculated as above.

Discrete approximations of the masks are used for computation, in both Gaussian smoothing and Sobel edge detection.

2.2 Feature Extraction and Classification

After prepocessing is complete, feature extraction and classification was carried out. CNN is used as a classifier and our convolutional layers have filter maps with increasing filter count, resulting in 32, 64 and 128 filter map sizes, respectively.

The main structure of our model is based on multi-channel CNN architecture as shown in Fig. 1. The normal cropped images have been given to the first channel. And in the second channel, on cropped, median filtered pictures, Sobel edge detection with a specific threshold value has been applied and fed. Median filtering and Gaussian smoothing has been applied and those images are fed into the third channel. The fully connected layers of these three channels are then concatenated. Thus, in this manner the features extracted by three channels are merged. Now, the information rich, multi-channel network is trained, using the ADAM optimiser.

3 Experimental Results and Discussion

In this section we discuss the effectiveness of the proposed approach and the reason for its efficiency. The method has been evaluated on three datasets, namely JAFFE [12], CK+ [11] and Oulu-CASIA dataset [22].

Japanese Female Facial Expression database (JAFFE), contains 213 grayscale images of 10 female Japanese subjects. The Extended Cohn Kanade (CK+) database has 593 image sequences, of 123 subjects. Both JAFFE and CK+ have seven expressionss, namely: anger, happiness, sadness, surprise, sadness, disgust and neutral. The third dataset, used was Oulu-CASIA database. The dataset is classified according to the illumination in which the video was taken for weak, strong and dark. Under the category of strong, for each of the 80 subjects, from a number of images of varying degree of a specific expressions, we chose the last seven images of peak expression, for a particular person. Also, the images could be classified into six basic expressions; all of the above mentioned expressions, except neutral.

In order to determine the number of optimum CNN channels to be used, the normal images of the three chosen datasets, were passed through a 1 channel, 2-channel, 3-channel and a 4-channel CNN model. For the simplicity of comparison, normal images, without any preprocessing are taken as inputs to the respective channels. As evident by the results shown in the graph, a contrast in accuracy was observed. The optimum number of channels were thus chosen to be three, as for more than three channels, overfitting reduced the accuracy and for less than three channels, the models learning efficiency and classification accuracy,

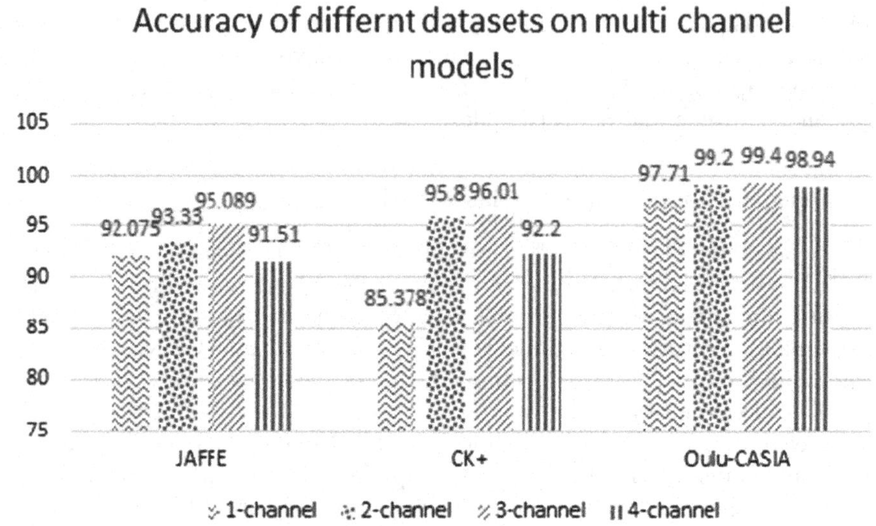

Fig. 2. The graphs are showing comparison of performance accuracy of models having different number of channels on the three chosen datasets.

did not improve much, as the model was not able to learn the features in a comprehensive manner (Fig. 2).

To improve the performance of the three channel model, the quality of the images were refined, by various preprocessing methods [5], as mentioned in the above section. According to our requirements, as we wanted better feature extraction, that would lead to an increased accuracy of Facial Expression Recognition, we explored different ways of information reduction i.e. removing uncorrelated information and keeping only necessary information and noise reduction in images.

For noise reduction, different types of filters were taken into consideration. We used median filtering because it gave a better performance, when compared to mean filtering. Unlike mean which can have any value, median value is always from the values present among the pixel values of the neighbourhood, making the image more representative and less blurry. Hence in the second and third channel prior to applying any other filter, we applied median filtering.

After applying Median filter in the second channel, for edge detection, we experimented with three major methods: Sobel edge detection, Canny edge detection [2] and Prewitt [14] edge detection. The main aim of edge detection was information reduction, and thus we looked for the most clear and defined edges, which were observed in Sobel edge detection. Canny had too many edges and thus did not accomplish our goal of uncorrelated information reduction and Prewitt had too less edges, and as a result was not suitable for a better feature extraction.

In the third channel, in addition to median filtering, on applying a Gaussian smoothing filter, we got an image with gentler smoothing and preserved edges.

For example, black spots, extra strands of hair etc. were almost removed, and thus a better image that lead to an efficient expressions recognition was acquired (Fig. 3).

Table 1. Comparison of performance of our proposed approach and various state-of-the-art methods.

Method	JAFFE	CK+	Oulu-CASIA
Weighted mixture double channel CNN [20]	92.20	97.00	92.30
Fiducial Point detection and NN [16]	93.80	99.00	84.70
Aly et al. [1]	87.32	88.14	84.21
Rivera et al. [15]	88.75	91.51	85.18
Lopes et al. [10]	88.73	93.68	86.42
Ours	**95.53**	**97.27**	**100**

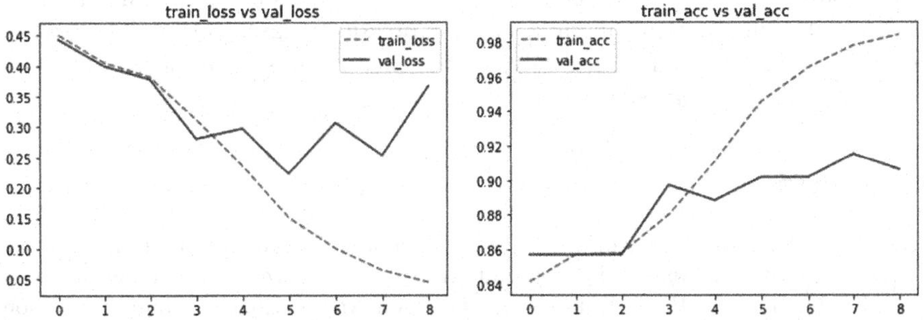

Fig. 3. As shown in the graphs above, for JAFFE dataset, the training and validation accuracy increases, and the training and validation loss decreases for increasing number of epochs.

In regard to various other methods of FER, from Table 1, it can clearly be seen that the proposed method performs better than the traditional hand-crafted methods of expressions recognition, due to the benefits of deep learning models [8]. In addition to this, in comparison to various deep learning models also, the proposed model performs fairly well. The multi-channel CNN model outperforms various deep learning single channel models, as shown in the table, due to simply its core concept [9,17,19].

The idea of multi-channel model is to connect the extracted features from different channels before the final output, so that the model can use richer information obtained from different channels. This is done by concatenating the fully connected layers of each channel and form one fully connected layer. This fully connected layer can now be used further, for classification. This process leads to a more superior feature extraction, which in turn provides a boost to the performance. It is more flexible as different types of input data can be merged, in an easier and simpler way. While training, the weights of each channel, get updated independently, and thus richer information is fed, while classification. And therefore, the proposed model is more robust to noise and errors, because of the multiple sources of independent information extraction. Fine tuning of images by the preprocessing stage, along with the innovative model suggested by us, gave the model immunity against noise, which ultimately lead to a better performance accuracy for FER.

4 Conclusion

In this paper, we have proposed an efficient method, for classification of facial expressions recognition. Various preprocessing steps like face cropping and Sobel edge detection help in focusing on the information content only. Noise reduction methods like median filtering and Gaussian smoothing improve the saturation, contrast and brightness of the image. Finally, lwhen the preprocessed images, were passed through our proposed model, consisting of a multi-channel CNN architecture, the features learned through the three channels in tandem, gave a boost to the feature learning and facial expressions classification. The testing accuracy, as quoted, is demonstrative of the effectiveness of proposed approach over many existing approaches.

References

1. Aly, S., Abbott, A.L., Torki, M.: A multi-modal feature fusion framework for kinect-based facial expression recognition using dual kernel discriminant analysis (DKDA). In: 2016 IEEE Winter Conference on Applications of Computer Vision (WACV), pp. 1–10. IEEE (2016)
2. Chen, X., Cheng, W.: Facial expression recognition based on edge detection. Int. J. Comput. Sci. Eng. Surv. **6**(2), 1 (2015)
3. Cuimei, L., Zhiliang, Q., Nan, J., Jianhua, W.: Human face detection algorithm via Haar cascade classifier combined with three additional classifiers. In: 2017 13th IEEE International Conference on Electronic Measurement & Instruments (ICEMI), pp. 483–487. IEEE (2017)
4. Deng, G., Cahill, L.: An adaptive Gaussian filter for noise reduction and edge detection. In: 1993 IEEE Conference Record Nuclear Science Symposium and Medical Imaging Conference, pp. 1615–1619. IEEE (1993)
5. Dharavath, K., Talukdar, F.A., Laskar, R.H.: Improving face recognition rate with image preprocessing. Indian J. Sci. Technol. **7**(8), 1170–1175 (2014)
6. Jung, H., Lee, S., Yim, J., Park, S., Kim, J.: Joint fine-tuning in deep neural networks for facial expression recognition. In: Proceedings of the IEEE International Conference on Computer Vision, pp. 2983–2991 (2015)
7. Kanopoulos, N., Vasanthavada, N., Baker, R.L.: Design of an image edge detection filter using the sobel operator. IEEE J. Solid-State Circuits **23**(2), 358–367 (1988)
8. Ko, B.: A brief review of facial emotion recognition based on visual information. Sensors **18**(2), 401 (2018)

9. Liu, S., Liu, Z.: Multi-channel CNN-based object detection for enhanced situation awareness. arXiv preprint arXiv:1712.00075 (2017)
10. Lopes, A.T., de Aguiar, E., De Souza, A.F., Oliveira-Santos, T.: Facial expression recognition with convolutional neural networks: coping with few data and the training sample order. Pattern Recognit. **61**, 610–628 (2017)
11. Lucey, P., Cohn, J.F., Kanade, T., Saragih, J., Ambadar, Z., Matthews, I.: The extended Cohn-Kanade dataset (CK+): a complete dataset for action unit and emotion-specified expression. In: 2010 IEEE Computer Society Conference on Computer Vision and Pattern Recognition-Workshops, pp. 94–101. IEEE (2010)
12. Lyons, M.J., Akamatsu, S., Kamachi, M., Gyoba, J., Budynek, J.: The Japanese female facial expression (JAFFE) database. In: Proceedings of Third International Conference on Automatic Face and Gesture Recognition, pp. 14–16 (1998)
13. Nagu, M., Shanker, N.V.: Image de-noising by using median filter and Weiner filter. Int. J. Innov. Res. Comput. Commun. Eng. **2**(9), 5641–5649 (2014)
14. Prewitt, J.M.: Object enhancement and extraction. Pict. Process. Psychopictorics **10**(1), 15–19 (1970)
15. Rivera, A.R., Castillo, J.R., Chae, O.O.: Local directional number pattern for face analysis: face and expression recognition. IEEE Trans. Image Process. **22**(5), 1740–1752 (2013)
16. Salmam, F.Z., Madani, A., Kissi, M.: Emotion recognition from facial expression based on fiducial points detection and using neural network. Int. J. Electr. Comput. Eng. **8**(1), 52 (2018)
17. Shi, H., Ushio, T., Endo, M., Yamagami, K., Horii, N.: A multichannel convolutional neural network for cross-language dialog state tracking. In: 2016 IEEE Spoken Language Technology Workshop (SLT), pp. 559–564. IEEE (2016)
18. Song, S., Xiao, J.: Deep sliding shapes for Amodal 3D object detection in RGB-D images. In: Proceedings of the IEEE Conference on Computer Vision and Pattern Recognition, pp. 808–816 (2016)
19. Sun, Y., Zhu, L., Wang, G., Zhao, F.: Multi-input convolutional neural network for flower grading. J. Electr. Comput. Eng. **2017** (2017)
20. Yang, B., Cao, J., Ni, R., Zhang, Y.: Facial expression recognition using weighted mixture deep neural network based on double-channel facial images. IEEE Access **6**, 4630–4640 (2018)
21. Yu, Z., Zhang, C.: Image based static facial expression recognition with multiple deep network learning. In: Proceedings of the 2015 ACM on International Conference on Multimodal Interaction, pp. 435–442. ACM (2015)
22. Zhao, G., Huang, X., Taini, M., Li, S.Z., PietikäInen, M.: Facial expression recognition from near-infrared videos. Image Vis. Comput. **29**(9), 607–619 (2011)
23. Zhao, X., Shi, X., Zhang, S.: Facial expression recognition via deep learning. IETE Tech. Rev. **32**(5), 347–355 (2015)

Data Driven Sensing for Action Recognition Using Deep Convolutional Neural Networks

Ronak Gupta[1(✉)], Prashant Anand[1], Vinay Kaushik[1], Santanu Chaudhury[1,2], and Brejesh Lall[1]

[1] Department of Electrical Engineering, Indian Institute of Technology Delhi, New Delhi 110016, India
ronakgupta143@gmail.com

[2] Indian Institute of Technology Jodhpur, Jodhpur 342037, India

Abstract. Tasks such as action recognition requires high quality features for accurate inference. But the use of high resolution and large volume of video data poses a significant challenge for inference in terms of storage and computational complexity. In addition, compressive sensing as a potential solution to the aforementioned problems has been shown to recover signals at higher compression ratios with loss in information. Hence, a framework is required that performs good quality action recognition on compressively sensed data. In this paper, we present data-driven sensing for spatial multiplexers trained with combined mean square error (MSE) and perceptual loss using Deep convolutional neural networks. We employ subpixel convolutional layers with the 2D Convolutional Encoder-Decoder model, that learns the downscaling filters to bring the input from higher dimension to lower dimension in encoder and learns the reverse, i.e. upscaling filters in the decoder. We stack this Encoder with Inflated 3D ConvNet and train the cascaded network with cross-entropy loss for Action recognition. After encoding data and undersampling it by over 100 times (10×10) from the input size, we obtain 75.05% accuracy on UCF-101 and 50.39% accuracy on HMDB-51 with our proposed architecture setting the baseline for reconstruction free action recognition with data-driven sensing using deep learning. We experimentally infer that the encoded information from such spatial multiplexers can directly be used for action recognition.

Keywords: Data driven compressive sensing (CS) · 3D Deep Convolutional Neural Networks (DCNN) · Perceptual compression · Reconstruction-free action recognition

1 Introduction

Action recognition is a fundamental task in computer vision community with widespread applications in video surveillance, unmanned aerial vehicles (UAV)

B. Deka et al. (Eds.): PReMI 2019, LNCS 11941, pp. 250–259, 2019.
https://doi.org/10.1007/978-3-030-34869-4_28

to name a few. In such applications, instead of using expensive methods for video compression and transmission implemented at transmitter end, compressive sensing (CS) can be applied to encode the data. The encoded data can then be decompressed and processed for high-level inference (action recognition). However, at higher compression ratio the reconstruction based action recognition framework will provide low quality results as they are not optimized to be used over compressed measurements or reconstructed data.

Recent works in deep learning have motivated us to perform data-driven CS for encoding the data and directly performing high-level inference on a large-scale dataset [1,2]. One of the key advantages data-driven encoder-decoder approach offers for CS based recognition is that it allows network to learn more complex patterns from data that may not be easily expressible in a model-based approach [3]. Thus, *this paper proposes a novel Data-driven sensing framework using Deep convolutional neural networks trained with joint MSE and perceptual loss in the context of reconstruction-free action recognition.*

One of the key challenges to perform reconstruction free action recognition using Deep convolutional neural networks(DCNN) is dimensionality reduction in convolutional layers. Usually, downsampling is performed in DCNN using max pooling, average pooling, stochastic pooling or spatial pyramid pooling [1]. Since the above methods are handcrafted which are designed for achieving translation invariance or fixed-length feature representation, they do not optimally preserve signal information while reducing dimensionality [1]. Therefore, learning dimensionality reduction from data itself is desired so as to preserve more information compared to handcrafted ways that do not use training data for downsampling. We have used sub-pixel convolutional layer for performing downsampling that uniformly distributes the samples into multiple dimensions thereby reducing their scale while increasing the number of channels of the data. This makes the data more suitable for dimensionality reduction using convolutional encoder where the higher dimension signal is brought to lower dimension by aggregating feature maps and then reducing the channels in the data. The signal can be recovered from undersampled measurements by using convolutional layers and a sub-pixel convolution layer which learns the upscaling filters and brings the signal to original dimension [4].

To train the network, we propose to use perceptual loss and mean squared error loss. The intuition of using perceptual loss is that the reconstructed output would be similar to the input image as it preserves structural information. In several works [5,6], it has also been shown that good quality images can be generated using perceptual loss functions based on differences between high-level percepts extracted from pre-trained convolutional neural networks, for classification, instead of based on per-pixel differences.

Outline of the paper is as follows: Sect. 2 revisits the existing work in this area. The proposed methodology is explained in Sect. 3. Section 4 presents experimental results to show the effectiveness of the framework and Sect. 5 concludes the paper.

2 Prior Work

Mousavi et al. [7] introduced first data driven sensing and recovery using deep learning framework. They applied stacked denoising autoencoder (SDA) as an unsupervised learning approach to capture statistical dependencies between the linear and non-linear measurements and improve the signal recovery compared to the conventional CS approach. The limitation of this work was that it consisted of a fully connected network. Kulkarni et al. [8] proposed a novel class of CNN architecture called ReconNet as decoder which takes in CS measurements of an image block as input and outputs reconstructed image block. Further the reconstructed image blocks are arranged appropriately and fed into an off-the-shelf denoiser to remove the artifacts. Mousavi et al. [1] proposed a deep convolutional neural network to learn a transformation from the original signals/images to a near-optimal number of undersampled measurements instead of using conventional random linear measurements obtained through a fixed sensing matrix and learns the inverse transformation for recovery from measurements to original signals/images. Learning undersampled measurements from original signal preserves more information. However their results were limited to 1D signals.

Early research focussed on non-deep learning architectures for recognition. Kulkarni et al. [9] presented a method for quantifying the geometric properties of high-dimensional video data in terms of recurrence textures for performing activity recognition at low data rates. In [10] Kulkarni et al. proposed a correlation-based framework in compressed domain and avoids reconstruction process. They showed that Action MACH (Maximum Average Correlation Height) correlation filters can be implemented in compressed domain to find correlation with compressed measurements using the concept of smashed filtering in the space-time domain. Recently, researchers have showed focus on deep learning based recognition on undersampled measurements. In [11] Adler et al. presented an end-to-end deep learning approach for Compressed Learning for classification of image in which the training jointly optimizes the sensing matrix and the inference operator. In [12] Lohit et al. shows that convolutional neural networks can be employed to extract discriminative non-linear features directly from data-independent random CS measurements. They project the CS measurements to the image space by a fixed projection matrix and later apply CNN to the intermediate projection to classify images. In [13], Zisselman et al. presented that optimizing the sensing matrix jointly with a nonlinear inference operator using neural networks, improved upon the methods which used a standard linear projection such as random sensing, PCA etc. In their experiments, the signals were reshaped to full dimension prior entering inference stage, since they used redesigned networks and added compression-decompression layers. In [3] Lohit et al. designed a three-stage training algorithm that allows learning the measurement operator and the reconstruction/inference network jointly such that the system can operate with adaptive measurement rates.

The proposed approach, shown in Fig. 1, does data-driven sensing and learns undersampled measurements with perceptual loss for action recognition. More-

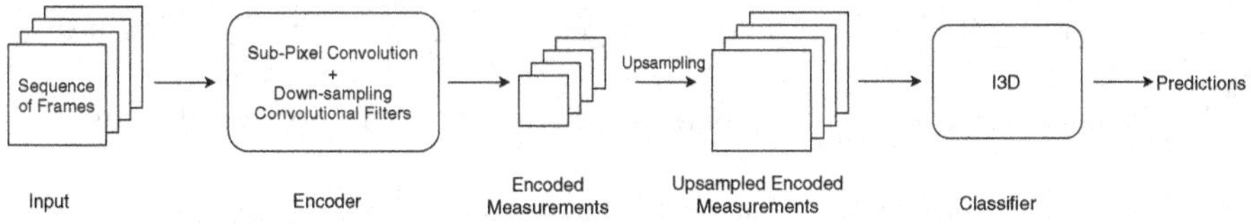

Fig. 1. Proposed framework for data-driven sensing and reconstruction free action recognition

over, the approach directly works on 2D signals which gives us good performance over compressively sensed UCF-101 and HMDB-51 action recognition datasets.

3 Methodology

3.1 Data-Driven Sensing Through Convolutional Autoencoder (CAE)

Our data-driven compression involves CAE (Fig. 2) with sub-pixel convolutional layers and optionally, VGG 16 pre-trained network. The input to the Encoder of CAE is the original image $x \epsilon \mathbb{R}^{H*W*3}$ which is multi-channel and 2-dimensional. The first layer is sub-pixel convolutional layer that learns to generate aggregated feature maps of reduced dimension. The value of reduced dimension is the height(H) and width(W) divided by r, which is the undersampling factor. So the output of first layer is $(H/r) \times (W/r) \times 3r^2$. It means the first sub-pixel convolutional layer divides the length of output feature map by a factor r^2 and increases the number of channels in output feature map by a factor of r^2. Mathematically, sub-pixel convolutional layer (inverse of pixel shuffle [4]) here can be described as:

$$\hat{x}(x,r)_{i,j,c} = x_{i*r+floor(\frac{mod(c,r^2)}{r}),\, j*r+mod(mod(c,r^2),r), floor(\frac{c}{r^2})} \quad (1)$$

The sub-pixel convolutional layer reduces the input image scale, however to undersample the input in Encoder, we employ several blocks of convolutional layers same as Inception module (Fig. 3) used in [14] to reduce the dimensionality by decreasing the total number of feature maps such that dimension of the output equals to $(H/r) \times (W/r) \times 3$. This reduces the total number of measurements by a factor of $r \times r$ and thus the undersampling ratio is $r^2(r \times r)$. Such sub-sampling ensures that the channels preserve the structural information of input image.

Once the undersampled measurements are obtained from Encoder, we now employ several blocks of convolutional layers to extract feature maps in Decoder. Hence in our architecture we learn to encode and reconstruct images directly to the measurement domain. Now, the output of encoder lies in $\mathbb{R}^{(H/r)*(W/r)*3}$ domain, however we want to recover the input signal which lies in $\mathbb{R}^{(H)*(W)*3}$ domain. Using the block of convolutional layers in Decoder would boost the dimensionality to $\mathbb{R}^{(H/r)*(W/r)*3r^2}$. In last we apply sub-pixel convolution layer to rearrange the feature maps and generate the output, $\hat{x}$, which lies in $\mathbb{R}^{(H)*(W)*3}$.

Mathematically the last layer(pixel shuffle layer [4])can be described as:

$$\hat{x}(x,r)_{i,j,c} = x_{floor(i,r),floor(j,r),c*r^2+r*mod(i,r)+mod(j,r)} \tag{2}$$

We train our CAE with MSE, perceptual and joint loss, weighted combination of MSE and perceptual loss. The perceptual loss we used here is defined in [5], that measures high-level perceptual and semantic differences between images. Their perceptual loss is function of deep convolutional neural networks (VGG 16) pre-trained for image classification. Instead of per-pixel loss between the output image $\hat{x}$ and original image x, we use the similar feature representations as computed by the loss network ϕ. Let $\phi_j(x)$ be the jth layer activations of network ϕ when processing the image x, if j is a convolutional layer then we get a feature map of size $C_j \times H_j \times W_j$ as $\phi_j(x)$. The perceptual loss is the sum of normalized Euclidean distance between feature representations of corresponding convolutional layers as showed in Eq. 3. That means the perceptual loss is minimum, when the classification output of pre-trained convolutional neural network (VGG 16 or VGG 19 [15]) for the reconstructed image would be same as that for the original image.

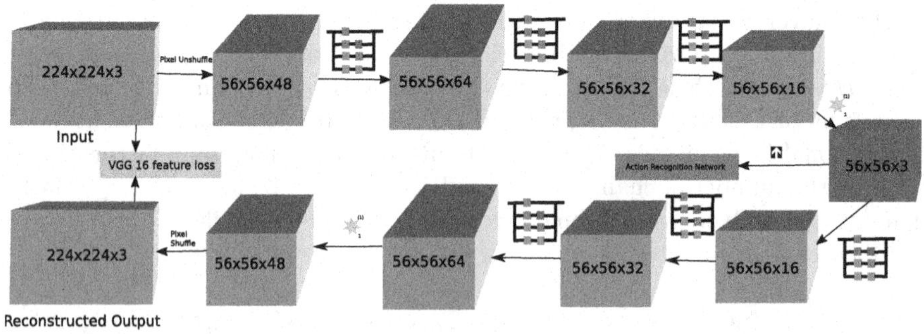

Fig. 2. Data driven Sensing framework for learning undersampled measurements (dimensions shown for undersampling ratio of 4×4)

$$l_{feat}^{\phi,j}(\hat{x},x) = \sum_j \frac{1}{C_j H_j W_j} ||\phi_j(\hat{x}) - \phi_j(x)||^2 \tag{3}$$

The image content and structural information is preserved but color, texture and exact shape are not preserved while using perceptual loss [5]. Hence we train our Autoencoder with joint loss (weighted) of MSE loss and perceptual loss as shown in Eq. 4.

$$L_{joint}(\hat{x},x) = \alpha\left(\frac{1}{s}\sum_{i=1}^{s}||\hat{x}-x||_2^2\right) + \beta\left(l_{feat}^{\phi,j}(\hat{x},x)\right) \tag{4}$$

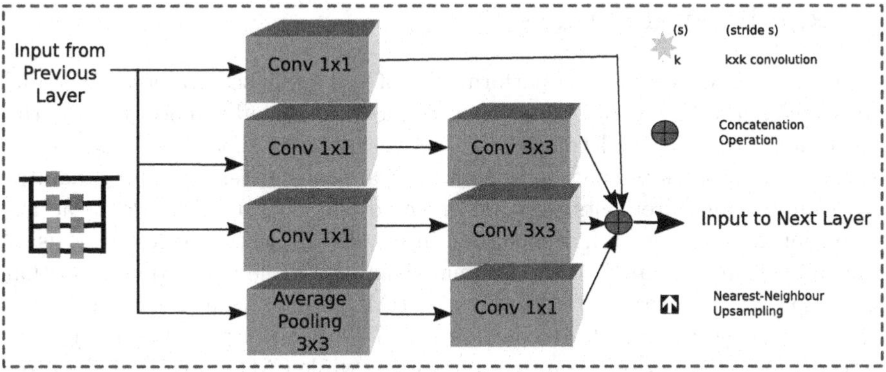

Fig. 3. Inception module in our CAE architecture

3.2 Learning Action Recognition in Compressed Domain

The convolutional encoder of the reconstruction network is now stacked with the Inflated 3D (I3D) convolutional neural network [16]. Since the structural information encoded in the Convolutional Encoder lies in $\mathbb{R}^{(H/r)*(W/r)*3}$, we upsample the encoded information by $r \times r$ times using a nearest neighbour upsampling layer. The upsampling of encoded feature maps is not equivalent to reconstruction. The encoder is initialized using learned weights from reconstruction network. The 3D convolutional network is initialized using pretrained weights from Imagenet and Kinetics dataset. The classification architecture is shown in Fig. 4. Further, this action recognition architecture is then trained with the cross-entropy classification loss using SGD optimizer.

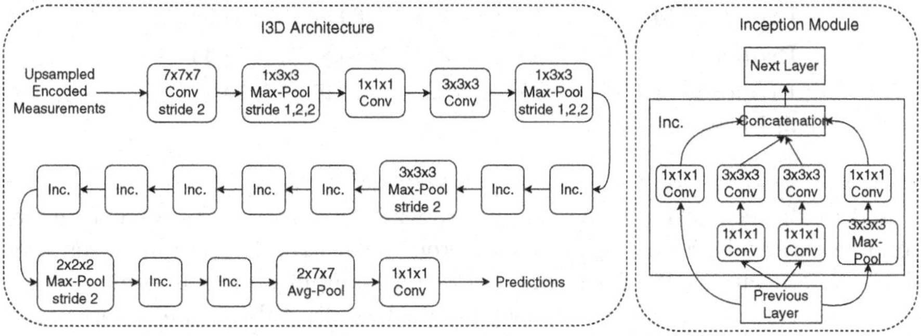

Fig. 4. Action recognition architecture

4 Experimental Results

In this section, we report the performance of our proposed framework at different undersampling ratios. Once our framework is end-to-end trained, the convolutional encoder and trained I3D network can be separated. The convolutional encoder does sensing on unseen data at transmitter end and generates undersampled measurements that are given to trained I3D network for action recognition at receiver end. Hence, our input size for action recognition is of lower dimension compared to the original signal. In Table 1, we compare action recognition results over UCF-101 [17] and HMDB-51 [18] datasets of different techniques. Here, STSF [10], Recon [19] + IDT [20] and our proposed approach are techniques that perform compressive sensing while C3D+ [21], RGB-I3D [16] are deep learning based action recognition over original signal. MB motion vectors+3DConvNet [22] shows action recognition results over macroblock motion vectors of H.264 compressed signal.

Table 1. Performances of different action recognition framework in compressed domain

Framework	Input size	Accuracy (%)	
		UCF-101	HMDB51
STSF [10]	FBI[a]	–	22.5
Recon [19]+IDT [20]	FBI	–	57.2
MB motion vectors	$24^2 \times 2 \times 160$	77.5	49.5
+3D ConvNet [22]			
C3D+ [21]	$112^2 \times 3 \times 16$	82.3	–
RGB-I3D [16]	$224^2 \times 3 \times 64$	95.6	74.8
Proposed (2 × 2)	$112^2 \times 3 \times 32$	91.25	66.66
Proposed (4 × 4)	$56^2 \times 3 \times 32$	89.53	65.88
Proposed (8 × 8)	$28^2 \times 3 \times 32$	80.84	56.33
Proposed (10 × 10)	$22^2 \times 3 \times 32$	75.05	50.39

[a]Full Blown Images

In Table 2, recognition results of our proposed approach with different undersampling ratios of original signal for different reconstruction loss has been presented. The accuracy for 2×2 and 4×4 undersampling ratio is similar, since there is not much change in Encoder parameters. In Table 2, we observe that the accuracy decreases in small amount at higher undersampling ratio and conclude that the information is correctly captured by the Encoder parameters. We present the same for both datasets displaying the efficiency of our proposed approach. Table 2 shows that when Encoder is trained with joint loss, the performance of action recognition pipeline increases as perceptual loss makes sure that the encoder captures more structural information. Table 3 compares the performance of sensing network with different number of Inception submodules

Table 2. Avg. Accuracy on UCF101 splits

Undersampling ratio	CAE RMSE loss	CAE perceptual loss	CAE joint loss	Encoder complexity (W)
4 (2×2)	87.89	90.85	**91.25**	3,651
16 (4×4)	89.40	88.79	**89.53**	14,894
64 (8×8)	80.78	76.50	**80.84**	179,408
100 (10×10)	73.62	73.96	**75.05**	404,278

in CAE, using reconstruction loss and action recognition accuracy as an evaluation metric over UCF-101 dataset splits at compression ratio of 4×4. The performance in Table 3 infers that optimal sensing is obtained with 3 Inception submodules in data-driven sensing encoder.

4.1 Implementation Details

For training CAE, we use ADAM optimizer with initial learning rate set to 10^{-3} which is reduced by 10^{-1} when validation loss gets saturated. While training CAE with joint loss function as shown in Eq 4, $\alpha = 1$ and $\beta = 0.04$ gives us the best results. To train the stacked network of Encoder and I3D for classification, we employ standard SGD with momentum set to 0.9 and initial learning rate set to 10^{-2}. All the networks were implemented in TensorFlow [23] and ran on nvidia-docker [24] for Tensorflow on NVIDIA DGX-1.

Table 3. Performance with respect to number of Inception modules in Proposed data-driven sensing encoder

No. of Inception submodules	Joint loss (CAE)	Accuracy on UCF-101 (%)	Encoder complexity (W)
1	7.5649	89.21	1,252
2	4.0129	85.94	4,322
3	4.2220	89.53	14,894
4	4.5798	87.84	53,766

5 Conclusion

A data-driven CS framework for reconstruction free action recognition is presented in the paper. Our proposed architecture preserves more structural information utilizing the joint loss function, based on perceptual loss and MSE loss, that provides better performance as compared to individual losses. Experimental results on UCF-101 and HMDB-51 are presented to show the effectiveness of the framework at various undersampling ratios. For future work, our undersampling ratio specific architecture can be modified to a generic architecture which works with different undersampling ratios.

Acknowledgment. The NVIDIA DGX-1 for experiments was provided by CSIR-CEERI, Pilani, India

References

1. Mousavi, A., Dasarathy, G., Baraniuk, R.G.: DeepCodec: adaptive sensing and recovery via deep convolutional neural networks. arXiv preprint arXiv:1707.03386 (2017)
2. Xu, K., Ren, F.: CSVideoNet: a real-time end-to-end learning framework for high-frame-rate video compressive sensing. In: 2018 IEEE Winter Conference on Applications of Computer Vision (WACV), pp. 1680–1688. IEEE (2018)
3. Lohit, S., Singh, R., Kulkarni, K., Turaga, P.: Rate-adaptive neural networks for spatial multiplexers. arXiv preprint arXiv:1809.02850 (2018)
4. Shi, W., et al.: Real-time single image and video super-resolution using an efficient sub-pixel convolutional neural network. In: Proceedings of the IEEE Conference on Computer Vision and Pattern Recognition, pp. 1874–1883 (2016)
5. Johnson, J., Alahi, A., Fei-Fei, L.: Perceptual losses for real-time style transfer and super-resolution. In: Leibe, B., Matas, J., Sebe, N., Welling, M. (eds.) ECCV 2016. LNCS, vol. 9906, pp. 694–711. Springer, Cham (2016). https://doi.org/10.1007/978-3-319-46475-6_43
6. Blau, Y., Mechrez, R., Timofte, R., Michaeli, T., Zelnik-Manor, L.: The 2018 PIRM challenge on perceptual image super-resolution. In: Leal-Taixé, L., Roth, S. (eds.) ECCV 2018. LNCS, vol. 11133, pp. 334–355. Springer, Cham (2019). https://doi.org/10.1007/978-3-030-11021-5_21
7. Mousavi, A., Patel, A.B., Baraniuk, R.G.: A deep learning approach to structured signal recovery. In: 2015 53rd Annual Allerton Conference on Communication, Control, and Computing (Allerton), pp. 1336–1343. IEEE (2015)
8. Kulkarni, K., Lohit, S., Turaga, P., Kerviche, R., Ashok, A.: Reconnet: non-iterative reconstruction of images from compressively sensed measurements. In: Proceedings of the IEEE Conference on Computer Vision and Pattern Recognition, pp. 449–458 (2016)
9. Kulkarni, K., Turaga, P.: Recurrence textures for human activity recognition from compressive cameras. In: 2012 19th IEEE International Conference on Image Processing (ICIP), pp. 1417–1420. IEEE (2012)
10. Kulkarni, K., Turaga, P.: Reconstruction-free action inference from compressive imagers. IEEE Trans. Pattern Anal. Mach. Intell. **38**(4), 772–784 (2016)
11. Adler, A., Elad, M., Zibulevsky, M.: Compressed learning: a deep neural network approach. arXiv preprint arXiv:1610.09615 (2016)
12. Lohit, S., Kulkarni, K., Turaga, P.: Direct inference on compressive measurements using convolutional neural networks. In: 2016 IEEE International Conference on Image Processing (ICIP), pp. 1913–1917. IEEE (2016)
13. Zisselman, E., Adler, A., Elad, M.: Compressed learning for image classification: a deep neural network approach. Process. Anal. Learn. Images Shapes Forms **19**, 1 (2018)

14. Szegedy, C., et al.: Going deeper with convolutions. In: Proceedings of the IEEE Conference on Computer Vision and Pattern Recognition, pp. 1–9 (2015)
15. Simonyan, K., Zisserman, A.: Very deep convolutional networks for large-scale image recognition. arXiv preprint arXiv:1409.1556 (2014)
16. Carreira, J., Zisserman, A.: Quo vadis, action recognition? A new model and the kinetics dataset. In: 2017 IEEE Conference on Computer Vision and Pattern Recognition (CVPR), pp. 4724–4733. IEEE (2017)
17. Soomro, K., Zamir, A.R., Shah, M.: Ucf101: a dataset of 101 human actions classes from videos in the wild. arXiv preprint arXiv:1212.0402 (2012)
18. Kuehne, H., Jhuang, H., Garrote, E., Poggio, T., Serre, T.: HMDB: a large video database for human motion recognition. In: 2011 IEEE International Conference on Computer Vision (ICCV), pp. 2556–2563. IEEE (2011)
19. Needell, D., Tropp, J.A.: Cosamp: Iterative signal recovery from incomplete and inaccurate samples. Appl. Comput. Harmon. Anal. **26**(3), 301–321 (2009)
20. Wang, H., Schmid, C.: Action recognition with improved trajectories. In: Proceedings of the IEEE International Conference on Computer Vision, pp. 3551–3558 (2013)
21. Tran, D., Bourdev, L., Fergus, R., Torresani, L., Paluri, M.: Learning spatiotemporal features with 3D convolutional networks. In: Proceedings of the IEEE International Conference on Computer Vision, pp. 4489–4497 (2015)
22. Chadha, A., Abbas, A., Andreopoulos, Y.: Compressed-domain video classification with deep neural networks: there's way too much information to decode the matrix. In: 2017 IEEE International Conference on Image Processing (ICIP), pp. 1832–1836. IEEE (2017)
23. Abadi, M., et al.: TensorFlow: large-scale machine learning on heterogeneous systems (2015). http://tensorflow.org/, software available from tensorflow.org
24. Nvidia gpu cloud tensorflow. nVIDIA offers GPU accelerated containers via NVIDIA GPU Cloud (NGC) for use on DGX systems. https://ngc.nvidia.com/catalog/containers/nvidia:tensorflow

Gradually Growing Residual and Self-attention Based Dense Deep Back Projection Network for Large Scale Super-Resolution of Image

Manoj Sharma[1,2](✉), Avinash Upadhyay[1](✉), Ajay Pratap Singh[1](✉), Megh Makwana[3](✉), Swati Bhugra[2], Brejesh Lall[2], Santanu Chaudhury[2], Deepak[1], and Anil Saini[1]

[1] CSIR-Central Electronics Engineering Research Institute, Pilani, Rajasthan, India
mksnith@gmail.com, avinres@gmail.com, singhajay518@gmail.com
[2] Department of Electrical Engineering, IIT Delhi, Delhi, India
[3] CCS Computer Pvt. Ltd., Delhi, India
meghmak95@gmail.com

Abstract. Due to the strong capacity of deep learning in handling unstructured data, it has been utilized for the task of single image super-resolution (SISR). These algorithms have shown promising results for small scale super-resolution but are not robust to large scale super-resolution. In addition, these algorithms are computationally complex and require high-end computational devices. Developing large-scale super-resolution framework finds its application in smart-phones as these devices have limited computational power. In this context, we present a novel light-weight architecture-Gradually growing Residual and self-Attention based Dense Deep Back Projection Network (GRAD-DBPN) for large scale image super-resolution (SR). The network is made of cascaded self-Attention based Residual Dense Deep Back Projection Network (ARD-DBPN) blocks to perform super-resolution gradually. Where each block performs 2X super-resolution and fine tuned in an end to end manner. The residual architecture facilitates the faster convergence of network and overcomes the issue of vanishing gradient. Experimental results on different benchmark data-set have been presented to compare the efficacy and effectiveness of the architecture.

Keywords: Large Scale Super-Resolution · Gradual · Residual · Dense · Deep Back Projection Network (DBPN) · Self-attention · Spectral normalization

1 Introduction

Single image super-resolution (SISR) is a challenging and an ill-posed task in computer vision since the objective is to recover the high resolution (HR) image

M. Sharma, A. Upadhyay, A. P. Singh and M. Makwana—Equal Contribution.

B. Deka et al. (Eds.): PReMI 2019, LNCS 11941, pp. 260–268, 2019.
https://doi.org/10.1007/978-3-030-34869-4_29

from its low resolution (LR) counterpart. Difficulty in recovering HR image grows gradually with the increase in super-resolution (SR) ratio since the probability space for the solution increases. Introduction of deep learning based frameworks have revolutionized this field with the distinctive improvement in peak signal-to-noise ratio (PSNR) as compared to the conventional approaches.

In 2014, Dong et al. [2] introduced the first deep-learning based solution for SISR which had only three convolution layers and showed significant improvement over empirical methods. Kim et al. [4] proposed a gradient clipping and skip connections based deeper convolutional network to increase the SR performance. In the aforementioned methods, the LR input was first super-resolved using interpolation techniques that resulted in the addition of spurious noise and artifacts to the LR image. Furthermore, it added an additional computational overhead to the process. To overcome this, Ledig et al. [6] proposed a skip connection-based model SRResNet to prevent the gradient diminishing phenomenon in very deep networks. SRResNet takes LR image as input (with no interpolation step) and up-samples the image within the network. They utilized this network as a generator in generative adversarial network SRGAN. Zhang et al. [14] and Tong et al. [12] proposed a residue and skip connection based dense networks respectively to address the gradient diminishing and low converging speed thus facilitating the flow of information through each layer with fast convergence. Recently, Dense DBPN [3] network has shown state-of-the-art performance for SISR. This network consists of consecutive up-sampling and down-sampling blocks attached in series that helps in generating variation rich HR feature and feeding it back in the LR space to enhance the LR features thereby effectively enhancing the super-resolution process. Although these aforementioned state-of-the-art networks performed instant up-sampling to the desired scale in one step their performance diminished with an increase in the scaling factor. This is due to the fact that at higher scales, there is a huge information gap between the LR and HR image. With instant up-sampling in a single step, the extracted features from the LR image lacks enough information for proper reconstruction of the image. Thus, the models fail to converge adequately. To address this problem, Lai et al. [5] and Zhao et al. [15] showed gradual up-scaling approach that perform large-scale super-resolution in multi-levels. However, in gradual up-scaling the artefacts produced by the lower scaling level of the model gets super-resolved in the consecutive higher scaling levels thereby decreasing the overall performance of the network. Manoj et al. [10] presented a residual gradual upscaling network to perform effective SR on larger scales by using residual architectures and end-to-end training of all levels in the model. This facilitated the fast convergence and reuse of the weights from preceding layers. End-to-end optimization of gradual network removed the artefacts created by the up-sampling network at each level.

Motivated by this paper, we present Gradually growing Residual and self-Attention based Dense Deep Back Projection Network (GRAD-DBPN) to super-resolve LR images for higher magnification scales. The network consists of self-attention based Residual-Dense-Deep Back-Projection Network(ARD-DBPN) block which performs 2X SR at every level. We employed self-attention blocks in each ARD-DBPN blocks to extract robust features. ARD-DBPN blocks also have residual connections between the first and last layer of the block. Here, residual architecture facilitated the fast convergence and reuse of the features from preceding layers. These blocks are then repeated to achieve the required magnification scale. The number of blocks required to get the respective scale can be obtained using the formula $log_2(X)$, where X represents the magnification factor. After reaching the desired scale the concatenated network is then fine-tuned in end-to-end manner by passing the error from the last layer of the last ARD-DBPN block to the first layer of the first ARD-DBPN block.

2 Related Work

In recent years, many SISR algorithms based on Convolutional neural network (CNN) have been proposed. Dong et al. [2] introduced the first primitive CNN based approach consisting of only three convolutional layers. It was later upgraded into a very deep CNN architecture by Kim et al. [4]. In the pursuit to improve SISR, many different approaches such as recursive convolutional network [11], deep residual network [7,11], merged shallow and deep CNN [11], sparse convolutional framework [11] and bi-directional recurrent convolutional network [11] have been proposed. The previously mentioned techniques follow two types of approach for the up-sampling step. For example, [2,4] incorporate a conventional method such as interpolation to increase the scale of input image first and then recreate the HR image using the up-scaled image as input. Wheras, in second type of approach the scale is increased by convolutional layers [6]. LapSRN [5], GUN [15] and IRGUN [10] employ gradual up-scaling technique to reconstruct HR image from LR image gradually.

3 Contribution

The major contributions of our work are: 1. A novel Gradually growing Residual and self-Attention based Dense Deep Back Projection Network for large scale SR. 2. Utilization of self-attention based model along with batch normalization and spectral normalization for effective Large scale SR. 3. Experimental study of the proposed network on different benchmark data-sets.

4 Methodology

GRAD-DBPN: The proposed framework GRAD-DBPN shown in Fig. 2 consists of three ARD-DBPN blocks connected back to back as shown in Fig. 1. Each ARD-DBPN block is responsible for 2X upscaling. These blocks consist of 4 stages, feature extraction, self-attention block, deep-back-projection and reconstruction. They are trained for their respective scales. The number of blocks that are required to achieve the specific SR scale X will be $log_2(X)$. These

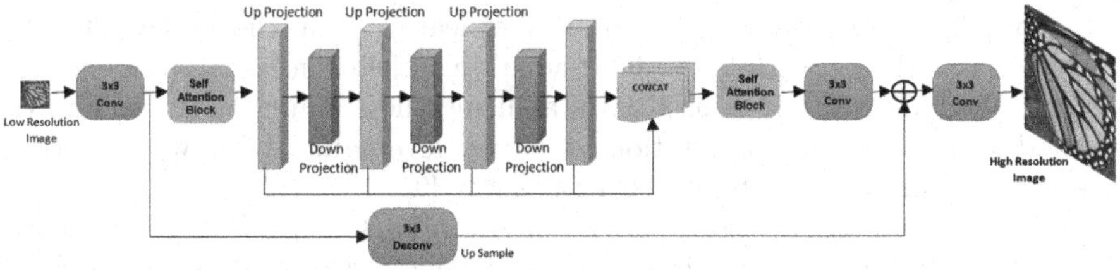

Fig. 1. Block diagram for ARD-DBPN for 2X SR

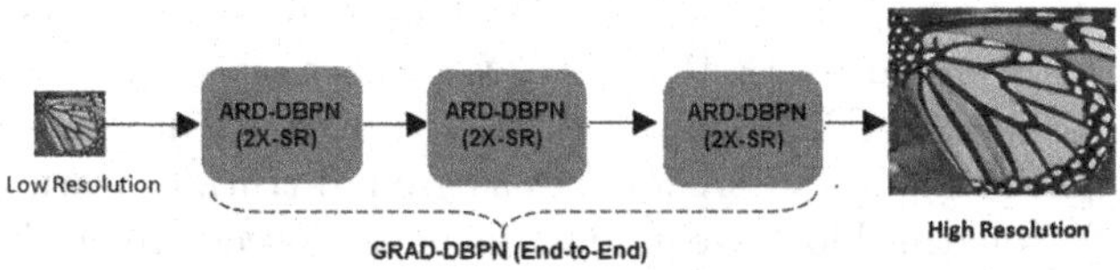

Fig. 2. Block diagram for GRAD-DBPN for 8X SR

ARD-DPBN Blocks are then cascaded to get required magnification factor. The weights from the previous ARD-DBPN block is used in the consecutive blocks instead of random weights except for the first block. Subsequent to training independently, all blocks are cascaded and then fine-tuned in end-to-end manner such that the optimization error from the last block of GRAD-DBPN is utilized for optimizing every ARD-DBPN block simultaneously. ARD-DBPN: An ARD-DBPN block represented in the Fig. 1 consists of following different parts performing respective tasks. 1. Feature Extraction: The first convolution layer L1 with 3X3 kernel size and 64 feature maps extracts the features from the LR image. Next convolution layer L2 with 1X1 kernel size and 32 feature map is used to reduce the dimension of the feature map extracted from the first convolution layer. 2. Self-Attention: The self-attention block [13] is depicted in Fig. 3. These blocks are introduced after the convolutional layer and after the concatenation operation of first and last up-projection network in ARD-DBPN as shown in the Fig. 1. These blocks helps in generating high fidelity natural images using information from all feature location and long range dependencies instead of only depending on spatially local points. 3. Back Projection: The back to back upscaling and downscaling network is used to extract the HR features and projecting it back to LR space to enhance the features. Further, the features of all the upscaling block of the back projection network are concatenated to gather all the enhanced features together. 4. Reconstruction. Another convolution layer L3 with 1X1 kernel size and 64 feature map is used to increase the dimension of the feature map to the dimension of the feature map created by the first layer. This is then passed to another convolution layer L4 of kernel size 3X3 and feature map of 64. Residue from the L1 layer is added here and passed to another convolution layer L6 with kernel size 3X3 and feature size 1 to reconstruct the image. 5. Residue: The features from the L1 layer are upscaled and added to the output of the L4 layer. These combined features are then passed to L5 layer for reconstruction. The passing of feature from the L1 layer to the L5 layer makes the network learn the residue which helps in fast convergence. 6. Spectral Normalization: It is used for stabilizing the training and it facilitates the model to use small computational time by leveraging the power iteration trick resulting in better stability during training [9].

Training: The training of the model is done on RGB image. First, the patches of size 128X128 are extracted from the HR image and termed as 8X-Patches. These patches are used as the ground truth for the end-to-end training of the GRAD-DBPN Model and training the last ARD-DBPN Block. These patches are downscaled to half, quarter and one-eighth of its original size using bi-cubic interpolation and termed as 4X-Patches, 2X-Patches, and 1X-Patches. For 8X SR we used three ARD-DBPN blocks. The first ARD-DBPN block is trained using 1X-Patches as input and 2X-Patches as ground-truth to learn 2X SR. The learned weights from this block are used in the next consecutive block. This block with the learned weights of the previous block is fine-tuned with input as 2X-Patches and ground truth as 4X-Patches. The process is repeated for the third block. After the individual training of all three blocks, they are connected in a cascaded manner and then fine-tuned end-to-end using 1X-Patches as input and 8X-Patches as ground truth. We have used leaky-ReLU as activation function for each convolutional and deconvolutional layer.

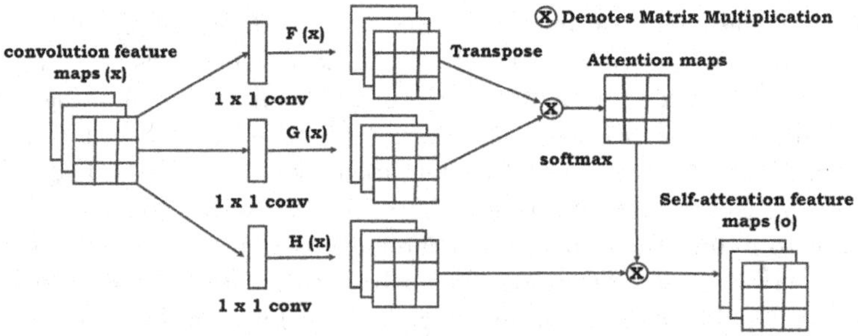

Fig. 3. Self-attention mechanism [13]

Testing: Testing of the model is done on the complete LR image, instead of patches to avoid the framing effect. Any dimension of LR input can be fed as an input to the model.

Model Specifications: We have used 8 Up-projection and Down-projection block layers. Each convolutional layer of Up-projection and down-projection blocks have 3X3 kernel size and 64 feature maps. LeakyRelu activation function is used. The learning rate of 0.00004 is used for optimization.

Table 1. Comparison of average SSIM and PSNR for various Image SR algorithms at 8X scale for benchmark datasets. Values in red are highest while values in blue are second highest.

Dataset	Scale	Bicubic	-	GUN	-	RDN	-	IRGUN	-	D-DBPN	-	GRAD-DBPN	-
-	-	PSNR	SSIM	PSNR	SSIM	PSNR	SSIM	PSNR	SSIM	PSNR	SSIM	PSNR	SSIM
Set5	8	24.39	0.657	25.99	0.713	26.10	0.730	26.28	0.740	**26.31**	**0.741**	**26.53**	**0.746**
Set14	8	23.19	0.568	24.23	0.610	24.39	0.614	24.53	0.641	**24.58**	**0.644**	**24.88**	**0.710**
BSD100	8	23.67	0.547	24.42	0.579	24.53	0.584	24.61	0.591	**24.78**	**0.596**	**25.23**	**0.701**
URBAN100	8	20.74	0.515	21.66	0.565	21.70	0.577	21.89	0.594	**21.94**	**0.598**	**22.33**	**0.629**
MANGA109	8	21.47	0.649	23.00	0.717	23.28	0.728	23.43	0.748	**23.46**	**0.749**	**23.85**	**0.792**
DIV2K Validation Dataset	8	23.82	0.549	24.36	0.609	24.47	0.628	24.99	0.652	**24.99**	**0.655**	**25.39**	**0.685**

Table 2. Comparison of average PSNR for various Image SR algorithms at 4X scale for benchmark datasets. Values in red are highest while values in blue are second highest.

Dataset	Scale	Bicubic	VDSR	GUN	RDN	EDSR	IRGUN	D-DBPN	GRAD-DBPN
Set5	4	28.42	31.35	31.50	31.58	32.62	32.65	**32.68**	**33.24**
Set14	4	26.10	28.03	28.04	28.29	28.94	28.98	**29.09**	**29.42**
BSD100	4	25.96	27.29	27.44	27.55	27.79	28.01	**28.24**	**28.74**
URBAN100	4	23.15	25.18	25.24	25.44	26.86	25.48	**25.67**	**26.32**
MANGA109	4	24.92	28.82	28.97	29.09	29.12	29.22	**29.23**	**29.62**
DIV2K Validation Dataset	4	27.32	28.22	28.67	28.92	28.99	29.1	**29.14**	**29.56**

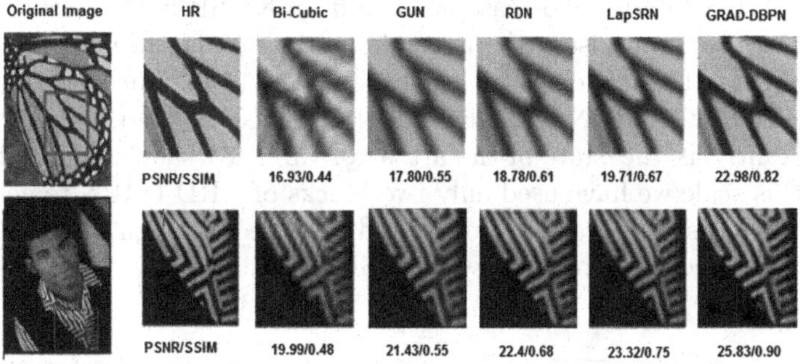

Fig. 4. Comparison at 8X SR on Set5 and BSD100 Dataset

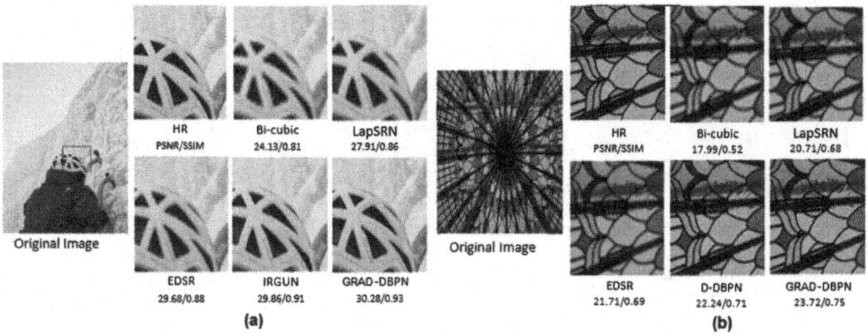

Fig. 5. Comparison at 8X SR on DIV2K Validation Dataset

5 Experiments

We have carried out numerous experiments to show the efficacy of proposed framework. We have achieved better performance from existing modern algorithms with less time complexity. We used popular datasets for all our experiments. Training is done using 50,000 images from the ImageNet dataset. To compare our findings with the existing state-of-the-art SR methods we have used **URBAN100**[10]. **BSD100** [8], **Set5** [2], **Set14** [2], **DIV2K** [1] and **Manga109** [10] datasets. We have used Intel Core i7 processor having clock speed 3.6 GHz and RAM of 128 GB with Nvidia GTX 1080 GPU for all our experiments.

5.1 Comparisons with Other State-Of-The-Art Methods

We have shown the comparison of the performance for proposed framework with other modern algorithms on higher scales(8X) in Table 1. Algorithms such as GUN [15], LapSRN [5], IRGUN [10] and D-DBPN [3] have shown state-of-the-art performance for 8X SR. We have gathered the source codes for these algorithms available publicly and trained them for 8X SR alongside our network. IRGUN and D-DBPN have performed reasonably well for 8X SR in the past. In Table 2 we have shown a comparison of our results with the algorithms which showed the state-of-the-art result on lower resolution scale (4X). These algorithms are VDSR [4], EDSR [7], RDN [11], IRGUN [10], GUN [15] and D-DBPN [3]. D-DBPN currently is the state-of-the-art algorithm. To show the result of our model on this scale we have used only two blocks of ARD-DBPN to get 4X SR. Since our model is deliberately trained and designed for high scale SR, we are not showing any comparison for low-scale SR (2X).

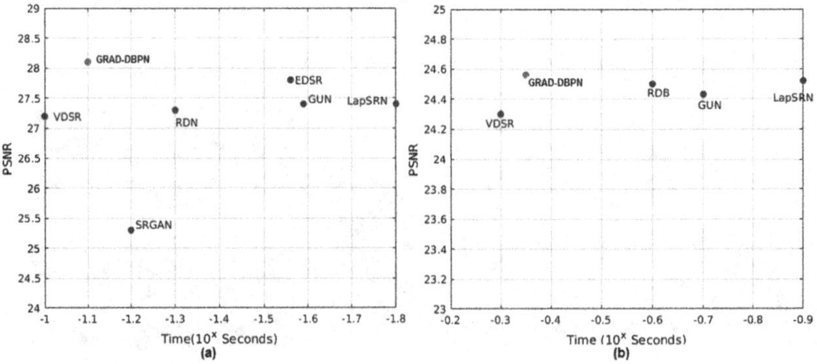

Fig. 6. Run-time performance comparison of various frameworks for (a) 4X SR on BSD100 dataset (b) 8X SR on BSD100 dataset

5.2 Result Analysis

As shown in Table 2, the proposed GRAD-DBPN network outperforms all other architectures in terms of PSNR and SSIM. Our framework have shown an average improvement of 0.36 dB in PSNR and 0.047 in SSIM over the current state-of-the-art framework D-DBPN for 8X. It has also shown a moderate improvement of 0.47 dB in PSNR over DBPN for 4X scale. In the Fig. 6 we have shown the average testing time on 4X and 8X scales for BSD100 Dataset. It is evident that our model gave good trade off between PSNR and time taken for testing in comparision of other frameworks. Visual results of our model is also depicted in Figs. 5 and 4 along with the PSNR and SSIM values for DIV2K and BSD100 datasets. The proposed framework is faster than the present state-of-the-art models without making any compromise over PSNR performance. This makes it suitable for smartphone applications. The model is light-weight and take less than 2 MB of storage space.

6 Conclusion

In this work, we presented a novel Gradually growing Residual and self-Attention based Dense Deep Back Projection Network (GRAD-DBPN) that showed significant improvement in terms of PSNR and SSIM metrics for single image super-resolution (SISR) as compared to the existing algorithms for large magnification ratio. Usage of spectral norm facilitated quicker and improved convergence of the error. Self attention and gradual growing improves the perceptual and objective quality while using less computational resources thus making it a light-weight architecture.

References

1. Agustsson, E., Timofte, R.: Ntire 2017 challenge on single image super-resolution: dataset and study. In: Proceedings of the IEEE Conference on Computer Vision and Pattern Recognition Workshops, pp. 126–135 (2017)
2. Dong, C., Loy, C.C., He, K., Tang, X.: Image super-resolution using deep convolutional networks. IEEE Trans. Pattern Anal. Mach. Intell. **38**(2), 295–307 (2015)
3. Haris, M., Shakhnarovich, G., Ukita, N.: Deep back-projection networks for super-resolution. In: Proceedings of the IEEE Conference on Computer Vision and Pattern Recognition, pp. 1664–1673 (2018)
4. Kim, J., Kwon Lee, J., Mu Lee, K.: Accurate image super-resolution using very deep convolutional networks. In: Proceedings of the IEEE Conference on Computer Vision and Pattern Recognition, pp. 1646–1654 (2016)
5. Lai, W.S., Huang, J.B., Ahuja, N., Yang, M.H.: Deep Laplacian pyramid networks for fast and accurate super-resolution. In: Proceedings of the IEEE Conference on Computer Vision and Pattern Recognition, pp. 624–632 (2017)
6. Ledig, C., et al.: Photo-realistic single image super-resolution using a generative adversarial network. In: Proceedings of the IEEE Conference on Computer Vision and Pattern Recognition, pp. 4681–4690 (2017)

7. Lim, B., Son, S., Kim, H., Nah, S., Mu Lee, K.: Enhanced deep residual networks for single image super-resolution. In: Proceedings of the IEEE Conference on Computer Vision and Pattern Recognition Workshops, pp. 136–144 (2017)
8. Martin, D., Fowlkes, C., Tal, D., Malik, J., et al.: A database of human segmented natural images and its application to evaluating segmentation algorithms and measuring ecological statistics. ICCV Vancouver (2001)
9. Miyato, T., Kataoka, T., Koyama, M., Yoshida, Y.: Spectral normalization for generative adversarial networks. arXiv preprint arXiv:1802.05957 (2018)
10. Sharma, M., Mukhopadhyay, R., Upadhyay, A., Koundinya, S., Shukla, A., Chaudhury, S.: Irgun: improved residue based gradual up-scaling network for single image super resolution. In: Proceedings of the IEEE Conference on Computer Vision and Pattern Recognition Workshops, pp. 834–843 (2018)
11. Timofte, R., Agustsson, E., Van Gool, L., Yang, M.H., Zhang, L.: Ntire 2017 challenge on single image super-resolution: methods and results. In: Proceedings of the IEEE Conference on Computer Vision and Pattern Recognition Workshops, pp. 114–125 (2017)
12. Tong, T., Li, G., Liu, X., Gao, Q.: Image super-resolution using dense skip connections. In: Proceedings of the IEEE International Conference on Computer Vision, pp. 4799–4807 (2017)
13. Zhang, H., Goodfellow, I., Metaxas, D., Odena, A.: Self-attention generative adversarial networks. arXiv preprint arXiv:1805.08318 (2018)
14. Zhang, Y., Tian, Y., Kong, Y., Zhong, B., Fu, Y.: Residual dense network for image super-resolution. In: Proceedings of the IEEE Conference on Computer Vision and Pattern Recognition, pp. 2472–2481 (2018)
15. Zhao, Y., Li, G., Xie, W., Jia, W., Min, H., Liu, X.: GUN: gradual upsampling network for single image super-resolution. IEEE Access **6**, 39363–39374 (2018)

3D-GCNN - 3D Object Classification Using 3D Grid Convolutional Neural Networks

Rishabh Tigadoli(✉), Ramesh Ashok Tabib, Adarsh Jamadandi, and Uma Mudenagudi

K.L.E. Technological University, Hubli 580032, Karnataka, India
rtatgml@gmail.com, {ramesh_t,adarsh.jamadandi,uma}@kletech.ac.in

Abstract. In this paper we propose to solve the problem of 3D Object Classification on point cloud data. We propose a 3D CNN architecture which we call 3D-GCNN that consumes point cloud data directly and performs classification. We present a novel method to represent the point cloud data using a margin based density occupancy grids which creates a minimum volume bounding box around the point cloud data. This information is fed to our proposed 3D CNN model which has far lesser trainable parameters and ensures convergence. We demonstrate our results on the ModelNet 10 and ModelNet 40 models and show we achieve better or comparable performance compared to other methods.

Keywords: 3D object classification · Occupancy grids · Point cloud data

1 Introduction

In this paper, we propose a 3D CNN framework to classify 3D objects. Our framework is able to consume point cloud data directly and doesn't require conversion to other 3D representations. We propose a novel method to represent the point cloud data based on a margin-based method that constructs a minimum volume bounding box around the point cloud which is then fed to our model. Human beings are generally better at classifying and recognizing objects, however it is much harder task for computers to do it algorithmically. With the advent of affordable 3D scanners, 3D representation of data has become ubiquitous, however indexing of this data requires semantic understanding of the objects involved. Traditional methods use various geometric handcrafted features to tackle the problem of 3D object classification, the results vary according to the features that are selected. Deep learning techniques have become pervasive and have found immense applications in the field of image classification, image-caption generation, video classification, action recognition in videos etc. These methods and frameworks have been extended to 3D data as well, some frameworks consider 3D CAD models, while some use meshed representations

B. Deka et al. (Eds.): PReMI 2019, LNCS 11941, pp. 269–276, 2019.
https://doi.org/10.1007/978-3-030-34869-4_30

as their input data. We propose a deep learning framework based on 3D CNNs to classify 3D objects directly based on point cloud data, towards this we make the following contributions:

1. We propose 3D-GCNN, a 3D CNN based deep learning model that consumes point clouds and converts them into novel margin-based occupancy grid representation.
2. We propose a 3D CNN architecture which has less trainable parameters but still achieves convergence and high accuracy.
3. We demonstrate our proposed framework on ModelNet10 and ModelNet40 dataset and show that we achieve better or comparable results to existing state of the art methods.

2 Related Work

The task of 3D object classification can be divided into two categories - techniques that use hand-crafted features and techniques that employ deep learning techniques. For the hand-crafted techniques many approaches have been proposed, they mainly utilize various geometric features like spin images [4], 3D shape context [1], point-feature histograms [7] and viewpoint feature histograms [8]. Authors in [2] propose to use Metric tensors and Christoffel symbols as novel geometric features and use a simple SVM classifier to perform 3D object categorization. On the contrary, deep learning techniques do not need any hand-crafted features, authors in [11] propose a deep learning architecture where the meshed models are represented as a binary tensors with 1 representing the mesh inside a voxel grid and 0 representing its absence. Authors in literature [9] propose an architecture where 3D models are represented as panoramic images and a variation of CNN is proposed that learns from such representations. All the works mentioned in the literature either work on 3D meshes, CAD models or involves representing 3D shapes in more manageable ways and then feeding them to deep learning frameworks. However, with the advent of inexpensive 2.5D scanners (*Microsoft Kinect*) the most direct representation of 3D models is point cloud data, following this line of thought, authors in [6] propose a deep learning framework that consumes point clouds directly for 3D object recognition and segmentation. Taking this approach further authors in [3] propose a novel method to represent point clouds using density occupancy grid representation. Inspired by these works we propose a refined approach to construct density occupancy grids and use this information to feed to our deep learning framework. Extensive experiments are performed on the ModelNet 10 and ModelNet 40 datasets to show how our framework outperforms/performs comparable to existing state-of-the-art methods.

3 Proposed Methodology

In this section we explain in detail the proposed methodology for 3D object classification. The proposed framework is depicted in Fig. 1.

3.1 Preprocessing Input Data

The proposed framework is experimented on the famous ModelNet dataset provided by [11]. The dataset consists of 662 object categories with a total of 127,915 models. A subset of these models called ModelNet10 and ModelNet 40 are benchmark datasets for 3D object classification and recognition tasks. The dataset consists of CAD models in mesh representation, we propose a pre-processing step to convert the CAD models into point clouds. The CAD models are sampled to obtain a sparse point cloud representation which can directly be fed to our 3D-GCNN Model. The conversion process involves performing a random mesh sampling which uses barycentric co-ordinate system for the triangle meshes. The idea is to generate point clouds inside the triangle using its vertices and a linear combination of three random numbers. Consider the three vertices of the triangle P, Q and R, we generate three random numbers u, v, w whose value lies in between $[0, 1]$ and is given by,

$$w = 1 - (u + v) \tag{1}$$

The value obtained from this equation is multiplied with the coordinates of the vertices of the triangle. In this manner the CAD models are sampled to have 2048 points.

3.2 Margin Based Occupancy Grid

In this section we explain the refined occupancy grid approach that we have formulated to represent the input point cloud data. Occupancy grids are a type of data structure [10] that allows us to obtain an efficient and compact representation of the volumetric space [3]. The occupancy grids provide good 3D shape cues that can be learned for various tasks. The occupancy grids can be formed using two approaches -

1. Binary Grids - In this approach a binary tensor is defined which has either 1 or 0 based on whether the grid is occupied.
2. Density Grids - In this approach the grids are defined by the density of the points present in the voxel.

In our formulation we use the density grids but unlike in the work [3] where the density grids are fixed with fixed leaf and grid size, we propose a margin based density grid formulation with variable voxel leaf size. The procedure to form occupancy grids can be summarized as follows:

1. We calculate the centroid of the 3D model and draw a margin based on the farthest and nearest points with respect to the centroid. All the points are translated about this margin to form the bounding boxes.
2. The 3 principal axes are equally segmented with respect to the bounding box limits. This allows for a variable voxel leaf-size which can be set based on the largest axis segmented.

3. The binning of the points is done based on the search and sort technique [5]. Each point is mapped to a voxel and voxels beyond the specific grid size are clipped to maintain uniformity.
   The size of the voxel is heuristically set to $15 \times 15 \times 15$ as opposed to $30 \times 30 \times 30$ as in [3]. This modification is in accordance to our cascaded 3D CNN architecture that accepts low resolution point cloud and can be trained on a CPU. Figure 1 shows conversion of CAD models to occupancy grids.

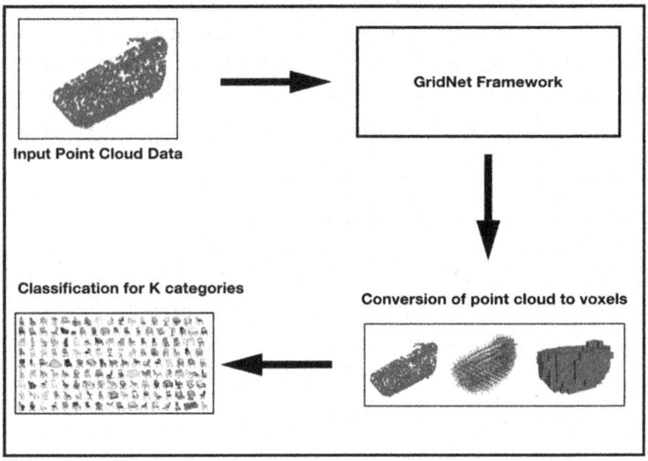

Fig. 1. An overview of the proposed 3D classification framework.

3.3 Network Architecture

In this section we describe our 3D CNN architecture that consumes low resolution point cloud data in the form of density occupancy grids and performs classification. CNNs have had tremendous success in various image-related tasks. The same idea has been extended for 3D object recognition and classification [3,6].The architecture is shown in Fig. 2. Our network architecture which we refer to as 3D-GCNN consists of 2 convolutional layers, a max-pooling layer, a dropout layer and finally dense layers and softmax to obtain classification scores. The 3D CNN architecture is designed to have lesser trainable parameters and to accept low-resolution point cloud so that it can be trained/tested on a local machine with no GPU compute while still achieving comparable performance to state-of-the-art work. The main layers along with the parameters are described below:

1. **Input Layer -** The input layer of the 3D-GCNN accepts low resolution point cloud and outputs the density occupancy grid based voxels of size $15 \times 15 \times 15$. This acts as an input layer to the first convolutional layer.

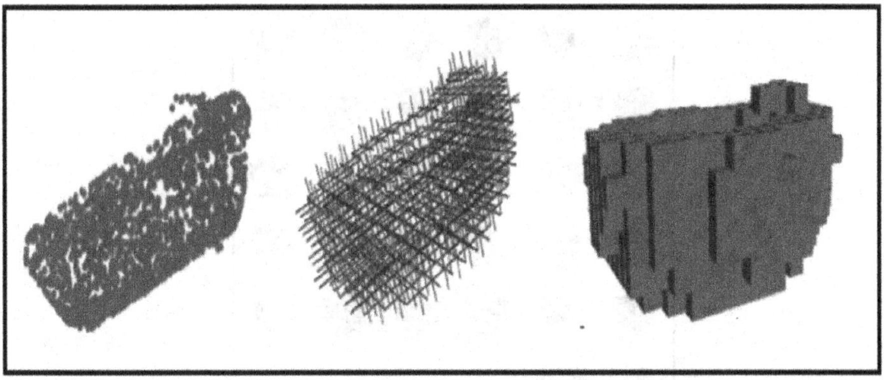

Fig. 2. The CAD models which are in mesh representation are first converted to low resolution point cloud data using random mesh sampling technique (Left). The point clouds are later mapped to voxels (Right) of variable leaf size by forming density occupancy grids (Middle).

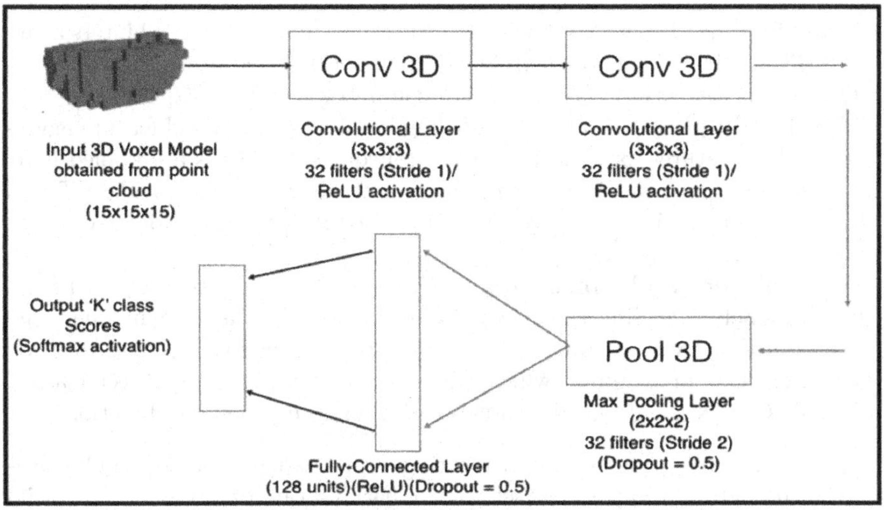

Fig. 3. Our network architecture which we call 3D-GCNN consists of cascaded 3D CNNs with less trainable parameters and can be trained on local CPU machines with comparable performance to existing state-of-the-art methods.

2. **Convolutional Layer 1 -** The first convolutional layer has 32 filters each of size $3 \times 3 \times 3$ and stride 1. The output of this layer is $13 \times 13 \times 13 \times 32$. This layer is followed by a ReLU layer given by $f(x_i) = max(0, x_i)$. The output activation maps are computed as follows,

$$f1 = max(0, W_1^T * x_i + B_1) \tag{2}$$

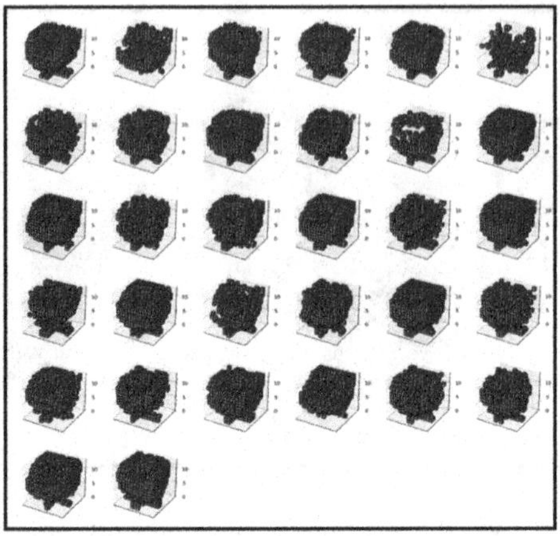

Fig. 4. A visualization of intermediate activation maps learning the features of chair model.

3. **Convolutional Layer 2 -** The second convolutional layer has 32 filters again of $3 \times 3 \times 3$ size and stride of 1. The output of this layer is $11 \times 11 \times 11 \times 32$. The output activation maps are again computed using the Eq. 2.
4. **Max-Pooling Layer -** The max-pooling layer acts as a global feature aggregator. The features extracted from the convolutional layers are pooled here. The kernel size of the max-pooling layer is $2 \times 2 \times 2$ with a stride of 2 and the output being $5 \times 5 \times 5 \times 32$. We use a dropout of 0.5 to avoid overfitting after the max-pooling layer.
5. **Dense Layers and Softmax Classification -** The activation maps are flattened before feeding them to the dense layers. We use 128 hidden units with ReLU activation function. The output layer is softmax classification which consists of K units, where K represents the different object classes. The 3D-GCNN is subjected to a categorical cross-entropy loss function.

The weights are refined by experimenting with Adadelta and Adam Optimizers. The Adadelta optimizer was set with a learning rate of 1.0 and momentum of 0.95 with no decay rate for the learning rate, while with the Adam optimizer the learning rate was set to 0.001 with bias correction of 0.9 and 0.999 respectively. From our experiments it was found that Adadelta generalized well while using Adam allowed the network to converge quickly.

4 Experiments and Results

In this section we describe the various experiments conducted on ModelNet dataset. We use the same train/test split as given by the ModelNet. We first divide the training set into a train/cross-validation set which is 80% and 20% respectively. This is true for both ModelNet 10 and ModelNet 40 datasets. The network was trained for 120 epochs on both datasets before the model converged.

Table 1. A comparison with state-of-the-art methods suggest that our framework performs comparable to existing methods.

Method	Input	ModelNet 10	ModelNet 40
Qi et al. [6]	Point Cloud	-	**89.2%**
Garcia et al. [3]	Point Cloud	77.6%	-
3D Shapnets [11]	Volume	83.5%	77%
DeepPano [9]	Image	85.45%	77.63%
3D-GCNN (Ours)	Point Cloud	**89.1%**	**83.5%**

The experiments done on a local machine with no GPU support took about 2 h 40 min to fully train for both the datasets, while the same model migrated to a Nvidia Quadro K5000 GPU with Ubuntu OS 16.04, 64 GB RAM and Intel Xeon processor took about 20 and 40 min for ModelNet 10 and ModelNet 40 datasets respectively. The per-category accuracy of our model is still comparable to state-of-the-art methods even without data augmentation. Our 3D-GCNN architecture achieves a per-category test accuracy of 89.1% and 83.5% on ModelNet 10 and ModelNet 40 datasets respectively. Table 1 summarizes the comparison of our model with existing methods. From the table it is clear that our proposed framework outperforms the methods in [3,9,11] and is comparable to the work of [6]. The lower accuracy can be attributed to the way CNNs extract features and perform classification, the presence of similar looking objects like desk and table affect the performance of the model. Authors in [3] show that making the network deeper or employing a balanced dataset led to any increase in the performance, thus our 3D-GCNN model with relatively lesser trainable parameters 541,994 performs better than deeper network models.

5 Conclusion

In this paper we propose 3D-GCNN, a 3D CNN based object classification framework that uses margin based density occupancy grids with variable leaf size to directly consume point clouds as input data and perform 3D object classification. Extensive experiments performed on standard datasets show that our framework performs comparable to existing state-of-the-art methods.

Acknowledgement. This work was partly carried out under Department of Science and Technology (DST) sponsored Indian Heritage in Digital Space (IHDS).

References

1. Frome, A., Huber, D., Kolluri, R., Bülow, T., Malik, J.: Recognizing objects in range data using regional point descriptors. In: Pajdla, T., Matas, J. (eds.) ECCV 2004. LNCS, vol. 3023, pp. 224–237. Springer, Heidelberg (2004). https://doi.org/10.1007/978-3-540-24672-5_18
2. Ganihar, S.A., Joshi, S., Shetty, S., Mudenagudi, U.: Metric tensor and Christoffel symbols based 3D object categorization. In: ACM SIGGRAPH 2014 Posters, SIGGRAPH 2014, pp. 38:1–38:1. ACM, New York (2014). https://doi.org/10.1145/2614217.2630582
3. Garcia-Garcia, A., Gomez-Donoso, F., Rodrguez, J., Orts, S., Cazorla, M., Azorin-Lopez, J.: PointNet: a 3D convolutional neural network for real-time object class recognition, pp. 1578–1584, July 2016. https://doi.org/10.1109/IJCNN.2016.7727386
4. Johnson, A.E., Hebert, M.: Using spin images for efficient object recognition in cluttered 3D scenes. IEEE Trans. Pattern Anal. Mach. Intell. **21**(5), 433–449 (1999). https://doi.org/10.1109/34.765655
5. Leimer, K.: External sorting of point clouds, September 2013. https://www.cg.tuwien.ac.at/research/publications/2013/leimer-2013-esopc/
6. Qi, C.R., Su, H., Mo, K., Guibas, L.J.: PointNet: deep learning on point sets for 3D classification and segmentation. CoRR abs/1612.00593 (2016). http://arxiv.org/abs/1612.00593
7. Rusu, R.B., Blodow, N., Beetz, M.: Fast point feature histograms (FPFH) for 3D registration. In: Proceedings of the 2009 IEEE International Conference on Robotics and Automation, ICRA 2009, Piscataway, NJ, USA, pp. 1848–1853. IEEE Press (2009). http://dl.acm.org/citation.cfm?id=1703435.1703733
8. Rusu, R.B., Bradski, G.R., Thibaux, R., Hsu, J.M.: Fast 3D recognition and pose using the viewpoint feature histogram. In: 2010 IEEE/RSJ International Conference on Intelligent Robots and Systems, pp. 2155–2162 (2010)
9. Shi, B., Bai, S., Zhou, Z., Bai, X.: DeepPano: deep panoramic representation for 3-D shape recognition. IEEE Signal Process. Lett. **22**(12), 2339–2343 (2015). https://doi.org/10.1109/LSP.2015.2480802
10. Thrun, S.: Learning occupancy grid maps with forward sensor models. Auton. Robot. **15**(2), 111–127 (2003). https://doi.org/10.1023/A:1025584807625
11. Wu, Z., et al.: 3D ShapeNets: a deep representation for volumetric shapes. In: 2015 IEEE Conference on Computer Vision and Pattern Recognition (CVPR), pp. 1912–1920, June 2015. https://doi.org/10.1109/CVPR.2015.7298801

A New Steganalysis Method Using Densely Connected ConvNets

Brijesh Singh[1(✉)], Prasen Kumar Sharma[1], Rupal Saxena[2], Arijit Sur[1], and Pinaki Mitra[1]

[1] Department of Computer Science and Engineering, Indian Institute of Technology Guwahati, Guwahati, India
{brijesh.singh,kumar176101005,arijit,pinaki}@iitg.ac.in

[2] Department of Chemistry, Indian Institute of Technology Guwahati, Guwahati, India
rupal.saxena@iitg.ac.in

Abstract. Steganography is an ancient art of communicating a secret message through an innocent-looking image. On the other hand, steganalysis is the counter process of the steganography, which targets to detect hidden trace within a given image. In this paper, a new approach to steganalysis is presented to learn prominent features and avoid loss of stego signals. The proposed model uses diverse sized filters to capture all useful steganalytic features through a densely connected convolutional network. Moreover, there is no fully connected network in the proposed model, which allows testing any size of images regardless of the image size used for training. To justify the applicability of the proposed scheme, it has been shown experimentally that the proposed scheme outperforms most of the related state-of-the-art methods.

Keywords: Steganography · Steganalysis · Convolutional Neural Network · DenseNet

1 Introduction

Steganography is the process of concealing secret information within an ordinary image. The image which is used for hiding the secret information is called a *cover* image. The image after embedding a secret message is known as a *stego* image. Steganography can be categorized into two types- (1) Spatial domain and (2) Transform domain steganography. The spatial domain steganography hides the secret by modifying the pixels of the cover image. The DCT domain steganography conceals the secret message within the DCT coefficients [1] of the image. One of the traditional steganography schemes such as LSB replacement [20] hides the secret information in the least significant bits (LSB) of the image pixels. Modern steganography schemes such as HUGO [15], wow [5], S-UNIWARD [6] hides a secret message by minimizing some heuristically defined distortion function. The distortion function assigns a high cost when embedding in the smooth regions in the image whereas a low cost to the noisy areas of the image.

B. Deka et al. (Eds.): PReMI 2019, LNCS 11941, pp. 277–285, 2019.
https://doi.org/10.1007/978-3-030-34869-4_31

Steganalysis is the process of detecting the trace of the hidden message in the given image. Steganalysis can be broadly divided into two types: (1) Blind and (2) Targeted steganalysis. *Blind steganalysis* detects the embedding in the image without knowing the steganographic algorithm used for embedding. The *Targated steganalysis* utilizes the knowledge of the steganographic scheme used for embedding. Steganalysis methods work in two stages. In the first stage, features are extracted using some tools, and in the second stage, classification is done based on the extracted features. The distortion function based recent steganographic schemes are more likely to distribute the secret message in the noisy or high textured area of the image than the flat areas. Therefore, the steganalysis schemes assume that the steganographic noise lies in the high-frequency components of the source image, thereby, strive to capture these feature for steganalysis.

2 Related Works

A lot of steganalysis works have been reported in the literature. These works can be broadly categorized into two types: (1) Handcrafted feature based and (2) Deep feature based steganalysis.

The conventional handcrafted feature based methods use some fixed handcrafted filters to extract the steganalytic features which are used for steganalytic classification. Pevny et al. introduced the Subtractive Pixel Adjacency Matrix (SPAM) [14], which utilized the fact that the steganographic noise alters the dependencies between the neighboring pixels of an image. Using higher-order Markov chain [19] these dependencies are captured. The transition probability matrix of the Markov chain is used as features to train an SVM classifier [4] for steganalysis. The Spatial Rich Model (SRM) [3] is proposed by Fridrich and Kodovsky, which uses several linear and non-linear filters to compute noise residual, followed by 106 different submodels to capture diverse kinds of relationships between neighboring pixels of noise residual. The submodels are used to train the Ensemble Classifier (EC) [10] for steganalytic classification. SRM [3] showed considerable improvement in detecting the trace of steganographic embedding in images over SPAM [14]. The performance of classifiers depends on the quality of features supplied to the classifiers. Handcrafted feature based schemes such as SRM [3] and SPAM [14] which rely on several fixed handcrafted filters, may be suboptimal in extracting all the precise steganalytic features.

Convolutional Neural Networks (CNN) are known for best automatic feature extractors which mitigate the problems of handcrafted feature extraction. Recently, many steganalysis works have been reported in the literature; some of them are as follows: Qian et al. proposed GNCNN [16], a CNN based model for steganalysis. The GNCNN [16] comprise of a fixed preprocessing layer and five convolution layers for feature extraction, followed by three fully connected layers for classification. GNCNN used a Gaussian activation to capture stego and cover signals more precisely. The preprocessing layer has a fixed high-pass filter, which exposes the stego noise and suppresses the image component. GNCNN reported a comparable results with SRM [3] on S-UNIWARD [6], HUGO [15], and WOW [5]. Xu et al. proposed XuNet [21] comprised of a preprocessing layer with a high-pass filter and five groups of layers for feature extraction followed by a fully-connected network for classification. Each group consists of a convolution layer followed by an average pooling and Batch Normalization (BN) [8]. The first group used an ABS layer to capture all the values of noise residual (negative as well as positive values) which might be discarded by some of the activation functions, followed by a convolution layer. Authors claimed a considerable performance over SRM with EC when detecting HILL [11] and S-UNIWARD [6]. Ye et al. [22] proposed a CNN based framework which initializes the first layer of the model with the filters of SRM [3] to better capture the noise residual. They also introduced an activation function named truncated linear unit to capture noise residual with low SNR. Authors reported better performance as compared to SRM [3] for WOW [5], S-UNIWARD [6] and HILL [11] embedding. Tian and Li [18] proposed a CNN based steganalysis using transfer learning. The model used a Gaussian high-pass filter for the preprocessing of images followed by the pre-trained Inception-V3 [17] model for steganalytic classification.

It has been observed from the literature that most of the existing CNN based steganalysis schemes: (i) Sharply increase the feature space by using a sequence of kernels in subsequent layers. (ii) Use some fixed-size kernels (with less variation) which may not be much expressive in learning the stego features since the stego signal is weak and sparse in nature. A kernel with lower spatial dimension may not learn, and a kernel with higher spatial dimension may lead to overfitting. (iii) Use a fully connected layer at the end for classification. The use of fully-connected layers imposes a constraint that the training and testing must be carried out on the images with the same spatial dimension. In order to use images of different sizes, due to the restriction mentioned above, the images must be resized before testing. However, resizing may lead to loss of stego signals, conceptually similar to pooling.

In this paper, considering the shortcomings mentioned above, a densely connected convolution network for steganalysis has been proposed for steganalysis. In contrast to the existing schemes, the proposed scheme makes the following contributions:

- A densely connected convolutional network without pooling layers is proposed, which progressively captures the steganalytic features at different scales.
- The fully connected layers are removed, which allows the model to be tested on any size of images regardless of the size of images used for training.

The proposed scheme is trained and tested on BOSSBase 1.0 [2] dataset and the steganalytic performance is compared with SRM [3], SPAM [14] against S-UNIWARD [6], HUGO [15], WOW [5] and HILL [11]. The performance of the proposed scheme is also compared with a recently proposed scheme of Tian and Li [18] against WOW [5] and S-UNIWARD [6].

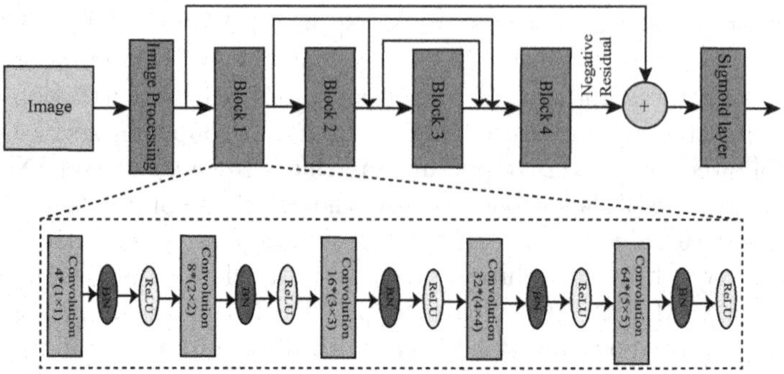

Fig. 1. The proposed model architecture. The architecture of each block is similar; one of the blocks (Block 1) is also shown in dotted box. Block consists of $4 \times (Conv \rightarrow BN \rightarrow ReLu)$ with sizes indicated for each convolution block.

3 Proposed Work

This section presents the proposed scheme for *targeted* steganalysis. The proposed model is inspired by DenseNet [7]. The model architecture of the proposed scheme is shown in Fig. 1. The proposed model comprises of an image processing layer followed by four densely connected convolution blocks and a sigmoid layer at the end for classification. Since the steganalytic classifiers are trained on the noise residual instead of the image components, a fixed high-pass filter (HPF) given in Eq. (1) has been used in the image processing layer.

$$HPF = \frac{1}{12}\begin{pmatrix} -1 & 2 & -2 & 2 & -1 \\ 2 & -6 & 8 & -6 & 2 \\ -2 & 8 & -12 & 8 & -2 \\ 2 & -6 & 8 & -6 & 2 \\ -1 & 2 & -2 & 2 & -1 \end{pmatrix} \tag{1}$$

The kernel of the image processing layer is kept fixed and is not updated while training. The noise residual extracted from the image processing layer is used as input to the subsequent dense blocks. The densely connected blocks are used to avoid the problem of vanishing gradients and stego features. Each block is connected to all its subsequent blocks. Consequently, all the blocks receive the feature map from all their preceding blocks. Each block comprises of five convolutional layers. The details of the layers used in each block are given in Table 1. All the blocks have the same configuration except for the last block (Block 4), where the output feature size is $1 \times 512 \times 512$. Convolutional layers in each block are followed by the Batch Normalization [8] for faster convergence and the ReLU [12] activation. Pooling layer has not been used since the use of pooling may result in loss of the stego noise. The number of convolutional filters progressively increase as $4, 8, 16, 32$ and 64, and the kernel size also increases gradually from 1×1 to 5×5 as each block slowly increases the scope of the convolution operator. The different sized kernels help to learn the features at different scales,

Table 1. Details of the layers in each dense block

Layer #	Input feature size	# of filters	Filter size	Output feature size
1	$1 \times 512 \times 512$	4	1×1	$4 \times 512 \times 512$
2	$4 \times 512 \times 512$	8	2×2	$8 \times 512 \times 512$
3	$8 \times 512 \times 512$	16	3×3	$16 \times 512 \times 512$
4	$16 \times 512 \times 512$	32	4×4	$32 \times 512 \times 512$
5	$32 \times 512 \times 512$	64	5×5	$64 \times 512 \times 512$

thereby avoiding the loss of stego signal and capturing more prominent features. The output of the densely connected blocks is a negative residual map which is pixel-wise added to noise residual extracted by the image processing layer to boost the noise components. The resulting output is used as input to the classification layer (sigmoid layer). The classification layer determines whether the input image is a stego or cover image by using the mean sigmoid over entire pixels. The whole framework is trained by minimizing the cross-entropy loss given in Eq. (2).

$$L = -\sum_{\forall x} p(x).\log(q(x)) \tag{2}$$

where $p(x)$ and $q(x)$ denotes the true and estimated distributions respectively, over a discrete variable x.

4 Implementation Details and Results

4.1 Experimental Setup

The experiments are carried out on BOSSBase v1.0 dataset [2]. The dataset consists of 10,000 cover images of size 512×512. The steganographic embedding algorithms[1] S-UNIWARD [6], HUGO [15], WOW [5] and HILL [11] are used to obtain stego images. Further, 10000 cover-stego pairs of images are divided into training: 5000, validation: 1000 and testing: 4000 cover-stego pairs. To compare the performance with the proposed model SRM [3] and SPAM [14] are implemented along with Ensemble Classifier v.2.0 [10], with the same split (5000 pairs for training and 5000 pairs for testing) as for proposed model. The proposed model is trained using Pytorch [13] on a standard workstation having NVIDIA Quadro M-4000 GPU (8 GB) for 90 epochs. The learning rate is initially set to 0.001 and decays by a factor of 10 every 30 epochs. The batch size is empirically kept as 8 (4 cover and 4 stego). Adam Optimizer [9] is used to optimize the proposed network parameters when training.

[1] Steganographic algorithms, feature extractors such as SRM, SPAM and Ensemble classifier can be found at: http://dde.binghamton.edu/download/.

Table 2. Steganalytic classification accuracy (in %) of the proposed scheme is compared to SRM [3] with Emsemble classifier [10] and SPAM [14] with Ensemble classifier against S-UNIWARD [6], HUGO [15], WOW [5] and HILL [11].

Scheme	Payload (bpp)	Proposed scheme	SRM with EC	SPAM with EC
S-UNIWARD	0.1	66.50%	59.05%	54.24%
	0.2	68.50%	62.17%	58.92%
	0.3	70.75%	65.80%	63.44%
	0.4	75.25%	73.70%	67.51%
HUGO	0.1	63.50%	60.03%	52.36%
	0.2	70.25%	67.98%	56.00%
	0.3	74.00%	74.46%	60.03%
	0.4	77.25%	78.30%	63.98%
WOW	0.1	67.25%	60.97%	52.46%
	0.2	70.75%	65.77%	55.82%
	0.3	74.00%	69.26%	58.98%
	0.4	76.50%	75.12%	62.33%
HILL	0.1	61.50%	55.45%	52.28%
	0.2	65.75%	61.37%	55.26%
	0.3	69.50%	67.11%	58.41%
	0.4	75.00%	72.58%	61.37%

Table 3. Comparison of the proposed scheme with Tian and Li [18] in terms of steganalytic classification accuracy (in %) against WOW [5] and S-UNIWARD [6].

Payload	WOW		S-UNIWARD	
bpp	Proposed scheme	Tian and Li [18]	Proposed	Tian and Li [18]
0.1	67.25%	67.90%	66.50%	65.10%
0.3	74.00%	69.00%	70.75%	67.20%
0.4	76.50%	71.4%	75.25%	69.80%

4.2 Results

The quantitative results for the proposed model are given in Table 2 when compared to the SRM with EC [3] and SPAM [14] with EC [10] against S-UNIWARD [6], HUGO [15], WOW [5], and HILL [11] steganographic schemes with different embedding rates. The results are measured in terms of percentage (%) classification accuracy. The best result is shown in the red color, and the blue color represents the second best result. A series of graphs are also given in Fig. 2 for a visual presentation where the proposed scheme is shown in red color, SRM with EC [3] is shown in green color and SPAM with EC [14] is shown in blue color. Results are evident that the proposed scheme outperformed SRM [3]

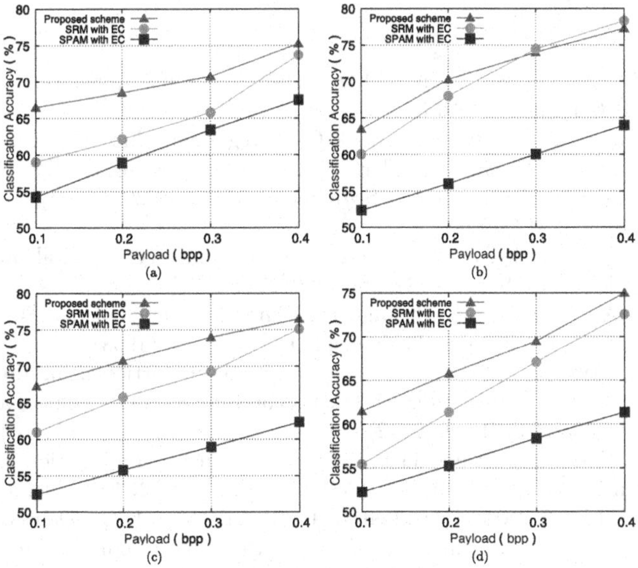

Fig. 2. Steganalytic performance comparison of the proposed scheme (Red) with SRM with EC (Green) and SPAM with EC (Blue) against: (a) S-UNIWARD (b) HUGO (c) WOW and (d) HILL steganography on embedding rates - {0.1, 0.2, 0.3, 0.4} bpp (Color figure online)

as well as SPAM [14] for most of the steganographic algorithms. The steganalytic performance of the proposed scheme is also compared with a recent work by Tian and Li [18], which has the same experimental setup in their work as the proposed scheme. The comparison is done against WOW [5] and S-UNIWARD [6] on embedding rates - {0.1, 0.3, 04} bits per pixel (bpp). The results are given in Table 3, the best result is shown in red color, and the next best is shown in blue color. The proposed scheme has comparable performance against WOW [5] on 0.1 bpp, and for the rest of steganographic embedding and payloads, the proposed scheme clearly outperformed Tian and Li [18].

5 Conclusion

In this paper, a densely connected convolution network based steganalysis is presented. The proposed model captures complex dependencies that are more appropriate for steganalysis, and the learned features avoid the loss of stego signals. The proposed model has no fully connected layer which adds advantage that the model can be tested on any size of the image unlike with fully-connected layers where the image size used for training and testing must be same. The steganalytic performance of the proposed scheme is compared with SRM, SPAM with Ensemble Classifier and a recent scheme by Tian and Li against different steganographic algorithm on different embedding rates. The proposed model outperforms the existing schemes with a considerable margin.

References

1. Ahmed, N., Natarajan, T., Rao, K.R.: Discrete cosine transform. IEEE Trans. Comput. **100**(1), 90–93 (1974)
2. Bas, P., Filler, T., Pevný, T.: "Break Our Steganographic System": the ins and outs of organizing BOSS. In: Filler, T., Pevný, T., Craver, S., Ker, A. (eds.) IH 2011. LNCS, vol. 6958, pp. 59–70. Springer, Heidelberg (2011). https://doi.org/10.1007/978-3-642-24178-9_5
3. Fridrich, J., Kodovsky, J.: Rich models for steganalysis of digital images. IEEE Trans. Inf. Forensics Secur. **7**(3), 868–882 (2012)
4. Hearst, M.A., Dumais, S.T., Osuna, E., Platt, J., Scholkopf, B.: Support vector machines. IEEE Intell. Syst. Their Appl. **13**(4), 18–28 (1998)
5. Holub, V., Fridrich, J.: Designing steganographic distortion using directional filters. In: 2012 IEEE International Workshop on Information Forensics and Security (WIFS), pp. 234–239. IEEE (2012)
6. Holub, V., Fridrich, J., Denemark, T.: Universal distortion function for steganography in an arbitrary domain. EURASIP J. Inf. Secur. **2014**(1), 1 (2014)
7. Huang, G., Liu, Z., Van Der Maaten, L., Weinberger, K.Q.: Densely connected convolutional networks. In: Proceedings of the IEEE Conference on Computer Vision and Pattern Recognition, pp. 4700–4708 (2017)
8. Ioffe, S., Szegedy, C.: Batch normalization: accelerating deep network training by reducing internal covariate shift. arXiv preprint arXiv:1502.03167 (2015)
9. Kingma, D.P., Ba, J.: Adam: a method for stochastic optimization. arXiv preprint arXiv:1412.6980 (2014)
10. Kodovsky, J., Fridrich, J., Holub, V.: Ensemble classifiers for steganalysis of digital media. IEEE Trans. Inf. Forensics Secur. **7**(2), 432–444 (2011)
11. Li, B., Wang, M., Huang, J., Li, X.: A new cost function for spatial image steganography. In: 2014 IEEE International Conference on Image Processing (ICIP), pp. 4206–4210. IEEE (2014)
12. Nair, V., Hinton, G.E.: Rectified linear units improve restricted Boltzmann machines. In: Proceedings of the 27th International Conference on Machine Learning (ICML 2010), pp. 807–814 (2010)
13. Paszke, A., Gross, S., Chintala, S., Chanan, G.: Pytorch: tensors and dynamic neural networks in Python with strong GPU acceleration. PyTorch: tensors and dynamic neural networks in Python with strong GPU acceleration 6 (2017)
14. Pevny, T., Bas, P., Fridrich, J.: Steganalysis by subtractive pixel adjacency matrix. IEEE Trans. Inf. Forensics Secur. **5**(2), 215–224 (2010)
15. Pevný, T., Filler, T., Bas, P.: Using high-dimensional image models to perform highly undetectable steganography. In: Böhme, R., Fong, P.W.L., Safavi-Naini, R. (eds.) IH 2010. LNCS, vol. 6387, pp. 161–177. Springer, Heidelberg (2010). https://doi.org/10.1007/978-3-642-16435-4_13
16. Qian, Y., Dong, J., Wang, W., Tan, T.: Deep learning for steganalysis via convolutional neural networks. In: Media Watermarking, Security, and Forensics 2015, vol. 9409, p. 94090J. International Society for Optics and Photonics (2015)
17. Szegedy, C., Vanhoucke, V., Ioffe, S., Shlens, J., Wojna, Z.: Rethinking the inception architecture for computer vision. In: Proceedings of the IEEE Conference on Computer Vision and Pattern Recognition, pp. 2818–2826 (2016)
18. Tian, J., Li, Y.: Convolutional neural networks for steganalysis via transfer learning. Int. J. Pattern Recognit Artif Intell. **33**(02), 1959006 (2019)

19. Tijms, H.C., Tijms, H.C.: Stochastic Models: An Algorithmic Approach, vol. 994. Wiley, Chichester (1994)
20. Wu, H.C., Wu, N.I., Tsai, C.S., Hwang, M.S.: Image steganographic scheme based on pixel-value differencing and LSB replacement methods. IEEE Proc.-Vis. Image Signal Process. **152**(5), 611–615 (2005)
21. Xu, G., Wu, H.Z., Shi, Y.Q.: Structural design of convolutional neural networks for steganalysis. IEEE Signal Process. Lett. **23**(5), 708–712 (2016)
22. Ye, J., Ni, J., Yi, Y.: Deep learning hierarchical representations for image steganalysis. IEEE Trans. Inf. Forensics Secur. **12**(11), 2545–2557 (2017)

Semi-supervised Multi-category Classification with Generative Adversarial Networks

Reshma Rastogi(✉) and Ritesh Gangnani

South Asian University, New Delhi 110021, India
reshma.khemchandani@sau.ac.in, ritesh.gangnani@gmail.com

Abstract. For training robust deep neural architectures to generate complex samples across varied domains, Generative Adversarial Networks (GANs) have shown promising performance in recent years. In previous works, the effectiveness of GANs in transforming images from the laseled source domain to the unlabeled target domain has shown high potential. In this paper, we outline a generalized semi-supervised learning framework where proposed 'Semi-supervised Multi-category Classification with Generative Adversarial Networks (SMC-GAN)' model, first, maps the data in the source domain to target domain to generate target-like source images and, then, learns to discriminate the target domain data using semi-supervised classifier. Extensive experimental evaluations on standard cross-domain datasets show that the proposed model is an efficient classifier and allows faster convergence than a conventional GAN approach for digit classification tasks.

Keywords: Domain adaptation · Adversarial learning · GAN · Semi-supervised learning

1 Introduction

Deep Convolutional Neural Networks (CNN) have shown amazing results for a variety of representation based learning tasks [1], but the success of these models purely rely on the availability of the substantial amount of labeled data, due to an event known as domain shift or dataset-bias [2]. However, in many real-world scenarios, acquiring the training labels can be expensive, tedious, or simply boring at times. One of the promising solution to overcome this problem is through domain adaptation, that aims to transfer the knowledge of labeled source domain to the unlabeled target domain and fine-tune underlying networks according to task-specific datasets. As the domains are of different distributions, the domain shift problem is a major concern. Thus, the goal of domain adaptation is to make the model generalise well on the target domain.

Adversarial networks are employed to generalise across different domains when a substantial amount of labeled data is available to train a deep CNN for source domain and no annotated data is available for training of target domain.

B. Deka et al. (Eds.): PReMI 2019, LNCS 11941, pp. 286–294, 2019.
https://doi.org/10.1007/978-3-030-34869-4_32

Many distinct models have been proposed to tackle this problem, primarily concentrating on reducing the distribution shift among the source and target domains to perform classification on the target domain. The alternative methods include transforming the image representations by mapping the source domain images to the target domain or by learning a generator model to reconstruct new target images from the source data [3]. Several recent Generative Adversarial Networks (GANs) have been proposed in sync with these approaches to adapt different data distributions [4–6] in the uni-directional manner. However, in [7], the authors proposed a symmetric bi-directional adaptation GAN model, that introduces a symmetric mapping between source domain to the target domain, and, additionally mapping target domain to the source domain, followed by training of two classifiers in both directions. In this model, both the classifiers are trained in a supervised manner on the labeled source data and generated target-like source data, respectively. However, authors did not considered target domain images at all while training the target classifier.

Although the above-mentioned classifiers only rely on supervised learning framework ignoring the available unlabeled data information while learning the classifier for the target data, it is intuitive that adversarial learning on both the source and target domain should take advantage of labeled and unlabeled information. Hence, we propose a framework termed as Semi-supervised Multi-category Classification with Generative Adversarial Networks (SMC-GAN), where the ultimate task is to learn a semi-supervised classifier for the unlabeled target data. As illustrated in Fig. 1, we first perform unsupervised domain adaptation that maps the labeled source images to the target domain using GANs and generates the target-like source images from the labeled source images employing domain-adversarial loss. Subsequently, we train the classifier on the generated target-like source images and then assign pseudo-labels to the unlabeled target images using the trained classifier by assigning the class labels based on maximum predicted probability. Finally, we re-train the classifier with these target data annotations resulting in the semi-supervised learning of the classifier model.

The rest of the paper is organized as follows: Sect. 2 reviews the related work in the Generative Adversarial Networks. Section 3 details out the problem Semi-supervised Multi-category Classification approach based on GAN. Subsequently, Sect. 4 summarizes the results and finally, Sect. 5 concludes the paper.

2 Related Work

Unsupervised domain adaptation has been a challenging research work in recent years, both theoretically and practically. Since a lot of prior work exists, our literature review primarily focuses on approaches using CNN for domain adaptation due to their firm empirical supremacy over the problem.

Generative Adversarial Networks (GAN) [8] consists of two components, a generator and a discriminator. The generator model generate samples approximately matching the distribution of the real data, whereas the discriminator model distinguishes between the generated samples drawn from the generator and real samples from training data. GANs have been applied to different domains successfully, such as imitating images of digit dataset and faces. They have been extended in several ways for unsupervised domain adaptation to tackle the domain shift. The conditional GAN [9] is a type of GAN, that employs the class annotations as additional input to both generator and discriminator model resulting in better learning of the networks when compared to baseline GAN models. In Adversarial Discriminative Domain Adaptation (ADDA) [10], author proposed a framework for unsupervised domain adaptation based on adversarial as well as discriminative learning. A discriminative mapping of target images to source feature space is learnt and the target test images are mapped to the same space for classification. Authors in [11] proposed CycleGAN which employed a round-trip mapping approach using two GANs for image translation i.e. source to target to source, which implies mapping source samples to target samples followed by mapping target back to source samples should return to their ground truth values, introducing a cycle-consistency loss. Use of this approach generates high quality transformed images to the target domain. In CyCADA [12], the model adapt representations at both pixel-level and feature-level, enforcing the cycle-consistency loss. Similarly, SBADAGAN [7] also encourage the entire network to generate target-like source images along with the source-like target images by employing two generative adversarial losses. Further, it simultaneously minimizes both the classification losses collectively to produce final annotations for the target images.

In [13], authors have used pseudo-label for unlabeled data, which corresponds to the maximum predicted probability for the semi-supervised learning framework. Thus, learning the model simultaneously on supervised as well as unsupervised dataset belonging to the same distribution. Similarly, in another approach [14], the semi-supervised model is trained by simultaneously minimizing the sum of supervised and unsupervised loss functions. To the best of our knowledge, the approach to combine CNN with semi-supervised learning has been explored in some recent works, however, the semi-supervised learning with adversarial domain adaptation has not been explored yet.

In the adversarial approaches discussed above, after successful feature mapping of either the source domain to target domain, and/or learning of target to source domain, the classifier is trained completely in a supervised manner on the labeled source data, and used for the labels prediction of target data. In our model, instead of training the classifier solely on, source data, we annotate, target data with pseudo-labels and train the classifier in a semi-supervised manner by using target-like source data (labeled) and actual target data (unlabeled).

3 Proposed Model

Our model is primarily focused on the semi-supervised classification task, predicting the correct labels of the provided unsupervised target domain in an efficient and well-generalised manner. In unsupervised adaptation, we assume access to two related datasets, i.e. a labeled source dataset $X_s = \{p_s^i, q_s^i\}_{i=1}^{N_s}$ drawn from a source domain S containing N_s samples, and an unlabeled set of

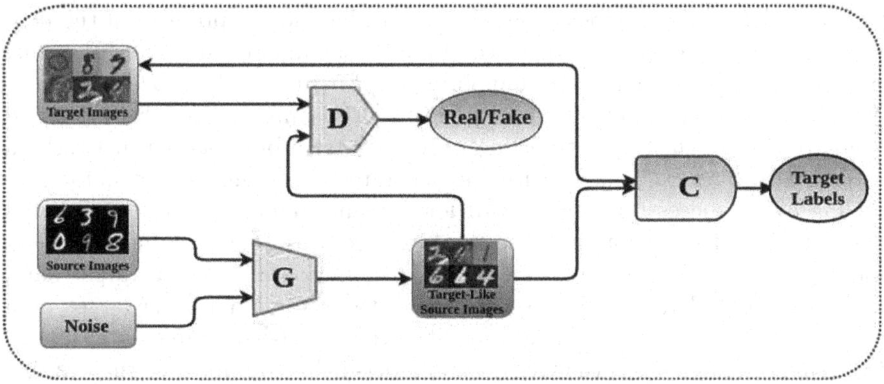

Fig. 1. An overview of Semi-supervised Multi-category Classification with Generative Adversarial Networks (SMC-GAN).

target images given as $X_t = \{p_t^i\}_{i=1}^{N_t}$ drawn from a target domain T with different distribution containing N_t samples, where no annotations are available for training. The ultimate task is to build a model to predict the class labels of the target dataset using the knowledge available in the source dataset. The structure of our GAN model is diagrammed in Fig. 1. Since there is a domain shift/gap among the datasets, thus, initially to bridge the domain shift/gap, we propose a framework that adapts the knowledge of source images after mapping them to the distribution of the target images in X_t. To perform the domain adaptation task, we use GANs to identify the mapping of a random noise vector z together with X_s to achieve the corresponding generated image. Thus, we train a generator network G that learns to map the source samples p_s^i to its target like version $p_{st}^i = G(p_s^i)_{i=1}^{N_s}$ defining the set $X_{st} = \{p_{st}^i, q_s^i\}_{i=1}^{N_s}$. The network is also extended with an adversarial discriminator, D, along with a semi-supervised classifier, C. The discriminator input takes X_t as well as generated target-like images X_{st} and recognizes them as two distinct sets, adversarially learning G. Thus, the generator is aimed to learn the distribution of real data, while the discriminator is aimed to correctly classify whether the input data is from the real data or the generator. The following part, i.e., classifier network, takes generated images as input and assigns the task-specific labels q_s^i to the generated target-like source images.

After the successful generation of the target-like source data from the generator, we train the semi-supervised classifier network on the generated labeled source data (i.e., from the generator) and the unlabeled target data. Subsequently, we propose a simple approach to predict the labels of the unsupervised target domain data using the knowledge of the data with a similar distribution. To achieve this, along with the GAN network, we simultaneously train a classifier using the labeled target-like generated images, that also benefits the generator model to improve through backpropagation. With a motivation to learn the classifier in a semi-supervised manner, we also require to train the classifier utilizing the knowledge in the unlabeled target data. To handle the problem of the non-availability of labels for images in the target domain, we use the self-labeling approach [13]. Consequently, we annotate each target sample p_t^i using the classifier model trained on generated target-like source samples X_{st}. These assigned pseudo-labels are then used transductively to re-train our classifier network on the original target data along with the generated target-like source data. Self-labeling has a successful track record for domain adaptation problems and is proved to be effective for modern deep architectures. The exactness of these assigned pseudo-labels is validated as the distribution of the real target data, and the generated target-like source data is the same. Thus, the target data should belong to one of the classes present in the target-like source data. In case of the slight domain shift, the true pseudo-labels helps to improvise the learning of the model. Whereas, for significant domain shift, the efficiency of the model is unaffected by the possible mislabeled samples as the distribution of both the labeled and unlabeled target samples is same.

3.1 Proposed Formulation

This section formalizes our proposed model and specify how the model is optimized using the loss functions.

We first describe the optimization of discriminator D given generator G. We input noise vector z belonging to uniform distribution along with the source images to the generator model. This allows an additional degree of freedom to model external variations in the dataset. We introduce a mapping from source to target through Generator G and train it to produce target-like samples that can fool Discriminator D. Adversarial discriminator attempts to classify the domains of real data and fake data. Thus, rather than robust binary cross-entropy which is used in common practice, we propose least square loss function for optimizing the discriminator parameters:

$$\min_G \max_D \mathcal{L}_D(D,G) = \mathbb{E}_{\boldsymbol{p}_t \sim T}[(D(\boldsymbol{p}_t) - 1)^2] + \mathbb{E}_{\boldsymbol{p}_s \sim S\,,\, \boldsymbol{z} \sim noise}[(D(G(\boldsymbol{p}_s, \boldsymbol{z})))^2], \quad (1)$$

where, p_s is sampled from source data distribution S, p_t is sampled from target data distribution T, z is sampled from the prior distribution $P_z(z)$ such as uniform distribution, and $E(\cdot)$ represents the expectation.

As the semi-supervised classifier network is trained considering the labeled as well as unlabeled data, the corresponding terms consists of two parts in the loss function and is given by:

$$\mathcal{L}_C = \alpha_1 \mathcal{L}_{supervised} + \alpha_2 \mathcal{L}_{unsupervised}, \quad (2)$$

where, $\mathcal{L}_{supervised}$ is the loss for the labeled target-like source data and $\mathcal{L}_{unsupervised}$ is the loss for unlabeled target data which is the self-labeling loss. For the semi-supervised classifier evaluated on the transformed source images as well as target images, corresponding loss $\mathcal{L}_C$ is a standard softmax cross-entropy, given as:

$$\mathcal{L}_{supervised} = \mathcal{L}_C(G, C) = \mathbb{E}_{\{\boldsymbol{p}_s, \boldsymbol{q}_s\} \sim \mathcal{S},\ \boldsymbol{z}_s \sim noise}[-\boldsymbol{q}_s \cdot \log(\hat{\boldsymbol{q}}_s)], \quad (3)$$

where, $\hat{\boldsymbol{q}}_s = (C(G(p_s, z_s)))$ and $\boldsymbol{q}_s$ is the one-hot encoding of the labels assigned to the target-like source data.

The loss for assignment of annotations to the X_t i.e. self-loss is a simple classification softmax cross-entropy, given as:

$$\mathcal{L}_{unsupervised} = \mathcal{L}_{self}(G, C) = \mathbb{E}_{\{\boldsymbol{p}_t, \boldsymbol{q}_{t_{self}}\} \sim T,\ z_t \sim noise}[-\boldsymbol{q}_{t_{self}} \cdot \log(\hat{\boldsymbol{q}}_{t_{self}})], \quad (4)$$

where, $\hat{\boldsymbol{q}}_{t_{self}} = (C(p_t))$ and $\boldsymbol{q}_{t_{self}}$ is the one-hot encoding of the assigned target labels. This loss is backpropagated to the generator G, that encourages the network to preserve the annotated category of the target images.

By collecting all the loss functions discussed above, we conclude SMC-GAN with the overall loss, given as:

$$\mathcal{L}_{SMC-GAN}(G, D, C) = \min_{G,C} \max_{D} \beta \mathcal{L}_D(D, G) + \alpha_1 \mathcal{L}_C(G, C) + \alpha_2 \mathcal{L}_{self}(G, C). \quad (5)$$

Here, $(\alpha_1, \alpha_2, \beta) \geq 0$ are weights to manipulate the interaction among the loss terms.

4 Experimental Results

We evaluate SMC-GAN for unsupervised domain adaptation across four different domain shifts. We explore four digits dataset of varying distributions, namely, MNIST, MNIST-M, USPS, SVHN, each consisting of 10 classes. The description of these datasets can be found in [7]. We compare our model against multiple state-of-the-art approaches, all based upon domain adversarial learning objectives.

4.1 Implementation Details

The model is implemented in python and experiments are performed using the Keras framework [19]. Our model architecture is analogous to that used in [20]. We used the ADAM [21] optimizer with the learning rate 10^{-4} for both the generator network as well as discriminator network. The model is trained for 500 epochs with the batch size of 32, resulting in no case of overfitting. The parameter α_1 defined in Eq. (5) is set to 1, whereas β is set to 10 to prevent the generator from indirectly switching labels. The training starts with the self-labeling loss and hence, α_2 is set to zero as this loss hinders the convergence of the generator model in the initial stage. After the generator starts to converge, α_2 is switched to 1 in order to increase the performance of the model.

Table 1. Comparison against existing works on unsupervised domain adaptation. SMC-GAN reports the resulting accuracy obtained by the proposed semi-supervised classifier.

	MNIST → USPS	MNIST → MNIST-M	SVHN → MNIST	USPS → MNIST
DANN [4]	85.1	77.4	73.9	73.0 ± 2.0
CoGAN [15]	91.2	62.0	not conv	89.1
ADDA [10]	89.4 ± 0.2	–	76.0 ± 1.8	90.1 ± 0.8
PixelDA [16]	95.9	98.2	–	–
DTN [5]	-	-	84.4	–
DIRT-T [17]	-	98.7	99.4	–
DA_{ass} [18]	–	89.5	97.6	–
CyCADA [12]	95.6 ± 0.2	–	90.4 ± 0.4	96.5 ± 0.1
SBADA-GAN [7]	97.6	99.4	76.1	95.0
SMC-GAN (Proposed)	91.74	98.4	75.43	98.07

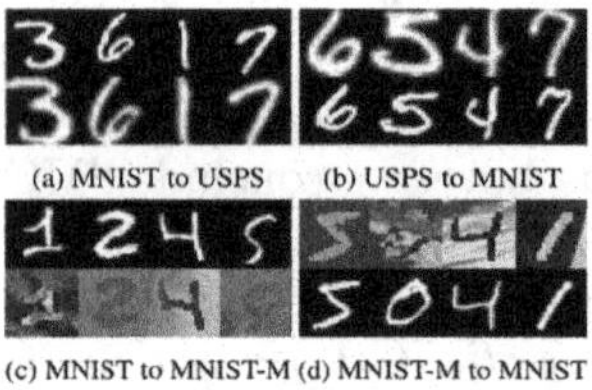

Fig. 2. Examples of generated digits. The top row represents the original samples, whereas the bottom row represents the corresponding generated image.

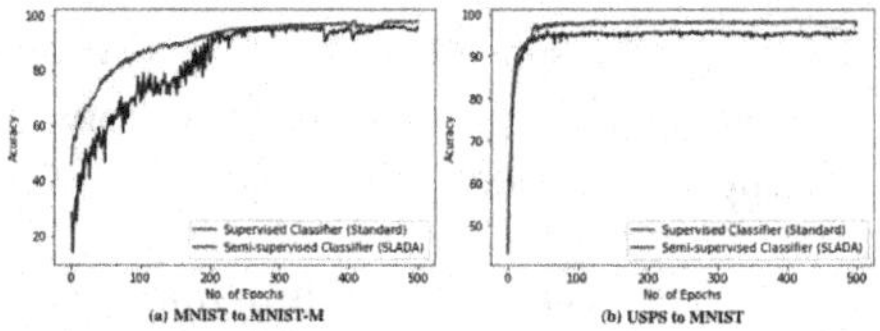

Fig. 3. Comparisons of accuracies obtained by supervised and semi-supervised classification framework as a function of number of epochs.

4.2 Results Discussion

Table 1 shows the results in above-mentioned evaluation settings. It can be seen that the proposed SMC-GAN model achieves competitive results and performs significantly better on most of the tasks when compared to previous unidirectional approaches [4,10,15] and few cyclic approaches [7,12]. In comparison to previous approaches, our model learns to generalise better for the target discriminative task under the guidance of the generator, thus, directly learning a target discriminative model through adversarial adaptation losses. The ultimate task of our model is to learn a semi-supervised classifier which has been shown to perform well when compared to supervised classifiers approaches used in earlier frameworks [7,9,10], and has a faster convergence rate of the generator model due to backpropagation by the classifier network. Moreover, we also experimented with MNIST-M (source) → MNIST (target) pair and achieved 99.2% accuracy for the classification task on the target dataset using proposed SMC-GAN. However, we have not included this setting in Table 1 as it is not reported by other referred papers. Furthermore, in Fig. 2, we show the generated target-like source images using the source samples.

Figure 3 shows the iterative comparison of accuracies as obtained by the standard supervised classifier and proposed semi-supervised classification framework.

It is evident that SMC-GAN framework achieves better performance in lesser number of epochs, which validates the claim that the proposed framework results into a faster and more generalizable model.

5 Conclusions

In this paper, we proposed a semi-supervised learning framework for efficient prediction of unlabeled samples in the target domain. We utilized unsupervised domain adaptation techniques based on adversarial objectives followed by a semi-supervised classification. Our method maps the source samples into the target domain to tackle the problem of distribution shifts between source and target data and learns a semi-supervised classifier for classifying the test patterns in target domain. We utilized the self-labelling approach on target samples and used these along with the generated target-like source images for the fine-tuning the resulting classification model. The proposed framework boosts the performance of the model significantly in terms of prediction performance and training time when compared to related models. In future, we would like to explore the semi-supervised methodology for cross-domain adaptation.

References

1. Yosinski, J., Clune, J., Bengio, Y., Lipson, H.: How transferable are features in deep neural networks? In: Advances in Neural Information Processing Systems, pp. 3320–3328 (2014)
2. Gretton, A., Smola, A., Huang, J., Schmittfull, M., Borgwardt, K., Schölkopf, B.: Covariate Shift and Local Learning by Distribution Matching, pp. 131–160. MIT Press, Cambridge (2009)
3. Ghifary, M., Kleijn, W.B., Zhang, M., Balduzzi, D., Li, W.: Deep reconstruction-classification networks for unsupervised domain adaptation. In: Leibe, B., Matas, J., Sebe, N., Welling, M. (eds.) ECCV 2016. LNCS, vol. 9908, pp. 597–613. Springer, Cham (2016). https://doi.org/10.1007/978-3-319-46493-0_36
4. Ganin, Y., et al.: Domain-adversarial training of neural networks. J. Mach. Learn. Res. **17**(1), 2030–2096 (2016)
5. Zhang, X., Yu, F.X., Chang, S.-F., Wang, S.: Deep transfer network: unsupervised domain adaptation. CoRR, abs/1503.00591 (2015)
6. Ganin, Y., Lempitsky, V.: Unsupervised domain adaptation by backpropagation. In: International Conference on Machine Learning, pp. 1180–1189 (2015)
7. Russo, P., Carlucci, F.M., Tommasi, T., Caputo, B.: From source to target and back: symmetric bi-directional adaptive GAN. In: Proceedings of the IEEE Conference on Computer Vision and Pattern Recognition, pp. 8099–8108 (2018)
8. Goodfellow, I., et al.: Generative adversarial nets. In: Advances in Neural Information Processing Systems, pp. 2672–2680 (2014)
9. Mirza, M., Osindero, S.: Conditional generative adversarial nets. CoRR, abs/1411.1784 (2014). http://arxiv.org/abs/1411.1784
10. Tzeng, E., Hoffman, J., Saenko, K., Darrell, T.: Adversarial discriminative domain adaptation. In: Proceedings of the IEEE Conference on Computer Vision and Pattern Recognition, pp. 7167–7176 (2017)

11. Zhu, J.-Y., Park, T., Isola, P., Efros, A.A.: Unpaired image-to-image translation using cycle-consistent adversarial networks. In: Proceedings of the IEEE International Conference on Computer Vision, pp. 2223–2232 (2017)
12. Hoffman, J., et al.: CyCADA: cycle-consistent adversarial domain adaptation. In:, Dy, J., Krause, A. (eds.) Proceedings of the 35th International Conference on Machine Learning, Proceedings of Machine Learning Research, vol. 80. Stockholmsmssan. PMLR, Stockholm Sweden, 10–15 July 2018, pp. 1989–1998 (2018)
13. Lee, D.-H.: Pseudo-label: the simple and efficient semi-supervised learning method for deep neural networks. In: Workshop on Challenges in Representation Learning. ICML, vol. 3, p. 2 (2013)
14. Rasmus, A., Berglund, M., Honkala, M., Valpola, H., Raiko, T.: Semi-supervised learning with ladder networks. In: Advances in Neural Information Processing Systems, pp. 3546–3554 (2015)
15. Liu, M.-Y., Tuzel, O.: Coupled generative adversarial networks. In: Advances in Neural Information Processing Systems, pp. 469–477 (2016)
16. Bousmalis, K., Silberman, N., Dohan, D., Erhan, D., Krishnan, D.: Unsupervised pixel-level domain adaptation with generative adversarial networks. In: 2017 IEEE Conference on Computer Vision and Pattern Recognition (CVPR), pp. 95–104 (2017)
17. Shu, R., Bui, H.H., Narui, H., Ermon, S.: A DIRT-T approach to unsupervised domain adaptation. In: International Conference on Learning Representations (2018). https://openreview.net/forum?id=H1q-TM-AW
18. Haeusser, P., Frerix, T., Mordvintsev, A., Cremers, D.: Associative domain adaptation. In: Proceedings of the IEEE International Conference on Computer Vision, pp. 2765–2773 (2017)
19. Chollet, F., et al.: Keras (2015)
20. Radford, A., Metz, L., Chintala, S.: Unsupervised representation learning with deep convolutional generative adversarial networks. CoRR, abs/1511.06434 (2016)
21. Kingma, D.P., Ba, J.: Adam: a method for stochastic optimization. CoRR, vol. abs/1412.6980 (2015)

Soft and Evolutionary Computing

A Soft Computing Approach for Optimal Design of a DC-DC Buck Converter

Barnam Jyoti Saharia[1(✉)] and Nabin Sarmah[2]

[1] Department of Electrical Engineering, Tezpur University, Tezpur 784028, Assam, India
bjsaharia@gmail.com

[2] Department of Energy, Tezpur University, Tezpur 784028, Assam, India
nabin@tezu.ernet.in

Abstract. The state space dynamic model of a DC-DC buck converter used for optimal designing of the converter to minimize the overall losses is presented in this paper. The optimum design criterion involves the selection of converter switching frequency, inductance and capacitance values for continuous conduction mode (CCM) of operation. The ripple in current, ripple in voltage and bandwidth are considered as constraints along with the criterion for CCM. Optimizing algorithms, namely Particle Swarm Optimization (PSO), Simulated Annealing (SA) and Firefly Algorithm (FA) are used to generate the solution to the optimal design problem. The comparative investigation of the algorithms reveals that PSO outperforms the FA and SA in terms of computational effort, convergence time, and also the most efficient design having minimum losses.

Keywords: Buck converter · Optimization techniques · Particle Swarm Optimization · Simulated Annealing · Firefly Algorithm

1 Introduction

The proper design of a DC-DC converter manually often is observed to be a process involving time and cost. This has led to applying optimization methods in an attempt to ease the burden of the design process. DC-DC buck converters have improved in performance over the years and some recent advances are due to the progress in the energy storing elements, which have reduced in size and are characterized by lower associated losses. Power electric switches have also been refined, in terms of better blocking voltage, low on-state resistance, and ability to withstand transient stresses. But selecting optimal converter design parameters that match the dimension constraints and gives better efficiency still is an optimization problem [1].

Available literature suggests the use of different methods to optimally select the converter designing parameters. They vary in terms of the objective functions as well as the constraints, in addition to the varying methodologies to implement the optimization process. Balachandran and Lee [2] discuss a practical optimization approach which involves selection of a working design which meets the power circuit performance parameters and concurrently optimizes the weight or total circuit losses. The technique enables a cost-effective design. Additionally, the computer-aided method enables the

B. Deka et al. (Eds.): PReMI 2019, LNCS 11941, pp. 297–305, 2019.
https://doi.org/10.1007/978-3-030-34869-4_33

designer the options in the solution to a tradeoff in weight-efficiency, investigation of the impacts of component characteristics and the converter requirements for the desired design, and optimal configuration in the system power. An design approach for a monolithic DC-DC buck converter is presented in [3] where the criterion for selection involved the decision variables which included voltage swing in MOSFET gate driver, switching frequency and the current ripple minimization. In addition, electromagnetic interference (EMI) minimization is also addressed along with the efficiency of the converters [4, 5]. While authors in [6] make use of Particle Swarm Optimization (PSO) for the control of DC-DC converters. In most cases, the optimization approach involves only one parameter. Other methods involve dividing the process into several stages, considering only one or two parameters at each stage. This doesn't allow for the selection of a number of constraints for the optimization problem and hence cannot accurately allow for all the possible constraint selection in the design procedure.

Seeman and Sanders [7] applied the Lagrangian function to optimize a switched capacitor converter while the augmented Lagrangian method for optimization of a half bridge DC-DC converter was discussed Wu et al. [8]. Quadratic programming [9] also finds its application in the designing of DC-DC converters. The above mentioned methods all have the drawback of solving the problem for local optima, which is dependent on the initial starting point and hence doesn't give the global optima. In order to ensure global optimum results are obtained, the current work involves the design of an optimal DC-DC buck converter using soft computing techniques. A state-space dynamic model of the converter is considered along with the design parameters. The converter is optimized using Particle Swarm Optimization (PSO), Simulated Annealing (SA) and Firefly Algorithm (FA) optimization algorithms respectively and a comparison on the performance of the algorithms is also presented.

Following sections present a brief introduction of the algorithms considered, followed by the state space model of the DC-DC buck converter used to optimally select the design parameters for efficient performance. The optimization process is presented which is followed the results and discussion. Conclusion and future work summarizes and draws a conclusion to the work carried out.

2 Optimization Algorithms

The following sub-sections cover in brief the theory of the three optimization algorithms, namely PSO, SA and FA. The underlying governing principle is discussed, which forms the basis for optimal design of dc-dc buck converter parameters used in the current work.

2.1 Particle Swarm Optimization (PSO)

Kennedy and Eberhart introduced Particle Swarm Optimization in 1995 [10]. The algorithm mimics the natural behavior of bird and fish as they make use of collective or swarm intelligence. Optimization problems that have possible multiple solutions adopt PSO as a solution approach to the optimization problem to arrive at the desired optimum. The algorithm involves movement of the swarm or the particles of the swarm to locate the overall best value of velocity and position in the collective swarm in a well-defined or bounded search space. The iterative algorithm involves the evaluation of the position and velocity of each particle of the swarm in each step to arrive at the global optimized value. They are then updated for the position and velocity of the individual components of the swarm is given as [11]:

$$x_i^{k+1} = x_i^k + v_i^{k+1} \quad (1)$$

$$v_i^{k+1} = wv_i^k + c_1 r_1 P_{besti} + c_2 r_2 g_{besti} \quad (2)$$

Where, v_i^{k+1} represents the swarm velocity of the ith particle in $k + 1^{th}$ iteration, P_{besti} gives the particles best value while the global best value of the swarm is represented by g_{besti}. For the learning factor is defined as w, and the position constants are represented by c_1, c_2 while r_1, r_2 have a random value in the range of (0, 1) [11]. Figure 1 gives a representation of the movement of the particles in the swarms.

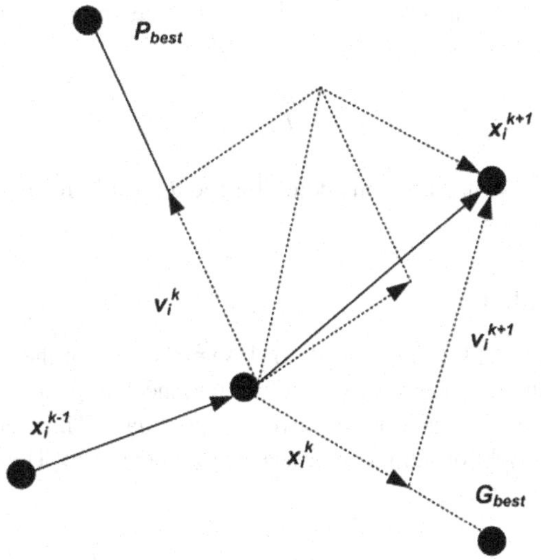

Fig. 1. Movement of particles in a swarm in the PSO algorithm [11]

2.2 Simulated Annealing (SA)

Simulated Annealing (SA) is used in the solutions for global optimization problems and is a random search technique in nature. The algorithm covers the imitation of the annealing phenomenon where a metal cools and freezes to form the crystalline state with the minimum of energy and the large crystals are formed to reduce the defects in the metal crystals. The application of SA into optimization process was initiated by Kirkpatrick, Gellat and Vecchi in 1983 [12]. Unlike gradient-based methods and deterministic search methods which are limited by their property of being trapped into the local minima, SA is able to avoid getting stuck at the local minima.

SA makes use of random search in terms of Markov Chain, which functions to accept improvement in the objective function and at the same time keeps some changes which are not ideal. For example, if we consider the case of the minimization problem, ideally we want to keep the changes in the iteration process that leads to an overall decrease in the value of the objective function f. However the algorithm also keeps solutions or values that lead to an increase in the value of f with a probability p called transitional probability which is defined as:

$$p = e^{-\frac{\Delta E}{k_B T}} \tag{3}$$

Where k_B is the Boltzmann's constant and for simplicity, we use k = 1. T gives the value for the Temperature that regulates the annealing process. Change in energy levels is given by ΔE. The transition probability which is based on Boltzmann distribution and links ΔE to Δf, i.e. the change of objective function is given by:

$$\Delta E = \gamma \Delta f \tag{4}$$

Where γ is a real constant. We assume its value to be unity for ease of computation. Thus the probability thus becomes

$$p(\Delta f, T) = e^{-\frac{\Delta f}{T}} \tag{5}$$

A random number is usually considered for the threshold regardless of whether the change is accepted or not.

2.3 Firefly Algorithm

Developed by Yang [13], Firefly Algorithm (FA) makes use of the attractiveness of the firefly by their brightness. Light absorption experiences exponential decay and light variation with distance is related by an inverse square law. The algorithm models this intensity variation of light or attractiveness as a non-linear term. The FA can be equated for the solution vector x_i as:

$$x_i^{t+1} = x_i^t + \beta_0 e^{-\gamma r_{ij}^2}(x_j^t - x_i^t) + \alpha \varepsilon_i^t \tag{6}$$

Where α represents a scaling factor which controls the step sizes for the randomized walks, γ is the parameter that controls the visibility of the fireflies the search mode, β_0 being the attractiveness constant for fireflies with zero distance between them. r_{ij} which denotes the distance between firefly i and firefly j and is represented in-terms of their Cartesian co-ordinates. To speed up the overall algorithm convergence the degree of randomness is usually reduced gradually using:

$$\alpha = \alpha_0 \theta^t \tag{7}$$

3 State Space Modeling of DC-DC Buck Converter

DC-DC buck converter performs well in the tracking of photovoltaic (PV) systems with power point tracking under low radiation and temperatures [14, 15]. The state-space dynamic modeling of the DC-DC buck converter considering the system losses, current and voltage ripples and other constraints that are intrinsic to the process of optimal converter design is presented in this section. A buck converter circuit is shown in Fig. 2.

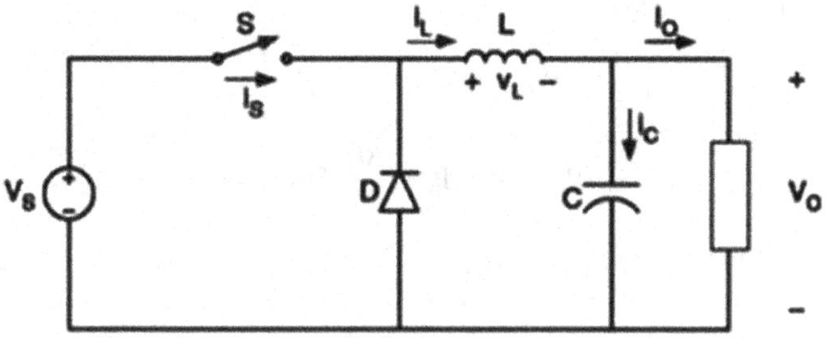

Fig. 2. Electrical circuit of a buck converter [14]

Using a state vector approach for the converter we have [16, 17]:

$$x = \begin{pmatrix} i_1 \\ v_0 \end{pmatrix} \tag{8}$$

$$\frac{di}{dt} = \frac{-V_0}{L} + \frac{V_i}{L}u \tag{9}$$

$$\frac{dv_0}{dt} = \frac{i_1}{C} - \frac{V_0}{RC} \tag{10}$$

Where i_1 represents inductor current while output voltage is given by v_0. The equations are dependent on the initial state determined by u which acts as the control signal for the switching device in the ON (u = 1) or OFF (u = 0) state. The inductor, the capacitor and the load resistance values are given by L, C and R and V_i represents the input voltage. For the design of the converter the current ripples, the voltage ripple, size of inductor for continuous conduction mode (CCM) and the constraints imposed by the Bandwidth (BW) are to be determined along with the state space dynamic model. For the buck converter we have:

$$\Delta i = \frac{V_0}{Lf_s}(1 - D) \tag{11}$$

$$\Delta v_o = \frac{V_0}{8Lf_s^2C}(1 - D) \tag{12}$$

$$Lf_s \geq \frac{V_0}{2I_0}(1 - D) \tag{13}$$

$$BW > 2\pi(10\%f_s) \tag{14}$$

Where V_O gives the output voltage; $f_s = 1/T_s$ is the switching frequency. The power calculations are important for the optimal design of the converter. This involves the losses due to parasitic resistances, switching losses caused by parasitic capacitance. The total power loss is given as:

$$P_{Q1} = P_{ONQ1} + P_{SWQ1} \tag{15}$$

$$P_{ONQ1} = (\mathrm{I}_0^2 + \frac{\Delta i_1^2}{12})\mathrm{DR}_{DS} \tag{16}$$

$$P_{SWQ1} = V_i I_0 (T_{swON} + T_{swOFF}) f_s \tag{17}$$

$$P_{IND} = (\mathrm{I}_0^2 + \frac{\Delta i_1^2}{12})R_L \tag{18}$$

$$P_{CAP} = (\frac{\Delta i_1^2}{12})R_C \tag{19}$$

$$P_{BUCK} = P_{onQ1} + P_{SW} + P_{IND} + P_{CAP} \tag{20}$$

The efficiency of the converter is given by:

$$\eta = \frac{P_{Load}}{P_{Load} + P_{Buck}} \tag{21}$$

T_{SWON} and T_{SWOFF} represents the transition time to ON and OFF the device, I_O is the average output current, R_{DS} is the on-state resistance of the MOSFET. R_L represents the loss component in the inductor while RC represents the equivalent series resistance in the capacitor. P_{Load} gives the averaged power at the load.

4 Optimal Design of DC-DC Buck Converter

To set up the optimization problem, we need to frame the objective function subject to certain constraints. The current work follows the optimal design of the converter by reducing the total power loss in the converter, i.e. to minimize P_{buck}, such that the efficiency of the converter is maximum. The considered constraints are the maximum and minimum size of the design variables, the admissible ripples, the CCM criterion (Eqs. 11–13) and the BW (Eq. 14). Mathematically it can be represented as:

Minimize P_{Buck} for

$$L_{min} \leq L \leq L_{max}$$

$$C_{min} \leq C \leq C_{max}$$

$$f_{min} \leq f_s \leq f_{max}$$

$$D_{min} \leq D \leq D_{max}$$

$$\Delta i_l \leq a\% I_O$$

$$\Delta v_O \leq b\% V_O$$

Where a and b limit the averaged magnitude of current and voltage in percentage respectively. In order to design a converter the parameters considered are presented in Table 1. These include the values of the converter parameters required for the design as well as the range of the design variables considered for the optimization process.

Table 1. Design parameters considered for optimization.

Parameter	Value	Parameter	Value
V_{in}	15 V	L_{min}	0.1 μH
V_o	5 V	L_{max}	100 mH
I_o	15 A	C_{min}	0.1 μF
R_{DS}	5.2 mΩ	C_{max}	100 μF
T_{swON}	10^{-8} s	f_{smin}	10 kHz
T_{swOFF}	10^{-8} s	f_{smax}	100 kHz
D_{min}	0.02	a	15% of I_O
D_{max}	0.2	b	15% of V_O

5 Result and Discussion

In order to optimize the buck converter, the optimization algorithms are formulated in m-files in the MATLAB software interface. The algorithms are implemented on a PC with 64 bit Windows 8 operating system having a Intel® core processor (i5) with 4.00 GB RAM. As the algorithms are inherently random in nature each algorithm is run for 100 iterations. Each iteration result is recorded and the results for best value, average and standard deviation obtained for each of the algorithms is summarized in Table 2.

Table 2. Optimized design parameters using PSO, SA and HAS

Comparison parameter	Algorithm	D	L (μH)	C (μF)	Fs (kHz)	P_{BUCK} (W)	η (%)
Best	**PSO**	**0.101**	**180**	**1.25**	**10.062**	**1.289**	**94.58**
	SA	0.100	78.6	820	10.68	1.292	94.57
	FA	0.101	148	574	24.01	1.295	94.56
Average	PSO	**0.101**	**148**	**108.2**	**10.06**	**1.292**	**94.57**
	SA	0.100	107	63.7	56.80	1.498	93.76
	FA	0.100	98.7	63	51.74	1.481	93.83
Standard deviation	**PSO**	**0.000242**	**42**	**334**	**0.063**	**0.013**	**0.051**
	SA	0.000265	45.3	362	26	0.117	0.456
	FA	0.000277	45	372	25.69	0.109	0.425

Table 2 shows the results of the optimization process for the PSO, SA and FA algorithm. The execution times for the three algorithms for 100 iterations were 1.56 s, 26.86 s, and 1.88 s for PSO, SA and FA algorithms respectively.

From the optimization results it is evident that the best result for designing a optimized buck converter, with the most efficient performance and minimum losses is obtained by the PSO algorithm. The PSO algorithm also takes less computational time to execute the 100 iterations and the variation in average values for power loss are 15% and 16% when compared to FA and SA respectively. The variation in average efficiency is within 1% with both SA and FA. Thus the PSO algorithm outperforms both SA and FA in-terms of computational time and to best optimize designed configuration at the minimum power loss.

6 Conclusion and Future Prospects

The paper discusses the problem of the optimal design of a DC-DC buck converter. Comparison is drawn on the performance of three popular optimization algorithms, namely PSO, SA and FA with respect to their performance in the design optimization problem. It has been found that PSO outperforms both SA and FA in arriving at the best results of converter design at minimal power loss. Additionally, the PSO algorithm gives average results with better design parameters at minimum losses which is an improvement by 15% and 16% when compared to the average results obtained using SA and FA respectively.

In recent times, many new optimization algorithms have been developed. The results of the current work can be compared to algorithms such as Biogeography based Optimization (BBO), Gravitational Search Algorithm (GSA), Cuckoo Search Algorithm(CS), Bat Algorithm(BA), which have shown good results in solving design optimization problems. Additionally, in order to find the statistically significant results, Wilcoxon Rank sum test can also be performed to compare the algorithms.

References

1. Leyva, R., Ribes-Mallada, U., Garces, P., Reynaud, J.F.: Design and optimization of buck and double buck converters by means of geometric programming. Math. Comput. Simul. **82**, 1516–1530 (2012)
2. Balachandran, S., Lee, F.C.Y.: Algorithms for power converter design optimization. IEEE Trans. Aerosp. Electron. Syst. **AES-17**, 422–432 (1981)
3. Kursun, V., Narendra, S.G., De, V.K., Friedman, E.G.: Low-voltage-swing monolithic DC-DC conversion. IEEE Trans. Circuits Syst. II Express Briefs **51**, 241–248 (2004)
4. Ray, R.N., Chatterjee, D., Goswami, S.K.: Reduction of voltage harmonics using optimisation-based combined approach. IET Power Electron. **3**, 334–344 (2010)
5. Yuan, X., Li, Y., Wang, C.: Objective optimisation for multilevel neutral-point-clamped converters with zero-sequence signal control. IET Power Electronics, vol. 3, pp. 755–763. Institution of Engineering and Technology (2010)
6. Fermeiro, J.B.L., Pombo, J.A.N., Calado, M.R.A., Mariano, S.J.P.S.: A new controller for DC-DC converters based on particle swarm optimization. Appl. Soft Comput. **52**, 418–434 (2017)
7. Seeman, M.D., Sanders, S.R.: Analysis and optimization of switched-capacitor DC–DC converters. IEEE Trans. Power Electron. **23**, 841–851 (2008)
8. Wu, C.J., Lee, F.C., Balachandran, S., Goin, H.L.: Design optimization for a half-bridge DC-DC converter. In: 1980 IEEE Power Electronics Specialists Conference, pp. 57–67 (1980)
9. Busquets-Monge, S., et al.: Design of a boost power factor correction converter using optimization techniques. IEEE Trans. Power Electron. **19**, 1388–1396 (2004)
10. Kennedy, J., Eberhart, R.: Particle swarm optimization. In: IEEE International Conference on Neural Networks, Proceedings, vol. 1944, pp. 1942–1948 (1995)
11. Saharia, B.J., Brahma, H., Sarmah, N.: A review of algorithms for control and optimization for energy management of hybrid renewable energy systems. J. Renew. Sustain. Energy **10**, 053502 (2018)
12. Kirkpatrick, S., Gelatt, C.D., Vecchi, M.P.: Optimization by simulated annealing. Science **220**, 671–680 (1983)
13. Yang, X.-S.: Firefly algorithm, stochastic test functions and design optimisation. Int. J. Bio-Inspired Comput. **2**, 78–84 (2010)
14. Bhattacharjee, S., Saharia, B.J.: A comparative study on converter topologies for maximum power point tracking application in photovoltaic generation. J. Renew. Sustain. Energy **6**, 053140 (2014)
15. Saharia, B.J., Manas, M., Sen, S.: Comparative study on buck and buck-boost DC-DC converters for MPP tracking for photovoltaic power systems. In: 2016 Second International Conference on Computational Intelligence & Communication Technology (CICT), pp. 382–387 (2016)
16. Rashid, M.H.: Power Electronics: Circuits, Devices, and Applications. Pearson Education India (2004)
17. Umanand, L.: Power Electronics Essentials and Applications, 1st edn. Wiley, New York (2009)

MR_IMQRA: An Efficient MapReduce Based Approach for Fuzzy Decision Reduct Computation

Kiran Bandagar, Pandu Sowkuntla(✉), Salman Abdul Moiz, and P. S. V. S. Sai Prasad

School of Computer and Information Sciences, University of Hyderabad, Hyderabad 500046, Telangana, India
bandagar.kiran30@gmail.com, pandu.sowkuntla@uohyd.ac.in, salman.abdul.moiz@gmail.com, saics@uohyd.ernet.in

Abstract. Fuzzy-rough set theory, an extension to classical rough set theory, is effectively used for attribute reduction in hybrid decision systems. However, it's applicability is restricted to smaller size datasets because of higher space and time complexities. In this work, an algorithm MR_IMQRA is developed as a MapReduce based distributed/parallel approach for standalone fuzzy-rough attribute reduction algorithm IMQRA. This algorithm uses a vertical partitioning technique to distribute the input data in the cluster environment of the MapReduce framework. Owing to the vertical partitioning, the proposed algorithm is scalable in attribute space and is relevant for scalable attribute reduction in the areas of Bioinformatics and document classification. This technique reduces the complexity of movement of data in shuffle and sort phase of MapReduce framework. A comparative and performance analysis is conducted on larger attribute space (high dimensional) hybrid decision systems. The comparative experimental results demonstrated that the proposed MR_IMQRA algorithm obtained good sizeup/speedup measures and induced classifiers achieving better classification accuracy.

Keywords: Fuzzy-rough sets · Hybrid decision systems · Attribute reduction · Iterative MapReduce · Apache Spark · Vertical partitioning

1 Introduction

The decision system with different types of attributes (e.g., categorical, real-valued, set-valued, and boolean) is called as Hybrid Decision System (HDS). Traditional approaches like rough sets [7] require discretization of numeric attributes to perform attribute reduction, which can result in significant information loss [4]. Extensions were proposed to classical rough set theory to overcome this problem. Dubois and Prade [2] developed fuzzy-rough sets and rough-fuzzy sets, as hybrid approaches combining strengths of fuzzy sets and rough sets together.

B. Deka et al. (Eds.): PReMI 2019, LNCS 11941, pp. 306–316, 2019.
https://doi.org/10.1007/978-3-030-34869-4_34

Out of these, fuzzy-rough sets have evolved as a standard approach for feature subset selection in hybrid decision systems.

Jensen et al. [5], proposed new approaches for fuzzy-rough attribute reduction, where, different algorithms were designed based on *attribute dependency degree measure* and *discernibility matrix* methods. Cornelis [1] proposed a selection of the subset of features with fuzzy decision reducts and designed a Modified Quick Reduct Algorithm (MQRA). Sai Prasad et al. [8] proposed an efficient approach IMQRA (Improved Modified Quick Reduct Algorithm) for fuzzy decision reduct computation based on MQRA [1] by incorporating a simplified computational model and positive region removal.

All the existing fuzzy-rough reduct computation algorithms are sequential and can only handle smaller size datasets. A little attention has been paid on parallel/distributed techniques for fuzzy-rough attribute reduction to deal with large-scale datasets, particularly high dimensional datasets. Therefore, it is the need of the hour to research the issue of fuzzy-rough set based attribute reduction in parallel/distributed approach.

With the objective of scalable fuzzy-rough set feature selection, in this paper, a novel MapReduce based fuzzy-rough Improved Quick Reduct Algorithm (MR_IMQRA) is proposed. It is implemented on iterative MapReduce framework of *Apache Spark* [12]. Existing classical rough set based MapReduce approaches for attribute reduction [11] use object space partitioning (horizontal partitioning technique) of the input data to the nodes of the cluster. This technique results in complicated shuffle and sort phase for the datasets having the larger attribute space (high dimensionality). In contrast, proposed MR_IMQRA is attribute space (vertical partitioning technique) partitioning based approach suitable for datasets of larger attribute space prevalent in the areas of Bioinformatics and document classification.

The rest of this paper is organized as follows. The related details of fuzzy-rough attribute reduction and the existing IMQRA algorithm are given in Sect. 2. The proposed MR_IMQRA algorithm is discussed in Sect. 3, along with MapReduce based implementation details. Comparative experimental results and analysis are provided in Sect. 4. Finally, the conclusion of this paper is given in Sect. 5.

2 Related Work

This section provides related definitions, terminology and concepts for fuzzy-rough attribute reduction based on [2,5,9] and presents the existing work of Improved Modified Quick Reduct Algorithm (IMQRA) [8].

2.1 Fuzzy-Rough Attribute Reduction

Let $HDT = (U, C^h = C^s \cup C^r, \{d\})$ be a Hybrid Decision Table. Here U represents the set of objects, C^s is set of symbolic (categorical) attributes, C^r is set of numerical (real valued) attributes, and d is the symbolic decision attribute. In fuzzy rough sets [2,5,9], a fuzzy similarity relation is defined on objects for measuring the graded indiscernibility based on numeric attribute. For a numeric attribute $a \in C^r$, R_a represents fuzzy similarity relation, where $R_a(i,j)$, $\forall(i,j) \in U \times U$ gives fuzzy similarity for any pair of objects i, j. It is to be noted that, if an attribute is qualitative (categorical), then the classical indiscernibility relation is adopted, hence $a \in C^s$, $R_a(i,j)$ is taken as either 1 (if the object values are equal) or 0 (if the object values are not equal). The fuzzy similarity relation R can be extended for a set of attributes $P \subseteq C^h$ by using a specified t-norm Γ as given,

$$R_P(i,j) = \Gamma\left(R_a(i,j)\right) \; \forall i,j \in U \; and \; \forall a \in P \tag{1}$$

Many approaches are existed in the literature to construct similarity relation. In the proposed design, the following procedure is used to build similarity relation.

$$R_a(i,j) = max\left(min\left(\frac{a(i) - a(j) + \sigma(a)}{\sigma(a)}, \frac{a(j) - a(i) + \sigma(a)}{\sigma(a)}\right), 0\right) \tag{2}$$

Here, $\sigma(a)$ is standard deviation of attribute a. From Radzikowska-Kerry's fuzzy-rough set model [9], the fuzzy-rough *lower approximation* of a fuzzy set A on U can be defined by using fuzzy similarity relation R in U.

$$R \downarrow A(j) = \underset{i \in U}{inf} \; I(R(i,j), A(i)) \tag{3}$$

where I is fuzzy implicator. From the *Lemma 1* of [8], the above (3) is simplified using the natural negation N_I of I for obtaining fuzzy-rough positive region based on $P \subseteq C^h$ as,

$$POS_P(j) = R_P \downarrow R_{d,j}(j) = \begin{cases} \underset{i \in U_2(j)}{min} (N_I(R_P(i,j))) & if \;\; U_2(j) \neq \phi \\ 1 & otherwise \end{cases} \tag{4}$$

Here, for an object, $j \in U$, the $U_1(j)$ represents the set of objects which belongs to the decision class of j and $U_2(j)$ represents the rest of the objects which belong to other decision classes. The resulting dependency degree measure is given as,

$$\gamma_P(\{d\}) = \frac{\sum_{i \in U} POS_P(i)}{|U|} \tag{5}$$

A fuzzy-rough *reduct* R is defined as minimal subset of attributes satisfying $\gamma_R(\{d\}) = \gamma_{C^h}(\{d\})$. The reduct generation can be done by using two control strategies, (i) *Sequential Forward Selection (SFS)*, and (ii) *Sequential Backward Elimination (SBE)*. In SFS strategy, reduct generation starts with an empty set, and attributes are incrementally added. It is possible in SFS strategy that the computed reduct may have some redundant attributes resulting as a super set of reduct (*superreduct*). In SBE strategy, reduct generation starts with whole attributes, and redundant attributes are removed one by one that results in minimal reduct. Even though SBE generates minimal reduct, the computational efficiency of the SFS strategy is more. In contrast to classical rough set approaches the redundancy in SFS reduct is very less owing to graded indiscernibility. Hence, the proposed algorithm is developed based on the attribute dependency degree measure approach that follows the SFS control strategy of the reduct generation, which has a less possibility of resulting in superreduct.

2.2 Improved Modified Quick Reduct Algorithm (IMQRA)

Sai Prasad et al. [8] proposed IMQRA algorithm based on the MQRA (Modified Quick Reduct Algorithm) [1]. A brief description of this algorithm is given below. Detailed theoretical and experimental description can be found in [8].

According to this algorithm, the fuzzy similarity relation for attribute $a \in C^h \cup \{d\}$ is represented as a symmetric similarity matrix with dimensions $U \times U$ and having entries $R_a(i,j)$, $\forall i,j \in U$. IMQRA starts with reduct set P initialized to an empty set, and in each iteration, attribute inducing maximum gamma gain is included into P. Objects achieving lower approximation membership of 1 are named as $ABSOLUTE\_POS_P$. It is proved in [8] that, removal of $ABSOLUTE\_POS_P$ does not affect the subsequent computations while resulting in significant space and time complexity gains. The algorithm terminates when P satisfies the reduct properties.

3 Proposed Work

The proposed MR_IMQRA algorithm is a scalable distributed/parallel version of IMQRA [8]. This section describes the proposed algorithm (given in Algorithm 1), along with its features. The proposed MR_IMQRA algorithm consists of two steps, namely, (i) Computation of distributed fuzzy-rough similarity matrix, and (ii) Fuzzy-rough reduct computation.

Algorithm 1. MR_IMQRA

Input: *HDT:* (U, $C^h = C^s \cup C^r$, {d}), R_{sim}*:* Fuzzy similarity relation, *N:* Fuzzy Negation, *Γ*: t-Norm.
Output: Fuzzy superreduct B
Procedure:
$AttrRdd\langle attr, attrData\rangle \leftarrow readAsRdd(HDT)$
$Dpartition \leftarrow U/\{d\}$
$simMatRdd\langle attr, R_{attr}\rangle \leftarrow AttrRdd.$**map**$\{\langle attr, attrData\rangle =>$
  Construct matrix R_{attr} from $attrData$ using R_{sim}on each pair of objects
  EMIT $\langle attr, R_{attr}\rangle$
}
$B \leftarrow \{\}$, $R_B \leftarrow \{\}$
$\gamma_B(\{d\}) \leftarrow 0$, $\gamma_{old} \leftarrow -1.0$
$posRegSum \leftarrow 0$
while $\gamma_B\{d\} > \gamma_{old}$ AND $\gamma_B\{d\} \neq 1$ AND $|nonAbsPos| > 0$ **do**
  broadcast(DPartition), **broadcast**(R_B), **broadcast**(B)
  $\gamma_{old} \leftarrow \gamma_B(\{d\}$
  PosRdd$\langle attr, |POS_{B\cup\{attr\}}(\{d\}|\rangle \leftarrow simRdd.$**map**$\{\langle attr, R_{attr}\rangle =>$
  if $attr \in B$ **then**
    EMIT $\langle attr, -1\rangle$
  else
    $R_{B\cup\{attr\}} = \Gamma(R_B, R_{attr})$
    Compute $POS_{B\cup\{attr\}}(\{d\}$
    EMIT $\langle attr, |POS_{B\cup\{attr\}}(\{d\}|\rangle$
  end if
  $\langle bA, |POS_{B\cup\{bA\}}|\rangle = PosRdd.$**reduce**$\{(\langle a1, |POS_{B\cup\{a1\}}|\rangle, \langle a2, |POS_{B\cup\{a2\}}|\rangle) =>$

  if $|POS_{B\cup\{a1\}}| > |POS_{B\cup\{a2\}}|$ **then**
    EMIT $|POS_{B\cup\{a1\}}|$
  else
    EMIT $|POS_{B\cup\{a2\}}|$
  end if
  }
  $B \leftarrow B \cup \{bA\}$
  $posRegSum \leftarrow |POS_{B\cup\{bA\}}|$
  $R_{bA} = simMatRdd.$**filter**$\{\langle attr, R_{attr}\rangle => (attrNo == bA)\}$.**map**(_._2)
  $R_B \leftarrow \Gamma(R_B, R_{bA})$
  $\langle nonAbsPos, absPos\rangle$ =getAbsolute(R_B))
  $$\gamma_B = \frac{|absPos| + posRegSum}{|U|}$$
  $simMatRdd\langle attr, R_{attr}\rangle = simMatRdd.$**filter**$\{\langle attr, R_{attr}\rangle => (attr! = bA)\}$
  Restrict DPartition and U to nonPos objects
end while
return B

3.1 Computation of Distributed Fuzzy-Rough Similarity Matrix

As mentioned earlier, the proposed algorithm uses the vertical partitioning technique to distribute the input data to the nodes of the cluster. To realize this technique, a necessary preprocessing step is to be done on the input dataset. The input dataset is converted into the form, such that the rows correspond to the attributes, and the column corresponds to the objects. Each row is prefixed with an attribute number for preserving the attribute identity in the partitioning of the dataset. Algorithm receives input data in two portions of conditional attributes data and decision attribute data.

The portion containing the information of conditional attributes is read as in *RDD* form $AttrRdd\langle attr, attrData\rangle$. Here, the key *attr* corresponds to the attribute number, and the value *attrData* corresponds to the object information of the attributes. (Note: An *RDD* in Apache Spark represents a *Resilient Distributed dataset* for performing parallel operations over several partitions of data in the cluster. The notation, $RDD < key, value >$ represents the structure of each object of *RDD* in the pair of *key* and *value*). As the entire attribute information is available within a single partition, the requisite similarity matrices for all the conditional attributes can be computed in parallel using a single *map()* operation. Here, for each record of *AttrRdd*, the corresponding similarity matrix is constructed using Eq. (2) and a new transformed $RDD : simMatRdd\langle attr, R_{attr}\rangle$ is constructed, where the value R_{attr} corresponds to the similarity matrix of *attr*.

3.2 Fuzzy-Rough Reduct Computation

The fuzzy-rough similarity matrices, computed in the earlier section, acts as the input for this fuzzy-rough reduct computation. Initially, the reduct set B and the associated similarity matrix R_B is set to $NULL$, the *gamma* value of the previous iteration γ_{old} is set to -1.0, *gamma* value of current iteration $\gamma_B(\{d\})$ is set to zero. In each iteration, decision equivalence classes *Dpartition*, current reduct set B, and reduct similarity matrix R_B are broadcasted to all the nodes of the cluster, as every partition requires this information for further computations. The computation in an iteration of MR_ IMQRA requires computation of $POS_{B\cup\{attr\}}(\{d\})$ for all $attr \in C^h - B$ and inclusion of best attribute into B.

In an iteration of the proposed algorithm, if an attribute is already in B, then a dummy $\langle key, value\rangle$ pair is generated as $\langle attr, -1\rangle$, so that it is not considered subsequently into the reduct. For every attribute $attr \in C^h - B$, the computation of $R_{B\cup\{attr\}}$ is done using *t-norm* operation. The creation of $R_{B\cup\{attr\}}$ is done locally and the corresponding memory is removed after computation of $POS_{B\cup\{attr\}}(\{d\})$. Then a key-value pair $\langle attr, |POS_{B\cup\{attr\}}(\{d\}|\rangle$ is generated. Through the *reduce()* operation, the global best attribute bA is selected and added to the reduct set B. The *reduce()* operation of Apache Spark involves local *reduce()* followed by global *reduce()*. Therefore, in every partition the local best attribute is selected and only it's corresponding key-value pair is communicated to the global Reducer. Hence, the proposed vertical partitioning based approach has a minimum data transfer across shuffle and sort phase in an iteration.

In the Driver, we need to update R_B as B is included with bA; this requires the availability of R_{bA} in the Driver. Hence a *filter()* operation is applied on *simMatRdd* to select a record corresponding to bA, and it's associated similarity matrix is fetched to driver and updation of R_B is done using *t-norm* operation. In this way MR_IMQRA algorithm continues till γ_B value reaches to 1 or γ_B remains unchanged for the last m number of iterations (hence indicating that it can not get a better *gamma* measure by further adding more *attributes*) or *nonAbsPos* has become zero. If the 2nd terminating condition meets, then it removes m lastly added *attributes* from B. At the end it returns the final reduct set B.

3.3 Absolute Positive Region Removal in MR_IMQRA

The absolute positive region objects are those objects which achieve the total positive region membership of 1 [8]. The removal of such objects does not affect the computations of remaining iterations and reduces the space complexity of the algorithm efficiently. As an RDD is immutable, the removal of these objects from the respective similarity matrices will become complex and requires the creation of a new RDD. Therefore, in MR_IMQRA, the removal of absolute positive region objects is done only from *Dpartition* and U. In the driver, using *getAbsolute* function on R_B, *nonPos* and *absPos* objects are determined and *Dpartition* and U are restricted to *nonPos* objects. Hence, the rest of the computations are restricted to only non-positive region objects in mappers. In this way, MR_IMQRA becomes a real implementation of IMQRA algorithm by incorporating the absolute positive region removal aspect that gives computational advantages.

4 Experimental Results and Analysis

In this section, experiments are conducted to illustrate the utility of the proposed MR_IMQRA algorithm for scalable fuzzy-rough set based attribute reduction.

4.1 Experimental Setup

The experiments are conducted on a cluster of five nodes, out of which one node is master (driver), and the rest of the nodes are workers (slaves). Every machine has Intel Core i5-7500 Processor with a clock frequency of 3.4 GHz, having 8 GB of RAM and all the nodes are installed with Ubuntu 18.04 LTS, java 1.8.0_191, Apache Spark 2.3.1, and Scala 2.11.8.

As mentioned in earlier sections, the proposed algorithm is suitable for the datasets having moderate object space and larger attribute space (i.e., high dimensional datasets). Accordingly, the datasets are chosen and downloaded from GitHub [6]. The description of the datasets is given in Table 1.

Table 1. Datasets used in the experiments

Dataset	Objects	Features	Classes
Ovarian	253	15156	2
Yeoh	248	12625	6
Chin	118	22215	2
Buyczynski	127	22283	3

Table 2. Comparative results of MR_IMQRA with MR_MDLP_IQRA

Dataset	MR_IMQRA		MR_MDLP_IQRA	
	Computational Time(s)	Reduct length	Computational Time(s)	Reduct length
Ovarian	15.69	3	816.01	2
Yeoh	19.35	5	862.20	4
Chin	11.34	4	801.38	4
Burczynski	14.03	5	794.198	5

4.2 Comparison of MR_IMQRA and MR_MDLP_IQRA

In the literature, it is observed that no significant work is done in MapReduce based fuzzy-rough set attribute reduction. Hence, to assess the importance of the vertical partitioning technique in MR_IMQRA algorithm, a fusion of two approaches MR_MDLP [10] (for scalable discretization of numerical attributes with MapReduce) and MR_IQRA_IG [11] (for computation of reduct on categorical dataset obtained from MR_MDLP) are done and represented as MR_MDLP_IQRA. The source code of the MR_MDLP is made available in GitHub [3].

Table 3. Classification accuracy of MR_IMQRA, and MR_MDLP_IQRA (in %)

Dataset	MR_IMQRA		MR_MDLP_IQRA	
	SVM	Random forest	SVM	Random forest
Ovarian	98.68	98.68	68.42	31
Yeoh	74.67	72.00	28.00	24.00
Chin	80.56	75.00	61.11	69.00
Burczynski	53.85	69.23	58.97	48.71

Experiments are conducted on algorithms, MR_IMQRA, and MR_MDLP_IQRA for the given datasets. The obtained computational time (in seconds) and reduct length are given in Table 2. From the results, it can be observed that MR_IMQRA algorithm is taking considerably less computational time

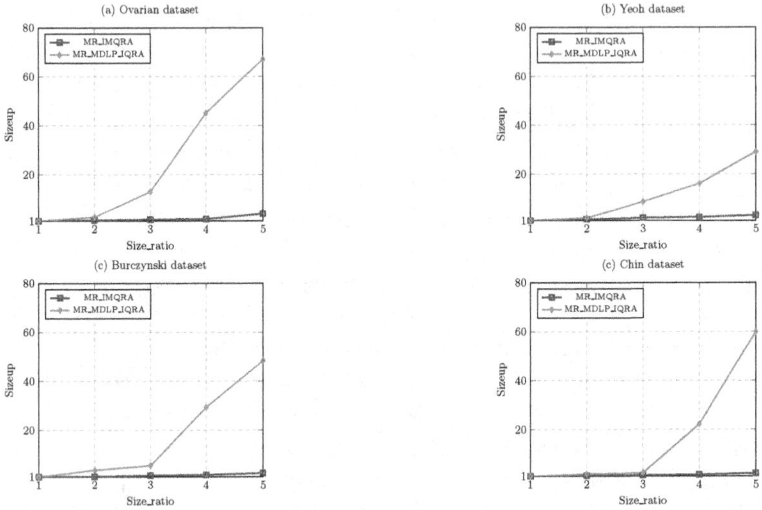

Fig. 1. Sizeup of MR_IMQRA and MR_MDLP_IQRA for different datasets

and almost giving similar reduct length like MR_MDLP_IQRA for all the datasets. The less computational times of MR_IMQRA are contrary to expectation as MR_IMQRA has a theoretical time complexity of $O(|C^h|^2|U|^2)$, where as MR_MDLP_IQRA has a time complexity of $O(|C^h|^2|U|log|U|)$. This phenomenon occurred because of vertical partitioning in MR_IMQRA leading to simplified shuffle and sort phase. In contrast, the horizontal partitioning in MR_MDLP_IQRA results in complex shuffle and sort phase, especially for high dimensional datasets.

Classification accuracy results using SVM, and Random forest classifiers for both algorithms, MR_IMQRA, and MR_MDLP_IQRA are given in Table 3 using 70% training data and 30% testing data. From the table, it is observed that MR_IMQRA achieved significantly higher classification accuracies than MR_MDLP_IQRA in both classifiers. It is observed that, both approaches are resulting in unrelated reducts. The classification analysis establishes that, MR_IMQRA has better potential in selection of relevant attributes in comparison to MR_MDLP_IQRA in which information loss due to discretization is affecting the selection of relevant reduct.

4.3 Performance Evaluation

Sizeup and speedup are the metrics used to asses the performance of the parallel algorithms. The sizeup experiments are conducted for different sizes of the datasets on the same cluster and are represented as follows,

$$Sizeup = \frac{Time\ taken\ by\ a\ dataset\ with\ corresponding\ size\_ratio}{time\ taken\ by\ a\ dataset\ of\ base\ size}.$$

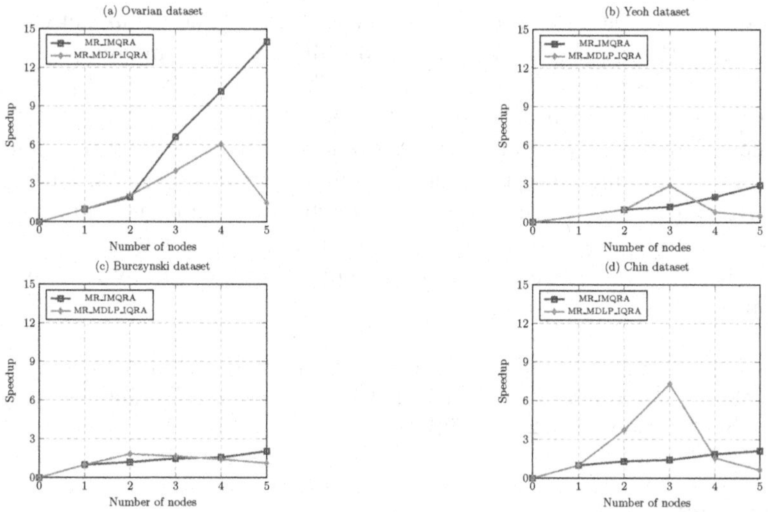

Fig. 2. Speedup of MR_IMQRA and MR_MDLP_IQRA for different datasets

Where, $Size\_ratio = \frac{H_{size}}{H_{base\_size}}$.
Here, H_{base_size} represents base dataset size and H_{size} represents current dataset size. The number of computers kept as five nodes. Each dataset size is increased with 20%, 40%, 60%, 80%, and 100% of attributes in the dataset. Figure 1 shows the sizeup performance results of MR_IMQRA and MR_MDLP_IG algorithms for different datasets with varying sizes of attribute space. Sizeup results shown in Fig. 1 establish that MR_IMQRA obtained a sub-linear sizeup measures in contrast to quadratic sizeup measures in MR_MDLP_IQRA.

Speedup experiments are conducted for the same datasets on different sizes of the cluster and it is represented as follows,

$$Speedup(n) = \frac{Computational\ time\ taken\ by\ a\ single\ node}{Computational\ time\ taken\ by\ a\ cluster\ of\ n\ nodes}.$$

Figure 2 shows the speedup results of MR_IMQRA and MR_MDLP_IQRA algorithms for different datasets with varied nodes from 1 to 5. MR_IMQRA has obtained the best speedup values in Ovarian dataset. In all the datasets MR_IMQRA has a steady increase in speedup measure values with increase in number of nodes, where in oscillations are observed in the results of MR_MDLP_IQRA. The results emperitically establish that proposed MR_IMQRA is recommended as a scalable solution for fuzzy-rough set reduct computation in high dimensional datasets.

5 Conclusion

The proposed work introduces a MapReduce based MR_IMQRA algorithm for attribute reduction in datasets of lesser object space and larger attribute space (high dimensional datasets) prevalent in Bioinformatics and document classification. MR_IMQRA is a distributed version of IMQRA algorithm and uses vertical partitioning to distribute the input dataset. The impact of vertical partitioning technique and the removal of the absolute positive region is shown vividly in the experimental analysis by obtaining reduct in lesser computational time and with reasonable sizeup and speedup values in comparison to horizontal partitioning based MR_MDLP_IQRA. The proposed algorithm also induced significantly better classifiers. In future, a MapReduce based SBE approach will be augmented to MR_IMQRA to remove existence of redundant attributes, if any resulting from SFS approach.

Acknowledgement. This work is supported by Department of Science and Technology (DST), Government of India under ICPS project [grant number: File DST/ICPS/CPS-Individual/2018/579(G)].

References

1. Cornelis, C., Jensen, R., Martn, G.H., Slezak, D.: Attribute selection with fuzzy decision reducts. Inf. Sci. **180**(2), 209–224 (2010)
2. Dubois, D., Prade, H.: Rough fuzzy sets and fuzzy rough sets*. Int. J. Gen. Syst. **17**(2–3), 191–209 (1990)
3. Lin, H.: MDLP-discretization (2017). https://github.com/hlin117/mdlp-discretization. Accessed 21 Nov 2017
4. Hu, Q., Yu, D., Xie, Z.: Information-preserving hybrid data reduction based on fuzzy-rough techniques. Pattern Recogn. Lett. **27**(5), 414–423 (2006)
5. Jensen, R., Shen, Q.: New approaches to fuzzy-rough feature selection. IEEE Trans. Fuzzy Syst. **17**(4), 824–838 (2009)
6. Ramey, J.: Datamicroarray (2016). https://github.com/ramhiser/datamicroarray/tree/master/data. Accessed 11 Jan 2016
7. Pawlak, Z.: Rough sets. Int. J. Comput. Inf. Sci. **11**(5), 341–356 (1982)
8. Prasad, P.S.V.S.S., Rao, C.R.: An efficient approach for fuzzy decision reduct computation Trans. Rough Sets **17**, 82–108 (2014). https://doi.org/10.1007/978-3-642-54756-0_5
9. Radzikowska, A.M., Kerre, E.E.: A comparative study of fuzzy rough sets. Fuzzy Sets Syst. **126**(2), 137–155 (2002)
10. Ramrez-Gallego, S., et al.: Data discretization: taxonomy and big data challenge. Wiley Interdisciplinary Rev.: Data Min. Knowl. Discov. **6**, 5–21 (2015)
11. Sai Prasad, P.S.V.S., Bala Subrahmanyam, H., Singh, P.K.: Scalable IQRA_IG algorithm: an iterative mapreduce approach for reduct computation. In: Krishnan, P., Radha Krishna, P., Parida, L. (eds.) ICDCIT 2017. LNCS, vol. 10109, pp. 58–69. Springer, Cham (2017). https://doi.org/10.1007/978-3-319-50472-8_5
12. Zaharia, M., et al.: Apache spark: a unified engine for big data processing. Commun. ACM **59**, 56–65 (2016)

Finding Optimal Rough Set Reduct with A^* Search Algorithm

Abhimanyu Bar, Anil Kumar, and P. S. V. S. Sai Prasad(✉)

School of Computer and Information Sciences, University of Hyderabad, Hyderabad 500046, India
{abhi16,anilhcu}@uohyd.ac.in, saics@uohyd.ernet.in

Abstract. Feature subset selection or reduct computation is a prominent domain for the classical rough set theory, which can preserve the most predictive features of a decision system. A given decision system has several reducts. Computation of all possible reducts was achieved through the computing prime implicants of the discernibility function. Currently, an optimal reduct based on any optimality criteria can only be achieved post-generation of all possible reducts. Indeed, it is an NP-hard problem. Several researchers have extended the alternative aspects with search strategies such as Genetic Algorithm, Ant Colony Optimization, Simulated Annealing, etc., for obtaining near-optimal reducts. In this paper, we propose an admissible and consistent heuristic for computing the optimal reduct having least number of induced equivalence classes or granules. A^*RSOR reduct computation algorithm is developed using the proposed consistent heuristic in A^* search. The proposed approach is validated both theoretically and experimentally. The comparative results establish the relevance of the proposed optimality criterion as the achieved optimal reduct has obtained significantly better accuracies with different classifiers.

Keywords: Rough sets · Attribute reduction · Feature subset selection · Optimal reduct · A^* search · Consistent heuristic

1 Introduction

The task of feature subset selection (FS) is a necessary preprocessing step for building learning models that increases predictive accuracy and model comprehensibility. Finding informative features is the most challenging task under inconsistent and imprecise information. The classical rough set theory (RST), proposed by Pawlak [14], is an essential mathematical method for dealing imprecise information without additional information about data. RST is only applicable to categorical decision systems [14] thus, it requires a prior discretization for numerical decision systems.

A given decision system has many reducts, and even a single reduct has the adequate capability for inducing a reliable classification model. The generalizability of classifier performance induced by the reduct varies from one reduct to

B. Deka et al. (Eds.): PReMI 2019, LNCS 11941, pp. 317–327, 2019.
https://doi.org/10.1007/978-3-030-34869-4_35

others, and there is no guarantee that the preferred single reduct can impose the best performance. Hence, researchers are interested in computing the best (optimal) reduct out of all possible reducts. Several heuristics approaches based on dependency measure, and discernibility matrix for single reduct computation are proposed in literature [8,15,17]. Even though these approaches are computationally efficient with polynomial time complexity, they can't assure the computation of optimal reduct.

In 1992, Skowron et al. [17] introduced boolean reasoning based approach for all reduct computation using crisp discernibility matrix. An optimal reduct can thus be selected by introducing the optimality criterion on all possible reducts. At present, this is the only way to obtain an optimal reduct. Though, in RST computing minimal/all reducts is an NP-hard problem [17]. Subsequently, several aspects has been investigated using the evolutionary algorithm such as genetic algorithm (GA) [19], ant colony optimization (ACO) [9], simulated annealing [11], particle swarm optimization (PSO) [18] etc., for near-optimal reduct computation. In this context, Wroblewski [19] proposed three approaches by combining GA with the greedy heuristics to generate minimal reduct. Jensen et al. [9] adopted a stochastic approach based on ACO to create a near-optimal reduct. While Wang et al. [18] proposed a reduct computation approach through PSO, Chen et al. [2] incorporated fish swarm optimization with rough sets for finding the reduct. Jensen et al. [11] proposed a mechanism for feature selection that combined the simulated annealing algorithm with the rough set theory.

However, most of the existing optimal/near-optimal reduct computation approaches are formulated with the optimality criteria of reduct having a minimum number of features (shortest length reduct). The dependency measures, such as gamma measure [8,20], conditional information entropy measure [10], discernibility based measure [1,17] etc., are used in reduct computation algorithms favouring those attributes with larger domain cardinality as the resultant finer granular space achieves better dependency measure value and can result in reduct with the smallest size. There is a correspondence between reduct and rule induction. Rules are induced from the granules (equivalence classes) obtained using reduct attributes [4]. In a finer granular space, the cardinality of rules will be larger, and the strength of the rule will be smaller, which can affect the generalizability of reduct induced classifiers. In contrast, coarser granular space will generate a smaller size rule set with higher strength and can induce better classifiers. Hence, in this work the optimality criterion is taken as minimizing the cardinality of equivalence classes or granules induced by reduct attributes. Based on this criterion, the resulting optimal reduct will produce coarsest granular space.

In this work, a consistent heuristic is proposed based on the considered optimality criterion. An optimal reduct computation algorithm (A^*RSOR) is developed by using the proposed consistent heuristic in A^* search. The resulting approach acquires a significance as an optimal reduct can be computed without generation of all possible reducts. The significance of A^*RSOR is validated both theoretically as well as through experimental analysis.

The remaining part of this paper is structured in the following order. The theoretical background of the classical rough sets along with the preliminaries of relative dependency measure and A^* search strategy are discussed in Sect. 2. In Sect. 3, the detailed theoretical explanation of the proposed optimal rough set reduct computation algorithm A^*RSOR is explored. Section 4 covers the comparative experimental evaluation and analysis of results. Lastly, in Sect. 5, the paper ends with the remarks and the future possibility of the proposed method.

2 Theoretical Background

2.1 Rough Set Theory

The decision system is represented as $DS = (\mathbb{U}, \mathbb{C} \cup \mathbb{D})$, where $\mathbb{U}$ represents the non-empty finite set of objects called universe, $\mathbb{C}$ is the set of conditional attributes such that $a : \mathbb{U} \rightarrow V_a$, $\forall a \in \mathbb{C}$ where V_a is the set of domain values with respect to 'a'. $\mathbb{D}$ is a set of decision attributes or response variables, usually $\mathbb{D} = \{d\}$ containing a single decision attribute. For any arbitrary subset $\mathbb{B} \subseteq \mathbb{C}$, there exist an associated equivalence relation called indiscernibility relation $IND(\mathbb{B})$ defined as:

$$IND(\mathbb{B}) = \{(u_i, u_j) \in \mathbb{U}^2 \mid \forall a \in \mathbb{B}, a(u_i) = a(u_j)\} \tag{1}$$

The collection of all equivalence classes(granules) of $IND(\mathbb{B})$ is represented as $\mathbb{U}/IND(\mathbb{B})$ or $\mathbb{U}/\mathbb{B}$, and is defined as: $\mathbb{U}/IND(\mathbb{B}) = \otimes\{a \in \mathbb{B} : \mathbb{U}/IND(\{a\})\}$, where $\otimes$ is the refinement operator. In the rest of the paper granules of $IND(\mathbb{B})$ are represented as $\mathbb{U}/\mathbb{B}$.

The set of granules $\mathbb{U}/\mathbb{B}$ constitute the granular space through which rough set based approximations are defined. Let $X \subseteq \mathbb{U}$ be the concept to be approximated, then the $\mathbb{B}$-lower and $\mathbb{B}$-upper approximations of X are computed as follows:

$$\underline{\mathbb{B}}X = \{u \in \mathbb{U} \mid [u]_{\mathbb{B}} \subseteq X\} \ and \ \overline{\mathbb{B}}X = \{u \in \mathbb{U} \mid [u]_{\mathbb{B}} \cap X \neq \emptyset\} \tag{2}$$

where $[u]_{\mathbb{B}}$ is the equivalence class of 'u'.

The positive region $POS_{\mathbb{B}}(\mathbb{D})$ is the collection of all objects that are certainly belongs to the concept $\mathbb{D}$, defined as:

$$POS_{\mathbb{B}}(\mathbb{D}) = \bigcup_{X \in \mathbb{U}/\mathbb{D}} \underline{\mathbb{B}}X \tag{3}$$

Consistent Decision System: A decision system DS is said to be a consistent decision system(CDS), if and only if $POS_{\mathbb{C}}(\mathbb{D}) = \mathbb{U}$. In case the given decision system is inconsistent, then the generalized decision operator has given in [5] is applied to convert DS into CDS. Hence in the rest of the paper, DS is assumed to be consistent.

2.2 Relative Dependency Measure

Han et al. [6] defined a dependency measure named as relative dependency. Let $\mathbb{B} \subseteq \mathbb{C}$, then the degree of relative attribute dependency, denoted as $\kappa_{\mathbb{B}}(\mathbb{D})$, of $\mathbb{D}$ over $\mathbb{B}$, is defined as:

$$\kappa_{\mathbb{B}}(\mathbb{D}) = \frac{|U/\mathbb{B}|}{|U/(\mathbb{B} \cup \mathbb{D})|} \tag{4}$$

$\mathbb{B}$ is a reduct of CDS, if only if, $\kappa_{\mathbb{B}}(\mathbb{D}) = \kappa_{\mathbb{C}}(\mathbb{D}) = 1$ and $\forall Q \subset \mathbb{B}, \kappa_Q(\mathbb{D}) \neq \kappa_{\mathbb{C}}(\mathbb{D})$. Here $|S|$ for any set S, represents the cardinality of S.

2.3 A^* Search Algorithm

In several domains, the solution to a problem can be formulated as a state-space search, represented as a graph, which contains a source node and several possible goal nodes. The solution is given in the form of a path or state according to the nature of the problem. A^* is a popular search strategy [7], introduced as part of Shakey Project [13]. This is a complete and optimal search algorithm, which was formulated by combining the Dijkstra's Shortest Path and Best-Fist Search mechanisms. Here along with the path cost, there is a prediction heuristic cost for reaching the goal node from a particular state 'n' is associated. Hence, the total cost to reach the goal state can be estimated as:

$$f(n) = g(n) + h(n) \tag{5}$$

where $g(n)$ denotes the path cost from the source node to the current node 'n' and $h(n)$ is the prediction heuristic which estimates the cost to reach the goal from node 'n'. Any node with $h(n)$ as zero is a candidate goal node.

Generally, A^* uses the openlist, which is defined as a priority queue to conduct the repeated selection of the least cost nodes to explore and closelist defined as the collection of explored nodes. The process is started by placing the source node into the *openlist*. At each iteration, the node with the lowest $f(n)$ value is removed from the *openlist* queue and placed in *closelist* queue, and their corresponding successor nodes with updated f values are added to the *openlist* queue. If the node selected for exploration from openlist has the heuristic values of zero, then A^* algorithm stops and returns the solution associated with the node.

A^* search is optimal, if $h(n)$ is admissible and consistent. A heuristic is admissible, if the value of $h(n)$ never overestimate the actual cost. The heuristic is consistent, if $h(n) \leq c(n, n') + h(n')$ provided that n' is the child node of n in the search space graph and $c(n, n')$ denotes the corresponding edge cost. In general, every consistent heuristic is admissible [7].

3 A^*RSOR Search Algorithm

The purpose of the A^*RSOR method is to find the reduct with the coarsest granular space for inducing least number of rules. Let $RED(CDS)$ represent set of all possible reducts for CDS, then an optimal reduct $\mathbb{B}^*$ with the coarsest granular space must satisfy the property:

$$|\mathbb{U}/\mathbb{B}^*| = \min_{\mathbb{B} \in RED(CDS)} |\mathbb{U}/\mathbb{B}| \tag{6}$$

The computation of reduct is a search problem in the search space of the power set of $\mathbb{C}$. The traditional SBE and SFS control strategies are the example of hill climbing approaches in this search space. The goal state corresponds to a reduct in $RED(CDS)$.In formulating A^* reduct computation algorithm, a heuristic function is needed for evaluating the cost of reaching any goal state. We have devised a heuristic function for states in the search space of reduct computation in the next subsection.

3.1 Partition Refinement Heuristic (PR-Heuristic)

The Kappa measure $\kappa_{\mathbb{B}}(\mathbb{D})$ is a measure of the purity of granular space induced by $\mathbb{B}$. A granule $g \in \mathbb{U}/\mathbb{B}$ is pure if 'g' contains objects of single decision class. A granular space is said to be a pure granular space if all of its granules are pure. For a pure granule, there is no further refinement takes place with the inclusion of decision attribute. As every reduct of CDS induces a pure granular space, we have $\kappa_{\mathbb{B}}(\mathbb{D}) = 1, \forall\ \mathbb{B} \in RED(CDS)$ as $|\mathbb{U}/\mathbb{B}| = |\mathbb{U}/(\mathbb{B} \cup \mathbb{D})|$.

If an attribute collection $\mathbb{B}$ is almost pure, then many of the granules in $\mathbb{U}/\mathbb{B}$ are pure, that is $|\mathbb{U}/(\mathbb{B} \cup \mathbb{D})|$ will be slightly higher than $|\mathbb{U}/\mathbb{B}|$ as only the remaining granules will participate in the refinement resulting in $\kappa_{\mathbb{B}}(\mathbb{D})$ near to one. If $\mathbb{B}$ induces almost impure granular space, then $|\mathbb{U}/(\mathbb{B} \cup \mathbb{D})|$ is much higher than $|\mathbb{U}/\mathbb{B}|$ resulting in $\kappa_{\mathbb{B}}$ which is near to zero. Hence, the number of splits additionally occurring through the refinement of $IND(\mathbb{B})$ into $IND(\mathbb{B} \cup \mathbb{D})$ estimates the furtherness of the current attribute set $\mathbb{B}$ in becoming reduct. This motivated us to formulate a heuristic named as partition refinement heuristic (PR-Heuristic) to estimate the cost attributes set $\mathbb{B}$ in becoming the reduct. For $\mathbb{B} \subseteq \mathbb{C}$, PR-heuristic h_{PR} is given by

$$h_{PR}(\mathbb{B}) = |\mathbb{U}/(\mathbb{B} \cup \mathbb{D})| - |\mathbb{U}/\mathbb{B}| \tag{7}$$

Here $h_{PR}(\mathbb{B})$ is the number of split (refinements) in $\mathbb{U}/\mathbb{B}$ occurring with inclusion of $\mathbb{D}$. In the reduct computation search space, let the child node of $\mathbb{B}$ is $\mathbb{B}' = \mathbb{B} \cup \{a\}$ for any $a \in \mathbb{C} - \mathbb{B}$. The cost of refinement from $\mathbb{B}$ to $\mathbb{B}'$ is $c(\mathbb{B}, \mathbb{B}') = |\mathbb{U}/\mathbb{B}'| - |\mathbb{U}/\mathbb{B}|$. h_{PR} is said to be consistent heuristic, if and only if $h_{PR}(\mathbb{B}) \leq h_{PR}(\mathbb{B}') + c(\mathbb{B}, \mathbb{B}')$. Theorem 1 gives the proof for establishing that h_{PR} is a consistent heuristic.

Theorem 1. *The partition refinement heuristic h_{PR} for state space of reduct computation in CDS is a consistent heuristic.*

Proof. Let h_{PR} denote the PR-heuristic. Let $\mathbb{B} \subseteq \mathbb{C}$ denotes a node being explored in A^* algorithm, and for any $a \in \mathbb{C} - \mathbb{B}$, a new node $\mathbb{B}' = \mathbb{B} \cup \{a\}$ is generated. Using Eq. 7, it follows as:

$h_{PR}(\mathbb{B}) = |\mathbb{U}/(\mathbb{B} \cup \mathbb{D})| - |\mathbb{U}/\mathbb{B}|$ and $h_{PR}(\mathbb{B}^{'}) = |\mathbb{U}/(\mathbb{B} \cup \{a\} \cup \mathbb{D})| - |\mathbb{U}/(\mathbb{B} \cup \{a\})|$ The edge cost $c(\mathbb{B}, \mathbb{B}^{'}) = |\mathbb{U}/(\mathbb{B} \cup \{a\})| - |\mathbb{U}/\mathbb{B}|$, then consider, $h_{PR}(\mathbb{B}^{'}) + c(\mathbb{B}, \mathbb{B}^{'}) = |\mathbb{U}/(\mathbb{B} \cup \{a\} \cup \mathbb{D})| - |\mathbb{U}/\mathbb{B}| \geq |\mathbb{U}/(\mathbb{B} \cup \mathbb{D})| - |\mathbb{U}/\mathbb{B}| = h_{PR}(\mathbb{B})$ (Since $IND(\mathbb{B} \cup \{a\} \cup \mathbb{D})$ is a refinement of $IND(\mathbb{B} \cup \mathbb{D})$). Hence, it is proved that h_{PR} is a consistent heuristic.

3.2 A^* Algorithm with PR-Heuristic

An A^* search based algorithm is formulated using PR-heuristic as A^* rough set optimal reduct (A^*RSOR) algorithm, which is presented in Algorithm 1. For a set of attributes $\mathbb{B} \subseteq \mathbb{C}$, path cost $g(\mathbb{B})$ represents cardinality of granular space induced by $\mathbb{B}$. i.e., $g(\mathbb{B}) = |\mathbb{U}/\mathbb{B}|$. Hence, the total cost $f(\mathbb{B})$ becomes $f(\mathbb{B}) = g(\mathbb{B}) + h_{PR}(\mathbb{B}) = |\mathbb{U}/(\mathbb{B} \cup \mathbb{D})|$.

Algorithm 1. A^* Rough Sets Optimal Reduct (A^*RSOR) Algorithm

```
Input : CDS: Consistent decision system, C: Set of conditional attributes, openlist:
        Priority queue over f, closelist: List of explored states.
Output: Optimal Reduct.
openlist = φ ;
closelist = φ ;
for every x in C do
    N= Create a node corresponding to {x}
    Insert(openlist, N)
end
while openlist is not empty do
    CN = openList(1);
    remove(openlist, CN);
    Insert(closelist, CN);
    if h_PR(CN) == 0 then
        Opt_Reduct = CN;
        Return(Opt_Reduct);
    end
    Cr = C − CN ;
    for every r in Cr do
        CS = CN ∪ {r};
        if CS is in closelist then
            Continue;
        end
        if CS is in openlist then
            Continue;
        else
            if SuperReductCheck(CS, openlist) == TRUE;
            then
                Continue;
            end
        end
        N=Create a node CS;
        Insert(openlist, N);
    end
end
```

In A^*RSOR algorithm, openlist represents a priority queue of frontier nodes in the increasing order of 'f' values. Initially, openlist is inserted with nodes corresponding to individual attributes of $\mathbb{C}$. In each iteration, the least 'f' value node CN is removed from openlist and inserted in closelist. If $h_{PR}(CN) = 0$ then we have identified the optimal reduct and the attribute set in CN is return as optimal reduct. Otherwise, a child node is generated for each attribute addition into CN, which is not already included in CN. The resulting node CS is inserted in the openlist, if and only if it is not in either the openlist or the closelist. In case CS corresponds to attribute set which is a superset to a candidate reduct node in the openlist (a node with $h_{PR} = 0$) then, it is not included in the openlist as it results in the superset of reduct. This verification is represented as a function $SuperReductCheck(CS, openlist)$.

4 Empirical Results and Observations

The proposed algorithm A^*RSOR is implemented in Matlab-2017a environment, and comparative experiments are conducted in the system with the following configuration: 3.40 GHz × 4-Intel(R) Core i5-7500 processor, 8GB DDR4 RAM, Ubuntu-16.04.1 LTS 64-bit operating system. Eight categorical benchmark datasets are used from UCI-machine learning repository [3], as shown Table 1. Additionally, the Wine, Sahart and Zoo datasets are discretized using "mdlp" discretization method [12] to transform into the categorical datasets. The verification of optimal reduct computation by A^*RSOR through ranking experiment and 10-fold cross-validation based induced classifier performance analysis is conducted. Comparative experimental studies are conducted with simulated annealing based near-optimal reduct computation algorithm $SimRSAR$ [11] and hill-climbing search based greedy reduct computational algorithm $IQRA\_IG$ [15].

Table 1. The description of experimental datasets

Sl.No	Data Sets	No. of objects	No. of variables	No. of classes
1	Austra	690	15	2
2	Breastcancer	699	10	2
3	Diab	768	9	2
4	Heart	294	14	2
5	Lymphography	148	19	7
6	Sahart	462	9	2
7	Wine	178	14	3
8	Zoo	101	16	7

4.1 Ranking Experiments

The correctness of the proposed method and implementation is verified in ranking experiment by checking the satisfiability of optimality criteria in the computed reduct. Towards this objective for each dataset, all reducts are computed using rough set exploration system ($RSES$) [16] and are ranked in the increasing order of optimality criteria ($f(\mathbb{R}) = |\mathbb{U}/IND(\mathbb{R})|$). The ranks obtained by reducts from A^*RSOR, $SimRSAR$, and $IQRA\_IG$ approaches are reported along with obtained 'f' values in Table 2. Table 2 also reported reduct length obtained by the respective algorithms. Total number reducts, obtained by $RSES$ tool, are reported under the column $|RSESReducts|$.

Table 2. Results of ranking experiment with total granular space and the corresponding length of the optimal reduct.

Datsets	$\|RSESReducts\|$	$SimRSAR$ Reduct			$IQRA\_IG$ Reduct			A^*RSOR Reduct		
		Rank	$f(R)$	Length	Rank	$f(R)$	Length	Rank	$f(R)$	Length
$Austra^*$	44	43	689	3	43	689	3	1	674	4
Breastcancer	20	4	299	4	5	308	4	1	288	4
$Diab^*$	28	20	768	3	20	768	3	1	764	3
$Heart^*$	20	18	290	6	1	278	7	1	278	7
Lymphography	17	5	145	11	5	145	11	1	143	12
Sahart	27	5	460	3	1	458	4	1	458	4
$Wine^*$	81	80	178	3	67	176	3	1	152	4
Zoo^*	33	18	27	5	4	22	5	1	20	5

Analysis of Results: The results demonstrate that the implemented A^*RSOR algorithm achieves the optimal reduct in all the datasets by obtaining rank one. The compared approaches $SimRSAR$, and $IQRA\_IG$ have varying rankings across the datasets. A significant difference in ranking order is observed in Austra, Diab, Heart, Wine, and Zoo datasets and are marked by $*$ symbol. Out of these only in Austra and Wine datasets, there is a significant variation in $f(\mathbb{R})$ values.

4.2 Ten-Fold Experiments

In this section, we conducted the 10-fold cross-validation experiments on given benchmark datasets for assessing the relevance of reduct in inducing different classifiers (Naive Bayes (NB), CART and Random Forest (RF)). In all the classifiers, default options are used while treating all attributes as categorical. Table 3 depicts the mean and standard deviation of reduct length and computational time (in seconds) resulted in ten-fold experiments. Table 4 demonstrates the resulting mean and standard deviation ($\mu \pm \sigma$) of classification accuracies obtained in ten-fold experiments. In Table 4, the column *ALLAttributes* refers to results from un-reduced training data.

Table 3. Reduct length and computational time in ten-fold cross-validation experiment

Datsets	Reduct length			Time		
	SimRSAR	*IQRA_IG*	A^*RSOR	SimRSAR	*IQRA_IG*	A^*RSOR
Austra	3 ± 0	3 ± 0	5.60 ± 1.07	157.05 ± 7.79	0.11 ± 0.01	234.31 ± 10.13
Breastcancer	4 ± 0	4 ± 0	4 ± 0	7.14 ± 0.31	0.03 ± 0.00	3.69 ± 0.17
Diab	2.80 ± 0.42	2.80 ± 0.42	3 ± 0	250.24 ± 24.64	0.06 ± 0.007	3.60 ± 0.19
Heart	5.90 ± 0.56	6.70 ± 0.67	6.80 ± 0.42	4.74 ± 0.27	0.06 ± 0.01	64.02 ± 3.11
Lymphography	10.30 ± 0.82	10.80 ± 0.63	10.70 ± 0.82	2.41 ± 0.12	0.11 ± 0.01	1.72 ± 0.05
Sahart	3 ± 0	4 ± 0	4 ± 0	36.58 ± 1.93	0.07 ± 0.009	3.28 ± 0.14
Wine	2.70 ± 0.48	2.70 ± 0.48	5.30 ± 0.82	9.34 ± 0.97	0.02 ± 0.007	13.42 ± 0.83
Zoo	4.90 ± 0.31	5.30 ± 0.67	5.10 ± 0.73	0.31 ± 0.02	0.02 ± 0.005	26.09 ± 7.39

Student t-test is conducted on classification accuracies for analysis of results between the proposed algorithm A^*RSOR with *IQRA_IG*, *SimRSAR* and *ALLattributes*, and the results of the t-test are depicted in Table 4 along with classification accuracies. Table 4 entries show the four significance levels indicated as (+/-)*, (+/-)**, (+/-)***, # as per p-value in the t-test. The significance level for experimental results based on t-test are divided as statistically significant and are indicated as * ($p - value \leq 0.05$), statistically highly significant as ** ($p - value \leq 0.01$), statistically extremely significant as *** ($p - value \leq 0.001$) and no statistical difference indicated as #. The prefix of '+' denotes the A^*RSOR has performed better than the compared algorithm and '-' denotes that it has underperformed.

Analysis of Results: In Table 3, *IQRA_IG* algorithm has obtained much lesser computational time than A^*RSOR and *SimRSAR* algorithms. This is in correlation with the theoretically lesser complexity of a greedy hill-climbing search algorithm in comparison to multiple subspace search algorithms of A^*RSOR and SimRSAR. The practical time complexity of A^*RSOR search algorithm depends on the depth at which the optimal reduct is formed. For instance, in Diabetes and Sahart datasets, the proposed A^*RSOR algorithm incurred only 3.6, 3.2 s on average, whereas *SimRSAR* incurred 250, 36.5 s. In contrast, *SimRSAR* obtained significantly lesser computational time in Austra, Heart, wine, and Zoo datasets. It is to be noted that out of these three algorithms, the worst cases time complexity of A^*RSOR is exponential while the other two algorithms are having polynomial time complexity. Both SimRSAR and *IQRA_IG* are formulated towards obtaining shorter length reduct. Hence, the reduct lengths of A^*RSOR are slightly higher or equal than other algorithms.

Table 4. Ten-fold cross-validation results for classification

Datsets	*AllAttributes*			*SimRSAR*			*IQRA_IG*			*A*RSOR*		
	NB	CART	RF	NB	CART	RF	NB	CART	RF	NB	CART	RF
Austra	62.89 ± 6.66 +*	81.30 ± 3.08 #	80 ± 4.77 #	58.40 ± 5.79 +***	55.36 ± 8.79 +***	57.10 ± 8.20 +***	56.37 ± 6.08 +***	51.88 ± 4.96 +***	53.33 ± 6.17 +***	69.85 ± 5.90	78.84 ± 8.002	77.39 ± 7.61
Breastcancer	97.13 ± 1.67 #	93.40 ± 3.09 #	96.84 ± 1.91 #	96.39 ± 2.30 #	93.23 ± 3.78 #	95.86 ± 2.64 #	96.29 ± 2.22 #	93.27 ± 3.34 #	95.25 ± 3.63 #	96.58 ± 1.61	93.41 ± 2.98	95.44 ± 2.15
Diab	65.41 ± .60 #	59.23 ± 6.38 #	64.22 ± 4.86 #	64.48 ± 6.67 #	61.22 ± 5.20 #	62.53 ± 6.85 #	64.24 ± 5.08 #	62.011 ± 5.66 #	61.76 ± 4.99 #	62.80 ± 5.14	58.07 ± 6.90	60.18 ± 4.67
Heart	82.14 ± 8.60 #	79.64 ± 9.68 #	81.10 ± 9.27 #	77.05 ± 9.36 #	73.90 ± 11.88 #	75.67 ± 9.85 #	80.33 ± 7.28 #	74.94 ± 11.04 #	79.38 ± 10.20 #	81.10 ± 8.21	77.31 ± 10.99	79.72 ± 10.25
Lymphography	38.83 ± 14.47 #	35.51 ± 12.49 #	43.11 ± 13.06 #	39.61 ± 13.74 #	35.12 ± 10.16 #	37.72 ± 9.70 #	36.75 ± 13.90 #	33.37 ± 12.92 #	38.37 ± 12.63 #	39.80 ± 12.19	30.84 ± 16.08	36.29 ± 12.44
Sahart	57.38 ± 7.68 #	52.21 ± 7.51 #	55.41 ± 6.50 #	53.03 ± 3.72 +*	50.45 ± 7.29 #	52.61 ± 9.14 #	54.52 ± 6.65 #	54.13 ± 5.64 #	53.48 ± 7.53 #	60.39 ± 8.13	52.83 ± 5.41	55 ± 7.80
Wine	59.74 ± 9.23 +***	57.97 ± 11.63 +***	94.30 ± 4.83 -**	55.22 ± 14.41 +***	53.45 ± 13.36 +***	70.94 ± 17.48 +*	52.87 ± 11.01 +***	54.04 ± 12.88 +***	67.03 ± 17.55 +**	88.02 ± 6.01	83.10 ± 8.52	83.88 ± 7.96
Zoo	91.00 ± 11.00 #	87.18 ± 14.86 #	94.18 ± 8.11 #	88.00 ± 10.32 #	89.09 ± 5.70 #	93 ± 6.74 #	84.09 ± 9.72 #	89.09 ± 8.77 #	95 ± 8.49 #	86.18 ± 14.22	88.18 ± 9.03	96.18 ± 6.53

In NB, CART and RF classifiers, A^*RSOR obtained statistically similar in classification accuracies with compared algorithms in Breastcancer, Diab, Lymphography and Zoo datasets. In Sahart dataset, A^*RSOR approach performed statistically significant than *SimRSAR* in NB classifiers. In Austra and Wine datasets, A^*RSOR algorithm achieved statistically extremely significant accuracies than compared algorithms using all three classifiers. Also, in Wine dataset, A^*RSOR algorithm performed extremely significant than *AllAttributes* in both NB & CART classifiers. Likewise, in Austra dataset, A^*RSOR algorithm performed statistically significant accuracy than *AllAttributes* using NB classifier. Results also demonstrate the additional advantages for RF in achieving better accuracies with *AllAttributes*, as RF is an ensemble classifier with bagging over attribute space. It is further noted that A^*RSOR achieves statistically similar results with *AllAttributes* in RF for all datasets except Wine dataset.

The datasets in which A^*RSOR obtained better accuracies i.e., Austra and Wine, are also the datasets in which *IQRA_IG*, *SimRSAR* algorithm obtained higher rank reducts along with significantly higher $f(R)$ value. This establishes that the chosen optimality criterion of the coarser granular space is relevant in obtaining reducts with greater potential in building better classification models.

5 Conclusion

Many approaches of rough set based reduct algorithm are aiming towards computing shortest length reduct both optimally and near optimally. In this work, the need for an alternative optimal criterion for reduct computation is identified, and A^*RSOR algorithm is developed for the computation of optimal reduct using A^* search. Partition refinement heuristic is introduced and proved to be a consistent heuristic. Comparative experimental results validated the utility of proposed optimality criteria. In the future, scalable algorithms for proposed A^*RSOR will be developed for enhancing the applicability to large scale decision systems.

References

1. Chen, J., Lin, Y., Li, J., Ma, Z., Tan, A.: A rough set method for the minimum vertex cover problem of graphs. Appl. Soft Comput. **42**, 360–367 (2016)
2. Chen, Y., Zhu, Q., Xu, H.: Finding rough set reducts with fish swarm algorithm. Knowl.-Based Syst. **81**, 22–29 (2015)
3. Dua, D., Karra Taniskidou, E.: UCI machine learning repository (2017). http://archive.ics.uci.edu/ml
4. Grzymala-Busse, J.W.: Rule induction. In: Maimon, O., Rokach, L. (eds.) Data Mining and Knowledge Discovery Handbook, pp. 249–265. Springer, Boston (2009). https://doi.org/10.1007/978-0-387-09823-4_13
5. Hassanien, A.E., Suraj, Z., Slezak, D., Lingras, P.: Rough Computing: Theories, Technologies and Applications. IGI Global, Hershey (2007)
6. Han, J., Hu, X., Lin, T.Y.: Feature subset selection based on relative dependency between attributes. In: Tsumoto, S., Słowiński, R., Komorowski, J., Grzymała-Busse, J.W. (eds.) RSCTC 2004. LNCS (LNAI), vol. 3066, pp. 176–185. Springer, Heidelberg (2004). https://doi.org/10.1007/978-3-540-25929-9_20
7. Hart, P.E., Nilsson, N.J., Raphael, B.: A formal basis for the heuristic determination of minimum cost paths. IEEE Trans. Syst. Sci. Cybern. **4**(2), 100–107 (1968)
8. Jensen, R.: Rough set-based feature selection: a review. In: Rough Computing: Theories, Technologies and Applications, pp. 70–107. IGI Global (2008)
9. Jensen, R., Shen, Q.: Finding rough set reducts with ant colony optimization. In: Proceedings of UKCI-2003, vol. 1, pp. 15–22 (2003)
10. Jensen, R., Shen, Q.: Fuzzy-rough attribute reduction with application to web categorization. Fuzzy Sets Syst. **141**(3), 469–485 (2004)
11. Jensen, R., Shen, Q.: Semantics-preserving dimensionality reduction: rough and fuzzy-rough-based approaches. IEEE Trans. Knowl. Data Eng. **16**(12), 1457–1471 (2004)
12. Kim, H.J.: MDLP: discretization using the minimum description length principle (2010). https://rdrr.io/cran/discretization/
13. Kuipers, B., Feigenbaum, E.A., Hart, P.E., Nilsson, N.J.: Shakey: from conception to history. AI Mag. **38**, 88–103 (2017)
14. Pawalk, Z.: Rough Sets: Theoretical Aspects of Reasoning About Data. Kluwer Academic Publishers, Dordrecht (1991)
15. Sai Prasad, P.S.V.S., Raghavendra Rao, C.: Extensions to IQuickReduct. In: Sombattheera, C., Agarwal, A., Udgata, S.K., Lavangnananda, K. (eds.) MIWAI 2011. LNCS (LNAI), vol. 7080, pp. 351–362. Springer, Heidelberg (2011). https://doi.org/10.1007/978-3-642-25725-4_31
16. Skowron, A., et al.: Rough Set Exploration System (RSES). Warsaw University, Warsaw (2005). https://www.mimuw.edu.pl/~szczuka/rses/
17. Skowron, A., Rauszer, C.: The discernibility matrices and functions in information systems. In: Słowiński, R. (ed.) Intelligent Decision Support. Theory and Decision Library (Series D: System Theory, Knowledge Engineering and Problem Solving), vol. 11, pp. 331–362. Springer, Dordrecht (1992). https://doi.org/10.1007/978-94-015-7975-9_21
18. Wang, X., Yang, J., Teng, X., Xia, W., Jensen, R.: Feature selection based on rough sets and PSO. Pattern Recogn. Lett. **28**(4), 459–471 (2007)
19. Wroblewski, J.: Finding minimal reducts using genetic algorithms. In: Proccedings of the Second Annual Join Conference on Infromation Science, vol. 2, pp. 186–189 (1995)
20. Zheng, K., Hu, J., Zhan, Z., Ma, J., Qi, J.: An enhancement for heuristic attribute reduction algorithm in rough set. Expert Syst. Appl. **41**(15), 6748–6754 (2014)

Analytical Study and Empirical Validations on the Impact of Scale Factor Parameter of Differential Evolution Algorithm

Dhanya M. Dhanalakshmy, G. Jeyakumar(✉), and C. Shunmuga Velayutham

Department of Computer Science and Engineering, Amrita School of Engineering, Amrita Vishwa Vidyapeetham, Coimbatore, India
{md_dhanya, g_jeyakumar, cs_velayutham}@cb.amrita.edu

Abstract. Differential Evolution (*DE*) is a popular optimization algorithm in the repository of Evolutionary Algorithm (*EAs*). The *DE* algorithm is known for its simple algorithmic structure, which has minimal number (only three) of control parameters. A propitious avenue for enhancement of *DE's* performance is making it a self-adaptive algorithm. There exist many algorithms for self-adapting one or more of *DE* parameters. The self-adaptiveness of any parameter needs critical analysis on the impact of that parameter. This paper analyzes and presents the impact of the parameter - mutation scale factor (*F*) of *DE*. Including empirical evidences for understanding the effect of *F* on the nature of convergence of *DE* at solving a problem is the novelty of this paper. The experiment includes implementing a set of benchmark functions, with diversified features, using different variants of *DE,* in order to critically analyze the role of *F*.

Keywords: Differential Evolution · Premature convergence · Stagnation · Mutation step size · Parameter adaptation

1 Introduction

Evolutionary Algorithms (*EAs*) are population based metaheuristic optimization procedures that play a vital role leading to solution of diverse sets of optimization problems. The Differential Evolution (*DE*) algorithm [1–3] is a recent addition to the family of *EA*s, whose conceptual simplicity and ease of implementation draw unabated attention of research scholars, soon after its introduction to the *EA* literature in 1995. Its simplicity lies in the fact that it has a minimal set of control parameters. Major parameters of *DE* that need to be set for a given optimization problem are scale factor (*F*), crossover rate (*CR*), population size (*NP*), type of crossover (T_c), type of mutation (T_m) and maximum number of generations (*Max_Gen*).

In order to motivate the research folks to readily incorporate *DE* among their preferred investigative tools, the *DE* control parameters are actively being worked on to be self-configurable or self-adaptive [4]. At the get-go, most of the published literatures apropos *DE* have been focused on the *F* and *CR* parameters. Few studies are keen on appraising the viability of self-adapting *NP* alone. Self-configurability of all or a

B. Deka et al. (Eds.): PReMI 2019, LNCS 11941, pp. 328–336, 2019.
https://doi.org/10.1007/978-3-030-34869-4_36

coherent subset of the *DE* control parameters poses real challenges. Research and any contribution towards self-adapting *DE* will greatly facilitate the successful adoption of the *DE* algorithm by the industry.

Prior to embellishment of *DE's* self-adaptive traits, it is imperative to attain a systematic understanding of the effect of each parameter on the performance of the algorithm. Besides, it is vital to be cognizant of any interactive effects among the parameters, culminating in the development of complete tuning-free *DE* algorithms. These bear testimony to the undiminished curiosity within the research community, oriented to accomplishment of tuning-free *DE* algorithms. This paper describes the experiments carried out to delve into the effects of the *F* parameter in population dynamics of *DE* algorithm.

This paper is organized into five sections. Section 2 reviews ongoing works related to effect of *F* on *DE* algorithm. The design of experiment (to gauge the *F* parameter's characteristics) is presented in Sect. 3. The observed results are summarized in Sect. 4, with closing remarks in Sect. 5.

2 Related Works

This section highlights recent research works on adaptability of parameters of *DE* algorithms, with special attention to probes the significance of the *F* parameter. Numerous *DE* adaptation strategies have been proposed in the literature. A classification of different adaptation and/or tuning strategies of *DE,* proposed by various researchers, is presented in [5] and [6]. The authors grouped those strategies into four categories, based on their algorithmic approach, and provided insights into the different tuning methods for the *F* and *CR* parameters. Adaptive *DE* variants are reviewed by Al-Dabbagh et al. [4], pointing out the advantages and disadvantages of each variant.

Since this paper aims at providing insight about the effect of scale factor (*F*) parameter on *DE* algorithm, the minimal research that has been published in this direction are described below. Storm [2] suggested that value of *F* can be chosen from the set of [0.5, 1.0], which has been acceptable to most of the researchers engaged in development of adaptation strategies for *F*. Gamperle et al. [7] inquired into the effect of *DE* parameters, and shared some insights on setting relevant values for these parameters, compatible with four benchmark functions and four *DE* variants. The competitive *DE* algorithm, proposed by Trvdik [8, 9] experimented with four variants of competitive *DE,* for various combinations of *F* and *CR,* on six benchmark functions to derive an optimal setup. Ali and Torn [10] performed empirical analysis to deduce a minimum value for *F,* in *DEPD* - differential evolution using pre-calculated differential (a modified *DE* algorithm); they proposed that minimum value for *F* should belong to the set of [0.4, 0.5].

Brest et al. [11] analyzed the effects of *F* and *CR,* in their pursuit of developing a self-adaptive *DE*. Lampinen and Liu [12] done experiments to assess the consequences of *F* in the range of [0, −2] and *CR* in the range [0, −1], on the failure to converge to optima. Jeyakumar and Shunmuga Velayutham [13, 14] marshaled a bootstrap test, proposed by Mezura et al. [15], for selection of appropriate *CR* values for both the classical *DE* and Distributed *DE*. Akhila et al. [16] studied the impact of *F* value on

Population Diversity. Opara and Arabas [17] furnished a method to calculate the critical region of *F*, below which premature convergence may occur.

Collective review of the reported probes reveals that the impact of the *DE* parameters on its performance is influenced by various factors. However, presentation of an analytical report on the impact of all the *DE* parameters, in a single article, is a cumbersome task. Hence, this paper focuses on the critical analysis of the impact of *DE*'s mutation scale factor *(F)* parameter, with strong empirical evidences derived through the results obtained from an extensive experimental set up.

3 Design of Experiments

This work is zeroed in on the impact of *DE's* Scale Factor *F* on the convergent nature of *DE,* apposite to solutions of different classes of optimization problems. The empirical set up is constituted of 4 benchmarking functions, 8 *DE* variants, 11 values of *F,* and two values of *CR*. Four types of benchmark functions (differing in terms of the number of optima and separability) used in the experiments are f_1 - Sphere Function (Unimodal, Separable), f_2 - Schwefel's Function 1.2 (Unimodal, non-separable), f_3 - Generalized Rastrigin's Function (Multimodal, Separable) and f_4 - Ackley's Function (Multimodal, non-separable). All these are minimization functions, with the global optimum at 0 (zero).

Except *F*, other parameters are set as follows: *NP* – 60, Problem Dimension *D* – 30, Total number of runs – 100 (T_r) and Maximum number of generations (*Max_Gen*) – 2000. Two strategies are used for setting values of *CR*. In Strategy 1, *CR* is set as 0.3, randomly. In strategy 2, it is set based on a statistical bootstrap testing. For *F*, the possible values are kept in the range of [0.1 to 1.0] as mentioned in the *DE* literature and two strategies are followed: $Unique_F$ and $Varying_F$. In $Unique_F$, *F* values are chosen at random in a single instance for each run. In $Varying_F$, it is done for each generation of a run. The population comprised of *NP* candidates, is randomly initialized, in the beginning of each run. *DE* has many variants based on the type of base vector, number of chromosome pairs used for the mutation, and type of crossover. The notation adopted to symbolize *DE* variants is *DE/m/n/c.* Here, *m* is a string representing type of base vector, *n* is the number of chromosome pairs used for mutation and *c* is the string representing type of crossover. Among the popular variants available, the eight variants chosen for the experiment are *DE/rand/1/bin, DE/rand/1/exp, DE/best/1/bin, DE/best/1/exp, DE/rand/2/bin, DE/rand/2/exp, DE/best/2/bin* and *DE/best/2/exp.*

Each variant of *DE* is experimented in solving above four test functions. Each Function-Variant combination used two setups for *CR* and eleven setups for *F* (ten $Unique_F$ and one $Varying_F$). Hence, a total of 704 experiments are carried out. 100 runs are performed for each experiment, and each run had 2000 generations.

4 Results and Discussions

Mean Objective Value (*MOV*), Mean Function Evaluations (*MFE*) and Percentage of Success (P_s) are the performance metrics used in the experiment. The *MOV* is calculated as the average of best objective values of all runs. The best objective Value (*BOV*) of each run is the minimum objective value over *Max_Gen* generations. The *MOV* is used as an indicator to derive the accuracy of the solution obtained by *DE*. The *MOV* is computed as shown in Eqs. (1) and (2).

$$MOV = \frac{1}{tr} * \sum_{i=1}^{tr} BOV_i \tag{1}$$

$$BOV_i = \min_j(OV_{ij}), j = 1, 2, \ldots, Max\_Gen \tag{2}$$

where, OV_{ij} is the best objective function value in the ith run and j^{th} generation.

A run gets terminated either at when *DE* reaches the global minimum or at the number of generations equal to *Max_Gen*. The maximum number of function evaluations (*MaxFE*) in a run is 120000 (*Max_Gen* * *NP*). If a run gets terminated before *Max_Gen*, the number of function evaluations (*NFE*) is less than *MaxFE*. The *MFE* is the average of *NFEs* over T_r runs. The *MFE* indicates speed of convergence of *DE*, and is calculated by Eq. (3). Among 100 runs of a combination of 'function-variant-parameter's value', the count of successful runs (C_{SR}) that reach the global optimum is used to calculate the value for P_s, as shown in Eq. (4).

$$MFE = \frac{1}{Tr} \sum_{i=1}^{Tr} NFE_i \tag{3}$$

$$P_s = \frac{\mathrm{C_{SR}}}{Tr} \tag{4}$$

The preliminary inferences drawn, from comparison of the empirical results, based on *MOV*, *MFE* and P_s are shown in Tables 1 and 2, for *DE/rand/*/** and *DE/best/*/** variants, respectively. The *DE/rand/*/** and *DE/best/*/** includes variants with random and best vector as base vector, respectively.

These preliminary findings are inadequate to formulate tangible conclusion about the role played by F, on evolutionary search of DE. Hence, the study is extended by adding experiments that analyze the convergent nature of DE variants. This experiment calculates the average of BOV at each generation, across 100 runs for each combination of variant-function, with different values of F, and two strategies of CR. The BOV for each generation in every run is stored, and is used for calculating the generation-wise average objective function values.

Based on *BOV* measure, nature of convergence of the *DE* variants is analyzed. If the *DE* variant converges to the global optimum, it is Successful Convergence (*SuC*). If it gets stuck at the nearest local optima, it is Premature Convergence (*PrC*). In some cases, the *DE* variant will not converge locally or globally; this is Stagnation (*STAG*).

88 experiments are conducted for each *DE* variant, from which the number of *SuC* (*SuC#*), *PrC* (*PrC#*) and *STAG* (*STAG#*), aroused by the *DE* variants are calculated. This detail is shown in Table 3. It shows that all *DE/*/*/exp* variants fall in more number of *STAGs,* comparing to *DE/*/*/bin* variants, and achieved minimal number of *SuCs*. However, it is interesting to note that the *DE/*/*/exp* variants yielded less number of *PrCs*; it is zero for *DE/rand/2/exp* and *DE/best/2/exp*. Among all *DE/*/*/bin* variants, *DE/rand/1/bin* turned out more number of *SuCs*. Other *DE/*/*/bin* variants seem to stagnate with more number of runs.

Table 1. Inferences on effect of *F* for *DE/rand/*/** variants

	Inferences
DE/rand//bin*	$F = 0.5$ & 0.6 has provided good results for all the functions
	$F = 0.3$ to $F = 0.7$ have provided good results for two functions with fixed *CR*
	$Varying_F$ strategy provided good results for two functions with Bootstrap *CR* and for one function with $CR = 0.3$
	Successful convergence occurred for f_1 and f_3 with Bootstrap *CR* at low values of *F*
DE/rand//exp*	Successful convergence occurred only for f_1
	Except for f_1, $Varying_F$ strategy does not give good results
	$Varying_F$ strategy gave non-zero P_s for f_1 with *CR* by Bootstrap
	Minimum *BOVs* obtained are at lower values of *F* in all cases ($F \leq 0.4$)

Table 2. Inferences on effect of *F* for *DE/best/*/** variants

	Inferences
DE/best// bin*	Only f_1 and f_3 have successful convergence. The convergence is at high *F* values
	Smallest *MOVs* occur for same *F* values in both Bootstrap and Fixed *CR* cases
	For f_2, lowest *MOVs* are obtained with Lower *F* value (*F* in the range of (0.1–0.4)) in both cases of *CR*
	$Varying_F$ strategy showed non-zero P_s value for f_1, f_2 and f_3
*DE/best/ */exp*	Only Unimodal functions are converging to optimum value. Both of them converges only for bootstrap *CR* values at $F = 0.7$
	For Unimodal functions, *VaryingF* strategy works in par or better than Fixed *F* values with Bootstrap *CR*
	Successful convergence occurred only for f_2
	MFE < *MaxFE* for f_1 with Bootstrap *CR* and $Varying_F$ strategy
	$P_s > 0$ for 3 cases, excluding $Varying_F$ strategy, where *F* values are either 0.5 or 0.6

To attain more precision, the counts on *PrC*, *SuC* and *STAG* are re-computed based on different Function groups, mutation type and crossover type (*bin* or *exp*); the results are shown in Tables 4, 5 and 6 respectively. The results showed that *PrC#* is less in *DE/rand/*/** comparing to *DE/best/*/** except for multimodal non-separable functions. *SuC#* is more for *DE/rand/*/** than *DE/best/*/**. In case of *STAG*, both variants are almost similar. The *DE/*/*/bin* variants are observed to carry more *PrC* and *SuC*, whereas *DE/*/*/exp* variants returned more *STAG* cases.

Next, the *F* values are categorized into: Low (0.1–0.3), Medium (0.4–0.7) and High (0.8–1.0). Then the relationship between nature of convergence and category of *F* is analyzed. Results are depicted in Tables 7 and 8. *PrCs* are found to occur with Low values of *F* in most cases. The *SuCs* occurred mainly with medium *F* values. In some variants, *SuC* occurred for $F = 0.8$ also. Then the *STAG* cases mostly occurred with Medium (0.6 and 0.7), and High *F* values. Based on the mutation type (Table 7), for *DE/rand/*/** variants the *PrCs* occur at *Low* values of *F*, *SuCs* occur mostly with *Medium* and *High* values of *F* and *STAGs* at all ranges of *F*. For *DE/best/*/** variants, *PrC* occur at *Low* and *Medium* values of *F*, *SuC* occur in m*edium* and *High* values of *F* and *STAG* at all ranges of *F*. Based on the crossover type (Table 8), *PrC* happens with *Low* values of *F* in most combinations of the function-variant, and with *Medium F* values in some combinations for *DE/*/*/bin* variants. *SuC* occur at *Medium* and *High* values of *F* for the *DE/*/*/bin* variants. *STAG* cases are found to occur across all three classes of *F* values for both *DE/*/*/bin* as well as *DE/*/*/exp* variants. It is worth noting that *DE/*/*/exp* variants gave very less *SuC* and *PrC*. The number of *STAGs* cases of *DE/*/*/exp* variants are more than the *DE/*/*/bin* variants.

Table 3. Nature of convergence vs. Variants

Variant	DE/rand/1/bin	DE/rand/1/exp	DE/best/1/bin	DE/best/1/exp	DE/rand/2/bin	DE/rand/2/exp	DE/best/2/bin	DE/best/2/exp
PrC#	31	3	31	3	8	0	17	0
SuC#	38	1	8	1	11	1	6	3
STAG#	19	84	49	84	69	87	65	85

Table 4. Nature of convergence based on mutation type

	f_1		f_2		f_3		f_4	
	rand	*best*	*rand*	*best*	*rand*	*best*	*rand*	*best*
PrC#	10	18	6	6	9	24	16	6
SuC#	22	16	0	1	10	1	0	0
STAG#	56	54	82	81	69	63	72	82

Table 5. Nature of convergence vs. Function

Function	f_1	f_2	f_3	f_4
PrC#	28	12	33	22
SuC#	38	1	11	0
STAG#	110	163	132	154

Table 6. Nature of convergence based on crossover type

	f_1		f_2		f_3		f_4	
	bin	*exp*	*bin*	*exp*	*bin*	*exp*	*bin*	*exp*
PrC#	25	3	11	1	28	5	22	0
SuC#	33	5	0	1	11	0	0	0
STAG#	30	80	77	86	49	83	66	88

Table 7. Effect of *F* values on convergence for *DE/rand/*/** and *DE/best/*/** variants

	f_1		f_2	
	rand	*best*	*rand*	*best*
PrC	Low	Low, Medium	Low	Low, Medium
SuC	Medium	Medium, High	–	With *VaryingF*
STAG	High	High, Low	Medium	Medium
	f_3		f_4	
	rand	*best*	*rand*	*best*
PrC	Low	Low, Medium	Low, Medium	Low
SuC	Medium, High	High	–	–
STAG	Medium	Medium	Medium	Medium

Table 8. Effect of *F* values on convergence for *DE/*/*/bin* and *DE/*/*/exp* variants

	f_1		f_2	
	bin	*Exp*	*bin*	*exp*
PrC	Low	Low	Low	Low
SuC	Medium	Medium	–	Medium
STAG	High	High, Medium	Medium	Medium
	f_3		f_4	
	bin	*exp*	*bin*	*exp*
PrC	Low	Low	Low, Medium	–
SuC	Medium, High	–	–	–
STAG	Medium	Medium	Medium	Medium

The cases depicting the occurrences of *PrC, SuC* and *STAG* for the benchmarking function with different *F* values are shown in Fig. 1. The graph shows the evolution *BOV* of a random run from generation 1 to generation 2000, in the unit step of 200. To highlight the significances in differences between the evolution of *BOVs* among the runs with *PrC, SuC* and *STAG*, the Wilcoxon Signed-Ranks Test for Paired Samples is performed. The parameters involved in this test are α - level of significance, *n* – number of samples, *T* - test statistic and *T-Crit* - critical value for *T*. The pairs considered for study are *PrC* Vs *STAG*, *PrC* Vs *SuC* and *STAG* Vs *SuC*. The value for α = 0.05 and *n* = 21 for all the pairs. The results obtained for these pairs by the Wilcoxon test is shown in Table 9.

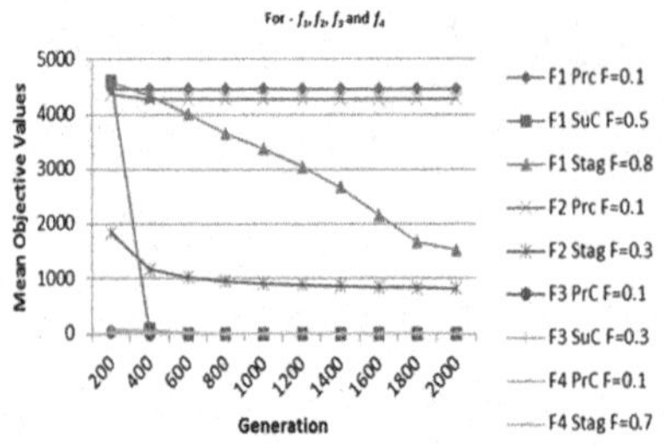

Fig. 1. Convergence graph for all function

The results show that there are significant differences between the *BOVs* of the runs with *SuC* Vs *PrC* and *SuC* Vs *STAG*. However, among *PrC* and *STAG*, the difference between *BOVs* is not statistically significant. The '+' in the table indicates that the difference is significant, and '-' indicates it is not.

The key observations from this empirical as well as comparative analytical study are summarized as follows: (i) The *PrCs* occur mostly at *Low* values of *F* irrespective function type, mutation type and Crossover type. (ii) The *SuCs* occur mostly at *Medium* *F* values. For *DE/best/*/** variants, *SuC* occurred at *High F* values also. (iii) It is interesting to note that the *STAG* occurs in all the ranges of *F*.

Table 9. Statistical results - wilcoxon signed-ranks test

Parameters	*PrC Vs STAG*	*PrC Vs SuC*	*STAG Vs SuC*
T	90	21	21
T-Crit	58	58	58
Significance	No	Yes	Yes
	–	+	+

5 Conclusions

This paper explained the observations and inferences obtained from experiments carried out to study the effect of Scale Factor (*F*) on *DE* algorithm. Generally, it is observed that Low values of *F* tend to produce Premature Convergence irrespective of *DE* variants used. The medium and high values of *F* produce successful convergence for *DE/rand/*/** and *DE/best/*/** variants, respectively. Stagnation is not associated with any particular value of *F*, but it is observed that *DE/*/*/exp* variants fall in stagnation often irrespective of the *F* values.

In future work, this experiment can be extended to include the *DE* algorithm with the Crossover operator is removed [18], to focus only on mutation. Also a visualization system to visualize the types of convergences may support for more clear understanding on the effect of *F*. Extending this work for multi-objective *DE* algorithm is also in the interest of future work [19].

References

1. Storn, R., Price, K.: Differential evolution—a simple and efficient adaptive scheme for global optimization over continuous spaces. Technical report - TR-95-012, ICSI (1995)
2. Storn, R.: On the usage of differential evolution for function optimization. In: Biennial Conference of The North American Fuzzy Information Processing Society, pp. 519–523 (1996)
3. Storn, R., Price, K.: Differential evolution—a simple and efficient heuristic strategy for global optimization and continuous spaces. J. Glob. Optim. **11**(4), 341–359 (1997)
4. Al-Dabbagh, R.D., Neri, F., Idris, N., Baba, M.S.: Algorithmic design issues in adaptive differential evolution schemes: Review and taxonomy. In: Swarm and Evolutionary Computation, vol. 43, pp. 284–311 (2018)
5. Pranav, P., Jeyakumar, G.: Control parameter adaptation strategies for mutation and Crossover rates of differential evolution algorithm–an Insight. In: Proceedings of IEEE International Conference on Computational Intelligence and Computing Research, pp. 563–568 (2015)

6. Dhanalakshmy, D.M., Pranav, P., Jeyakumar, G.: A survey on adaptation strategies for mutation and crossover rates of differential evolution algorithm. Int. J. Adv. Sci. Eng. Inf. Technol. **6**(5), 613–623 (2016)
7. Gamperle, R., Muller, S.D., Koumoutsakos, P.: A parameter study for differential evolution. In: Proceedings of WSEAS International Conference on Advances in Intelligent Systems, Fuzzy Systems, Evolutionary Computation, pp. 293–298 (2002)
8. Trvdik, J.: Competitive differential evolution. In: Proceedings MENDEL, pp. 7–12 (2006)
9. Trvdik, J.: Differential evolution with competitive setting of control parameters. Task Q. **10** (4), 1001–1011 (2007)
10. Ali, M.M., Törn, A.: Population set-based global optimization algorithms: some modifications and numerical studies. Comput. Oper. Res. **31**(10), 1703–1725 (2004)
11. Brest, J., Greiner, S., Boskovic, B., Mernik, M., Zumer, V.: Self-adapting control parameters in differential evolution: a comparative study on numerical benchmark problems. IEEE Trans. Evol. Comput. **10**(6), 646–657 (2006)
12. Liu, J., Lampinen, J.: A fuzzy adaptive differential evolution algorithm. Springer-Soft Comput. **9**, 448–462 (2005)
13. Jeyakumar, G., Shunmuga Velayutham, C.: Distributed mixed variant differential evolution algorithms for unconstrained global optimization. Memetic Comput. **5**(4), 275–293 (2013)
14. Jeyakumar, G., Shunmuga Velayutham, C.: Distributed heterogeneous mixing of differential and dynamic differential evolution variants for unconstrained global optimization. Springer-Soft Comput. **18**(10), 1949–1965 (2014)
15. Mezura-Montes, E., Velazquez-Reyes, J., Coello, C.A.: A comparative study on differential evolution variants for global optimization. In: Proceedings of the 8th Annual Conference on Genetic and Evolutionary Computation, pp. 485–492 (2006)
16. Akhila, M.S., Vidhya, C.R., Jeyakumar, G.: Population diversity measurement methods to analyze the behavior of differential evolution algorithm. Int. J. Control Theor. Appl. **8**(5), 1709–1717 (2016)
17. Opara, K.R., Arabas, J.: Differential evolution: a survey of theoretical analyses. Swarm Evol. Comput. **44**, 546–558 (2019)
18. Dhanalakshmy, D.M., Jeyakumar, G., Shunmuga Velayutham, C.: Crossover-free differential evolution algorithm to study the impact of mutation scale factor parameter. Int. J. Recent Technol. Eng. **7**(6), 1728–1737 (2019)
19. Shinde, S.S., Devika, K., Thangavelu, S., Jeyakumar, G.: Multi-objective evolutionary algorithm based approach for solving RFID reader placement problem using weight-vector approach with opposition-based learning method. Int. J. Recent Technol. Eng. **7**(5), 177–184 (2019)

Image Registration Using Single Swarm PSO with Refined Search Space Exploration

P. N. Maddaiah and P. N. Pournami(✉)

National Institute of Technology Calicut, Kozhikode 673601, Kerala, India
maddaiah@nitc.ac.in, pournamipn@nitc.ac.in

Abstract. Image registration is an elementary task in Computer Vision, which geometrically aligns multiple images of a scene, captured at different times, from various viewpoints, or by heterogeneous sensors. The optimisation strategy we employ for achieving the optimal set of transformation vectors is a major factor that determines the success and effectiveness of an automatic registration procedure. This paper discusses a scheme to modify the conventional Particle Swarm Optimisation (PSO) algorithm for better search space exploration and for faster convergence. While PSO is running, after half of the total number of iterations, find the particle which is in worst position in space, then reposition that particle by mean value of its current position and the global solution. It is observed that re-positioning the worst particle in space helps that particle from premature convergence to a local optimum solution and motivates the particle to generate unique search directions, which increased the possibility of finding the globally best solution. An image registration algorithm using this modified PSO method is also presented. From the experimental results presented here, it is visible that the proposed algorithm guarantees superior results in terms of registration accuracy and reduced execution time, even in the case of large deformations between the reference and float images.

Keywords: Single swarm PSO · Image registration · RMSE

1 Introduction

In a typical problem instance, an image registration system considers two images: *a reference image, denoted by* I_R and *a float image, denoted by* I_F. The outcome of the registration algorithm is a transformation T such that when this geometrical transformation is applied on the float image, the reference image I_R and the transformed float image $T(I_F)$ are in perfect alignment with each other [18]. Thus the aim is to find a transformation T^* such that T^* gives maximum similarity between the reference image and the float image, with proper guidance from an optimisation technique. A prominent category of algorithms fall within the class of automatic or global registration methods, which use the full image data to derive an appropriate transformation for aligning the input images.

B. Deka et al. (Eds.): PReMI 2019, LNCS 11941, pp. 337–346, 2019.
https://doi.org/10.1007/978-3-030-34869-4_37

In an automatic framework, the registration process is formulated as a mathematical problem by constructing a cost function defined over the similarity or dissimilarity between the input images. Then search for a single transformation, T^*, imposed on the whole image, which maximizes or minimizes the cost function by searching the parameter space of all allowable transformations. A more robust approach is to register images using feature points identified on both the images. A mapping is found based on these control points which is directly used for achieving correspondence between the reference and the float images.

From an operational point of view, image registration methods choose an intelligent combination of the following four coupled components [8,18].

1. *Feature detection:* Salient features in the input images are manually marked or automatically detected.
2. *Feature matching:* A mapping between the features detected in the float image and those detected in the reference image is determined.
3. *Transform model estimation:* The attributes of the mapping functions are estimated in order to align the float image with the reference image.
4. *Image resampling and transformation:* The float image is transformed using the mapping functions and the intensity values at non-integer coordinates are estimated by some interpolation technique.

The registration problem can be reduced to determining parameters of the transformation in the search space that provides the highest similarity. This is an optimisation problem in multidimensional space. The number of parameters in the transformation model is the dimensional size of this problem. Hence, it is necessary to keep the number of parameters in the model as small as possible for a faster registration process. The number of sub-processes involved makes the whole process of image registration very complex and poses many challenges in the registration of digital images. In general, image registration has huge computational cost and the overall performance of a registration algorithm rely upon the stability of the optimisation strategy. Although considerable research has gone into developing efficient optimisation techniques, more attention has to be paid to formulate improved strategies for finding the optimal match between the reference and the float images.

2 Review of Literature

The word *registration* was first coined by Becker in the year 1900 in a US patent [4]. In the literature, several reports on image registration methods can be found. Probably, the most comprehensive survey of the universal image registration methods was published by Brown in 1992 [5]. She provided a framework for understanding the existing registration techniques and also a methodology for supporting the selection of appropriate technology for a particular registration problem. Comprehensive surveys of classic image registration methods were tabled by Zitova [18] in 2003.

The optimisation technique used for obtaining optimal transformation parameters has a pivotal role in determining the success and effectiveness of any registration method. Swarm intelligence, Tabu Search, Simulated Annealing, Genetic Algorithms, SVMs and ANNs are well-known examples of intelligent algorithms that use clever simplifications and methods to solve computationally complex problems [14]. These global optimisation methods guarantee improved performance when applied on standard problems such as Traveling Salesmen Problem, Knapsack Problem, Searching and Sorting etc.

Seixas et al. [15] makes use of genetic algorithm to address image registration based on point matching. The approach uses nearest neighbor point matching and mapping was performed using affine transformation. Due to the use of genetic algorithm, pre-alignment between images are not required to guarantee good results. However, accuracy depends on the random choice of initial population. In many cases, the hybrid particle swarm technique provided better registration performances than the evolutionary techniques providing comparable convergence. It is observed that hybrid approach employing crossover operator improves accuracy. However, in few cases convergence was adversely affected due to the prevention of particle from moving to global optimum.

Valsecchi et al. [16] reported a real-coded genetic algorithm based image registration approach using intensity-based technique. Genetic Algorithm (GA) makes use of cross-over and random mutation. Registration is performed through multiple stages following a multi-resolution strategy, where complexity increases with stages, and incorporation of provisions for restart of optimisation. The restart procedure improved reliability. The authors report that the convergence issues experienced was properly addressed after the initial investigation stage. Chen et al. [6] proposed particle swarm optimisation for medical image registration. Rigid transformation is used for global transformation of image and non-rigid for local transformation of images by cubic B-spline curves.

Another PSO-based technique is proposed in [12]. Here, a hybrid particle swarm optimisation is developed for multi modal 3D medical image registration, by including sub-population and crossover from GA techniques into traditional PSO. A dynamic brain image registration is proposed by Li et al. in [11]. The algorithm combines PSO and inheritance idea. The algorithm inherits information from reference image to guide registration process. The algorithm has a complexity of $\mathcal{O}(n * m * t)$, where t is the time taken to compute fitness value of a particle, m denotes the population size, and n is the number of iterations.

Yet another PSO based registration approach is that by Wachowiak et al. [17] for multimodal images. Here, a single slice 3D biomedial image registration is done using PSO. A new hybrid PSO approach incorporating initial user guidance is employed. Performance of the optimisation approach was compared with mutual information metric under different evolutionary strategies. Another work is the non-rigid approach of registration for medical images proposed by Anna et al. [3] to enhance the quality of registration using DTCWT (Dual-Tree Complex Wavelet Transform) and Niche PSO (NPSO). The approach employs multi-scale key-points as features, and DCTWT is used to detect features, and Hausdorff distance metric is employed as the metric to compute image similarity. NPSO is used to compute optimal affine parameters. The authors claim better robustness and accuracy in noisy images. Krusienski and Jenkins raise many suggestions and solutions for overcoming the limitations of the conventional PSO [9]. There are modifications of the conventional PSO which assert conditions either on convergence time or on search capacity of the algorithm.

The literature survey carried out reveals that a major issue in automatic image registration is the need for efficient and faster optimisation algorithms for determining the parameters of registration. Exhaustive or brute force approaches are computationally impractical. Simulated Annealing (SA), Particle Swarm optimisation (PSO) and Genetic algorithm (GA) are some of the popular global search algorithms employing heuristics to arrive at faster solutions. It is also observed that when applied to image registration, these algorithms can offer some clever formulations which minimize the overall computational overhead. Research in this area continues to be motivated by the need to exploit new mechanisms and principles to boost the performance of the conventional PSO for a variety of problems in practice, including image registration.

3 Single Swarm PSO with Refined Search Space Exploration

Swarm Intelligence is the emergent collective intelligence of groups of social organisms called swarms. Swarms consist of large number of individuals coordinated by indirect communication, self-organization and decentralized control. Each individual is named a particle of the population. Computational swarm intelligence analyses the behavior of natural swarms, such as fish schools, bird flocks and ant colonies, and translates the learned theories into an algorithm. When a solution to a problem is desired, the population of individuals iteratively evolve by combined efforts and competition among the individuals. Particle Swarm Optimisation (PSO) is introduced by Kennedy and Eberhart, which is a population-based metaheuristic [7].

In the iterative PSO algorithm, each particle is initialized with uniform random values in the given D-dimensional search space S. Particle's motion is based on its velocity and current position and these values are updated at each iteration. The ith particle of the group is defined by a velocity vector v^i=(v^{i1}, v^{i2},...v^{id},...v^{iD}) and a position vector x^i=(x^{i1}, x^{i2},...x^{id},...x^{iD}) in the D-dimensional search space S. In the nth iteration, each particle i has a position x_n^{id} in the search space S and moves with a velocity v_n^{id}. The velocity of a particle is updated in every iteration according to Eq. 1.

$$v_{n+1}^{id} = w * v_n^{id} + c_1 * r_{1n}(p_{best^n}^{id} - x_n^{id}) + c_2 * r_{2n}(g_{bestn} - x_n^{id}) \quad (1)$$

Subsequently, the new particle position can be determined in terms of Eq. 2.

$$x_{n+1}^{id} = x_n^{id} + v_{n+1}^{id} \quad (2)$$

In these equations, N stands for the total number of iterations and n =1,2,3... N. c_1 and c_2 are the acceleration coefficients known as the cognitive learning rate and the social learning rate, respectively. r_1 and r_2 are random numbers between 0 and 1. w is the inertia coefficient which will vary according to the Eq. 3.

$$w = (w_{max} - w_{min})[-V_{max}, V_{max}] * \frac{n}{N} \tag{3}$$

The velocity of particles should be in the range of $[-V_{max}, V_{max}]$ to ensure that the particle does not exit from the allowed search space. The particles accelerate toward the other particles which have better fitness values. To establish the local and global attractor, the fitness or objective function $f : S \subseteq R^n \rightarrow R$ is used. The velocity and position adaptation is done iteratively for all particles of the swarm until a specified termination criterion is reached. Once all the particles find their best solution, $p^{id}_{best^n}$, i=1, 2, ... N, the best solution is again calculated from these N values. The best found position by the swarm is the result and thereby the return value of the algorithm. This best value is globally accepted as the final solution, which is represented as g_{bestn}.

We propose the following modification to the conventional PSO algorithm as a tool to solve image registration problem to arrive at faster solutions. ***The proposal is to re-position the worst particle in space to help that particle from premature convergence to a local optimum solution. The re-randomization is performed after half of the iterations. This motivates the stray particles to generate unique search directions, which increases their possibility of faster convergence to the global solution.***

3.1 Validation Using Benchmark Functions

The optimization ability of the proposed technique is validated using a set of standard benchmark functions [10]. There are unimodal and multimodal functions. A function with only one local minimum is called unimodal function, whereas a multimodal function has more than one local minima. As the dimension of the function increases, the number of local minima will also increase. Parameters for the experiment include dimension of the optimization function D, number of particles M, total number of iterations N, inertia coefficient ω and acceleration coefficients c_1 and c_2. The proposed PSO algorithm was run for various dimensions D = 2, 3, 4, 5 and 10. The inertia coefficient ω is set to 0.2 and both the acceleration coefficients c_1 and c_2 are set to 2.0. The other important parameters number of particles M, total number of iterations N are set to 200 and 5000 respectively.

Table 1 reports the best parameter setting of the proposed PSO, for each of these functions, to converge to the global optimum value. This table was prepared after the proposed algorithm was tested for 30 independent runs. NFE is the number of function evaluations, which is the most important characteristic of an optimisation algorithm.

Table 1. Performance of proposed PSO on benchmark functions

Function	D = 2		D = 3		D = 4		D = 5		D = 10	
	NFE	Min. value	NFE	Min. value	NFE	Min. value	NFE	Min. value	NFE	Min. value
Ackley	99	8.88E−16	113	8.88E−16	120	8.88E−16	297	8.88E−16	170	4.44E−15
Beale	118	0	118	0	136	0	124	0	128	0
Bohachevsky	66	0	64	0	64	0	64	0	64	0
Booth	108	0	110	0	105	0	109	0	106	0
Branin	14	0.39789	1	0.39789	4	0.39789	14	0.39789	12	0.39789
Easom	18	−1	22	−1	25	−1	24	−1	24	−1
Goldstein Price	13	3	17	3	17	3	18	3	17	3
Griewank	148	0	135	0	160	0	79	0	4500	0.0295
Levy	95	1.5E−32	92	1.5E−32	93	1.5E−32	97	1.5E−32	165	1.5E−32
Matyas	919	0	898	0	906	0	936	0	929	0
Rastrigin	58	0	73	0	71	0	116	0	118	1.9899
Shubert	23	−186.7309	26	−186.7309	26	−186.7309	22	−186.7309	24	−186.7309
Sphere	544	0	546	0	546	0	546	0	1290	0

4 Results and Discussion

The problem under consideration is the global, rigid registration of two 2D images. A rigid-body transformation in two dimensions is defined by four parameters: two translations in x and y directions and two rotations in x an y directions. The objective function to be maximized is the Mutual Information (MI) between the input images. MI [13] is the most celebrated similarity measure employed for image registration. The experiment environment is Intel Core i5 (1.3 GHz) processor with 4GB primary memory and the algorithm is run on 64-bit Mac OSX Yosemite system. The proposed technique is implemented using MATLAB v7.10. The first set of experiments use images from CMU House sequence [1]. The CMU house data set contains 111 two dimensional gray scale images of a toy house. Each image is of size 480 × 512 and is taken at different angles of camera position. Next set of experiment uses NewYork Data set [2], where the images are taken by a rotating camera. This data set contains 35 images of 512 × 512 size. Figure 1 is a set of sample images from these data sets.

Fig. 1. Sample images from CMU house sequence and NewYork datasets

4.1 Translation

As the first step, the modified PSO is tested using CMU House Sequence. Here the first image is taken and is translated by various levels. The reference image is translated by various units, in x and y directions, to create the set of float images. A total of $M = 20$ particles are initialized uniformally in the search space. The parameters $c1$ and $c2$ are set to 2. Mutual Information (MI) between the input image is used as the cost function. Registration is performed using both the traditional PSO and the proposed PSO algorithms. In order to analyze the registration accuracy, Root Mean Square Error (RMSE) between the reference image $I_R(x, y)$ and the registered float image $I_F(T(x, y))$ is calculated using Eq. 4.

$$RMSE = \sqrt{\frac{1}{N}\sum_{i=1}^{N}(I_R(x, y) - I_F(T(x, y)))^2} \quad (4)$$

Table 2 summarizes the results for both these algorithms over 20 independent runs for the given set of input images. Total number of iterations for both the algorithms are set to $N = 200$.

Table 2. Translation

Translation[x,y]	Basic PSO			Proposed PSO		
	TIME (sec)	MI	RMSE	TIME (sec)	MI	RMSE
[10 10]	1002.18	5.5327	0	954.33	5.5327	0
[−10 10]	1025.62	5.543	0.0509	834.88	5.543	0.0509
[10 − 10]	879.69	5.5052	0.0261	818.85	5.5052	0.0261
[−10 − 10]	881.09	5.5056	0.0576	824.146	5.5056	0.0576
[15 15]	862.91	5.4396	0	817.46	5.4396	0
[−15 15]	945.87	5.4692	0.0897	822.95	5.4692	0.0897
[15 − 15]	873.54	5.4392	0	818.20	5.4392	0.0365
[−15 − 15]	870.96	5.4486	0.0966	871.09	5.4486	0.0966
[20 20]	860.94	5.3418	0	823.46	5.3418	0
[−20 20]	926.94	5.3946	0.1076	872.21	5.3946	0.1076
[20 − 20]	935.48	5.3713	0.0546	881.24	5.3713	0.0546
[−20 − 20]	1014.63	5.3915	0.1207	944.83	5.3915	0.1207

4.2 Rotation

The second set of experiments were designed to analyze the effectiveness of the modified PSO for images with varying degrees of rotations. Images from NewYork data set are utilized for this. The first image in this dataset is fixed as the reference image and the remaining 34 images are used to create the float image set. Registration is performed using both the traditional PSO and the modified PSO. 20 independent runs were carried out for each input image set and the best values are selected. Here for images from 1 to 10, the parameter setting for the experiments was the following. A total of $M = 20$ particles are initialized uniformally in the search space, $c1$ and $c2$ are set to 2, $N = 200$. Mutual Information (MI) between the input images is used as the cost function. Preliminary results are included in Fig. 2 and the results are summarized in Table 3.

Table 3. Rotation

[Image1, Image2]	Basic PSO			Proposed PSO		
	TIME (sec)	MI	RMSE	TIME (sec)	MI	RMSE
[1, 2]	364.08	1.1529	9.073	307.75	1.4167	5.8136
[1, 4]	367.52	0.3596	31.4317	314.99	1.1991	6.4673
[1, 6]	379.47	1.2789	4.9321	345.96	1.2789	4.9321
[1, 8]	1227.01	0.4099	32.4321	1566.16	1.1318	7.3927
[1, 10]	1326.13	1.3233	7.6751	1062.70	1.3234	7.6751
[1, 12]	357.99	0.3763	32.2158	338.92	0.4117	31.7973
[1, 14]	539.85	0.4008	34.0943	364.4	0.3847	31.7818
[1, 16]	417.74	0.4002	34.5227	340.92	0.3986	34.7637
[1, 18]	367.14	0.3452	35.2884	365.42	0.3519	35.2025
[1, 20]	433.02	0.3827	35.509	389.27	0.3892	37.6945
[1, 22]	482.09	0.3662	37.0449	576.28	0.389	35.5047
[1, 24]	467.68	0.3773	36.1851	356.44	0.3751	35.2837
[1, 26]	365.42	0.5181	22.3919	322.8	1.0687	13.114
[1, 28]	369.35	0.4352	24.7789	319.34	1.1792	9.874
[1, 30]	357.96	0.5945	18.2654	304.57	1.2322	7.5635
[1, 32]	332.73	0.4408	23.7133	279.05	1.2205	6.752
[1, 34]	386.96	1.3382	6.1607	321.923929	1.3382	6.1607

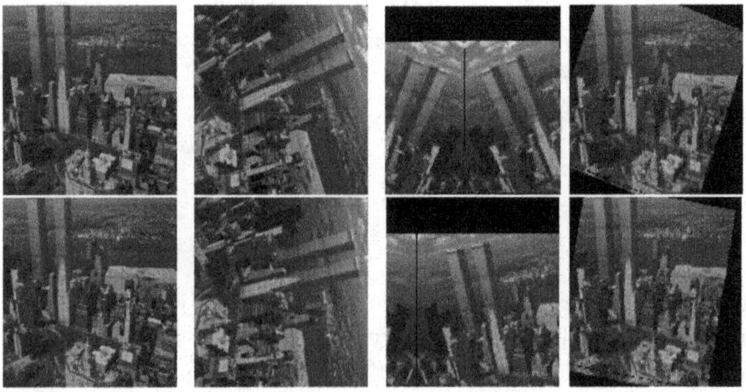

Fig. 2. Reference Image, Float Image, PSO output, Proposed PSO Output

4.3 Translation and Rotation

The final set of experiments were designed to analyze the effectiveness of the modified PSO for images with varying degrees of translations and rotations using images from NewYork data set. Float images are created by translating each image by +5, +10 and +15 units in both x and y directions. Relevant parts of the results are compiled in Table 4.

Table 4. Rotation and Translation +15 in x and +15 y direction

[Image1, Image2]	Basic PSO			Proposed PSO		
	TIME (sec)	MI	RMSE	TIME (sec)	MI	RMSE
[1, 2]	383.54	0.449	34.3989	304.68	1.293	34.3757
[1, 3]	358.39	0.39	33.9644	288.34	1.1724	34.1937
[1, 4]	363.34	0.3934	36.0159	302.27	1.1075	34.3546
[1, 5]	339.53	0.3641	33.6465	290.05	1.0706	33.9921
[1, 6]	356.68	0.4313	36.2148	306.04	1.4221	34.0194
[1, 7]	352.7	0.3853	35.1371	337.67	1.3286	33.4782
[1, 8]	344.88	0.4329	35.4526	298.92	1.2242	32.8412
[1, 9]	365.99	0.3638	36.0035	305.91	1.406	33.7141
[1, 10]	355.86	0.4068	34.1809	311.95	1.2622	34.2951

5 Summary

It is quite visible that the proposed PSO algorithm could optimize the benchmark functions for different dimensions. Various experiments were designed to demonstrate the efficiency of the modified PSO in achieving maximum similarity between the reference and float images; thus achieving registered images. The proposed PSO algorithm shows better average RMSE value and it acquired better similarity between the input images. It is evident from the convergence characteristics that the improved PSO attains better MI in fewer iterations and the best values for transformation parameters are sought in the remaining iterations. Generally area-based image registration methods mostly survive when the distortion between the input images is small. Here, the proposed algorithm guarantees highly acceptable solutions even in the case of large deformations between the reference and float images.

References

1. CMU House Sequence. http://vasc.ri.cmu.edu/idb/html/motion. Accessed Dec 2018
2. Oxford Dataset. http://www.robots.ox.ac.uk/~vgg/research/affine/. Accessed Dec 2018

3. Anna, W., Tingjun, W., Jinjin, Z., Silin, X.: A novel method of medical image registration based on DTCWT and NPSO. In: 2009 Fifth International Conference on Natural Computation, vol. 5, pp. 23–27, August 2009
4. Becker, J.: Focusing-camera, 4 April 1916. http://www.google.co.in/patents/US1178475
5. Brown, L.G.: A survey of image registration techniques. ACM Comput. Surv. **24**, 325–376 (1992)
6. Chen, Y.W., Lin, C.L., Mimori, A.: Multimodal medical image registration using particle swarm optimization. In: 2008 Eighth International Conference on Intelligent Systems Design and Applications, vol. 3, pp. 127–131, November 2008. https://doi.org/10.1109/ISDA.2008.321
7. Eberhart, R., Kennedy, J.: A new optimizer using particle swarm theory. In: 1995 Proceedings of the Sixth International Symposium on Micro Machine and Human Science, MHS 1995, pp. 39–43. IEEE (1995)
8. Goshtasby, A.A.: 2DD and 3-D Image Registration: For Medical, Remote Sensing, and Industrial Applications. Wiley-Interscience, Hoboken (2005)
9. Krusienski, D.J., Jenkins, W.K.: A modified particle swarm optimization algorithm for adaptive filtering. In: 2006 IEEE International Symposium on Circuits and Systems, pp. 4–140, May 2006. https://doi.org/10.1109/ISCAS.2006.1692541
10. Li, X., Tang, K., Omidvar, M.N., Yang, Z., Qin, K.: Benchmark functions for the CEC'2013 special session and competition on large-scale global optimization, January 2013
11. Li, Y., Lai, H., Lu, L., Gao, Y., Wang, P.: Dynamic brain magnetic resonance image registration based on inheritance idea and PSO. In: 2011 4th International Conference on Biomedical Engineering and Informatics (BMEI), vol. 1, pp. 263–267, October 2011. https://doi.org/10.1109/BMEI.2011.6098345
12. Lin, C.L., Mimori, A., Chen, Y.W.: Hybrid particle swarm optimization and its application to multimodal 3d medical image registration. Intell. Neurosci. **2012**, 6:6 (2012). https://doi.org/10.1155/2012/561406
13. Maes, F., Collignon, A., Vandermeulen, D., Marchal, G., Suetens, P.: Multimodality image registration by maximization of mutual information. IEEE Trans. Med. Imag. **16**(2), 187–198 (1997)
14. Pham, D., Karaboga, D.: Intelligent Optimisation Techniques: Genetic Algorithms, Tabu Search. Simulated Annealing and Neural Networks. Springer, Heidelberg (2012)
15. Seixas, F.L., Ochi, L.S., Conci, A., Saade, D.M.: Image registration using genetic algorithms. In: Proceedings of the 10th Annual Conference on Genetic and Evolutionary Computation, pp. 1145–1146. ACM (2008)
16. Valsecchi, A., Damas, S., Santamaría, J., Marrakchi-Kacem, L.: Genetic algorithms for voxel-based medical image registration. In: 2013 Fourth International Workshop on Computational Intelligence in Medical Imaging (CIMI), pp. 22–29 (2013)
17. Wachowiak, M.P., Smolikova, R., Zheng, Y., Zurada, J.M., Elmaghraby, A.S.: An approach to multimodal biomedical image registration utilizing particle swarm optimization. IEEE Trans. Evol. Comput. **8**(3), 289–301 (2004). https://doi.org/10.1109/TEVC.2004.826068
18. Zitova, B., Flusser, J.: Image registration methods: a survey. Image Vis. Comput. **21**(11), 977–1000 (2003)

Shuffled Particle Swarm Optimization for Energy Efficiency Using Novel Fitness Function in WSN

Amruta Lipare(✉) and Damodar Reddy Edla

National Institute of Technology Goa, Farmagudi, India
amruta.lipare@gmail.com, dr.reddy@nitgoa.ac.in

Abstract. Clustering of sensor nodes in Wireless Sensor Networks (WSN) is a critical task in order to minimize energy consumption as well as prolonging the lifetime of the network. In cluster based WSN, the gateway from the cluster collects, aggregates and sends data to the base station. To perform these operations, the energy consumption depends on two significant factors; load on the gateway and the data transmission distance. The improper clustering of sensor nodes may consume more energy, and the network may die soon. Therefore, load balancing of gateways according to the transmission distance is necessary to increase the network lifetime. We applied the shuffling strategy from Shuffled Frog Leaping Algorithm (SFLA) to Particle Swarm Optimization (PSO) to implement an algorithm called shuffled PSO (SPSO) for improving energy efficiency in WSN. The quality of the solution is measured by the novel fitness function concerning the lifetime of gateways and average cluster distance. The extensive simulations are performed with some of the existing algorithms. From the experimental analysis, it is observed that the proposed clustering approach gives effective results.

Keywords: Clustering · Energy efficiency · Particle swarm optimization · Shuffled frog leaping algorithm · Wireless sensor networks

1 Introduction

Wireless sensor networks (WSNs) has wide variety of applications like automation, health-care systems, defense systems etc. In most of the applications, the sensor nodes are battery operated. Sensor nodes sense the flactuations in environmental conditions such as variation in temperature, pressure change. Nodes collect the sensed data to the base station (BS) [1]. Such operations led to the power drop of sensor nodes. Battery charging or replacing is a difficult task in such WSN deployment. Therefore, efficient energy consumption is essential to increase the network lifetime [1]. Researchers applied different energy efficient algorithms such as clustering, routing, load balancing techniques etc. In the network with clusters of sensor nodes, the data sensed from the environment by

B. Deka et al. (Eds.): PReMI 2019, LNCS 11941, pp. 347–355, 2019.
https://doi.org/10.1007/978-3-030-34869-4_38

the cluster member nodes gets collected to the cluster head (CH). CH combines overall received data and transfers to the BS. BS is a device connected to the internet to get updates from the targeted area. In this paper, we used the concept of gateways. Gateway is an instrument having high battery and works similar to the CH [9]. We adopt two bio-inspired algorithms namely shuffled frog leaping algorithm (SFLA) [4] and particle swarm optimization (PSO) [6]. The shuffling policy from SFLA is applied on PSO. Therefore, named as shuffled PSO (SPSO) algorithm for energy efficiency in WSN.

2 Literature Survey

Some of the energy efficient algorithms used in WSN are discussed. Low-Energy Adaptive Clustering Hierarchy (LEACH) [5] is a famous clustering method implemented to WSNs. In LEACH algorithm, the CHs are selected dynamically; so load on the network gets balanced. But, the CH selection is probabilistic. Therefore, CH with low power may get selected for several times leading to its death in early stages. To avoid this conditions, researchers have come up with improved LEACH [10], by selecting the CH according to its residual energy, its distance from BS, etc. Many researchers adopted number of the bio-inspired algorithms to resolve different issues in WSN. Kuila et al. have used a genetic algorithm (GA) for load balancing [7] and PSO for energy efficiency of the network [8]. Here, they used standard deviation of the load as a fitness function. Edla et al. used shuffled complex evolution (SCE) method and improved SFLA for clustering in WSN [2,3], where they used two novel fitness functions considering heavy and under loaded gateways.

3 Fundamental Methods

In this section, we address some of the fundamental methods in the literature required for implementing the proposed SPSO algorithm.

3.1 Outline of Particle Swarm Optimization

The flocking nature of birds inspires Particle Swarm Optimization [6]. Birds always travel in a group to search for food and shelter. Birds fly with the same velocity and follow the fitter members of the swarm [6]. PSO is composed of a group of particles moving around the search space according to their best particle location. By using Eq. 1, at each iteration, the velocity of the particle is updated.

$$v_i(t+1) = v_i(t)+(c_1\times rand()\times(p_i^{best}-p_i(t)))+(c_2\times rand()\times(p_g^{best}-p_i(t))) \quad (1)$$

Where $v_i(t+1)$ is the updated velocity of the ith particle, c_1 and c_2 are the weighted coefficients of local best and global best particles respectively. $p_i(t)$ is the position of particle at time t. p_i^{best} is the local best position of ith particle in the population throughout the iterations. p_g^{best} is the global best position of particle [6]. $rand()$ function generates a random number from 0 to 1 uniformly. The position of particle is updated using Eq. 2.

$$p_i(t+1) = p_i(t) + v_i(t) \tag{2}$$

3.2 Outline of Shuffled Frog Leaping Algorithm

SFLA [4] is a memetic algorithm and uses shuffling mechanism within the set of solutions. Following steps describe SFLA in details.

Step 1. **Initialization:** Initialize the set of parameters such as frog population size, number of memeplexes and sub-memeplexes, the maximum number of iteration in local exploration and algorithmic iterations.

Step 2. **Fitness evaluation and sorting:** A fitness function is used to evaluate each frog (solution). According to the fitness value the solutions are sorted in the descending order.

Step 3. **Formation of memeplexes and sub-memeplexes:** Solutions are partitioned into m memeplexes $M_1, M_2, \ldots, M_m$, such that each memeplex contain k solutions and first solution is assigned to first memeplex, second solution is assigned to second memeplex, m^{th} solution is assigned to m^{th} memeplex and $(m+1)^{th}$ solution is assigned to first memeplex and so on. Later, each memeplex is partitioned into n sub-memeplexes.

Step 4. **Evaluation of each sub-memeplex:** The worst solution in each sub-memeplex is evolved as described in the following steps.
   (a) **Off-spring generation phase:** Information of worst frogs changes to best frog using single point crossover.
   (b) **Censorship:** In this step, the fitness value of the new offspring is tested. If it is not better, crossover worst frog (solution) from each sub-memeplex according to these three steps serially and stop when fitness value improve; (i) crossover with the best frog of memeplex. (ii) crossover with the best frog among overall population. (iii) generate a new solution.

Step 5. **Convergence checking phase:** The algorithm terminates after reaching to the maximum algorithmic iterations.

3.3 Energy Model

In this work, we used radio model from [7]. If the transmission distance between sender and receiver is greater than d_0, then multi-path channel is used, else free-space channel is used. Let E_{elec} be an amount of energy used by the electronic circuitry at the sender node and receiver node. Therefore, the energy consumption to send the l-bit of data to the destination over the distance d is expressed in Eq. 3.

$$E_T(l,d) = \begin{cases} l * E_{elec} + l * \epsilon_{fs} * d^2, & d < d_0 \\ l * E_{elec} + l * \epsilon_{mp} * d^4, & d \geq d_0 \end{cases} \tag{3}$$

Similarly, the energy consumption to receive l-bit of data by the receiving sensor node is expressed in Eq. 4.

$$E_R(l) = l * E_{elec} \tag{4}$$

4 Shuffled PSO for Energy Efficiency in WSN

We applied an adaptive learning strategy from PSO [6] and the theory of shuffle policy from SFLA [4]. Our contributions in this algorithms are listed below.

(i) We restricted the assignment of sensor nodes and gateway in the initial population phase of PSO. Instead of assigning gateways randomly, assignment is restricted within the communication range. This increases the chances of generating a better solution in the first phase itself.
(ii) In the phase of information change from SFLA, only the worst solution is replaced instead of replacing both the parent solutions. This modification helps to save the best solution among the population.
(iii) A new phase is added after the phase of information change from SFLA called 'Relocation Phase', in which the allocation of the sensor node which is connected to farther gateway is changed to its nearest gateway. This operation decreases an amount of energy consumption of the sensor node.
(iv) Shuffle strategy from SFLA is applied on the best P solutions from PSO.

The following subsections describe the proposed shuffled PSO.

4.1 Initial Population Generation

An initial population phase from PSO is carried out with randomly generated solutions. The solution represents the allocation of sensor nodes and gateways.

4.2 Fitness Function Evaluation

One of the contributions in SPSO is, a novel fitness function designed to evaluate the solution. The energy consumption required to do the intra-cluster activities is expressed in Eq. 5. Let n_i be the number of sensor nodes connected to the gateway g_i in the network. Hence the energy consumption by gateway g_i can be formulated as in Eq. 5.

$$E_{gateway} = n_i \times E_r + n_i \times E_{da} + n_i \times E_t(g_i, BS) \tag{5}$$

where E_r is the energy required to receive the data. E_{da} is an amount of energy needed for data aggregation and $E_t(g_i, BS)$ is the energy needed for data transmission from g_i to BS. Let $E_{remaining}$ be the residual energy of gateway after number of rounds. The lifetime of the gateway [8] is framed in Eq. 6.

$$L(g_i) = \frac{E_{remaining}}{E_{gateway}} \tag{6}$$

The standard deviation of lifetime of gateways ($\sigma_{(L_i)}$) will determine the even distribution of gateway life. $\sigma_{(L_i)}$ is calculated as shown in Eq. 7. μ_L is the average of lifetime of n gateways and formulated in Eq. 8.

$$\sigma_{(L_i)} = \sqrt{\frac{\sum_{i=1}^{n}(\mu_L - (L_i))}{n}} \tag{7}$$

$$\mu_L = \frac{\sum_{i=1}^{n}(L_i)}{n} \tag{8}$$

The energy consumption also influenced by transmission distance between sender and receiver [8]. Therefore, we tried to minimize the average cluster distance (AvgClustDist) to increase the lifetime of the network. So, fitness function including the $\sigma_{(L_i)}$ and (AvgClustDist) is formulated in Eq. 9

$$Fitness \propto \frac{1}{\sigma_{(L_i)}} \times \frac{1}{AvgClustDist} \tag{9}$$

We put proportionality constant K = 1, so, fitness function is as in Eq. 10

$$Fitness = \frac{1}{\sigma_{(L_i)} \times AvgClustDist} \tag{10}$$

4.3 Velocity and Position Update

After evaluating the solutions, fitness values for each solution are obtained. The best solution among overall population is identified. All solutions update their velocity using Eq. 1 and position using Eq. 2. Finally, the best solution among updated solutions is fetched.

4.4 Partitioning the Solutions

The best P solutions from the set of updated PSO population are selected to undergo shuffling procedure. The shuffling procedure starts here. Initially, the solutions are sorted in descending order. The sorted solutions are partitioned as described in the step 3 of Sect. 3.2. The memeplexes are divided vertically to form sub-memeplexes.

4.5 Information Change and Relocation Phase

Two parent solutions are selected for information change. There are three conditions while selecting parents.

Step 1. Initially select best and worst solutions from a sub-memeplex. Randomly select a point 'x' from both the solutions. Change the information from the worst solution after point x. Test the new solution with a fitness function. If the fitness value improves then replace the worst solution with new solution else; go to Step 2.

Step 2. Select the best solution from own memeplex and follow the same procedure as maintained in step 1. If the fitness value improves then replace the worst solution with new solution else; go to Step 3.
Step 3. Select the best solution as the global best solution and follow the same procedure as maintained in step 1. If the fitness value improves then replace the worst solution with new solution else; go to Step 4.
Step 4. Generate a new solution and replace it with the worst solution.

In relocation phase, the assignment of the sensor node connected to farther gateway is changed to its nearest gateway. Thus, the sensor node saves its energy.

4.6 Sorting and Shuffling

As the worst solutions from each sub-memeplexes get replaced by better solutions. According to the updated fitness values, all the solutions from the population get sorted. This sorting operation causes shuffling in the solutions.

5 Results and Discussions

For the implementation of SPSO algorithm Matlab 2015a is used on a Windows 7 operating system having 2GB RAM with intel core i3 processor. For the experimentation, the targeted area $50 \times 50\,\mathrm{m}^2$ with 50 sensor nodes and 8 gateways is considered. The BS is positioned at the side of the area i.e. (45, 25). Figure 1 shows the network structure. PSO-based clustering approach [8] and SFLA-based clustering approach [3] are implemented for comparison purpose. Sensor nodes and gateways are positioned at the targeted area in a random manner and fixed their positions once they get deployed. The initial energies of sensor nodes and gateways are 1 J and 10 J respectively. The remaining tuning parameters are described in the Table 1. The SPSO based clustering algorithm

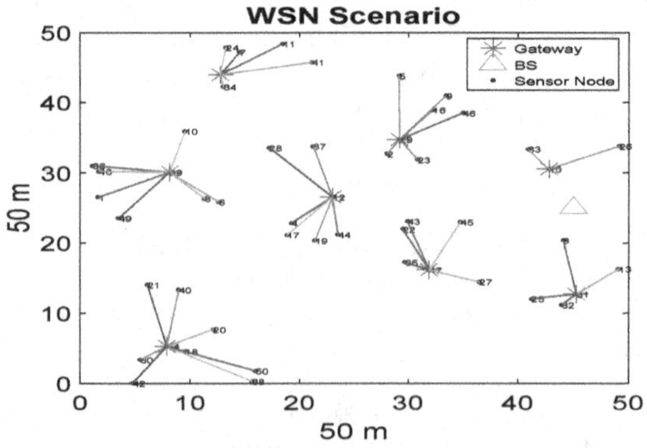

Fig. 1. Network structure

Table 1. Tuning parameters

Parameter	Value	Parameter	Value	Parameter	Value
Communication range	10 m	E_{elec}	50 nJ/bit	*Number of maximum iterations*	100
AggregationRatio	10%	ϵ_{fs}	10 pJ/bit/m^2	ϵ_{mp}	0.001 pJ/bit/m^4

is experimented to examine the network lifetime against PSO and SFLA based clustering algorithms. All three algorithms are reviewed with the constant and variable load on the sensor nodes. It is observed that in both the cases, in SPSO the first gateway dies (FGD) after longer rounds than other compared algorithms as shown in Tables 2 and 3. This is due to the deviation in the lifetime of gateways. The round when half of the gateways are alive (HGA) is examined. The difference between the FGD and HGA defines the network stability. Higher the variance longer is the network lifetime. SPSO performs superior to compared algorithms as given in Tables 4 and 5.

Table 2. Results of the first gateway dies for **constant** load on the nodes

Algorithm	PSO	SFLA	**SPSO**
Round number when FGD	3599	3146	**4251**

Table 3. Results of the first gateway for **variable** load on the nodes

Algorithm	PSO	SFLA	**SPSO**
Round number when FGD	3084	2667	**3676**

Table 4. Results for the half of the gateways die for **constant** load on the nodes

Algorithm	PSO	SFLA	**SPSO**
Round number when HGA	8365	7954	**9304**

Table 5. Results for the half of the gateways for **variable** load on the nodes

Algorithm	PSO	SFLA	**SPSO**
Round Number when HGA	7815	7041	**8592**

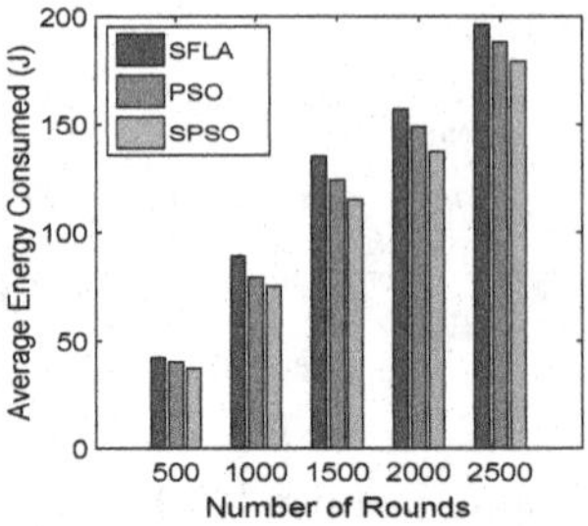

(a) Constant load

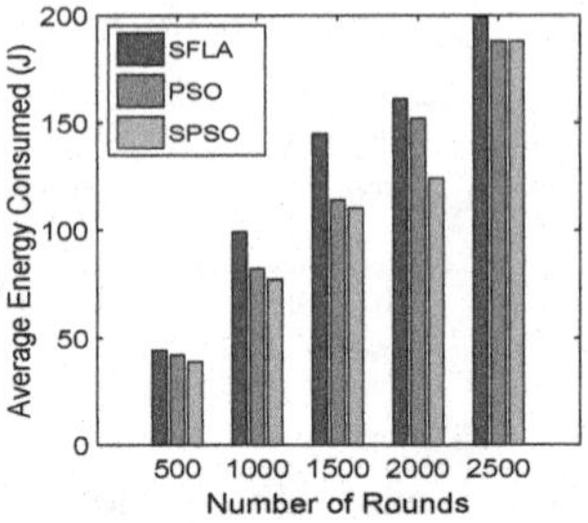

(b) Variable load

Fig. 2. Plot of number of rounds v/s energy consumption

The energy efficiency is one of the most critical factor in WSN. We calculated the overall energy consumption of gateways according to the energy model in Sect. 3.3. After 500, 1000, 1500, 2000 and 2500 rounds, the average energy consumption is noted. Even though the first gateway in the network dies after long time, energy consumption is less in SPSO for both constant and variable load as shown in Fig. 2. This is due to the average distance factor in the fitness function and the deviation in the network lifetime. The data is sent over the optimal distance, leading to less energy consumption of the network.

6 Conclusion

In this paper, we adopted PSO and SFLA for clustering of the sensor nodes in WSN. A novel fitness function is designed including two significant parameters, the standard deviation of the lifetime of gateways and average data transmission distance in the network. The best solutions from PSO are fetched and the shuffled strategy of SFLA is applied on the selected best solutions. Finally, the fittest solution among the population is selected as the optimum solution. We experimented with some of the existing algorithms namely PSO based clustering approach and SFLA based clustering approach in WSN. It is observed that SPSO outperforms compared algorithms for the network lifetime parameters and energy consumption.

References

1. Akyildiz, I., Su, W., Sankarasubramaniam, Y., Cayirci, E.: Wireless sensor networks: a survey. Comput. Netw. **38**(4), 393–422 (2002). https://doi.org/10.1016/S1389-1286(01)00302-4
2. Edla, D.R., Lipare, A., Cheruku, R.: Shued complex evolution approach for load balancing of gateways in wireless sensor networks. Wirel. Pers. Commun. **98**(4), 3455–3476 (2018). https://doi.org/10.1007/s11277-017-5024-3

3. Edla, D.R., Lipare, A., Cheruku, R., Kuppili, V.: An efficient load balancing of gateways using improved shuffled frog leaping algorithm and novel fitness function for WSNs. IEEE Sens. J. **17**(20), 6724–6733 (2017). https://doi.org/10.1109/JSEN.2017.2750696
4. Eusuff, M., Lansey, K., Pasha, F.: Shuffled frog-leaping algorithm: a memetic metaheuristic for discrete optimization. Eng. Optim. **38**(2), 129–154 (2006). https://doi.org/10.1080/03052150500384759
5. Handy, M., Haase, M., Timmermann, D.: Low energy adaptive clustering hierarchy with deterministic cluster-head selection. In: 4th International Workshop on Mobile and Wireless Communications Network, pp. 368–372. IEEE (2002). https://doi.org/10.1109/MWCN.2002.1045790
6. Kennedy, J.: Particle swarm optimization. Encyclopedia of machine learning, pp. 760–766 (2010)
7. Kuila, P., Gupta, S.K., Jana, P.K.: A novel evolutionary approach for load balanced clustering problem for wireless sensor networks. Swarm Evol. Comput. **12**, 48–56 (2013). https://doi.org/10.1016/J.SWEVO.2013.04.002
8. Kuila, P., Jana, P.K.: Energy efficient clustering and routing algorithms for wireless sensor networks: particle swarm optimization approach. Eng. Appl. Artif. Intell. **33**, 127–140 (2014). https://doi.org/10.1016/.ENGAPPAI.2014.04.009
9. Lipare, A., Edla, D.R.: Novel fitness function for SCE algorithm based energy efficiency in WSN. In: 2018 9th International Conference on Computing, Communication and Networking Technologies (ICCCNT), pp. 1–7. IEEE, July 2018. https://doi.org/10.1109/ICCCNT.2018.8494185
10. Xiangning, F., Yulin, S.: Improvement on LEACH protocol of wireless sensor network. In: 2007 International Conference on Sensor Technologies and Applications (SENSORCOMM 2007), pp. 260–264. IEEE, October 2007. https://doi.org/10.1109/SENSORCOMM.2007.4394931

An Empirical Analysis of Genetic Algorithm with Different Mutation and Crossover Operators for Solving Sudoku

D. Srivatsa, T. P. V. Krishna Teja, Ilam Prathyusha, and G. Jeyakumar(✉)

Department of Computer Science and Engineering, Amrita School of Engineering, Amrita Vishwa Vidyapeetham, Coimbatore, India
{cb.en.u4cse17459, cb.en.u4cse17465, cb.en.u4cse17424}@cb.amrita.students.edu, g_jeyakumar@cb.amrita.edu

Abstract. Prospective optimization tools such as Evolutionary Algorithms (*EAs*), are widely used to tackle optimization problems in the real world. Genetic Algorithm (*GA*), one of the instances of *EAs*, has potential research avenues of testing its applicability in real-world problems and improving its performance. This paper presents a study on the capability of the Genetic Algorithm (*GA*) to solve the classical Sudoku problem. The investigation includes various mutations and crossover schemes to unravel the Sudoku problem. A comparative study on the performance of *GA* with these schemes was conducted involving Sudoku. The findings reveal that *GA* is ineffective to deal with the Sudoku problem, as compared to other classical algorithms, as it often fails to disengage itself from some local optimum condition. On a positive note, *GA* was able to solve the Sudoku problems much faster, only the Sudoku had very few unfilled elements. A critical appraisal of the observed behavior of *GA* is presented in this paper, covering combinations of two mutations and three crossovers schemes.

Keywords: Genetic algorithm · Mutations · Crossovers · Puzzle · Sudoku · Local minima

1 Introduction

In the field of Computer Science, under Artificial Intelligence, Evolutionary Computing (*EC*) is a subfield of Soft Computing. *EC* has a family of algorithms called Evolutionary Algorithms (*EAs*) to resolve global optimization problems. Technically, *EAs* belong to the families of population-based, trial and error metaheuristic problem-solvers involving stochastic computation. *EAs* use biological evolutions such as reproduction, recombination, mutation, and selection. Natural selection (survival of the fittest) is espoused by individuals that consider fitness score provided by the fitness functions. The best individuals are selected for the iterative evolution process until a final solution is achieved with an appropriate fitness score.

Genetic Algorithm (*GA*), a component of *EA*, is used to generate high-quality accurate solutions to optimization and search problems [1, 2]. Parametric values, for the

B. Deka et al. (Eds.): PReMI 2019, LNCS 11941, pp. 356–364, 2019.
https://doi.org/10.1007/978-3-030-34869-4_39

Genetic algorithm, -several initial solutions, initial population, the maximum number of generations, mutation type, and crossover-type - can be initialized or provided by the user. *GA* encodes a population of solutions, embracing the natural evolution process, to arrive at an optimal solution in the population, with the help of the fitness function. The *GA* routine is executed until either the appropriate solution is achieved, or limited by the permissible maximum number of generations.

This paper aims at testing the suitability of applying *GA* to solve an *NP*-complete problem. The Sudoku problem was selected as a test case to solve Sudoku using *GA*.

A Sudoku puzzle is defined as a logic-based, number-placement puzzle. It can be of different sizes but the Sudoku used in this paper s 9×9 square, in which it has 9 rows, 9 columns, and 9 3×3 grids. Initially, some cells of a Sudoku puzzle are provided with numerical digits in the range of 1 to 9. The puzzle is said to be solved, when each of the rows, columns, and grids, contain only single instances of numbers in the range of 1 to 9, such that no digit is repeated within the same row or column or grid. For a traditional Sudoku, the sum of digits in each row, column or grid should be equal to 45.

The remaining part of the paper is organized as follows: Sect. 2 discusses related works, while Sect. 3 presents the design of the experiment. Section 4 reports on the results and the observations, followed by closing remarks in Sect. 5.

2 Related Works

An insight into the assessment of *EA* in tackling Sudoku was presented in [3], which suggested that random mutation may have a significant impact on the performance of *EAs*. Coming to grips with Sudoku, using *GA,* by considering each 3×3 grid as a block was reported in [4]. The paper's authors treated each sub-block as a problem, and applied uniform crossover was to the particular block, in which the crossover positions were limited to the links between sub-blocks. Swap mutation technique was adopted such that it avoids duplication of numerals within a sub-block. This work also did a comparative study of disparity hypothesis and gene duplication using child. It was found The Sudoku was solved with a high degree of accuracy.

Code written in *C++*, to solve Sudoku, using *GA,* was explained by Weiss in [5]. He reported that *GA's* performance in Sudoku was impaired by slow convergence and inability to escape from local minima. In [6], descriptions on solving Sudoku, using different types of *GA* were delineated. A novel hybrid genetic algorithm (*HGA*) was generated, and the workings of *HGA*, interactive genetic algorithm (*IGA*), and classical *GA* were compared. He concluded that *HGA* gives better results than other *GAs*. However, the *HGA* is not that suitable for solving difficult Sudoku puzzles (the puzzles with more unfilled elements).

The authors in [7] described solving Sudoku, using *GA* mutation and crossover strategies. The Pencil and pen algorithm is one of the frequently proposed methods in the literature, for solutions of the Sudoku puzzle [7]. In *GA*, used in [7], to solve Sudoku puzzle one crossover type and two mutation types are used. The first mutation operation is applied to a gene (a cell in the matrix) of the chromosome. The second mutation is applied to each of the sub-square rows.

[8] discussed 3 objectives: (1) to check if the *GA* is efficient in solving the Sudoku problem. (2) Can *GA* be used to efficiently generate new puzzles for Sudoku? (3) Can *GA* be used to check the difficulty level of Sudoku problem?

The authors in [9] proposed a modified form of GA, Retrievable Genetic Algorithm (Ret-GA). Ret-GA was applied such that initially it creates a Blueprint Matrix (of the same size as the Sudoku puzzle), wherein '1' is assigned to a particular entry, only if the corresponding entry in the original Sudoku puzzle had an entered value (i.e. non-blank), otherwise that Sudoku location is assigned with value '0'. The initial population generated is subjected to Row-wise Uniform Crossover.

[10] described diverse algorithms - Cultural Genetic Algorithm (*CGA*), Repulsive Particle Swarm Optimization (*RPSO*), Quantum Simulated Annealing (*QSA*), and the Hybrid method that combines *GA* with Simulated Annealing (*HGASA*), which are used to solve the Sudoku puzzle. This paper concluded that *QSA* and *HGASA* are successful in solving the Sudoku puzzle. [11] solved the Sudoku puzzle using a novel multistage genetic algorithm (*MGA*). To solve a given puzzle, initially, a group table was constructed, with an initial random population. In every cycle, *GA* worked to find a better solution, and at each iteration, the group table was updated. Swap mutation technique was used, when mutation probability is satisfied.

Dealing with Sudoku through Variable Neighborhood Search (*VNS*) was shown in [12]. The basic idea is to explore a set of predefined neighborhoods, to provide a better diversification of successful solutions. A particular generation was selected based on the fitness function, invert an exchange strategy was undertaken to obtain the solutions. The performance was compared with Harmony Search, revealed that Harmony search is not as efficient as *VNS*. Finally, the paper concluded that *VNS* can produce competitive results at an easy level, and promising results in harder levels.

Authors in [13] introduced a heuristic to grapple with Sudoku, adopting modified crossover and mutation operators of *GA*. [14]. described a teaching strategy, engaging in-class exercise to introduce *GA* in Microsoft Excel to solve the Sudoku puzzle A permutation and row-crossover operators were designed for *GA* to solve Sudoku in [15]; the proposed algorithm was tested on different instances of Sudoku.

On a close review of different versions of *GA* available in the literature to solve Sudoku problems, the work presented in this paper proposes a study of the effect of using *GA* for solving Sudoku with combinations of different mutation and crossover operations.

3 Design of Experiments

As noted previously, a Sudoku puzzle is said to be solved, when each of the rows, columns, and grids, contain only single instances of numbers in the range of 1 to 9, such that no digit is repeated within the same row or column or grid. For a traditional Sudoku, the sum of digits in each row, column or grid should be equal to 45. The components of *GA* are designed as follows.

Population for the GA - The population for the GA to solve is the set of candidates each mean in Sudoku. A two-dimensional array was adopted to represent a Sudoku. Thus, a set of two-dimensional arrays in the population are used in this study.

Fitness of Algorithm - The fitness of each candidate has calculated an algorithm. This algorithm follows the following steps

(1) Calculate the number of occurrences of each number row-wise.
(2) Calculate the number of occurrences of each number column-wise.
(3) Calculate the number of occurrences of each number in all nine 3 × 3 grid.
(4) If the row-wise, column-wise, and 3 × 3 grid count is equal to 81 the puzzle is solved.

Mutations - In this experiment, two mutations were considered – Random Resetting and Swap Mutation.

- Random Resetting - If the value selected in each row is a fixed value then don't randomize it, otherwise insert a random value into it.
- Swap Mutation - Randomly select two genes, row-wise, and if both the elements do not belong to the set of fixed elements of the puzzle, then swap them; otherwise, continue, until both the selected elements are not fixed elements. The constraint is that the puzzle must contain at least two nonfixed elements in each row.

Crossovers - The three crossover operators deliberated in this study were a one-point crossover, two-point crossover, and uniform crossover.

- One-point crossover - In this procedure, a point is selected at random, which is said to be the crossover point; by dividing the chromosome, and swapping the genes of the chromosome concerning crossover point, new off-springs are produced.
- Two-point crossover - In this procedure, two points are selected at random, which are said to be the crossover points; genes of the parents between the two crossover points are swapped, and the genes, before the first crossover point, will be from the first parent, and genes from second crossover point, will be from the second parent. By combining these, new off-springs are generated.
- Uniform Crossover - In this procedure, the crossover points are randomly generated, and the genes of the parents are exchanged, at that particular point. No division of chromosome is done in this process, but the genes are replaced at the crossover points. There can be more than one crossover point in a chromosome.

4 Results and Discussions

The results of all the six combinations and mutations and crossovers used with *GA* to solve the Sudoku are explained in this section. The combinations of mutations and crossover are as follows: Random Resetting Mutation and One-Point Crossover, Random Resetting Mutation and Two-Point Crossover, Random Resetting Mutation and Uniform Crossover, Swap Mutation and One-Point Crossover, Swap Mutation and Two-Point Crossover and Swap Mutation and Uniform Crossover.

The optimal score: Each element in 9 × 9 Sudoku is to be counted one time and since there are 9 numbers in 9 rows, 9 columns, 9 grids, the optimal fitness score of Sudoku, of size 9 × 9, considered in the experiment, is 81, row-wise and column-wise. The optimal fitness score for the 3 × 3 grid is also 81. Summing up all the three fitness scores, a total value of 243 is the optimum score for the solution of Sudoku.

The *GA,* with each of the above-noted combinations of mutation and crossover, was applied to solve Sudoku puzzles, for 10 different runs and the average performance of *GA* is presented and discussed.

The results presented in Table 1 consist of the run number, whether or not the Sudoku puzzle was solved, Number of generations, Optimum score of each execution and the Time taken to complete execution. Table 1 depicts the performance of *GA* in solving Sudoku puzzles when Random Resetting Mutation and One-point crossover were used. An average number of generations obtained was 1091.6; the average optimum score was 244.40, and the average execution time was 54.71 s. The success ratio is 7:10 i.e., 7 out of 10 times the algorithm solved Sudoku perfectly.

Table 1. Sudoku results - *GA* with random resetting mutation and one-point crossover.

Run no	Puzzle solved?	Generation	Optimum score	Time (sec)
1	NO	2000	247	93.40
2	YES	255	243	20.08
3	YES	529	243	29.14
4	YES	182	243	16.73
5	YES	217	243	12.60
6	NO	2000	249	80.97
7	YES	663	243	37.2
8	YES	1565	243	82.01
9	NO	2000	247	97.83
10	YES	1505	243	77.21
Average		1091.60	244.40	54.72

The results, using *GA* with Random resetting mutation and two-point crossover, are shown in Table 2. An average number of generations obtained was 1149.8, the average optimum score was 234.90, and average execution time was 47.15 s. The success ratio was 4:10, i.e. 4 out of 10 times, the *GA* algorithm solved Sudoku puzzles, successfully.

Table 2. Sudoku results - *GA* with random resetting mutation and two-point crossover.

Run no	Puzzle solved?	Generation	Optimum score	Time (sec)
1	YES	487	243	21.23
2	YES	133	243	13.89
3	NO	2000	253	71.86
4	NO	2000	249	82.25
5	YES	241	243	14.12
6	NO	2000	249	73.30
7	NO	2000	253	68
8	YES	512	243	35.30
9	NO	2000	249	81.39
10	NO	125	124	10.16
Average		1149.80	234.90	47.15

Table 3 depicts the performance of *GA* when Random mutation and Uniform crossover was used. An average number of generations obtained was 1096.80, the average optimum score was 244, and the average execution time is 47.47 s. The success ratio was 8:10 implies that this combination of mutation and, crossover with *GA* was able to solve Sudoku in 8 out of 10 runs. The performance in solving Sudoku with Swap mutation and One-Point crossover is presented in Table 4. As seen from the results, the average number of generations obtained was 2000, the average optimum score was 263.20, and average execution time was 96.33 s. The success ratio was 0:10. It is worth noting that this combination of mutation and crossover does not support *GA* to solve the Sudoku problem in all cases of runs. This gives an insight that the impact of this combination to be studied further to understand the reason for such performance.

Table 3. Sudoku results - *GA* with random resetting mutation and uniform crossover.

Run no	Puzzle solved?	Generation	Optimum score	Time (sec)
1	YES	727	243	31.46
2	YES	405	243	19.84
3	NO	2000	247	80.65
4	NO	2000	249	93.90
5	YES	539	243	28.01
6	YES	1817	243	70.24
7	YES	414	243	19.85
8	YES	308	243	15.66
9	YES	1774	243	70.21
10	YES	984	243	44.19
Average		1096.80	244	47.40

Table 4. Sudoku results - *GA* with random swap mutation and one-point crossover.

Run no	Puzzle solved?	Generation	Optimum score	Time (sec)
1	NO	2000	263	83.1
2	NO	2000	261	75.41
3	NO	2000	263	85.54
4	NO	2000	265	81.69
5	NO	2000	261	88.25
6	NO	2000	259	87.99
7	NO	2000	269	118.3
8	NO	2000	265	116.3
9	NO	2000	255	106.5
10	NO	2000	271	120.2
Average		2000	263.20	96.33

Table 5 displays the performance of *GA,* with Swap mutation and Two-Point crossover. Results indicate that in all cases, *GA* uses the maximum generation of 2000, without reaching the solution. The average optimum score it obtained was 256, and average execution time was 95.06 s. The success ratio is 0:10. Table 6 delineates that the *GA* with Swap mutation and Uniform crossover also failed to attain the solution for Sudoku puzzle, in any of the runs, even with a maximum number of generations. The average optimum score was 260.66, and average execution time was 112.12 s.

The experimental results have established that the Swap Mutation routine is inadequate for *GA* to solve the Sudoku problem, irrespective of the associated crossover study. Hence, the comparative study was extended to probe the Random Resetting mutation, with all the three types of crossovers. The results were compared by the Success Ratio (*SR*), an average number of generations taken for successful runs (*AnG#*), and the Execution Time. The results are presented in Table 7.

Table 5. Sudoku results - *GA* with random swap mutation and two-point crossover.

Run no.	Puzzle solved?	Generation	Optimum score	Time (sec)
1	NO	2000	263	127.50
2	NO	2000	263	110.90
3	NO	2000	259	88.24
4	NO	2000	249	66.50
5	NO	2000	257	86.50
6	NO	2000	255	91.24
7	NO	2000	261	85.27
8	NO	2000	259	101.7
9	NO	2000	235	102.10
10	NO	2000	259	90.60
Average		2000	256	95.06

Table 6. Sudoku results - *GA* with random swap mutation and two-point crossover.

Run no	Puzzle solved?	Generation	Optimum score	Time (sec)
1	NO	2000	261	123.50
2	NO	2000	265	126.80
3	NO	2000	259	94.54
4	NO	2000	261	103.90
5	NO	2000	261	112.70
6	NO	2000	255	105.80
7	NO	2000	255	103
8	NO	2000	261	125.70
9	NO	2000	263	112.20
10	NO	2000	265	113.10
Average		2000	260.60	112.12

Table 7. Comparison of random resetting mutation.

Crossover	*SR*	*AnG#*	*Time (Sec)*
One-point	7	702.29	39.28
Two-point	4	299.61	18.94
Uniform	8	871	37.43

Test observations show that Random Setting mutation with Uniform crossover captured the top position, with higher *SR,* by yielding more number of successful runs. However, the Two-point crossover led the top position, for its speed, by achieving the solutions, with a minimal number of generations. Although its *SR* was 4, its *AnG#* was 299.61. These conflicting performances of Uniform and Two-point crossover need to be investigated further.

5 Conclusions

This paper presented evaluations of solving Sudoku problems by the Genetic Algorithm, using various combinations of Random mutations and crossover operations. Each blend of mutation and crossover was found to furnish different performance results. Empirical evidence of solving Sudoku problems, by various mixes of mutation and crossovers, has shown that deployment of Random mutation with Uniform crossover turned out more successful runs, whereas the same commingled with Two-Point crossover was effective in the speed of convergence.

References

1. Janani, N., Shiva Jegan, R.D., Prakash, P.: Optimization of virtual machine placement in cloud environment using genetic algorithm. Res. J. Appl. Sci. Eng. Technol. **10**(3), 274–287 (2015)
2. Raju, D.K.A., Velayutham, C.S.: A study on GA based video abstraction system. In: Proceedings of World Congress on Nature Biologically Inspired Computing (2009)
3. Mcgerty, S., Moisiadis, F.: Are evolutionary algorithms required to solve sudoku problems? In: Fourth International Conference on Computer Science and Information Technology, pp. 365–377 (2014). https://doi.org/10.5121/csit.2014.4231
4. Sato, Y., Inoue, H.: Solving sudoku with genetic operations that preserve building blocks. In: Proceedings of the 2010 IEEE Conference on Computational Intelligence and Games (2010). https://doi.org/10.1109/itw.2010.5593375
5. Weiss, J.M.: Genetic algorithms and sudoku. In: Midwest Instruction and Computing Symposium (MICS 2009), pp. 1–9 (2009)
6. Deng, X.Q., Da Li, Y.: A novel hybrid genetic algorithm for solving Sudoku puzzles. Optim. Lett. **7**, 241–257 (2013)
7. Kazemi, S., Fatemi, B.: A retrievable genetic algorithm for efficient solving of sudoku puzzles. Int. J. Comput. Inf. Eng. **8**(5) (2014)
8. Mantere, T., Koljonen, J.: Solving, rating and generating Sudoku puzzles with GA. In: IEEE Congress on Evolutionary Computation (2007)
9. Das, K.N., Bhatia, S., Puri, S., Deep, K.: A retrievable GA for solving sudoku puzzles. Technical report, Citeseer (2012)
10. Perez, M., Marwala, T.: Stochastic optimization approaches for solving sudoku. arXiv:0805.0697v1 (2008)
11. Chel, H., Mylavarapu, D., Sharma, S.: A novel multistage genetic algorithm approach for solving sudoku puzzle. In: Proceedings of 2016 International Conference on Electrical, Electronics, and Optimization Techniques (ICEEOT) (2016)
12. Hamza, K.A., Sevkli, A.Z.: A variable neighborhood search for solving sudoku puzzles. In: Proceedings of the International Joint Conference on Computational Intelligence, vol. 1, pp. 326–331 (2014)
13. Gerges, F., Azar, D., Zoueii, G.: Genetic algorithms with local optima handling to solve sudoku puzzles. In: Proceedings of the 2018 International Conference on Computing and Artificial Intelligence (2018)
14. Ernstberger, K.W., Venkataramanan, M.A.: Announcing the engagement of sudoku: an in-class genetic algorithm game. J. Innov. Educ. **16**(3), 185–196 (2018)
15. Rodriguez Vasquez, K.: GA and entropy objective function for solving Sudoku puzzle. In: Proceedings of Genetic and Evolutionary Computation Conference Companion (2018)

A Preliminary Investigation on a Graph Model of Differential Evolution Algorithm

M. T. Indu[1], K. Somasundaram[2], and C. Shunmuga Velayutham[1](✉)

[1] Department of Computer Science and Engineering, Amrita School of Engineering, Amrita Vishwa Vidyapeetham, Coimbatore, India
cb.en.d.cse17010@cb.students.amrita.edu, cs_velayutham@cb.amrita.edu
[2] Department of Mathematics, Amrita School of Engineering, Amrita Vishwa Vidyapeetham, Coimbatore, India
s_sundaram@cb.amrita.edu

Abstract. Graph based analysis of Evolutionary Algorithms (EAs), though having received little attention, is a propitious method of analysis for the understanding of EA behavior. However, apart from mere proposals in literature, the graph model(s) of EAs have not been aptly demonstrated of their full potential. This paper presents a critical analysis of an existing prominent graph model of EAs. The comprehensive analysis involved modelling as graphs, the interactions within Differential Evolution (DE) algorithms (*DE/rand/1/bin* and *DE/best/1/bin* by way of example), while solving CEC2005 benchmark functions. Subsequently, the graph-theoretic measures- degree centrality, clustering coefficient, betweenness centrality and closeness centrality- were deployed to investigate the potential of graph model in providing insights about EA dynamics. However, the model falls short in providing useful insights about the DE runs. The observed insights were qualitative in nature and not competitive compared to empirical analysis. This indicates the need for considerable research attention in designing robust graph models.

Keywords: Evolutionary algorithm dynamics · Graph-based analysis · Differential evolution · Graph model

1 Introduction

Evolutionary Computation (EC) is primarily concerned with the design and analysis of robust search algorithms, called Evolutionary Algorithms (EAs), modelled after natural evolution. Besides robustness, EAs stochastically exhibit complex population dynamics, which renders them difficult to comprehend and be harnessed. Empirical [5], theoretical [6] and visual [7] analyses are the means by which research efforts are being undertaken to comprehend the dynamics and behavior of EAs. Few research works have attempted to represent and visualize interactions within EAs as graphs. However, these works have merely proposed several graph-theoretic measures to analyze EA dynamics, without appropriate validation.

B. Deka et al. (Eds.): PReMI 2019, LNCS 11941, pp. 365–374, 2019.
https://doi.org/10.1007/978-3-030-34869-4_40

The first research on graph-based analysis of EAs was presented in [12]. Ivan Zelinka et al. [15] modelled Differential Evolution (DE) and Self-Organizing Migrating Algorithm (SOMA) as graphs and this idea was extended, with little or no modification, to Differential Evolution (DE) [2,8,9] and Particle Swarm Optimization (PSO) [8]. Various properties of graph that could provide insights into EA dynamics were suggested in [2,8], and are summarized in Table 1. In all these works, several graph-theoretic measures were mapped to some EA dynamics/property, but lacked appropriate validation of these mappings.

Table 1. Graph measures and possible insights about EAs [2,8]

Graph measures	Insights about EA dynamics
Degree centrality	To identify premature convergence or stagnation in an EA run
Clustering coefficient	To show the population diversity
Betweenness centrality	Incidentally, the node having the highest betweenness and the best fitness are the same
Closeness centrality	A measure of rate of distribution of information in the graph (as well in population)
Clique	For general subgraph analysis
k-Club and k-Clan	To isolate individuals which are redundant

This paper presents critical analysis of a prominent graph model, proposed in [15], and applied in [2,8,9,14]. The in-depth analysis involves modelling the interactions within *DE/rand/1/bin* and *DE/best/1/bin* algorithms as graphs, while solving CEC2005 benchmark functions [10]. Subsequent to modelling, the graph-theoretic measures- degree centrality, clustering coefficient, betweenness centrality and closeness centrality- are employed to investigate the potential of the graph model in understanding EA dynamics. The remaining sections of this paper are structured as follows: Sect. 2 outlines the prominent graph model from the literature, in the context of DE algorithm. Section 3 explains the experimental design. Section 4 presents the simulation results and analysis. Section 5 concludes this work with suggestions on future studies.

2 Graph Model for DE

Differential Evolution is a simple, robust evolutionary algorithm, primarily applied for continuous optimization problems. For each target vector in the DE population, three random solutions, in the population, interact to create a mutant vector, which in turn undergoes crossover with the corresponding target vector to create a trial vector. In graph creation, the identified graph model focuses on these individuals, except the transitory mutant vector. The graph creation has been explained in Algorithm 1 (for *DE/rand/1/bin*) and Fig. 1 shows an example for the same.

Since the number of vertices is always the population size (N), a new trial vector in the next generation gets represented by the same vertex that used to represent its corresponding target vector in a previous generation. Thus, vertices in the graph cannot be related directly to individuals, instead can be interpreted as place holders, for individuals in newer evolving populations. The graph model, described in this section, attempts to effectively capture the interactions within a typical DE algorithm. However the potential of this graph model has not been amply demonstrated in respective literature.

Algorithm 1. Algorithm for creating graph of a DE run

Initialize the population with N randomly generated solutions (called target vectors): X_1, X_2, .. X_N, each with dimension, D
Calculate the fitness of the N solutions
for $gen = 1$ to $Maxgen$ **do**
    for $i = 1$ to N **do**
        Select 3 distinct random vectors X_{r_1}, X_{r_2}, X_{r_3} for X_i
        Create mutant vector V_i using differential mutation
        Create trial vector U_i using binomial crossover between X_i and V_i
        if $fitness(U_i)$ is better than $fitness(X_i)$ **then**
            Create vertices for X_{r_1}, X_{r_2}, X_{r_3}, X_i if they do not exist
            Create edges from X_{r_1}, X_{r_2}, X_{r_3} to X_i with weight 1 if they do not exist
            Add 1 to the edge weights of edges from X_{r_1}, X_{r_2}, X_{r_3} to X_i if they already exist
        end if
    end for
    for $i = 1$ to N **do**
        Replace X_i with U_i if $fitness(U_i) \leq fitness(X_i)$
        Represent U_i by the same vertex that represented X_i
    end for
end for

3 Experimental Design

The critical evaluation of the graph model, outlined in the previous section, calls for a rigorous, systematic empirical analysis. Towards this, *DE/rand/1/bin* and *DE/best/1/bin* algorithms were used to optimize 25 functions from the CEC2005 benchmark suite [10]. The following parameters were used during the simulations: $N = 50$, $D = 10$, $F = 0.4$, $Cr = 0.5$ and $Maxgen = 5000$. 30 runs were executed, for each function in the benchmark suite, using both the algorithms. The graph creation algorithm, described in Algorithm 1, was used to create directed weighted graphs, for each run-function-algorithm in the benchmark suite. The following four prominent graph theoretic measures (refer Table 1), proposed in the literature, have been employed for graph analysis.

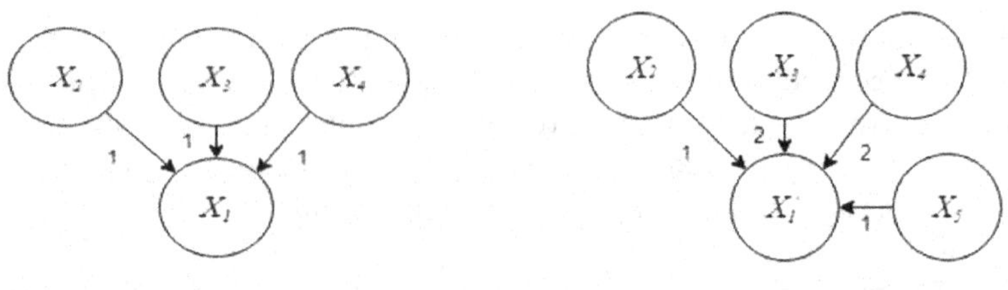

(a) In generation 1, vectors X_2, X_3, X_4 improves X_1. Let the better offspring be $X_1^{'}$

(b) In generation 2, $X_1^{'}$ occupies the position of X_1. Vectors X_3, X_4, X_5 improves $X_1^{'}$

Fig. 1. An example for graph creation during DE run with the considered graph model

1. Degree Centrality: Since the average indegree of vertices in a graph is equal to the average outdegree, we limit our focus to the indegree of vertices. Indegree of a vertex in a directed weighted graph is the sum of edge weights coming into that vertex [13].
$$indeg(u) = \sum_{v \in V - \{u\}} w_{vu} \tag{1}$$
where V is the set of vertices, and w_{vu} is the weight of the edge vu representing the number of times the individual at vertex v helped to improve the individual at vertex u.
2. Clustering Coefficient: The local clustering coefficient of a vertex, in a weighted directed graph, is computed using the following equation [3]:
$$cc(i) = \frac{1/2 \sum_j \sum_k (w_{ij}^{\frac{1}{3}} + w_{ji}^{\frac{1}{3}})(w_{ik}^{\frac{1}{3}} + w_{ki}^{\frac{1}{3}})(w_{jk}^{\frac{1}{3}} + w_{kj}^{\frac{1}{3}})}{d_i^{tot}(d_i^{tot} - 1) - (2 \times d_i^{bidir})} \tag{2}$$
where, i, j and k are three distinct vertices, d_i^{tot} is the total degree of vertex i, d_i^{bidir} is the number of bidirectional edges at vertex i.
3. Betweenness Centrality: The betweenness centrality of a vertex, in a weighted directed graph, is calculated by the following equation [4]:
$$betw(i) = (\sum \frac{\sigma_{uv}^{i}}{\sigma_{uv}}) \times \frac{1}{(N-1)(N-2)} \tag{3}$$
where, u, v and i are three distinct vertices, σ_{uv}^{i} is the count of shortest paths between u and v through i, σ_{uv} is the count of shortest paths between u and v, and N is the number of vertices in the graph.
4. Closeness Centrality: The closeness centrality of a vertex, in a weighted directed graph, is computed by the following equation [1]:
$$close(i) = \frac{|R(i)|}{N-1} \times \frac{|R(i)|}{\sum_{u \in R(i)} d(i,u)} \tag{4}$$

where, $R(i)$ is the set of vertices reachable from vertex i, N is the number of vertices in the graph, $\frac{|R(i)|}{N-1}$ is the normalization for closeness centrality, and $d(i,u)$ is the shortest distance from i to u.

4 Simulation Results and Analysis

The four identified graph theoretic measures were computed, for every graph generated per run of *DE/best/1/bin* and *DE/rand/1/bin*, while solving CEC2005 functions. The results were plotted across the number of generations. Based on these plotted results, the analysis, are presented below, for each of the four graph measures. In all the following figures, showing the plots for 30 runs, red color solid lines represent convergence, cyan color dashed lines represent premature convergence and magenta color dashed-dotted lines represent stagnation. Due to space limitations, the plots in this section have been organized to show few patterns of growth for each graph measure. It is also worth pointing out that though each pattern of growth is shown for only one function, the remaining functions also displayed patterns similar to the ones presented here.

4.1 Degree Centrality

The degree centrality analysis has been proposed, in the literature, to observe the premature convergence or stagnation, in an evolutionary run (refer Table 1). Figure 2 shows runs of F15 with *DE/rand/1/bin*, which resulted in convergence and premature convergence. Figure 3 shows runs of F25, which resulted in premature convergence and stagnation. We can see that the state of the DE run (whether it is convergence/premature convergence/stagnation) is not completely distinguishable from the figures. As it happens with a typical EA run, initial generations are marked by faster fitness improvements followed by a long slower improvement phase. As long as there are fitness improvements in the population, edges are formed between vertices contributing to indegree growth. The indegree growth stops if the population gets into convergence/premature convergence as a single (global/local) solution dominates and takes over the population making it to lose diversity in subsequent generations. The indegree growth may also stop if there is stagnation where the population is still diverse but unable to evolve further. Figure 4 shows the average fitness (of single run) plotted against indegree growth, for the function F15. A quick comparison reveals that degree centrality based graph analysis do not offer advantage over average fitness and observing the latter in few past generations is suffice between the two to sense premature convergence/stagnation.

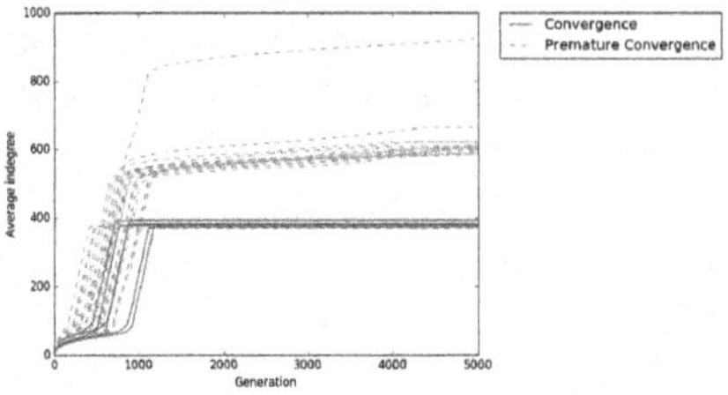

Fig. 2. Average indegree for 30 runs of F15 for *DE/rand/1/bin*

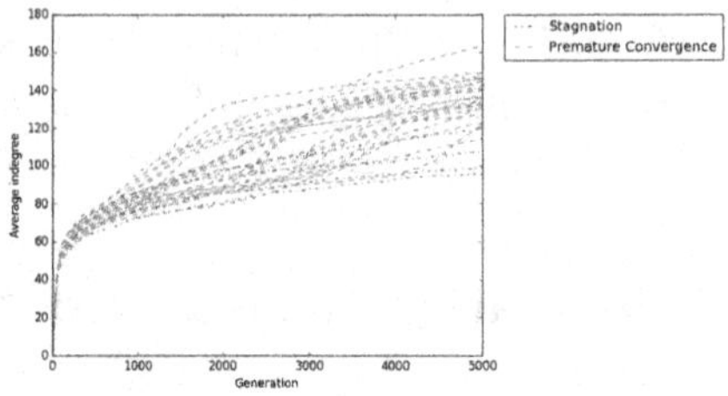

Fig. 3. Average indegree for 30 runs of F25 for *DE/rand/1/bin*

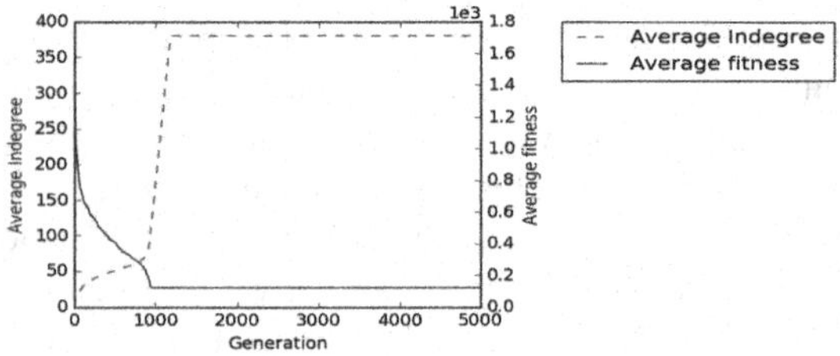

Fig. 4. Average indegree and Average fitness of F15 for *DE/rand/1/bin*

4.2 Clustering Coefficient

It is suggested in the literature (refer Table 1) that clustering coefficient shall be used to get insights about population diversity. Figure 5 shows the average weighted clustering coefficient for 30 runs of F14 plotted against generations for *DE/best/1/bin*. As the population evolves over generations, the edge weights in the neighborhood of nodes increases, thereby increasing the average weighted clustering coefficient. The successful evolution of a population also correlates with gradual loss of its diversity. This inverse relationship can be observed in Fig. 6 (diversity calculated using P-measure [11]). The population diversity, when measured directly and used effectively, is potential enough to serve as an effective means to understand the dynamics of an EA run.

4.3 Betweenness Centrality

It has been commented in the literature (refer Table 1) that the node having the highest betweenness incidentally has the best fitness. We observed that quite often there was no such correlation. In case of the underlying graph model a node is more like a place holder! The increased contribution of a node, towards the improvement of another node, is captured by increasing the weight of the edge between them. On the contrary, the betweenness centrality, which calculates shortest path, favors the nodes with lesser edge weights. The undertaken graph model does not address this conflict. The above observation extends to the closeness centrality measure too. Since this work is intended to analyze the graph

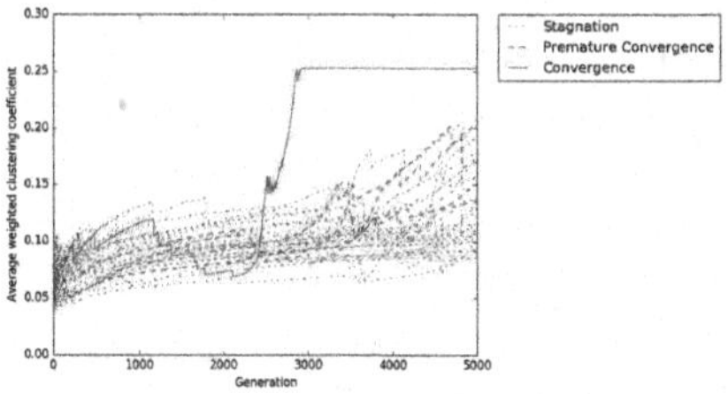

Fig. 5. Average weighted clustering coefficient of F14 for *DE/best/1/bin*

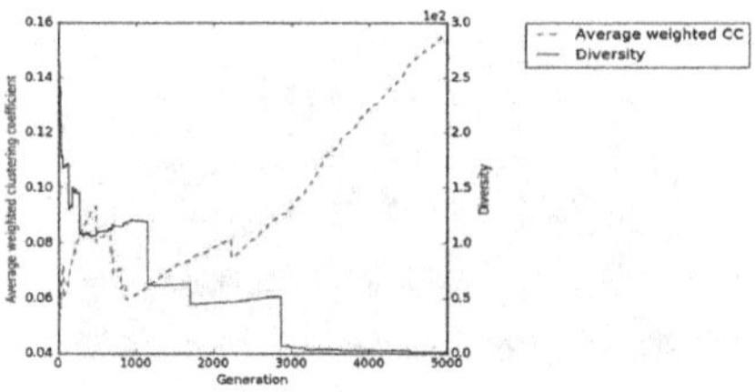

Fig. 6. Average weighted clustering coefficient and population diversity of F14 for *DE/best/1/bin*

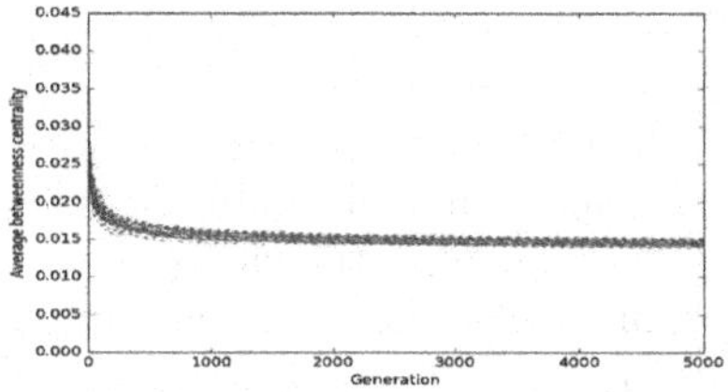

Fig. 7. Average betweenness centrality for 30 runs of F8 for *DE/rand/1/bin*

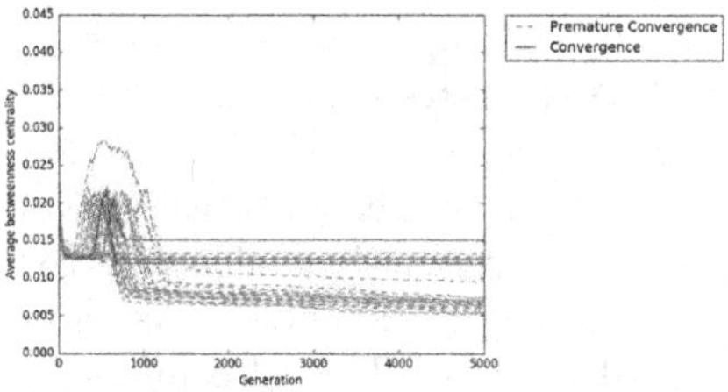

Fig. 8. Average betweenness centrality for 30 runs of F15 for *DE/rand/1/bin*

model as is, we proceeded with betweenness centrality and closeness centrality calculations. Figures 7 and 8 show the average betweenness centrality plotted against generations for F8 and F15 solved by *DE/rand/1/bin*. At the beginning, edges are added between nodes leading to increased betweenness values. When the graph becomes more and more complete during evolution, the betweenness centrality reduces and then stabilizes.

4.4 Closeness Centrality

Closeness centrality, in literature, has been proposed as a measure of rate of distribution of information in the population (refer Table 1). Figure 9 shows as an example, the cumulative closeness centrality for each node in the graph, at the end of 5000 generations for F1 solved by *DE/best/1/bin*. It can be observed from the figures that there are nodes which have displayed very active contribution in evolution while there are also quite a few nodes which have not contributed much in the improvement of other nodes. The concept of nodes as place holder makes this observation not so worthwhile. The behavior of cumulative closeness of all the N nodes over 5000 generations is shown in Fig. 10. Being cumulative, it just increases over generations.

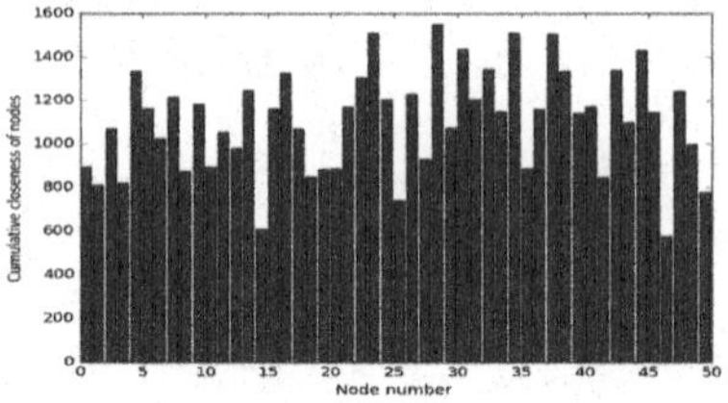

Fig. 9. Cumulative closeness of each node at the end of one run of F1 for *DE/best/1/bin*

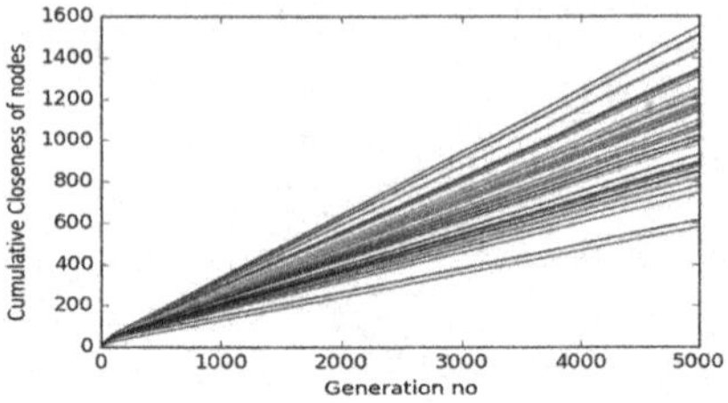

Fig. 10. Cumulative closeness of all nodes over generations during one run of F1 for *DE/best/1/bin*

5 Conclusion

This paper presented a critical analysis of an existing prominent graph model of EAs. Graph theoretic analysis of EAs could be used to verify the empirical and theoretical analyses' about EAs and also to complement them in understanding those algorithms. This also offers the prospect of controlling the dynamics of EAs, so as to improve the efficacy of EAs leading to better and robust designs. However the potential of graph based analysis of EAs largely rests with the underlying graph model. The graph model, analyzed in this paper, represents N individuals in the population as N vertices in the graph. Since the number of vertices remain the same, even as the population evolves, the vertices represent the descendants of each individual in the initial population. Nodes in the graph serve as place holders for individuals. While this modelling abstraction attempts to capture interactions within DE, in a cumulative sense, it certainly limits analysis, because apart from interactions, the evolution of population is also a critical factor in understanding EAs.

In case of degree centrality, the analysis did seem to provide only a coarse insight about the convergence of EA runs. Statistical analysis of DE runs could very well provide better insights into convergence. Since the clustering coefficient of the modelled graph displayed an inverse relationship with population diversity, the latter could serve as an effective means to understand the dynamics of an EA run. The betweenness centrality and closeness centrality measures did not seem to provide any valuable insights about DE runs. It is felt that the inability of prominent graph properties to provide any worthwhile insight is by virtue of the adopted graph model, whose potential is largely determined by its very modelling process. Since graph based analysis of EAs is a prospective method of analysis, complementing empirical and theoretical analyses, designing effective graph model of EAs are perceived to be the major attention for future investigations.

References

1. Applied social network analysis in python. https://www.coursera.org/learn/python-social-network-analysis/lecture/noB1S/degree-and-closeness-centrality. Accessed 03 Oct 2018
2. Davendra, D., Zelinka, I., Metlicka, M., Senkerik, R., Pluhacek, M.: Complex network analysis of differential evolution algorithm applied to flowshop with no-wait problem. In: 2014 IEEE Symposium on Differential Evolution (SDE), pp. 1–8. IEEE (2014). https://doi.org/10.1109/SDE.2014.7031536
3. Fagiolo, G.: Clustering in complex directed networks. Phys. Rev. E **76**(2), 026107 (2007). https://doi.org/10.1103/PhysRevE.76.026107
4. Hagberg, A., Schult, D., Swart, P.: Networkx reference (2012). https://networkx.github.io/documentation/latest/_downloads/networkx_reference.pdf. Accessed 27 Nov 2018
5. Jeyakumar, G., Velayutham, C.S.: An empirical comparison of differential evolution variants on different classes of unconstrained global optimization problems. In: 2009 World Congress on Nature & Biologically Inspired Computing (NaBIC), pp. 866–871. IEEE (2009). https://doi.org/10.1109/NABIC.2009.5393495
6. Qi, X., Palmieri, F.: Theoretical analysis of evolutionary algorithms with an infinite population size in continuous space. part I: basic properties of selection and mutation. IEEE Trans. Neural Netw. **5**(1), 102–119 (1994). https://doi.org/10.1109/72.265965
7. Sathyajit, B.P., Velayutham, C.S.: Visual analysis of genetic algorithms while solving 0-1 knapsack problem. In: Hemanth, D.J., Smys, S. (eds.) Computational Vision and Bio Inspired Computing. LNCVB, vol. 28, pp. 68–78. Springer, Cham (2018). https://doi.org/10.1007/978-3-319-71767-8_6
8. Šenkerık, R., Pluhácek, M., Viktorin, A., Janoštık, J.: On the application of complex network analysis for metaheuristics. In: 7th BIOMA Conference, pp. 201–213 (2016)
9. Skanderova, L., Fabian, T.: Differential evolution dynamics analysis by complex networks. Soft Comput. **21**(7), 1817–1831 (2017). https://doi.org/10.1007/s00500-015-1883-2
10. Suganthan, P.N., et al.: Problem definitions and evaluation criteria for the cec 2005 special session on real-parameter optimization. Technical report, Nanyang Technological University Singapore and KanGAL Report Number 2005005 (2005)
11. Vitaliy, F.: Exploration and exploitation. In: Differential evolution-in search of solutions, vol. 4, pp. 69–81. Springer, Heidelberg (2006). https://doi.org/10.1007/978-0-387-36896-2_4
12. Walczak, Z.: Graph based analysis of evolutionary algorithm. In: Kłopotek, M.A., Wierzchoń, S.T., Trojanowski, K. (eds.) Intelligent Information Processing and Web Mining, vol. 31, pp. 329–338. Springer, Heidelberg (2005). https://doi.org/10.1007/3-540-32392-9_34
13. Yang, Y., Xie, G., Xie, J.: Mining important nodes in directed weighted complex networks. Discrete Dyn. Nat. Soc. **2017** (2017)

14. Zelinka, I.: On analysis and performance improvement of evolutionary algorithms based on its complex network structure. In: Sidorov, G., Galicia-Haro, S.N. (eds.) MICAI 2015. LNCS (LNAI), vol. 9413, pp. 389–400. Springer, Cham (2015). https://doi.org/10.1007/978-3-319-27060-9_32
15. Zelinka, I., Davendra, D., Snášel, V., Jašek, R., Šenkeřík, R., Oplatková, Z.: Preliminary investigation on relations between complex networks and evolutionary algorithms dynamics. In: 2010 International Conference on Computer Information Systems and Industrial Management Applications (CISIM), pp. 148–153. IEEE (2010). https://doi.org/10.1109/CISIM.2010.5643674

An Evolutionary Approach to Multi-point Relays Selection in Mobile Ad Hoc Networks

Alok Singh(✉) and Wilson Naik Bhukya

School of Computer and Information Sciences, University of Hyderabad, Hyderabad 500 046, Telangana, India
{alokcs,naikcs}@uohyd.ernet.in

Abstract. Multi-point relaying is a technique to carry out the flooding of broadcast messages in the mobile ad hoc network (MANET) in a highly efficient manner. In this technique, each node v chooses a subset S of nodes among its one-hop neighbors, and only the nodes belonging to S are allowed to re-transmit the broadcast messages received from v. The nodes in S are called multi-point relays. By limiting the privilege to re-transmit the broadcast messages received from v to only multi-point relays allows for significant reduction in redundant broadcast messages. In this paper, we have proposed a steady-state genetic algorithm based approach for the selection of multi-point relays. Unlike the previously proposed approaches, which use bit-vector encoding to represent a solution, we have used the integer encoding where the multi-point relay nodes are represented as an ordered list. For dense networks, where the number of nodes in multi-point relays is only a small fraction of one-hop neighbors, a significant savings is achieved not only in memory usage, but also in computation time as the efficiency of genetic operators depends on the length of the chromosome. Computational results show the effectiveness of our approach.

Keywords: Genetic algorithm · Multi-point relaying technique · Mobile ad hoc network

1 Introduction

A mobile ad hoc network (MANET) has challenges of its own due to the mobility of its nodes and the use of wireless mode of communication. Due to the mobility of nodes, topological information needs to be updated regularly throughout the network. Further, higher the mobility of the nodes, more frequently the topological information needs to be updated. Therefore, control traffic increase with increase in mobility. On the other hand, limited bandwidth and higher transmission error rate associated with wireless mode of communication call for the

This work is supported by the research grant no. MTR/2017/000391of SERB, Government of India.

B. Deka et al. (Eds.): PReMI 2019, LNCS 11941, pp. 375–384, 2019.
https://doi.org/10.1007/978-3-030-34869-4_41

judicious use of available bandwidth. Further, the communication range of each node is limited, and, as a result, multi-hop communication is the norm. Hence, mobility needs to be managed by consuming as little network bandwidth as possible. This will have positive implication on energy consumption also.

The information that a node needs to share over the network to manage the mobility consists of its relative movement, its new position, its new neighborhood etc. Depending on the design of a MANET, this information may be of use only in the vicinity of the node sharing this information or in the entire network. The latter case generates a huge volume of control traffic as each topological change needs to be propagated or broadcasted to the entire network. A simple and easy to implement technique for broadcasting is called flooding where a node re-transmits an incoming message received by it for the first time to all links other than the link through which the message is received. Flooding ensures that each node in the network receives the broadcast message provided all the nodes are connected to the network. However, it consumes lot of network bandwidth as it generates redundant network traffic consisting of nodes receiving the same message over multiple links. Further, redundant traffic is undesirable from the perspective of energy efficiency also.

Multi-point relaying (MPR) technique was introduced by Quayyum et al. [8] to reduce the volume of redundant network traffic while forwarding a broadcast message. In this technique, each node v selects a subset S of nodes among its one-hop neighbors, and only the nodes in S are allowed to re-transmit the broadcast messages received from v. The nodes in S are called multi-point relays, and are so chosen that all two-hop neighbors of v are one-hop neighbors of the nodes in S. This ensures that all two-hop neighbors of v receive the broadcast message. By restricting the privilege to re-transmit the broadcast messages received from v to only multi-point relays allows for significant reduction in redundant broadcast messages. Multi-point relaying technique works in a distributed manner, where each node independently maintains its own set of multi-point relays which is continuously updated as per the topological changes occurring in its neighborhood. The optimized link state routing protocol (OLSR) [4] utilizes MPR technique. However, the problem of finding the multi-point relays of minimum cardinality for a node is $\mathcal{NP}$-hard as the problem of finding the minimum dominating set in a graph reduces to it. Quayyum et al. [8] proposed a greedy heuristic to compute the multi-point relays of minimum cardinality for a give node. This greedy heuristic is described in detail in the next section. This greedy heuristic works well for sparse networks. However, for dense networks, where number of neighbors for each node is high, its performance deteriorates. This is due to large number of common two-hop neighbors (of v) between multi-point relay nodes [2].

Several techniques exist in the literature (e.g. [1,5–7,9]) to improve the multi-point relay nodes selection mechanism based on various characteristics of MANET. Most of these techniques make use of the same greedy heuristic as proposed in [8]. However, the degree of preference of a node changes from one technique to the other as per the characteristics of MANET being tackled.

Chizari et al. [2] proposed an elitist generational genetic algorithm for multi-point relay node selection that works well even for dense networks. This genetic algorithm used bit vector encoding to represent a solution, and, employed roulette wheel selection method, uniform crossover (called scattered crossover method in [2]) and adaptive feasible mutation. Instead of minimizing the number of multi-point relay nodes for a node v, it computes for each multi-point relay node u, the set of nodes that are common between two-hop neighbors of v and one-hop neighbors of u and minimizes the sumtotal of cardinalities of these sets. This objective function aims at minimizing the redundant entries in the routing table stored at v, which, in turn, aids in further reducing the redundant messages. Instead of repairing the infeasible solutions, this genetic algorithm uses a penalty term in the fitness function to deal with them. In comparison to the greedy heuristic of [8], this genetic algorithm obtained better results in general, and this difference in solution quality widens as the density of the network increases. Later Chizari et al. [3] proposed a genetic algorithm, a tabu search approach, a simulated annealing approach and a recursive hill-climbing approach for selecting multi-point relays with the objective of minimizing the cardinality of the set of multi-point relay nodes. Except for few minute differences, this genetic algorithm is the same as the one proposed in [2].

In this paper, we have proposed a steady-state genetic algorithm for selecting multi-point relays with the same objective function as used in [2]. However, our genetic algorithm differs entirely from the one proposed in [2]. Our genetic algorithm uses the integer encoding, where a solution is encoded as an ordered list of multi-point relay nodes. For dense networks, where the number of nodes in a multi-point relay set is only a small fraction of one-hop neighbors, a significant savings, not only in terms of memory usage, but also in terms of computation time is achieved by using this encoding instead of bit-vector encoding as the efficiency of genetic operators depends on the length of the chromosome. However, when using this encoding, we have to invariably deal with variable length chromosomes. Our crossover operator copies the nodes common to two solutions to the child chromosomes, whereas the mutation operator is a ruin-and-recreate operator. We have used the probabilistic binary tournament selection to select the parent solutions for crossover and mutation. We have compared the results of our approach with the heuristic of [8] and the genetic algorithm of [2]. Computational results show the superiority of our approach as our approach is able to find as good as or better solutions in comparison to these two approaches.

The remaining part of this paper is organized as follows: Sect. 2 formally defines the multi-point relay nodes selection problem tackled in this paper. It also describes the greedy heuristic of [8] as it is the most commonly used approach for this problem. The steady-state genetic algorithm developed by us is presented in Sect. 3. Section 4 reports the computational results of our approach, and compares them with the previously proposed approaches. Finally, Sect. 5 concludes the paper by outlining the contributions made and few directions for future research.

2 Problem Formulation and the Greedy Heuristic

2.1 Problem Formulation

Let us assume that we need to select the multi-point relays for the node x. We will refer to node x for which the multi-point relays are being chosen as the base node. Let us denote the set of one-hop neighbors of x by $N(x)$, and the set of two-hop neighbors of x by $N^2(x)$. Clearly, $N^2(x) \subset \cup_{y \in N(x)} N(y)$. Let $I_y = N(y) \cap N^2(x) \forall y \in N(x)$ and $|I_y|$ the cardinality of I_y.

The multi-point relay node selection problem considered in this paper seeks a subset $S_{MPR} \subseteq N(x)$ such that $N^2(x) \subset \cup_{s \in_{MPR}} N(s)$ and the $\sum_{s \in S_{MPR}} |I_s|$ is minimized. This is the same objective function as used in [2], and aids in reducing the redundant messages further in case of dense networks in comparison to the objective of directly minimizing the cardinality of S_{MPR}. Hereafter, this problem will be refereed to as *P-MPR*.

2.2 The Greedy Heuristic

Quayyum et al. [8] proposed a greedy heuristic for selecting multi-point relay nodes. To describe this heuristic, we will use the notational conventions introduced in the previous section. Let S_{MPR} be the set of multi-point relay nodes to be computed for node x. Initially, S_{MPR} is empty. This heuristic begins by adding those nodes of $N(x)$ to S_{MPR}, which are the sole neighbors of some nodes in $N^2(x)$, and, then an iterative process ensues. The nodes in $N^2(x)$, which are neighbors of some node in S_{MPR} are said to be covered, whereas all other nodes are uncovered. During each iteration, a node of $N(x)$, which is not yet included in S_{MPR} and which is the neighbor of maximum number of uncovered nodes in $N^2(x)$ is added to S_{MPR}. This process is repeated till all the nodes in S_{MPR} is covered. Algorithm 1 provides the pseudo-code of this greedy heuristic. Quayyum et al. [8] also proved that the cardinality of the set of multi-point relay nodes computed by this heuristic is always within $\log n$ times the cardinality of the optimal set, where n is the number of nodes in the network.

3 The Steady-State Genetic Algorithm for *P-MPR*

We have developed a steady-state genetic algorithm based approach for *P-MPR*. Hereafter, this approach will be referred to as SSGA. Subsequent subsections describe the salient features of SSGA.

3.1 Solution Encoding

We have represented a solution directly by the set of multi-point relay nodes. An ordered list is used to represent this set. This encoding results in substantive savings in memory usage and computation time over the bit-vector encoding for dense networks as the number of nodes in a multi-point relay set is only

Algorithm 1: Pseudo code of the greedy heuristic of [8]

Input: A network and a base node x
Output: Set of multi-point relay nodes S_{MPR} for node x
$S_{MPR} := \emptyset$;
$S_{sole} := \{y : y \in N(x) \wedge (\exists z : z \in N^2(x) \wedge (N(z) \cap N(x) = \{y\}))\}$;
$M(x) := N(x) \setminus S_{sole}$;
$S_{MPR} := S_{MPR} \cup S_{sole}$;
$M^2(x) := N^2(x) \setminus \{z : z \in N^2(x) \wedge (N(z) \cap S_{MPR} \neq \emptyset)\}$;
while $(M^2(x) \neq \emptyset)$ **do**
  $y_m := \arg\max_{y \in M(x)} |N(y) \cup M^2(x)|$;
  $S_{MPR} := S_{MPR} \cup \{y_m\}$;
  $M(x) := M(x) \setminus \{y_m\}$;
  $M^2(x) := M^2(x) \setminus \{z : z \in M^2(x) \wedge y_m \in N(z)\}$;
return S_{MPR};

a small fraction of one-hop neighbors. The crossover and mutation operators need to scan the entire chromosome one or more time to produce offspring, and, as a result, efficiency of these operators depend on the length of the chromosome. Though, there is an overhead of sorting in maintaining the ordered list, it allows for efficient implementation of crossover operator as discussed subsequently. Please note that our encoding yields variable length chromosomes. However, the way, we have designed our crossover and mutation operators, the variable length chromosomes do not pose any additional challenges.

To illustrate the advantage of our encoding, suppose a node x has 10 nodes numbered 1 to 10 as one-hop neighbors, and, out of these 10 nodes, nodes 2, 5 & 6 form a set of multi-point relay nodes for node x. In our encoding, this set is represented by an ordered list 2 5 6, which has length 3. On the other hand, in case of bit-vector encoding, this set is represented as 0 1 0 0 1 1 0 0 0 0, which has length 10. Hence, in this particular case our encoding results in 70% savings in memory usage.

3.2 Fitness

Objective function as described in Sect. 2 is used as the fitness function. As *P-MPR* is a minimization problem, so a lesser value of the objective function indicates a more fit solution.

3.3 Selection

Probabilistic binary tournament selection method is used to select the two parents for crossover and a single parent for mutation, where p_{bt} is the probability of selecting the better of the two individuals in the tournament.

3.4 Crossover

Our crossover operator begins by copying the nodes common to both the parents to the child. The reason for doing this is based on assertion that if a node occurs in both the selected parents, then the chances of this node occurring in other good solutions are high. Owing to our encoding, nodes common to both the parents P_1 and P_2 can be computed in $\mathcal{O}(min(|P_1|, |P_2|))$ instead of $\mathcal{O}(|P_1| \times |P_2|)$. To make the child thus obtained a multi-point relay set, other nodes are added to the child through an iterative process, where during each iteration, a node from $N(x)$ is added in a quasi-random manner. During each iteration, with probability p_g, a node which is not yet included in the child and which is the neighbor of the maximum number of uncovered nodes in $N^2(x)$ is added to the child. Otherwise, a node is selected for addition at random from among nodes which are not yet included in child and which are neighbors of at least one uncovered node in $N^2(x)$. This process is repeated till all the nodes in $N^2(x)$ get covered, i.e., child becomes a multi-node relay set.

In a bid to further improve the child solution, we look for redundant nodes in the child. These are the nodes whose deletion from child solution will not make any node in $N^2(x)$ uncovered. The nodes in the child solution are checked one-by-one in some random order for redundancy, and, if a node is found redundant, it is deleted immediately before checking another node for redundancy.

3.5 Mutation

Mutation operator copies each node of the parent to the child with probability p_m. If a node of the parent is the sole neighbor of some nodes in $N^2(x)$, then it is always copied. The child thus obtained is transformed into a multi-point relay set, and, then redundant nodes are removed using the same procedures as used in the crossover operator.

We have used crossover and mutation in a mutually exclusive manner. With probability p_c crossover is used, or else mutation is used. Actually, our crossover operator preserves the nodes common to both parents in a bid to create even better solutions utilizing these nodes. Using mutation after crossover can remove some of these nodes.

3.6 Population Replacement Model

Steady-state population replacement model is used in our genetic algorithm, i.e., during each generation a single child is produced, which replaces the population member having the worst fitness provided it is different from all the current population members. In case, the child matches any current population member, it is discarded.

3.7 Initial Solution Generation

Each initial solution is generated in the same manner as in the greedy heuristic except for one difference, i.e., the statement $y_m := random(\{y : y \in M(x) \wedge$

Algorithm 2: Pseudo-code of SSGA for *P-MPR*

```
Input: A network and a base node x
Output: The best set of multi-point relay nodes S_best for node x found by
        SSGA
Generate ps initial solutions S1, S2, ..., Sps; // Section 3.7
S_best := Best solution among S1, S2, ..., Sps;
while (termination condition is not satisfied) do
    if (u01 < pc) then
        P1 := Selection(S1, S2, ..., Sps); // Section 3.3
        repeat
            P2 := Selection(S1, S2, ..., Sps); // Section 3.3
        until (P1 ≠ P2);
        SC := Crossover(P1, P2); // Section 3.4
    else
        P1 := Selection(S1, S2, ..., Sps); // Section 3.3
        SC := Mutation(P1); // Section 3.5
    if (SC is better than S_best) then
        S_best := SC;
    Apply replacement policy to include SC in population; // Section 3.6
return S_best;
```

$(N(y) \cap M^2(x) \neq \emptyset)\}$ is used in place of the first statement inside the while loop ($y_m := \arg\max_{y \in M(x)} |N(y) \cup M^2(x)|$) in Algorithm 1. Here, *random*(S) is a function that takes as input a set S and returns one element of that set uniformly at random. Redundant nodes are removed from the solution thus obtained in the same manner as in our crossover operator. This solution is included into the initial population only when it is different from all the initial population members generated so far. Otherwise, this solution is discarded. This uniqueness checking is done to conform with steady state population replacement model.

The pseudo-code for SSGA has been provided in Algorithm 2. In this pseudo-code, *Selection*(), *Crossover*() and *Mutation*() are three functions that implement probabilistic binary tournament selection method (Sect. 3.3), crossover operator (Sect. 3.4) and mutation operator (Sect. 3.5) respectively. Further, ps is the population size and $u01$ is a uniform random variate in [0, 1].

4 Computational Results

SSGA has been coded in C and executed on an Intel Core i5-3470S processor based system with 4 GB RAM running at 2.90 GHz under 64-bit Debian 6.0.9 operating system. We have used a population of 20 individuals ($ps = 20$). However, if during initial population generation, we fail to generate i^{th} solution different from all $i - 1$ current population members even after 10 consecutive attempts, then initial population generation process is stopped, and ps is set to $i - 1$. We have used $p_c = 0.5$, $p_g = 0.1$, $p_m = 0.66$ and $p_{bt} = 0.8$. All these parameter values have been chosen empirically. These values are so chosen that they provide good results on most of the instances. However, they are in no way optimal parameter values. SSGA terminates when best solution has not improved over 500 iterations.

To test our approach, we have performed the experiments in the same manner as in [2]. In all the experiments, the network is assumed to be a grid with varying size ranging from 10×10 to 20×20. For all the experiments, the number of nodes are taken to be 100. Nodes are assumed to be placed on a grid of specific size randomly such that no two nodes are at the same position. Please note that for 10×10 grid, all 100 positions are occupied and randomness has no role to play. All experiments are done under following assumptions [2]:

- All nodes have the same transmission range.
- All transmissions between nodes are error-free
- All transmission links are bi-directional
- The movement of nodes during the procedure of finding the MPR nodes is negligible and can be ignored

Two sets of experiments were performed in [2]. In the first set of experiments, network is assumed to be a grid of 10×10 with 100 nodes and transmission range of nodes is varied from 1 to 6. In the second set of experiments, transmission range was fixed to 6 and grid size was varied from 10×10 to 20×20 in increments of 2 along each dimension, i.e., experiments were performed with grid sizes 10×10, 12×12, 14×14, 16×16, 18×18, 20×20.

We have compared the performance of SSGA with genetic algorithm (GA) of [2] and greedy heuristic (GH) of [8]. For comparison, we have re-implemented GH. So we can compare SSGA with GH under any scenario. However, the results of GA are taken from [2]. So comparison between SSGA and GA is possible for the first set of experiments only, because a single scenario only exists with 100 nodes placed on a 10×10 grid. However, so many scenarios exist for grid sizes larger than 10×10 with respect to allocation of 100 nodes to grid locations, and hence, the results of SSGA are not comparable with those of GA for second set of experiments.

The results of first set of experiments are reported in Table 1. For these experiments, for each of the value of the transmission range, we have computed the set of multi-point relay nodes for each of the 100 nodes by taking each of them as base node one-by-one, and reported the minimum, maximum and average objective function values over these 100 nodes obtained by each of GH, GA and SSGA approaches under columns *Min*, *Max* and *Avg* respectively. In addition, we have reported the average percentage improvement obtained by GA and SSGA over GH for each value of transmission range under the column *%-imp*. We have observed that the results obtained by GH differ from those reported in [2] for some transmission ranges, and hence, the results obtained by our implementation of GH have been used to compute the average percentage improvement values. These results clearly show the superiority of SSGA over GH and GA. Further, it can be clearly observed that the difference in performance of SSGA over GH and GA widens as transmission range increases, i.e., as the

Table 1. Results of GH, GA and SSGA for different transmission ranges over the 100 node 10 × 10 grid network

Range	GH			GA				SSGA			
	Min	Max	Avg	Min	Max	Avg	%-imp	Min	Max	Avg	%-imp
6	5	135	61.17	5	129	50.96	16.69	5	98	42.96	29.77
5	25	168	90.01	23	146	78.24	13.07	21	142	70.57	21.60
4	57	108	89.95	59	116	87.65	2.56	57	102	84.89	5.63
3	36	92	59.40	36	114	58.53	1.46	36	92	57.24	3.64
2	10	32	21.19	10	32	21.16	0.14	10	32	21.12	0.33
1	4	12	9.04	4	12	9.04	0.00	4	12	9.04	0.00

Table 2. Results of GH and SSGA for 100 node network with transmission range of 6 and different grid sizes

Grid Size	GH			SSGA			
	Min	Max	Avg	Min	Max	Avg	%-imp
10 × 10	5.00	135.00	61.17	5.00	98.00	42.96	29.77
12 × 12	24.90	187.80	88.58	20.30	179.70	74.24	16.19
14 × 14	45.30	161.80	91.02	41.30	154.00	83.55	8.20
16 × 16	40.70	129.60	80.07	39.80	125.20	75.42	5.81
18 × 18	29.30	120.60	67.44	29.30	111.80	64.73	4.02
20 × 20	20.10	105.60	52.88	20.10	102.70	51.00	3.56

density of the network increases. We have not reported the execution times as they are negligible for SSGA (less than 0.01 s).

For second set of experiments, for grid sizes larger than 10 × 10, nodes can be allocated to grid locations in multiple ways, and hence, for each grid size, we have taken 10 scenarios and reported the average results of 10 scenarios for SSGA and GH in Table 2. Various column names under GH and SSGA in this table have the same meaning as in the previous table. As the grid size increases, the density of the network decreases. Here also, it can be observed that the difference in performance of SSGA over GH widens as density of the network increases. Though we can not compare the performance of SSGA with GA of [2] for second set of experiments, it is clear from Tables 1 and 2, that relative performance of algorithms depends on the density of the network, and therefore, it is expected that relative performance similar to Table 1 will be observed for second set of experiments too between SSGA and GA.

5 Conclusions

In this paper, we have proposed a steady-state genetic algorithm for *P-MPR*. A key feature of our genetic algorithm is integer encoding based on ordered list to represent a solution. We have also incorporated a redundant node removal strategy, which helps in further improving the child solution obtained through crossover and mutation. Computational results show the effectiveness of our approach in comparison to previously proposed approaches.

We intend to extend our approach to the version of multi-point relay nodes selection problem having the objective of minimizing the number of multi-point relay nodes. Similar approaches can be developed for other related problems also such as dominating set problem, uni-cost set covering problem, router selection problem etc. Another possible future work is to work on a bi-objective version of multi-point relay node selection problem, where one objective can be minimizing the number of multi-point relay nodes, and, the other objective can be minimizing the size of the resulting routing table.

References

1. Chang, Y.K., Ting, Y.W., Su, S.C.: Power-efficient and path-stable broadcasting scheme for wireless ad hoc networks. In: Proceedings of the 21st International Conference on Advanced Information Networking and Applications Workshops (AINAW 2007), pp. 707–712. IEEE (2007)
2. Chizari, H., Hosseini, M., Razak, S.A.: Multipoint relay selection using GA. In: Proceedings of the 2009 IEEE Symposium on Industrial Electronics and Application (ISIEA 2009), pp. 957–962. IEEE (2009)
3. Chizari, H., Hosseini, M., Salleh, S., Razak, S.A., Abdullah, A.H.: EF-MPR, a new energy efficient multi-point relay selection algorithm for manet. J. Supercomput. **59**, 744–761 (2012). https://doi.org/10.1007/s11227-010-0470-7
4. Jacquet, P., Mühlethaler, P., Clausen, T., Louiti, A., Qayyum, A., Viennot, L.: Optimized link state routing protocol for ad hoc networks. In: Proceedings of the 2000 IEEE International Multi Topic Conference (INMIC 2001), pp. 62–68. IEEE (2001)
5. Khan, A.Y., Rashid, S., Iqbal, A.: Mobility vs. predictive MPR selection for mobile ad hoc networks using OSLR. In: Proceedings of the 2005 IEEE International Conference on Emerging Technologies (IET 2005), pp. 52–57. IEEE (2005)
6. Liang, Q., Sekercioglu, Y.A., Mani, N.: Gateway multipoint relays - an MPR-based broadcast algorithm for ad hoc networks. In: Proceedings of the 10th IEEE Singapore International Conference on Communication Systems (ICC 2006), pp. 1–6. IEEE (2006)
7. Nguyen, D., Minet, P.: Analysis of MPR selection in the OLSR protocol. In: Proceedings of the 21st International Conference on Advanced Information Networking and Applications Workshops (AINAW 2007), pp. 887–892. IEEE (2007)
8. Qayyum, A., Viennot, L., Laouiti, A.: Multipoint relaying: an efficient technique for flooding in mobile wireless networks. Technical Report 3898, Institut National De Recherche En Informatique Et En Automatique, INRIA, France (2000)
9. Yawut, C., Paillassa, B., Dhaou, R.: Mobility versus density metric for OLSR enhancement. In: Fdida, S., Sugiura, K. (eds.) AINTEC 2007. LNCS, vol. 4866, pp. 2–17. Springer, Heidelberg (2007). https://doi.org/10.1007/978-3-540-76809-8_2

An Aggregated Rank Removal Heuristic Based Adaptive Large Neighborhood Search for Work-over Rig Scheduling Problem

Naveen Shaji(✉), Cheruvu Syama Sundar, Bhushan Jagyasi, and Sushmita Dutta

Accenture Advanced Technology Centers in India, Mumbai, India
{naveen.shaji,syama.sundar.cheruvu,bhushan.jagyasi, sushmita.dutta}@accenture.com

Abstract. Work-over Rig Scheduling Problem (WRSP) is a well known challenge in oil & gas industry. Given the limited number of work-over rigs to cater to the maintenance needs of a large number of wells, the challenge lies in planning an optimum schedule that minimizes the overall production loss. In this work, we propose a new Aggregated Rank Removal Heuristic (*ARRH*) applied to Adaptive Large Neighborhood Search to solve WRSP. The proposed approach results in more efficient searches as compared to existing heuristics - Genetic Algorithm, Variable Neighborhood Search and Adaptive Large Neighborhood Search.

1 Introduction

Oil wells require various maintenance services from time to time. Work-over rigs are specialized mobile units that are designed to carry out these maintenance activities. However rigs are expensive units and hence limited in number. The wells that require maintenance, wait in queue for a rig to visit them and carry-out the needed work-over, all the while remaining idle and resulting in production loss. Hence, there is a need to generate an optimum maintenance schedule that minimizes the production loss. The problem of optimizing schedule for work-over rigs is combinatorial optimization problem with constraints and belongs to the NP-Hard difficulty. As the number of wells to be serviced increases, the size of the problem increases exponentially which can cause the execution times of these searches to go beyond an acceptable time.

The Work-over Rig Scheduling Problem (WRSP) can be seen as a type of the Vehicle Routing Problem (VRP) [3,10] but with the objective to minimize the waiting time for the wells multiplied by the production loss. There have been several solutions proposed for solving the WRSP using various heuristics which has been summarized in [7]. Aloise et al. [1] employed Variable Neighborhood Search (VNS) heuristic to solve WRSP. Ribeiro et al. [8] compares between three meta heuristics, Iterated local search (ILS), Clustering search (CS) and Adaptive

B. Deka et al. (Eds.): PReMI 2019, LNCS 11941, pp. 385–394, 2019.
https://doi.org/10.1007/978-3-030-34869-4_42

large neighborhood search (ALNS). Clustering search proceeds by assigning the solutions obtained by the search to various clusters based on minimum hamming distance. The authors claim that the best results were obtained with ALNS.

Several recent works [4–6,9] deals with Capacitated Cumulative Vehicle Routing Problem (CCVRP) which shares a close analogy with WRSP. Ngueveua et al. [6] is the first to formulate CCVRP, draw insights into why CCVRP is tougher than CVRP and uses Memetic Algorithm (Genetic Algorithm (GA) supported with local search) to solve it. Lysgaard et al. [5] uses the Branch-and-Cut-and-Price algorithm to solve CCVRP. Sze et al. [9] and Liu et al. [4] proposes hybrids of the Large Neighborhood Search to solve CCVRP and highlights the robustness and efficiency of the approach.

Since ALNS is one of the state of the art algorithms for solving WRSP, our work focuses on improving ALNS for achieving better solutions in a quicker time. The key contribution of this paper is a new Aggregated Rank Removal Heuristic (*ARRH*) based ALNS to solve the WRSP. We compare the proposed *ARRH* based ALNS with existing ALNS to show the performance benefits in terms of production loss.

2 Problem Definition

In this section, the work-over rig scheduling problem (WRSP) and its constraints is formulated mathematically. Different wells require different levels of maintenance which can only be serviced by a rig having a equal or higher level work-over capability. To model WRSP, it is assumed that we know before hand the wells that require work-over, their production loss and various travel times between the wells. A good model of the problem and its constraints has been provided by Ribeiro et al. [8], which has been used in our work.

Let W be the set of wells to be scheduled and K be the set of available rigs. Every schedule s is an association of each rig $k \in K$ to a directed graph, $G^k = (V^k, A^k)$, where V^k is the set of nodes and A^k is the set of arcs. Let W^k be the set of wells assigned to rig k then $V^k = W^k \cup \{o_k, z_k\}$, where o_k is the node representing the starting position of rig k, $k \in K$, z_k is the node representing the ending position of rig k, $k \in K$. Every arc $(i, j) \in A^k$ is characterized by a time duration $c_{ij} = e_{ij} + d_j$, where e_{ij} is the travel time between nodes i and j , $i, j \in W \cup_{k \in K} \{o_k\}$ and d_j is the duration of the maintenance (work-over time) required by node j, $j \in W$. In case of missing paths between any two nodes, travel time will be calculated as the minimum spanning time with one or more intermediate nodes.

We assume dummy arcs (i, z_k), $i \in W^k$ with zero duration. Also $i \in W^k \cup \{o_k\}$ & $j \in W^k$ $\forall (i, j) \in A^k$.

To denote the presence or absence of an edge (i, j) in schedule of rig k we use a binary variable y_{ij}^k as given by Eq. 1.

$$y_{ij}^k = \begin{cases} 1 & if\ (i,j) \in A^k, \\ 0 & otherwise. \end{cases} \tag{1}$$

We now formulate $x_j(s)$, the time waited by the node j in the queue for the maintenance to begin in schedule s, using Eq. 2.

$$for\ \ s\ s.t\ \ j \in A^k,\ \ x_j(s) = \begin{cases} e_{o_k j} & if\ y^k_{o_k j} = 1 \\ x_i(s) + d_i + e_{ij} & if\ y^k_{o_k j} = 0\ \ and\ \ y^k_{ij} = 1 \end{cases} \tag{2}$$

Objective Function: The objective function Ψ to be minimized is given by Eq. 3.

$$\Psi(s \mid s \vdash G^k,\ k \in K) = \sum_{k \in K} \sum_{j \in W^k} (x_j + d_j) p_j \tag{3}$$

where, p_j is the oil production rate of node j, $j \in W$.

Constraints

1. All wells demanding maintenance must be present in the combined schedule of all rigs. That is $W = \sum_{k \in K} \cup W^k$.
2. Each well is visited exactly once in the combined schedule of all rigs.

$$\sum_{k \in K} \sum_{j:(i,j) \in A^k} y^k_{ij} = 1,\ \ i \in W \tag{4}$$

3. The classic network flow constraints are modeled as.

$$\sum_{j:(o_k,j) \in A^k} y^k_{o_k j} = \sum_{i:(i,z_k) \in A^k} y^k_{i z_k} = 1, k \in K \tag{5}$$

$$\sum_{j:(i,j) \in A^k} y^k_{ij} - \sum_{j:(j,i) \in A^k} y^k_{ji} = 0,\ \ k \in K,\ i \in W^k \tag{6}$$

4. Level Constraint: Level constraint is modeled to make sure that wells are visited by those rigs that are equipped to serve the level of maintenance demanded.

$$q_k \geq l_j\ \ \forall\, j : (j,i) \in A^k,\ \ k \in K \tag{7}$$

where, l_j is the level of maintenance request at node j, $j \in W$ and q_k is the maximum level of maintenance request that can be attended by rig k.

3 Algorithms

3.1 Genetic Algorithm Supported Variable Neighborhood Search (VNS+GA)

We provide below a brief description of the implemented GA and VNS Algorithms along with the hybrid GA supported VNS.

Genetic Algorithm. The initial population was created by means of iteration through each rig multiple times appending a random well (not yet assigned), to its schedule till all requests for maintenance were met by the schedule.

Crossover- In this work, to cross between two schedules the following approach was used.
Let PA, PB denote the individuals to be crossed.

1. All wells that belong outside crossover point is removed from PA to form PA'.
2. All wells that are in PA' are eliminated from PB forms PB'.
3. Concatenation of PB' to PA' retaining the structure gives child C.

Mutation- In this work, mutation implements two variation in the child. A random interchange of wells within the schedule of a rig and a random well being shifted from one rig's schedule to another. The number of children to be mutated are given by a user defined parameter τ. Mating Pool- In this work the method of Roulette wheel selection [2] is used. The number of children is given by a user defined parameter n_{child}.

Variable Neighborhood Search (VNS). Our attempt to solve the WRSP using VNS was inspired and extends the work of Aloise et al. [1] which employed VNS heuristic to solve WRSP. The VNS framework adopts a recursive search involving an exhaustive search of local neighborhoods and a random search of global neighborhoods (a neighborhood is a set of possible schedules that are similar to a given schedule in some way).

For a given schedule its neighborhoods are generated using the following operations

1. Swap Wells from Same Work-over rig ($SWSW$): In the given schedule, swap two wells assigned to same rig.
2. Swap Wells from Different Work-over rigs (SWDW): In the given schedule, two wells assigned to two different rigs respectively are interchanged.
3. Add Drop (AD): In the given schedule, a well assignment to one rig is dropped and it the well is assigned to another rig.

The local neighborhoods considered are SWSW, SWDW and AD. The global neighborhoods are SWSW done twice and thrice, SWDW done twice and thrice and the AD done twice and thrice.

Genetic Algorithm Supported Variable Neighborhood Search. In Genetic Algorithm Supported Variable Neighborhood Search a GA is employed by the VNS heuristic to carry out its local search routine. From the current schedule several solutions in the neighborhood defined by the VNS are created and these solutions become the initial population of GA. The idea here is to remove the in-efficient exhaustive search and replace it by the much faster GA.

3.2 Proposed *ARRH* Based ALNS

In this section, we first briefly describe the existing ALNS implementation, and then present our proposed *ARRH* based ALNS.

Adaptive Large Neighbourhood Search (ALNS). Our work is inspired from and extends the work of Ribeiro et al. [8] which highlights the superiority of ALNS. Below we provide a brief description of ALNS, a detailed view is available in Ribeiro et al. [8]. ALNS proceeds by destroying a part of the existing solution and recreating it. In each iteration it removes a certain number wells from the schedule and then inserts it back into better positions. For this purpose, it employs various insertion and removal heuristics. The probability of selection of a removal or insertion heuristics depends on its past performance in the search.

Objective Function- The objective function in ALNS is modified to add an additional term λ for the broken level constraints. Here λ is a user defined parameter.

The Proposed Aggregated Rank Removal Heuristic Based ALNS. Aggregated Rank Removal Heuristic (*ARRH*) which is proposed to replace the Worst Removal Heuristic (*WRH*) in ALNS. The general idea here is that although ALNS is selecting wells based on travel times and production losses using separate methods, such methods lack the power to make unbiased identification of opportunities that arise due to a combination of these factors. The *ARRH* is designed to solve two shortcomings of the *WRH* as given below.

1. Bias Factor - For optimization, the general idea behind removal heuristics is to remove those wells that has an opportunity to be placed in a better position. However, evaluating the wells solely based on loss has a tendency to give more preference to those wells that have higher production rate. Similarly large wait times can also occur from wells with large work-over time or a well isolated from others, inducing a large travel time. Hence larger production loss doesn't always imply the presence of an opportunity for the well to be placed in a better position.
2. Masking Factor - The *WRH* takes into consideration of the contribution to production loss only. A well's contribution to over all production loss is driven not only by its own maintenance requirement but also because of the previous wells that make the well under consideration to wait for its turn. However these other wells' contributions are masked in *WRH*.

Proposed ARRH- The proposed approach, Aggregated Rank Removal Heuristic (*ARRH*) works on ranked values of (a) production loss, (b) travel times and (c) work-over time. Ranked values corresponding to each metric is created by ordering the wells in the increasing order of the metric. Then a discrete set of ordered values ranging from 0 to 1 of equal spacing is created and assigned to the wells as their ranked value of the metric. Ranking the metric in such fashion will help identify opportunities for a better solution without bias.

Henceforth, in this paper, the ranked values of travel times, work-over time, production rate and ratio of production rate to work-over time will be represented by $\hat{e}_{i,j}$, $\hat{d}_j$, $\hat{p}_j$ and $\hat{r}_j$ respectively. *Scoring* - Scoring the well based on multiple criteria can help pinpoint the cause of the increased production loss. *ARRH* removes wells with the highest scores. In this work, three criteria have been used to determine the goodness of a solution.

- Suffocation - If any rig is overloaded with many wells with high production losses and large work-over times no matter the ordering, the wells assigned to the rig will face large production losses, indicating that some of the wells assigned to the rig needs to be removed and placed in the schedule of another rig. The suffocation coefficient for a rig is calculated by Eq. 8.

$$\Gamma(k) = \sum_{j \in W^k} \hat{p}_j + \hat{d}_j \tag{8}$$

- Ordering - Ordering criterion measures how the wells are ordered in the schedule of a rig. It follows the same criteria as described in Bassi et al. [11]. The ordering coefficient for a rig is calculated by Eq. 9.

$$\zeta(k) = \frac{4}{n(n+1)} \sum_{j \in W^k} i\,\hat{r}_j \tag{9}$$

  Where i = 0, . . ., n is the numbering of wells in schedule of rig r.
- Travel-time - The suffocation and ordering criteria deals with the distribution and arrangement of rigs based on $\hat{p_j}$ and $\hat{d}_j$. An optimization based only on these criteria alone may lead to coupling between wells that are far apart and induce large travel times rendering the entire procedure in-feasible. The travel time criterion mitigates this by evaluating the travel times at the edges of each well. The travel time coefficient is given by the Eq. 10.

$$\xi(i) = \begin{cases} 2e_{i,\widehat{i+1}} & if\ y_{o_k i} = 1 \\ 2e_{i,\widehat{i-1}} & if\ y_{i z_k} = 1 \\ e_{i,\widehat{i+1}} + e_{i,\widehat{i-1}} & Otherwise \end{cases} \tag{10}$$

  Where $i \in W^k$ and $(i, i+1)$, $(i-1, i) \in A^k$

In order to give an equal importance to all criteria, the coefficients by the design of the equations are set to vary between 0 and 2.

The criteria mentioned above reflects on some trends we want to see in the schedule however, they are not mutually exclusive. A schedule that is suitable to one criteria need not suit another. Therefore there is a need to strike a balance between them using stochastic aggregation.

In *ARRH* wells are removed based on a score given by Eq. 11.

$$s(i) = x\Gamma(k|i \in k) + y\xi(i) + z\zeta(k|i \in k) \tag{11}$$

where i is the well,
r is the rig to which i belongs in the current solution,
x, y, z are random numbers between 0.8 and 1.

A time complexity comparison between $ARRH$ and WRH shows that they both have similar average time complexities $O(|W|^2)$, where $|W|$ is the number of wells that needs to be serviced. Note that that time complexity for both the heuristics is not dependent on the number of rigs.

4 Results

4.1 Experimental Setup

The data used to create different instances of the problem were created using random simulations. All algorithms were implemented using python 3, on a Windows 10 Enterprise 64 Bit system with Intel(R)Core(TM) i7-8650U CPU @ 1.90 GHz and available physical memory of 9.82 GB. For readability and ease of comparison all losses reported are scaled down by a factor of 5000.

The parameters were created as:

- Production loss (p_j)- Production loss of each well is sampled from a uniform distribution ranging from 20 to 60.
- Work-over times (d_j)- Work-over times were sampled from a uniform distribution ranging from 20 to 80.
- Travel times ($e_{i,j}$)- x and y co-ordinates were sampled from a uniform distribution ranging from 0 to 100. Travel times were calculated as the euclidean distance between two points.

4.2 A Brief Comparison of GA, VNS+GA and ALNS

All three algorithms were tried to solve an instance where VNS failed to deliver results in a feasible time. That is, for 50 wells and 10 Rigs, VNS gave a loss of 78.8956 for an epoch taking 9138 s. Hence we had an estimated 130 h for 50 epochs. For 5 runs, ALNS, VNS+GA and GA gave 70.04, 78.23, 110.51 average losses respectively, for a run time of 2000 seconds. The parameters used for the algorithm are given in Table. 1. The Fig. 1(a) compares the production loss with respect to the run time of these three algorithms.

4.3 Comparison of ALNS Vs *ARRH* Based ALNS

ALNS and *ARRH* based ALNS were compared for 18 instances and the results are tabulated in Table. 2. All instances were run 5 times repeatedly to decrease the uncertainty in the result. The entries made into the table are:

- Instance: Specifies the instance, the n/m notation stands for n wells and m rigs.

Table 1. User defined parameters

Heuristic	Symbol	Meaning	Value
GA	n_{pop}	Population Size	700
	n_{child}	Number of Children	600
	τ	Rate of Mutation	13
VNS+GA	n_{pop}	Population Size	20
	n_{child}	Number of Children	16
	τ	Rate of Mutation	2
ALNS and M_ALNS	ϵ	Controls Number of wells removed	0.4
	ϕ	Number of iterations in a segment	50
	λ	Penalizes broken level constraints	15
	ν	Avoids determinism in Shaw removal heuristics	2
	Φ	Avoids determinism in Route removal heuristic	3
	ρ	Avoids determinism in Worst removal heuristic	2
	T_{start}	Starting temperature in Simulated Annealing	15
	c	Temperature reduction factor	0.995

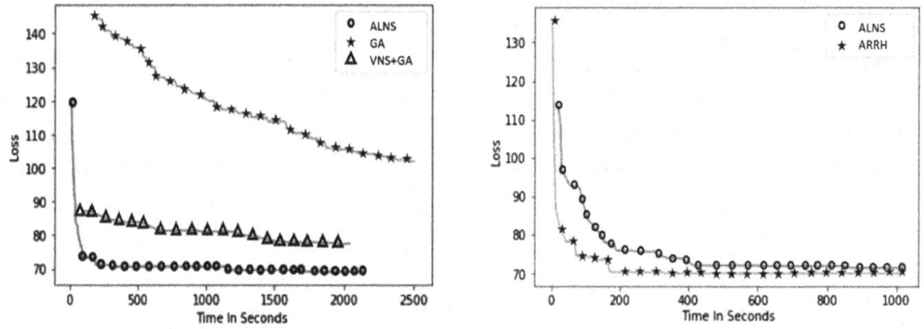

(a) Comparison between ALNS, GA supported VNS and GA.

(b) Comparison between ALNS and *ARRH* based ALNS

Fig. 1. Loss vs time comparison

- Time: To be fair both algorithms were run for the same time for an instance and is mentioned in the table in seconds.
- Average Loss: Gives the average loss over 5 times.
- Best: Gives the best-found solution from the 5 times.
- Deviation: Here deviation is defined as $\frac{AverageLoss-Best}{Best} \times 100$.

Figure 1 provides a comparison between ALNS and proposed *ARRH* based ALNS. As seen from the graph, proposed *ARRH* based ALNS results in improvement in production losses as compared to ALNS for the same run time. Further if there is a benchmark of production losses *ARRH* based ALNS is able to achieve it faster.

Table 2. ALNS Vs *ARRH* based ALNS

Instance	Time (s)	ALNS			*ARRH* based ALNS		
		Average	Deviation	Best	Average	Deviation	Best
50/5(1)	250	123.44	1.59	121.47	119.53	1.69	117.50
50/5(2)	250	126.12	3.53	121.66	123.46	2.10	120.86
50/5(3)	250	139.07	1.86	136.47	132.12	5.22	125.22
50/10(1)	250	74.91	3.63	72.19	73.48	1.97	72.03
50/10(2)	250	78.23	3.50	75.49	74.65	1.55	73.49
50/10(3)	250	77.45	4.78	73.75	73.95	2.71	71.94
100/5(1)	3600	494.75	1.31	488.24	472.08	2.39	460.78
100/5(2)	3600	450.97	2.22	440.94	442.14	1.19	436.86
100/5(3)	3600	395.22	3.18	382.63	390.52	3.51	376.81
100/10(1)	3600	246.63	1.80	242.17	236.45	1.53	232.83
100/10(2)	3600	260.58	0.73	258.65	257.27	0.99	254.70
100/10(3)	3600	236.34	1.17	233.57	231.89	0.68	230.29

ALNS and *ARRH* based ALNS were also compared using Manhattan distance to calculate distance between the wells. The results were collected for 5 instances of 20 and 30 wells each and *ARRH* based ALNS was found to give an average improvement of 2.14 % over ALNS. This insight is similar to results obtained from euclidean distance, hence the approach is robust to various distance metrics.

5 Conclusion

The Workover Rig Scheduling Problem (WRSP) has been addressed to improve the production losses of the well. We presented the implementation of the Genetic Algorithm based Variable Neighborhood Search algorithm for the same problem. This combination had a significant improvement than both GA and VNS implementations. However, GA based VNS was not found to be superior than existing Adaptive Large Neighborhood based (ALNS) approach. In this paper, a new heuristic - Aggregated Rank Removal Heuristic (ARRH) had been proposed to be implemented with the existing Adaptive Large Neighborhood Search (ALNS) algorithm for WRSP. The proposed ARRH is based on the utility of rank-based methods to make unbiased judgments about the schedules. The results indicates that this proposed approach results in the lower production losses than ALNS approach for equal run times. In this work, a removal heuristic was designed and used on the basis of ranking. In future we aim to introduce a rank based insertion heuristic as well.

References

1. Aloise, D.J., Aloise, D., Rocha, C.T., Ribeiro, C.C., Filho, J.C.R., Moura, L.S.: Scheduling workover rigs for onshore oil production. Discrete Appl. Math. **154**(5), 695–702 (2006). https://doi.org/10.1016/j.dam.2004.09.021. http://www.sciencedirect.com/science/article/pii/S0166218X05003008. IV ALIO/EURO Workshop on Applied Combinatorial Optimization
2. Goldberg, D.E.: Genetic Algorithm in Search Optimization and Machine Learning. Addison-Wesley, Reading (1989)
3. Li, F., Golden, B., Wasil, E.: The open vehicle routing problem: algorithms, large-scale test problems, and computational results. Comput. Oper. Res. **34**(10), 2918–2930 (2007). https://doi.org/10.1016/j.cor.2005.11.018. http://www.sciencedirect.com/science/article/pii/S0305054805003515
4. Liu, R., Jiang, Z.: A hybrid large-neighborhood search algorithm for the cumulative capacitated vehicle routing problem with time-window constraints. Appl. Soft Comput. **80**, 18–30 (2019). https://doi.org/10.1016/j.asoc.2019.03.008. http://www.sciencedirect.com/science/article/pii/S1568494619301267
5. Lysgaard, J., Wøhlk, S.: A branch-and-cut-and-price algorithm for the cumulative capacitated vehicle routing problem. Eur. J. Oper. Res. **236**, 800–810 (2014). https://doi.org/10.1016/j.ejor.2013.08.032
6. Ngueveu, S.U., Prins, C., Calvo, R.W.: An effective memetic algorithm for the cumulative capacitated vehicle routing problem. Comput. Ope. Res. **37**(11), 1877–1885 (2010). https://doi.org/10.1016/j.cor.2009.06.014. http://www.sciencedirect.com/science/article/pii/S0305054809001725. metaheuristics for Logistics and Vehicle Routing
7. Ribeiro, G., Mauri, G., Lorena, L.: A simple and robust simulated annealing algorithm for scheduling workover rigs on onshore oil fields. Comput. Ind. Eng. **60**, 519–526 (2011). https://doi.org/10.1016/j.cie.2010.12.006
8. Ribeiro, G.M., Laporte, G., Mauri, G.R.: A comparison of three metaheuristics for the workover rig routing problem. Eur. J. Oper. Res. **220**(1), 28–36 (2012). https://doi.org/10.1016/j.ejor.2012.01.031. http://www.sciencedirect.com/science/article/pii/S0377221712000665
9. Sze, J.F., Salhi, S., Wassan, N.: The cumulative capacitated vehicle routing problem with min-sum and min-max objectives: an effective hybridisation of adaptive variable neighbourhood search and large neighbourhood search. Transp. Res. Part B Methodol. **101**, 162–184 (2017). https://doi.org/10.1016/j.trb.2017.04.003. http://www.sciencedirect.com/science/article/pii/S0191261516308396
10. Vidal, T., Crainic, T.G., Gendreau, M., Prins, C.: Heuristics for multi-attribute vehicle routing problems: A survey and synthesis. Eur. J. Oper. Res. **231**(1), 1–21 (2013). https://doi.org/10.1016/j.ejor.2013.02.053. http://www.sciencedirect.com/science/article/pii/S0377221713002026
11. Bassi, H.V., Ferreira Filho, V.J.M., Bahiense, L.: Planning and scheduling a fleet of rigs using simulation-optimization. Comput. Ind. Eng. **63**, 1074–1088 (2012). https://doi.org/10.1016/j.cie.2012.08.001

Image Processing

Unsupervised Detection of Active, New, and Closed Coal Mines with Reclamation Activity from Landsat 8 OLI/TIRS Images

Jit Mukherjee[1(✉)], Jayanta Mukherjee[2], Debashish Chakravarty[3], and Subhas Aikat[2]

[1] Advance Technology Development Centre, Indian Institute of Technology, Kharagpur, Kharagpur, West Bengal, India
jit.mukherjee@iitkgp.ac.in
[2] Department of Computer Science and Engineering, Indian Institute of Technology, Kharagpur, Kharagpur, West Bengal, India
jay@cse.iitkgp.ac.in, subhas.aikat@iitkgp.ac.in
[3] Department of Mining Engineering, Indian Institute of Technology, Kharagpur, Kharagpur, West Bengal, India
dc@mining.iitkgp.ernet.in

Abstract. Monitoring and classification of surface coal mine regions have several research aspects, as they have huge impacts on the eco-environment of a region. Mining regions change over time immensely. New potential mines get opened. Whereas, few mines stay active and many of them get closed. Closed mine regions are reclaimed to environment through plantation. In the past, semi-supervised and supervised techniques have been used to detect mine classes and assess the changes of land use and land cover classes. In this work, mine regions are detected in an adaptive manner from satellite images unlike the techniques in the literature. Further, a change detection technique is used to detect active, new, and closed surface coal mines in a region. Detected closed mines are further analysed to evaluate reclamation of that region. Average precision and recall of active, new, and closed mining regions of the proposed technique are found to be [84.7%, 62.8%], [74.2%, 64.5%], and [70.1%, 58.2%], respectively.

Keywords: Clay mineral ratio · Coal mine index · Surface coal mine · Change detection · Mine classes · Landsat 8

1 Introduction

Surface mining is a widely used mining technique, and has several advantages. It creates irreversible damage to the ecosystem, serious land degradation, loss of vegetation, contamination, etc. [4,7]. Large scale surface mining directly impacts soil fertility, desertification factors, biodiversity, etc. [4,8]. Coal surface mining

B. Deka et al. (Eds.): PReMI 2019, LNCS 11941, pp. 397–404, 2019.
https://doi.org/10.1007/978-3-030-34869-4_43

has severe environmental adversities because of coal seam fires. In surface mining, a shallow ore deposit is extracted by removing the overlaying soil and creating a pit. Extracted minerals and waste materials are dumped in nearby regions. Such land use and land cover changes due to mining show the paramount correlation of human activities to environment [13]. Therefore, quantification of land use and land cover changes in mining regions have several impacts and various research challenges. Detection and classification of surface mining regions have various research aspects such as mine waste water detection and treatment [9,11], assessment of reclamation success [13,14], detection of surface mining land classes [13], coal seam fire detection and management [6], etc. Supervised support vector machine based classification has been used in [13] to detect changes due to surface mining and reclamation in the island of Milos in Greece and to detect various land classes and to quantify reclamation success [7]. In [4], various vegetation indexes are considered to detect mine wasteland. Vegetation recovery on reclaimed coal surface mining areas has been measured through field validation in south western Virginia, USA [5]. In [3], two different classification approaches are used to detect coal surface mine land cover classes such as object and spectral based approaches. Most of these works detect several land classes in mine region through supervised and semi-supervised techniques and subsequently use these land classes to assess reclamation success through change detection. Clay mineral ratio detects the volume of hydrothermally altered rock in clay [2]. It has been found resourceful to detect various surface mine land classes in adaptive fashion [9,11]. In [10], coal mine regions are detected using a spectral index over *SWIR-I* and *SWIR-II* bands, called coal mine index (*CMI*). Further, it is also used to detect coal mine regions adaptively through hierarchical K-Means clustering [12]. In this work, the concept presented in [12] has been used to detect coal mine regions over multi-temporal Landsat images. These regions are further analyzed to check whether they are active, closed or newly created. The closed down mines are further checked whether they are reclaimed.

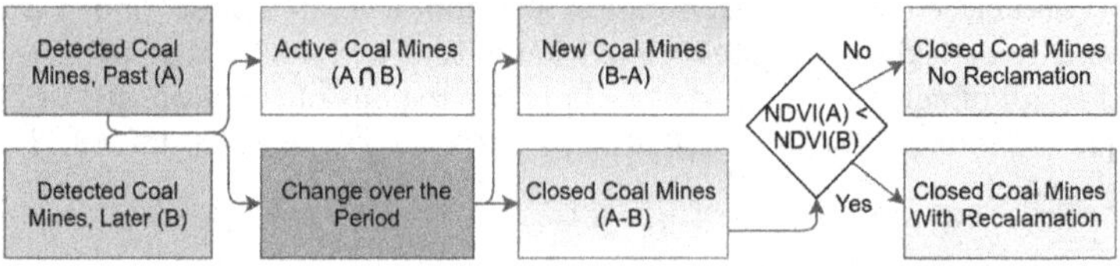

Fig. 1. Flow diagram of the proposed technique

2 Methodology

In this work, Landsat 8 multi-temporal images are used. The flow diagram of the proposed method is shown in Fig. 1. As shown in Fig. 1(A) and (B), coal mine regions are detected in unsupervised fashion from Landsat 8 images, where acquisition time of A is before the acquisition time of B. These images are further analysed by a change detection technique. It produces two images showing the unchanged and the changed regions, where land cover changes have occurred. The latter image is further analysed to detect the closed and new coal mines. Thereafter, Normalized difference vegetation index ($NDVI$) values are examined over the closed mine regions to detect reclamation regions.

2.1 Adaptive Detection of Coal Mine Regions

A spectral index, namely coal mine index (*CMI*), has been proposed as $\phi(SWIR-I, SWIR-II)$, i.e. $\frac{\lambda_{SWIR-I}-\lambda_{SWIR-II}}{\lambda_{SWIR-I}+\lambda_{SWIR-II}}$, where λ_{SWIR-I}, and $\lambda_{SWIR-II}$ represent reflectance values in short wave infra-red one, and short wave infra-red two band, respectively [10]. It is detected as a coal mine region if $\phi(SWIR-I, SWIR-II) < \tau$ [10], where τ is an empirical threshold. In [12], a novel method has been proposed to detect coal mine regions in an adaptive manner using *CMI*. It has been found that lower values of *CMI* have higher probability of being a coal mine region [10,12]. In [12], *CMI* values are hierarchically clustered using K-Means Clustering, where $k = 2$. The cluster associated with the higher *CMI* values is discarded, and that with lower CMI values is preserved. The process is repeated at a number of levels of hierarchy. It has been found that coal mine regions are detected at level three for Jharia coal mine region [12]. In this work, the proposed method uses the technique in [12] to detect active, new and closed mine using temporal analysis.

2.2 Adaptive Detection of Active, New and Closed Coal Mines with Reclamation

Active, new, and closed coal mine regions are detected using change detection technique. Let, A, and B be the detected coal mining regions of the same geographical region over a period, where acquisition time of A is before the acquisition time of B. A, and B have several regions in common. These regions are detected as active mine regions ($ActiveMines = A \cap B$). Mines, which are detected in A but not in B, are denoted as closed mines ($ClosedMines = A - B$). Whereas, mines, which are detected in B but not in A, are denoted as new mines ($NewMines = B - A$).

Further, reclamation impacts are computed in closed mine regions. In reclamation, trees are planted in a closed mine area to control its impact on environment. Therefore, vegetation index values of the closed mines are further analysed. Here, the $NDVI$ is used to compute the vegetation values. It is denoted by $\phi(NIR, Red)$ [4]. Here, *Red*, and *NIR* are the reflectance values of Red and Near Infra-red bands. Higher values of $NDVI$ denote vegetation in a region [15]. A closed mine is detected as reclamation region if $NDVI(A) < NDVI(B)$. Otherwise, the regions are denoted as closed coal mine regions without reclamation. The proposed method is applicable for multispectral remote sensing images with *Red*, *NIR*, *SWIR-I*, and *SWIR-II* bands.

3 Data and Study Area

In this work radiometrically corrected, geo-referenced, multi-temporal Landsat 8 $L1$ images are used over the Jharia Coal Field (JCF) regions from $USGS$ archive. It has an additional criteria of $< 10\%$ cloud cover. The JCF region is situated in the Gondwana basin of the Damodar valley in the state of Jharkhand, India ranging between $23°38'$ N and $23°50'$ N and longitudes $86°07'E$ and $86°30'E$. The JCF is a sickle shaped coal field, which resides among other prominent coal fields of Raniganj, Bokaro, Karanpura, Ramgarh and Hutar in the eastern part of Damodar valley. It has diverse range of land cover classes such as mine regions, dense forest, urban settlement, fresh water bodies, rivers, river bed, grassland, dry crop lands, etc. Landsat 8 provides nine multi-spectral and two thermal image bands with the temporal resolution of 16 days. They have spatial resolution of 30 meter except the panchromatic band. $L1$ data products provide top of atmosphere (TOA) reflectance values. In this work, Landsat images of 2013, and 2018 have been used for experimentation. High resolution Google Earth™ historical images of this period are obtained for ground truth generation. In this work, coal quarry and coal dump regions are treated as target locations. These regions are marked from historical and recent Google Earth™ images through visual inspection and experts' opinion. Further, active, new, and closed mine locations are also marked for validation of the proposed technique.

4 Results

In this work, TOA reflectance values are computed from multi-temporal Landsat 8 images as per [1] to detect active, new, closed coal surface mines of a region. The proposed technique has a pipeline of processes. First, coal mine regions are detected from Landsat 8 $L1$ images in 2013, and 2018. The coal mine regions are detected by an adaptive technique using *CMI* [10], as described in Sect. 2.1. It has been found that *CMI* can significantly differentiate coal mine regions from other land classes. The manual thresholding technique over *CMI*

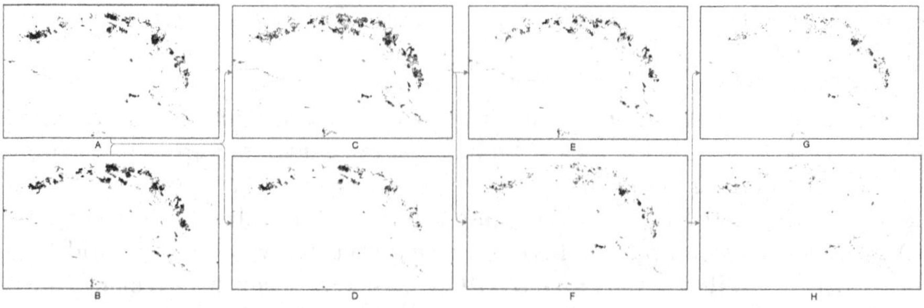

Fig. 2. A. Detected mines in 2013, B. Detected mines in 2018, C. Change in mine locations during 2013–2018, D. Active mines, E. New mines in 2018, F. Closed mines in 2018, G. Closed mines without reclamation, H. Reclaimed closed mines.

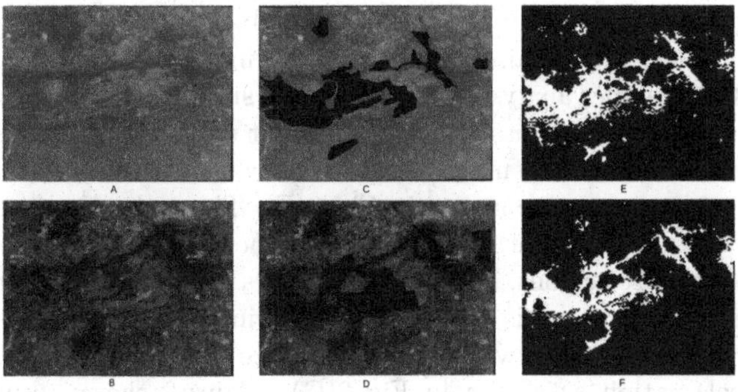

Fig. 3. (A) Google earth™ image in 2013, (B) Google earth™ image in 2018, (C) Marked mining region of (A) in black, (D) Marked mining region of (B) in black, (E) Detected mining region in 2013, (F) Detected mining region in 2018.

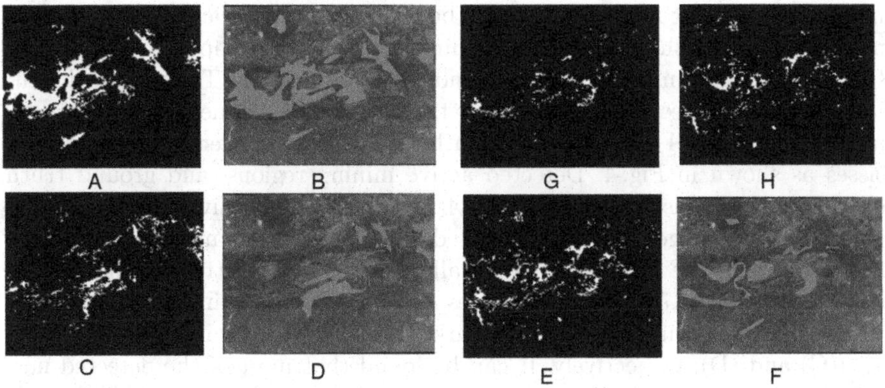

Fig. 4. (A) Active mines (B) Ground truth active mines marked in red (C) New mines (D) Ground truth new mines in red (E) Closed mines (F) Ground truth closed mines in red (G) Non-reclaimed closed mine (H) Reclaimed closed mines.

provides average accuracy[1] of 86.24% [10]. The automated detection of coal mine region has average precision, and recall of 76.43%, and 62.75%, respectively [12]. Detected coal mine regions of *JCF* in 2013, and 2018 are shown in Fig. 2(A) and (B), respectively. The black pixels in Fig. 2(A)–(H) shows various detected land cover classes of mining regions. Next, these two images with detected coal mine regions are further analysed to detect the changes of land cover and land use over the period as discussed in Sect. 2.2. Coal mine regions, which remain unchanged during this period, are treated as the active mine regions as shown in Fig. 2(D). Figure 2(C) shows the land use changes in mine regions from 2013

[1] Precision, Recall, and Accuracy are defined as $t_p/(t_p + f_p)$, $t_p/(t_p+f_n)$, and $(t_p + t_n)/(t_p + t_n + f_p + f_n)$, where t_p, f_p, f_n, and t_n are true positive, false positive, false negative, and true negative, respectively.

to 2018. Figure 2(C) is composed of various sub classes such as, new mines, closed mines, which are reclaimed, and closed mines, which are not reclaimed. Next, Fig. 2(C) is further analysed to detect these sub classes. Regions, which are detected as non mine land classes in Fig. 2(A) but detected as mining regions in Fig. 2(B), are marked as new mining regions. It is shown in Fig. 2(E). Detected closed down mine regions are shown in Fig. 2(F). These closed mines may have reclaimed to the environment. Hence, it is further analysed by the change of $NDVI$ values in these regions. If the $NDVI$ values are found to be greater than previous image, the regions is considered to be in the process of reclamation, as shown in Fig. 2(H). Otherwise, the regions are considered to be not in the process of reclamation as shown in Fig. 2(G). Figure 2 shows result over the whole Jharia region. For better understanding, results of an area of interest with ground truth images in 2013 and 2018 are shown in Fig. 3. The black portion of Fig. 3(C) and (D) represent the ground truth surface mining region in 2013, and 2018 images, respectively. It can be observed from ground truth and output images in Fig. 3 that mine regions have been changed significantly between 2013 and 2018. Further, the changes of the mining regions are computed. Let, A, and B be the processed images from 2013, and 2018, respectively. The proposed technique computes three images, such as active mines ($A \cap B$), new mines ($B - A$), and closed mines ($A - B$). Ground truths are generated over these three land classes as shown in Fig. 4. Detected active mining regions, and ground truth active mining regions are shown in Fig. 4(A) and (B), respectively. It is observed from these two images that the proposed technique detects many of the active mining regions. The precision, and recall of detection of active mining regions are found to be 84.7%, and 62.8%, respectively as shown in Table 1 (Right). Detected new mining regions, and marked ground truth image are shown in Fig. 4(C) and (D), respectively. It can be found that most of the detected new mining regions are similar with the ground truth new mining regions. Whereas, Fig. 4(C) also falsely detect some regions as new mining regions. The precision, and recall of new mining regions are found to be 74.2%, and 64.5, respectively. Similarly, detected, and ground truth closed mines are shown in Fig. 4(E) and (F), respectively. Precision, and recall of closed mining regions are found as 70.1%, and 58.2%, respectively. Precision and recall of closed and new mining regions are less than active mining land classes. The perimeter of a mining region changes over time. The walls of the mining pits are detected as mining region by the technique discussed in [10,12]. It is observed from Fig. 4(E), some of the detected regions are the walls of the mining pit. These affect the accuracy of the proposed technique. In the past, supervised and semi-supervised techniques are used to detect such mine land classes. In [7], a supervised support vector machine has been used in multi-temporal Landsat images to classify various land cover classes with average accuracy of 89.4%, and 96% in Landsat 5, and 8 images, respectively. Thereafter, reclamation is analyzed over these land classes. In [16], land reclamation monitoring has been done using pixel and object based classification in ArcGIS 9.0 platform. It has been found that 52% of closed mining areas are reclaimed, whereas field study shows 79% areas are reclaimed [16].

In [13], various land cover classes in mining region are classified using supervised support vector machine with average accuracy of 94% to detect land cover changes. The proposed method, though providing a lower accuracy, is an unsupervised technique to detect such land classes, and does not require labeled data set. Further, closed mining regions are analyzed for reclamation. Closed mining regions without reclamation and with reclamation are shown in Fig. 4(G) and (H), respectively. The *NDVI* detects the portion of vegetation in a region. In reclamation, mining regions are taken back to environment through plantation. Therefore, *NDVI* is used here to separate closed mining region with reclamation from other mine land classes. It has been observed that most of the regions, which are detected as possible reclamation regions in Fig. 4(H), have visible vegetation in the ground truth images. Further, for validation, t-test has been performed over the null hypothesis $\mu_{Reclamation} = \mu_{mlc}$, where $\mu_{Reclamation}$, and μ_{mlc}, represent mean of ground truth reclamation areas, and mean of other mine land cover classes, respectively. Here, ground truth coal quarry, coal dump, mine water bodies, and coal overburden areas are considered as other mine land cover classes. $S_{Reclamation} \neq S_{mlc}$ is considered here, where $S_{Reclamation}$, and S_{mlc} represent variance of ground truth reclamation regions, and other mine land classes, respectively. t-test result with null, and alternative hypothesis as $\mu_{Reclamation} = \mu_{mlc}$, and $\mu_{Reclamation} \neq \mu_{mlc}$, respectively, are shown in Table 1 (Left). As shown in Table 1 (Left), the null hypothesis is rejected. Therefore, the *NDVI* can distinguish reclamation regions from other mine land classes. Another advantage of the proposed technique is that use of CMI makes it applicable over the season [10,12].

Table 1. (Left) T test results with mine reclamation regions and other mine land classes. (Right) Precision and recall of various mine land classes

T test results					Accuracy computation			
	Dump	Quarry	Overburden	Mine swamp		New	Active	Closed
t_0	7.46	8.74	3.15	12.99	Precision	74.2	84.7	70.1
df	180.5	183.1	183.2	99	Recall	61.5	62.8	58.2
P value	<0.00001	<0.00001	0.0019	<0.00001	F_1 score	67.25	72.12	63.59

5 Conclusion

In this work, an unsupervised method has been proposed to detect active, new, and closed coal mine regions in an adaptive manner. First, coal mine regions are detected without supervision using *CMI*. A change detection technique during 2013–2018 is used over to detect the mentioned land classes. Closed mining regions are further analyzed to detect reclamation regions. In the past, semi-supervised and supervised techniques have been used to detect such classes. The proposed method has average precision and recall of [84.7%, 62.8%], [74.2%, 64.5%], and [70.1%, 58.2%], for active, new and closed mining regions, respectively. Changes of perimeter of the pit walls are detected by the proposed technique as closed or new mining regions, which affects the accuracy. Further experimentation in this regard is considered as a future direction. The method detects only those reclamation regions which are effective in this period.

References

1. Using the usgs landsat8 product. https://landsat.usgs.gov/using-usgs-landsat-8-product. Accessed 29 03 2018
2. Drury, S.A.: Image Interpretation in Geology. No. 551.0285 D796 1993, Chapman and Hall, London (1993)
3. Gao, Y., Kerle, N., Mas, J.F.: Object-based image analysis for coal fire-related land cover mapping in coal mining areas. Geocarto Int. **24**(1), 25–36 (2009)
4. Han, Y., Li, M., Li, D.: Vegetation index analysis of multi-source remote sensing data in coal mine wasteland. New Zealand J. Agric. Res. **50**(5), 1243–1248 (2007)
5. Holl, K.D.: Long-term vegetation recovery on reclaimed coal surface mines in the eastern USA. J. Appl. Ecol. **39**(6), 960–970 (2002)
6. Huo, H., et al.: A study of coal fire propagation with remotely sensed thermal infrared data. Rem. Sens. **7**(3), 3088–3113 (2015)
7. Karan, S.K., Samadder, S.R., Maiti, S.K.: Assessment of the capability of remote sensing and gis techniques for monitoring reclamation success in coal mine degraded lands. J. Environ. Manag. **182**, 272–283 (2016)
8. Lima, A.T., Mitchell, K., O'Connell, D.W., Verhoeven, J., Van Cappellen, P.: The legacy of surface mining: remediation, restoration, reclamation and rehabilitation. Environ. Sci. Pol. **66**, 227–233 (2016)
9. Mukherjee, J., Mukherjee, J., Chakravarty, D.: Automated seasonal separation of mine and non mine water bodies from landsat 8 OLI/TIRS using clay mineral and iron oxide ratio. IEEE JSTARS **12**(7), 2550–2556 (2019)
10. Mukherjee, J., Mukherjee, J., Chakravarty, D., Aikat, S.: A novel index to detect opencast coal mine areas from landsat 8 OLI/TIRS. IEEE JSTARS **12**(3), 891–897 (2019)
11. Mukherjee, J., Mukhopadhyay, J., Chakravarty, D.: Investigation of seasonal separation in mine and non mine water bodies using local feature analysis of landsat 8 OLI/TIRS images. In: IEEE IGARSS 2018, Valencia, Spain, 22–27 July 2018, pp. 8961–8964 (2018)
12. Mukherjee, J., Mukhopadhyay, J., Chakravarty, D., Aikat, S.: Automated seasonal detection of coal surface mine regions from landsat 8 oli images. In: IEEE IGARSS 2019, Yokoham, Japan, 27 July–3 August 2019 (2019)
13. Petropoulos, G.P., Partsinevelos, P., Mitraka, Z.: Change detection of surface mining activity and reclamation based on a machine learning approach of multi-temporal landsat TM imagery. Geocarto Int. **28**(4), 323–342 (2013)
14. Raval, S., Merton, R., Laurence, D.: Satellite based mine rehabilitation monitoring using worldview-2 imagery. Min. Technol. **122**(4), 200–207 (2013)
15. Silleos, N.G., Alexandridis, T.K., Gitas, I.Z., Perakis, K.: Vegetation indices: advances made in biomass estimation and vegetation monitoring in the last 30 years. Geocarto Int. **21**(4), 21–28 (2006)
16. Singh, N., Gupta, V., Singh, A.: Geospatial technology for land reclamation monitoring of open cast coal mines in India. In: Proceedings of ISPRS, pp. 1–4 (2011)

Extraction and Identification of Manipuri and Mizo Texts from Scene and Document Images

Loitongbam Sanayai Meetei(✉), Thoudam Doren Singh, and Sivaji Bandyopadhyay

Department of Computer Science and Engineering, National Institute of Technology Silchar, Silchar, Assam, India
loisanayai@gmail.com, thoudam.doren@gmail.com, sivaji.cse.ju@gmail.com

Abstract. The content inside an image is exceptionally compelling. As such, text within an image can be of special interest and compared to other semantic contents, it tends to be effectively extracted. Text detection within an image is the task of detecting and localizing the portion of an image that contains the text information. Manipuri and Mizo are respectively the lingua francas of two neighboring northeastern states of Manipur and Mizoram in India. While Manipuri, is currently written using Meetei Mayek script and Bengali script, Mizo is written in Roman script with circumflex accent added to the vowels. In this work, we report the task of text detection in natural scene images and document images in Manipuri and Mizo. We made a comparative study between Maximally Stable Extremal Regions (MSER) coupled with Stroke Width Transform (SWT) and Efficient and Accurate Scene Text Detector (EAST) for the text detection. The detected text portion of both the languages is subjected to Optical Character Recognition (OCR) and a post OCR processing of spelling correction. In our experiment of the text detection, EAST outperformed the other method.

Keywords: Text detection · SWT · MSER · EAST · OCR · Manipuri · Mizo

1 Introduction

Manipuri and Mizo are the official languages of two neighboring northeastern states of India, namely Manipur and Mizoram. Manipuri is considered to be a Tibeto-Burman language though a clear cut boundary is not drawn and Mizo is a Kuki-Chin family language. Further, both Tibeto-Burman and Kuki-Chin languages are considered to be the sub-family of Sino-Tibetan languages. Manipuri is written using Meetei Mayek (also known as Meitei Mayek) along with Bengali script in the field of academics and online news websites. In the case of Mizo,

B. Deka et al. (Eds.): PReMI 2019, LNCS 11941, pp. 405–414, 2019.
https://doi.org/10.1007/978-3-030-34869-4_44

25 letters of the Roman script are used (excluding the letter 'q') along with circumflex accent added to the vowels. Also, a character Ṭ (minuscule: ṭ) is used in the orthography of Mizo which is pronounced as 'tr'. The usage of Manipuri and Mizo can be found in both the states and other neighboring northeastern states like Assam and Tripura. Their usage can also be found in the neighboring countries: Myanmar and Bangladesh. The number of speakers for Manipuri is around 3 million and that of Mizo is around 1 million.

The amount of natural language text available in the digital or electronic format is increasing day by day but mostly unstructured. While the data available in the textual format are somewhat consumable and ready for processing, the text in the image or video frames need further processing to be made ready for use. In whatever form the text appear in any multimedia format, text wants to be found or noticed. The text is displayed in a printed or any multimedia format for readability, meaning the contrast between the text and background is high and the strokes are regular enough that any normal person can detect them. The text can provide a great amount of useful information such as describing the theme of the image or other useful information e.g. name of a location, license plate number, etc. The extracted text information can be used for a variety of applications such as text-based image indexing, keyword-based image search, etc. The overarching goal is to convert the text data appearing in an image into high-quality information.

The type of dataset for performing text detection can be of a wide variety. Some of them are as follows: 1. Document images in the gray-scale format and multicolor format. This includes single or multiple column text from a book or news articles, textbook covers, etc. 2. Images with the caption, the text could be overlaid on the background or inside a frame for better contrast. Such images are mostly found in video frames, newspaper, etc. 3. Scene text, where the text appears on the image captured by an electronic device in an environment. Unlike the text in the document images, text in scene images tends to suffer from variations in skew, perspective, blur, illumination, alignment, etc.

The general steps for text detection from a multimedia document can be grouped into the following four stages: (1) Text detection, where the presence of text content is checked in the input data. If text content is detected the system proceeds to the next step of localization. (2) In text localization step, the area where the text appears in an image is localized. It can be carried out in either texture based or connected component based approach. (3) In the third step, after tracking the text area present in the image, the particular segment is extracted. The extracted portion is then enhanced for better visibility. (4) Finally, the segmented image portion of the text area from step 3 is processed for character recognition. The process is also known as optical character recognition (OCR) where the recognized characters are converted to machine-encoded text.

Detection of text from multimedia document images has a wide variety of applications in different domains. With smartphone devices becoming ubiquitous and playing a vital role in our daily routine, a consumer can use it to fetch the information about a product by sending the image captured from their smartphones to a remote server [16]. It can also be helpful in minimizing laborious work such as building book inventories [3]. The application of text detection in a scene images can provide aid to a visually impaired person in their daily commute [5,8]. Globally, the number of vehicles is increasing steeply. The detection of a vehicle license plate from a natural scene image [2,10,11] can be utilized in monitoring the traffic system, building an automated parking and other law enforcement activities. To the best of our knowledge, there is no report on text detection of Mizo from an image. While the research work of text detection in Manipuri is very recent activity. The main objective of our work is to detect the text written in Manipuri and Mizo in natural scene image and document image. And further, carry out the process of OCR on the segmented portion of the image containing text. The recognized text is again subjected to the post-OCR processing step of spelling correction. The work can be helpful in empowering various research area such as categorization of images, text mining, etc. The remaining of this paper is structured as follows. Section 2 describes the related works, we present our approach model in Sect. 3. The experimental results are discussed in Sect. 4, with Sect. 5 summarizing the conclusion and future work.

2 Related Works

A work on automatically detecting text in complex color images was reported by Zhong et al. [17]. The authors proposed two methods for detecting the text portion: (1) By segmenting the image into connected components with homogeneous color or gray value, and then applying several heuristics, the area which is likely to contain text are identified. This method cannot identify characters with blurred boundaries and touching characters such as cursive printed text. (2) The second method was devised for the horizontally aligned handwritten text by locating the area with high variance after the computation of the local spatial variation in the gray-scale image. A combination of the above two methods was reported to perform better on detecting text on a set of test image from a video frame, CD and book covers.

Epshtein et al. [7] proposed a method called Stroke Width Transform (SWT) for text detection in natural images. The unique feature of text i.e. uniformly thick strokes separates it from other elements in an image. However, the method does not require a letter to have a constant stroke width. After the implementation of Canny edge detection [1] on the input image, it computes the width of each stroke that forms an object in the image. With F-measure of 0.66, the method was reported to perform better than other previously proposed methods while experimenting on the database of ICDAR 2003 and 2005.

A text detection method on natural scene images using neural network was proposed by Zhou et al. [18]. The model consists of two stages, initially, a fully convolutional network is used to predict a word or text-line. Then, the predicted output is fed into a Non-Maximum Suppression to produce the final output. Although the model is reported to get an F-score of 0.78 on the ICDAR 2015 [9] dataset, it has some limitations on its capability to detect long text lines and vertically oriented texts.

A work on Meetei Mayek script text detection is reported by Devi et al. [6]. MSER (Maximally Stable Extremal Regions) [13] features were used to detect the text area appearing in an image, however, because of its poor performance in a blurred image, Edge Enhanced MSER [4] was applied to overcome the drawbacks of MSER. To exclude the non-text element in the output of Edge Enhanced MSER, geometric and SWT filtering was used. The system is reported to achieved F-measure of 0.69.

3 Architecture

We employ two models wherein the first model, we use MSER for feature extraction and used the extracted features for finding the text segment in the form of boxes using SWT. In the second model, EAST [18] is used for text detection. The results obtained from both the model is then fed into the OCR system for text recognition. Figure 1 illustrates the working of our models. The following section describes the approach of our models.

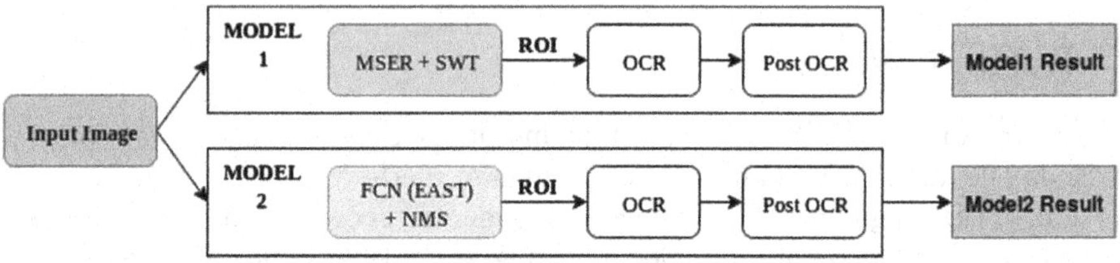

Fig. 1. Framework of our model.

3.1 MSER+ SWT

Maximally Stable Extremal Regions (MSER). One of the blob (a portion of an image with some similar properties) detection method, MSER, proposed by Matas et al. [13], is used as a feature detection technique in our experiment. The MSER extracts a set of co-variant extremal regions from an image by segmenting the image into a number of portions that are maximally stable. This feature tends to identify blobs that segregate themselves from their surroundings.

Stroke Width Transform (SWT). SWT, devised by Epshtein et al. [7], tries to detect strokes inside an image by computing per pixel the width of the most likely stroke containing the pixel. Stroke is defined as a contiguous element of an image with finite width bounded by two roughly parallel edges forming a constant width band. The transformation begins by detecting high contrast edges appearing inside the image. Traversing on each edge pixel in the orthogonal direction, the method attempt to detect a parallel edge that indicates a stroke. Each complete stroke is identified by the connecting adjacent cross-sections with a similar width.

Steps for First Model (Model1): The following steps are carried out in our first model:

1. The input image is first converted into a grayscale image (grayImg).
2. MSER is applied on the grayImg for feature extraction.
3. Apply Canny edge detection in the MSER regions and used it as input to SWT to get the final output: region of interest (ROI) in the form of a box.

3.2 EAST

EAST, an Efficient and Accuracy Scene Text, proposed by Zhou et al. [18] is a text detection technique build using a deep learning model.

Steps for Second Model (Model2): The following steps are carried out in our second model:

1. Resize the input image into a multiple of 32.
2. The resized image is then fed to the pre-trained model of Fully Convolutional Network (FCN) implementation of EAST.
3. Apply non-maxima suppression to suppress the weak and overlapping bounding boxes to get the final ROI.

The ROI obtained are the coordinates of bounding boxes in the form of a rectangle, detected as a region containing the text by the model.

The above two models are implemented using OpenCV (Open Source Computer Vision Library)[1], Python and its libraries. For the implementation of text detection, in Model1, the hyperparameters described in [15] for geometric and SWT elimination is used and for Model2, a pre-trained EAST model[2] is used.

3.3 OCR and Post-processing

OCR allows us to extract the recognized text from a region in the form of a machine-readable Unicode format. After getting the ROI from the above model, further steps of pre-processing are carried as follows:

1. Add padding on the surrounding of ROI by computing the dimensions of ROI.
2. Convert the region obtained from step 1 to grayscale format.
3. The grayscale image is then converted to a binary image format using Otsu's binarization method [14].
4. The output from step 3 is then resized by increasing its dimensions which is the final input to the OCR module.

[1] https://opencv.org/.
[2] https://github.com/argman/EAST.

The OCR is implemented by using an open-source tool: Tesseract v4's LSTM deep learning text recognition algorithm[3]. After getting the list of characters or word(s) from the OCR module, the text obtained is again subjected to a spelling correction process. The spelling correction is implemented using Symspell[4], a language-dependent spell correction module. The post OCR processing, spelling correction module, requires a dictionary of tokens along with its frequency generated from a corpus. After the collection of the corpus, special symbols are removed by tokenizing to build the word frequency dictionary module. For the word frequency dictionary, we have collected the dataset from news articles and a module to generate word frequency dictionary using Java is developed inhouse.

4 Experimental Result and Discussion

The proposed models are employed on a minimum number of input images: natural scene image[5] and images captured from the textbook, containing the text in Manipuri and Mizo separately. To evaluate our model, the ROI containing the Manipuri text and Mizo text on the input images are identified manually for our reference as ground truth. In Model1, we observed that certain non-text areas were also detected as the ROI. The Model2 is able to capture the ROI with better precision than the Model1. Sample outputs for each of the image containing text in Manipuri and Mizo are shown in Fig. 2, natural scene image in Manipuri (NSMn) and Mizo (NSMz), and Fig. 3, document image in Manipuri (DCMn) and Mizo (DCMz). Figures 2a and b shows the ROI detected by Model1 and Model2 in NSMn, while Fig. 2c and d shows the ROI detected by Model1 and Model2 in NSMz respectively. For the document image output, Fig. 3a and b shows the ROI detected by Model1 and Model2 in DCMn, while Fig. 3c and d shows the ROI detected by Model1 and Model2 in DCMz respectively.

From Figs. 2 and 3, it is observed that Model1 detects the image segment that does not contain text as the ROI. While the ROI detected by Model2 is the image segment where the text appears. To evaluate our model, the measurement system in [12] is used to compute the precision(P) and recall(R). F-score is calculated as the harmonic mean of P and R as follows:

$$F - score = \frac{2 \times P \times R}{P + R} \tag{1}$$

Table 1 shows the summarizes evaluation of our two models for text detection in the sample images. In Table 1, M1P, M1R, and M1F represent the precision, recall and F-score of Model1 respectively, and M2P, M2R, and M2F represent the precision, recall and F-score of Model2 respectively. The values in Table 1 are in terms of percentage. From Table 1, it can be observed that Model2 achieve better precision and F-score than the Model1.

[3] https://github.com/tesseract-ocr/tesseract.
[4] https://github.com/wolfgarbe/SymSpell.
[5] Natural scene image source: http://e-pao.net; https://iias.asia/.

(a) NSMn-Model1 (b) NSMn-Model2

(c) NSMz-Model1 (d) NSMz-Model2

Fig. 2. ROI detected by Model1 and Model2 in the sample natural scene images

(a) DCMn-Model1 (b) DCMn-Model2 (c) DCMz-Model1 (d) DCMz-Model2

Fig. 3. ROI detected by Model1 and Model2 in the sample document images

Table 1. Evaluation of Model1 and Model2 on the sample input images

Image Ids	M1P	M1R	M1F	M2P	M2R	M2F
NSMn	38.71	100	**55.8**	100	91.67	**95.6**
NSMz	36.8	100	**53.8**	100	78.5	**87.9**
DCMn	60	100	**75**	75	100	**85.7**
DCMz	35	85	**49**	85	85	**85**

After the completion of the text detection task, the ROI obtained is subjected to the OCR module. To get a better result in OCR, we added padding on the bounding box of ROI, especially on the height of the ROI detected. We performed the OCR for the Meetei Mayek script of Manipuri (NSMn and DCMn) and the Roman scripts of Mizo (NSMz and DCMz). The OCR result using the ROI detected from Model1 contains a lot of noisy character(s) as compared to the one from Model2 because of the detection of the non-text image segment as the final ROI. After the application of OCR, a post OCR processing of spelling correction is carried out. Spelling correction module requires a word frequency dictionary for each of the languages. For building the word frequency dictionary of Manipuri, a dataset of 93000 words with 18500 unique word forms is collected from a local newspaper[6] and for the Mizo, we collected the dataset from the Mizoram local newspaper[7] from April 2013 to February 2019. The Mizo text corpus consists of 21 million words with 191736 unique word forms. As the text in images is less likely to contain any special characters, any such characters present in the OCR result is removed before applying the spelling correction. The OCR result of the ROI detected by the Model2 in Figs. 2 and 3 and the post OCR processing of spelling correction is shown in Table 2. The ground truth of the above sample is given below:

Table 2. OCR and spelling correction result of sample images.

Image Ids	OCR result	Spelling correction result
NSMn	[illegible]	[illegible]
NSMz	;Âm upate zah thiam Mizo. Zemawli MIZORAM PENSIONERS ASSOCIATION : ZARKAWT — T2013	*em* upate zah thiam Mizo *Zemawi* MIZORAM PENSIONERS ASSOCIATION ZARKAWT *2013*
DCMn	[illegible]	[illegible]
DCMz	NAZARET JSUA Thawkrimte Thian Rev. /. R. Lalthanmawia	NAZARET *ISUA* Thawkrimte Thian Rev R Lalthanmawia

1. NSMn: [illegible]
2. NSMz: Aia upate zah thiam Mizo Zemawi
MIZORAM CIVIL PENSIONERS ASSOCIATION ZARKAWT 2013
3. DCMn: [illegible]
4. DCMz: NAZARET ISUA Thawkrimte Ṭhian Rev R Lalthanmawia

The final results obtained in our experiment are close to the ground truth except for the case of document image with text in Meetei Mayek (DCMn).

[6] http://hueiyenlanpao.com/.
[7] https://www.vanglaini.org/.

5 Conclusion and Future Work

In our work, we have created a prototype model for detecting and identifying the text in Manipuri and Mizo. We experimented with two types of models on different images (natural scene and document). The second model, EAST detect the text with better precision in all the cases. The techniques used in both the languages show that the performance of these systems doesn't depend on the language family but it's largely on the script used to represent them. The grapheme complexity of Meetei Mayek font makes it harder to achieve a quality result. The detection of the text for both languages are close in terms of F-score. However, the OCR made a difference in the overall performance. The spelling correction module of both languages is found to be effective in our experiment. In the future, work on enhancing the OCR module for better result and handling of mixed language texts of different scripts can be carried out. Also, increasing the corpus size of both languages will enhance the performance of the spelling correction module.

Acknowledgments. This work is supported by Scheme for Promotion of Academic and Research Collaboration (SPARC) Project Code: P995 of No: SPARC/2018-2019/119/SL(IN) under MHRD, Govt of India.

References

1. Canny, J.: A computational approach to edge detection. IEEE Trans. Pattern Anal. Mach. Intell. PAMI **8**(6), 679–698 (1986)
2. Cano, J., Pérez-Cortés, J.-C.: Vehicle license plate segmentation in natural images. In: Perales, F.J., Campilho, A.J.C., de la Blanca, N.P., Sanfeliu, A. (eds.) IbPRIA 2003. LNCS, vol. 2652, pp. 142–149. Springer, Heidelberg (2003). https://doi.org/10.1007/978-3-540-44871-6_17
3. Chen, D.M., Tsai, S.S., Girod, B., Hsu, C.H., Kim, K.H., Singh, J.P.: Building book inventories using smartphones. In: Proceedings of the 18th ACM international conference on Multimedia, pp. 651–654. ACM (2010)
4. Chen, H., Tsai, S.S., Schroth, G., Chen, D.M., Grzeszczuk, R., Girod, B.: Robust text detection in natural images with edge-enhanced maximally stable extremal regions. In: 2011 18th IEEE International Conference on Image Processing, pp. 2609–2612. IEEE (2011)
5. Chen, X., Yuille, A.L.: Detecting and reading text in natural scenes. In: Proceedings of the 2004 IEEE Computer Society Conference on Computer Vision and Pattern Recognition, 2004, CVPR 2004, vol. 2, pp. II–II. IEEE (2004)
6. Devi, C.N., Devi, H.M., Das, D.: Text detection from natural scene images for manipuri meetei mayek script. In: 2015 IEEE International Conference on Computer Graphics, Vision and Information Security (CGVIS), pp. 248–251. IEEE (2015)
7. Epshtein, B., Ofek, E., Wexler, Y.: Detecting text in natural scenes with stroke width transform. In: 2010 IEEE Computer Society Conference on Computer Vision and Pattern Recognition, pp. 2963–2970. IEEE (2010)

8. Ezaki, N., Bulacu, M., Schomaker, L.: Text detection from natural scene images: towards a system for visually impaired persons. In: Proceedings of the 17th International Conference on Pattern Recognition, 2004, ICPR 2004, vol. 2, pp. 683–686. IEEE (2004)
9. Karatzas, D., et al.: ICDAR 2015 competition on robust reading. In: 2015 13th International Conference on Document Analysis and Recognition (ICDAR), pp. 1156–1160. IEEE (2015)
10. Kim, K.I., Jung, K., Kim, J.H.: Color Texture-based object detection: an application to license plate localization. In: Lee, S.-W., Verri, A. (eds.) SVM 2002. LNCS, vol. 2388, pp. 293–309. Springer, Heidelberg (2002). https://doi.org/10.1007/3-540-45665-1_23
11. Kim, S.K., Kim, D.W., Kim, H.J.: A recognition of vehicle license plate using a genetic algorithm based segmentation. In: Proceedings of 3rd IEEE International Conference on Image Processing, vol. 2, pp. 661–664. IEEE (1996)
12. Lucas, S.M., Panaretos, A., Sosa, L., Tang, A., Wong, S., Young, R.: ICDAR 2003 robust reading competitions. In: Seventh International Conference on Document Analysis and Recognition, 2003, Proceedings, pp. 682–687. Citeseer (2003)
13. Matas, J., Chum, O., Urban, M., Pajdla, T.: Robust wide-baseline stereo from maximally stable extremal regions. Image Vis. Comput. **22**(10), 761–767 (2004)
14. Otsu, N.: A threshold selection method from gray-level histograms. IEEE Trans. Syst. Man Cybern. **9**(1), 62–66 (1979)
15. Özgen, A.C., Fasounaki, M., Ekenel, H.K.: Text detection in natural and computer-generated images. In: 2018 26th Signal Processing and Communications Applications Conference (SIU), pp. 1–4. IEEE (2018)
16. Tsai, S.S., et al.: Mobile product recognition. In: Proceedings of the 18th ACM International Conference on Multimedia, pp. 1587–1590. ACM (2010)
17. Zhong, Y., Karu, K., Jain, A.K.: Locating text in complex color images. Pattern Recogn. **28**(10), 1523–1535 (1995)
18. Zhou, X., et al.: East: an efficient and accurate scene text detector. In: Proceedings of the IEEE Conference on Computer Vision and Pattern Recognition, pp. 5551–5560 (2017)

Fast Geometric Surface Based Segmentation of Point Cloud from Lidar Data

Aritra Mukherjee[1(✉)], Sourya Dipta Das[2], Jasorsi Ghosh[2], Ananda S. Chowdhury[2], and Sanjoy Kumar Saha[1]

[1] Department of Computer Science and Engineering, Jadavpur University, Kolkata, India
kalpurush1601@gmail.com, sks_ju@yahoo.co.in

[2] Department of Electronics and Telecommunication Engineering, Jadavpur University, Kolkata, India
dipta.math@gmail.com, jasorsi13@gmail.com, ananda.chowdhury@gmail.com

Abstract. Mapping the environment has been an important task for robot navigation and Simultaneous Localization And Mapping (SLAM). LIDAR provides a fast and accurate 3D point cloud map of the environment which helps in map building. However, processing millions of points in the point cloud becomes a computationally expensive task. In this paper, a methodology is presented to generate the segmented surfaces in real time and these can be used in modeling the 3D objects. At first an algorithm is proposed for efficient map building from single shot data of spinning Lidar. It is based on fast meshing and subsampling. It exploits the physical design and the working principle of the spinning Lidar sensor. The generated mesh surfaces are then segmented by estimating the normal and considering their homogeneity. The segmented surfaces can be used as proposals for predicting geometrically accurate model of objects in the robots activity environment. The proposed methodology is compared with some popular point cloud segmentation methods to highlight the efficacy in terms of accuracy and speed.

Keywords: Unsupervised surface segmentation · 3D point cloud processing · Lidar data · Meshing

1 Introduction

Mapping of environment in 3D is a sub-task for many robotic applications and is a primary part of Simultaneous Localization And Mapping (SLAM). For 3D structural data sensing, popular sensors are stereo vision, structured light, TOF (Time Of Flight) cameras and Lidar. Stereo vision can extract RGBD data but the accuracy is highly dependent on the presence of textural variance in the scene. Structured light and TOF cameras can extract depth information and

B. Deka et al. (Eds.): PReMI 2019, LNCS 11941, pp. 415–423, 2019.
https://doi.org/10.1007/978-3-030-34869-4_45

RGBD data respectively in an accurate fashion. But these are mostly suitable for indoor uses and in a low range. Lidar is the primary choice for sensing accurate depth in outdoor scenarios over long range. Our focus in this work is on the unsupervised geometric surface segmentation in three dimensions based on Lidar data. We would like to emphasize that models built from such segmentation can be very useful for tasks like autonomous vehicle driving.

According to Nguyen *et al.* [13], the classic approaches in point cloud segmentation can be grouped into edge based methods [3], region based methods [1,9,11,17], attributes based methods [4,6,8,19], model based methods [16], and graph based methods [2,10]. Vo *et al.* [17] proposed a new octree-based region growing method with refinement of segments and Bassier *et al.*. [1] improved it with Conditional Random Field. In [9,11], variants of region growing methods with range image generated from 3D point cloud are reported. Ioannou *et al.* [8] used Difference of Normals (DoN) as a multiscale saliency feature used in a segmentation of unorganized point clouds. Himmelsbach *et al.* [7] treated the point clouds in cylindrical coordinates and used the distribution of the points for line fitting to the point cloud in segments. Ground surface was recognized by thresholding the slope of those line segments. In an attempt to recognize the ground surface, Moosmann *et al.* [12] built an undirected graph and characterized slope changes by mutual comparison of local plane normals. Zermas *et al.* [18] presented a fast instance level LIDAR Point cloud segmentation algorithm which consisting of deterministic iterative multiple plane fitting technique for the fast extraction of the ground points, followed by a point cloud clustering methodology named Scan Line Run (SLR). On the other hand, in supervised methods, PointNet [14] takes the sliding window approach to segment large clouds with the assumption that a small window can express the contextual information of the segment that it belongs to. Landrieu *et al.* [10] introduced superpoint graphs, a novel point cloud representation with rich edge features encoding the contextual relationship between object parts in 3D point clouds. This is followed by deep learning on large-scale point clouds without major sacrifice in fine details.

In most of the works, neighbourhood of a point used to estimate normal is determined by tree based search. This search is time consuming and the resulting accuracy is also limited for sparse point cloud as provided by a Lidar in robotic application. This observation acts as the motivation for us to focus on developing a fast mesh generation procedure that will provide the near accurate normal in sparse point cloud. Moreover, processing Lidar data in real-time requires considerable computational resources limiting its deployability on small outdoor robots. Hence a fast method is also in demand.

2 Proposed Methodology

The overall process consists of three steps as shown in Fig. 1. First, the point cloud is sensed by the Lidar, subsampled and the mesh is created simultaneously. Second, surface normal is calculated for node points using the mesh. Finally, surface segmentation is done by labelling the points on the basis of spatial continuity of normal with a smooth distribution.

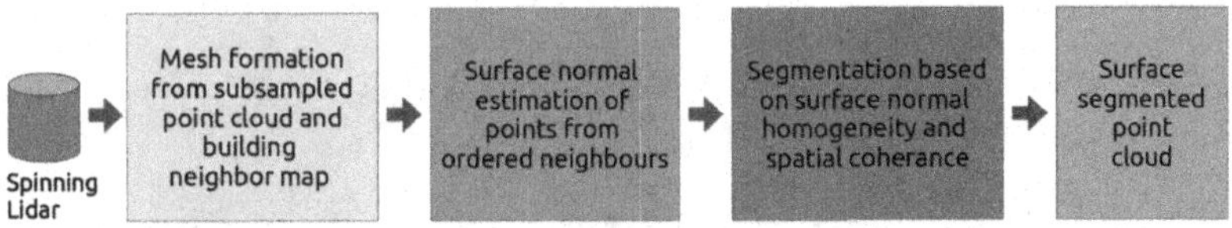

Fig. 1. Block diagram of the entire system, the first stage can be merged with Lidar scanning by improvising the Lidar firmware

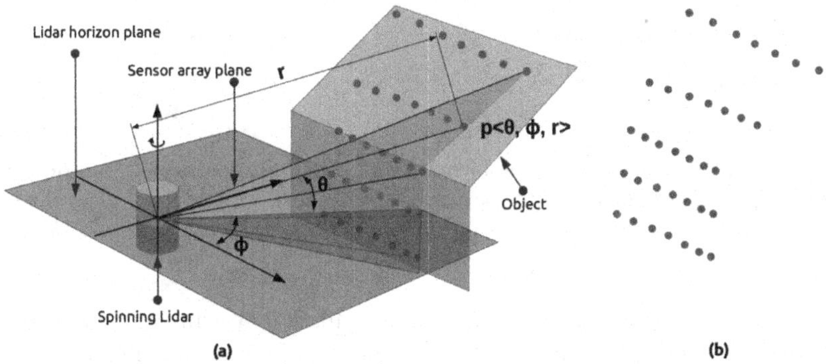

Fig. 2. (a) A schematic showing the formation of point cloud by Lidar and (b) the resultant point cloud

The proposed methodology segments surface from point cloud obtained by spinning Lidars only. Spinning Lidars work on the principle of spinning a vertical array of divergent laser distance sensors and thus extracts point cloud in spherical coordinates. The point cloud consists of a set of coordinates $P = \{p(\theta, \phi, r)\}$ where θ is the fixed vertical angle of a sensor from the plane perpendicular to the spinning axis, ϕ is the variable horizontal angle due to spinning of the array and r is the distance measured by the laser sensor. This form of representation is exploited by our methodology to structure the data in an ordered form the only caveat being running it for a single spin. By varying the factor of sub-sampling of ϕ, the horizontal density of the point cloud can be varied. Figure 2 shows the operational procedure of a spinning Lidar and the resultant point cloud for an object with multiple surfaces. Please note that not every point during the sweep is considered for mesh construction as noisy points too close to each other horizontally produces erroneous normal. Sub-sampling is done to rectify this error by skipping points uniformly during the spin.

Mesh Construction: The significant novelty of this work is the fast mesh generation process that enables quick realization of subsequent steps. The mesh construction is a simultaneous process during the data acquisition stage. The connections are done during sampling in the following manner. Let a point be denoted as $p(\theta, \phi, r)$. Let the range of θ be $[\theta_0, \theta_n]$ which corresponds to $n + 1$ vertical sensors in the array; and the range of ϕ be $[\phi_0, \phi_m]$ where $m + 1$ is the

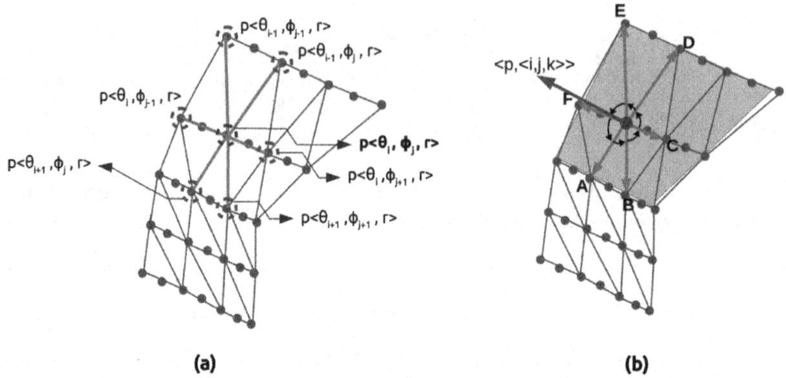

Fig. 3. (a) A schematic showing the formation of mesh on subsampled (factor 2) cloud with the neighbour definition of a point and (b) the normal formation from the neighbours

number of times the sensor array is sampled uniformly during a single spin. The corresponding distance of the point is r from the Lidar sensor. Let the topmost sensor in the array corresponds to angle θ_0 and the count proceeds from top to bottom the vertically spinning array. Further, let first shot during the spin corresponds to ϕ_0. The mesh is constructed by joining neighbouring points in the following manner: $p(\theta_i, \phi_j, r)$ is joined with $p(\theta_{i+1}, \phi_j, r)$, $p(\theta_{i+1}, \phi_{j+1}, r)$ and $p(\theta_i, \phi_{j+1}, r)$ for all points within range of $[\theta_0, \theta_{n-1}]$ and $[\phi_0, \phi_{m-1}]$. The points from the last vertical sensor, *i.e.*, corresponding to θ_n, are joined with the immediate horizontal neighbour. Thus, $p(\theta_n, \phi_j, r)$ is joined with $p(\theta_n, \phi_{j+1}, r)$ where j varies from 0 to $m-1$. For points corresponding to ϕ_m, $p(\theta_i, \phi_m, r)$ is joined with $p(\theta_i, \phi_0, r)$, $p(\theta_{i+1}, \phi_0, r)$ and $p(\theta_{i+1}, \phi_m, r)$. The point $p(\theta_n, \phi_m, r)$ is joined with $p(\theta_n, \phi_0, r)$ to create the whole cylindrical connected mesh. For all pairs the joining is done if both the points have an r that is within range of the Lidar. If all the neighbours of a point is present then six of them are connected by the meshing technique instead of all eight. This is done to ensure there are no overlapping surface triangles. Figure 3(a) shows the connectivity a point which have six valid neighbours on the mesh. The mesh is stored in a map of vectors $M = \{< p, v >| p \in P, v = \{q_n \mid q_n \in P, n \leq 6, \;\; and \;\; q_n \;\; is \;\; a \;\; neighbour \;\; of \;\; p\}\}$ where each point is mapped to the vector v containing its existent neighbors in an ordered fashion. The computational complexity of the meshing stage in $O(n_{sp})$ where n_{sp} is the number of sub-sampled points. As the meshing is performed on the fly with the sensor spinning the absolute time depends on the angular frequency and sub-sampling factor.

Normal Estimation: The structured mesh created in the previous step helps toward a fast computation of normal at a point. A point forms vectors with its neighbour. Pair of vectors are taken in an ordered fashion. Normal is estimated for that point by averaging the resultant vectors formed by cross multiplication of those pairs. The ordering is performed during the mesh construction stage only. For a point $p(\theta_i, \phi_j, r)$, vectors are formed with the existing neighbours as stored in M in an anti-clockwise fashion. A normal can be estimated for p if its corresponding v has $|v| \geq 2$. From the neighbour vector v of p obtained from M, let $p(\theta_i, \phi_j, r)$ forms $\boldsymbol{A}$ by joining with $p(\theta_{i+1}, \phi_j, r)$, $\boldsymbol{B}$ by joining with $p(\theta_{i+1}, \phi_{j+1}, r)$, and so on until it forms $\boldsymbol{F}$ by joining with $p(\theta_i, \phi_{j-1}, r)$. Then cross- multiplication of existing consecutive vectors is performed. In general, if every point exists, then $\boldsymbol{A} \times \boldsymbol{B}$, $\boldsymbol{B} \times \boldsymbol{C}$ etc. are computed ending with $\boldsymbol{F} \times \boldsymbol{A}$ to complete the circle. This arrangement is illustrated in Fig. 3(b). The normal is estimated by averaging all the $\hat{i}, \hat{j}, \hat{k}$ components of the resultant vectors individually. The normals are stored in the map $N = \{< p, n >|\ p \in P, n = \{\hat{i_p}, \hat{j_p}, \hat{k_p}\}\}$. Due to the inherent nature of the meshing technique, sometime points from disconnected objects get connected to the mesh. To mitigate this effect of a different surface contributing to the normal estimation, weighted average is used. The weight of a vector formed by cross multiplication of an ordered pair is inversely proportional to the sum of lengths of the vectors in the pair.

Algorithm 1. Surface segmentation on normal distribution

$L = \{< p, l >|\ p \in P, l = 0\}$;
label $\leftarrow 1$, *stack* $S \leftarrow \{\}$;
for *each* $p \in M$ **do**
  if $l = 0$ *for* p *in* L **then**
    $L \leftarrow< p, l \leftarrow label >$;
    S.push(p);
    while $S \neq \{\}$ **do**
      $p \leftarrow$ S.pop();
      for *each* q *in* v *corresponding to* $p \in M$ **do**
        if $l = 0$ *for* q *in* L **then**
          get $\hat{i_q}, \hat{j_q}, \hat{k_q}$ for q from N;
          if $|\hat{i_p} - \hat{i_q}| < I$, $|\hat{j_p} - \hat{j_q}| < J$, $|\hat{k_p} - \hat{k_q}| < K$ **then**
            $L \leftarrow< q, l \leftarrow label >$;
            S.push(q);
          end
        end
      end
    end
  else
    search next p in M
  end
end
Result: L

Segmentation by Surface Homogeneity: Based on the normal at a point, as computed in the previous step, we now propagate the surface label. A label map $L = \{< p, l >|\ p \in P, l = 0\}$ is used for this purpose. This label map stores the label of each point p by assigning a label l. If for any point p, its $l = 0$ denotes the point is yet to be labelled. The criteria of assigning the label of p to its neighbour q depends on the absolute difference of their normal components. Three thresholds I, J, K are empirically set depending on the type of environment. Segment labelling is propagated by a depth first search approach as described in algorithm 1. Two neighbouring points will have same label provided the absolute difference of corresponding components of their normals are within component-wise threshold. Computations of normals and mesh, as discussed earlier, generates the normal map N and the mesh M respectively. Subsequently algorithm 1 uses N and M to label the whole sub-sampled point cloud in an inductive fashion. Due to sub-sampling, all points in P will not get a label. This issue is resolved by assigning the label of its nearest labelled point along the horizontal sweep. An optional post-processing may be arranged by eliminating segments with too few points.

3 Experiments Results and Analysis

The proposed methodology is implemented with C++ on a linux based system with DDR4 8 GB RAM, 7th generation i5 Intel Processor. Experiments were performed on a synthetic dataset which simulates Lidar point clouds. The methodology is compared with the standard region growing algorithm used in point cloud library [15] and a region growing algorithm combined with merging for organized point cloud data [19] and it is observed that it excels in terms of both speed and accuracy.

We have used the software "Blensor" [5] to simulate the output of the Velodyne 32E Lidar in a designed environment. Blensor can roughly model the scenery with the desired sensor parameters. BlenSor also provides an exact measurement ground truth annotated over the model rather than the point cloud. We have included primitive 3D model structures such as cylinder, sphere, cube, cone and combined objects placed in different physically possible orientations in the scene to simulate the real environment. Different percentage of occlusion and crowding levels is included in the model environment to test out the property of scene complexity independence, of the proposed solution. We have a total of 32 different environments with increasing order of different types of objects, occlusion, complexity of geometry and pose, and crowding levels. We have used Gaussian noise as our noise model for Lidar with zero mean and variance of 0.01. With Velodyne 32E Lidar the total number of sensors in the spinning array is 32 and a resolution of 0.2 degree is set for horizontal sweep resulting in 1800 sensor firing in one spin. Thus for our dataset the range of θ and ϕ are $[\theta_0, \theta_{31}]$ and $[\phi_0, \phi_{1799}]$. The output at different stages of our methodology is shown in Fig. 4.

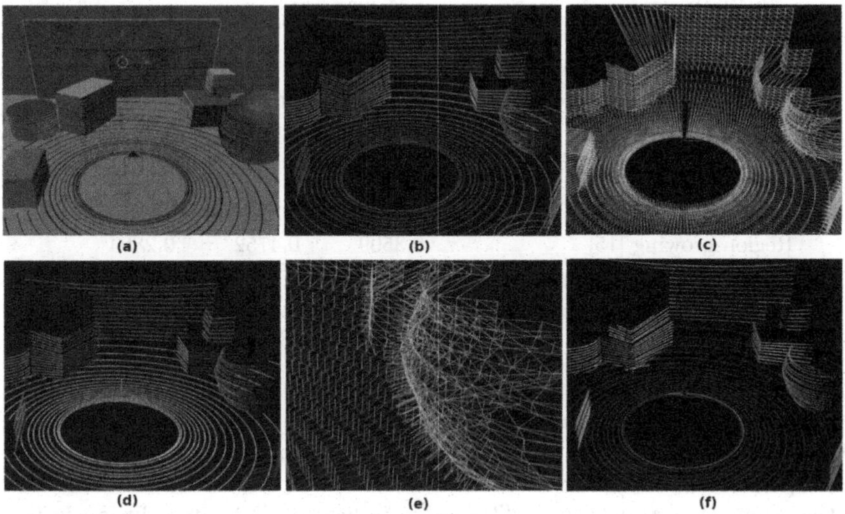

Fig. 4. (a) Synthetic scene with scanned point cloud overlayed (b) Point cloud with distance color coded from blue(least) to red(highest) (c) Mesh and normals with subsampling factor of 5 (d) Point cloud segment surface ground truth (e) A detailed look at the mesh and normals (f) Segmented point cloud by proposed methodology (Color figure online)

Table 1. Comparison of execution times (all units in milliseconds) of different competing methods.

Method	Sampling interval	Average time (in ms)	Max time (in ms)	Min time (in ms)
Proposed method	5	54.33	63	41
	10	35.06	48	28
	15	25.53	32	20
Region growing [15]		275.13	1507	134
Region growing with merging [19]		368.46	1691	129

Performance of Proposed Methodology: The input scene and point cloud along with the output at different stages are given in Fig. 4. It should be mentioned that the different colors of the mesh are due to separation of ground plane on the basis of normal and is rendered for better visualization only. Performance is evaluated using the precision-recall and f1 score metric. As the segments may lack semantic labels, edge based comparisons were performed with overlapping of dilated edge points with ground-truths. We vary the sampling interval from 5 to 15 in steps of 5, as a sampling interval of 5 corresponds to 1 degree of sweep of the Lidar. The thresholds for checking normal homogeneity are kept at $<\hat{i},\hat{j},\hat{k}> = <0.2, 0.2, 0.2>$. These values are determined empirically for scenes

Table 2. Comparison of accuracy of different competing methods

Method	Sampling interval	Average F1 score	Average precision	Average recall
Proposed method	5	0.7406	0.7616	0.7224
	10	0.7147	0.7330	0.6998
	15	0.6910	0.7190	0.6673
Region growing [15]		0.3509	0.4752	0.2804
Region growing with merging [19]		0.3614	0.3849	0.3430

with standard objects on a flat surface. Our methodology is compared with [15] and [19]. The tuning parameters of the methods are kept at default settings. Table 1 shows the execution times for different methods and clearly reveals that the proposed method is much faster. In Table 2, we present the accuracy values for the competing approaches. This table indicates that our solution is much more accurate. Overall, the results demonstrate that we are successful in providing a fast yet accurate solution to this complex problem.

4 Conclusion

In this work we have presented an unsupervised surface segmentation algorithm which is fast, accurate and robust to noise, occlusion and different orientations of the surface with respect to the Lidar. This work serves as the first step for mapping environments with geometric primitive modelling in SLAM applications for unmanned ground vehicles. In future, supervised classifier can be utilized for segment formation on data collected by Lidar on a real environment. Thereafter, the surface segments will enable the model generation of 3D objects.

References

1. Bassier, M., Bonduel, M., Van Genechten, B., Vergauwen, M.: Segmentation of large unstructured point clouds using octree-based region growing and conditional random fields. Int. Arch. Photogram. Rem. Sens. Spat. Inf. Sci. **42**(2W8), 25–30 (2017)
2. Ben-Shabat, Y., Avraham, T., Lindenbaum, M., Fischer, A.: Graph based over-segmentation methods for 3d point clouds. Comput. Vis. Image Underst. **174**, 12–23 (2018)
3. Bhanu, B., Lee, S., Ho, C.C., Henderson, T.: Range data processing: representation of surfaces by edges. In: Proceedings of the Eighth International Conference on Pattern Recognition, pp. 236–238. IEEE Computer Society Press (1986)
4. Feng, C., Taguchi, Y., Kamat, V.R.: Fast plane extraction in organized point clouds using agglomerative hierarchical clustering. In: 2014 IEEE International Conference on Robotics and Automation (ICRA), pp. 6218–6225. IEEE (2014)

5. Gschwandtner, M., Kwitt, R., Uhl, A., Pree, W.: BlenSor: blender sensor simulation toolbox. In: Bebis, G., et al. (eds.) ISVC 2011. LNCS, vol. 6939, pp. 199–208. Springer, Heidelberg (2011). https://doi.org/10.1007/978-3-642-24031-7_20
6. Hackel, T., Wegner, J.D., Schindler, K.: Fast semantic segmentation of 3d point clouds with strongly varying density. ISPRS Ann. Photogram. Rem. Sens. Spat. Inf. Sci. **3**(3), 177–184 (2016)
7. Himmelsbach, M., Hundelshausen, F.V., Wuensche, H.J.: Fast segmentation of 3d point clouds for ground vehicles. In: 2010 IEEE Intelligent Vehicles Symposium, pp. 560–565. IEEE (2010)
8. Ioannou, Y., Taati, B., Harrap, R., Greenspan, M.: Difference of normals as a multi-scale operator in unorganized point clouds. In: 2012 Second International Conference on 3D Imaging, Modeling, Processing, Visualization & Transmission, pp. 501–508. IEEE (2012)
9. Jiang, X.Y., Meier, U., Bunke, H.: Fast range image segmentation using high-level segmentation primitives. In: Proceedings Third IEEE Workshop on Applications of Computer Vision. WACV 1996, pp. 83–88. IEEE (1996)
10. Landrieu, L., Simonovsky, M.: Large-scale point cloud semantic segmentation with superpoint graphs. In: Proceedings of the IEEE Conference on Computer Vision and Pattern Recognition, pp. 4558–4567 (2018)
11. Li, M., Yin, D.: A fast segmentation method of sparse point clouds. In: 2017 29th Chinese Control and Decision Conference (CCDC), pp. 3561–3565. IEEE (2017)
12. Moosmann, F., Pink, O., Stiller, C.: Segmentation of 3d lidar data in non-flat urban environments using a local convexity criterion. In: 2009 IEEE Intelligent Vehicles Symposium, pp. 215–220. IEEE (2009)
13. Nguyen, A., Le, B.: 3d point cloud segmentation: a survey. In: 2013 6th IEEE Conference on Robotics, Automation and Mechatronics (RAM), pp. 225–230. IEEE (2013)
14. Qi, C.R., Su, H., Mo, K., Guibas, L.J.: Pointnet: deep learning on point sets for 3d classification and segmentation. In: Proceedings of the IEEE Conference on Computer Vision and Pattern Recognition, pp. 652–660 (2017)
15. Rusu, R.B., Cousins, S.: 3d is here: point cloud library (pcl). In: 2011 IEEE International Conference on Robotics and Automation, pp. 1–4. IEEE (2011)
16. Tarsha-Kurdi, F., Landes, T., Grussenmeyer, P.: Hough-transform and extended ransac algorithms for automatic detection of 3d building roof planes from lidar data. In: ISPRS Workshop on Laser Scanning 2007 and SilviLaser 2007, vol. 36, pp. 407–412 (2007)
17. Vo, A.V., Truong-Hong, L., Laefer, D.F., Bertolotto, M.: Octree-based region growing for point cloud segmentation. ISPRS J. Photogram. Rem. Sens. **104**, 88–100 (2015)
18. Zermas, D., Izzat, I., Papanikolopoulos, N.: Fast segmentation of 3d point clouds: a paradigm on lidar data for autonomous vehicle applications. In: 2017 IEEE International Conference on Robotics and Automation (ICRA), pp. 5067–5073. IEEE (2017)
19. Zhan, Q., Liang, Y., Xiao, Y.: Color-based segmentation of point clouds. Laser Scan. **38**(3), 155–161 (2009)

Class Similarity Based Orthogonal Neighborhood Preserving Projections for Image Recognition

Purvi A. Koringa(✉) and Suman K. Mitra

Dhirubhai Ambani Institute of Information and Communication Technology, Gandhinagar 382007, India
201321010@daiict.ac.in, suman_mitra@daiict.ac.in

Abstract. Recent Dimensionality reduction methods like Locality Preserving Projections and Neighborhood Preserving Projections learn local neighborhood characteristics and try to preserve these characteristics in the lower dimensional space. In supervised settings, conventional Orthogonal Neighborhood Preserving Projections (ONPP) uses knowledge of class label to identify the neighbors of data points. When data points are closely placed or the classes are overlapping, such hard decision rule may not help finding a good low dimensional representation. To overcome this limitation, we are proposing a novel neighborhood selection rule, where low dimensional representation is used with Logistic Regression to find the probability of every data point to belong into a particular class. Based on these probabilities a new distance measure - Class Similarity based distance is used to define a local neighborhood of data points. It is observed that class similarity based ONPP very well represents the relationship of neighbors in low dimensions. The proposed scheme is used to recognize face images and handwritten numerals images. The proposed Class Similarity based neighborhood scheme achieves same recognition performance with significantly less number of subspace dimensions.

Keywords: Dimensionality reduction · Image recognition · Subspace learning

1 Introduction

In recent years, the visual information surge has led to the widespread application of image recognition or identification in the field of pattern recognition where a test image is to be identified based on the samples in the training data. Many of the recognition algorithms are appearance based where an image is considered as a large array of intensity values. While working with these methods, an image of $m \times n$ size is represented as an mn-dimensional vector which demands high computation and often results in slower computation, also known as *curse of dimensionality.* Though images are represented as high dimensional data, natural images have only limited degrees of freedom thus often lie in comparatively very

B. Deka et al. (Eds.): PReMI 2019, LNCS 11941, pp. 424–432, 2019.
https://doi.org/10.1007/978-3-030-34869-4_46

low dimensional linear or non-linear manifold. This led to the development of Dimensionality Reduction (DR) techniques that reduce redundant information present in higher dimensional space and remove curse of dimensionality.

The elementary task of DR techniques is to search for a linear or non-linear mapping of the data from a high dimensional space to a lower dimensional subspace which preserves some specific information of high dimensional space that leads to favorable outcomes for recognition while reducing the computational load. Some of the most used linear DR techniques are Principal Component Analysis (PCA) [15] and Linear Discriminant Analysis (LDA) [8]. These methods search low dimensional representation of data assuming it lies on a linear manifold but this assumption is usually not true in complex data such as images. To overcome this shortcoming, nonlinear DR methods were developed to search underlying low dimensional manifold. Some of the well-known nonlinear DR are Locally Linear Embeddings (LLE) [12] and Laplacian Eigenmaps (LE) [1]. These methods suffer from the limitation of embedding out-of-sample data because of the absence of explicit mapping from high dimensional space to low dimensional subspace. To overcome it, linear extensions of LLE is proposed in Orthogonal Neighborhood Preserving Projection (ONPP) [4].

The neighborhood selection and the distance used to define it have the paramount effect on the learned manifold. Most DR methods use Euclidean distance to define neighborhood of a data point, which generally do not match the classification properties. It is possible that data points from different classes may be considered neighbors based on their small Euclidean distance, on the other hand, data points from same class may not be considered neighbors because of large Euclidean distance. One approach to handle this problem is incorporating class information in neighborhood selection, data points from the same class are considered neighbors of each other. In image recognition, where the knowledge of the class label is available, neighbors are selected based on their class label only. These methods are known as supervised methods but when data points from different classes are overlapping or closely placed, such hard neighborhood selection rule does not capture the data geometry efficiently.

Various approaches to define this neighborhood is attempted in the past. It is proved in [7] that combining different distance matrices or dissimilarity representations can often increase the performance of individual ones. In [11], authors have proposed Supervised LLE (SLLE) that uses knowledge of class label to modify the pair-wise distances that define neighborhood. It blindly adds a constant to the Euclidean distance of the data points belonging to different classes. In [10], k-means clustering based approach is proposed to find neighbors of a data point which is unsupervised approach that does not consider class label knowledge to define the neighborhood. Work documented in [16] proposes an adaptive neighborhood of varying size based on the local linearity.

In this article we are addressing the problem with neighborhood selection based on either class label or a distance measure only. The main contribution of this paper is the computation of class similarity based distance between two data points and a new neighborhood selection rule based on this class similarity. The proposed neighborhood selection rule merges the effect of class knowledge and Euclidean distance. The new neighborhood selection rule is used in conventional ONPP for image recognition. We report the results of recognition experiments on two kinds of data: Face image data with very high dimensions, and handwritten numerals data with a large number of samples. The proposed Class Similarity based ONPP outperforms the ONPP with significantly less number of subspace dimensions. The paper also compares the recognition performance of a Modified variant of ONPP, namely MONPP [5] with proposed neighborhood rule.

The article is organized as follows, Sect. 2 explains ONPP in detail, Sect. 3 proposes modified distance based on Class Similarity. Section 4 documents recognition experiments with conventional ONPP approach and Class Similarity based ONPP approach. Section 5 concludes the proposed work.

2 Orthogonal Neighborhood Preserving Projections

ONPP achieves lower dimensional subspace in three steps. First, a local neighborhood of each data point is defined, each data point is then expressed as a linear combination of its neighbors. The third step seeks subspace bases that preserve this linear relationship among neighbors in the lower dimensional representation of data through a minimization problem. In applications involving images, Let $\mathbf{X} = [\mathbf{x_1}, \mathbf{x_2}, ...\mathbf{x_N}]$ be the data matrix of vectorized images so that an image $\mathbf{x_i}$ constitutes a point in mn dimensional space. The goal of subspace based DR methods is to find an orthogonal/non-orthogonal subspace bases $\mathbf{V} \in \mathcal{R}^{mn \times d}$ such that the low dimensional representation $\mathbf{Y} = \mathbf{V}^T\mathbf{X}$.

Step 1: Finding Neighborhood: For each data point $\mathbf{x_i}$, a local neighborhood is defined using simple clustering methods such as *k-Nearest Neighbor (k-NN)* or ε - *Neighbors (ε-NN)*. Let $\mathcal{N}_{x_i}$ be the set of k neighbors of $\mathbf{x_i}$.

Step 2: Calculating Reconstruction Weights: ONPP assumes that the neighborhood lies on a locally linear manifold, thus each data point can be expressed as a linear combination of its neighbors. For $\mathbf{x_i}$, the linear combination can be denoted as $\sum_{j=1}^{k} w_{ij}\mathbf{x_j}$ where, $\mathbf{x_j} \in \mathcal{N}_{x_i}$. The linear weights w_{ij} for each $\mathbf{x_i}$ can be computed by posing the problem as minimization of the reconstruction error

$$\arg\min_{\mathbf{W}} \sum_{i=1}^{N} \| \mathbf{x_i} - \sum_{j=1}^{k} w_{ij}\mathbf{x_j} \|^2 \qquad \text{s.t.} \sum_{j=1}^{k} w_{ij} = 1 \tag{1}$$

An improved variant of ONPP is proposed as Modified Orthogonal Neighborhood Preserving Projections (MONPP) in [5]. MONPP stresses on the fact that for larger neighborhoods, the local linearity assumption may not hold true and the neighborhood assumed to be a linear patch may have some inherent non-linearity. To take this non-linearity into account, MONPP uses nonlinear weights in place of linear weights to improve recognition performance. In this article, ONPP as well as MONPP is used to compare the performance of proposed Class Similarity based neighborhood selection rule.

Step 3: Finding Subspace: This step is dimensionality reduction by finding the bases $\mathbf{V} \in \mathcal{R}^{mn \times d}$ of low dimensional subspace which preserves the linear relationship of each $\mathbf{x_i}$ with its neighbors $\mathbf{x_j}$ with corresponding reconstruction weights w_{ij} in each projection $\mathbf{y_i}$ and its neighbors $\mathbf{y_j}$ with same weights w_{ij}. Such embedding is obtained by minimizing the reconstruction errors in the subspace. Hence, the minimization problem is defined as

$$\arg\min_{\mathbf{Y}} \sum_{i=1}^{N} \| \mathbf{y_i} - \sum_{j=1}^{k} w_{ij}\mathbf{y_j} \|^2 \qquad \text{s.t. } \mathbf{V}^T\mathbf{V} = \mathbf{I} \tag{2}$$

The bases $\mathbf{V}$ turns out to be the eigen-vectors of $\mathbf{X}(\mathbf{I}-\mathbf{W})(\mathbf{I}-\mathbf{W})^T\mathbf{X}^T$ corresponding to the smallest d eigen-values ($d \ll mn$). For recognition tasks, an out-of-sample data point $\mathbf{x_l}$ can now be projected to subspace as $\mathbf{y_l} = \mathbf{V}^T\mathbf{x_l}$.

DR methods can be implemented in supervised mode when class label knowledge is available. It is proved in [4] that incorporating class information in neighborhood selection improves recognition, but it is not always a good idea to ignore Euclidean distance entirely. Moreover, in tasks like image recognition, the number of samples N available to learn a projection space is less than dimension mn (known as small-sample size problem). To overcome this limitation all dimensionality reduction algorithms apply PCA as preprocessing to achieve an intermediate low dimensional space to learn bases. Based on these two observations, in the next section, we propose a new class similarity based distance that uses preprocessed data to define a new neighborhood selection rule.

3 Class Similarity Based ONPP (CS-ONPP)

In conventional ONPP, the neighbors of data point $\mathbf{x_i}$ are selected based on euclidean distance (in unsupervised settings) or based on class label information (in supervised settings). To incorporate underlying similarity between data points along with their class label many works have been done, in [17] authors have proposed an Enhanced Supervised LLE (ESLLE) where the Euclidean distance is simply modified by adding a constant increment for the pairs of data that belongs to different class, keeping the distance of intra-class data point pairs unchanged. The scheme does not consider similarity between intra-class data or inter-class data in any way. When data points are very similar, they are closely placed in the high dimensional space and the classes are overlapping. In such cases, hard decision rule based on class label may not help finding a good low dimensional representation. To overcome this limitation of neighborhood finding rule, we are proposing a novel neighborhood rule inspired by [18].

Instead of claiming a data point $\mathbf{x_i}$ belonging to an unique class and modifying distance accordingly, let us define a C-dimensional class probability vector $\mathbf{p}(\mathbf{x_i}) = [p_1(\mathbf{x_i}), p_2(\mathbf{x_i}), ..., p_C(\mathbf{x_i})]^T$. Here, C is number of classes in data. The cth element of the vector $p_c(\mathbf{x_i})$ represents the probability of a data point $\mathbf{x_i}$ belonging to c^{th} class.

For given data matrix $\mathbf{X}$ with known class label, the probability of each data point $\mathbf{x_i}$ belonging to class c can be computed using Logistic Regression (LR).

The LR assumes that the logit of the probability $\pi(\mathbf{x_i})$ is a linear combination of features of $\mathbf{x_i}$, that can be given by

$$\log\left(\frac{\pi(\mathbf{x_i})}{1-\pi(\mathbf{x_i})}\right) = \alpha + \beta^T \mathbf{x_i}$$

Specifically for a class c,

$$\pi(\mathbf{x_i};\alpha_c,\beta_c) = \frac{\exp(\alpha_c + {\beta_c}^T \mathbf{x_i})}{1+\exp(\alpha_c + {\beta_c}^T \mathbf{x_i})}, \quad c = 1, ..., C. \tag{3}$$

Where $\alpha_c \in \mathcal{R}$ and $\beta_c \in \mathcal{R}^{mn}$ are parameters for class c learned on training data with class knowledge using maximum likelihood estimation.

Performing LR on high dimensional data causes huge computational burden, thus we take advantage of pre-processing performed in ONPP and use lower dimensional representation sought using PCA to find these class probabilities for each data point $\mathbf{x_i}$. Let $\mathbf{z_i}$ be a lower dimensional PCA representation of $\mathbf{x_i}$ to find probability for class c. The Eq. (3) becomes

$$\pi(\mathbf{x_i}) = \pi(\mathbf{z_i};\alpha_c,\beta_c) = \frac{\exp(\alpha_c + {\beta_c}^T \mathbf{z_i})}{1+\exp(\alpha_c + {\beta_c}^T \mathbf{z_i})} \tag{4}$$

To find probability vector $\mathbf{p}(\mathbf{x_i})$, each entry is the probability $p_c(\mathbf{x_i})$ for class c can be computed by

$$p_c(\mathbf{x_i}) = \frac{\pi(\mathbf{z_i};\alpha_c,\beta_c)}{\sum_{c=1}^{C} \pi(\mathbf{z_i};\alpha_c,\beta_c)} \tag{5}$$

Note that, PCA representation $[\mathbf{z_1}, \mathbf{z_2}, ...\mathbf{z_N}]$ carries class information form corresponding data points $[\mathbf{x_1}, \mathbf{x_2}, ...\mathbf{x_N}]$. For a pair of data points $\mathbf{x_i}$ and $\mathbf{x_j}$,class-similarity $\mathcal{S}(i,j)$ is proposed as (6) to define a new distance measure Δ',

$$\mathcal{S}(i,j) = \begin{cases} 1; & \mathbf{x_i} = \mathbf{x_j} \\ \mathbf{p}(\mathbf{x_i})^T \mathbf{p}(\mathbf{x_j}); & \mathbf{x_i} \neq \mathbf{x_j} \end{cases} \tag{6}$$

$$\Delta'(\mathbf{x_i}, \mathbf{x_j}) = \|\mathbf{x_i} - \mathbf{x_j}\| + \alpha \max(\Delta)(1 - \mathcal{S}(i,j)) \tag{7}$$

The new distance formula modifies Euclidean distance based on the similarity value $\mathcal{S}(i,j)$ of two data points. Similarity of a data point with itself is defined as 1 to make the distance of the data point with itself 0. For two distinct data points $\mathbf{x_i}$ and $\mathbf{x_j}$, the similarity is defined as an inner product of class probability vectors. If $\mathbf{x_i}$ and $\mathbf{x_j}$ belong to different classes, the inner product of class probability vectors is expected to be smaller, thus a smaller similarity $\mathcal{S}(i,j)$ resulting into larger $\Delta'(\mathbf{x_i}, \mathbf{x_j})$ compared to Euclidean distance.

Based on this new distance $\Delta'(\mathbf{x_i}, \mathbf{x_j})$, neighbors for data point $\mathbf{x_i}$ will be selected, which incorporates class information as well as similarity among neighbors. The rest of the method of finding subspace is similar to ONPP. Table 1 gives algorithm to find Class-Similarity based ONPP subspace.

Table 1. Class-Similarity based ONPP Algorithm

Input: Dataset $\mathbf{X} \in \mathcal{R}^{mn \times N}$ and subspace dimensions d
Output: Subspace bases $\mathbf{V} \in \mathcal{R}^{mn \times d}$, Lower dimension projections $\mathbf{Y} \in \mathcal{R}^{d \times N}$
1: Find low dimensional representation $\mathbf{z_i}$ of data point $\mathbf{x_i}$ by projecting on d_{pca} dimensional space using PCA ($\mathbf{z_i} = V_{pca}^T \mathbf{x_i}$)
2: Use Logistic Regression on $\mathbf{z_i}$ to find class probability vector $\mathbf{p_i}(\mathbf{x_i})$ corresponding to data point $\mathbf{x_i}$
3: Calculate modified distance for all data point pairs $\Delta'(\mathbf{x_i}, \mathbf{x_j})$ using Eq. (7)
4: Search NN $\mathcal{N}_{\mathbf{x_i}}$ with modified distance $\Delta'(\mathbf{x_i}, \mathbf{x_j})$
5: Compute the weight W for each neighbor data point $\mathbf{x_j} \in \mathcal{N}_{\mathbf{x_i}}$ by solving (1)
6: Compute Projection matrix $\mathbf{V} \in \mathcal{R}^{mn \times d}$
7: Compute lower dimensional projection of data $\mathbf{X}$ by $\mathbf{Y} = \mathbf{V}^T \mathbf{X}$

4 Experiments and Results

Class-similarity based neighborhood selection approach is applied to ONPP and MONPP, now onwards denoted as CS-ONPP and CS-MONPP. The recognition performance and dimensionality reduction performance of Conventional ONPP and Modified ONPP are compared with that of the CS-ONPP and CS-MONPP on some benchmark face databases and handwritten numerals databases.

The face databases used are ORL [13], UMIST [3] and CMU-PIE [14] having nearly 400, 564 and 1596 images respectively showing variations in terms of pose, lighting, occlusions and expressions. For uniformity, all images are resized to 40×40, out of which 50% images are used for training. The Handwritten numerals databases used are MNIST [6], Gujarati [9] and Devanagari [2] having nearly 68000, 13000 and 18000 images respectively showing large variations in stroke width, orientation, shape etc. All images were resized to 30×30. For each database, randomly 1000 images were used for training.

To analyze the behavior of proposed method with respect to parameters PCA dimensions d_{PCA}, the tuning parameter α and ONPP subspace dimension d, experiments are repeated with various set of (d_{PCA}, α, d), where values of $d_{PCA} \in \{2, 4, 6, 8, 10\}$ $\alpha \in \{0.25, 0.50, 0.75\}$ and ONPP subspace dimensions are considered from $\{5, 10, \dots\}$. To achieve unbiased results, such 20 randomization for all set of (d_{PCA}, α, d) were performed on all databases. Best recognition (in %) results achieved with conventional ONPP and MONPP (column 3) with corresponding subspace dimensions (column 4) are reported in Table 2 with subspace dimensions required, corresponding PCA dimension and α (column 5,6,7) for CS based approaches to achieve the same recognition accuracy.

For all databases, it is observed that Class similarity based approaches achieve better recognition at less number of subspace dimensions compared to conventional approaches. For ORL, CS-ONPP and CS-MONPP needs average 55 and 62 lesser subspace dimensions respectively. For UMIST, proposed methods needs on average 100 and 85 lesser dimensions. In CMU-PIE, CS-ONPP

Table 2. Best Recognition Accuracy (%) achieved using ONPP and MONPP with corresponding subspace dimensions d. To achieve same recognition accuracy, subspace dimensions required in CS-ONPP and CS-MONPP are reported with corresponding tuning parameter α and PCA dimension d_{PCA}

1	2	3	4	5	6	7
Database	Method	Conventional approach		Class similarity based approach		
		Best recognition (%)	Subspace dimensions	Subspace dimensions	α	d_{PCA}
ORL	ONPP	94.00	155	95	0.25	4
	MONPP	94.50	150	85	0.25	4
UMIST	ONPP	100	145	40	0.50	4
	MONPP	100	105	20	0.50	4
CMU-PIE	ONPP	93.86	955	945	0.50	4
	MONPP	93.91	865	245	0.25	6
MNIST	ONPP	86.52	90	40	0.50	4
	MONPP	87.56	70	30	0.50	4
Devanagari	ONPP	86.34	50	25	0.25	8
	MONPP	87.94	40	25	0.50	4
Gujarati	ONPP	91.76	60	40	0.50	6
	MONPP	92.08	60	30	0.50	6

Table 3. Best Recognition Accuracy(%) of proposed CS-ONPP and CS-MONPP with parameters subspace dimensions (d), tuning parameter α and PCA dimensions d_{PCA}

	CS-ONPP		CS-MONPP	
Database	Accuracy (%)	[d, α, d_{PCA}]	Accuracy (%)	[d, α, d_{PCA}]
ORL	97.50	55, 0.25, 4	98.50	30, 0.50, 4
UMIST	100	40, 0.50, 4	100	20, 0.50, 4
CMU-PIE	93.91	955, 0.75, 4	93.91	245, 0.75, 4
MNIST	88.42	30, 0.50, 4	90.21	40, 0.50, 4
Devanagari	88.89	35, 0.75, 4	90.01	35, 0.75, 2
Gujarati	92.24	55, 0.50, 6	92.42	45, 0.50, 6

improved dimension reduction with only a small margin, but CS-MONPP needs comparatively 700 less dimensions to achieve best recognition. For MNIST, the best recognition can be achieved with average 45 lesser dimensions using both CS-ONPP and CS-MONPP. In Devanagari data, to achieve best recognition, CS-ONPP needs on average 20 less dimensions, where as CS-MONPP needs average 15 less dimensions. In Gujarati, to reach best recognition of ONPP, CS-ONPP needs average 30 less dimensions, where as CS-MONPP needs average 27 less dimensions. It is also observed that the proposed neighborhood rule increases the overall recognition performance in terms of accuracy. Table 3 reports best recognition accuracy for proposed methods CS-ONPP and CS-MONPP along with parameters subspace dimensions (d), tuning parameter (α) and PCA dimensions (d_{PCA}).

5 Conclusion

Conventional ONPP selects neighbors based on Euclidean distance or the class knowledge, which may not be the best rule when data distribution is highly overlapping. We propose a new probability based neighborhood selection rule which incorporates both the information - Euclidean Distance and Class Similarity between two data points. Logistic Regression is used to compute the probability. Experiments performed on Face data and Handwritten numerals data, Class Similarity based approaches CS-ONPP and CS-MONPP outperforms conventional algorithms in recognition and achieves superior recognition rates with comparatively less number of subspace dimensions. In the future, it will be an interesting work to observe whether this neighborhood rule improves the performance of class of DR methods that are based on local neighborhood.

References

1. Belkin, M., Niyogi, P.: Laplacian eigenmaps for dimensionality reduction and data representation. Neural Comput. **15**(6), 1373–1396 (2003)
2. Bhattacharya, U., Chaudhuri, B.: Databases for research on recognition of handwritten characters of Indian scripts. In: Proceedings of 8th International Conference on Document Analysis and Recognition, pp. 789–793. IEEE (2005)
3. Graham, D.B., Allinson, N.M.: Characterising virtual eigensignatures for general purpose face recognition. In: Face Recognition, pp. 446–456. Springer, Berlin (1998). https://doi.org/10.1007/978-3-642-72201-1_25
4. Kokiopoulou, E., Saad, Y.: Orthogonal neighborhood preserving projections: a projection-based dimensionality reduction technique. IEEE Trans. Pattern Anal. Mach. Intell. **29**(12), 2143–2156 (2007)
5. Koringa, P., Shikkenawis, G., Mitra, S.K., Parulkar, S.K.: Modified orthogonal neighborhood preserving projection for face recognition. In: Kryszkiewicz, M., Bandyopadhyay, S., Rybinski, H., Pal, S.K. (eds.) PReMI 2015. LNCS, vol. 9124, pp. 225–235. Springer, Cham (2015). https://doi.org/10.1007/978-3-319-19941-2_22
6. Lecun, Y., Cortes, C.: The MNIST database of handwritten digits (1999). http://yann.lecun.com/exdb/mnist/
7. Lee, W.J., Duin, R.P.W., Ibba, A., Loog, M.: An experimental study on combining Euclidean distances. In: 2010 2nd International Workshop on Cognitive Information Processing, pp. 304–309 (June 2010). https://doi.org/10.1109/CIP.2010.5604238
8. Lu, J., Plataniotis, K.N., Venetsanopoulos, A.N.: Face recognition using LDA-based algorithms. IEEE Trans. Neural Netw. **14**(1), 195–200 (2003)
9. Nagar, R., Mitra, S.K.: Feature extraction based on stroke orientation estimation technique for handwritten numeral. In: 8th International Conference on Advances in Pattern Recognition (ICAPR), pp. 1–6 (2015)

10. Nie, F., Wang, X., Huang, H.: Clustering and projected clustering with adaptive neighbors. In: Proceedings of the 20th ACM SIGKDD International Conference on Knowledge Discovery and Data Mining, pp. 977–986. ACM (2014)
11. de Ridder, D., Kouropteva, O., Okun, O., Pietikäinen, M., Duin, R.P.W.: Supervised locally linear embedding. In: Kaynak, O., Alpaydin, E., Oja, E., Xu, L. (eds.) ICANN/ICONIP -2003. LNCS, vol. 2714, pp. 333–341. Springer, Heidelberg (2003). https://doi.org/10.1007/3-540-44989-2_40
12. Roweis, S.T., Saul, L.K.: Nonlinear dimensionality reduction by locally linear embedding. Science **290**(5500), 2323–2326 (2000)
13. Samaria, F., Harter, A.: Parameterisation of a stochastic model for human face identification. In: Proceedings of 2nd IEEE Workshop on Applications of Computer Vision. AT&T Laboratories Cambridge (December 1994)
14. Sim, T., Baker, S., Bsat, M.: The CMU pose, illumination, and expression database. In: Proceedings of Fifth IEEE International Conference on Automatic Face and Gesture Recognition, pp. 46–51. IEEE (2002)
15. Turk, M., Pentland, A.: Eigenfaces for recognition. J. Cogn. Neurosci. **3**(1), 71–86 (1991)
16. Wen, G., Jiang, L., Wen, J., Shadbolt, N.R.: Performing locally linear embedding with adaptable neighborhood size on manifold. In: Yang, Q., Webb, G. (eds.) PRICAI 2006. LNCS (LNAI), vol. 4099, pp. 985–989. Springer, Heidelberg (2006). https://doi.org/10.1007/978-3-540-36668-3_119
17. Zhang, S.Q.: Enhanced supervised locally linear embedding. Pattern Recogn. Lett. **30**(13), 1208–1218 (2009)
18. Zhao, L., Zhang, Z.: Supervised locally linear embedding with probability-based distance for classification. Comput. Math. Appl. **57**(6), 919–926 (2009)

Identification of Articulated Components in 3D Digital Objects Using Curve Skeleton

Sharmistha Mondal[1], Nilanjana Karmakar[2(✉)], and Arindam Biswas[1]

[1] Department of Information Technology, Indian Institute of Engineering Science and Technology, Shibpur, India
sharmistha28101990@gmail.com, barindam@gmail.com

[2] Department of Information Technology, St. Thomas' College of Engineering and Technology, Kolkata, India
nilanjana.nk@gmail.com

Abstract. A novel combinatorial algorithm for the segmentation of the articulated components of a 3D digital object is presented in the paper. As a preprocessing step, the single voxel thick 3D curve skeleton of the object is extracted by a topological technique. The curve skeleton is segmented by exploiting the concept of antipodal points in 3D orthogonal domain and calculating the isothetic distance of each voxel to a leaf voxel. The isothetic distance of the voxels on the object surface to the voxels on the segmented curve skeleton is used to separate the articulated components of the object into segments. The segmentation of the curve skeleton is based on the object geometry rather than the topology of the skeleton. The effectiveness of the algorithm has been demonstrated by the experimental results on a variety of articulated objects.

Keywords: Articulated components · Antipodal voxels · 3D curve skeleton · Image segmentation

1 Introduction

Image segmentation is a widely used concept of considerable importance in the field of computer vision. It is applicable in the field of content-based image retrieval, medical imaging, object detection, recognition tasks, etc. In the current paper the articulated components in a 3D digital object are identified by segmentation. There are four main approaches of segmentation, namely thresholding, boundary detection, region-based and hybrid methods [7]. A survey on different segmentation and partitioning techniques of boundary meshes has been presented in [10], where the segmentation problem has been formulated as an optimization problem. Similarly, [13] provides a wide overview over the common

Partially supported by Visvesvaraya PhD Scheme for Electronics & IT, Ministry of Electronics & Information Technology (MeitY), Government of India.

B. Deka et al. (Eds.): PReMI 2019, LNCS 11941, pp. 433–441, 2019.
https://doi.org/10.1007/978-3-030-34869-4_47

binarization and segmentation methods used in 3D image processing. Further, a comparative study of some of the mesh segmentation algorithms and their results have been provided in [2].

A work on hierarchical segmentation of articulated shapes involving eigen functions of the Laplace-Beltrami operator and persistent homology has been reported in [9]. A volume-based shape-function called the shape-diameter-function (SDF) has been used in [11] to construct partitioning of skeletons which remain consistent across a family of objects. A novel hierarchical pose-invariant mesh segmentation algorithm has been proposed in [4] which helps in the extraction of important feature points and of the core component of the mesh using a spherical mirroring operation. A semantic oriented 3D mesh hierarchical segmentation problem in [12] uses enhanced topological skeleton to identify junction areas and obtain a fine segmentation of the object. An interactive and automatic method of model segmentation based on random walks has been demonstrated in [6].

2 Definitions and Preliminaries

- **Digital Object :** A *digital object* $\mathcal{A}$ is defined as a finite subset of $\mathbb{Z}^3$, with all its constituent points (i.e., voxels) having integer coordinates and connected in 26-neighborhood. Each voxel is equivalent to a *3-cell* [5] centered at the concerned integer point. The *isothetic distance* between two points $p(x_1, y_1, z_1)$ and $q(x_2, y_2, z_2)$ is defined as the Minkowski norm L_∞ given by $d_\top(p,q) = \max\{|x_1 - x_2|, |y_1 - y_2|, |z_1 - z_2|\}$. The (isothetic) distance of a point p from an object $\mathcal{A}$ is $d_\top(p, \mathcal{A}) = \min\{d_\top(p,q) : q \in \mathcal{A}\}$, and the distance between two connected components $\mathcal{A}_1$ and $\mathcal{A}_2$ is $d_\top(\mathcal{A}_1, \mathcal{A}_2) = \min\{d_\top(p,q) : p \in \mathcal{A}_1, q \in \mathcal{A}_2\}$.
- **Antipodal points :** Two points are antipodal (i.e., each is the antipode of the other) if they are diametrically opposite [1]. In mathematics, the concept of antipodal points is generalized to spheres of any dimension: two points on the sphere are antipodal if they are opposite through the centre. In the orthogonal domain, two voxels v_1 and v_1' are termed as antipodal voxels w.r.t another voxel v_2 if the following conditions are satisfied.
  - v_1 and v_1' are located at diametrically opposite positions w.r.t v_2.
  - $|d_\top(v_1, v_2) - d_\top(v_1', v_2)| < t$, where $d_\top(v_1, v_2)$ and $d_\top(v_1', v_2)$ denote the isothetic distances between the corresponding pair of voxels and t is a threshold.
- **Articulated Components :** A 3D object is said to be articulated if it consists of two or more flexibly connected rigid 3D object-components [8]. The components are attached through joints such that they can move w.r.t one another. The 3D model of a human body is an example of articulated object where the hands, legs, and head are the articulated components.

3 Proposed Work

Let us consider a 3D digital object A provided as a triangulated data set such that exactly two triangles are incident on each edge of the triangulation. Let the

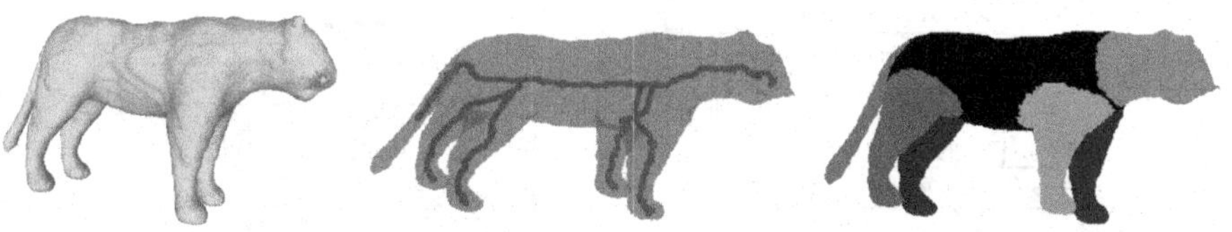

Fig. 1. A digital object `Tiger` (Left), its 3D curve skeleton (Middle), and identified articulated components (Right) (Color figure online).

object be embedded on a 3D digital grid represented as a set of unit grid cubes (UGCs) each of length g. Our objective is to identify the articulated components of the object by exploiting its curve skeleton. The curve skeleton of an object is defined as a 3D curve which is single voxel thick, connected and is centered w.r.t to the object thereby capturing the object shape [3]. Figure 1(Right) shows the segmented articulated components of the digital object `Tiger` (Fig. 1(Left)) extracted by using its 3D curve skeleton in Fig. 1(Middle).

3.1 Preprocessing

The 3D curve skeleton has been extracted using the 3D isothetic inner cover $\underline{P}_{\mathbb{G}}(A)$. The 3D isothetic inner cover is defined as the 3D polyhedron of maximum volume defined w.r.t an underlying grid $\mathbb{G}$ having surfaces parallel to the coordinate planes and inscribing the entire object [3]. If a voxel p is intersected by one or more triangles on the object surface, then it is considered as an object voxel. A UGC is considered as partially object-occupied if at least one of its constituent voxels is a background voxel. Hence, a UGC intersected by one or more triangles is a partially object-occupied UGC. All object voxels within partially occupied UGCs constitute the set of boundary voxels $\mathcal{B}$.

The combinatorial algorithm in [3] has been used to extract the 3D curve skeleton of a digital object. The voxels at the boundary of the inner cover is represented in a topological space. The object voxels enclosed by the inner cover and satisfying certain conditions along the three coordinate planes are expressed in another topological space. The topological spaces are related by exploiting the concepts of homotopy equivalence and homology computation. Homotopy equivalence of the two topological spaces indicates that the 3D curve skeleton accurately represents the shape of the object. The nth homology groups of the two topological spaces are computed to be isomorphic to demonstrate that they are topologically identical in n-dimensional space. The 3D curve skeleton, thus obtained, is single voxel thick, connected, and centered w.r.t the object.

3.2 Segmentation of the 3D Curve Skeleton

Let $S(A)$ be the set of voxels representing the 3D curve skeleton of the digital object A. If a voxel belonging to the curve skeleton is such that exactly one of the voxels in its 26-neighborhood belongs to the curve skeleton, then the voxel

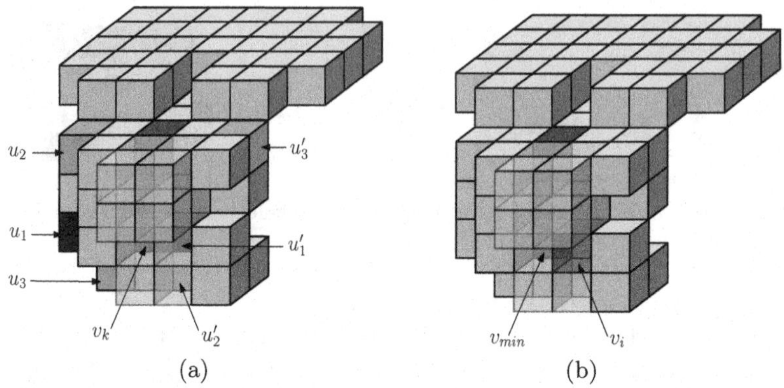

Fig. 2. (a) A sample 3D curve skeleton (red), a leaf voxel v_k, and pairs of antipodal voxels $\{u_1, u'_1\}$ (blue), $\{u_2, u'_2\}$ (green), and $\{u_3, u'_3\}$ (orange) w.r.t v_k. (b) The boundary voxel v_i is assigned the color of its nearest skeleton voxel v_{min} (red) (Color figure online).

is termed as *leaf voxel.* Figure 2(a) shows a part of a digital object where the voxels representing the skeleton and the leaf voxel v_k are shown (in red).

Let L denote the list of leaf voxels in $S(A)$. Let the voxel v_k belong to the curve skeleton such that $v_k \in L$. Starting from v_k, the object voxels may be traversed in 26 directions so that a boundary voxel is reached at the end of traversal in each direction. Hence we have 26 boundary voxels w.r.t v_k. The 26 boundary voxels form 13 pairs of antipodal voxels (Sect. 2). Figure 2(b) shows three such pairs of antipodal voxels, $\{u_1, u'_1\}$ (blue), $\{u_2, u'_2\}$ (green), and $\{u_3, u'_3\}$ (orange).

Let (u_j, u'_j) denote a pair of antipodal voxels w.r.t v_k. Let d_j and d'_j be the distances in terms of voxels from v_k to u_j and u'_j respectively. If the difference between d_j and d'_j lies within a threshold say, t, then $d_m = max(d_j, d'_j)$ is calculated. A pair of antipodal voxels is said to be at more or less equal distance from v_k if the above threshold criterion is satisfied. Let $dmin_k$ be the minimum value of d_m out of the thirteen values of d_m corresponding to the available pairs of antipodal voxels w.r.t v_k. A segment number is assigned to v_k. Let E denote the list of segmented skeleton voxels. v_k along with its segment number is inserted into E. As v_k has been traversed, it is marked as visited.

Since the 3D curve skeleton is single voxel thick and connected, a single unvisited neighbor is always available for a skeleton voxel. Let v_i be the unvisited neighbor of v_k. The procedure of finding the minimum value of d_m is repeated with v_i. Let $dmin_i$ be the minimum value of d_m obtained w.r.t v_i. If the difference between $dmin_k$ and $dmin_i$ lies within a threshold say, t_1, then v_i is assigned with the same segment number as v_k and is inserted into E. The procedure is repeated as long as the threshold criterion with threshold t_1 is satisfied. Every time a voxel is traversed, it is marked as visited. Once this condition fails, we start with the next leaf voxel in L and a new segment number is assigned to it. The entire procedure is repeated for every leaf voxel in L. The list E obtained at the end contains all the segmented skeleton voxels.

3.3 Identification of Articulated Components

Initially, the triangulated data set representing the 3D digital object is provided. $\mathcal{B}$ is the set of boundary voxels representing the object surface. Let us consider the segmented skeleton in the form of the list E. For a boundary voxel $v_i \in \mathcal{B}$, let v_{min} be the nearest (in terms of number of voxels) skeleton voxel (Fig. 2 (b)). Let d_{min} be the distance between v_{min} and v_i. Now w.r.t v_{min}, we consider the 13 pairs of antipodal voxels. Let (u_j, u'_j) denote a pair of such antipodal voxels. Let d_j and d'_j be the distances in terms of voxels from v_{min} to u_j and u'_j respectively. For a pair of antipodal voxels, if the difference between d_j and d'_j lies within a threshold t, then $d_m = max(d_j, d'_j)$ is computed. Out of the thirteen available pairs of antipodal voxels w.r.t v_{min}, let d_{max} be the maximum value of d_m. If d_{min} is less or equal to $(t_2 \times d_{max})$, then the segment number of v_{min} is assigned to the boundary voxel v_i. In Fig. 2(b), the color of the nearest skeleton voxel v_{min} is assigned to v_i. The same procedure is followed for each and every boundary voxel belonging to $\mathcal{B}$. The assignment of respective segment numbers to the boundary voxels leads to the proper identification of the articulated components of the digital object.

4 Algorithm SEGMENTATION3D

The algorithm for the proposed work has been given in Fig. 3. Given the set of voxels $S(A)$ which belongs to the 3D curve skeleton of the digital object A and the set of boundary voxels $\mathcal{B}$ (voxels intersected by the triangulated surface of A), all the voxels of $S(A)$ are initially marked unvisited (Steps 1 and 2 of SEGMENTATION3D). A leaf voxel v_k of the 3D curve skeleton is considered. For each pair of antipodal voxels (u_j, u'_j), Steps 2–8 of procedure FINDDIST are carried out and the resultant distance is returned (Step 12). The procedure FINDDIST calculates the minimum distance among the 13 pairs of antipodal voxels which are situated at more or less equal distance w.r.t v_k. This minimum distance is stored in $dmin_k$ (Step 5 of SEGMENTATION3D). The leaf voxel v_k is then assigned a segment number, is marked as visited and is inserted into a list E along with its segment number (Steps 6–9 of SEGMENTATION3D). Let v_i be an unvisited neighbor of v_k (Steps 10, 16, and 17). The procedure FINDDIST is called for v_i and the calculated distance is stored in $dmin_i$ (Step 18). If the difference between $dmin_k$ and $dmin_i$ is within a threshold t_1 (where t_1 is a threshold based on some factors), then the segment number of v_k is assigned to v_i (Step 20). v_i is marked as visited and is inserted into E along with its segment number (Steps 21–22). Steps 13–23 is repeated for the unvisited neighbor of each v_i. This procedure is continued until the condition in Step 19 fails. Once this condition fails, the entire procedure is repeated for the next leaf voxel. Steps 5–23 is carried out for every leaf voxel.

Algorithm
SEGMENTATION3D$(S(A), \mathcal{B})$

1. **for** each voxel $v_i \in S(A)$
2. $visited[v_i] \leftarrow 0$
3. $SegmentNum \leftarrow 0$
4. **for** each leaf voxel $v_k \in L$ $\triangleright$ L is the list of leaf voxels
5. $dmin_k \leftarrow$ FINDDIST $(v_k, 1)$
6. $SegmentNum \leftarrow SegmentNum + 1$
7. $segment[v_k] \leftarrow SegmentNum$
8. $E \leftarrow E \cup (v_k, segment[v_k])$
9. $visited[v_k] \leftarrow 1$
10. Let $v_i \leftarrow v_k$, where $i = 1$
11. $flag \leftarrow 1$
12. **while** $(flag = 1)$
13. $i \leftarrow i + 1$
14. $flag \leftarrow 0$
15. $j \leftarrow 1$
16. $v_i \leftarrow N(v_{i-1})$
17. **if** $(visited[v_i] = 0)$
18. $dmin_i \leftarrow$ FINDDIST $(v_i, 1)$
19. **if** $(|dmin_i - dmin_k| \leq t_1)$
20. $segment[v_i] \leftarrow segment[v_k]$
21. $E \leftarrow E \cup (v_i, segment[v_k])$
22. $visited[v_i] \leftarrow 1$
23. $flag \leftarrow 1$
24. SURFACE-SEG $(E, \mathcal{B})$

Procedure FINDDIST (v, f)

1. $\widehat{d} \leftarrow \infty$, $\widetilde{d} \leftarrow 0$
2. **for** each pair of antipodal voxels (u_j, u'_j) $\triangleright$ $1 \leq j \leq 13$
3. $d_j \leftarrow dist(v, u_j)$
4. $d'_j \leftarrow dist(v, u'_j)$
5. **if** $(|d_j - d'_j| \leq t)$
6. $d_m \leftarrow max(u_j, u'_j)$
7. **if** $(f = 1$ **and** $d_m < \widehat{d})$
8. $\widehat{d} \leftarrow d_m$
9. **else if** $(f = 2$ **and** $d_m > \widetilde{d})$
10. $\widetilde{d} \leftarrow d_m$
11. **if** $(f = 1)$
12. return $\widehat{d}$
13. **if** $(f = 2)$
14. return $\widetilde{d}$

Procedure SURFACE-SEG $(E, \mathcal{B})$

1. **for** each voxel $v_i \in \mathcal{B}$
2. $d_{min} \leftarrow \infty$
3. **for** each voxel $v_j \in E$
4. **if** $(d_\top(v_i, v_j) < d_{min})$
5. $d_{min} \leftarrow d_\top(v_i, v_j)$
6. $v_{min} \leftarrow v_j$
7. $d_{max} \leftarrow$ FINDDIST $(v_{min}, 2)$
8. **if** $(d_{min} \leq (t_2 \times d_{max}))$
9. $segment[v_i] \leftarrow segment[v_{min}]$

Fig. 3. The algorithm for the segmentation of articulated components (Color figure online).

Now that the 3D curve skeleton is segmented, the SURFACE-SEG procedure is used to segment the surface of the digital object A which leads to the identification of the articulated components of A. For each boundary voxel v_i, the nearest skeleton voxel v_{min} is determined. The isothetic distance from v_i to v_{min} is stored in d_{min} (Steps 3–6 of SURFACE-SEG). All the voxels from v_i through v_{min} that constitutes the isothetic distance are object voxels. The maximum distance (d_{max}) among the 13 pairs of antipodal voxels which are situated at more or less equal distance w.r.t v_{min} is determined in Step 7. If d_{min} is less than or equal to $(t_2 \times d_{max})$ (where t_2 is a threshold based on some factors), then the segment number of v_{min} is assigned to v_i (Steps 8–9). The process is repeated for each boundary voxel in Steps 2–9. Thus, a segment number is assigned to all the voxels on the object surface belonging to each articulated component. Hence, the surface of the object is segmented.

5 Time Complexity

Let n be the number of voxels on the object surface and m be the number of voxels belonging to the 3D curve skeleton where $m = O(n^{3/2})$. The leaf voxels are detected by traversing the curve skeleton in $O(m)$ time. Starting from each leaf voxel the curve skeleton is segmented by traversing the voxels of the curve skeleton exactly once in $O(m)$ time. For each skeleton voxel the procedure FindDist is executed in constant time, i.e., $O(1)$.

For each voxel on the surface, the closest skeleton voxel is identified in $O(m)$ time. Assignment of segment number to the voxels on the object surface takes $O(1)$ time (procedure FindDist). Hence, the Surface-Seg procedure is executed in $O(n) \times O(m) = O(mn)$ time. The following improvement may, however, be suggested to reduce the complexity. While executing the procedure FindDist w.r.t each skeleton voxel, the skeleton voxel closest to each voxel on the surface may be identified. Hence, the procedure Surface-Seg will be executed in

Fig. 4. Results for identification of articulated components in some digital objects (Color figure online).

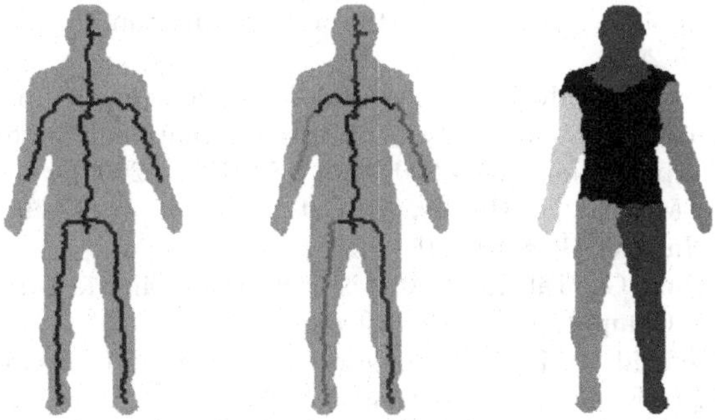

Fig. 5. The 3D curve skeleton (black) of `Human` with the leaf voxels (red) (Left), the segmented curve skeleton (Middle) and the identified articulated components (Right) (Color figure online).

$O(n) \times O(1) = O(n)$ time. Therefore, the total complexity of the segmentation algorithm is given by $O(m) + O(m) + O(n) = O(n^{3/2}) + O(n^{3/2}) + O(n) \simeq O(n^{3/2})$.

6 Experimental Results

The implementation of the proposed algorithm has been done in C in Ubuntu 16.04, Intel Core i5-7500 CPU @ 3.40GHz × 4. The segmented articulated components of the digital objects `Hand`, `Ant`, `Horse`, and `Turtle` has been demonstrated by the experimental results in Fig. 4. The step-by-step approach of the proposed method has been illustrated for the digital object `Human` in Fig. 5. It is observed that the algorithm can separate the articulated portions of the objects upto a high degree of accuracy.

7 Conclusion

The speciality of the proposed algorithm lies in considering all the three orthogonal directions along the coordinate axes (x, y, and z-axes) together for the operations instead of considering the coordinate planes (yz-, zx-, and xy-plane) separately. Since the process takes place at the voxel level, the accuracy of segmentation is independent of the grid resolution of the objects. As the segmentation of the curve skeleton is dependent on the geometry of the object rather than the topology of the curve skeleton, there have been instances of over-segmentation in some cases (Fig. 4, `Turtle`). Improvements based on topological techniques may be attempted in future to ensure natural segmentation. The robustness of the algorithm may be tested in future in terms of the rotation invariance and pose invariance.

References

1. Antipodal Points. http://mathworld.wolfram.com/AntipodalPoints.html. Accessed June 5 2018
2. Attene, M., Katz, S., Mortara, M., Patané, G., Spagnuolo, M., Tal, A.: Mesh segmentation-a comparative study. In: IEEE International Conference on Shape Modeling and Applications, SMI 2006, pp. 7–7. IEEE (2006)
3. Karmakar, N., Mondal, S., Biswas, A.: Determination of 3D curve skeleton of a digital object. Inf. Sci. **499**, 84–101 (2019)
4. Katz, S., Leifman, G., Tal, A.: Mesh segmentation using feature point and core extraction. Vis. Comput. **21**(8–10), 649–658 (2005)
5. Klette, R., Rosenfeld, A.: Digital Geometry: Geometric Methods for Digital Picture Analysis. Morgan Kaufmann, San Francisco (2004)
6. Lai, Y.K., Hu, S.M., Martin, R.R., Rosin, P.L.: Rapid and effective segmentation of 3D models using random walks. Comput. Aided Geom. Des. **26**(6), 665–679 (2009)

7. Lin, Z., Jin, J., Talbot, H.: Unseeded region growing for 3D image segmentation. In: Selected papers from the Pan-Sydney workshop on Visualisation, vol. 2, pp. 31–37. Australian Computer Society, Inc. (2000)
8. Martínez, G.: Shape estimation of articulated 3D objects for object-based analysis-synthesis coding (OBASC). Signal Process. Image Commun. **9**(3), 175–199 (1997)
9. Reuter, M.: Hierarchical shape segmentation and registration via topological features of Laplace-Beltrami eigenfunctions. Int. J. Comput. Vis. **89**(2–3), 287–308 (2010)
10. Shamir, A.: A survey on mesh segmentation techniques. In: Computer Graphics Forum, vol. 27, pp. 1539–1556. Wiley Online Library (2008)
11. Shapira, L., Shamir, A., Cohen-Or, D.: Consistent mesh partitioning and skeletonisation using the shape diameter function. Vis. Comput. **24**(4), 249 (2008)
12. Tierny, J., Vandeborre, J.P., Daoudi, M.: Topology driven 3D mesh hierarchical segmentation. In: IEEE International Conference on Shape Modeling and Applications 2007, SMI 2007, pp. 215–220. IEEE (2007)
13. Wirjadi, O.: Survey of 3D image segmentation methods (2007)

EAI-NET: Effective and Accurate Iris Segmentation Network

Sanyam Rajpal[1], Debanjan Sadhya[2(✉)], Kanjar De[3], Partha Pratim Roy[4], and Balasubramanian Raman[4]

[1] Indiana University Bloomington, Bloomington, IN, USA
srajpal@iu.edu
[2] ABV-Indian Institute of Information Technology and Management Gwalior, Gwalior, India
debanjan@iiitm.ac.in
[3] NTNU Norwegian University of Science and Technology, Gjovik, Norway
kanjar.de@ntnu.no
[4] Indian Institute of Technology Roorkee, Roorkee, India
{proy.fcs,balarfcs}@iitr.ac.in

Abstract. In iris-based biometric models, segmentation of the iris region from the rest of the eye is a crucial step. The quality of the segmented region directly affects the extracted iris features, which subsequently determines the overall recognition accuracy of the model. In this work, we propose *EAI-Net*, which is an effective and accurate iris segmentation network based on the U-Net architecture. In comparison to the previous works, we treat the segmentation process as a 3-class problem wherein the pupil, iris and the rest of the image are treated as separate classes. Furthermore, we have increased the complexity degree of our model by encoding the complex regions of the iris more efficiently. We have conducted both qualitative and quantitative assessments of our results over four benchmark iris databases - UBIRISv2, IITD, CASIAv4-Interval, and CASIAv4-Thousand. The obtained results demonstrate the superiority of our model over the other state-of-the-art deep-learning based approaches in solving the problem of iris segmentation in both the visible (VIS) and near-infrared (NIR) spectrum.

Keywords: Iris · Segmentation · U-Net

1 Introduction

Iris is the annular region in the eye which is present between the sclera and the pupil. It primarily consists of complex texture patterns which are unique to an individual. Biometric recognition systems which are operationally based on this particular trait are considered to be one of the most secure forms for entity authentication [3]. Furthermore, the advent of *mobile biometrics* has proliferated the use of these models in large-scale government and semi-government projects. Due to all these reasons, the development of accurate and robust iris-based

B. Deka et al. (Eds.): PReMI 2019, LNCS 11941, pp. 442–451, 2019.
https://doi.org/10.1007/978-3-030-34869-4_48

recognition systems which can work in unconstrained environments is an active area of research.

Segmentation of the iris region is arguably the most crucial stage in the entire recognition process. This important phase involves detecting and subsequently isolating the iris region from the corresponding input image. Importantly, the quality of the features extracted from the segmented area heavily relies on the accuracy of the associated segmentation procedure. As such, inaccurate iris segmentation results in the largest source of error for iris-based authentication models [5,17]. The main factors which affect the segmentation process are: (i) occlusions caused due to eyelids and eyelashes, (ii) specular reflections and non-uniform illumination, (iii) imaging distance and, (iv) noise from the acquisition device (sensor) [17].

Our work in this paper proposes *EAI-Net*, which is an end-to-end deep-learning based segmentation model for non-ideal iris images that are characterized with real-world covariates such as variable imaging distances, subject perspectives and non-uniform lighting conditions. Our proposed model utilizes the U-Net architecture [18] for segmenting the iris region from their corresponding images. Importantly, this architecture can work with relatively few training images while yielding precise segmented regions. We have tested our model on four benchmark iris databases, for which our model comprehensively outperforms other deep-learning based studies.

2 Related Work

With the advent of deep neural networks, highly challenging problems in computer vision like object detection and object classification have shown excellent results. Some of the earliest works involved the use of Fully convolutional networks (FCN) [13] and Densely connected convolutional networks (DenseNet) [8] for performing the task of semantic segmentation. The use of deep-learning based models for iris segmentation was initially studied by Liu et al. [12] wherein Hierarchical convolutional neural network (HCNN) and Multi-scale fully convolutional network (MFCN) were introduced. Other deep models such as fully convolutional encoder-decoder networks [7] and a domain adaption technique for CNN based iris segmentation [6] were subsequently used in later works. Most recent works have utilized the design of Fully convolutional deep neural network (FCDNN) [1] and Generative adversarial networks (GAN) [2] for segmenting lower quality iris images which are obtained in the visible spectrum. The U-Net architecture has also been used in some previous works [11,14]. However in our work, we have demonstrated that this architecture can give more accurate results when the iris and pupil sections of the eye are segregated. In such a scenario, the pupil is treated as a separate class and is not included with the background class. This feature facilitates the EAI-Net model in encoding the complex boundary of the iris region more accurately.

3 The EAI-Net Model

In this section, we describe in details the proposed EAI-Net model along-with the underlying U-Net architecture.

3.1 U-Net Architecture

In this paper, we have used U-Net to effectively learn the features from different regions of the eye. U-Net is one of the most popular architectures of convolutional neural networks which deals with the problem of end-to-end image segmentation. The initial U-Net model was successfully used for the segmentation of bio-medical images [18]. This architecture is basically an encoder-decoder model which consists of a contracting path (which works as an encoder) and an expanding path (which works as a decoder). Most of the operations in U-Net include convolution, which is followed by a non-linear activation function. In the contracting path, max-pooling operations are present for reducing the size of the feature maps. The expansion path consists of a sequence of up-convolutions in combination with the concatenation of high-resolution features from the contracting path. Each level in the U-Net architecture has four layer depth for extracting higher-level features from the iris. In each level of the U-Net, there is a convolution operation from a 3×3 kernel, which is followed by the ReLU activation function and batch normalization. Each max-pooling operation is performed by a factor of 2 for finding out the features at different scales. To avoid any information and content loss due to convolutions, skip connections are added. Similar to the contracting path, up-sampling with a factor of 2 is done in the expanding path for generating the upscaled maps. In each level of the expanding path, 3×3 kernel convolution operations are performed. This process is followed by the non-linear ReLU activation function and batch normalization (similar to the contracting path). We use a soft-max layer after the last convolutional operation in the expanding path for generating the final output segmentation mask. The implemented U-Net based architecture is illustrated in Fig. 1.

3.2 Pre-processing of Ground-Truth

The iris segmentation problem is generally treated as a 2-class problem where the iris is considered as the foreground and the rest of the image is considered as the background. The main issue in adopting such an approach is that the iris and pupil have similar visual appearances, for which their exact discrimination becomes very difficult. To address this problem, we modify the problem into a 3-class problem where the pupil and iris are treated as separate classes. This process enables the deep-neural network to learn distinguishing features between the iris and the pupil, which subsequently results in a more accurate segmentation of the iris region. We achieve this particular objective in our work by using elements from computational geometry. Specifically speaking, we convert the binary problem into a 3-class problem using a combination of convex hulls, fitting contours and morphological operations. Furthermore, we had to use a combination of the

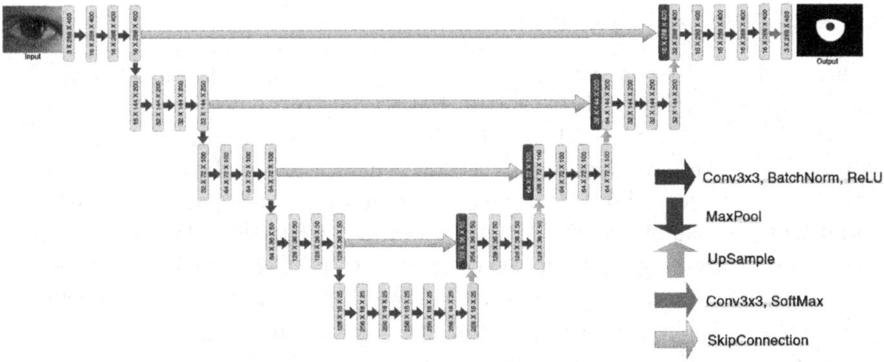

Fig. 1. The proposed *EAI-Net* model for iris segmentation.

convex hull with concave hull [15] and the morphological closing operation for generating the augmented ground-truth for the CASIAv4-T database. The reason for using these additional pre-processing operations was due to the presence of some poorly labeled noisy samples in this particular database. The process of generating the 3-class ground-truth where the classes are labeled as 0 (for background), 1 (for iris), and 2 (for pupil) is presented in Algorithm 1.

Algorithm 1. Generation of augmented Ground-truth

Input: $M \in \{0,1\}^{M \times N}$
Output: $GT \in \{0,1,2\}^{M \times N}$

$GT \leftarrow 0$
$contour \leftarrow findContour(M)$ using [19]
$Area\_List \leftarrow []$
for $i \leftarrow 1$ to $|contour|$ **do**
  $Area\_List.append(contour[i], i)$
$sorted\_List, index \leftarrow SORT(Area\_List)$
for $j \leftarrow 1$ to 2 **do**
  $Hull\_Points \leftarrow CONVEX\_HULL(contour(index[j]))$ using [4]
  $GT \leftarrow Fill\_Contour(Hull\_Points, GT, j)$
return GT

4 Experimental Setup

In this section, we describe the experimental datasets and associated quantitative measures. We also elaborate on the network training process.

4.1 Database Description

We have performed extensive experiments on the following four publicly available benchmark iris databases: IITD-1 [10], UBIRISv2 [16], CASIAv4-Interval (further referred to as CASIAv4-I) and CASIAv4-Thousand (further referred to as CASIAv4-T)[1]. We have specifically selected these four databases for validating our work due to the variability of both image quality and quantity in them. The ground-truth masks of the IITD, CASIAv4-I and UBIRISv2 database are provided by the University of Salzburg via their *IRISSEG-EP* package [5][2]. Alternatively, the ground-truth masks for the CASIAv4-T database are distributed by Bezerra et al. [2]. However, it should be noted that the ground-truths corresponding to all the images of the respective databases are not provided. For instance, the total number of available annotations for UBIRISv2 and CASIAv4-T are 2250 and 1000 respectively.

4.2 Evaluation Protocol and Metrics

To evaluate the performance of EAI-Net, we use the following statistical quantities: **NICE-I**, **NICE-II** [7], and **F1-Score**. The NICE-I and NICE-II scores represent the overall segmentation errors between the segmentation mask (obtained from the network) and the corresponding ground-truth mask. The NICE-I score estimates the segmentation error by computing the proportion of the disagreeing pixels between the two masks, whereas the NICE-II score is intended to balance the disproportion between the prior probabilities of iris and non-iris pixels in the images. The F1-Score is a standard measure of the segmentation accuracy. It represents the harmonic mean of the corresponding precision and recall values. All these three metrics are bounded in the range $[0, 1]$.

4.3 Model Training Details

The entire framework for supervised iris segmentation has been implemented in Pytorch. Information like the number of channels, the number of filters, the type of connection and activation functions are visually depicted in Fig. 1. The receptive field has been kept identical for implementation in the different datasets. The batch size for training was kept at 4. All the experiments were conducted on a computer having Intel Xeon E5 processor with NVIDIA Quadro K620 2GB RAM graphics card. The model takes around 25 epochs to converge. We have used the *Adam Optimizer* [9] for conducting all the experiments. The hyper-parameters associated with this optimizer include *learning rate* $= 0.0001$, $\beta_1 = 0.9$, and $\beta_2 = 0.999$. The learning rate was multiplied with 0.5 every time the validation loss did not decrease (validation was done after every 150 iterations). For training the U-Net, we have chosen *Categorical cross entropy* as the loss function.

[1] http://biometrics.idealtest.org/.
[2] http://www.wavelab.at/sources/Hofbauer14b/.

5 Results and Discussions

Now we present and analyze all of our obtained results. In accordance with the previous works, we perform both quantitative and qualitative assessment of our results.

5.1 Ablation Study

We initially perform an ablation study by comparing the traditional 2-class segmentation problem with the 3-class problem. As presented in Table 1, some improvements in performance can be immediately noticed when the iris, the pupil and the background were considered as separate classes. Specifically speaking, both the NICE-I and NICE-II error scores were relatively lower and the F1 score was comparatively higher for the 3-class problem. This trend was consistently noted for all the four iris databases. Hence these results vindicate the importance of segmenting the entire eye image into three distinct classes (instead of two).

Table 1. Average values of the evaluation metrics while considering 2-class and 3-class segmentation problems.

Database	NICE-I		NICE-II		F1-Score	
	2-class	3class	2-class	3-class	2-class	3-class
CASIAv4-I	0.0156	0.0152	0.0193	0.0193	0.9813	0.9842
IITD	0.0192	0.0185	0.0242	0.0231	0.9684	0.9764
UBIRISv2	0.0082	0.0073	0.0331	0.0316	0.9692	0.9699
CASIAv4-T	0.007	0.0054	0.0227	0.0204	0.9586	0.9785

5.2 Quantitative Evaluation

We quantitatively compare the performance of EAI-Net with the other state-of-the-art deep-learning based iris segmentation techniques. For evaluation purpose, we have used the performance measures explained previously in Sect. 4.2. The mean (μ) and standard deviation (σ) of these measures are presented in Table 2.

As observable, the best F1 Score of 0.9842 was obtained for CASIAv4-I, which indicates the presence of high precision and recall values. Alternatively, the least F1 Score of 0.9699 was noticed for the UBIRISv2 databases, which denotes relatively poor segmentation of the iris regions. This result can be aptly justified due to the presence of off-angle noisy iris samples in this database. Interestingly, low NICE-I scores of 0.0054 and 0.0073 were noticed for the CASIAv4-T and UBIRISv2 databases respectively. This particular outcome can be attributed to the fact that the area of the iris region is comparatively much smaller in the samples of these datasets. This resulted in a lesser number of disagreeing pixels

Table 2. Mean (μ) and standard deviation (σ) values of the evaluation metrics.

Database	NICE-I		NICE-II		F1-Score	
	μ	σ	μ	σ	μ	σ
CASIAv4-I	0.0152	0.0059	0.0193	0.0088	0.9842	0.0098
IITD	0.0185	0.0074	0.0231	0.0116	0.9764	0.0129
UBIRISv2	0.0073	0.0082	0.0316	0.0341	0.9699	0.0572
CASIAv4-T	0.0054	0.0036	0.0204	0.0195	0.9785	0.0337

between the ground-truth and the corresponding predicted mask, which consequently produced low NICE-I scores. Another noticeable observation pertains to the CASIAv4-T database. Although this database is characterized by covariates such as specular reflection and non-uniform illumination (much like UBIRISv2), the corresponding F1 score of 0.9785 is relatively high. One possible reason for this result might relate to its associated spectral band. Since all of the images for this database were captured in NIR, the iris regions had more richly structured textural information which the EAI-Net exploited.

The superiority of our framework over the other deep-learning based techniques is demonstrated in Table 3. For all the iris databases, our model results in comparatively better values of NICE-I, NICE-II and F1 Score. The best improvement in the segmentation error corresponded to the UBIRISv2 database, wherein a decrease of approximately 18.88% over the next best (lowest) reported result [12] was noted. Considering the quality of the samples in this database, this is a considerable improvement over the previous results. The only anomaly was noticed for the IITD database, for which a smaller error score of 0.0133 was observed in the GAN model [2]. However, it should be noticed that our U-Net based model is relatively more efficient than GAN in terms of the required memory resources.

5.3 Qualitative Evaluation

Now we visually analyze a few instances of the iris segmentation results given by our model. Figure 2 illustrates sample results from the four databases used for our evaluation. As expected, the EAI-Net model gives excellent results for the CASIAv4-I and IITD datasets. Although both the UBRISv2 and CASIAv4-T are very challenging iris dataset, EAI-Net works well on them too. As understandable from Fig. 2, our model effectively handles samples from both the VIS and NIR spectrum. Important covariates such as imaging-distance and camera angle are also efficiently supervised by our model.

The segmentation errors for some noisy samples are illustrated in Fig. 3. The EAI-Net model is unable to accurately segment the iris regions when it is affected by strong reflections and drooping eyelashes. Due to this reason, pre-processing these iris samples for eliminating the effects of these covariates would potentially improve the segmentation accuracy of our network. Noticeably, the sample from

Table 3. Comparative analysis of the average segmentation scores for the four iris databases.

Database	Technique	NICE-I	NICE-II	F1-Score
CASIAv4-I	**Proposed EAI-Net**	**0.0152**	**0.0193**	**0.9842**
	FCEDN (Basic variant) [7]	0.033	0.0382	0.9375
	FCEDN (Bayesian variant) [7]	0.0412	0.0362	0.925
	Domain adaptation [6]	0.033	0.038	0.937
IITD	**Proposed EAI-Net**	0.0185	**0.0231**	**0.9764**
	FCEDN (Basic variant) [7]	0.0277	0.0322	0.951
	FCEDN (Bayesian variant) [7]	0.0292	0.0337	0.95
	FCN [2]	0.0148	-	0.9744
	GAN [2]	**0.0133**	-	0.9584
	Domain adaptation [6]	0.027	0.032	0.951
UBIRISv2	**Proposed EAI-Net**	**0.0073**	**0.0316**	**0.9699**
	FCEDN (Basic variant) [7]	0.0262	0.0687	0.79
	FCEDN (Bayesian variant) [7]	0.0187	0.0675	0.8625
	FCN [2]	0.01	-	0.882
	GAN [2]	0.03	-	0.9142
	MFCN [12]	0.009	-	-
	HCNN [12]	0.011	-	-
CASIAv4-T	**Proposed EAI-Net**	**0.0054**	**0.0204**	**0.9785**
	FCN [2]	0.0061	-	0.9442
	GAN [2]	0.014	-	0.9538

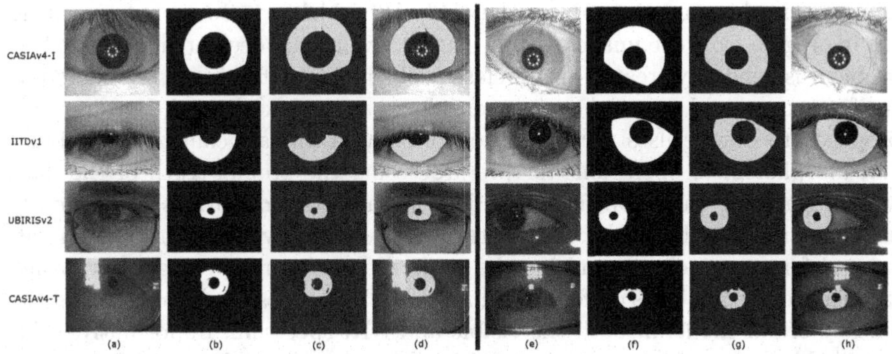

Fig. 2. Qualitative analysis of the segmentation results for some selected iris samples. Columns (a), (e) represent the original images, columns (b), (f) represent the available ground-truths, columns (c), (g) represent the predicted iris masks, and columns (d), (h) represent the segmented iris region (yellow section). All the images are scaled uniformly for representational purpose (Color figure online).

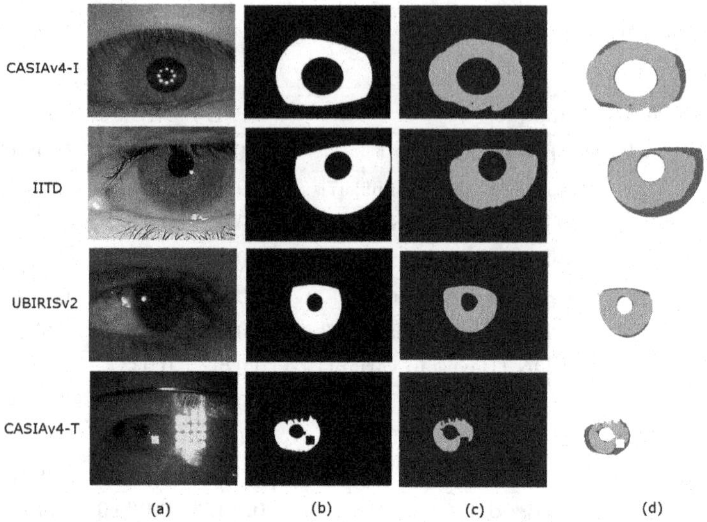

Fig. 3. Segmentation results for some selected noisy iris samples. Column (a) represents the original images, column (b) represents the corresponding ground-truths, column (c) represents the predicted iris masks, and column (d) shows the error in the predicted masks (red regions) (Color figure online).

the UBIRISv2 database is additionally characterized with low contrast since the entire UBIRISv2 database was collected in the VIS spectrum.

6 Conclusion

Our work in this paper introduces the *EAI-Net* model for accurately segmenting the iris region from eye images. While using conventional deep architectures, this problem is generally treated as a 2-class problem where the iris is considered as the foreground and rest of the eye is considered as the background. However, our proposed technique uses a combination of computational geometry techniques and morphological operations for pre-processing the ground-truth of the data while separating the pupil from iris. This 3-class ground-truth is subsequently used for training the U-Net architecture whose receptive fields have been calculated for accurately recognizing the structure of the iris. We have performed extensive empirical tests on four benchmark iris databases for demonstrating the efficacy of our model in both the visible and NIR spectrum. Importantly, EAI-Net is able to accurately segment the iris region for two of the most challenging iris databases, namely UBIRISv2 and CASIAv4-T. In the future extension of our work, we would investigate this model in combination with region proposal networks for extracting the iris region after initially localizing the eyes. Furthermore, we would like to focus on developing strategies that seek to optimize performance and computational aspects of the used architecture.

References

1. Bazrafkan, S., Thavalengal, S., Corcoran, P.: An end to end deep neural network for iris segmentation in unconstrained scenarios. Neural Netw. **106**, 79–95 (2018)
2. Bezerra, C.S., et al.: Robust iris segmentation based on fully convolutional networks and generative adversarial networks. CoRR. http://arxiv.org/abs/1809.00769 (2018)
3. Daugman, J.: Information theory and the iriscode. IEEE Trans. Inf. Forensics Secur. **11**(2), 400–409 (2016)
4. Graham, R.L., Yao, F.F.: Finding the convex hull of a simple polygon. J. Algorithms **4**(4), 324–331 (1983)
5. Hofbauer, H., Alonso-Fernandez, F., Wild, P., Bigun, J., Uhl, A.: A ground truth for iris segmentation. In: 2014 22nd International Conference on Pattern Recognition, pp. 527–532 (August 2014)
6. Jalilian, E., Uhl, A., Kwitt, R.: Domain adaptation for CNN based iris segmentation. In: 2017 International Conference of the Biometrics Special Interest Group (BIOSIG), pp. 1–6 (September 2017)
7. Jalilian, E., Uhl, A.: Iris segmentation using fully convolutional encoder–decoder networks. In: Bhanu, B., Kumar, A. (eds.) Deep Learning for Biometrics. ACVPR, pp. 133–155. Springer, Cham (2017). https://doi.org/10.1007/978-3-319-61657-5_6
8. Jégou, S., Drozdzal, M., Vazquez, D., Romero, A., Bengio, Y.: The one hundred layers tiramisu: fully convolutional densenets for semantic segmentation. In: 2017 IEEE Conference on Computer Vision and Pattern Recognition Workshops (CVPRW), pp. 1175–1183. IEEE (2017)
9. Kingma, D.P., Ba, J.: Adam: a method for stochastic optimization. arXiv preprint. arXiv:1412.6980 (2014)
10. Kumar, A., Passi, A.: Comparison and combination of iris matchers for reliable personal authentication. Pattern Recogn. **43**(3), 1016–1026 (2010)
11. Lian, S., Luo, Z., Zhong, Z., Lin, X., Su, S., Li, S.: Attention guided U-Net for accurate iris segmentation. J. Vis. Commun. Image Represent. **56**, 296–304 (2018)
12. Liu, N., Li, H., Zhang, M., Liu, J., Sun, Z., Tan, T.: Accurate iris segmentation in non-cooperative environments using fully convolutional networks. In: 2016 International Conference on Biometrics (ICB), pp. 1–8 (June 2016)
13. Long, J., Shelhamer, E., Darrell, T.: Fully convolutional networks for semantic segmentation. In: Proceedings of the IEEE Conference on Computer Vision and Pattern Recognition, pp. 3431–3440 (2015)
14. Lozej, J., Meden, B., Struc, V., Peer, P.: End-to-end iris segmentation using U-Net. In: 2018 IEEE International Work Conference on Bioinspired Intelligence (IWOBI), pp. 1–6 (July 2018)
15. Moreira, A., Santos, M.Y.: Concave hull: a k-nearest neighbours approach for the computation of the region occupied by a set of points (2007)
16. Proenca, H., Filipe, S., Santos, R., Oliveira, J., Alexandre, L.A.: The ubiris. V2: a database of visible wavelength iris images captured on-the-move and at-a-distance. IEEE Trans. Pattern Anal. Mach. Intell. **32**(8), 1529–1535 (2010)
17. Proenca, H., Alexandre, L.A.: Iris recognition: analysis of the error rates regarding the accuracy of the segmentation stage. Image Vis. Comput. **28**(1), 202–206 (2010)
18. Ronneberger, O., Fischer, P., Brox, T.: U-Net: convolutional networks for biomedical image segmentation. In: Navab, N., Hornegger, J., Wells, W.M., Frangi, A.F. (eds.) MICCAI 2015. LNCS, vol. 9351, pp. 234–241. Springer, Cham (2015). https://doi.org/10.1007/978-3-319-24574-4_28
19. Suzuki, S., et al.: Topological structural analysis of digitized binary images by border following. Comput. Vis. Graph. Image Process. **30**(1), 32–46 (1985)

Comparative Analysis of Artificial Neural Network and XGBoost Algorithm for PolSAR Image Classification

Nimrabanu Memon[1], Samir B. Patel[2](✉), and Dhruvesh P. Patel[1]

[1] Civil Engineering, School of Technology, Pandit Deendayal Petroleum Univerisity, Gandhinagar, Gujarat, India
{nimrabanu.hphd18,dhruvesh.patel}@sot.pdpu.ac.in

[2] Computer Science and Engineering, School of Technology, Pandit Deendayal Petroleum Univerisity, Gandhinagar, Gujarat, India
samir.patel@sot.pdpu.ac.in
http://www.pdpu.ac.in/

Abstract. Image classification has become an important area of research in remote sensing. In this paper, the algorithms Artificial Neural Network (ANN) and Extreme Gradient Boosting (XGBoost) are used to classify compact polarimetric (CP) RISAT-1 cFRS mode data for land cover categorisation over Mumbai region. After preprocessing, Raney decomposition technique was applied to obtain the R, G, B channels of the image. Hyperparameter tuning of ANN was also performed to get the optimal parameters for the classification. Comparative analysis showed that both the algorithms showed almost equal performance on the data in terms of accuracy. However, there was only 1% of the increment found in both the train and test the accuracy of XGBoost classifier. ANN method required tuning, and thus it requires more time for computation while XGBoost algorithm works well without any tuning and thus, XGBoost outperforms the image classification task for CP RISAT-1 data than ANN.

Keywords: ANN · XGBoost · CP · Machine learning · SAR · RISAT-1

1 Introduction

Recently, Compact Polarimetry (CP) is an important topic of research among researchers due to its potentiality and advantages over single, dual, as well as quad-pol data [14]. It's wide swath coverage and half transmitted power requirement is an added advantage [21]. So far it has proven its potentiality in various applications, one among them is the land cover (LC) classification which serves as a bases to different applications such as in hydrology, disaster management, agriculture monitoring, etc. On the other hand, machine learning is a powerful and popular tool, not only in the field of computer, and data science but also

B. Deka et al. (Eds.): PReMI 2019, LNCS 11941, pp. 452–460, 2019.
https://doi.org/10.1007/978-3-030-34869-4_49

in the field of remote sensing. Due to its self-learning property, it can easily handle the complexity of SAR data and help to gain more information from the imagery. In remote sensing, it is commonly used for image segmentation and classification. PolSAR image classification is generally performed using (1) statistical model [15,25], (2) scattering mechanism [15,24,27], or (3) through image processing [23,26]. Also, several studies are done using some combinations of the above three classification approaches [15,20].

Recently, there has been an addition in machine learning techniques to classify Polarimetric data, among which many advanced classifiers based on neural network had been excellent performers [5]. The most popular among them is an artificial neural network. There are also many complex neural network architecture that has increased the classification accuracy. However, besides the improvement in the accuracy, their complexity leads to time consumption. This ultimately makes them impractical to process the entire image (huge amount of data). Therefore, an efficient classifier would be the one with high accuracy and low time consuming in the practical use.

2 Why Artificial Neural Network and XGBoost?

2.1 Artificial Neural Network

Artificial neural network is a distribution free approach to image classification [7]. Many researchers so far have proven artificial neural network to be a powerful as well as self-adaptive method of pattern recognition as compared to traditional linear models [11,13]. From early 1990's, artificial neural networks are being applied to analyze the remotely sensing images with promising results [1]. The advantage of using ANN is

1. Its ability to learn complex functional relationships between input and output training data. It does this by employing a nonlinear response function, iterating in the network [16].
2. Generalization capability in the noisy environment makes it more robust in the presence of imprecise or missing data [10].
3. Ability to make use of a priori knowledge along with the realistic physical parameters for the analysis [6,8].

Thus, ANN perform tasks more accurately than any other statistical classification techniques, particularly on the complex feature space (having different statistical distribution) [2,22]. Comparatively, many studies have proved that ANN can classify remotely sensed data more precisely than supervised maximum likelihood classifier [3,4,9,12,19].

2.2 XGBoost (Extreme Gradient Boosting) [18]

XGBoost algorithm has become a dominating algorithm in the field of applied machine learning. It is used over other gradient boosting machines (GBMs) due

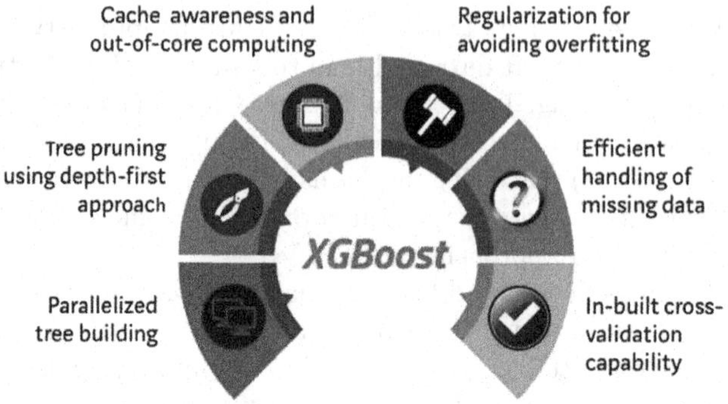

Fig. 1. Features in XGBoost for optimization [17]

to its fast execution speed and model performance. It can be run on c++, command line interface, Python, R, Julia, Java, and other JVM languages like scala. Figure 1 shows the special features of XGBoost that has improved the algorithm over other GBM frameworks. XGBoost includes parallel computation to construct trees using all the CPUs during training. Instead of traditional stopping criteria (i.e. criterion first), it make use of 'max_depth' parameter, and starts tree pruning from backward direction. This significantly improves the computational performance and speed of XGBoost over other GBM frameworks. Next, it can automatically learns best missing value depending on training loss, and thus, can handle different types of sparsity patterns in the input data more efficiently. In one study [17] compared various machine learning algorithms like random forest, logistic regression, standard gradient boosting along with XGBoost algorithm and found XGBoost to be most efficient of all the algorithms used in comparison. Both the classifiers have their potential in image classification and thus, in this paper a comparative study was made on RISAT-1 data between popular ANN and XGBoost for the LC classification of PolSAR data.

3 Study Area and Data Sources

The study is focused on Mumbai and its surrounding area, in Maharashtra, India. It's center is located at 19°13′14.99″N latitude and 72°55′58.03″E longitude. The major land features in this region are forest, mangrove forest, agriculture/fallow land, saltpans, urban, and water body. For this study, we have tried to covered all the major land cover classes for the level-1 classification of the study area and accuracy assessment of the obtained results is performed. Sensor specifications are provided in Table 1.

4 Methodology

SAR data provides high resolution images containing the scattering information, along with it is associated noise with salt and pepper texture, which may be the

Table 1. RISAT-1

Parameters	Description
Acquisition date	12 November, 2012
Incidence angle	35.986
Polarization	RH, RV
Pixel spacing	1.801
Line spacing	2.469
Altitude	541.366

result of interference of transmit and receive channels during data acquisition or may be due to some sensor error. So, it is the basic requirement of any SAR imagery to be preprocessed before classification. Thus, RISAT-1 imagery was first preprocessed & converted into coherency matrix. Next refined LEE filter with 5X5 window size was used for speckle removal and then 7:10 multilook was applied to convert the pixels in square pixels. After that, coherency matrix was decomposed into three scattering mechanism double, volume and odd bounce using Raney Decomposition [21]. Using double bounce band as red, volume as green and odd as blue band, an RGB image as shown in Fig. 2 was generated and selected machine learning techniques were applied for further segmentation and classification. Both the algorithms are supervised learning algorithms and thus ground truth data is required for training. Google Earth provided detailed level-1 information of land features and is sufficient for land cover classification and thus ground truth data (total of 46392 pixels) was prepared manually from google earth based on knowledge. Figure 3 shows the google earth image overlaid with ground truth pixels taken for training and testing the algorithms. Out of 46392 sample pixels 80% (37113 pixels) were randomly selected for training and remaining 20% (9279 pixels) were kept to test the performance of both the algorithms. The train-test size was chosen based on experiment and literature survey. Ground truth data was kept common for both the algorithms and both training and testing pixels were kept non overlapping to each other. The algorithms were implemented in python using tensorflow and keras libraries and ANN was tuned to find the optimal hyperparameters for the network with respect to our study. Finally, land cover maps were prepared with six LC classes (i.e. water body, urban, forest, saltpans, fallow, and mangrove forest). There may be other classes in the image, however, for initial investigation we have considered these six classes only. The classified maps were visually compared and accuracy assessment of each class was performed.

5 Results and Discussion

5.1 Training, Testing and Algorithm Tuning

There were seven hyperparameters to train the ANN with one hidden layer. One can manually search for these parameters or can employ advanced search

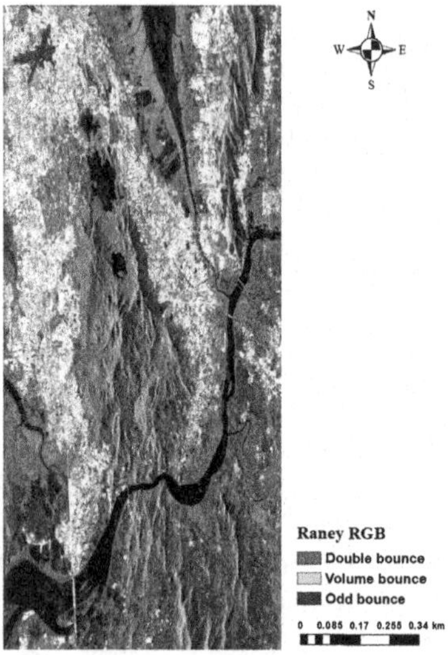

Fig. 2. Raney RGB (Color figure online)

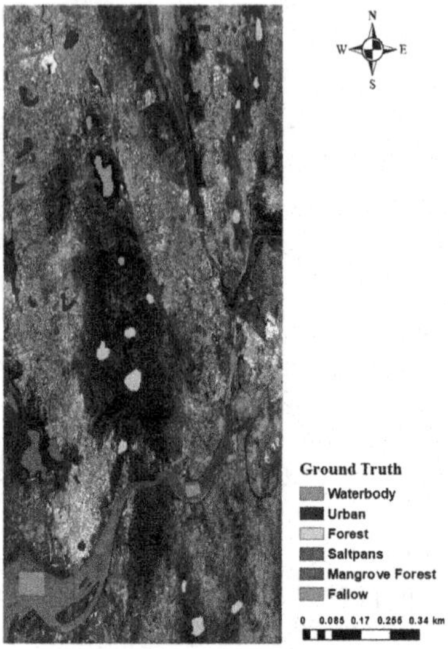

Fig. 3. Ground truth image

methods like GridSearchCV or RandomSearchCV method. In this paper we have used GridSearchCV method for tuning the network for four kernel initializers namely 'random uniform', 'uniform', 'Orthogonal', 'glorot normal'; two batch sizes (128,256); two optimizers namely 'sgd', 'adam'. Three activation techniques viz. 'relu', 'tanh', 'sigmoid' were used to check their potentiality for 100 epochs and 20 nodes (based on initial experiments) in both the input and the hidden layers. Generally, there is no thumb rule to choose number of epochs to train the classifier. The best way to choose epochs is to plot the accuracy vs epochs graph and select the saturation point for the accuracy. So we initially tested the algorithm for 50, and 100 epochs using 10-fold cross validation and through the analysis we decided to use 100 epochs for further training. Optimal parameters found through GridSearchCV is shown in Table 2. Figure 4 demonstrates one of the parameter combinations comparison obtained through GridSearchCV method. These parameters were used to train ANN.

Table 2. Optimal hyper-parameters obtained after tuning

Activation	Batch size	Epochs	Kernel initializer	Optimizer	Units (Neurons)
Relu	128	100	Uniform	Adam	20

5.2 Accuracy Assessment

The models were evaluated using k-fold cross validation technique with the default ten cross validation folds. K-fold cv accuracy of XGBoost algorithm was

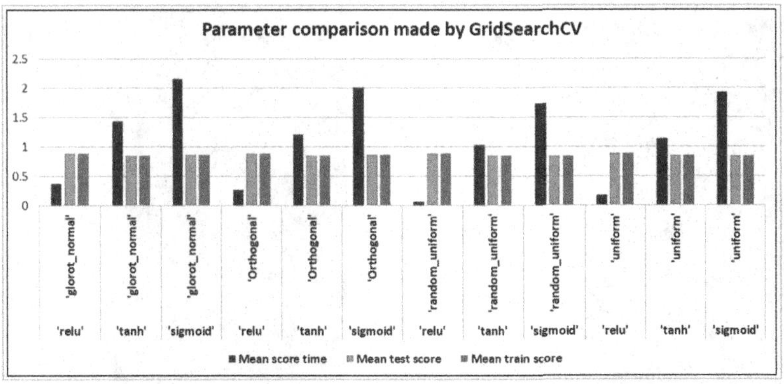

Fig. 4. GridSearchCV comparing parameters with batch size 128, 100 epochs and 20 neurons

found better than ANN and thus supports the superiority of XGBoost classifier (in terms of accuracy) than ANN on the given dataset. Accuracy of each class is shown in Fig. 5. It indicated that urban, water and fallow classes showed same number of correctly classified pixels in both the classifiers. Also, an increment in mangroves, & forest class and decrement in saltpans category was observed for XGBoost classifier. ANN achieved 91.92% of the training accuracy and 91.62% of the testing accuracy. Next the same train and test dataset were used for XGBoost. XGBoost achieved 92.25%, and 92.08% accuracy respectively. The classified images are shown in the Figs. 6 and 7. The algorithms were trained on high computational GPU then also ANN took almost 15 hrs to complete the tuning process of selected parameters through GridSearchCV method, while XGBoost algorithm took not more than 30 min to complete the entire classification process. XGBoost did not require external parameter tuning and gave almost equivalent results in terms of accuracy on the same data for ANN and thus was comparatively faster. XGBoost outperformed the comparative analysis.

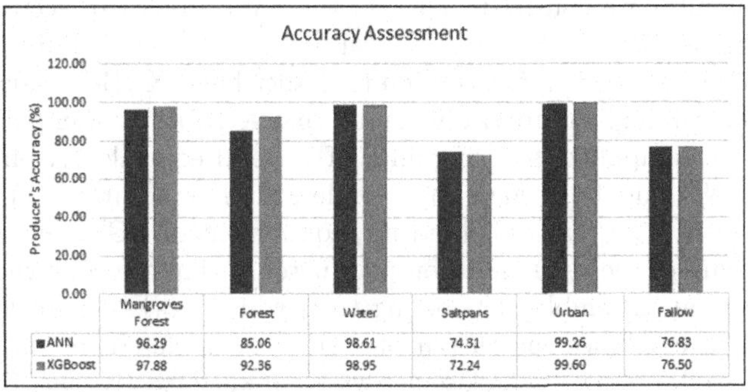

Fig. 5. Accuracy assessment

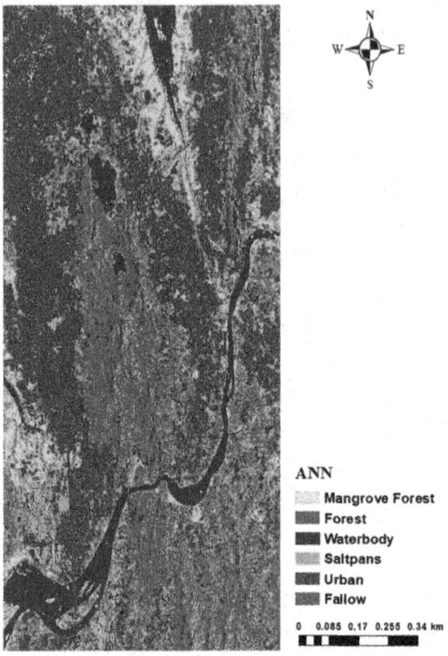

Fig. 6. ANN

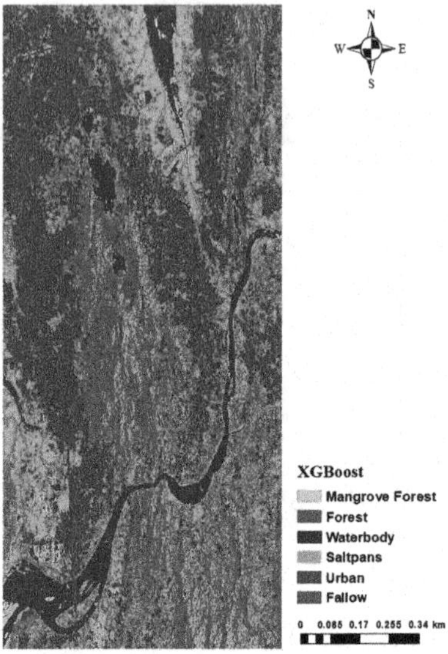

Fig. 7. XGBoost

6 Conclusion

PolSAR image classification is a very tedious task due to the complexity associated with the data. One can make use of polarimetric features using suitable decomposition techniques to make them more interpretable, and to construct RGB image. Machine learning has good potentiality to handle the complex data and thus is very helpful in classification of PolSAR data. In the present study hybrid polarimetric SAR data RISAT-1 has been used and Raney decomposition was applied in order to generate red, green and blue channels for the image. Gdal library was used in python, to create RGB image. Two supervised machine learning algorithms were used to classify the data and their accuracy assessment has been performed. Artificial Neural Network required tuning to find the optimal parameters for classification, on the other hand XGBoost algorithm did not require any tuning. Comparison showed that XGBoost is comparatively fast (due to parallel computations using all CPU's), and equivalently efficient algorithm than ANN and both gave more or less similar accuracy on both train and test datasets. The reason behind no requirement of XGBoost algorithm is that it takes care of the regularization parameter during the construction of the algorithm. It is an ensemble of boosting trees which works on weak classifiers. It means that it combines the output of various weak classifiers and only those classes are passed to next weak classifier which are incorrectly classified and further decision tree approach is applied on the passed inputs, this process is self repeated until all the input pixels are covered followed by majority voting. All these steps are automatically performed during the training phase and thus less tuning is needed. This makes it an ensemble of various weak classifiers(decision trees, like in random forest). The approach is similar to random forest but it uses gradient descent method to optimize the algorithm and thus it is found to be more efficient than ANN for this particular study.

Acknowledgement. Authors of this paper are thankful to SAC, Ahmedabad, for providing RISAT-1 dataset for the research. The research is a part of funded project "Land-Cover Classification of Polarimetric SAR Image/Data for Agricultural and Urban Region" by SAC, Ahmedabad under the grant "EPSA/3.1.1/2017". The authors have no conflict of interest regarding the publication of the research.

References

1. Atkinson, P.M., Tatnall, A.: Introduction neural networks in remote sensing. Int. J. Remote Sens. **18**(4), 699–709 (1997)
2. Benediktsson, J.A., Swain, P.H., Ersoy, O.K.: Conjugate-gradient neural networks in classification of multisource and very-high-dimensional remote sensing data. Int. J. Remote Sens. **14**(15), 2883–2903 (1993)
3. Chiuderi, A.: Multisource and multitemporal data in land cover classification tasks: the advantage offered by neural networks. In: 1997 IEEE International Geoscience and Remote Sensing. Remote Sensing-A Scientific Vision for Sustainable Development 1997, IGARSS 1997, vol. 4, pp. 1663–1665. IEEE (1997)
4. Civco, D.L.: Artificial neural networks for land-cover classification and mapping. Int. J. Geogr. Inf. Sci. **7**(2), 173–186 (1993)
5. Dong, H., Xu, X., Wang, L., Pu, F.: Gaofen-3 PolSAR image classification via XGBoost and polarimetric spatial information. Sensors **18**(2), 611 (2018)
6. Foody, G.M.: Land cover classification by an artificial neural network with ancillary information. Int. J. Geogr. Inf. Syst. **9**(5), 527–542 (1995)
7. Foody, G.M., McCulloch, M.B., Yates, W.B.: Classification of remotely sensed data by an artificial neural network: issues related to training data. Photogram. Eng. Remote Sens. **61**(4), 391–401 (1995)
8. Foody, G.: Using prior knowledge in artificial neural network classification with a minimal training set. Remote Sens. **16**(2), 301–312 (1995)
9. Gopal, S., Woodcock, C.E., Strahler, A.H.: Fuzzy neural network classification of global land cover from a 1 AVHRR data set. Remote Sens. Environ. **67**(2), 230–243 (1999)
10. Hewitson, B.C., Crane, R.G.: Neural Nets: Applications in Geography: Applications for Geography, vol. 29. Springer, Berlin (1994). https://doi.org/10.1007/978-94-011-1122-5
11. Jiang, D., Yang, X., Clinton, N., Wang, N.: An artificial neural network model for estimating crop yields using remotely sensed information. Int. J. Remote Sens. **25**(9), 1723–1732 (2004)
12. Kavzoglu, T., Mather, P.M.: Pruning artificial neural networks: an example using land cover classification of multi-sensor images. Int. J. Remote Sens. **20**(14), 2787–2803 (1999)
13. Keiner, L.E., Yan, X.H.: A neural network model for estimating sea surface chlorophyll and sediments from thematic mapper imagery. Remote Sens. Environ. **66**(2), 153–165 (1998). https://doi.org/10.1016/S0034-4257(98)00054-6. http://www.sciencedirect.com/science/article/pii/S0034425798000546

14. Kumar, V., Rao, Y.: Comparative analysis of RISAT-1 and simulated RADARSAT-2 hybrid polarimetric SAR data for different land features. Int. Arch. Photogramm. Remote Sens. Spat. Inf. Sci. **8** (2014)
15. Lee, J.S., Grunes, M.R., Ainsworth, T.L., Du, L.J., Schuler, D.L., Cloude, S.R.: Unsupervised classification using polarimetric decomposition and the complex Wishart classifier. IEEE Trans. Geosci. Remote Sens. **37**(5), 2249–2258 (1999)
16. Lek, S., Guégan, J.F.: Artificial neural networks as a tool in ecological modelling, an introduction. Ecol. Model. **120**(2–3), 65–73 (1999)
17. Morde, V.: XGBoost Algorithm: Long May She Reign! https://towardsdatascience.com/https-medium-com-vishalmorde-xgboost-algorithm-long-she-may-rein-edd9f99be63d. Accessed 20 Aug 2019
18. Norbert: Docker-Course-XGBoost (2019). https://github.com/ParrotPrediction/docker-course-xgboost. Accessed 5 May 2019
19. Paola, J.D., Schowengerdt, R.A.: A detailed comparison of backpropagation neural network and maximum-likelihood classifiers for urban land use classification. IEEE Trans. Geosci. Remote Sens. **33**(4), 981–996 (1995)
20. Pottier, E.: Unsupervised classification scheme of Polsar images based on the complex Wishart distribution and H/A/α polarimetric decomposition theorems. In: Proceedings of 3rd European Conference on Synthetic Aperture Radar, EUSAR 2000 (2000)
21. Raney, R.K.: Hybrid-polarity SAR architecture. IEEE Trans. Geosci. Remote Sens. **45**(11), 3397–3404 (2007)
22. Schalkoff, R.J.: Pattern Recognition. Wiley, Hoboken (1992)
23. Tan, C.P., Lim, K.S., Ewe, H.T.: Image processing in polarimetric SAR images using a hybrid entropy decomposition and maximum likelihood (EDML). In: 2007 5th International Symposium on Image and Signal Processing and Analysis, pp. 418–422. IEEE (2007)
24. Van Zyl, J.J.: Unsupervised classification of scattering behavior using radar polarimetry data. IEEE Trans. Geosci. Remote Sens. **27**(1), 36–45 (1989)
25. Wu, Y., Ji, K., Yu, W., Su, Y.: Region-based classification of polarimetric SAR images using Wishart MRF. IEEE Geosci. Remote Sens. Lett. **5**(4), 668–672 (2008)
26. Ye, Z., Lu, C.C.: Wavelet-based unsupervised SAR image segmentation using hidden Markov tree models. In: Object Recognition Supported by User Interaction for Service Robots, vol. 2, pp. 729–732. IEEE (2002)
27. Zou, T., Yang, W., Dai, D., Sun, H.: Polarimetric SAR image classification using multifeatures combination and extremely randomized clustering forests. EURASIP J. Adv. Signal Process. **2010**, 4 (2010)

Multi-scale Attention Aided Multi-Resolution Network for Human Pose Estimation

Srinika Selvam(✉) and Deepak Mishra(✉)

Indian Institute of Space Science and Technology, Thiruvananthapuram, Kerala, India
srinika0812@gmail.com, deepak.mishra@iist.ac.in

Abstract. In this paper, we propose attention maps at various scales on multi-resolution feature extractor baseline network for human pose estimation. The baseline network captures information across various scales with the help of repeated bottom-up and top-down approach using successive pooling and up-sampling. We propose a network named Refinement Net for regressing the predicted heatmaps to 2D joint locations to remove ambiguities in predicted position. We experiment with three levels of attention schemes - global, heatmap and multi-resolution. Attention masks helps in generating basin of attraction that helps the network on deciding where to "look". The proposed network performance is at par with the state-of-the-art two dimensional pose estimation methods on MPII dataset.

Keywords: Human pose estimation · Multi-resolution · Attention maps

1 Introduction

Pose estimation is a fundamental yet challenging problem in computer vision. Human pose estimation is task of identifying humans in an image and recovering their body poses. It is one of the longest serving problems in computer vision due to complex models involved in observing poses. People's poses are extensively used for pedestrian detection, human-robot interaction, sign-language understanding, virtual reality, etc. Pose estimation can be solved very efficiently by posing it as a deep learning problem. Hence, it has seen a lot of advancements over the recent years.

Convolutional neural networks (CNN) have seen an explosive success in the field of deep learning. Most state-of-the art pose estimation networks use CNN architectures to achieve outstanding performance on publicly available datasets such as MPII, Human3.6M, Leeds Sports, COCO, etc. In this paper, we propose a CNN architecture, called multi-resolution network (MRN) that has conv-deconv structure by successive pooling and up-sampling. We stack multiple MRNs to allow repeated refinement of prediction at consecutive stages. The prediction

B. Deka et al. (Eds.): PReMI 2019, LNCS 11941, pp. 461–472, 2019.
https://doi.org/10.1007/978-3-030-34869-4_50

maps are then fed to Refinement Network for regressing accurate 2D joint locations. Since the joint locations do not depend on the background region (non-human), these regions need to be eliminated before feeding into the pose estimation network to prevent the network from learning false features. Hence we propose attention schemes at various scales that allows the network to learn the required attention basin. Attention is a self-learning module and does not require any labelled data while training. The output of the attention module is a probability map that boosts near-human regions and suppresses other regions that are not useful for identifying the joint locations. Detailed explanation of the work is described in Sect. 3.

This work makes the following contributions:

- **Multi-Resolution Network (MRN)** - allows repeated bottom-up and top-down inference across various scales. Output of MRN is a single heatmap for all 16 joint locations.
- **Refinement Net** - Output of MRN is fed to Refinement Net for accurate regression of 2D joint locations from output heatmap.
- **Multi-scale Attention Modules** - To help the network learn faster and efficiently, we propose attention modules to boost regions of interest and mask unimportant regions.

Baseline network without attention achieves highest accuracy of 83.2%. However, with global attention, we achieve 3.5% increase in accuracy whereas with multi-resolution attention, we get 0.9% increase on MPII dataset.

2 Related Work

CNNs have revolutionized the task of human pose estimation by incorporating complex feature extracting network with reduced the number of parameters as compared to fully connected networks. Although it has significantly enhanced the performance, learning the exact (x, y) positions of body joints from an image is a complex task. To solve this problem, researchers turned to the heatmap, which is made by placing Gaussian blobs at every joint location. Plenty of methods were designed to regress the heatmap instead of (x, y) coordinates, such as Tompson et al. [16], Newell et al. [13], Wei et al. [18].

Pose Estimation. Introduction of Stacked Hour-Glass structure by [13] has certainly steered pose estimation to a new direction by focusing on multi-scale feature extraction. It uses CNN as their base and produces heatmap output for each joint location. Other major contributions to pose estimation include [1,9,10]. Numerous variations to [13] were also experimented by [4,6,8]. Cascaded pyramid network by [3] came up with a module named RefineNet for fine tuning of the heatmaps obtained from previous network module. RefineNet concatenates all the pyramid features rather than simply using the up-sampled features at the end of hourglass module. Different from RefineNet, we propose a module designed to eliminate any ambiguity that could arise due to spread spectrum of Gaussian blobs at each joint location or due to ordering of joint locations. This module regresses the heatmap output to corresponding joint locations.

Self Visual Attention. Visual attention concept is computationally efficient and effective for object detection [7,22], image recognition [2,17], caption generation [11,19,21], etc. In most cases, attention masks are usually learned from extra bounding box manual annotations. Learning maps from annotated data not only is time-consuming but also does not bring any progress to the area of deep learning. However, very few work has been done incorporating attention modules for human pose estimation, they include [5,15]. Attention aided pose estimation helps the network learn the regions of interest in the image and lead to faster convergence.

3 Network Architecture

In this section we elaborate the important constituents of the proposed human pose estimation architecture i.e. MRN, Refinement Net, and attention modules.

3.1 Multi-Resolution Network

The proposed network architecture is as described in Fig. 1. The architecture is inspired from Stacked Hourglass structure [13]. It was aimed at collecting information from various scales from the image. In the original implementation of the hourglass network, they make use of residual blocks that have skip connections within itself. We believe that the second tier of skip connection is redundant and does not help in learning better features required for pose estimation. Our network omits the implementation of residual blocks. The network starts with 7×7 convolutional layer with stride 2, followed by a residual module and a round of max pooling to bring the resolution down from 256 to 64. This is implemented in the Front Module. The network branches off to skip connections at every resolution to apply more convolutions at the pre-convolved layers as shown in Fig. 1. After reaching the final resolution of 4×4, the network does nearest neighbour interpolation with few more convolutional layers to bring back the network output to 64×64 resolution. At the lowest resolution, two consecutive rounds of 1×1 convolutional are performed. At each level of resolution, there are 256 feature vectors. At every up-sampled resolution, the feature vector is added with the skip connections coming from the corresponding resolution of bottom part of the hourglass. Local evidence helps in identifying faces, hands, etc whereas the consolidated features from various scales help in understanding the body pose. This feature is very critical to the network's performance over other networks. We extend our network architecture by stacking multiple MRNs back-to-back, feeding the output of one as input to the next. This provides the network a mechanism for repeated bottom-up, top-down inference allowing for reevaluation of initial estimates and features across the whole image. Multiple

MRNs help in refining the output prediction of previous MRNs. The output of this network is a single heatmap for all joint locations in the image. The resulting prediction heatmap value at a position gives the probability of any joint occurring at that location.

3.2 Refinement Network

Predictions from MRN does not provide any information about correspondence of prediction to joint locations. That is, we do not know which Gaussian peak corresponds to which joint in the human body. In order to regress body joint locations from the heatmap, we propose a module named "Refinement Net". Output of stacked MRNs is fed to Refinement Net. Refinement net helps in learning probability map to joint correspondence. The order of output of the Refinement network encodes the information about the sequence of body joint locations. Also, the heatmaps are obtained by placing Gaussian blobs at corresponding true values of the 2D pixel locations in the image. So, when going back from heatmap to joint locations, due to the spread spectrum of Gaussian blob, multiple peaks for a single joint location may occur. To avoid this complication, we redundantly predict 2D joint locations from the predicted heatmap from previous network. This module refines the prediction from MRN. Moreover, while training with heatmaps, the network (Front Module + MRN) gets stuck at a local minima corresponding to a heatmap with all joint prediction values 0.5. Refinement Network also helps in overcoming this saddle point and allows faster convergence of the model. Refinement network architecture is as described in Fig. 2 and the overall network architecture is as described in Fig. 3.

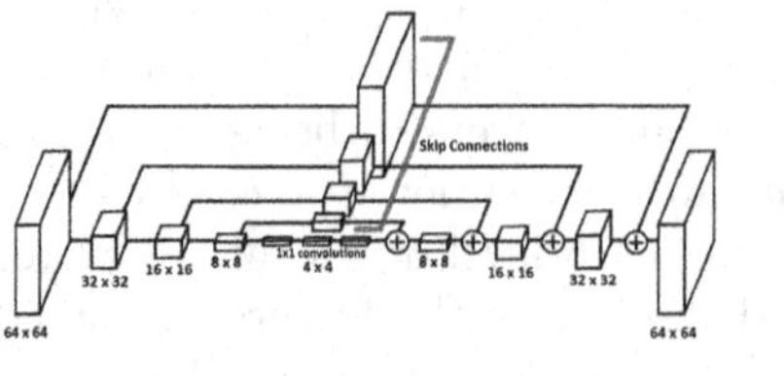

Fig. 1. Multi-resolution network is inspired from the popular Stacked hourglass network - aimed at capturing features at various resolution. Each block in MRN is a convolutional layer with 256 feature vectors - unlike stacked hourglass in which each block is a residual module.

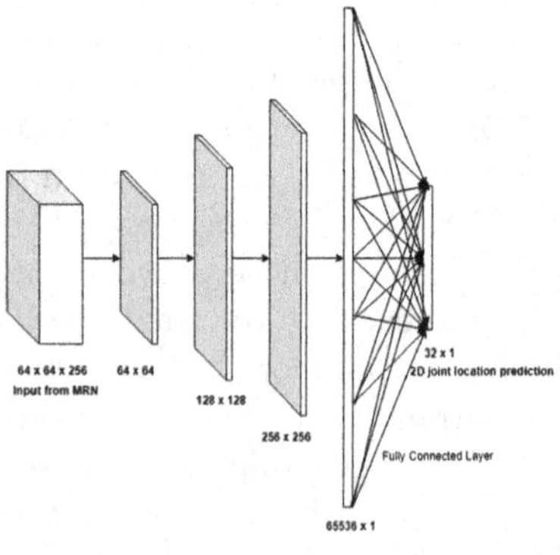

Fig. 2. Architecture of Refinement Net. Refinement Net helps in regressing joint positions from heatmaps that lacks information about correspondence between predictions and joint locations.

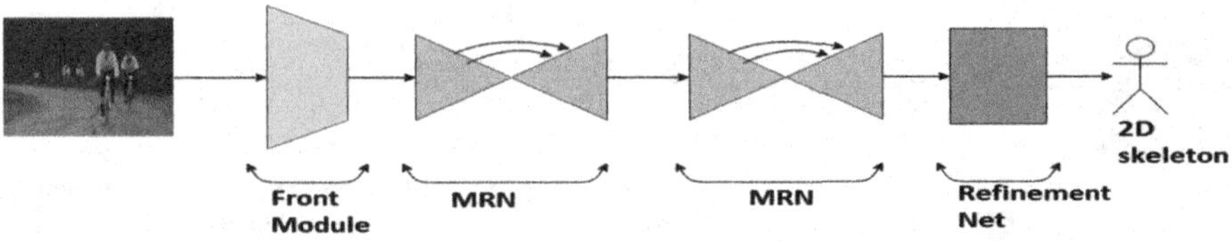

Fig. 3. Baseline network with MRN and Refinement Network

3.3 Attention Module

Attention schemes help the network in deciding where to look. The output of the attention module is a relevance map that boosts near-human regions and suppresses other regions that might not be useful for identifying the joint locations. Attention can be of two types: 1. Hard Attention and 2. Soft Attention. In hard attention, the attention mask has binary values - $\{0, 1\}$. 0 at a location indicates that information content at that position is irrelevant to the task of joint prediction whereas 1 indicates otherwise. In soft attention, each pixel position has values in the range [0, 1] - indicative of the relevance of the information at that position for output prediction. In our implementation, we extensively use soft attention masking since hard attention masks are difficult to back propagate. In this work, we propose attention schemes at three levels: **Global**, **Heatmap** and **Multi-Scale** Attention Schemes. Moreover, we do not require any manual annotations for attention masks.

Global Attention. Global attention is implemented at the input resolution. Every image is passed through global attention module and then to MRN and Refinement Net. The output of the module is an attention mask, same size as that of the image. The attention mask is element-wise multiplied (denoted by $\odot$) with all 3 channels of the image. This emphasizes the regions of interest and de-emphasizes unimportant regions in the original image. The output is then fed to MRN followed by Refinement Net. Global attention module also has hourglass structure with skip connections. It is aimed at capturing masking information from various scales across the image. Figure 4 shows the architecture of Global attention scheme.

Attention mask on the image is generated by passing the image through attention convolutional layers described in Eq. 1 where f denotes the attention transformation, I denotes the input image, W_{att} is the weights of the attention module, M is the attention mask and X is the input to stacked MRNs.

$$M = f_{att}(I; W_{att}) \tag{1}$$

$$X = I \odot M \tag{2}$$

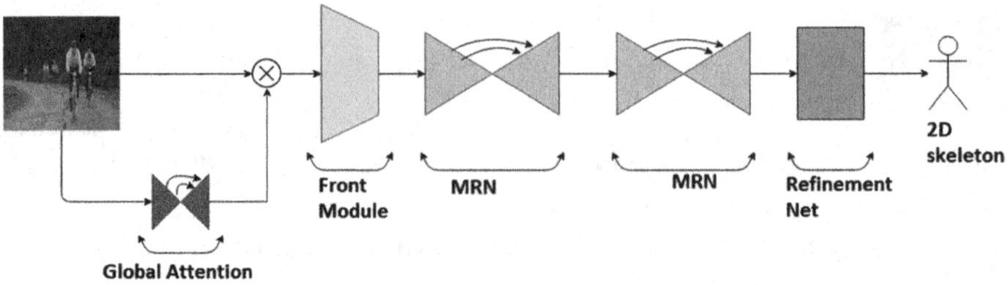

Fig. 4. Global Attention Scheme - Attention basins are generated on the input image. Unimportant regions are masked before passing through the network. Attention map dimension is as that of the input image.

Heat Map Attention. Heat map attention masks are implemented on the output of stacked MRNs. Attention module for heatmap predictions is an redundant layer that helps in filtering out wrong predictions before passing the predictions onto subsequent stages. This is important because, in our implementation of stacked MRN network, the subsequent stages learn from the previous stage's prediction. Applying attention onto each of the predicted output heatmaps will help the network learn faster and efficiently. Attention mask in each layer is element-wise multiplied with the predicted heatmap and passed to the next stage of the hourglass. Figure 5 shows the architecture of heatmap attention scheme.

Let X^i denote the output (heatmap) of ith MRN. To generate attention maps on X_i, we pass it through attention module (Eq. 3). This mask is element-wise multiplied with heatmap prediction and added with original prediction to enhance the regions of interest (Eq. 4). This is done in order to overcome the problem of vanishing gradient. Attention mask M^i is a relevance map that consists of values in the range [0, 1]. 1 implies high relevance and 0 implies low relevance. When this mask is multiplied with the predicted heatmap which is a probability map whose range is also [0, 1], the resulting values are infinitesimal that leads to vanishing gradient when propagated through multiple stacks of MRN. Hence, we add identity mapping to avoid this complication. Y^i is the input to *(i+1)*th MRN.

$$M^i = f_{att}(X^i; W^i_{att}) \tag{3}$$

$$Y^i = X^i \odot M^i + X^i \tag{4}$$

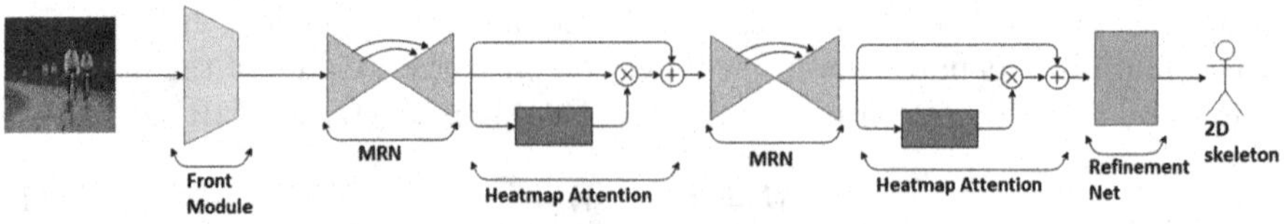

Fig. 5. Heatmap attention scheme - Attention basins are generated for each heatmap. Unwanted regions of the heatmap are masked before passing through subsequent stacks of MRN. Each attention map dimension is 64 × 64 - as that of MRN input/output.

Multi-Resolution Attention. Instead of building an explicit attention module, we can incorporate attention into MRNs itself. This attention scheme learns attention masks on all resolutions present in MRN. The attention masks run parallel to the skip connections in MRN. Each resolution attention mask is element-wise multiplied with corresponding resolution layer output and propagated to the subsequent layers in each hourglass and eventually to subsequent MRN stacks and Refinement net. Multi-resolution attention emphasize relevant information present at each scale. This improves the quality of each resolution layer output as well as predicted 2D joint positions. Figure 6 shows the architecture of multi-resolution attention scheme.

Let X_t^i denote the ith layer in the top half of the MRN and X_b^i denote its corresponding layer in the bottom half in a MRN. Attention maps are generated only on bottom half of MRN. Attention maps are learnt from X_b^i. X_b^i is passed through the attention module. This mask is element-wise multiplied with the corresponding feature vector for upsampled layer in the network and added with the original feature vector to enhance the regions of importance to overcome the problem of vanishing gradient as described in Subsect. 3.3. Y^i is added with the skip connections coming from the corresponding top section of the MRN. X_s^i denotes the skip connection coming from i^{th} top layer in MRN. Z^i is the input to $(i+1)^{th}$ layer in MRN. This is described in Eqs. 5, 6, and 7.

$$M^i = f_{att}^i(X_b^i; W_{att}^i) \tag{5}$$

$$Y^i = X_t^i \odot M^i + X_t^i \tag{6}$$

$$Z^i = Y^i + X_s^i \tag{7}$$

3.4 Training Details

Entire network was trained on MPII dataset with single person annotations with all visible body joints. At every convolutional layer, kernel and activity regularization are added to avoid over-fitting. We add batch normalization at every convolutional layers to speed up the training process, except in the 1 × 1 convolutional layers. Leaky ReLu activation is imposed for all convolutional layers, except the attention layers. We use sigmoid activation to generate the attention masks. Kernels of size (3, 3) and stride (1, 1) are used throughout MRNs. The network weights are all initialized with tensors of all ones. Adam optimizer is used to optimize the network on a single Titan X GPU. The network is trained on mini-batches of size 30 and shuffled between epochs. Network is trained using MSE as loss function between 16 predicted (x, y) joint positions

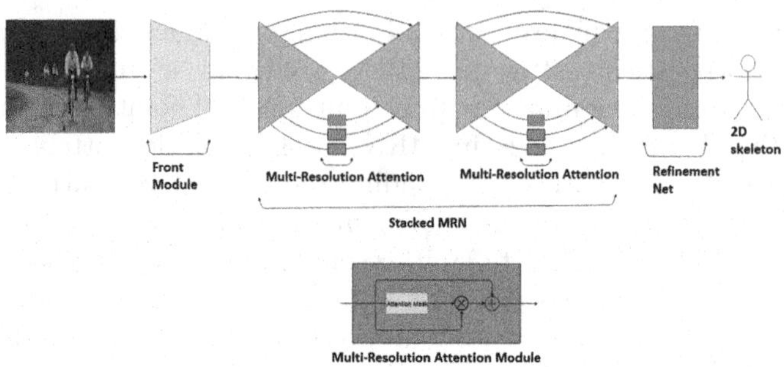

Fig. 6. Multi-resolution attention scheme - attention basins are generated on every scale in MRN. Unimportant regions are filtered out before passing through subsequent resolution levels in MRN. Each attention map has dimensions as that of each scale in MRN - 64 × 64, 32 × 32, 16 × 16, 8 × 8 and 4 × 4.

and ground truth as described in Eq. 8. For images with multiple people in it, output is predicted on a single person with all visible body joints.

$$Loss = Mse(P_{x,y}; G_{x,y}) = \sum_{i=1}^{16}(G^i_{x,y} - P^i_{x,y})^2 \tag{8}$$

Tasked Learning. The training method is derived from [15]. We represent body joint learning as task A and attention learning as task B. The result of task B has great influence on task A. For example, if task B filters out body regions, task A would never converge to a better solution, because the input feature maps do not contain much important information. So, firstly, we freeze the parameters of task B, and initialize all attention masks with all ones tensors and train task A. Then we freeze the parameters for task A, and train task B. Finally, we jointly train the two tasks. Training algorithm is as described in Algorithm 1. The learning rate is dropped by a factor 10 at the completion of each tasked training epoch.

4 Results

We compare our results with leading state-of-the-art results in Table 1. The metric used for comparison is average accuracy (in percent) over all body joints. Example results can be found in Fig. 7 for baseline network and in Fig. 9 for attention aided architecture. We also analyze the performance of the baseline network and attention aided network only various categories such as walking, dancing, exercising, standing, etc. Table 2 stages the results. Intermediate results of the baseline network are staged in Fig. 8.

Algorithm 1. Tasked Training Algorithm

```
procedure TRAIN(model, input, max_epochs)
    for i in range(max_epochs) do
        prediction ← model.predict(input)
        error ← (ground_truth − prediction)^2
        Back propagate error
Initialize model.attention_weights to ones
while convergence not attained do
    model.attention_weights.trainable ← False        ▷ Freeze attention weights
    TRAIN(model, input, max_epochs)
    model.attention_weights.trainable ← True         ▷ Release attention weights
    model.pose_network_weights.trainable ← False ▷ Freeze baseline network weights
    TRAIN(model, input, max_epochs)
    model.pose_network_weights.trainable ← True      ▷ Release baseline weights to
make the network end-to-end trainable
    TRAIN(model, input, max_epochs)
    Reduce learning_rate
                                                     ▷ while loop continues
```

Table 1. Comparison of results on MPII human pose dataset

Method	Average accuracy	Method	Average accuracy	Method	Average accuracy
[16]	82.0	[14]	82.4	[12]	85.0
[18]	88.5	[13]	90.9	[5]	91.5
[20]	92.0	[15]	92.0	Ours - Fig. 3	**83.2**
Ours - Fig. 4	**86.7**	Ours - Fig. 5	**78.6**	Ours - Fig. 6	**84.1**

Fig. 7. Example results of our baseline (MRN + Refinement Net) on MPII dataset

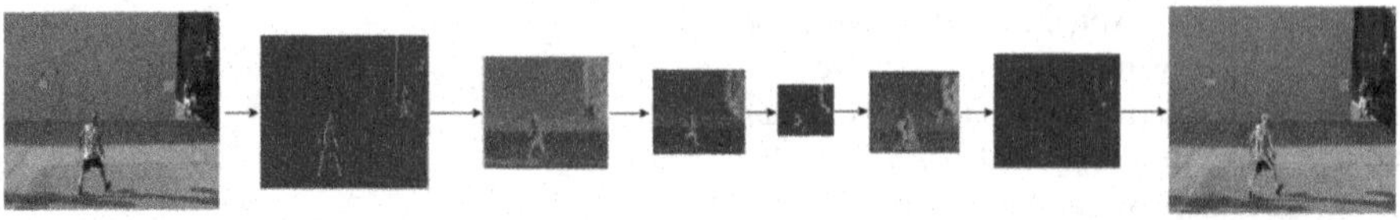

Fig. 8. Intermediate features extracted from the network. Bottom half of MRN does human localization whereas top half of the network derives higher level features from localized.

Fig. 9. Example results of our attention aided baseline network on MPII dataset. First column is the original input images. Second column is the corresponding attention masks generated for the input image. Third column is the predicted 2D joint locations of the network.

Table 2. Category-wise assessment of average accuracy (in percentage) on MPII dataset

Our method	Standing	Walking	Dancing	Skiing	Exercising	Bicycling	Mountain biking	Volleyball	Basketball
Fig. 3	81.93	**93.62**	76.67	84.29	82.82	**83.12**	**80.00**	60.2	78.42
Fig. 4	83.86	92.98	76.67	81.43	**87.00**	82.69	82.38	62.46	80.18
Fig. 5	81.93	89.36	76.67	80.00	79.90	79.89	73.31	60.82	78.42
Fig. 6	**85.44**	89.36	76.67	**84.29**	**87.00**	**83.12**	**80.00**	**67.38**	**81.93**

5 Conclusion

Our work is a significant contribution to attention-based human pose estimation since seldom researches have explored this field. Baseline network performs at an average accuracy of 83.2%. The reduced accuracy compared to [13] is due to omission of residual connections at the benefit of lesser number of parameters. With the use of attention schemes, the performance of the network (in terms of accuracy and time to convergence) has significantly improved. Global attention schemes gives an additional 3.5% increase in accuracy whereas multi-resolution attention gives an additional 0.9% improvement. However, heatmap attention modules decrease the performance of the baseline network by a significant 4.6%. Heatmaps have very concise and important information about the body joints. Imposing attention on heatmaps leads to disposal of important regions from the heatmap. Moreover when multiple MRNs are stacked, this leads to successive filtration of the heatmaps by the attention modules which leads to significant loss of information, reflecting in the reduced accuracy of the network.

References

1. Alp Güler, R., Neverova, N., Kokkinos, I.: DensePose: dense human pose estimation in the wild. In: CVPR (2018)
2. Chen, Y., Zhao, D., Lv, L., Li, C.: A visual attention based convolutional neural network for image classification. In: 2016 WCICA (2016)
3. Chen, Y., Wang, Z., Peng, Y., Zhang, Z., Yu, G., Sun, J.: Cascaded pyramid network for multi-person pose estimation. In: CVPR (2018)
4. Chou, C.J., Chien, J.T., Chen, H.T.: Self adversarial training for human pose estimation. In: 2018 APSIPA ASC (2018)
5. Chu, X., Yang, W., Ouyang, W., Ma, C., Yuille, A.L., Wang, X.: Multi-context attention for human pose estimation. In: CVPR (2017)
6. Guo, C., Du, W., Ying, N.: Multi-scale stacked hourglass network for human pose estimation (2018)
7. Hara, K., Liu, M.Y., Tuzel, O., Farahmand, A.M.: Attentional network for visual object detection. arXiv preprint arXiv:1702.01478 (2017)
8. Huang, F., Zeng, A., Liu, M., Qin, J., Xu, Q.: Structure-aware 3D hourglass network for hand pose estimation from single depth image. arXiv preprint arXiv:1812.10320 (2018)
9. Insafutdinov, E., et al.: Arttrack: articulated multi-person tracking in the wild. In: CVPR (2017)
10. Insafutdinov, E., Pishchulin, L., Andres, B., Andriluka, M., Schiele, B.: DeeperCut: a deeper, stronger, and faster multi-person pose estimation model. In: Leibe, B., Matas, J., Sebe, N., Welling, M. (eds.) ECCV 2016. LNCS, vol. 9910, pp. 34–50. Springer, Cham (2016). https://doi.org/10.1007/978-3-319-46466-4_3
11. Li, L., Tang, S., Deng, L., Zhang, Y., Tian, Q.: Image caption with global-local attention. In: Thirty-First AAAI Conference on Artificial Intelligence (2017)
12. Lifshitz, I., Fetaya, E., Ullman, S.: Human pose estimation using deep consensus voting. In: Leibe, B., Matas, J., Sebe, N., Welling, M. (eds.) ECCV 2016. LNCS, vol. 9906, pp. 246–260. Springer, Cham (2016). https://doi.org/10.1007/978-3-319-46475-6_16

13. Newell, A., Yang, K., Deng, J.: Stacked hourglass networks for human pose estimation. In: Leibe, B., Matas, J., Sebe, N., Welling, M. (eds.) ECCV 2016. LNCS, vol. 9912, pp. 483–499. Springer, Cham (2016). https://doi.org/10.1007/978-3-319-46484-8_29
14. Pishchulin, L., et al.: DeepCut: joint subset partition and labeling for multi person pose estimation. In: CVPR (2016)
15. Sun, G., Ye, C., Wang, K.: Focus on what's important: self-attention model for human pose estimation. arXiv preprint arXiv:1809.08371 (2018)
16. Tompson, J.J., Jain, A., LeCun, Y., Bregler, C.: Joint training of a convolutional network and a graphical model for human pose estimation (2014)
17. Wang, W., Shen, J.: Deep visual attention prediction. IEEE Trans. Image Process. **27**, 2368–2378 (2018)
18. Wei, S.E., Ramakrishna, V., Kanade, T., Sheikh, Y.: Convolutional pose machines. In: CVPR (2016)
19. Xu, K., et al.: Show, attend and tell: neural image caption generation with visual attention. In: ICML (2015)
20. Yang, W., Li, S., Ouyang, W., Li, H., Wang, X.: Learning feature pyramids for human pose estimation. In: ICCV (2017)
21. You, Q., Jin, H., Wang, Z., Fang, C., Luo, J.: Image captioning with semantic attention. In: CVPR (2016)
22. Zhang, D.Z., Liu, C.C.: A visual attention based object detection model beyond top-down and bottom-up mechanism. In: ITM Web of Conferences (2017)

Detection of Splicing Forgery Using CNN-Extracted Camera-Specific Features

Shiv Kumar Tiwari, Aniruddha Mazumdar(✉), and P. K. Bora

Department of Electronics and Electrical Engineering,
Indian Institute of Technology Guwahati, Guwahati 781039, Assam, India
kshiv0599@gmail.com, {m.aniruddha,prabin}@iitg.ac.in

Abstract. This paper proposes a convolutional neural network (CNN)-based method to detect forgeries in images, leveraging camera-specific features. The underlying principle is that all the pixels in a pristine image are captured by a single camera. On the other hand, the forged regions present in a spliced image are likely to be captured by cameras other than the one used to capture the pristine regions. The proposed method divides the test image into a number of non-overlapping patches, and extracts the camera-specific features from each patch using a CNN, which is trained for camera model identification. Once the features are obtained, they are clustered using the *hierarchical agglomerative clustering* technique for localizing the forgery. The test image is decided to be forged on the basis of the percentage of forged patches. The experimental results show the relative merits of the method over the state-of-the-art.

Keywords: Convolutional neural network · CNN · Deep learning · Image forensics · Camera model identificaction · Hierarchical agglomerative clustering

1 Introduction

Image forensics is a branch of study that focuses on the verification of the authenticity of images. Image forensics techniques can be broadly classified into two groups: *active* and *passive*. Active techniques rely on the insertion of digital signature or watermark into an image, which gets modified if the image is tampered to create forgeries. On the other hand, passive techniques do not require any prior knowledge about the test image, and can detect forgeries using the pixel intensity values only.

Image *splicing* is one of the popular image forgery techniques, which involves the creation of a composite image using parts from two or more images. Figure 1 shows the example of a splicing forgery, where parts from an image is copied and pasted onto a different image, to represent an event that has never happened. There are different approaches to detect splicing forgeries, such as compression-based [7], lighting condition-based [5], noise-based [6], camera sensor-based [2], [1] etc. In the compression-based approach, the properties of JPEG compression

B. Deka et al. (Eds.): PReMI 2019, LNCS 11941, pp. 473–481, 2019.
https://doi.org/10.1007/978-3-030-34869-4_51

are utilized to detect and localize the spliced regions. The lighting condition-based approach uses the mismatch in the lighting conditions estimated from different objects present in an image as a cue to detect the forgery. The inconsistencies in the noise profile is utilized for detecting the spliced parts in noise-based methods. In the camera sensor-based methods, different attributes of the imaging sensor, such as the color filter array (CFA) demosaicing technique and the photo response non-uniformity (PRNU) noise, are estimated and compared to reveal the presence of possible forgery.

This paper proposes a camera sensor-based method for the detection and localization of splicing forgery. The method is based on the intuition that the spliced region in an image is more likely to be captured by a camera which is different from the one used to capture the pristine regions. On the other hand, all the pixels in an authentic image are captured by a single camera only. Therefore, extracting camera-specific features from different parts of an image and checking their consistency may expose splicing forgery. The proposed method utilizes a convolutional neural network (CNN), trained for extracting camera-specific features for the camera model identification task. These features are further processed for detecting and localizing the spliced regions.

Fig. 1. An example of image splicing. (Photo tampering throughout history, http://www.fourandsix.com/photo-tampering-history.)

2 Related Work

Camera model identification finds the model of the source camera used to capture a given image. Motivated by the success of deep learning in various computer vision tasks, researchers have focused on developing CNN-based camera model identification techniques [1,9]. In these techniques, CNNs are trained on a large set of training images coming from different camera models. Once trained, the CNN learns to extract different features which maps the test images to their respective models.

Bondi *et al.* [2] proposed a method to detect and localize splicing forgeries using a CNN trained for camera model identification. While training the CNN, the authors considered only the patches having high texture contents and excluded the homogenous and the saturated patches. This is because the homogenous and the saturated patches do not contain enough statistical information to extract camera specific features [2]. Given an image under investigation, it is first divided into non-overlapping patches of size 64×64. An 18-dimentional feature vector is extracted from each patch using the pretrained CNN. The features are extracted from the last layer of the CNN, which is a softmax layer with 18 neurons. The output of each neuron in the softmax layer represents the probability of a patch being classified into one of the 18 camera models considered in the training stage. The features of all the patches present in the test image are clustered using the k-means clustering algorithm. If there are more than one clusters, the test image is decided as spliced, otherwise it is decided as authentic. In case of a spliced image, the patches corresponding to the features present in the cluster having the highest cardinality are considered as authentic patches and the rest are considered as spliced patches. The limitation of Bondi *et al.*'s method is that the features extracted from the softmax layer do not generalize well to unknown camera models. Since the output of the softmax layer is class probabilities, it can extract features from the patches coming from known camera models accurately. However, the patches coming from unknown camera models will also be classified into any of the known camera models. Therefore, the 18-dimensional feature may sometimes fail to differentiate between different camera models and hence may fail to detect splicing forgeries.

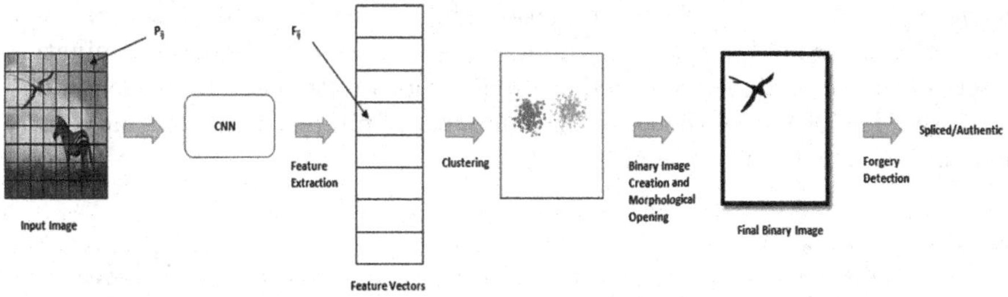

Fig. 2. Block diagram of the proposed method depicting the steps involved.

3 Proposed Method

The proposed method is based on the assumption that a spliced image is the composition of two or more images, which are more likely to be captured through different cameras. Therefore, extracting the camera-specific features from small patches of the images and clustering them may give information about the authenticity of the images. In case of an authentic image, since all the patches are captured by a single camera only, there should be only one cluster. On the other hand, if the features form more than one clusters in the feature space, then we can decide the test image to be spliced.

The main principles of the proposed method are as follows:

1. The method extracts the camera-specific feature from a layer before the softmax layer of a CNN, trained to identify camera models. Because of this, the extracted features are not directly dependant on the camera models present in the training stage. This in contrast to the features extracted by Bondi *et al.*, which are the probability scores extracted from the softmax layer of the CNN.
2. The extracted features are clustered using the *hierarchical agglomerative clustering (HAC)* technique [3,8] based on cosine distance. The cosine distance between two feature vectors extracted from two patches coming from the same camera is expected to be less than that coming from two different cameras.

Figure 2 shows the steps involved in the proposed method to localize and detect splicing forgeries. They are described below.

3.1 Feature Extraction Using CNN

3.1.1 CNN

The CNN architecture used in this work is similar to the one proposed by Bondi *et al.* [1]. It has 5 convolutional layers, 3 max-pooling layers and 2 fully connected layers. The input to the CNN is an image patch of size 64×64. There is a trade off between the camera model discrimination ability and the localization ability of the method. As the size of the patch increases, the camera model discrimination ability of the method increases and the splicing localization ability of the method decreases. The details of the kernel size, number of filters and the output feature map size of each layer are shown in Fig. 3.

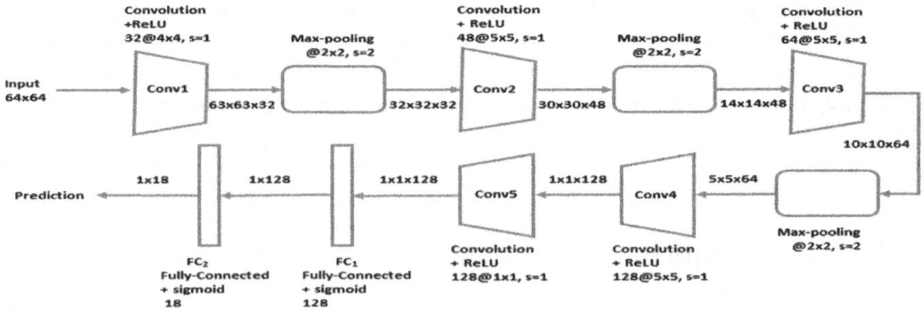

Fig. 3. The CNN architecture used in the proposed method.

3.1.2 Feature Extraction

Given an image under investigation I_T of size $W \times H$, it is first divided into a number of non-overlapping patches P_{ij} of size 64×64, where $i \in (1, \lfloor \frac{W}{64} \rfloor)$, $j \in (1, \lfloor \frac{H}{64} \rfloor)$ and $\lfloor . \rfloor$ denotes floor operator. Each image patch is fed to the CNN which is already trained to perform camera model identification. Since our motive is not to get the camera model scores, which is at the output of the last layer, i.e. the softmax layer, we propose to extract features from the first fully-connected layer as FC_1. As FC_1 has 128 neurons, the output features will be 128-dimensional vectors. The reasons for extracting features from FC_1 are as follows: Since the last layer classifies the patches to one of the camera model classes, the output of this layer will be dependent on the camera models present during training. Hence, the features extracted from this layer will not generalize to patches coming from cameras not present during the training stage. On the other hand, FC_1, does not learn to classify patches to any specific class. Rather, it extracts features which are more generic. Therefore, the features extracted from FC_1 will be more efficient in discriminating camera models not present in the training stage. In the feature space if feature vectors of two patches lie in close proximity they are considered to be from the same camera. If the feature vectors are far apart, they are considered to be captured from two different cameras.

3.1.3 Localization of Forged Regions using HAC

After obtaining the feature for all the patches, a binary image is estimated based on the proximity of patches in the feature space. Initially the CNN features for all the patches are clustered using HAC [3]. HAC combines feature vectors into small clusters, those clusters into larger clusters, and so forth, thus creating a hierarchy of clusters. Initially, all the CNN features are considered as a single-element cluster. At each step of the algorithm, the two clusters that are the most similar are combined into a new bigger cluster. This method is iterated till either all feature vectors are clustered into a single cluster or the distances between the clusters formed are greater than a certain threshold T_d. Therefore, the cluster which has the highest number of patches, is considered to be the one consisting of the authentic patches, and the other clusters are considered to be composed of either tampered regions or noise. In the case of a pristine image, all the features will be clustered into a single cluster. In the case of a spliced image, we are likely to get more than one cluster. This is because there is a high chance that the forged parts in a spliced image were captured by camera models different from the one used to capture the authentic region.

Let M_{ij} be the binary output image for the patch P_{ij}. If P_{ij} is an authentic patch, we assign all the pixels of M_{ij} as 0. Otherwise we assign 1 to all the pixels of the M_{ij}. A binary image M is created by putting together all the M_{ij}s in the appropriate image coordinates, i.e. $\mathrm{M} = [\mathrm{M}_{ij}]$, where $i \in (1, \lfloor \frac{W}{64} \rfloor)$ and $j \in (1, \lfloor \frac{H}{64} \rfloor)$. Because of noise, there may be a few false alarm patches scattered throughout M. To remove these false alarms, morphological opening is performed on M with structuring element of size 128×128 to get the binary image $\widehat{\mathrm{M}}$. This is based on the assumption that the smallest possible forgery can be of size 128×128. $\widehat{\mathrm{M}}$ localizes the potential forged patches of I_T.

3.1.4 Detection of the Forged Image using Thresholding

Because of the presence of noise, there may be some patches of an authentic image misidentified as forged patches. Therefore, there is a need to take image level decision about the test image, i.e. whether the image is pristine or spliced. For this, we check the percentage of forged patches present in I_T. Let T_f represent the percentage of forged patches present in I_T. We decide I_T to be pristine if the percentage of forged patches is less than a certain threshold T_p, i.e. if $T_f < T_p$. Otherwise, I_T is decided to be spliced (Fig. 4).

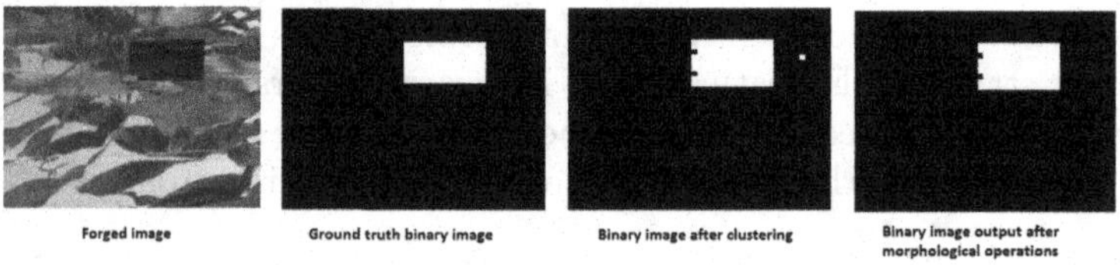

Fig. 4. An example of forged image along with the ground truth binary image, the binary output image obtained after clustering and the final binary output image obtained after the morphological post-processing operation.

4 Experimental Results

We use the Dresden Image Database [4] to train the CNN for performing the camera model identification task. The database contains 16,961 images captured using 28 different camera models. Out of these 28 models, we use 18 camera models for training the CNN. Since the CNN accepts images of size 64 × 64, we have divided the images present in all the 18 classes into patches of the same size. For testing the performance of splicing detection and localization, two tampered datasets are created. Dataset D_1 is created using images from the 18 known camera models (i.e. models present in the training stage) and dataset D_2 is created using 10 unknown camera models. Both the known and the unknown datasets contain 300 pristine and 300 spliced images along with their groundtruth binary image, representing the authentic and spliced parts. The tampered images are created as follows: Two images I_1 and I_2 are first selected randomly from the dataset. Then, a rectangular region from I_2 is copied randomly and pasted onto I_1 at a random location. The size of the rectangular region is varied from 128 × 128 to 512 × 512.

The CNN is first trained to perform the camera model identification task. Once the CNN is trained, we use it to extract features from the patches of size 64 × 64 present in the image under investigation. The features are later clustered to detect and localize the spliced regions.

As explained earlier, the proposed algorithm depends upon two thresholds T_d and T_p. The threshold T_d is the maximum distance between the means of

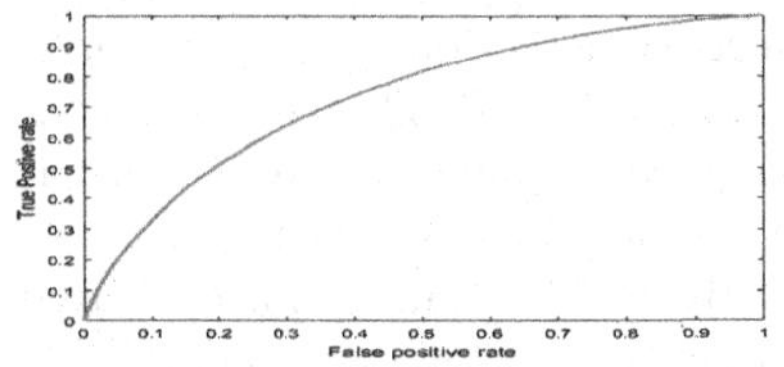

Fig. 5. ROC curve varying the threshold T_d.

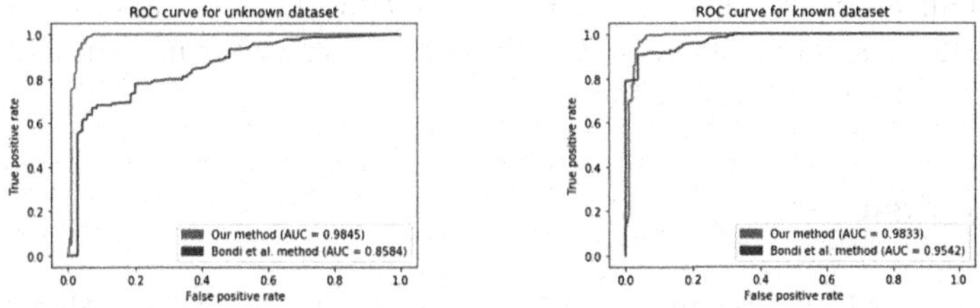

(a) For unknown dataset (D_2) (b) For known dataset (D_1)

Fig. 6. ROC curve for splicing detection

Table 1. Tampering detection results.

T_d	T_p	ACC	TPR	TNR
0.4	0	0.91	**0.985**	0.835
0.4	0.08	**0.94**	0.935	**0.945**
0.3	0	0.88	**0.942**	0.818
0.3	0.08	**0.90**	0.911	**0.889**

two clusters. To find the optimal value for T_d, we created 18,000 authentic feature pairs and 18,000 spliced feature pairs. The authentic pairs are obtained by taking two patches captured by same camera model. The spliced pairs are obtained by taking two patches from two different cluster of features which are captured using different camera models. The receiver operating characteristic curve (ROC) is then computed by plotting the true positive rate (TPR) against the false positive rate (FPR) at various threshold T_d as shown in Fig. 5. The optimal value of threshold T_d, is selected as a value which maximizing accuracy according to the ROC. The threshold T_p is used to take the image level decision of a test image. The optimal value of threshold T_p, is selected as a value which maximizes the accuracy according to ROC results shown in Fig. 6a and b.

The ROC is obtained by plotting the TPR against the FPR at various threshold T_p and fixing threshold $T_d = 0.4$ for the unknown dataset and known dataset as shown in Fig. 6a and b respectively. In Fig. 6a, area under the curve (AUC) is 0.9845 which shows that best forgery detection is achieved for $T_d = 0.4$ and AUC obtained using the Bondi *et al.* [2] method is 0.8584. In Fig. 6b, AUC is 0.9833 which shows that best forgery detection is achieved for $T_d = 0.4$. AUC obtained using the Bondi *et al.* [2] method is 0.9542. Also, on known and unknown datasets the proposed method achieves accuracies (ACCs) of 94.9% and 93.3% respectively. While, Bondi *et al.* achieves ACCs of 91.8% and 82% on known and unknown datasets respectively. Comparing both the ROC it is evident that for any unknown dataset, our method outperforms the existing method. Table 1 shows the results in terms of accuracy (ACC), TPR and TNR. For $T_p = 0$, an image is considered to be pristine only if all the patches are pristine so the TPR is maximum. From the table it is evident that for $T_d = 0.4$ and $T_p = 0.08$, we get the best result in terms of ACC. These values are later used in checking the authenticity of a given test image.

5 Conclusions

In this paper, we proposed an algorithm which exploits a CNN to extract camera-specific features from image patches. These features are further analysed in order to detect whether an image is forged or not. In the first phase, we train a CNN using images from the Dresden dataset to learn the camera-specific features for 18 camera models. In the second phase, we divide the test image into a number of non-overlapping patches and extract features using the trained CNN. After that, these features are clustered using the HAC technique for detecting and localizing splicing forgeries. The method can detect splicing present in images coming from known as well as unknown camera models. Experiments were conducted on 1200 known as well as unknown images coming from 28 camera models. The experimental results show that the proposed method outperforms the state-of-the-art for images captured by both known and unknown camera models. Our ongoing work aim at detecting splicing forgeries on post-processed image using this framework.

References

1. Bondi, L., Baroffio, L., Güera, D., Bestagini, P., Delp, E.J., Tubaro, S.: First steps toward camera model identification with convolutional neural networks. IEEE Signal Process. Lett. **24**(3), 259–263 (2016)
2. Bondi, L., Lameri, S., Güera, D., Bestagini, P., Delp, E.J., Tubaro, S.: Tampering detection and localization through clustering of camera-based CNN features. In: 2017 IEEE Conference on Computer Vision and Pattern Recognition Workshops (CVPRW), pp. 1855–1864. IEEE (2017)
3. Dubes, R.C., Jain, A.K.: Algorithms for clustering data (1988)
4. Gloe, T., Böhme, R.: The 'dresden image database' for benchmarking digital image forensics. In: Proceedings of the 2010 ACM Symposium on Applied Computing, pp. 1584–1590. ACM (2010)
5. Johnson, M.K., Farid, H.: Exposing digital forgeries in complex lighting environments. IEEE Trans. Inf. Forensics Secur. **2**(3), 450–461 (2007)
6. Lyu, S., Pan, X., Zhang, X.: Exposing region splicing forgeries with blind local noise estimation. Int. J. Comput. Vision **110**(2), 202–221 (2014)

7. Thai, T.H., Cogranne, R., Retraint, F., et al.: Jpeg quantization step estimation and its applications to digital image forensics. IEEE Trans. Inf. Forensics Secur. **12**(1), 123–133 (2016)
8. Xu, G., Shi, Y.Q.: Camera model identification using local binary patterns. In: 2012 IEEE International Conference on Multimedia and Expo, pp. 392–397. IEEE (2012)
9. Yang, P., Ni, R., Zhao, Y., Zhao, W.: Source camera identification based on content-adaptive fusion residual networks. Pattern Recogn. Lett. **119**, 195–204 (2019)

Multi-focus Image Fusion Using Sparse Representation and Modified Difference

Amit Vishwakarma, M. K. Bhuyan, Debajit Sarma(✉), and Kangkana Bora

Department of Electronics and Electrical Engineering,
Indian Institute of Technology (IIT) Guwahati, Guwahati 781039, Assam, India
{a.vishwakarma,mkb,s.debajit,kangkana.bora}@iitg.ac.in

Abstract. Multi-focus fusion technique is used to combine images obtained from single or different cameras with different focal distance, etc. In the proposed method, the non-subsampled shearlet transform (NSST) is employed to decompose the input image data into the low-frequency and high-frequency bands. These low-frequency and high-frequency bands are combined using sparse representation (SR) and modified difference based fusion rules, respectively. Then, inverse NSST is employed to get the fused image. Both qualitative and quantitative results confirm that the proposed approach yields a better performance as compared to state-of-the-art fusion schemes.

Keywords: Image fusion · K-SVD · Cosine bases · Multi-focus · Sparse representation · Dictionary learning

1 Introduction

The image fusion is an integration process of knowledge obtained from different or similar sources of images. Sometimes, it is impracticable to obtain an image having, all important focused objects due to the restricted depth of focus of an optical lens in digital cameras, optical microscopes, etc. Fusion techniques can be employed here to get a fused image, which consists of all the focus parts of source images. In our proposed approach, the non-subsampled shearlet transform (NSST) is employed for fusion over curvelet and contourlet. NSST have different characteristics, for example, better localization in both spatial and frequency domain, optimal sparse representation and fine detail sensitivity using parabolic scaling scheme [2].

In the last few years, sparse representation (SR)-based methods have arisen as an effective branch in image fusion research with many improved approaches [8,17]. In the proposed approach, a novel fusion criterion is proposed to combine high-frequency bands after source image decomposition using NSST. For fusion of low-frequency bands, SR-based technique is employed with a hybrid of learned and local cosine bases dictionaries [10,11]. Local cosine bases dictionary efficiently captures texture information of the input images, while

B. Deka et al. (Eds.): PReMI 2019, LNCS 11941, pp. 482–489, 2019.
https://doi.org/10.1007/978-3-030-34869-4_52

learned dictionary well approximates the fine details, low- and high-frequency features.

NSST [2–4] is employed to capture high-frequency details such as edges. contours, etc. NSST have good localized waveforms at various scales, positions, and angles. Moreover, it also minimizes the misregistration error because of the shift-invariant feature in the fused image.

Hybrid Dictionary Based on Learning and Local Cosine Bases: Local cosine dictionary obtained from the cosine modulated filter bank is efficient to represent the texture information of source images. K-SVD technique based learned dictionary efficient to describe fine features of the source image. Therefore, in the proposed fusion approach dictionaries obtained using local cosine bases and K-SVD method are concatenated to make a single dictionary to represent significant details of the input images.

2 The Proposed Method

The layout of the proposed fusion procedure is shown in Fig. 1. The presented approach first decomposes the input images into low and high-frequency coefficients using NSST. Later implementing proposed fusion schemes and taking inverse NSST fused version of the image is attained.

2.1 Modified Differences

To seize fine details from high-frequency coefficients generally, a variation of the center pixel to its horizontal and vertical pixels are estimated. This rule ignores the relationship of diagonal neighboring pixels with the center pixel. To overcome this limitation improved version of the operation is proposed, where the variation of the center point pixel to each neighboring pixels are estimated. We named this method as modified difference method [13] due to a large number of different operations in 3×3 window as shown in (1). Let $f(i, j)$ be an image of dimension $M \times N$ and kernel dimension is 3×3. Applying the modified difference method, we get an image $\mathbf{MDf}(i, j)$ considering $w_1 = w_2 = 3$. Note, (1) is applicable only for a window of size 3×3 [13].

$$\mathbf{MDf}(i, j) = \left\{ \sum_{w_1, w_2 = -1}^{1} |f(i, j) - f(i + w_1, j + w_2)| \right\}. \tag{1}$$

2.2 The Proposed Fusion Framework

Our presented method apply NSST for the source image decomposition upto 2-level. Input images $\mathbf{A}$ and $\mathbf{B}$ are initially decomposed into the low-frequency band and high-frequency coefficients. NSST decomposition of the two source images $\{\mathbf{I}_A, \mathbf{I}_B\}$ are carried out, which will give for each source images, one low-frequency bands $\{\mathbf{L}_A, \mathbf{L}_B\}$ and four high-frequency coefficients $\{\mathbf{H}_{k,l}^A(i, j), \mathbf{H}_{k,l}^B(i, j)\}$. Here, l and k represent directional band and the level of decomposition, respectively.

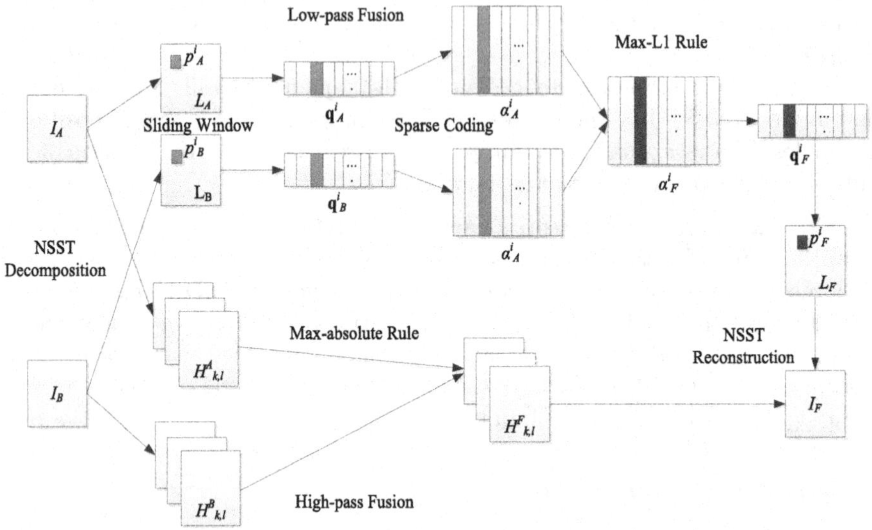

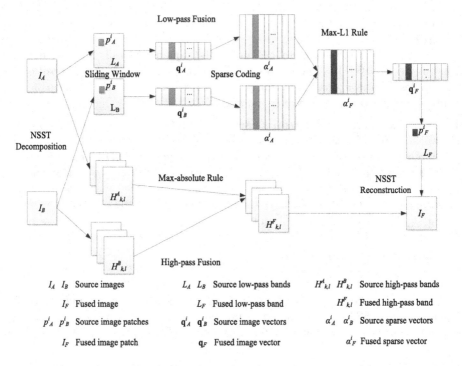

Fig. 1. Diagram of proposed image fusion approach [8].

Low-Frequency Coefficients Fusion Scheme: Low-frequency coefficients obtained using NSST decomposition are combined in five steps, given as [8]:

(I) Using the sliding window technique, images $\mathbf{L}_A$ and $\mathbf{L}_B$ are divided into $\sqrt{n}\times\sqrt{n}$ patch sizes initiated from upper left to down right with s pixel step length. In $\mathbf{L}_A$ and $\mathbf{L}_B$, T number of patches are represented as $\{p^i{}_A\}_{i=1}^{T}$ and $\{p^i{}_B\}_{i=1}^{T}$, respectively.

(II) Then we rearrange $\{p^i{}_A, p^i{}_B\}$ into column vectors $\{\mathbf{q}^i{}_A, \mathbf{q}^i{}_B\}$ for every position i in images. Then we attain normalized vectors $\{\hat{\mathbf{q}}^i_A, \hat{\mathbf{q}}^i_B\}$ as follows:

$$\begin{aligned} \hat{\mathbf{q}}^i_A &= \mathbf{q}^i{}_A - \bar{q}^i_A.\mathbf{1}, \\ \hat{\mathbf{q}}^i_B &= \mathbf{q}^i{}_B - \bar{q}^i_B.\mathbf{1}, \end{aligned} \tag{2}$$

where $\bar{q}^i_A$ and $\bar{q}^i_B$ are mean vectors of $\mathbf{q}^i{}_A$ and $\mathbf{q}^i{}_B$, respectively. **1** represent $n \times 1$ vector having all one entry.

(III) Now we perform sparse coding using orthogonal matching pursuit (OMP) algorithm to find sparse coefficient vectors $\{\alpha^i{}_A, \alpha^i{}_B\}$ from $\{\hat{\mathbf{q}}^i_A, \hat{\mathbf{q}}^i_B\}$ with dictionary $\mathbf{D}$ as follows [8]:

$$\begin{aligned} {\alpha^i}_A &= \arg\min_{\alpha} ||\alpha||_0 \quad such \quad that \quad ||\hat{\mathbf{q}}^i_A - \mathbf{D}\alpha||_2 < \varepsilon, \\ {\alpha^i}_B &= \arg\min_{\alpha} ||\alpha||_0 \quad such \quad that \quad ||\hat{\mathbf{q}}^i_B - \mathbf{D}\alpha||_2 < \varepsilon, \end{aligned} \tag{3}$$

(IV) Using max-L1 fusion rule ${\alpha^i}_A$ and ${\alpha^i}_B$ are fused to get the fused sparse vector given as [8]:

$$ {\alpha^i}_F = \begin{cases} {\alpha^i}_A & \text{if} \quad ||{\alpha^i}_A||_1 > ||{\alpha^i}_B||_1 \\ {\alpha^i}_B & \text{otherwise} \end{cases} \tag{4}$$

fused mean value $\bar{q}^i_F$ is obtained by

$$ \bar{q}^i_F = \begin{cases} \bar{q}^i_A & \text{if} \quad {\alpha^i}_F = {\alpha^i}_A \\ \bar{q}^i_B & \text{otherwise} \end{cases} \tag{5}$$

Now fused version of ${\mathbf{q}^i}_A$ and ${\mathbf{q}^i}_B$ is obtained as [8]:

$$ {\mathbf{q}^i}_F = \mathbf{D}{\alpha^i}_F + \bar{q}^i_F.\mathbf{1} \tag{6}$$

(V) Above four steps are iteratively applied for all image patches. For all patches $\{{p^i}_A\}_{i=1}^T$ and $\{{p^i}_B\}_{i=1}^T$ fused vectors $\{{\mathbf{q}^i}_F\}_{i=1}^T$ is obtained. Let $\mathbf{L}_F$ is low-pass fused image. Now for each ${\mathbf{q}^i}_F$ convert it by reshaping, into ${p^i}_F$ then put it into its original position in $\mathbf{L}_F$. Due to overlapped patches every pixel in $\mathbf{L}_F$ is averaged over number of times it is accumulated [8].

High-Frequency Coefficients Fusion Scheme: To combine high-frequency coefficients $\{\mathbf{H}^A_{k,l}(i,j)\}$ and $\{\mathbf{H}^B_{k,l}(i,j)\}$, modified difference fusion rule is proposed. (1) is applied on each high-frequency band. $\mathbf{H}^A_{k,l}(i,j)$, $\mathbf{H}^B_{k,l}(i,j)$ and $\mathbf{H}^f_{k,l}(i,j)$ are high-frequency coefficients at l^{th} direction sub-image at k^{th} level of decomposition and (i,j) location of input and fused images, respectively. Then applying modified difference in $\mathbf{H}^A_{k,l}(i,j)$ and $\mathbf{H}^B_{k,l}(i,j)$, modified difference features $\mathbf{MDH}^A_{k,l}(i,j)$ and $\mathbf{MDH}^B_{k,l}(i,j)$ are attained. Here fused coefficients *i.e.* $\mathbf{H}^f_{k,l}(i,j)$ are obtained as:

$$ \mathbf{H}^f_{k,l}(i,j) = \begin{cases} \mathbf{H}^A_{k,l}(i,j), & \mathbf{MDH}^A_{k,l}(i,j) \geq \mathbf{MDH}^B_{k,l}(i,j) \\ \mathbf{H}^B_{k,l}(i,j), & \text{otherwise.} \end{cases} \tag{7}$$

For fusion of high-frequency bands modified difference rule is used to enhance the significant high-frequency information. This rule enhances the corners, points, edges, contrast information in the final fused image. Moreover, these information improves the interpretation capability of the fused images. Now from fused low-frequency and high-frequency sub-images, inverse NSST is implemented to obtain a fused version of source images.

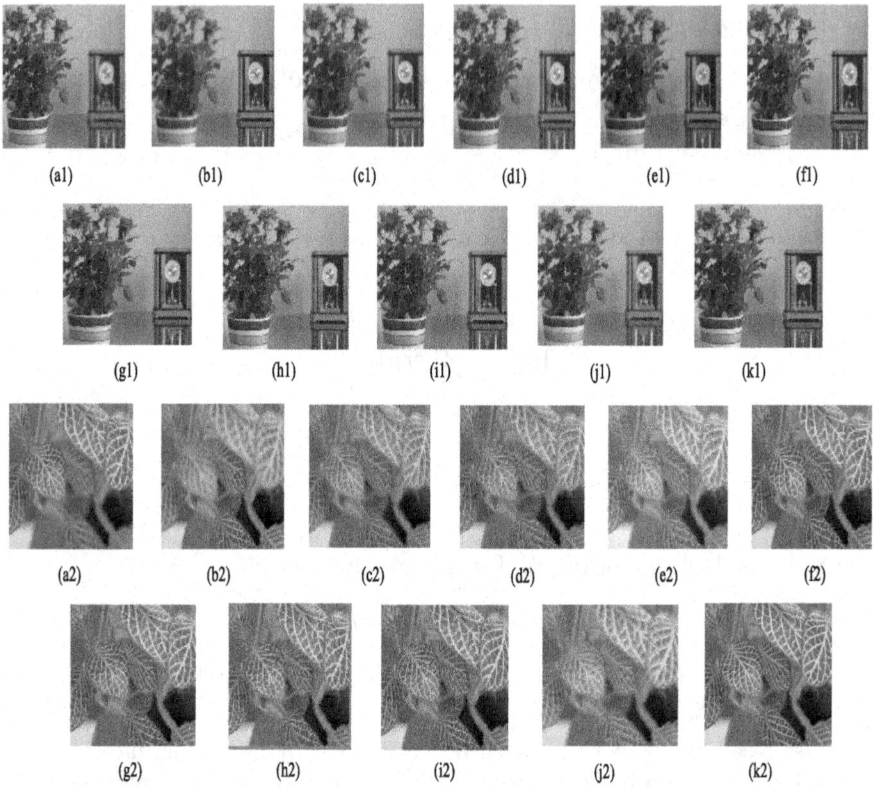

Fig. 2. Qualitative comparative study of state-of-the-art and proposed method of dataset-1. (a1)–(b1) and (a2)–(b2) are source image pairs. (c1)–(c2) GP [12], (d1)–(d2) DWT [5], (e1)–(e2) HOSVD [7], (f1)–(f2) GFF [6], (g1)–(g2) MST SR [8], (h1)–(h2) DCT SF [1], (i1)–(i2) DSIFT [9], (j1)–(j2) NSCT FAD [16], (k1)–(k2) Proposed.

3 Results and Comparison

To perform experiments database of multi-focus images are obtained from URLs given in [8]. The presented approach is compared with number of state-of-the-art approaches [1,5–9,12,16]. The proposed method is examined by employing three fusion metrics for example mutual information index (MI) [16], edge-based similarity index $Q^{AB/F}$ [14], and structure similarity measure Q_s [15]. Metric MI indicates the mutual information between the fused and source images. Metric $Q^{AB/F}$ indicates the edge preserving information in the fused image and metric Q_s indicates the structural similarity information between the fused and source images. For better fusion performance MI should be high, in addition, values of the $Q^{AB/F}$ and Q_s should be near to one.

Proposed fusion techniques are tested in two image pairs of two databases. Our approach for multi-focus modalities images is compared with the methods proposed using gradient pyramid (GP) [12], discrete wavelet transform

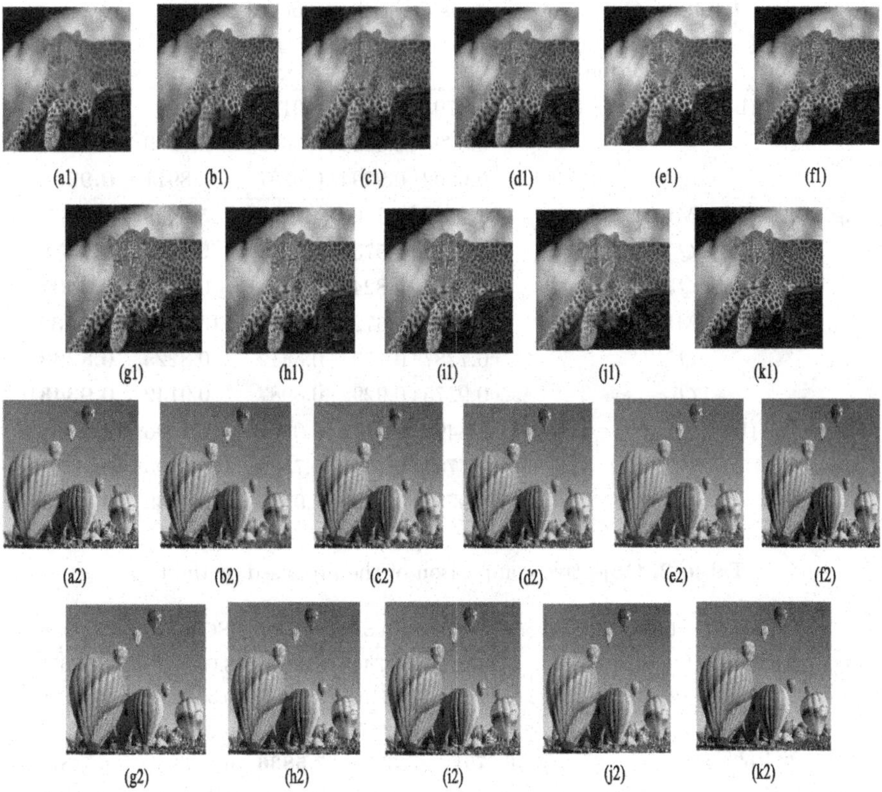

Fig. 3. Qualitative comparative study of state-of-the-art and proposed method of dataset-2. (a1)–(b1) and (a2)–(b2) are source image pairs. (c1)–(c2) GP [12], (d1)–(d2) DWT [5], (e1)–(e2) HOSVD [7], (f1)–(f2) GFF [6], (g1)–(g2) MST SR [8], (h1)–(h2) DCT SF [1], (i1)–(i2) DSIFT [9], (j1)–(j2) NSCT FAD [16], (k1)–(k2) Proposed.

(DWT) [5], high-order singular value decomposition (HOSVD) [7], guided filtering (GFF) [6], multi-scale transform and sparse representation (MST SR) [8], discrete cosine transform beside spatial frequency (DCT SF) [1], dense scale-invariant feature transform (DSIFT) [9] and NSCT focus area detection (NSCT FAD) [16]. In our proposed approach, the Max-L1 rule is used to select sparse vectors and coefficients of high-frequency bands are fused or combined using modified difference information.

In Figs. 2 and 3, qualitative comparison of various fusion schemes is presented. The main objective analysis of the presented approach is shown in Tables 1 and 2. From Tables 1 and 2, metrics values of our presented scheme are better as compared to state-of-the-art methods. Bold metric values indicate better results. If the input images are not registered perfectly, then both GP and DWT methods result in some ringing artifacts in the fused images due to shift variance of GP and DWT. The HOSVD scheme generates distorted edges and blur artifacts in

Table 1. Objective comparison of the proposed method

	Evaluation indices	GP	DWT	HOSVD	GFF	Proposed
Fig. 2(a1)–(b1)	MI	6.0704	6.057	6.0116	7.2995	6.9767
	$Q^{AB/F}$	0.579	0.5545	0.5678	0.7109	**0.7431**
	Q_s	0.9292	0.9294	0.9287	0.8911	**0.9296**
Fig. 2(a2)–(b2)	MI	4.4577	4.3506	4.1205	5.6563	6.7487
	$Q^{AB/F}$	0.6126	0.5731	0.4388	0.7177	**0.7376**
	Q_s	0.8305	0.8242	0.7801	0.8253	**0.8401**
Fig. 3(a1)–(b1)	MI	7.5316	8.6723	6.5946	10.6991	**10.9632**
	$Q^{AB/F}$	0.7787	0.7704	0.5812	0.8225	**0.8273**
	Q_s	0.9275	0.9292	0.8987	0.9142	**0.9343**
Fig. 3(a2)–(b2)	MI	8.8496	9.4177	9.0964	11.1296	**11.2063**
	$Q^{AB/F}$	0.7704	0.7606	0.7582	0.8195	0.8159
	Q_s	0.9758	0.9764	0.9799	0.9649	**0.9799**

Table 2. Objective comparison of the proposed method

	Evaluation indices	MST SR	DCT SF	DSIFT	NSCT FAD	Proposed
Fig. 2(a1)–(b1)	MI	7.2134	**8.3744**	8.3289	8.099	6.9767
	$Q^{AB/F}$	0.6969	0.7349	0.7338	0.5571	**0.7431**
	Q_s	0.8919	0.8867	0.8867	0.9033	**0.9296**
Fig. 2(a2)–(b2)	MI	6.7731	7.4374	**7.5836**	6.6193	6.7487
	$Q^{AB/F}$	0.7188	0.6978	0.7363	0.4968	**0.7376**
	Q_s	0.8238	0.8393	0.816	0.7783	**0.8401**
Fig. 3(a1)–(b1)	MI	10.7112	10.9357	10.9226	9.6359	**10.9632**
	$Q^{AB/F}$	0.8216	0.8258	0.8269	0.7204	**0.8273**
	Q_s	0.9142	0.9137	0.9137	0.5068	**0.9343**
Fig. 3(a2)–(b2)	MI	11.0353	11.1846	11.1847	10.2113	**11.2063**
	$Q^{AB/F}$	0.8181	0.8218	**0.8219**	0.7324	0.8159
	Q_s	0.9651	0.9648	0.9648	0.9688	**0.9799**

the fused/combined image. Fused image obtained through GFF approach usually loses some crucial texture information and have some blurring artifact. MST SR method affected from few noisy distortions because of the absolute value max select fusion scheme for high-frequency coefficients fusion.

DCT SF and DSIFT methods perform better than GFF and MST SR but underperformed as compared to our proposed method. DCT approach is inefficient to seize directional contents of the input images and have some blocking artifacts. Due to spatial frequency based fusion rule, the results of the DCT scheme is superior to GFF, MST SR and comparable with DSIFT. At last, NSCT FAD method unable to perform well because of inefficient sum-modified-Laplacian-based fusion rule for low-frequency bands.

4 Conclusion

In our proposed method, we describe NSST and SR-based fusion methods for multi-focus images. Our proposed method uses the modified difference based fusion rule to select high-frequency coefficients. Our proposed method gives better subjective and objective results in contrasted to existing multi-focus image fusion methods.

References

1. Cao, L., Jin, L., Tao, H., Li, G., Zhuang, Z., Zhang, Y.: Multi-focus image fusion based on spatial frequency in discrete cosine transform domain. IEEE Signal Process. Lett. **22**(2), 220–224 (2015)
2. Easley, G., Labate, D., Lim, W.Q.: Sparse directional image representations using the discrete shearlet transform. Appl. Comput. Harmon. Anal. **25**(1), 25–46 (2008)
3. Guo, K., Labate, D.: Optimally sparse multidimensional representation using shearlets. SIAM J. Math. Anal. **39**(1), 298–318 (2007)
4. Guorong, G., Luping, X., Dongzhu, F.: Multi-focus image fusion based on non-subsampled shearlet transform. IET Image Process. **7**(6), 633–639 (2013)
5. Li, H., Manjunath, B., Mitra, S.K.: Multisensor image fusion using the wavelet transform. Graph. Models Image Process. **57**(3), 235–245 (1995)
6. Li, S., Kang, X., Hu, J.: Image fusion with guided filtering. IEEE Trans. Image Process. **22**(7), 2864–2875 (2013)
7. Liang, J., He, Y., Liu, D., Zeng, X.: Image fusion using higher order singular value decomposition. IEEE Trans. Image Process. **21**(5), 2898–2909 (2012)
8. Liu, Y., Liu, S., Wang, Z.: A general framework for image fusion based on multi-scale transform and sparse representation. Inf. Fusion **24**, 147–164 (2015)
9. Liu, Y., Liu, S., Wang, Z.: Multi-focus image fusion with dense SIFT. Inf. Fusion **23**, 139–155 (2015)
10. Malvar, H.S.: Signal Processing with Lapped Transforms. Artech House, Norwood (1992)
11. Meyer, F.G.: Image compression with adaptive local cosines: a comparative study. IEEE Trans. Image Process. **11**(6), 616–629 (2002)
12. Petrovic, V.S., Xydeas, C.S.: Gradient-based multiresolution image fusion. IEEE Trans. Image Process. **13**(2), 228–237 (2004)
13. Vishwakarma, A., Bhuyan, M.K.: Image fusion using adjustable non-subsampled shearlet transform. IEEE Trans. Instrum. Measur. **68**(9), 1–12 (2018)
14. Xydeas, C., Petrovic, V.: Objective image fusion performance measure. Electron. Lett. **36**(4), 308–309 (2000)
15. Yang, C., Zhang, J.Q., Wang, X.R., Liu, X.: A novel similarity based quality metric for image fusion. Inf. Fusion **9**(2), 156–160 (2008)
16. Yang, Y., Tong, S., Huang, S., Lin, P.: Multifocus image fusion based on nsct and focused area detection. IEEE Sens. J. **15**(5), 2824–2838 (2015)
17. Yin, H., Li, S., Fang, L.: Simultaneous image fusion and super-resolution using sparse representation. Inf. Fusion **14**(3), 229–240 (2013)

Usability of Foldscope in Food Quality Assessment Device

Sumona Biswas(✉) and Shovan Barma(✉)

Indian Institute of Information Technology Guwahati (IIITG), Guwahati, India
{sumona, shovan}@iiitg.ac.in

Abstract. This work focuses on the quality assessment of agricultural product based on microscopic image, generated by Foldscope. Microscopic image-based food quality assessment always be an efficient method, but its system complexity, costly, bulk size and requirement of special expertise confines it usability. To encounter such issues, Foldscope which is small, lightweight, cheap and easy to use has been considered to verify its usability as food quality assessment device. In this purpose, measuring starch of potato has been selected to check its compatibility and microscopic images are taken from two image modalities—conventional microscope and Foldscope and the results have been compared. The image processing techniques including morphological filtering followed by Otsu's method has been employed to detect starch efficiently. In total, 20 images from each of the system have been captured. Following the experiment, the presence of starch (in %) estimated based on the image taken from microscope and Foldscope are 23.50 ± 0.79 and 24.29 ± 0.73 respectively, which is consistent. Such results reveal that the Foldscope can be used in food quality assessment system, which could make such devices simple, portable and handy.

Keywords: Foldscope · Microscopic image processing · Potato starch

1 Introduction

Food quality assessment is always being an important part at the time of purchasing or consuming any sort of foods. Now-a-days, it becomes more evident as fertilizer and pesticides in crops are being used enormously without maintaining the nutrition value to meet the huge demand of food for rapidly growing population. In addition, harmful chemicals are also being used in vegetables and fruits for immoral purposes; for instance, Copper Sulphate ($CuSO_4$) is used to look the vegetable fresh; Calcium Carbide (CaC_2) is used to ripe fruits artificially and so on. Besides, several synthetic colors are applied on vegetables to maintain their freshness. Such malpractices make the food quality assessment more evident. Generally, in market the quality of the agricultural products are evaluated based on their shape, size, colour etc. which are performed by means of vision, touch, smell, odour, taste, flavour etc. Indeed, all these methods are tedious, time consuming, very, much subjective, but not even very reliable. Therefore, it is very significant to develop a rapid, reliable, easy to use system to examine the agricultural products.

B. Deka et al. (Eds.): PReMI 2019, LNCS 11941, pp. 490–498, 2019.
https://doi.org/10.1007/978-3-030-34869-4_53

In this direction, comparing to other aforementioned manual techniques to assess the food quality, the visual perceptions (usually bare eyes) dominate the other means, which is not only inefficient but also unreliable due to subjectivity, inconsistency etc. However, such obscures can be tackled by the computer vision technology that corresponds to the human vision, in inspecting quality of fruits and vegetables. It captures images electronically, then interpret and recognize the characters/information to examine the quality, and grading of agricultural products. In this direction, several attempts have been made for grading of food including apples [1, 2], mango [2] sweet pepper [3], papaya [4], tomato [5, 6] etc. However, all the methods consider the traditional camera images, taken from the surface of the vegetables and fruits, which are failed to provide inside details such as nutrients contents of a products. Such details are the key components to quantify the food quality or grading.

Certainly, the microscopic images (resolution of 200 nm or smaller) provide insight information at cell level which could be appropriate to examine the food quality more precisely than the use of conventional camera images. Few works have been conducted in this direction based on such microscopic images to evaluate the quality of vegetable and fruits such as evaluation of potato and its starch [7], browning of apple due to storage [8], and effect of freezing in blueberries [9]. However, these images are usually generated by microscopes which are expensive, complex, bulky and need some sort of expertise to operate. In this regard, recently, a newly developed light weight ($\sim$9 gm), low cost ($\sim$Rupees 250), and 140X magnification powered origami based paper microscopic called Foldscope [10] could be a good solution to generate the microscopic images. Besides, the generated images can be captured directly by simple mobile camera. Thus, a combining Foldscope with mobile camera for capturing microscopic images and further employing computer vision and image processing techniques to analyze those images could be a very proper food quality assessment system. Therefore, in this work, microscopic images of potato has been considered to verify the usability of such device. The potato has been used as one of the most important crop consumed in India after rice, wheat and maize [11]. Usually, it contains water (80%) and dry matter (20%) which is well-known as starch [11]. Inevitability, these starch amount defines the quality of a potato [12].

In a microscopic image of potato, usually starches are glided on the surface of the cells which makes it challenging to distinguish from regular cells. In this purpose, staining method enhances either cell boundaries or starch, could be useful in such detection process. But, it requires not only involvement of additional arrangement (chemical process etc.), but also high-end microscope (like electron microscopes) [13]. Another way, use of image processing techniques such as morphological image enhancement on microscopic image directly could be very convenient to sidestep staining process [14]. Certainly, a well-established, simple and accurate image segmentation techniques such as Otsu's segmentation method [15] can be employed. It works with intensity of an image and considers thresholding technique to discriminate two set of classes of an image. Thus, it could be applicable to distinct starch from cell background. Nevertheless, such thresholding sometimes leads to improper segmentation that can be resolved easily by morphological filtering. Therefore, measuring the starch content using microscopic images which will be captured by above mentioned setup and deploying image processing technique is the main objective of this work.

The paper has been organized as follows: Sect. 2 provides a system overview of the proposed methods. In Sects. 3 and 4, the methodology and experiment have been elaborated respectively. Results and discussion have summarized in Sect. 5. Finally, conclusions have been drawn in Sect. 6.

2 System Overview

A system overview of the proposed method has been displayed in Fig. 1 which involves several stages including sample preparation, image acquisition, image processing, starch detection and estimation. Firstly, conventional methods for acquiring microscopic images such as sectioning of samples etc. are conducted to acquire microscopic image of potato. Next, images are generated using Foldscope as well as conventional microscopic. Further, image processing techniques such as morphological filtering followed by segmentation are conducted to detect the starch from those two sets of images. Finally, the percentage of starch has been estimated and the results are compared.

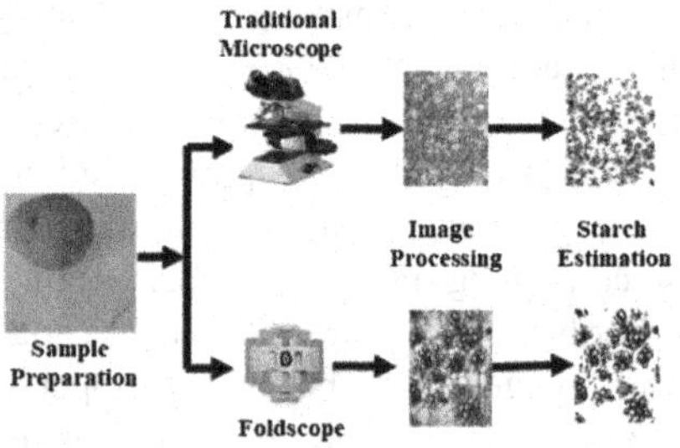

Fig. 1. System overview of the proposed method.

3 Methodology

Aim of the proposed work is to measure the starch content in potato directly (avoiding traditional staining process) from microscopic image which are acquired from two image acquisition modalities—Foldscope and traditional microscopic. Generated images from both of the modalities consist of starch spreading over regular cells and its boundaries. However, the intensity difference between them is relatively small which confines to employ simple thresholding technique to discriminate starch from regular cells—leads to under segmentation. Certainly, a popular segmentation method known as Otsu's method which performs the thresholding, based on statistical measures can be used. Nonetheless, direct use of Otsu's method sometimes results over segmentation which can be overcome by a morphological filtering. Therefore, a morphological filtering has been conducted followed by Otsu's algorithm to improve the system performances by eradicating uneven illumination and the cell boundaries. Further, to estimate the starch quantity efficiently binarization and morphological operation have been performed.

3.1 Morphological Filtering

Morphological operation analyses the geometrical structure of an image. It performs dilation, erosion, opening, closing operation with an image of definite shape and size (structuring element). Morphological filter can be constructed on the basis of morphological operations. It offers better result than the linear filtering as it deforms the image geometry. The morphological filter has been designed based on two morphological operations called top-hat and bottom hat transform by opening and closing operations respectively. The whole operation performs on grayscale image. The brightest parts of the images are enhanced by top-hat transform; whereas, the bottom-hat transform does the reverse process. Mathematically, for a grayscale image $g(x, y)$ with structuring element $a(x, y)$,

$$\text{Top - hat Transform}, T_t(g) = g - (g \circ a) \tag{1}$$

$$\text{Bottom - hat transform}, T_b(g) = (g \bullet a) - g \tag{2}$$

Top-hat transform followed by bottom-hat transform increases the contrast between cell boundaries and starch which further benefits to discriminate the starch distinctly. The contrast enhanced image (h) is obtained by adding the image itself with its top-hat transform as shown in (3); whereas, the starch boundaries (i) are further enhanced by subtracting the bottom-hat transformed image from contrast enhanced image (h) as displayed in (4).

$$\text{Contrast enhanced image}, h = g + T_t(g) \tag{3}$$

$$\text{Gap enhanced image}, i = h - T_b(g) \tag{4}$$

Furthermore, the starches have been discriminated from (3) and (4). These steps define morphological filtering of gray scale images.

3.2 Starch Detection

As mentioned, the segmentation operation has been performed by Otsu's method [15] which estimated the threshold values based on statistical measures. It separates the image into two classes based on threshold value t. Further, t can be optimized by either minimizing the weighted sum of within-class variance (σ_w) or maximizing the between-class variance (σ_b).

Two class probabilities $q_{c1}(t)$ and $q_{c2}(t)$ are calculated in terms of intensity probability $P(k)$ for different intensity k, from I bins of histogram expressed as,

$$q_{c1}(t) = \sum_{k=1}^{t} P(k) \tag{5}$$

$$q_{c2}(t) = \sum_{k=t+1}^{I} P(k) \tag{6}$$

and class means $\mu_{c1}(t)$ and $\mu_{c2}(t)$ expressed as

$$\mu_{c1}(t) = \sum_{k=1}^{t} \frac{kP(k)}{q_{c1}(t)} \tag{7}$$

$$\mu_{c2}(t) = \sum_{k=t+1}^{I} \frac{kP(k)}{q_{c2}(t)} \tag{8}$$

Therefore, the total class variance,$\sigma_{total(cv)}$ for any given threshold t in terms of class probabilities (q_c) and class mean σ can be expressed as

$$\sigma^2_{total(cv)} = \sigma^2_w(t) + q_{c1}(t)[1 - q_{c1}(t)][\mu_{c1}(t) - \mu_{c2}(t)]^2 \tag{9}$$

And between-class variance

$$\sigma^2_b = q_{c1}(t)[1 - q_{c1}(t)][\mu_{c1}(t) - \mu_{c2}(t)]^2 \tag{10}$$

The algorithm has been worked as follows:

Start	
Step 1	**Preprocessing** **Step 1.1** Input: Microscopic RGB image (*I*) **Step 1.2** Convert *I* to grayscale image (*J*)
Step 2	**Morphological Filtering** **Step 2.1** Perform top-hat (*K*) transform on *J* **Step 2.2** Perform bottom-hat (*L*) transform on *J* $M_1 = J + K$ $M_2 = M_1 - L$ Filtered image, $N = M_1 + M_2$
Step 3	**Segmentation by Otsu's method** **Step 3.1** Calculate threshold *T* of *N* **Step 3.2** Binary image *N* using *T*
Step 4	**Starch detection** **Step 4.1** Detection of starch from *N* **Step 4.2** Starch estimation (%) from *N*
Step 5	**End**

4 Experiment

4.1 Experimental Setup

The microscopic images are generated and captured by two modalities as mentioned earlier. One is the Foldscope which is an origami based optical microscope which can be assembled from a printed A4 size paper in 10 min. The dimension of the device is

70 × 20 × 2 mm^3 and of weight 8 gm. The images are generated by 140 × magnifications with 2micron resolution. Another, a traditional microscope Olympus CH20i using 10X objective lens has been used to generate images. Two set of images of same sample has been generated and captured by a smart phone camera (MotoG4 Play, 1280X720 HD, 294 ppi, 5 inches diagonal display, 8 MP). The simulation has been conducted using Matlab tool.

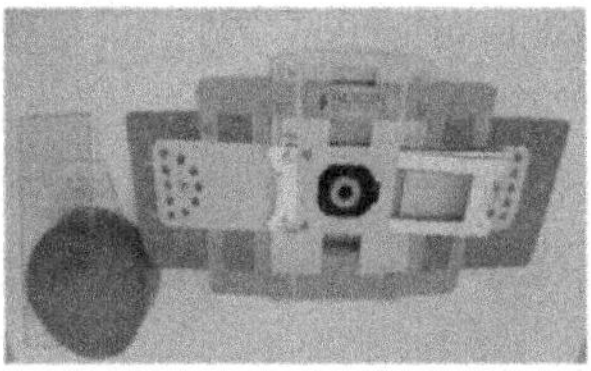

Fig. 2. Experimental set up.

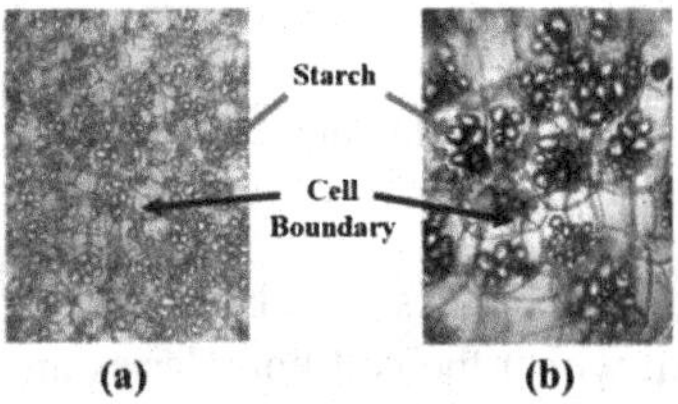

Fig. 3. Microscopic image of potato cells (a) Traditional microscope; (b) Foldscope

4.2 Data

For experimental validation, potato starch has been taken into account. Among different varieties of potatoes, the 'Jyoti' variety is been selected as it is largely consumed in India. The specification of the 'Jyoti' variety has been briefed in Table 1. To generate the microscopic images, thin sections of potato are placed under the Foldscope as well as the traditional microscope and consequently the images are captured. The images display the cell boundaries along with starch as indicated in Fig. 3(a) and (b) which are captured by traditional microscopic and Foldscope respectively. In total 20 images from both of the modalities are taken.

Table 1. Specification of potato sample

Particulates	Specifications
Common name	Alu, Alo, Aaloo
Season varieties	KufriJyoti (1968)
Family	Solanaceae
Botanical name	Solanumtuberosu L

5 Results and Discussions

The whole experiment performed on grayscale image and after morphological filtering the enhanced images has been displayed in Fig. 4(a) and (b) in which few cell boundaries can be observed; whereas starches are very prominent.

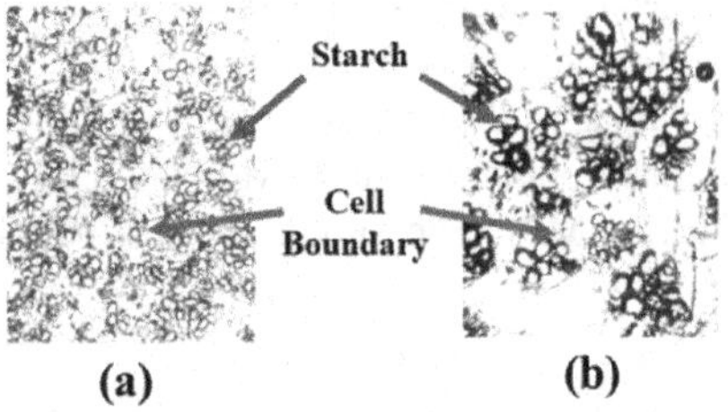

Fig. 4. Images after morphological filtering (a) microscope (b) Foldscope

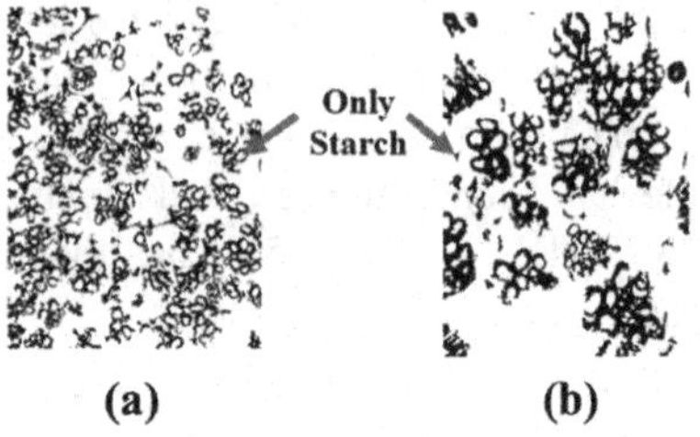

Fig. 5. Starch detection (a) microscope (b) Foldscope.

Next, the starches have been detected by employing Otsu's method as illustrated in Fig. 5, in which the cell boundaries are totally removed.

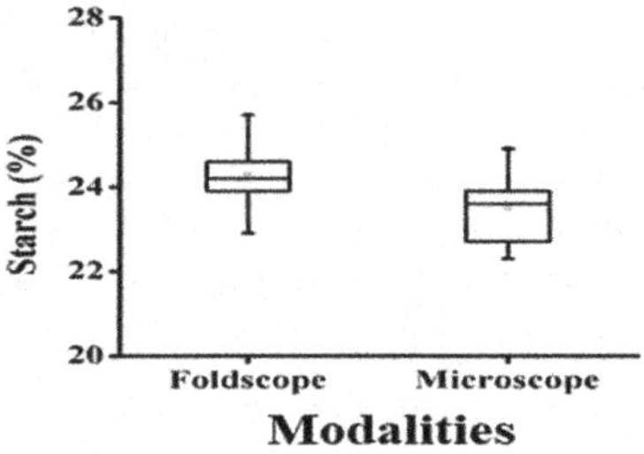

Fig. 6. Starch measurement using different microscopic image generation modalities.

Further, the starch present in potato has been calculated for both sets of images and their results are compared. The presence of starch has been calculated from the final image using (11)

$$\text{Starch} = \frac{BP_s}{P_s} \times 100\% \tag{11}$$

where, P_s and BP_s are the total no of pixels and black pixel (Starch) of the sample image respectively. The presence of starch (in %) estimated based on the image taken from microscope and Foldscope are 23.50 ± 0.79 and 24.29 ± 0.73 respectively and plotted in Fig. 6 in which the horizontal and vertical axes refers to measured starch and image generation modalities respectively. As seen the results are very much consistent.

6 Conclusion

In this wok, two set of microscopic images are generated and captured by conventional microscope and Foldscope which is a low cost, portable, paper microscope. Aiming its usability in food quality assessment system, estimation of starch in potato has been considered. The microscopic images are generated without staining by Foldscope as well as conventional microscope and captured by mobile camera. Next, morphological filtering followed by and Otsu's method has been employed. The starch of potato for both the image modalities are detected and measured successfully. The results for both of the cases are very consistent which offers that the Foldscope could be used in food quality assessment system. In future, more set of images along with several crops will be verified. In addition, deep learning-based recognition system will be incorporated to improve is efficiently.

Acknowledgement. The work has been supported by Department of Science and Technology, Govt. of India under IMRPINT-II with number IMP/2018/000538.

References

1. Cárdenas-Pérez, S., Chanona-Pérez, J., Méndez-Méndez, J.V., Calderón-Domínguez, G., López-Santiago, R., Perea-Flores, M.J.: Evaluation of the ripening stages of apple (Golden Delicious) by means of computer vision system. Biosyst. Eng. **159**, 46–58 (2017). https://doi.org/10.1016/j.biosystemseng.2017.04.009
2. Ali, M.A., Thai, K.W.: Automated fruit grading system. In: IEEE International Symposium in Robotics and Manufacturing Automation (ROMA), Kuala Lumpur, Malaysia, pp. 1–6, September 2017
3. Sa, I., Ge, Z., Dayoub, F., Upcroft, B., Perez, T., McCool, C.: Deepfruits: a fruit detection system using deep neural networks. Sensors **16**, 1222 (2016)
4. Pereira, L.F.S., Barbon Jr., S., Valous, N.A., Barbin, D.F.: Predicting the ripening of papaya fruit with digital imaging and random forests. Comput. Electron. Agric. **145**, 76–82 (2018)
5. Mehra, T., Kumar, V., Gupta, P.: Maturity and disease detection in tomato using computer vision. In: International Conference on Parallel, Distributed and Grid Computing (PDGC), Solan, India, pp. 399–403, December 2016
6. Arakeri, M.P.: Computer vision based fruit grading system for quality evaluation of tomato in agriculture industry. Procedia Comput. Sci. **79**, 426–433 (2016)
7. Liu, Q., Donner, E., Tarn, R., Singh, J., Chung, H.-J.: Advanced analytical techniques to evaluate the quality of potato and potato starch. In: Advances in Potato Chemistry and Technology, pp. 221–248. Elsevier (2009)
8. Cropotova, J., Tylewicz, U., Cocci, E., Romani, S., Dalla Rosa, M.: A novel fluorescence microscopy approach to estimate quality loss of stored fruit fillings as a result of browning. Food Chem. **194**, 175–183 (2016)
9. Allan-Wojtas, P., Goff, H., Stark, R., Carbyn, S.: The effect of freezing method and frozen storage conditions on the microstructure of wild blueberries as observed by cold-stage scanning electron microscopy. Scanning **21**, 334–347 (1999)
10. Foldscope. https://www.foldscope.com
11. Borah, S., Bowmick, B., Hazarika, C.: Production behaviour of potato in Assam-A critical analysis across zones and size groups of farms. Econ. Affairs **61**, 23 (2016)
12. Lutaladio, N., Castaldi, L.: Potato: the hidden treasure. J. Food Compos. Anal. **22**, 491–493 (2009)
13. Liu, Q., Zhang, L.J., Liu, X.P.: Microscopic image segmentation of Chinese herbal medicine based on region growing algorithm. In: Advanced Materials Research, pp. 4110–4115 (2013)

14. Ravi, S., Khan, A.: Morphological operations for image processing: understanding and its applications. In: Proceedings of 2nd National Conference on VLSI, Signal Processing & Communications NCVSComs, Guntur, India, December 2013
15. Otsu, N.: A threshold selection method from gray-level histograms. IEEE Trans. Syst. Man Cybern. **9**, 62–66 (1979)

Learning Based Image Selection for 3D Reconstruction of Heritage Sites

Ramesh Ashok Tabib(✉), Abhay Kagalkar, Abhijeet Ganapule, Ujwala Patil, and Uma Mudenagudi

KLE Technological University, Hubballi, India
ramesh_t@kletech.ac.in

Abstract. In this paper, we propose learning based pipeline with image clustering and image selection methods for 3D reconstruction of heritage site using cleaned internet sourced images. Cleaned internet sourced images means the images that do not contain an image with text, blur, occlusion, and shadow. 3D reconstruction of heritage sites is one of the emerging topics and is gaining importance as efforts are made to digitally preserve the heritage sites. 3D reconstruction using internet-sourced images is challenging as they often contain thousands of images taken from the same viewpoint. We propose to use autoencoders to extract robust features from images to cluster similar parts of heritage sites. We propose to use the image selection algorithm to select images from each cluster with the removal of redundant images. We demonstrate the proposed pipeline using available 3D reconstruction pipeline for a variety of heritage sites which contain one cluster to eight clusters and obtain better visual 3D reconstruction.

Keywords: Learning · Clustering · Image selection · 3D reconstruction

1 Introduction

In this paper, we propose learning based image clustering and selection methods for 3D reconstruction of heritage sites. We use cleaned images which do not have an image with watermark, blur, and occlusion as input to the proposed pipeline. When a particular search for a heritage site is given on the internet we get a few millions of images. For Example Hampi - a UNESCO world heritage site is spread over 16 sq mi and has about 1600 monuments. There are more than three million images on Google Image Search under the search label "Hampi". The search result contains various temples, fountains, stupas, paintings, sculptures etc. Most of these images are captured from hundreds or thousands of viewpoints and illumination conditions and may contain images with blur, occlusion, watermark etc. We use the method proposed in [11] to clean the images before giving into our proposed pipeline. This method proposed in [11] removes the images which contain blur, watermark, and occlusion. But even after the removal, the acquired

B. Deka et al. (Eds.): PReMI 2019, LNCS 11941, pp. 499–506, 2019.
https://doi.org/10.1007/978-3-030-34869-4_54

subset has redundant images which are not useful for 3D reconstruction and may affect the quality of 3D reconstruction as shown in Fig. 1. Typically heritage sites contain different parts like stupas (A hemispherical structure containing relics that is used as place of meditation), pillars, mandapas (It is a pillared outdoor hall), when 3D reconstruction of the whole space is carried without selecting images the reconstruction is not good due to a large number of outliers coming images with very near viewpoints which led to the selection of representative images for 3D reconstruction. For the selection process, we categorize the image into different parts of the heritage site and select a set of images from each part.

Most of the image clustering methods in literature talk about clustering as a tool to parallelize the sparse bundle adjustment step as including this step reduces the computational time. For large scale, 3D reconstruction efforts [1,8] are made to cluster the input image set with respect to viewpoints, but we propose to cluster the similar parts of the heritage site and a select subset of images for reconstruction. In [12] Thorsten Thormhlen et al., proposed an algorithm that selects the images having a large number of feature points by estimating the error of initial camera motion and object structure. In [3] Yasutaka Furukawa et al., proposed an approach for enabling existing multi-view stereo methods to operate on extremely large unstructured images and to remove an image if the conditions hold good even after the removal. By using these proposed approaches we propose to select images by removing redundant images that do not contribute to 3D reconstruction.

In this paper, we propose a pipeline with image clustering and selection modules to select images for 3D reconstruction. We summarize our contribution as follows:

- We propose learning based unsupervised image clustering to cluster similar-parts of the heritage site. An autoencoder is trained on the cleaned internet sourced images to extract robust features which are fed to a clustering algorithm to obtain image clusters containing similar parts of the heritage site.
- We propose an image selection module to get a subset of images from each cluster by removing redundant images. We use camera parameters and reprojection error factor in images to calculate the contributed value of the image for 3D reconstruction and by using a preset threshold we decide whether to retain or reject the image.
- We demonstrate the proposed pipeline using different heritage containing one to eight clusters and also compare with 3D models obtained using cleaned images.

In Sect. 2 we discuss the methodology of the proposed pipeline. In Sect. 3 we demonstrate the results of the proposed pipeline. In Sect. 4 we provide concluding remarks.

2 Learning Based Image Selection

We propose a pipeline as shown in Fig. 2, which consists of image clustering and selection modules. We assume our input images to be free from the text,

Fig. 1. Essence of image selection for 3D reconstruction

blur, occlusion etc. As it is evident from Fig. 1 even after the removal of these images the obtained point cloud from cleaned images is noisy and contain a large number of outliers. In the clustering module, images are clustered into respective similar-parts of the heritage site. Images from all clusters are given to the image selection algorithm, which removes redundant images. The selected set of images are then used to 3D reconstruct the heritage site. In what follows we explain in detail the image clustering and selection modules.

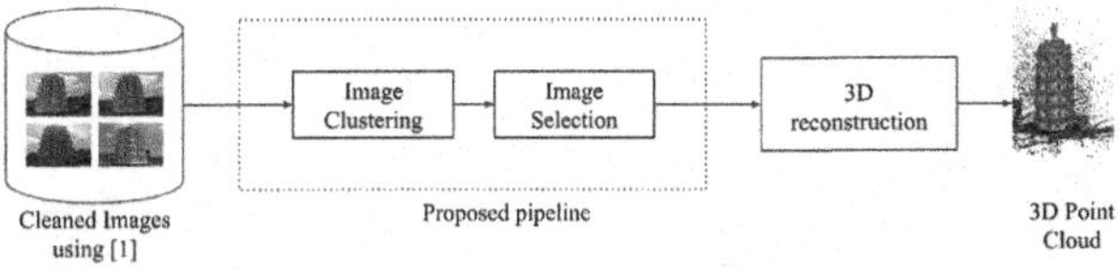

Fig. 2. Learning based image selection for 3D reconstruction

2.1 Image Clustering

Internet-sourced images contain all parts of the heritage site stored under a single label. Hence it makes clustering an important step before images are given to the image selection algorithm. In many clustering problems use handcrafted features like SIFT [5], SURF (Speeded-Up Robust Features) etc. But not all of the selected features are useful and relevant. In such a case choosing robust features leads to better performance. In unsupervised learning, class labels are not defined and choosing the right set of features becomes challenging because all features are not important. Redundant and irrelevant features may misguide clustering results. Hence we use autoencoders to extract robust features from input images.

Autoencoders [4,10,13] work by compressing the input into a latent-space representation and then reconstructing the output from this representation. It consists of an Encoder and a decoder as shown in Fig. 3. By pre-training an autoencoder network we except it to learn salient features of the input data. This feature vector is given as input to the clustering algorithm to obtain desired clusters.

We propose a network that consists of convolution layers stacked at the beginning to extract hierarchical features from the input images then flatten all units in the last convolution layer to form a vector followed by an up-sampled fully connected layer. Finally, it is down-sampled using subsequent fully connected layers with only 25 units which are called an embedded layer. The input image is thus transformed into a 25-dimensional feature vector as shown in Fig. 4 This obtained feature vector is given to the Mean-shift clustering algorithm [2] to obtain the clusters as shown in Fig. 5. In what follows we explain in detail the Image selection module.

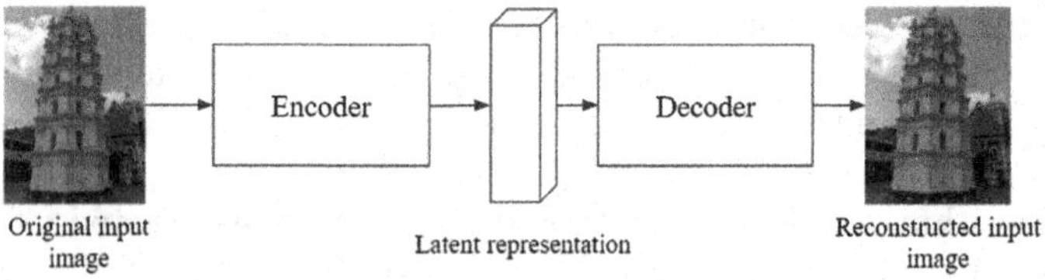

Fig. 3. Encoder-decoder network to learn robust features

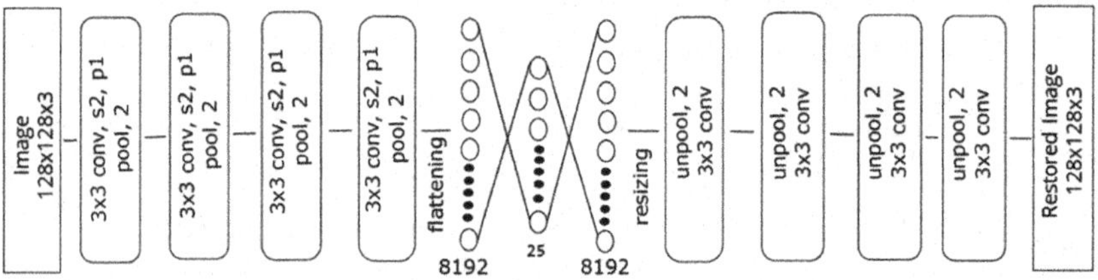

Fig. 4. Convolutional Autoencoder architecture trained for feature extraction in our proposed pipeline

2.2 Image Selection

Most of the image-based reconstruction for the generation of a detailed and informative 3D model relies on the images chosen. So image selection plays an important role in 3D reconstruction. Internet-sourced images are unstructured and redundant in nature. Many Multi-view Stereo algorithms aim to reconstruct a 3D model using all the images available [7]. Such an approach is not feasible as the number of images grows. Hence, it becomes an important step to select the right subset of images. After clustering, the clusters may contain redundant images that may not contribute to 3D reconstruction. So we use image selection to remove redundant images. The state of art Structure from Motion(SfM) [9] provides precise camera poses and 3D coordinates of the scene. A measure function proposed in [15] is used to estimate. The contributed value of two matching images to 3D reconstruction.

$$f_{jk} = fac\_angle_{jk} * fac\_error_{jk} \tag{1}$$

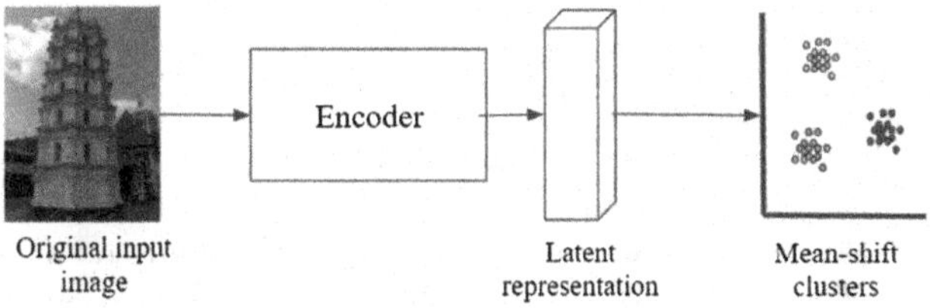

Fig. 5. Clustering using the features extracted from encoder

where f_{jk} represents the contributed value to 3D reconstruction of the combination of image j and image k. Let b be threshold and in our pipeline b is set to 7 heuristically. By using Algorithm 1 we remove the redundant images from acquired image subsets. The retained images are used for 3D reconstruction as shown in Fig. 6.

Algorithm 1. Image Selection

```
Input: Images i_1 ... i_n
for i ← 1 to n do
    for j ← 1 to n do
        f_ij += fac_angle_ij * fac_error_ij
for k ← 1 to n do
    calculate contributed value without k^th image;
    for p ← 1 to n do
        for q ← 1 to n do
            f_pq += fac_angle_pq * fac_error_pq
    if f_pq/f_ij less than 0.97 then
        Retain k^th image;
    else
        Remove k^th image;
```

3 Results and Discussions

In this section, we present the results of our proposed pipeline. The proposed pipeline is implemented on NVIDIA DGX-1 server with 500 GB RAM, 32 GB NVIDIA Tesla V100 graphics processor. We used internet sourced and passive crowd-sourced images to evaluate our proposed pipeline. We use OpenSfM to compute depth maps and generate the 3D point cloud.

Note that a heritage site containing more number of clusters or no viewpoint in common then the clusters are reconstructed separately. For the demonstration of results, we use three heritage site datasets. Shanta Durga Temple, Bankapura

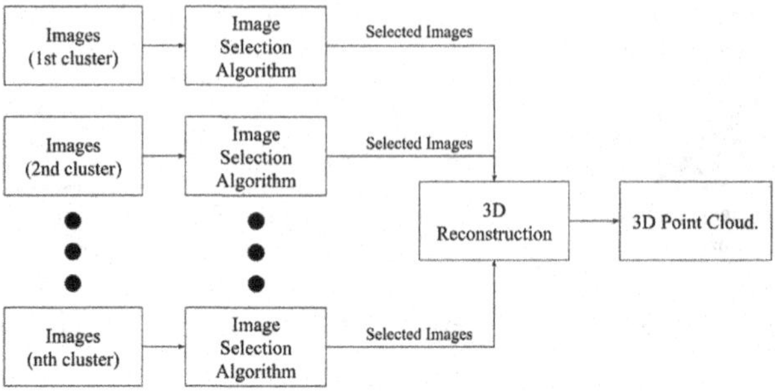

Fig. 6. Image selection pipeline

Table 1. Statistics

Dataset		Shantadurga temple	Bankapur Nageshwar	Pattadkal
No. of images	Input	611	703	236
	Cleaned	579	647	220
	After image selection	305	139	92
No. of clusters		2	1	2
No. of SFM points	Input	7,038,446	16,869,227	6,559,431
	Cleaned	5,879,238	10,788,004	4,401,547
	After image selection	3,007,136	3,033,418	1,544,580

(a) (b)

Fig. 7. 3D point cloud of (a) Shanta Durga Deepa Stambha from input images (579 Images). (b) Shanta Durga Deepa Stambha with our pipeline (305 Images).

Nageshwar Temple, Pattadkal are chosen for the demonstration. Figure 7 is 3D point cloud of Shanta Durga temple situated in Ponda (Tq), Goa, India. Figure 8 is the 3D point cloud of Bankapura Nageshwara temple situated in Haveri (Dist.), Karnataka, India. Figure 9 is the 3D point cloud of Pattadkal temple situated in Bagalkot (Dist.), Karnataka, India. Figure 7a illustrates 3D point cloud generated using 579 images. After image selection, 274 images were rejected. Note that the image selection discarded nearly 47.32% of images. We can observe that

(a) (b)

Fig. 8. 3D point cloud of (a) Bankapur Nageshwar Temple from original images (647 Images) (b) Bankapur Nageshwar Temple with our pipeline (139 Images).

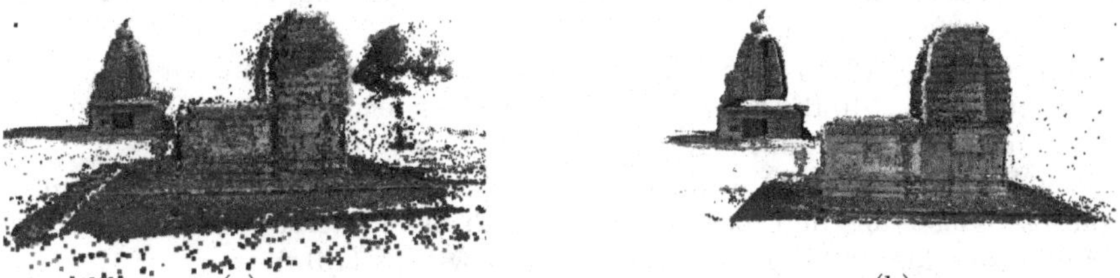

(a) (b)

Fig. 9. 3D point cloud of (a) Pattadkal Temple from original images (220 Images) (b) Pattadkal Temple with our pipeline (92 Images).

the 3D point cloud is noisy. Figure 7b illustrates 3D point cloud using 305 images after Image Selection. The obtained point cloud is free from noise and clear. This result indicates that Image selection is useful.

Similarly Figs. 8a and 9a illustrates 3D point cloud generated using 647 and 220 images respectively. And Figure. 8b and 9b illustrates the 3D point cloud generated using 139 and 92 images respectively after Image selection. Table 1 provides more information about number of clusters obtained and SFM points in generated 3D point clouds.

4 Conclusions

In this paper, we have proposed learning based image clustering and selection methods for 3D reconstruction of heritage site using cleaned internet sourced images. Cleaned internet sourced images means the images that do not contain an image with text, blur, occlusion, and shadow. 3D reconstruction of heritage sites is one of the emerging topics and is gaining importance as efforts are made to digitally preserve the heritage sites. 3D reconstruction using internet-sourced images is challenging as they often contain thousands of images taken from the same viewpoint. We have proposed to use autoencoders to extract robust features from images to cluster similar parts of heritage sites. We have proposed to use the image selection algorithm to select images from each cluster with the removal of redundant images. We have demonstrated the proposed pipeline using available 3D reconstruction pipeline for a variety of heritage sites which contain one cluster to eight clusters and obtain better visual 3D reconstruction.

Acknowledgement. This project is partly carried out under Department of Science and Technology (DST) sponsored project Indian Heritage in Digital Saces (IHDS).

References

1. Agarwal, S., Snavely, N., Simon, I., Sietz, S.M., Szeliski, R.: Building Rome in a day. In: Twelfth IEEE International Conference on Computer Vision (ICCV 2009). IEEE, September 2009
2. Carreira-Perpiñán, M.Á.: A review of mean-shift algorithms for clustering. CoRR abs/1503.00687 (2015)
3. Furukawa, Y., Curless, B., Seitz, S.M., Szeliski, R.: Towards internet-scale multi-view stereo. In: 2010 IEEE Computer Society Conference on Computer Vision and Pattern Recognition, pp. 1434–1441, June 2010
4. Jiang, Y., Yang, Z., Xu, Q., Cao, X., Huang, Q.: When to learn what: deep cognitive subspace clustering. In: Proceedings of the 26th ACM International Conference on Multimedia, MM 2018, pp. 718–726. ACM, New York (2018)
5. Lowe, D.G.: Distinctive image features from scale-invariant keypoints. Int. J. Comput. Vision **60**(2), 91–110 (2004)
6. Peng, X., Feng, J., Lu, J., Yau, W.Y., Yi, Z.: Cascade subspace clustering (2017). https://aaai.org/ocs/index.php/AAAI/AAAI17/paper/view/14442
7. Pons, J.P., Keriven, R., Faugeras, O.: Multi-view stereo reconstruction and scene flow estimation with a global image-based matching score. Int. J. Comput. Vision **72**(2), 179–193 (2007)
8. Sawyer, S.M., Ni, K., Bliss, N.T.: Cluster-based 3D reconstruction of aerial video. In: IEEE Conference on High Performance Ext Computing, pp. 1–6, September 2012
9. Snavely, N., Seitz, S.M., Szeliski, R.: Modeling the world from internet photo collections. Int. J. Comput. Vision **80**(2), 189–210 (2008)
10. Song, C., Liu, F., Huang, Y., Wang, L., Tan, T.: Auto-encoder based data clustering. In: Ruiz-Shulcloper, J., Sanniti di Baja, G. (eds.) CIARP 2013. LNCS, vol. 8258, pp. 117–124. Springer, Heidelberg (2013). https://doi.org/10.1007/978-3-642-41822-8_15
11. Sujay, A., Vaishnavi, A., Tabib, R., Mudenagudi, U.: Deep learning-based filtering of images for 3D reconstruction of heritage sites. In: 2018 11th Indian Conference on Computer Vision, Graphics and Image Processing, November 2018
12. Thormählen, T., Broszio, H., Weissenfeld, A.: Keyframe selection for camera motion and structure estimation from multiple views. In: Pajdla, T., Matas, J. (eds.) ECCV 2004. LNCS, vol. 3021, pp. 523–535. Springer, Heidelberg (2004). https://doi.org/10.1007/978-3-540-24670-1_40
13. Xie, J., Girshick, R.B., Farhadi, A.: Unsupervised deep embedding for clustering analysis. CoRR abs/1511.06335 (2015). http://arxiv.org/abs/1511.06335
14. Yang, B., Fu, X., Sidiropoulos, N.D., Hong, M.: Towards k-means-friendly spaces: simultaneous deep learning and clustering (2016). http://arxiv.org/abs/1610.04794
15. Yang, C., Zhou, F., Bai, X.: 3D reconstruction through measure based image selection. In: 2013 Ninth International Conference on Computational Intelligence and Security, pp. 377–381, December 2013

Local Contrast Phase Descriptor for Quality Assessment of Fingerprint Images

Ram Prakash Sharma(✉) and Somnath Dey

Indian Institute of Technology Indore, Indore, India
{phd1501201003,somnathd}@iiti.ac.in

Abstract. Fingerprint image quality is one of the main factors affecting the recognition performance of Automatic Fingerprint Identification System (AFIS). Therefore, analysis of fingerprint image quality is an important task during the acquisition. In this work, local contrast phase descriptor (LCPD) is used to analyze the texture quality of fingerprint images. Spatial and transform domain features computed using LCPD are fed to Support Vector Machine (SVM) classifier for fingerprint texture classification in wet, dry, and good class. Experimental evaluations performed on low-quality FVC 2004 DB1 dataset outperforms the current state-of-the-art methods. Therefore, utilizing the proposed method for quality control of fingerprint images during acquisition can help in improving the performance of fingerprint recognition system.

Keywords: Biometrics · Fingerprint quality · Local texture descriptor · Support vector machine

1 Introduction

The recognition of individuals using fingerprints is deployed in the areas of attendance system, personal identification, visa applications, airports, and border control etc. However, due to its widespread use, the demand for their performance improvement is also growing rapidly. The poor quality fingerprint images captured with the sensors due to the humidity, dust, temperature etc. causes degradation in the recognition performance. Therefore, identification of fingerprint quality is essential in order to process or enhance the poor quality fingerprint images before the recognition. The work presented in this paper provides a solution of identify the factors resulting in poor quality fingerprint images. This feedback will help in removing poor quality fingerprint images, quality enhancement, and updating or replacing poor samples [14].

In literature, various studies [1,11,22] have been proposed for the fingerprint quality assessment/classification. These methods either provide an overall quality score [16,21] or a quality label (i. e, good or bad) [6,15,17,20] to the fingerprint images. Identification of the factors resulting in poor quality fingerprint images can be useful to improve the recognition performance. Various studies related to fingerprint quality assessment [2,6,9,13,18,19] identifies the factors

B. Deka et al. (Eds.): PReMI 2019, LNCS 11941, pp. 507–514, 2019.
https://doi.org/10.1007/978-3-030-34869-4_55

which degrade the quality of fingerprint images during acquisition. These methods require multiple ridge features for fingerprint texture analysis. Therefore, we have proposed a novel method using single feature for the fingerprint texture assessment in this work.

In this work, a spatial and transform domain based local descriptor is proposed for fingerprint texture classification. The LCPD is designed specifically to analyze the texture variance of fingerprint images [5]. The fingerprint images are partitioned into 32 × 32 blocks to identify different quality region present in a fingerprint image. For each block, the local amplitude contrast and orientation information are extracted using spatial and transform domain features of LCPD. A 1D histogram is generated with the concatenation of bi-dimensional contrast-phase histogram. Using the 1D histogram, the SVM classifier is trained to classify the fingerprint blocks in appropriate quality classes (wet, dry, and good). Finally, an overall quality label is assigned to images of DB1 dataset using block t texture assessment method iteratively.

2 Related Works

The skin characteristics of the fingertip surface is the main cause which influences fingerprint recognition performance [1,22]. A literature review of some of these approaches are presented in the following.

In a recent approach, Tertychnyi et al. [18] utilizes the deep learning framework VGG16 to categorize the poor quality (wet, dry, etc.) fingerprint images. They have also tested their approach using SVM and Random Forest (RF) classifier. However, their results indicates that the performance of VGG16 deep learning model outperforms other two classifiers. A dry fingerprint detection method proposed by Wu et al. [19] utilizes various ridge-valley dependent features from fingerprint images of different resolution (500–1200 dpi). Thereafter, based on the extracted ridge features, classification of fingerprint images into normal and dry classes is done using SVM classifier. Sharma et al. [13] have utilized several ridge-valley distribution based features (i.e., mean, moisture, and ridge line count, etc) with Decision Tree (DT) classifier to train a model for quality classification of fingerprint images into dry, normal dry, good, normal wet, and wet classes. Awasthi et al. [2] proposed a block-based method which identifies the quality impairment (dry, wet, and good) information using various features related to orientation, consistency, pixel intensity, directional contrast of fingerprint images. The fingerprint images are also provided a quality score which depends on the number of dry, wet, and good quality blocks. The method proposed by Munir et al. [9] utilizes various frequency and spatial domain features and hierarchical k-means clustering for texture classification of fingerprint images. Lim et al. [6] have utilized features like directional strength, ridge/valley uniformity, etc. for fingerprint quality assessment. In their experiments, they have utilized self-organizing map (SOM), naive Bayes classifier, and radial basis function neural network (RBFNN) for quality classification of fingerprint blocks.

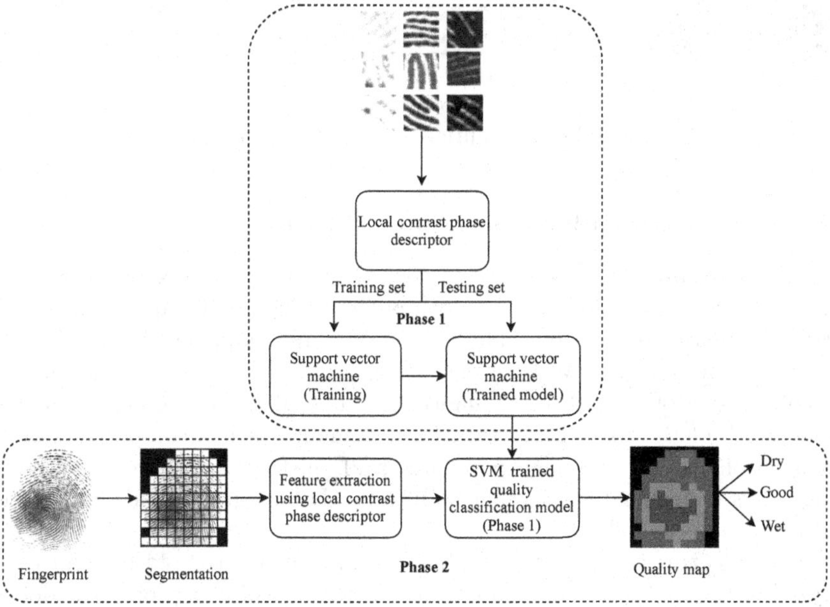

Fig. 1. Block diagram of the proposed method with all processing blocks

3 Proposed Method

Fingerprint quality analysis during acquisition can be helpful in acquiring better quality fingerprint images. This may lead to performance improvement of the fingerprint recognition system. The quality assessment of fingerprint images in the proposed work is done using the well-known LCPD. The block diagram indicating the various processing blocks of the proposed approach is provided in Fig. 1. The method proposed in this work operates in two different phases.

- **Block texture assessment:** In this phase, SVM classifier based model is designed to identify wet, dry, and good quality blocks.
- **Fingerprint texture assessment:** In this phase, the foreground blocks of fingerprint images are assigned a suitable quality class using the block texture assessment model. Finally, fingerprint images are assigned an overall quality class depending on maximum region of particular quality in the fingerprint image.

In this work, the manual labeling of fingerprint quality is done by three human experts. Some sample blocks of different qualities (dry, good, and wet) are given in Fig. 2.

In the work presented in this paper, a texture descriptor called LCPD is utilized for fingerprint quality assessment. The brief description of LCPD is provided in the following.

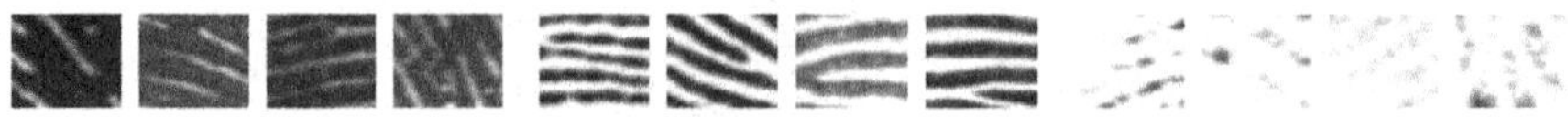

Fig. 2. Blocks of different qualities. (a) Wet (b) Good (c) Dry quality

3.1 Local Contrast Phase Descriptor (LCPD)

LCPD [5] is well-known local texture descriptor used for texture assessment of images. The LCPD features are computed using two components namely, spatial and phase components. The spatial domain component is based on the homologous component of Weber Local Descriptor (WLD) [4] while the rotation-invariant version of Local Phase Quantization (LPQ) [10] is used as phase component. Both of the spatial and transform domain features computed over the each pixel of fingerprint image populates a 2D histogram. Thereafter, the 2D histogram is concatenated together to form a 1D histogram. The computation of LCPD is done as follows:

The two features describing orientation and contrast are jointly defined and analyzed in the LCPD feature computation. The measure of contrast is defined in Eq. 1

$$\xi(x) = arctan\left[\frac{LoG[I(x)]}{I(x)}\right] \tag{1}$$

where $I(x)$ is input fingerprint image and LoG is the Laplacian of Gaussian operator.

In the LCPD feature computation, the rotation invariant version of the LPQ feature (LPQ^{ri}) [10] is used instead of angle of gradient. This LPQ feature is computed on a larger support in the order of 10×10 pixels which makes it more descriptive than the gradient angle and more resilient to random noise. The contrast ($\xi(x)$) and the LPQ^{ri} computed over the whole fingerprint image form a 2D histogram $LCPD(\xi(x), LPQ^{ri}(x))$ of dimension $S \times 256$. This 2D histogram is concatenated to form a 1D histogram feature vector.

For the descriptive LPQ feature, 256 bins are reserved contrary to WLD, and only a dozen for the contrast. Thereafter, the feature vector is passed through arctan non-linearity which results in a uniform quantization with N levels. When $N = 8$, the final feature becomes relatively long, i.e., 2048, and if the training set is relatively small, might result in over-fitting problems. So, a feature reduction technique is used to obtain more descriptive bins of the histogram.

4 Experimental Results

The performance evaluation of proposed approach is done on publically available quality sensitive FVC 2004 DB1 dataset [7]. The fingerprint images of DB1 dataset are acquired in different session by varying the fingertip surface conditions. This acquisition builds a quality based dataset which contain the fingerprint images of various qualities and makes it suitable for performance analysis

Table 1. Classification results obtained for the proposed block texture assessment method using LCPD features

		SVM			
		Dry	Good	Wet	Total
Subjective quality	Dry	**199**	1	0	200
	Good	3	**188**	9	200
	Wet	0	7	**193**	200
	Total	202	196	202	600
	Accuracy	99.50%	94.00%	96.50%	96.67%

Table 2. Comparative analysis of the results of block texture assessment

Methods	Classifier	Quality		
		Bad	Good	Overall
Proposed method	SVM	**392 (98.00%)**	**188 (94.00%)**	580 (**96.67%**)
Subjective quality	Manual	400	200	600
Lim et al. [6]	RBFNN	378 (94.50%)	186 (93%)	564 (94.00%)
Lim et al. [6]	SOM	359 (89.75%)	175 (87.50%)	534 (89.00%)
Lim et al. [6]	Naive Bayes	375 (93.75%)	181 (90.50%)	556(92.66%)
Sharma et al. [13]	DT	379 (94.75%)	191 (95.50%)	570 (95.00%)

of proposed quality assessment method. The DB1 dataset of FVC 2004 database contains 800 fingerprint images which are constituted from the eight samples of 100 fingers each.

4.1 Block Texture Assessment

A quality based dataset is constituted for performance evaluation of block texture assessment method using the 32×32 size foreground blocks of DB1 dataset. After this, three human experts having expertise in the field of fingerprint textures have marked 1000 blocks into wet, dry, and good quality classes. Thereafter, features are extracted using LCPD from the blocks. 5-fold cross-validation is used to partition the dataset in training and testing sets. Radial basis function (RBF) kernel is used for the SVM classifier [12] for performance evaluation of the proposed approach. Grid search algorithm [3] is used to find the optimal values of kernel (γ) and penalty (C) parameters. The optimal values obtained using this are $C = 2$ and $\gamma = 0.0625$ which are providing the best results. The classification performance of the LCPD features is reported in Table 1. The classification accuracies for the dry, good, and wet blocks are 99.50%, 94.00%, and 96.50%, respectively. Based on these accuracies, an overall classification accuracy of 96.67% is obtained for the proposed method. The comparative analysis of the results is provided in Table 2. Proposed method is compared with other

Table 3. Comparative analysis of the results obtained for the fingerprint texture assessment method

	Quality class			
	Wet	Good	Dry	Overall
Proposed method	**167 (97.09%)**	**403 (91.38%)**	**179 (95.72%)**	**749 (93.62%)**
Subjective quality	172	441	187	800
Awasthi et al. [2]	159 (92.44%)	405 (91.83%)	174 (93.04%)	738 (92.25%)
Munir et al. [9]	168 (97.67%)	393 (89.11%)	169 (90.37%)	730 (91.25%)
Terty et al. [18]	163 (94.76%)	387 (87.75%)	175 (93.58%)	725 (90.62%)

methods by combining wet and dry class to represent the bad quality class. The comparative evaluations indicate the higher performance of proposed method as compared with other methods.

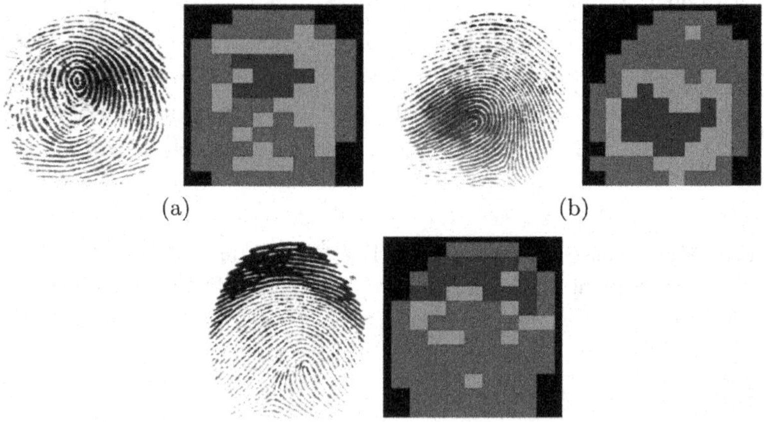

(a) (b) (c)

Fig. 3. Fingerprint images and their respective quality maps where blue, green, and red regions indicate wet, good, and dry texture, respectively. (Color figure online)

4.2 Fingerprint Texture Assessment

Fingerprint texture assessment results are reported in this section. Initially, a subjective quality classification is done by three human experts who classifies the fingerprint images into wet, good, and dry quality classes. Utilizing the block texture assessment method iteratively on the foreground blocks identified using [8], a decision on the overall quality of fingerprint images is made. The maximum number of blocks assigned to a particular quality class decides the overall quality class of a fingerprint image. The comparative analysis of the proposed approach is done with subjective quality, Awasthi et al. [2], Terty et al. [18], and Munir et al. [9] methods. The comparative analysis of the results is provided in Table 3. The comparative analysis of the results indicate that proposed method performs better (93.62%) as compared with the other quality assessment methods. The different quality regions contained in fingerprint images are shown in the quality maps in Fig. 3. The blue, green, and red colors show wet, good, and dry regions in the fingerprint images. The condition of the fingertip can be improved based on these maps, and an improved fingerprint image can be obtained during acquisition phase.

5 Conclusion

In this work, we have utilized LCPD for fingerprint quality assessment. The superior experimental results of the proposed approach demonstrates its capability as compared with other recent methods. Therefore, the proposed approach can be very effective to control the fingerprint quality during acquisition phase of fingerprint recognition system. Further, only a single feature is required by the proposed approach for texture assessment of fingerprint images which makes it a fast and simple approach. In future research directions of this work, texture quality of fingerprint images can be evaluated with some other texture based descriptors.

Acknowledgment. This research work has been carried out with the financial support provided from Science and Engineering Research Board (SERB), DST (ECR/2017/000027), Govt. of India.

References

1. Alonso-Fernandez, F., et al.: A comparative study of fingerprint image-quality estimation methods. IEEE Trans. Inf. Forensic Secur. **2**(4), 734–743 (2007)
2. Awasthi, A., Venkataramani, K., Nandini, A.: Image quality quantification for fingerprints using quality-impairment assessment. In: IEEE Workshop on Applications of Computer Vision (WACV), pp. 296–302 (2013)
3. Chang, C.C., Lin, C.J.: LIBSVM: a library for support vector machines. ACM Trans. Intell. Syst. Technol. **2**(3), 1–27 (2011)
4. Chen, J., Shan, S., Zhao, G., Chen, X., Gao, W., Pietikainen, M.: A robust descriptor based on Webers law. In: IEEE Conference on Computer Vision and Pattern Recognition, pp. 1–7 (2008)
5. Gragnaniello, D., Poggi, G., Sansone, C., Verdoliva, L.: Local contrast phase descriptor for fingerprint liveness detection. Pattern Recogn. **48**(4), 1050–1058 (2015)
6. Lim, E., Toh, K.A., Suganthan, P.N., Jiang, X., Yau, W.Y.: Fingerprint image quality analysis. In: International Conference on Image Processing (ICIP), vol. 2, pp. 1241–1244 (2004)
7. Maio, D., Maltoni, D., Cappelli, R., Wayman, J.L., Jain, A.K.: FVC2004: third fingerprint verification competition. In: Zhang, D., Jain, A.K. (eds.) ICBA 2004. LNCS, vol. 3072, pp. 1–7. Springer, Heidelberg (2004). https://doi.org/10.1007/978-3-540-25948-0_1

8. Mehtre, B.M.: Fingerprint image analysis for automatic identification. Mach. Vis. Appl. **6**(2), 124–139 (1993)
9. Munir, M.U., Javed, M.Y., Khan, S.A.: A hierarchical k-means clustering based fingerprint quality classification. Neurocomputing **85**, 62–67 (2012)
10. Ojansivu, V., Rahtu, E., Heikkila, J.: Rotation invariant local phase quantization for blur insensitive texture analysis. In: 2008 19th International Conference on Pattern Recognition, pp. 1–4 (2008)
11. Olsen, M.A., Smida, V., Busch, C.: Finger image quality assessment features: definitions and evaluation. IET Biometrics **5**(2), 47–64 (2016)
12. Schölkopf, B., Williamson, R., Smola, A., Shawe-Taylor, J., Platt, J.: Support vector method for novelty detection. In: Proceedings of the 12th International Conference on Neural Information Processing Systems, pp. 582–588 (1999)
13. Sharma, R.P., Dey, S.: Fingerprint image quality assessment and scoring. In: Ghosh, A., Pal, R., Prasath, R. (eds.) MIKE 2017. LNCS (LNAI), vol. 10682, pp. 156–167. Springer, Cham (2017). https://doi.org/10.1007/978-3-319-71928-3_16
14. Sharma, R.P., Dey, S.: Two-stage quality adaptive fingerprint image enhancement using fuzzy c-means clustering based fingerprint quality analysis. Image Vis. Comput. **83–84**, 1–16 (2019). https://doi.org/10.1016/j.imavis.2019.02.006
15. Shen, L.L., Kot, A., Koo, W.M.: Quality measures of fingerprint images. In: Bigun, J., Smeraldi, F. (eds.) AVBPA 2001. LNCS, vol. 2091, pp. 266–271. Springer, Heidelberg (2001). https://doi.org/10.1007/3-540-45344-X_39
16. Tabassi, E.: Development of NFIQ 2.0. NIST (2015). https://www.nist.gov/services-resources/software/development-nfiq-20
17. Tabassi, E., Wilson, C.L.: A novel approach to fingerprint image quality. In: International Conference on Image Processing, vol. 2, pp. 37–40. IEEE (2005)
18. Tertychnyi, P., Ozcinar, C., Anbarjafari, G.: Low-quality fingerprint classification using deep neural network. IET Biometrics **7**(6), 550–556 (2018)
19. Wu, C., Chiu, C.: Dry fingerprint detection for multiple image resolutions using ridge features. In: IEEE International Workshop on Signal Processing Systems (SiPS), pp. 1–5, October 2017
20. Yang, X.K., Luo, Y.: A classification method of fingerprint quality based on neural network. In: International Conference on Multimedia Technology, pp. 20–23 (2011)
21. Yao, Z., bars, J.L., Charrier, C., Rosenberger, C.: Quality assessment of fingerprints with minutiae delaunay triangulation. In: International Conference on Information Systems Security and Privacy (ICISSP), pp. 315–321 (2015)
22. Yao, Z., Bars, J.M.L., Charrier, C., Rosenberger, C.: Literature review of fingerprint quality assessment and its evaluation. IET Biometrics **5**(3), 243–251 (2016)

A Polynomial Surface Fit Algorithm for Filling Holes in Point Cloud Data

Vishwanath S. Teggihalli(✉), Ramesh Ashok Tabib, Adarsh Jamadandi, and Uma Mudenagudi

K.L.E. Technological University, Hubli 580032, Karnataka, India
vasantteggihalli30@gmail.com,
{ramesh_t,adarsh.jamadandi,uma}@kletech.ac.in

Abstract. In this paper, we propose a novel framework for detecting and filling missing regions in point cloud data. We propose to investigate the properties of point cloud data and develop a framework that can detect boundaries of intricate holes in the point cloud data of complex shapes and fill the hole consequently. The holes in point cloud data are caused owing to many reasons like reflectance, transparency, occlusions etc. Detecting holes in point cloud data is a non-trivial task, since point cloud data is unstructured and comes with no adjacency/connectivity information. We propose a Centroid-Shift algorithm that exploits the distance of cluster centroid from the member points to detect the boundaries of holes, further we propose a polynomial surface fit framework to accurately fill the missing regions/holes without losing the original shape attributes of the objects. We demonstrate our framework on popular 3D objects, we provide qualitative results and report RMS error as the evaluation metric to measure the effectiveness of the hole-filling algorithm.

Keywords: Hole filling algorithm · Hole detection · Polynomial Surface Fit

1 Introduction

In this paper[1]. We propose a novel framework to detect missing regions in point clouds and consequently fill them without losing the original shape attributes of the objects involved. Point cloud representations are becoming popular owing to rise of 3D scanners, which help us acquire data in an unstructured point cloud data format. These point clouds come equipped with certain properties like normals and colors, however the surface information is not explicitly known and has to be determined by inferring the relationship of these points with their neighborhood. Missing regions in point clouds are a result of infidel 3D scanners, reflectance, occlusions and/or transparency. Even with high-fidelity 3D scanners, it is possible to end up with missing regions in final integrated models

[1] The authors would like to thank the team at Samsung Research India, Bangalore for their valuable insights and discussions.

B. Deka et al. (Eds.): PReMI 2019, LNCS 11941, pp. 515–522, 2019.
https://doi.org/10.1007/978-3-030-34869-4_56

because of high grazing angles, missing regions in the original model, limited number of range scans from different viewing directions etc. While point clouds are easy to represent and manipulate compared to mesh representations, certain trivial operations often encountered in geometry modeling, like determining hole boundaries and filling holes become ill-posed problems. Hole detection and filling finds important applications in 3D shape representation and analysis, surface reconstruction, 3D model completion etc. One of the most direct methods of detecting hole boundaries and then filling the holes is to convert the point clouds into meshed representations, the techniques include Delaunay Triangulation [4] which uses the point clouds as vertices for forming a triangle mesh or by using surface reconstruction methods like Poisson [8]. The problem with these methods apart from being computationally expensive is that they don't work well in presence of large missing regions in the original point clouds, this prompts for a framework that can directly work on the original point clouds. In this work, we propose a new algorithm called Centroid-Shift that exploits the geometric properties like distance of point cloud cluster centroids from member points to detect the candidate hole boundaries in point clouds. We also propose a post-processing step that helps in filtering out the external boundary, unwanted edges and corners of the objects involved. Finally, we describe a Polynomial Surface Fit framework that is used to fill in the missing regions without the loss of original shape attributes of the objects involved. Towards this we make the following contributions:

1. We propose a framework, which we call the Centroid-Shift algorithm that helps in selecting the candidate boundary points of all potential holes/missing regions.
2. We propose a post-processing step based on underlying surface variations of the point clouds to eliminate pseudo-boundary points and unwanted edges.
3. Finally, we propose a Polynomial Surface Fit hole-filling algorithm that can be used to fill/complete missing regions in point clouds without losing the original shape attributes of the objects.

2 Related Work

Detection of holes and filling missing regions have been explicitly discussed in the context of mesh representations [1,7,10]. Recently with the rise of affordable 3D scanners, point cloud representations have become ubiquitous and various techniques to address the problems of detecting missing regions and hole-filling in point clouds have become apparent. While its easy for a human being to visually identify the boundaries of holes by just looking, the problem is non-trivial when it comes to algorithmically detecting the boundaries of missing regions in order to fill them. Several approaches that exploit various geometric properties have been proposed, in [6] the authors propose an angle criterion in which the point clouds are projected onto a tangent plane and are sorted radially and then the angle between the consecutive samples is measured, the larger the angle gap the more the probability of the points being boundary-points. Authors in [2] take this approach further and propose multiple criteria to detect the boundary points. They propose to use an angle criterion, a weighted-average criterion that assumes the boundary points are homeomorphic to half-disk and finally an ellipsoidal criterion that uses the eigenvalues which encode the shape of the ellipsoid. Finally, all these criteria are combined in a weighted average manner to detect boundary points. After detecting the boundaries of the missing regions, they have to be filled such that the original shape attributes of the object is not deformed. Since the underlying geometric shape is not apparent from just point clouds, it becomes difficult to ascertain what the missing region is. Authors in this survey paper [5] provide a detailed account of various hole-filling methods.

3 Proposed Methodology

In this section, we discuss our hole-boundary detection and hole-filling algorithm. Figure 1 outlines the proposed approach.

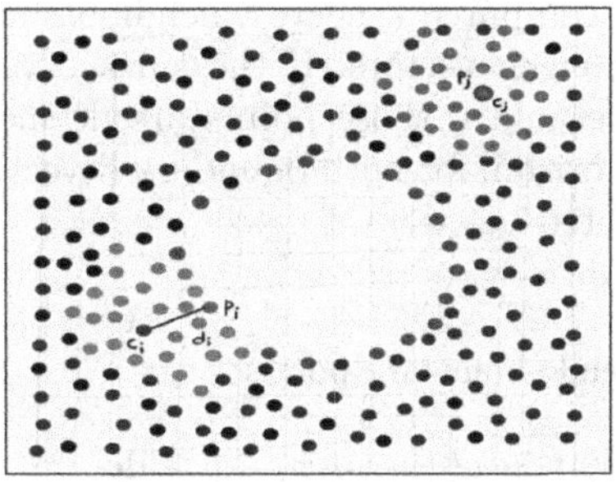

Fig. 1. In a certain point cloud with a large missing region, the euclidean distance d_i is calculated between a candidate boundary point P_i and the centroid C_i of a cluster. The points, which are farther from the centroid are possible boundary points. Such points are accumulated to detect the boundaries of the holes, later the pseudo-boundary points are discarded using a post-processing step.

3.1 Hole Boundary Detection

In this section, we describe the technique to detect the boundaries of missing regions, which we refer to as Centroid-Shift algorithm. This framework gives us a set of candidate boundary points, which could be plausible boundaries of missing regions, we further use a refining step to filter out pseudo-boundary points. Here, our main contribution lies in refining the detected boundaries and discarding the pseudo-boundaries and unwanted edges that are detected during the hole-boundary detection step. Consider a set of point clouds with a large missing region, we use the K-Nearest Neighbor (KNN) to find the neighbouring points of a query point, the K value is set heuristically based on the density of the point clouds, as a general rule we set the K value as 5% of the point cloud density, now for a query point P_i. We compute the centroid of these K neighbouring points. After computing the centroid of the sample, we compute the euclidean distance d_i between all the K neighbouring points and the cluster centroid. Similarly, we compute for all the points. The points whose distance is greater than the quarter of the mean of the distance of the neighbor points, given by the Eq. 1 are considered the candidate boundary points [3]. The proposed method is shown in Fig. 1.

$$d_i > \frac{\sum_1^k |p_i - c_i|}{4k} \tag{1}$$

3.2 Filtering Pseudo-Boundary Points

The centroid-shift algorithm detects the object boundaries and unwanted edges which satisfy the criteria mentioned above but are not actually holes/missing regions. The different types of pseudo-boundary holes detected are shown in Fig. 3. To discard the false boundary points, we consider surface variations. The process involves already computed k-nearest neighbors and then computing the surface variations using eigen vectors. If the surface variation is found to be greater than certain threshold t, those points are deemed as pseudo-boundary points. The proposed algorithm for both boundary detection and eliminating the pseudo points is highlighted below.

Result: Eliminate Pseudo-boundary points
initialization;
for *Each point* $p_i \in$ *point cloud* $\{p_1, p_2, p_3, ...p_n\}$ **do**
  Compute k-nearest neighbours points $N = \{n_1, n_2, n_3, ...n_k\}$ check for centroid shift criteria, if it is a boundary point then compute covariance matrix C for their respective neighbouring k points.
  Compute the eigen vectors for the covariance matrix C, which gives us eigen values $e_1(P_i) \geq e_2(P_i) \geq e_3(P_i)$ and then can calculate the surface variation SV as given by equation 2.
  if $SV > t$ **then**
    Discard the boundary point b_i;
  else
    Keep the boundary point as real hole boundary;
  end
end

Algorithm 1: Pseudo-boundary elimination algorithm via surface variation calculation.

The algorithm involves computing surface variation using the eigen vectors via the co-variance matrix of the neighborhood points. The co-variance matrix is calculated as given by the equation below,

$$cov(x,y) = \frac{\sum_1^K (N_{xmean} - n_{ix})(N_{ymean} - n_{iy})}{k-1} \tag{2}$$

The eigen vectors are used to calculate the surface variation as follows, and a surface variation greater than some threshold t is considered as pseudo-boundary.

$$SurfaceVariation = \frac{e_3(P_i)}{e_1(P_i) + e_2(P_i) + e_3(P_i)} \tag{3}$$

The value of k and t is heuristically set to achieve best possible scenario of objects whose pseudo-boundary points are not detected as actual holes.

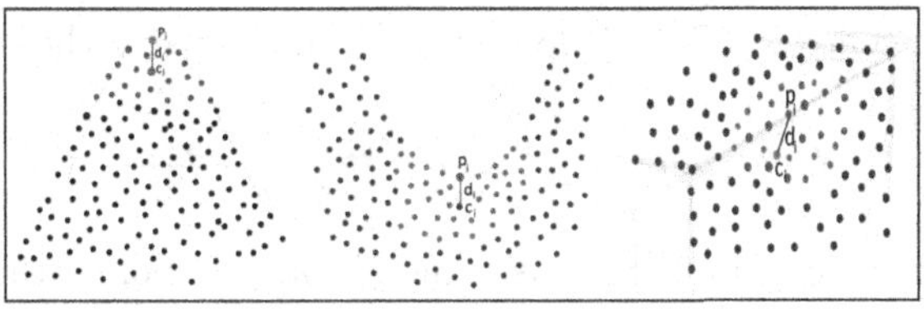

Fig. 2. The different cases of pseudo-boundary holes detected. In this first figure (from left) the query point is at a corner, in the second figure the point is located at a valley and in the last figure the query point is located at an edge.

In Fig. 3 we have experimented with varying k and t value to obtain object without pseudo-boundary points. In Fig. 3a, the k value and t value was set to 30 and 1.2 respectively, in Fig. 3b the values were set to k = 30 and t = 0.1, in Fig. 3c the values were k = 20 and t = 0.2 and for Fig. 3d the values were set to k = 20 and t = 0.1. By varying the k value and threshold value, we can obtain an optimum detection of all actual holes.

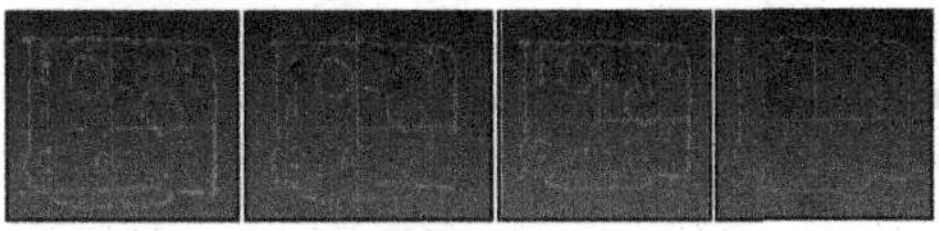

Fig. 3. We have experimented with varying k and t value to obtain object without pseudo-boundary points.

3.3 Hole Filling via Polynomial Surface Fit

In this section, we describe our proposed hole-filling algorithm. After detecting the candidate boundary points and eliminating the pseudo boundary points we aim to fill the detected missing regions to ensure better surface reconstruction. Our algorithm involves fitting an optimum polynomial surface patch and then sampling the neighborhood, most of the 3d objects can be approximated within

4th order polynomial fit. The holes in euclidean space doesn't give a proper sense of hole orientation and allows for overlapping of closely located holes making it difficult to preserve the sampling density of the original point cloud and to ascertain the patch necessary for filling the holes. Hence, we project the individual hole on to its euclidean plane, which in turn is computed via neighbouring points of boundary and thus reducing the problem of reconstructing holes in 3D to the simpler interpolation problem. Figure 5 shows projecting the holes onto eigen space as opposed to euclidean space. The steps involved in filling the holes/missing regions are outlined below:

1. For an identified true hole boundary consisting of m points, $\mathbf{B} = \{b_1, b_2, b_3, ...b_m\}$
2. Compute the centroid of the hole boundary points using the equation,

$$Centroid = (\frac{\sum_{i=1}^{m} x_i}{m}, \frac{\sum_{i=1}^{m} y_i}{m}, \frac{\sum_{i=1}^{m} z_i}{m}) \tag{4}$$

3. Compute a learning radius R consisting of learning points that will help us ascertain the order of the surface fit as follows,

$$LearningRadius(R) = 2(\frac{\sum_{i=1}^{m} |Centroid - b_i}{m}) \tag{5}$$

4. For the learning points $\mathbf{L} = \{L_1, L_2, L_3, ...L_n\}$ compute the covariance matrix and calculate the eigen vectors as outlined in Eq. 2.
5. Project all the learning points onto the eigen space to fit a higher order surface F(x,y). Projecting of learning points onto eigen space helps in efficient interpolation of higher order surface holes even though they have complex spatial orientations.
6. Interpolate the new points $\mathbf{I} = \{p_1, p_2, p_3, ...p_j\}$ using regression fit F(x,y) as $P_{iz} = F(P_{ix}, P_{iy})$. P_{ix} and P_{iy} are considered such that the interpolated points preserve the point density of original point cloud. i.e P_{ix} and P_{iy} should be uniformly distributed.
7. Finally the eigen-space projected points are reprojected back to the original space via inverse transformation using already computed eigen vectors.

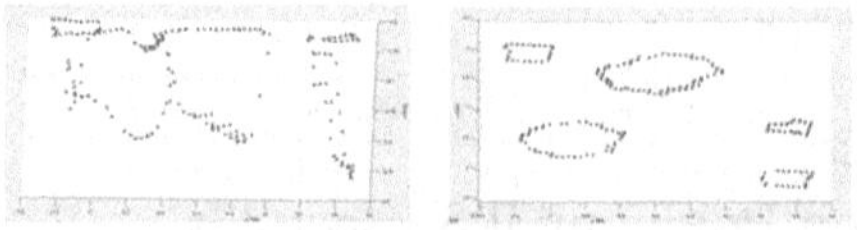

Fig. 4. Projecting the hole boundaries onto eigen-space provides a proper sense of spatial orientations of hole boundaries allowing us to efficiently fill them. In figure a, the hole boundaries are shown in euclidean space while in figure b the holes are projected onto eigen-space.

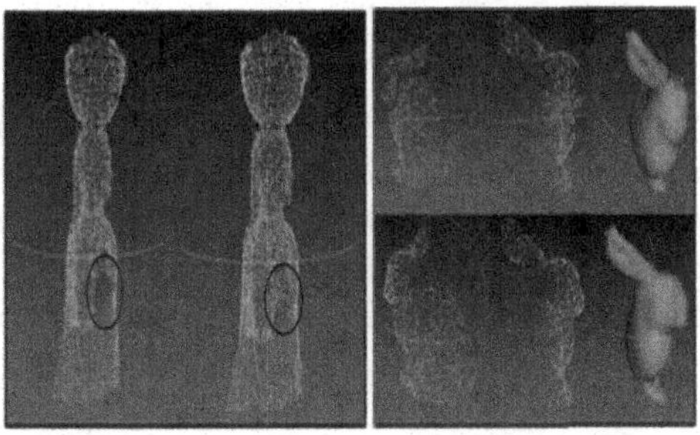

Fig. 5. After detecting the hole boundaries and eliminating the pseudo-boundaries, we apply our polynomial surface algorithm to fill the missing region. The figure shows the region is faithfully filled without distorting the original shape attributes of the model as reported by the RMS error in Table 1.

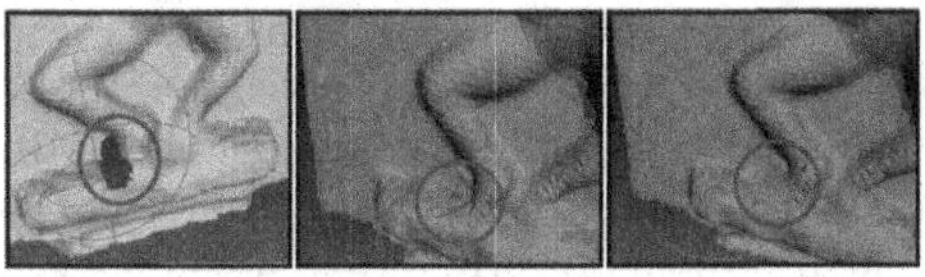

Fig. 6. Comparison of 3D reconstruction performed before and after hole filling.

4 Results

In this section, we present the results of our hole detection and filling framework. We perform various experiments on popular 3D models and report RMS error as a quantitative evaluation metric to assess the effectiveness of our hole filling algorithm. In Fig. 5 (left), we perform hole filling on a lady statue using our proposed framework. In Figs. 5 (right) and 6 we have compared surface reconstruction with and without hole filling, the top row in Fig. 5 shows the Sanford bunny model surface reconstructed without filling in missing regions and the below model shows surface reconstruction performed after hole filling. Clearly, filling the missing regions using our proposed framework yields better quality surface reconstruction. We report the RMS error for the filled patch in Table 1. As evident from the Table 1, the RMS error decreases as we fit higher order surfaces to fill the missing regions.

Table 1. RMS errors for Polynomial Surface Fit algorithm performed on various 3D models.

Model	First order fit	Second order fit	Third order fit	Fourth order fit
Stanford Bunny	0.910	0.0707	0.0601	0.0601
Girl statue	0.0904	0.0533	0.0495	0.0488

5 Conclusion

In this paper, we have proposed a novel algorithm to detect holes/large missing regions in point cloud data and consequently fill them without losing the original shape attributes of the model involved. Our proposed framework can successfully detect holes, filter out pseudo-boundaries and consequently help fill them. Experiments carried out on popular 3D models demonstrates the effectiveness of our framework.

References

1. Attene, M., Campen, M., Kobbelt, L.: Polygon mesh repairing: an application perspective. ACM Comput. Surv. **45**(2), 15:1–15:33 (2013). https://doi.org/10.1145/2431211.2431214
2. Bendels, G.H., Schnabel, R., Klein, R.: Detecting holes in point set surfaces. J. WSCG **14** (2006)
3. Chalmovianský, P., Jüttler, B.: Filling holes in point clouds. In: Wilson, M.J., Martin, R.R. (eds.) Mathematics of Surfaces. LNCS, vol. 2768, pp. 196–212. Springer, Heidelberg (2003). https://doi.org/10.1007/978-3-540-39422-8_14
4. Delaunay, B.N.: Sur la sphère vide. Bull. Acad. Sci. URSS **1934**(6), 793–800 (1934)
5. Guo, X., Xiao, J., Wang, Y.: A survey on algorithms of hole filling in 3D surface reconstruction. Vis. Comput. **34**(1), 93–103 (2018). https://doi.org/10.1007/s00371-016-1316-y
6. Hoppe, H., DeRose, T., Duchamp, T., McDonald, J., Stuetzle, W.: Surface reconstruction from unorganized points. In: Proceedings of the 19th Annual Conference on Computer Graphics and Interactive Techniques, SIGGRAPH 1992, pp. 71–78. ACM, New York (1992). https://doi.org/10.1145/133994.134011
7. Jun, Y.: A piecewise hole filling algorithm in reverse engineering. Comput.-Aided Des. **37**(2), 263–270 (2005). https://doi.org/10.1016/j.cad.2004.06.012, http://www.sciencedirect.com/science/article/pii/S0010448504001320
8. Kazhdan, M., Bolitho, M., Hoppe, H.: Poisson surface reconstruction. In: Proceedings of the Fourth Eurographics Symposium on Geometry Processing, SGP 2006, pp. 61–70. Eurographics Association, Aire-la-Ville, Switzerland, Switzerland (2006). http://dl.acm.org/citation.cfm?id=1281957.1281965
9. Setty, S., Ganihar, S.A., Mudenagudi, U.: Framework for 3D object hole filling. In: 2015 Fifth National Conference on Computer Vision, Pattern Recognition, Image Processing and Graphics (NCVPRIPG), pp. 1–4, December 2015. https://doi.org/10.1109/NCVPRIPG.2015.7490062
10. Wu, X.J., Wang, M.Y., Han, B.: An automatic hole-filling algorithm for polygon meshes. Comput.-Aided Des. Appl. **5**(6), 889–899 (2008). https://doi.org/10.3722/cadaps.2008.889-899

Sparsity Regularization Based Spatial-Spectral Super-Resolution of Multispectral Imagery

Helal Uddin Mullah[1], Bhabesh Deka[1(✉)], Trishna Barman[1], and A. V. V. Prasad[2]

[1] Department of Electronics and Communication Engineering, Tezpur University, Tezpur 784028, India
bdeka@tezu.ernet.in

[2] National Remote Sensing Centre, Hyderabad 500037, India

Abstract. Multispectral (MS) remote sensing image is composed of several spectral bands of distinct wavelengths. Most earth observation satellites provide MS images consisting several low-resolution (LR) bands together with a single high-resolution (HR) image. A single image super-resolution (SISR) method tries to produce a HR MS output from the given LR MS input using digital image processing algorithms. In this work, we present a patch-wise sparse representation based MS image SR using a coupled overcomplete trained dictionary. The dictionary learning is carried out from patches extracted from the given HR panchromatic (PAN) image itself. Experiments are carried out using test MS images from QuickBird satellites and results are compared with other state-of-the-art MS image SR and pan-sharpening methods.

Keywords: Super-resolution · Sparse representation · Dictionary learning · Pan-sharpening · Spectral information

1 Introduction

Remote sensing applications, like, geographical navigation, disaster management, agricultural monitoring, etc., require images with high spatial/spectral resolutions. The high spatial information allows an accurate geometric analysis while high spectral resolution allows a better thematic interpretation. Although the MS images usually contain high spectral information yet they have limitations in terms of low spatial resolution. There are numerous applications using HR MS remote sensing images where such LR images fail to provide high-quality visuals as well as a proper analysis of such images. So, reconstruction of high spatial/spectral MS images from available LR MS images becomes an important topic of research [5,10,14].

In remote sensing, several satellites, like, Landsat, QuickBird, SPOT, etc., provides a set of LR MS band images along with a corresponding high-resolution

B. Deka et al. (Eds.): PReMI 2019, LNCS 11941, pp. 523–531, 2019.
https://doi.org/10.1007/978-3-030-34869-4_57

panchromatic (HR PAN) image. For example, the data provided by QuickBird is composed of a HR PAN band of 0.65 m and several LR MS bands of 2.62 m resolution. Similarly, that of the Landsat-7 satellite is 15 m and 30 m, respectively. A MS color image is formed by combining the LR spectral bands while the PAN image is a single band gray-scale image only. Figure 1 shows an example of MS and PAN images of the same area from which the visual differences between the two images can be clearly viewed.

HR MS images can be produced by fusion of the HR PAN and the LR MS images. These methods are known as pan-sharpening as the resolution of MS images are made equal to that of the PAN images. A large number of works on fusion of MS and PAN image are explained in [2]. Some of the existing pan-sharpening methods which are very popular are named as intensity-hue-saturation method (IHS) [11], principal component analysis (PCA) method [9] and Brovery transform based method [13]. These are component substitution based methods where the MS image is first transformed into a color image and then an interpolated MS image is obtained to get a spatial resolution in the order of the PAN image's resolution. Next, the pixels of the luminance channel of the color transformed MS image is replaced with those of the PAN image and the final HR image is obtained through an inverse color transformation. One major limitation in this approach is that due to different statistical distribution of the pixels of the color image's luminance channel and the PAN image, the output images produced by such methods suffer from significant spectral distortions [10].

Fig. 1. Example of QuickBird Images of the same area: PAN (left) and MS (right).

The prime focus of this paper is to produce a HR MS image from the given LR MS image using the super-resolution (SR) method. Although, the pan-sharpening and SR methods have similarities in their approaches but they are different by the fact that pan-sharpening enhances the spatial information of MS images only while SR tries to estimate the target HR image's spatial as well as spectral information from the LR image itself.

The problem of SR image reconstruction from an available LR image is an inverse problem. This is also an ill-posed inverse problem as there may be many HR images which yield to the similar LR image. Literature studies refer that sparse representation theory in signal and image processing is popularly applied for image denoising and restoration problems [3,4]. Sparse representation is also used effectively in several recent single image super-resolution works [1,7,12]. These works are based on learning a pair of dictionaries (LR and HR) from a given dataset of HR RGB images and reconstruct the luminance channel of the LR input using patch-wise sparse representation technique. However, in case of MS image SR, a standard HR MS image dataset is usually not obtainable to learn the dictionary pairs and on the other hand, reconstruction from a transformed RGB image in pan-sharpening gives spectral distortions.

In this work, firstly, to overcome the dictionary learning issue, we focus on utilizing input HR PAN image to train the LR/HR dictionary pair as they contain high spatial details that is desired in the target HR image. A work by Zhu *et al.* [14] presents a similar pan-sharpening work where they consider the PAN image patches as dictionary atoms. We focus on training the dictionaries based on extraction of features, e.g. edges, etc., from the PAN image patches such that these features can be utilized for improved representation of an LR patches during reconstruction. Secondly, instead of reconstructing a RGB MS image, we apply reconstruction of each MS band separately and combine them to obtain the resulted HR MS image such that each band can maintain it's original spectral specifications. The major contributions of this paper are as stated below:

- A self-learning based K-singular value decomposition (K-SVD) coupled dictionary learning method is developed using the single HR PAN image.
- Reconstruction of a high spatial/spectral MS output image is made through band-wise super-resolution using sparsity regularization.
- Comparison of different methods are carried out using standard spatial/spectral quality and quantity evaluation metrics.

The remaining part of this paper is organized as follows: In Sect. 2, a background of the image acquisition process is discussed. The proposed method of MS image super-resolution is explained in Sect. 3. The details of experiments carried out and the simulations results obtained are given in Sect. 4. Finally, in Sect. 5, a conclusion is provided.

2 Background

The general image acquisition model can be expressed as [12]

$$\mathbf{X} = \mathbf{SHX}' + v, \tag{1}$$

where $\mathbf{X}$ represents the LR image, $\mathbf{X}'$ represent the HR image, and $\mathbf{H}$ and $\mathbf{S}$ are the blurring and down-sampling operators, respectively. Here, v is the additive noise which is ideally equal to zero.

Now, $\mathbf{X}'$ can be estimated by solving the inverse problem shown below.

$$\hat{\mathbf{X}}' = \arg\min \|\mathbf{X} - \mathbf{SHX}'\|_2^2 \tag{2}$$

The problem of Eq. 2 is an ill-posed inverse problem since many HR image $\mathbf{X}'$ may obey the above condition for any input LR image $\mathbf{X}$. Therefore, to resolve this issue, we regularize the problem using a local patch-wise sparsity prior based modeling of it. The detail working of the model proposed in this work is discussed in the following section.

3 Proposed MS Image Super-Resolution Method

The proposed SR algorithm considers the input PAN image $\mathbf{Y}$ for learning a pair of overcomplete dictionaries $\mathbf{D_H}$ and $\mathbf{D_L}$. Then, patch-based sparse representation of the input LR MS image $\mathbf{X}$ is carried out to generate the desired output HR MS image $\mathbf{X}'$. The work flow of the proposed SR algorithm is depicted in Fig. 2. It comprises of two main steps: dictionary training and SR reconstruction.

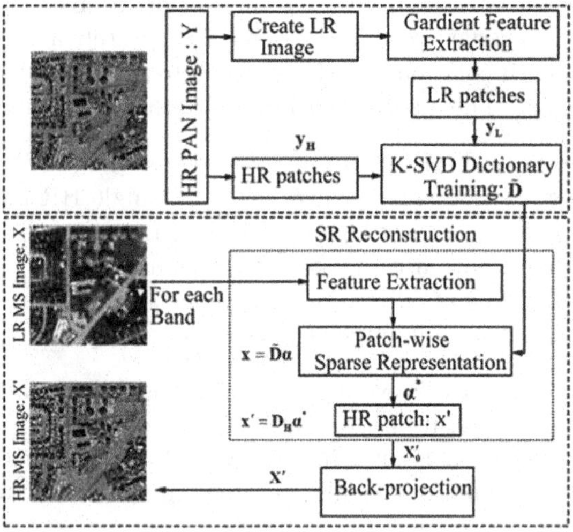

Fig. 2. Schematic of the proposed MS image SR algorithm.

3.1 Dictionary Training

For extraction of LR patches from the given PAN image $\mathbf{Y}$, it is blurred and down-sampled so that it can have the point spread function (PSF) identical to that of the input MS image. This converted LR PAN image is then passed through four 1D-filters of order 1 and 2, to extract both horizontal and vertical gradient features from it.

$$f_1 = [-1, 0, 1], \quad f_2 = [1, 0, -2, 0, 1], \quad f_3 = f_1^T, \quad f_4 = f_3^T, \tag{3}$$

The four feature vectors obtained from each patch after filtering are concatenated into a single vector that represents a LR patch. Again, a HR patch is extracted directly from the given HR images. Two vectors $\mathbf{y_H}$ and $\mathbf{y_L}$ are created which contain all the HR and LR patches. A sparse representation problem is formulated to train the coupled dictionary $\tilde{\mathbf{D}}$ from the combined patch vector of $\mathbf{y_H}$ and $\mathbf{y_L}$.

$$\min_{\{\mathbf{D_H},\mathbf{D_L},\mathbf{z}\}} \left\|\mathbf{y_C} - \tilde{\mathbf{D}}\mathbf{z}\right\|_2^2 + \lambda \left\|\mathbf{z}\right\|_1, \tag{4}$$

where $\tilde{\mathbf{D}} = \begin{bmatrix} \frac{1}{\sqrt{m}}\mathbf{D_H} \\ \frac{1}{\sqrt{n}}\mathbf{D_L} \end{bmatrix}$ and $\mathbf{y_C} = \begin{bmatrix} \frac{1}{\sqrt{m}}\mathbf{y_H} \\ \frac{1}{\sqrt{n}}\mathbf{y_L} \end{bmatrix}$; here, m, n represents the size of the HR and LR patches in vector form.

The least-square minimization based problem in Eq. 4 is solved by utilizing the optimized K-SVD training algorithm [8] to obtain the coupled trained overcomplete dictionary $\tilde{\mathbf{D}}$.

3.2 SR Reconstruction

The LR MS image is processed band-wise for SR reconstruction. A selected band image is first applied to the feature extraction stage to get the feature patch vectors as done in dictionary training. Next, for each feature patch vector $\mathbf{x}$ in LR MS image, a sparse representation problem is formulated using the dictionary $\tilde{\mathbf{D}}$ and it is given as:

$$\alpha^* = \arg\min_{\alpha} \left\{ \left\|\tilde{\mathbf{D}}\alpha - \mathbf{x}\right\|_2^2 + \lambda \left\|\alpha\right\|_1 \right\} \tag{5}$$

Equation 5 is a ℓ_1- ℓ_2 minimization problem. We estimate the sparse coefficient vector α^* by solving the feature-sign search based convex optimization algorithm [6].

Since, the patches of both LR and HR images share common sparse representation with their individual dictionaries, HR image patches can be reconstructed as follows:

$$\mathbf{x}' = \mathbf{D_H}\alpha^* \tag{6}$$

Tiling all the reconstructed patches in its corresponding channel yields to an intermediate HR image $\mathbf{X'_0}$. Finally, back-projection applies the global imaging model constraint of Eq. 2 on $\mathbf{X'_0}$ to obtain the final HR image $\mathbf{X'}^*$. Mathematically,

$$\mathbf{X'}^* = \arg\min_{\mathbf{X'}} \left\|\mathbf{X} - \mathbf{SHX'}\right\|_2^2 + c\left\|\mathbf{X'} - \mathbf{X'_0}\right\|_2^2 \tag{7}$$

Equation 7 is efficiently solved using gradient descent method.

4 Experimental Results

Experiments are carried out for SR reconstruction using the proposed method on two test MS images and comparison of results are shown with four other existing

MS image SR methods based on pan-sharpening, namely, IHS [11], PCA [9], Brovery [13], and SparseFI [14]. Datasets containing PAN and MS images of size 2048×2048 and 512×512 are acquired from a QuickBird image taken over the region India-The Sundarbans and captured on 02 November, 2002 from[1]. Simulations are carried out using MATLAB on a PC running Windows 7 OS and having 4 GB RAM.

In this experiment, to keep a HR reference image for quality assessment, the input PAN and MS images are downsampled by a factor 4. Thus, for both the datasets, the dimensions of the test LR MS image and trainable PAN image are resized to 128×128 and 512×512, respectively.

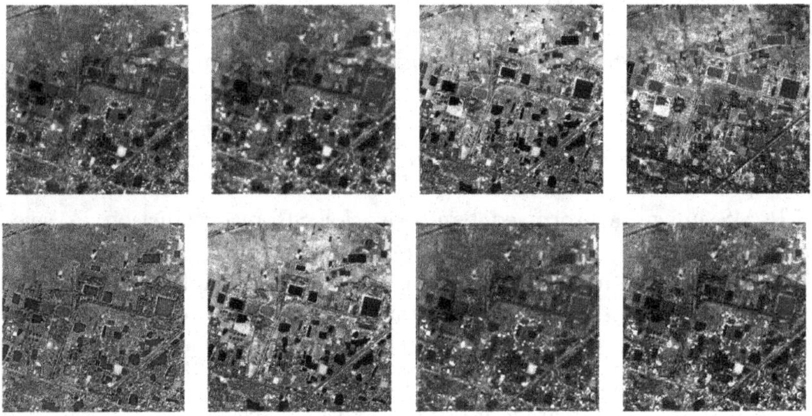

Fig. 3. Visuals for QuickBird first dataset using different methods. First row (from left): original MS, Down-sampled MS, Down-sampled PAN, IHS output; second row (from left): results by PCA, Brovery, SparseFI and Proposed method.

During the training phase, a coupled overcomplete dictionary consisting 1024 number of atoms is learned using 10000 sample patches taken from the input HR PAN image. Here, we consider extracting patches of size 7×7 which is preferable for better results as explained by Zhu *et al.* [14]. In this work, each LR MS band image is upscaled individually. So, during reconstruction, each feature patch of the input image is processed by sparse representation and an HR patch is obtained using the sparse coefficients and HR dictionary. Finally, the reconstructed HR images are combined to get the output HR MS image.

For quality assessment of the resulted images of different methods, the quantitative metrics computed are as follows: root mean-square error (RMSE), spatial correlation coefficient (CC), spectral distortion (SD), the universal image quality index (UIQI), spectral angle mapper (SAM), and erreur relative globale adimensionnelle de synthese (ERGAS).

[1] Global Land Cover Facility, (Accessed on 20 March, 2019). ftp://ftp.glcf.umd.edu/glcf/QuickBird/.

Table 1. Performance evaluation for QuickBird first dataset.

Parameter	IHS [11]	PCA [9]	Brovery [13]	SparseFI [14]	Proposed
RMSE	36.10	33.37	33.97	24.23	23.62
CC	0.589	0.738	0.715	0.769	0.813
SD	38.88	29.25	17.66	15.55	13.18
SAM	9.04	8.29	10.63	7.20	7.02
UIQI	0.742	0.842	0.878	0.805	0.860
ERGAS	22.15	12.82	18.24	8.24	8.11

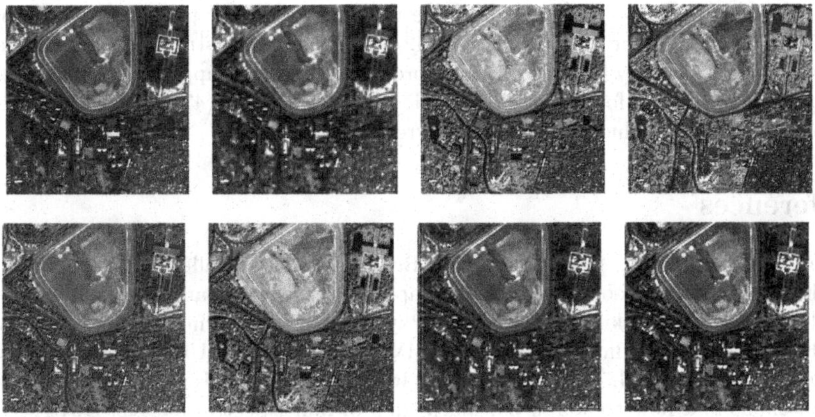

Fig. 4. Visuals for QuickBird second dataset using different methods. First row (from left): original MS, Down-sampled MS, Down-sampled PAN, IHS output; second row (from left): results by PCA, Brovery, SparseFI and Proposed method.

Table 2. Performance evaluation for QuickBird second dataset.

Parameter	IHS [11]	PCA [9]	Brovery [13]	SparseFI [14]	Proposed
RMSE	24.51	18.90	20.86	14.10	13.25
CC	0.714	0.807	0.817	0.828	0.839
SD	31.70	27.93	14.52	12.31	10.92
SAM	8.76	7.97	9.39	7.45	0.703
UIQI	0.773	0.813	0.818	0.845	0.894
ERGAS	25.14	13.57	20.14	9.06	8.76

The output images by different methods for the two test datasets are visually presented in Figs. 3 and 4. Moreover, their visual outputs are quantitatively validated using the above mentioned parameters in Tables 1 and 2. Results obtained indicate that, in case of the proposed method, the spatial information is better while the spectral distortion is less compared to others.

5 Conclusion

A new MS image SR method using sparse representation is presented in the paper. K-SVD technique based coupled overcomplete dictionary learning from input HR PAN image is also shown. HR output MS image is reconstructed based on patch-wise sparsity regularization along with back-projection. Simulations are carried out on two standard datasets. The proposed method performs superior while compared to other SR and pan-sharpening methods in terms of both quantitative and visual analysis. As a future work, more effective prior term based regularization can be considered for better spatial or spectral reconstructions.

Acknowledgements. Authors would like to thank ISRO for providing funds under the RESPOND project (ISRO/RES/4/642/17-18) and Ministry of Electronics and Information Technology (MeiTY), GoI for providing financial support under the Visvesvaraya Ph.D. Scheme for Electronics & IT (Ph.D./MLA/ 04(41)/2015-16/01) which helped in smooth conduction of the above research work.

References

1. Alvarez-Ramos, V., Ponomaryov, V., Reyes-Reyes, R., Gallegos-Funes, F.: Satellite image super-resolution using overlapping blocks via sparse representation. In: International Kharkiv Symposium on Physics and Engineering of Microwaves. Millimeter and Submillimeter Waves (MSMW 2016), Kharkiv, Ukraine, pp. 1–4 (2016)
2. Amro, I., Mateos, J., Vega, M., Molina, R., Katsaggelos, A.K.: A survey of classical methods and new trends in pansharpening of multispectral images. EURASIP J. Adv. Signal Process. **2011**(1), 1–22 (2011)
3. Deka, B., Datta, S.: Compressed Sensing Magnetic Resonance Image Reconstruction Algorithms. SSBN, vol. 9. Springer, Singapore (2019). https://doi.org/10.1007/978-981-13-3597-6
4. Elad, M.: Sparse and Redundant Representations: From Theory to Applications in Signal and Image Processing, 1st edn. Springer, New York (2010). https://doi.org/10.1007/978-1-4419-7011-4
5. Fernandez-Beltran, R., Latorre-Carmona, P., Pla, F.: Single-frame super-resolution in remote sensing: a practical overview. Int. J. Remote Sens. **38**(1), 314–354 (2017)
6. Lee, H., Battle, A., Raina, R., Ng, A.Y.: Efficient sparse coding algorithms. In: Schölkopf, B., Platt, J.C., Hoffman, T. (eds.) Advances in Neural Information Processing Systems 19 (NIPS-2006), pp. 801–808. MIT Press, Vancouver (2007)
7. Mullah, H.U., Deka, B.: A fast satellite image super-resolution technique using multicore processing. In: Abraham, A., Muhuri, P.K., Muda, A.K., Gandhi, N. (eds.) HIS 2017. AISC, vol. 734, pp. 51–60. Springer, Cham (2018). https://doi.org/10.1007/978-3-319-76351-4_6
8. Rubinstein, R., Zibulevsky, M., Elad, M.: Efficient implementation of the K-SVD algorithm using batch orthogonal matching pursuit. Technical report 40, Computer Science Departmen, Technion, Haifa, Israel (2008)
9. Shah, V.P., Younan, N.H., King, R.L.: An efficient pan-sharpening method via a combined adaptive PCA approach and contourlets. IEEE Trans. Geosci. Remote Sens. **46**(5), 1323–1335 (2008)
10. Shahdoosti, H.R., Ghassemian, H.: Fusion of MS and PAN images preserving spectral quality. IEEE Geosci. Remote Sens. Lett. **12**(3), 611–615 (2015)

11. Tu, T.-M., Su, S.C., Shyu, H.C., Huang, P.S.: A new look at IHS-like image fusion methods. Inf. Fusion **2**(3), 177–186 (2001)
12. Yang, J., Wright, J., Huang, T.S., Ma, Y.: Image super-resolution via sparse representation. IEEE Trans. Image Process. **19**(11), 2861–2873 (2010)
13. Zhang, Y.: Problems in the fusion of commercial high-resolution satelitte as well as landsat 7 images and initial solutions. Int. Arch. Photogram. Remote Sens. Spat. Inf. Sci. **34**, 587–592 (2002)
14. Zhu, X.X., Bamler, R.: A sparse image fusion algorithm with application to pansharpening. IEEE Trans. Geosci. Remote Sens. **51**(5), 2827–2836 (2013)

Quantum Image Edge Detection Based on Four Directional Sobel Operator

Rajib Chetia[1,2], S. M. B. Boruah[2], S. Roy[2], and P. P Sahu[2(✉)]

[1] Department of ECE, Central Institute of Technology (CIT) Kokrajhar, BTAD, Kokrajhar 783370, Assam, India

[2] Department of ECE, Tezpur University, Napaam, Tezpur 784028, Assam, India

ppstezu@gmail.com

Abstract. Sobel operator is used to detect the edge points of an image. Sobel operator is limited to only two directions but our proposed quantum image edge detection based on novel enhanced quantum representation (NEQR) is sensitive in four directions. As a result, it gives the accurate detection and reduces the loss of edge information of digital image. The computational complexity of our proposed algorithm is $\sim O(n^2 + q^2)$, which result into exponential speed up compare to all the conventional edge detection algorithms and existing Sobel quantum image edge detection method, especially when the image data is growing exponentially.

Keywords: Quantum image processing · Four directional Sobel operator · Edge extraction · NEQR · Computational complexity

1 Introduction

Quantum image processing (QIP) has been focused due to its ability to storing N bits classical information in only log 2^N quantum bits. The important properties of QIP such as entanglement, parallelism and exponential increase of quantum storage capacity make it unique [1]. In fact QIP is three-step technique-first step, representation of the classical digital image in a quantum image. The second step, quantum image is processed using quantum computation technologies and finally, images are extracted from processed quantum images. There are two broad groups of QIP - quantum inspired image processing and classically inspired quantum image processing [2–4]. In two groups, the primary task of QIP starts with the proper representation of conventional images in terms of quantum bits. In this direction many techniques such as quantum image model based on Qubit Lattice [5, 6], flexible representation of quantum image (FRQI) [7], novel enhanced quantum representation (NEQR) [8], Quantum log-polar images [9], normal arbitrary quantum superposition state [10] are already reported. Different QIP techniques such as quantum image scrambling [11–13], quantum image geometric transformation [14], and quantum image scaling [15], have been developed on the basis of quantum image representation (QIR) model.

B. Deka et al. (Eds.): PReMI 2019, LNCS 11941, pp. 532–540, 2019.
https://doi.org/10.1007/978-3-030-34869-4_58

The main three drawbacks which makes constraint in the application such as medical, pattern recognition, image features extraction, inaccurate retrieve of an original classical image, restricted quantum operations performed in QIP and limited to the representation of a large number of color using angle parameter of qubits. In spite of these drawbacks, Fan et al. [16, 17] have reported quantum image edge extraction based Sobel operator and Laplacian operators. Till the main issues in image edge extraction are high computational complexity and loss of edge information in a certain direction [14]. In comparison to classical image edge extraction, QIP can process all pixels of an image simultaneously by using unique superposition and entanglement properties of quantum mechanics. Both FRQI and NEQR have already been used for quantum image extraction [16, 18, 17]. In both cases, time complexity decreases exponentially but these schemes provide misclassification of noise pixels as edge points. Although Fan et al. [17] has proposed an enhanced quantum edge extraction scheme based on Laplacian filtering with NEQR for detection of image edge by smoothing. But till problem lies in accurate edge detection in all the directions.

Here we have proposed quantum image edge detection using four directional Sobel operator for gradient estimation to obtain more actual edge information. We use NEQR [8] method for QIR because it is extremely similar to classical image representation. NEQR utilizes the quantum states of superposition to store all the pixel of the image, therefore all the pixels data can be process simultaneously. The Mathematical modules of a proposed scheme has been developed to obtain an accuracy of edge extraction.

2 Related Work

The novel enhanced quantum representation (NEQR) [8]. In NEQR model [18], two entangled qubit sequences are used to store the grayscale value and positional information of all the pixels of image in superposition states [18]. Quantum image representation for $2^n \times 2^n$ image is expressed as

$$|I> = \frac{1}{2^n}\sum_{Y=0}^{2^n-1}\sum_{X=0}^{2^n-1}|C_{YX}>|Y>|X> = \frac{1}{2^n}\sum_{YX=0}^{2^{2n}-1}{}_{k=0}^{q-1}\otimes|C_{YX}^k>\otimes|YX> \tag{1}$$

where grayscale value of the pixel coordinate (Y, X),

$$C_{YX} = C_{YX}^{q-1}C_{YX}^{q-2}\ldots\ldots C_{YX}^{1}C_{YX}^{0};\ C_{YX}^{k}\in\{0,1\};\ C_{YX}\in[0,2^q-1]$$

The accurate classical image can be retrieved from NEQR method.

2.1 Classical Sobel Edge Detection Algorithm

Sobel operator is a discrete differential operator. It has two sets of 3×3 mask shown in Fig. 1(b) and (c) and mainly used for edge detection. If G_H and G_V represents the image gradient value of the original image for horizontal and vertical direction then the calculation of G^H & G^V is defined as

$$G^V = S(Y-1,\ X+1) + 2S(Y,\ X+1) + S(Y+1,\ X+1) - S(Y-1,\ X-1) - 2S(Y,\ X-1) - S(Y+1,\ X-1) \tag{2}$$

$$G^H = S(Y+1,\ X-1) + 2S(Y+1,\ X) + S(Y+1,\ X+1) - S(Y-1,\ X-1) - 2S(Y-1,\ X) - S(Y-1,\ X+1) \tag{3}$$

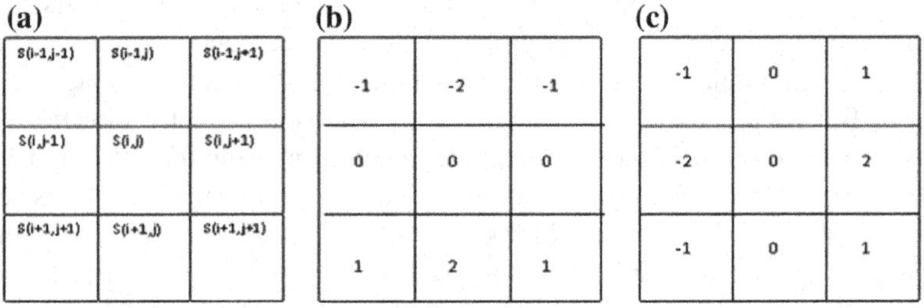

Fig. 1. (a) The 3 × 3 neighborhood templates, **(b)** and **(c)** are two Sobel masks.

The total gradient for each pixel is as follows

$$G \cong |G^V| + |G^H| \tag{4}$$

If G ≥ T (Threshold), the pixel will be the part of edge points.

2.2 Four Directional Sobel Edge Detection

Classical Sobel operator is limited to only vertical and horizontal direction. The performance of this operator can further be increased if the direction template are changed into four directions for gradient estimation (i.e. vertical, horizontal, 45° and 135° direction) is shown in Fig. 2(a) and (b). If G^{1d} and G^{2d} represents the gradient value of an image through 45° and 135° directions then gradient estimation can be expressed as follows

$$G^{D1} = S(Y,\ X+1) + 2S(Y+1,\ X+1) + S(Y+1,\ X) - S(Y-1,\ X) - 2S(Y-1,\ X-1) - S(Y,\ X-1) \tag{5}$$

$$G^{D2} = S(Y-1,\ X) + 2S(Y-1,\ X+1) + S(Y,\ X+1) - S(Y,\ X-1) - 2S(Y+1,\ X-1) - S(Y+1,\ X) \tag{6}$$

Total gradient of each pixel is calculated as follows

$$G \cong \left|G^{V}\right| + \left|G^{H}\right| + \left|G^{D1}\right| + \left|G^{D2}\right| \tag{7}$$

(a)

-2	-1	0
-1	0	1
0	1	2

(b)

0	1	2
-1	0	1
-2	-1	0

Fig. 2. (a) 45° direction. (b) 135° direction.

3 Quantum Image Edge Detection Based on Four Directional Sobel Operator

3.1 Algorithm for Edge Detection

I. Prepare the quantum image $|I\rangle$ from the original digital image. Here we use NEQR model for QIR.
II. Use cyclic shift transformations to obtain the shifted quantum image set.
III. Calculate the every pixel gradient for all the four directions (i.e. vertical, horizontal, 45° and 135°) using Sobel mask and record them as $|G>$.
IV. Compare all the gradients (i.e. $|G^{V}>$, $|G^{H}>$, $|G^{D1}>$ & $|G^{D2}>$) with a threshold value $|T_H>$ through four quantum comparators. If each of the gradient value is less than $|T_H>$ then it is defined as a non-edge point, otherwise it is an edge point.

Different steps of the algorithm can be describe as follows:

Step-I: The original digital image is represented into a quantum image using NEQR method and (2n + q) qubits are needed to store this quantum image. In order to store the eight pixels of a 3 × 3 neighborhood window, 8 extra qubits are required and these qubits are the tensor product with $|I\rangle$ i.e.

$$\begin{aligned} |0>^{\otimes 8q} \otimes |I> &= \frac{1}{2^n}\sum_{Y=0}^{2^n-1}\sum_{X=0}^{2^n-1}|0>^{\otimes 8q}|C_{YX}>|Y>|X> \\ &= \frac{1}{2^n}\sum_{Y=0}^{2^n-1}\sum_{X=0}^{2^n-1}|0>^{\otimes q}\ldots\ldots|0>^{\otimes q}|C_{YX}>|Y>|X> \end{aligned} \tag{8}$$

Step-II: The main operation of this stage is X-shift and Y-shift transformation of pixels. Any neighborhood eight pixels of $|I>$ are stored and encoded in the quantum states as follows

$$\frac{1}{2^n}\sum_{Y=0}^{2^n-1}\sum_{X=0}^{2^n-1}|C_{Y-1,X}>\otimes|C_{Y-1,X+1}>\otimes|C_{Y,X+1}>\otimes|C_{Y+1,X+1}>\otimes|C_{Y+1,X}>\otimes |C_{Y+1,X-1}>\otimes|C_{Y,X-1}>\otimes|C_{Y-1,X-1}>\otimes|C_{Y,X}>|Y>|X> \tag{9}$$

Step-III: Different quantum arithmetic operations are used to calculate the gradient of every pixel. The output quantum states $|G>$ for intensity gradient can be expressed as follows

$$|G> = \frac{1}{2^n}\sum_{Y=0}^{2^n-1}\sum_{X=0}^{2^n-1}|G^V_{Y,X}>|G^H_{Y,X}>|G^{D1}_{Y,X}>|G^{D2}_{Y,X}>|Y>|X> \tag{10}$$

The quantum state $|G>$ can be encoded as follows

$$|0>\otimes|G> = \frac{1}{2^n}\sum_{Y=0}^{2^n-1}\sum_{X=0}^{2^n-1}|0>|G^V_{Y,X}>|G^H_{Y,X}>|G^{D1}_{Y,X}>|G^{D2}_{Y,X}>|Y>|X> \tag{11}$$

The quantum circuit realization of gradient $|G>$ is shown in Figs. 3(a) and (b).

Step-IV: In order to detect the edge points of $|I>$, four quantum comparator are used to compare the intensity gradients of every pixel in the four directions with a given threshold $|T_H>$. All the gradients of $|G^V_{Y,X}>$, $|G^H_{Y,X}>$, $|G^{D1}_{Y,X}>$ & $|G^{D2}_{Y,X}>$ greater than equal to $|T_H>$ classified as an edge point. The final state for quantum edge extraction can be represented as follows

$$|D_E> = \frac{1}{2^n}\sum_{Y=0}^{2^n-1}\sum_{X=0}^{2^n-1}|D_{Y,X}>|Y>|X>, D_{Y,X}\in\{0,1\} \tag{12}$$

where $D_{Y,X}=0$, means no edge point and $D_{Y,X}=1$, means edge point. Circuit realization of quantum states $|D_E>$ is shown in Fig. 3(c). During the circuit realization, we ignored all the other garbage qubits of the output.

3.2 Quantum Circuit Realization for Edge Detection

The various quantum operations are required for image edge extraction. Here we use quantum image cyclic shift transformations [14, 19], parallel controlled-NOT operations, Quantum operation of multiplied by powers of 2, quantum ripple-carry adder (QRCA) [20], quantum absolute value (QAV) module [21, 22] and quantum comparator (QC) [23] to realize the quantum circuit for edge detection algorithm.

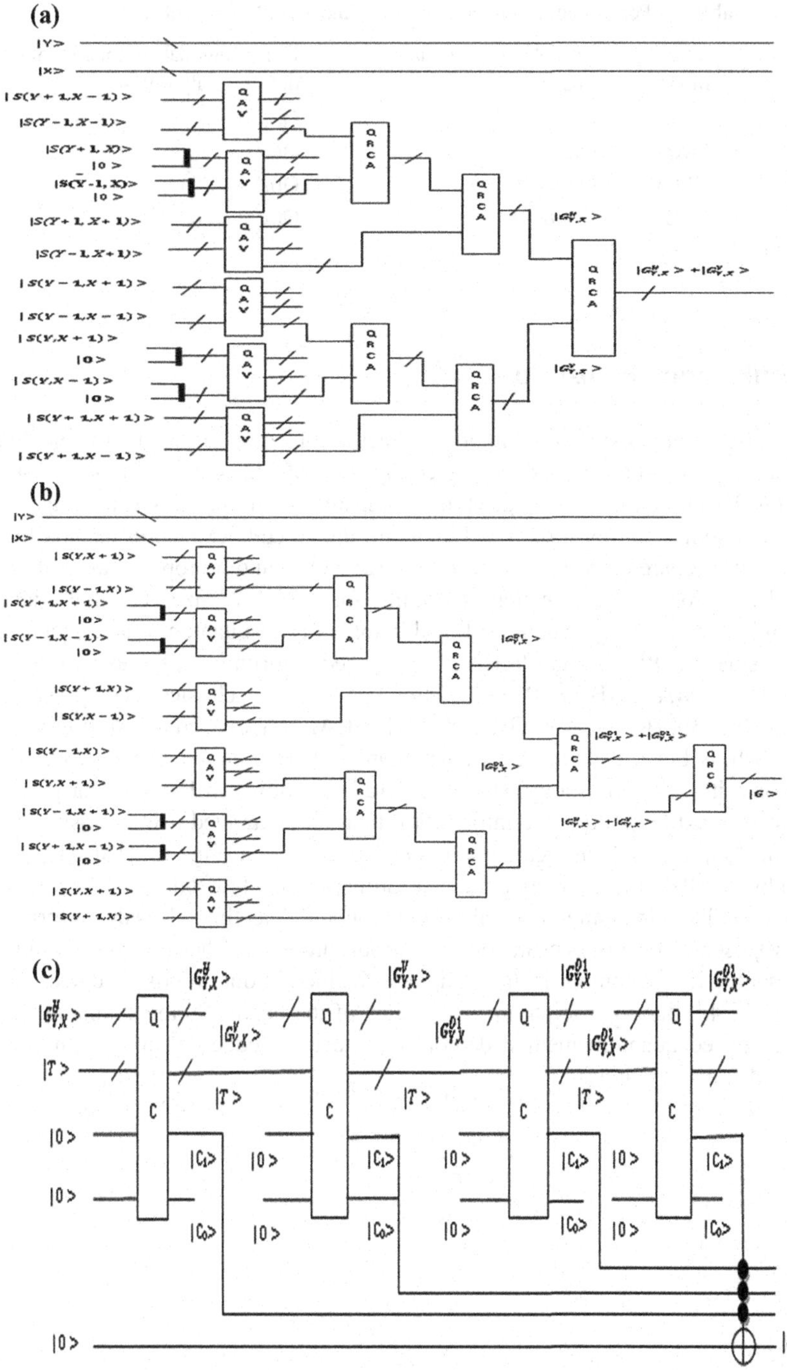

Fig. 3. (a) Circuit realization of quantum module for calculating the gradient of a quantum image through vertical and horizontal direction. (b) Circuit realization of quantum module for calculating the total gradient of each pixel of a quantum image. (c) Circuit realization of quantum module for final states of our proposed algorithm and stored in a quantum states $|D_E>$.

Table 1. Performance comparison of an image with a size of $2^n \times 2^n$.

Algorithm	Image model	Complexity of quantum image preparation	Computational complexity of edge detection algorithm
Sobel	–	–	$O(2^{2n})$
Zhang [18]	FRQI	$O(2^{4n})$	$O(n^2)$
Fan [16]	NEQR	$O(qn2^{2n})$	$O(n^2 + 2^{q+4})$
Our algorithm	NEQR	$O(qn2^{2n})$	$O(n^2 + q^2)$

4 Results and Discussion

Computational complexity is an important factor for comparison of different image edge detection techniques. The circuit complexity of our scheme is $\sim O(n^2 + q^2)$, which result into exponential speed up compare to all the classical image edge extraction schemes and existing Sobel quantum image edge extraction method [6, 16, 18]. We have prepared a performance comparison between our proposed algorithm and other existing Sobel edge detection algorithms for a $2^n \times 2^n$ size of digital image as shown in Table 1. However computational complexity of quantum image preparation is not consider in QIP. In order to test our proposed algorithm, all the experiments are simulated in MATLAB R2018a environment on a classical computer Intel (R) Core™i58300HCPU 2.30 GHz, 8.00 GB RAM. Three common original test images such as Lena, peppers and Cameraman [24] were chosen for edge extraction simulation with same threshold value ($T_H = 127$) as shown in Fig. 4a. Our proposed algorithm can extract more edge information than that extracted using classical Sobel algorithm (Fig. 4b) [25] and quantum Sobel edge extraction (Fig. 4c) [16]. This is due to bit to bit parallel processing and gradient estimation in all directions. However, there is less probability of missing of actual information and detecting false edge information in our proposed algorithm as estimation in all four directions. The simulation result also shows that the image contour is more brightly obtained in our proposed edge extracted images (Fig. 4d) due to absolute intensity value of the pixels. So the simulation result of our proposed quantum method demonstrate accurate edge detection with smooth edge in all four directions.

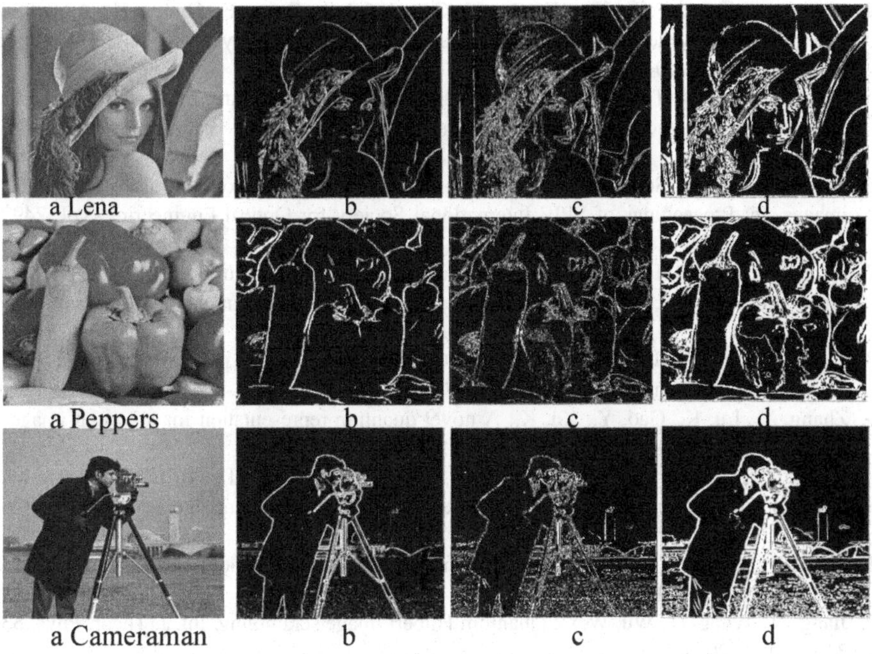

Fig. 4. Test images and their edge extraction (a) original image of Lena, Pepper and Cameraman [24]. (b) Classical Sobel edge extracted images. (c) Quantum Sobel edge extracted images. (d) Result images of our proposed algorithm.

5 Conclusion

Due to bit-level parallelism, the edge extracted process of QIP is much richer than the classical image processing (CIP). Compare to classical Sobel algorithm, it extracts more information and edges of the image is more specific. However, real time problem, discontinuity, roughness and other defects of Sobel edge detection algorithms can be reduced using our proposed algorithm. The proposed algorithm would be a clinically feasible practice for edge detection of medical images in future.

References

1. Yan, F., Iliyasu, A.M., Le, P.Q.: Quantum image processing: a review of advances in its security technologies. Int. J. Quant. Inf. **15**(03), 1730001 (2017)
2. Iliyasu, A.M.: Towards the realisation of secure and efficient image and video processing applications on quantum computers. Entropy **15**, 2874–2974 (2013)
3. Iliyasu, A.M.: Algorithmic frameworks to support the realisation of secure and efficient image-video processing applications on quantum computers. Ph.D. (Dr Eng.) Thesis, Tokyo Institute of Technology, Tokyo, Japan, 25 September 2012

4. Iliyasu, A.M., Le, P.Q., Yan, F., Bo, S., Garcia, J.A.S., Dong, F., Hirota, K.: A two-tier scheme for grayscale quantum image watermarking and recovery. Int. J. Innov. Comput. Appl. **5**, 85–101 (2013)
5. Tseng, C., Hwang, T.: Quantum digital image processing algorithms. In: Proceedings of the 16th IPPR Conference on Computer Vision, Graphics and Image Processing, pp. 827–834 (2003)
6. Fu, X., Ding, M., Sun, Y., et al.: A new quantum edge detection algorithm for medical images. In: Proceedings of SPIE International Society for Optical Engineering, vol. 7497, p. 749724 (2009)
7. Le, P.Q., Dong, F., Hirota, K.: A flexible representation of quantum images for polynomial preparation, image compression, and processing operations. Quantum Inf. Process. **10**(1), 63–84 (2011)
8. Zhang, Y., Lu, K., Gao, Y., Wang, M.: NEQR: a novel enhanced quantum representation of digital images. Quantum Inf. Process. **12**(8), 2833–2860 (2013)
9. Zhang, Y., Lu, K., Gao, Y., Xu, K.: A novel quantum representation for log-polar images. Quantum Inf. Process. **12**(9), 3103–3126 (2013)
10. Li, H., Zhu, Q., Zhou, R., Song, L., Yang, X.: Multi-dimensional color image storage and retrieval for a normal arbitrary quantum superposition state. Quant. Inf. Process. **13**, 991–1011 (2014)
11. Jiang, N., Wu, W.Y., Wang, L.: The quantum realization of Arnold and Fibonacci image scrambling. Quant. Inf. Process. **13**, 1223–1236 (2014)
12. Jiang, N., Wang, L., Wu, W.Y.: Quantum Hilbert image scrambling. Int. J. Theor. Phys. **53**, 2463–2484 (2014)
13. Zhou, R.G., Sun, Y.J., Fan, P.: Quantum image gray-code and bit-plane scrambling. Quant. Inf. Process. **14**, 1717–1734 (2015)
14. Le, P.Q., Iliyasu, A.M., Dong, F., et al.: Fast geometric transformations on quantum images. IAENG Int. J. Appl. Math. **40**(3), 113–123 (2010)
15. Zhou, R.-G., Hu, W., Fan, P., Ian, H.: Quantum realization of the bilinear interpolation method for NEQR. Sci. Rep. **7**(1), 2511 (2017)
16. Fan, P., Zhou, R.G., Hu, W., Jing, N.: Quantum image edge extraction based on classical Sobel operator for NEQR. Quant. Inf. Process. **18**, 24 (2019)
17. Fan, P., Zhou, R.G., Hu, W.W., Jing, N.: Quantum image edge extraction based on Laplacian operator and zero-cross method. Quant. Inf. Process. **18**, 27 (2019)
18. Zhang, Y., Lu, K., Gao, Y.: QSobel: a novel quantum image edge extraction algorithm. Sci. China Inf. Sci. **58**(1), 1–13 (2015)
19. Le, P.Q., Iliyasu, A.M., Dong, F., et al.: Strategies for designing geometric transformations on quantum images. Theoret. Comput. Sci. **412**, 1406–1418 (2011)
20. Cuccaro, S.A., Draper, T.G., Kutin, S.A., et al.: A new quantum ripple-carry addition circuit. arXiv:quant-ph/0410184 (2004)
21. Thapliyal, H., Ranganathan, N.: Design of efficient reversible binary subtractors based on a new reversible gate. In: Proceedings of the IEEE Computer Society Annual Symposium on VLSI, Tampa, Florida, pp. 229–234 (2009)
22. Thapliyal, H., Ranganathan, N.: A new design of the reversible subtractor circuit. Nanotechnology **117**, 1430–1435 (2011)
23. Wang, D., Liu, Z.H., Zhu, W.N., Li, S.Z.: Design of quantum comparator based on extended general Toffoli gates with multiple targets. Comput. Sci. **39**(9), 302–306 (2012)
24. http://www.imageprocessingplace.com/root_files_V3/image_databases.htm
25. Robinson, G.S.: Edge detection by compass gradient masks Comput. Graph. Image Process. **6**, 492–501 (1977)

Texture Image Retrieval Using Multiple Filters and Decoded Sparse Local Binary Pattern

Rakcinpha Hatibaruah(✉), Vijay Kumar Nath, and Deepika Hazarika

Department of Electronics and Communication Engineering, Tezpur University, Tezpur, Assam, India
{rakcinp,vknath,deepika}@tezu.ernet.in

Abstract. In local pattern based feature extraction, usually the raw spatial image provides limited information about the relationship between the pixels in a local neighborhood. A few recent methods address this issue by first filtering the images with bag of filters and then calculating local pattern over each filtered images. It is observed that the filtered images complement the discriminativeness of local pattern based features which enhances retrieval efficiency. Motivated by these approaches, a new approach based on multiple filters and decoded sparse local binary pattern (MF-DLBP) is proposed in this paper, wherein we first filter the raw spatial image with multiple filters to extract the low frequency and high frequency information. However, unlike previous approaches, we extract features from low pass and high pass filtered images adopting separate strategies, since characteristically they contain contrasting information. From each gradient filtered images, we compute sparse LBPs using LBP4hv (considering only horizontal and vertical neighbors) and LBP4d (considering only diagonal neighbors) techniques. To enhance the discriminativeness of the descriptor, two decoders are also used which compute the inter frequency relationship between the sparse local binary pattern maps of high pass filtered images only. The low pass filtered image is encoded with sign LBP (LBP_S) and magnitude LBP (LBP_M) to extract sign as well as magnitude information present in this image. The proposed approach is low dimensional and it shows highly competitive retrieval performance when tested on three benchmark texture databases which are Kylberg, Brodatz and STex.

Keywords: Local binary pattern · Decoder · Feature vector · Image retrieval · CBIR

1 Introduction

Texture analysis and retrieval has immense value in computer vision and pattern recognition and is also very useful in industrial identification applications. Both texture analysis and retrieval have gained ample attention among the researchers

B. Deka et al. (Eds.): PReMI 2019, LNCS 11941, pp. 541–550, 2019.
https://doi.org/10.1007/978-3-030-34869-4_59

for past few decades. The vast amount of works in this field can be categorized into several categories such as statistical, structural, model based, learning based approaches etc. [10]. Recently introduced deep leaning based approaches are also shown to be very efficient for retrieval purpose [11,16]. However, the deep learning networks need to be pre-trained with large volume of training data. Besides these methods are often time consuming and their memory requirements are very high. Recently, local pattern based texture classification is also gaining popularity among the researchers because of it's easy to understand and adaptive feature extraction architecture. Local binary pattern (LBP) is the pioneering work in this field [15]. To enhance the efficacy of LBP in retrieval, a number of new feature descriptors have also been proposed, which extracts the local texture information in more efficient manner [4,7,12,14,17,20]. High retrieval efficiencies have been reported for most of these approaches; however the feature dimension of most of these approaches are also very large. Some of these methods are suitable for only a particular kind of database. These limitations motivated us to develop a new feature descriptor for texture images, which has comparatively low feature dimension yet highly effective for retrieval purpose. Since, LBP encodes only the sign of the differences between the center pixel and the neighboring pixels and doesn't consider the magnitude of difference values, in [7], a completed modeling of LBP (CLBP) was proposed, considering both sign and magnitude information. The effect of adjacent neighbors in a local neighborhood on image texture is analyzed in [20]. Both these methods are shown to improve the retrieval performance of LBP in various image databases. The fusion of co-occurrence information with local pattern to enhance the effectiveness of the features is investigated in [8,9,13,19]. These approaches usually show state of art performance for retrieval of texture, face and bio medical images. However, one drawback of these approaches is their higher feature dimension, which may slow down the retrieval process [8,13,19]. In [4,5], feature extraction from the filtered images using local pattern is explored for retrieval. Both these approaches show that the discriminative capability of the local features computed from edge detected filtered images are superior to those computed from the original raw images.

The methods proposed in [4,5,7] inspired us to propose a new feature descriptor for retrieval of texture images. We first filter the raw images with one low pass and multiple high pass filters. Then unlike in [4,5], features are extracted from the low and high pass filtered images using different techniques. Since low pass filtered image contains more information than a high pass filtered image, we propose to encode it using sign LBP and magnitude LBP. In order to limit the feature dimension, we compute uniform histograms from the encoded low pass maps. However for encoding the multiple high pass filtered images we use sparse LBP in order to shorten the feature dimension. To further improve the discrimination capability of the descriptor the mutual inter-frequency information among the high pass LBP maps are computed using two decoders.

The paper is organized as follows: In Sect. 1, a brief review of some of the works related to texture analysis and retrieval is presented. Section 2 presents the proposed method. In Sect. 3, the experimental results and relevant discussions are provided. The paper is summarized and concluded in Sect. 4.

2 Proposed Method

The proposed method is depicted in a schematic block diagram in Fig. 1. The feature extraction procedure of the proposed descriptor can be summarized into following steps:

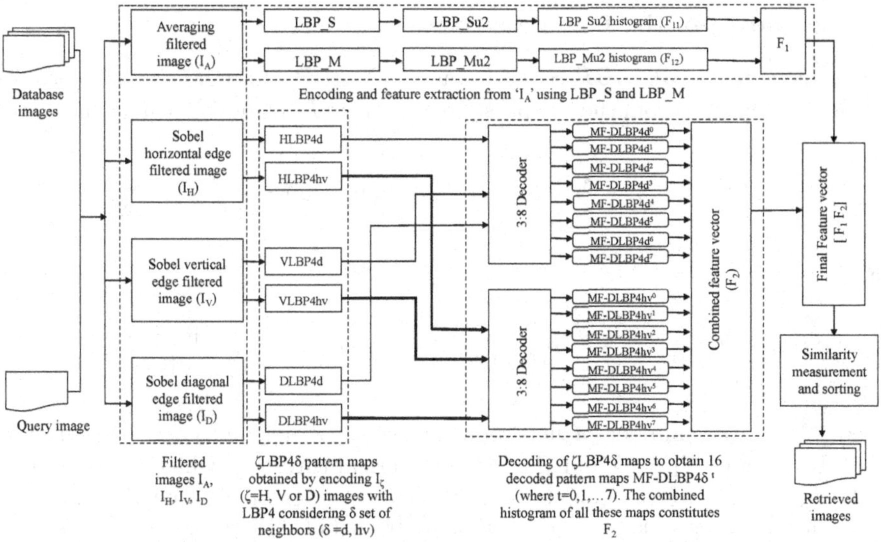

Fig. 1. Block diagram of the proposed MF-DLBP approach in a CBIR framework.

2.1 Filtering of the Image with Multiple Filters

In this step, the image is filtered with an averaging filter (F_A) and three high pass filters which are Sobel horizontal (F_H), Sobel vertical (F_V) and Sobel diagonal (F_D) edge filters [6]. All the four filtering masks considered here are shown in Fig. 2. The filtered images are obtained from these masks using following equation-

$$I_i = G * F_i \tag{1}$$

where, G is the input image, F_i is the filtering mask with $i \in [A, H, V, D]$, I_i is the filtered image corresponding to F_i and $*$ denotes the convolution operation on image G by F_i.

1/9	1/9	1/9
1/9	1/9	1/9
1/9	1/9	1/9

F_A

-1	-2	-1
0	0	0
1	2	1

F_H

-1	0	1
-2	0	2
-1	0	1

F_V

-2	-1	0
-1	0	1
0	1	2

F_D

Fig. 2. Filtering masks used in proposed technique; F_A: Averaging filter; F_H: Sobel horizontal edge filter; F_V: Sobel vertical edge filter; F_D: Sobel diagonal edge filter.

2.2 Encoding of the Low Pass Image

We employ sign LBP (LBP_S) and magnitude LBP (LBP_M) approaches to extract the features from the low pass filtered image I_A using following equations:

$$LBP\_S^{R,N_1}(I_A^c) = \sum_{i=0}^{N_1-1} 2^i \times f_1(I_A^i - I_A^c);\ f_1(x) = \begin{cases} 1, \text{ if } & x \geq 0 \\ 0, & \text{otherwise} \end{cases} \quad (2)$$

and

$$LBP\_M^{R,N_1}(I_A^c) = \sum_{i=0}^{N_1-1} 2^i \times f_2(|I_A^i - I_A^c|);\ f_2(x) = \begin{cases} 1, \text{ if } & x \geq Th \\ 0, & \text{otherwise} \end{cases} \quad (3)$$

where, N_1 is the number of neighbors in the local neighborhood of radius R (Here R = 1 and $N_1 = 8$) and 'Th' is an empirical threshold which is taken here as 25, I_A^i and I_A^c are the neighboring pixels and center pixel respectively.

Uniform histograms [18] are computed from both these pattern maps and are concatenated to construct the feature vector F_1.

$$F_1 = [F_{11}, F_{12}];\ \textit{where}\ \mathrm{F_{11}} = \mathrm{hist(LBP\_Su2(I_A))}\ \text{and}\ \mathrm{F_{12}} = \mathrm{hist(LBP\_Mu2(I_A))} \quad (4)$$

The dimension of F_1 alone is $2 \times 59 = 118$ features.

2.3 Encoding of the High Pass Filtered Images

The high pass filtered images which contain fine details are encoded with sparse LBP. In computation of sparse LBP, we first consider only four diagonal neighbors among the eight neighbors present in a local neighborhood of radius 1 as shown in Fig. 3(c). Then LBP map (LBP4d) is calculated considering only these four diagonal neighbors along with the center pixel as shown in Fig. 3(e). Similarly considering only the horizontal and vertical neighbors and the center pixel (Fig. 3(d)), the second LBP pattern map (LBP4hv) is formed as shown in Fig. 3(f).

LBP4d and LBP4hv are calculated from I_H, I_V and I_D images as shown in Fig. 3. After computing LBP4d and LBP4hv from I_H, I_V and I_D images, we get six binary pattern maps which are HLBP4d, HLBP4hv (corresponding to

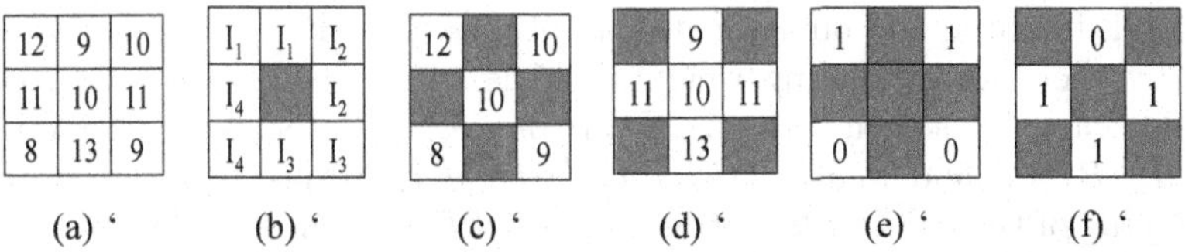

Fig. 3. An example of LBP4 computation. (a) sample image (b) position of the neighbors (c) diagonal neighbors (d) horizontal and vertical neighbors (e) diagonal LBP4 bit pattern map (*LBP4d*) (f) horizontal-vertical LBP4 bit pattern map (*LBP4hv*).

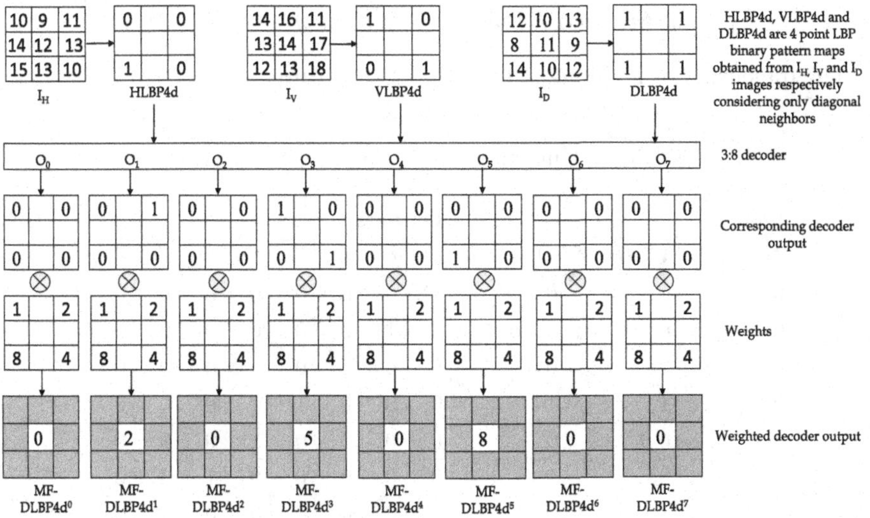

Fig. 4. An illustration of working of decoder with HLBP4d, VLBP4d and DLBP4d binary inputs.

I_H), VLBP4d, VLBP4hv (corresponding to I_V), DLBP4d and DLBP4hv (corresponding to I_D) as shown in Fig. 1. The advantage of this sparse approach is that it minimizes the feature dimension to a great extent by providing pattern values ranging between 0 to 15 only.

2.4 Computation of Inter Frequency Relationship Among the Encoded Pattern Maps of High Pass Images

To enhance the discrimination capability of the features, the inter frequency relationship among the LBP pattern maps of the high pass filtered images are computed with the help of two 3:8 decoders. Each of these decoders take three 4 bit LBP map as input from horizontal, vertical and diagonal filtered images respectively and then decode them into eight 4 bit LBP maps at the output. Therefore, *HLBP4d*, *VLBP4d* and *DLBP4d* maps are fed as input into the first decoder, while *HLBP4hv*, *VLBP4hv* and *DLBP4hv* are fed as input into the second decoder. The binary pattern outputs of both the decoders are multiplied by their corresponding weights and is summed up to a pattern value (the value ranges between 0 to 15). Therefore, a total of sixteen decoded maps $MF - DLBP4d^t$ and $MF - DLBP4hv^t$ ($t \epsilon [0,7]$) respectively are obtained. The working of the decoder to obtain $MF - DLBP4d^t$ maps is illustrated using an example in Fig. 4. The truth table for the decoders are given in Table 1.

The feature vector (F_2) is constructed by computing histograms of all $MF - DLBP4\delta^t$ ($\delta = d$ and $hv]$ and $t \in [0,7]$) maps and concatenating all of them as

$$F_2 = \begin{bmatrix} hist(MF - DLBP4d^0), \; ... \; , \; hist(MF - DLBP4d^7), \\ hist(MF - DLBP4hv^0), \; ... \; , \; hist(MF - DLBP4hv^7) \end{bmatrix} \tag{5}$$

The dimension of F_2 alone is $16 \times 16 = 256$ features.

Table 1. Truth table of the decoder ($\delta = d/hv$)

Inputs			Outputs							
$HLBP4\delta$	$VLBP4\delta$	$DLBP4\delta$	O_0	O_1	O_2	O_3	O_4	O_5	O_6	O_7
0	0	0	1	0	0	0	0	0	0	0
0	0	1	0	1	0	0	0	0	0	0
0	1	0	0	0	1	0	0	0	0	0
0	1	1	0	0	0	1	0	0	0	0
1	0	0	0	0	0	0	1	0	0	0
1	0	1	0	0	0	0	0	1	0	0
1	1	0	0	0	0	0	0	0	1	0
1	1	1	0	0	0	0	0	0	0	1

2.5 Final Feature Vector Formation

The final feature vector (FV) is constructed by concatenating the features extracted from low pass filter image i.e F_1 with the features of the high pass filtered images i.e F_2. Therefore,

$$FV = [F_1, \quad F_2] \tag{6}$$

The total feature vector length of proposed descriptor is $118 + 256 = 374$ features.

3 Results and Discussions

In order to evaluate retrieval performance of the proposed method, experiments are conducted on three benchmark texture image databases namely Kylberg [2], Brodatz [3] and STex [1]. The summary of these databases are presented in

Table 2. Summary of image databases

No	Database name	Total number of images	Number of images in each class	Total class	Dimension
1	Kylberg	640 (all grey scale images)	32	20	144 × 144
2	Brodatz	2800 (all grey scale images)	25	112	128 × 128
3	STex	7716 (color images)	16	476	128 × 128

Table 2. The retrieval performances are evaluated using two standard parameters namely Average Retrieval Precision (ARP) and Average Retrieval Recall (ARR) [19]. Higher values of them indicate better retrieval performance. In order to compute them, every image in the database is considered as query image once for each database and against each query, 'n' number of closely matched images are retrieved. The term 'n' is the number of top matches considered. Similarity between query image 'Q' and database images are computed using 'd_1' distance measure [20] for the proposed method. The retrieval performance of the proposed method is compared with six well known local pattern based methods: LBP (Dim: 256) [15], CSLBCoP (Dim: 1024) [19], BOF-LBP (Dim: 1280) [5], LTriDP (Dim: 768) [20], CoALTP (Dim: 2048) [13] and FDLBP (Dim: 4096) [4]. Except BOF-LBP which we have implemented ourselves, the retrieval results of all other techniques were obtained using matlab codes provided by their original authors.

The first experiment is conducted on Kylberg database [2]. To evaluate the retrieval performance, the images are retrieved in groups of 2, 4, 6..., 32 images and ARP and ARR values are calculated for each group of top matches. The plots for ARP and ARR are presented in Fig. 5(a–b). From the plots, it is observed that in most of the top matches, the proposed approach out performs all the other approaches. However in higher top matches (top matches of 28 or higher) however, CoALTP and CSLBCoP perform slightly better than the proposed approach. However, it should be noted that CoALTP and CSLBCoP have feature dimension much larger than the proposed approach. Overall, in this database, for 32 top matches, the proposed approach shows percentage improvement of 6.71% over LBP, −0.22% over CSLBCoP, 3.42% over BOF-LBP, 2.11% over LTriDP, −0.88% over CoALTP and 5.76% over FDLBP both in terms of ARP and ARR.

We conducted the second experiment on Brodatz database, [3]. In the experiment, the images are retrieved considering 5, 10, 15, 20 and 25 number of top matches. The ARP and ARR plots are shown in Fig. 5(c-d), which suggest that in this database, the proposed method consistently outperforms all the other methods both in terms of ARP and ARR. For 25 top matches, the proposed approach improves performance of LBP by 11.92%, of CSLBCoP by 4.40%, of

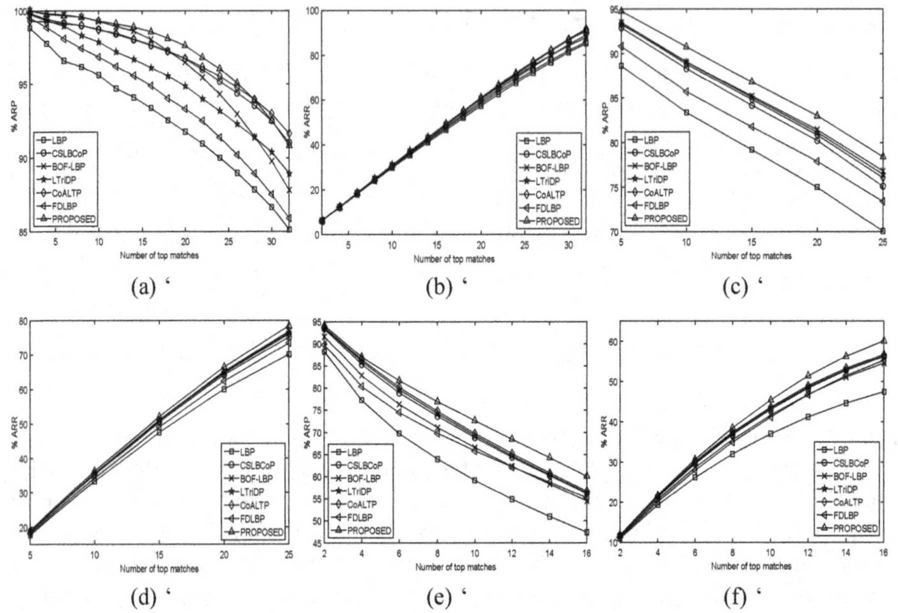

Fig. 5. Comparison of retrieval performance of the proposed approach with other approaches in terms of ARP and ARR respectively in (a–b) Kylberg database (c–d) Brodatz database (e–f) STex database.

BOF-LBP by 2.05%, of LTriDP by 2.57%, of CoALTP by 3.08% and of FDLBP by 6.82% both in terms of ARP and ARR in Brodatz database.

In our third experiment, Salzburg Texture Image Database (STex) is considered [1]. Since the images of this database are color images, as part of preprocessing, we first converted the images into gray scale images, before applying the feature descriptors. To evaluate the retrieval performance, we retrieved the images in groups of 2, 4, 6,...16 top matches. We present the ARP and ARR plots in Fig. 5(e–f) and from the plots it can be seen that as compared to all the techniques, the proposed approach achieves the best performance in STex database, at all the top matches. The proposed approach has shown percentage improvement of 26.76% over LBP, 6.97% over CSLBCoP, 10.30% over BOF-LBP, 6.67% over LTriDP, 6.01% over CoALTP and 8.87% over FDLBP both in terms of ARP and ARR, for 16 number of top matches.

The retrieval performance of LBP, CSLBCoP, BOF-LBP, LTriDP, CoALTP, FDLBP and MF-DLBP techniques for maximum number of top matches and their corresponding feature vector dimensions are presented in Table 3.

Table 3 shows only LBP has less feature dimension than the proposed method. The feature vector of the proposed method is 1.46 times higher than that of LBP, whereas the feature dimensions of CSLBCoP, BOF-LBP, LTriDP, CoALTP and FDLBP are 2.73, 3.42, 2.05, 5.47, 10.95 times higher than the

Table 3. Retrieval results comparison in terms of % ARP in Kylberg (for 32 number of top matches), Brodatz (for 25 number of top matches) and STex (for 16 number of top matches) databases

Method Name	LBP	CSLBCoP	BOF-LBP	LTriDP	CoALTP	FDLBP	MF-DLBP
Kylberg database	85.12	91.03	87.82	88.96	91.64	85.88	90.83
Brodatz database	70.06	75.11	76.83	76.45	76.07	73.40	78.41
STex database	47.43	56.20	54.51	56.36	56.71	55.22	60.12
Feature dimension	256	1024	1280	768	2048	4096	374

proposed method. However, the proposed method still outperforms all the other methods in most of the cases.

4 Conclusion

In this article, a new feature descriptor MF-DLBP is proposed for efficient texture retrieval. The descriptor first filter the image using one low pass and multiple high pass filters and then to extract the features, the low and high pass images are encoded using separate techniques. While the low pass image is encoded with LBP_S and LBP_M techniques to extract both sign and magnitude information of the image texture, the high pass images are encoded using a sparse LBP technique. The features are extracted from the encoded low pass images by calculating uniform histograms from them. On the other hand, unlike FDLBP, two decoders are used to compute the inter frequency relation among the encoded LBP maps of the high pass images only, which further enhances the discrimination capability of the extracted features. When conducted experiments on three bench mark texture databases, it is observed that the proposed approach with significantly smaller feature dimensions shows state of the art retrieval performance.

Acknowledgments. This work was supported by Digital India Corporation (formerly Media Lab Asia), Ministry of Electronics and Information Technology, Govt. of India, through Visvesvaraya Ph.D scheme.

References

1. Salzburg texture image database (STex). www.wavelab.at/sources/STex/. Accessed February 2019
2. Kylberg texture dataset. http://www.cb.uu.se/~gustaf/texture/. Accessed November 2018
3. Brodatz texture database. https://multibandtexture.recherche.usherbrooke.ca/original_brodatz.html. Accessed October 2018
4. Dubey, S.R.: Face retrieval using frequency decoded local descriptor. Multimed. Tools Appl., 1–21 (2018)

5. Dubey, S.R., Singh, S.K., Singh, R.K.: Boosting local binary pattern with bag-of-filters for content based image retrieval. In: 2015 IEEE UP Section Conference on Electrical Computer and Electronics (UPCON), pp. 1–6. IEEE (2015)
6. Gonzalez, R.C., Woods, R.E.: Digital Image Processing. Pearson Education, London (2002)
7. Guo, Z., Zhang, L., Zhang, D.: A completed modeling of local binary pattern operator for texture classification. IEEE Trans. Image Process. **19**(6), 1657–1663 (2010)
8. Hatibaruah, R., Nath, V., Saikia, K., Hazarika, D.: Elliptical local binary co-occurrence pattern for face image retrieval. J. Stat. Manag. Syst. **22**(2), 223–236 (2019)
9. Hatibaruah, R., Nath, V.K., Hazarika, D.: An effective texture descriptor for retrieval of biomedical and face images based on co-occurrence of similar center-symmetric local binary edges. Int. J. Comput. Appl., 1–12 (2019)
10. Humeau-Heurtier, A.: Texture feature extraction methods: a survey. IEEE Access **7**, 8975–9000 (2019)
11. Li, C., Huang, Y.: Deep decomposition of circularly symmetric Gabor wavelet for rotation-invariant texture image classification. In: 2017 IEEE International Conference on Image Processing (ICIP), pp. 2702–2706. IEEE (2017)
12. Murala, S., Maheshwari, R., Balasubramanian, R.: Local tetra patterns: a new feature descriptor for content-based image retrieval. IEEE Trans. Image Process. **21**(5), 2874–2886 (2012)
13. Naghashi, V.: Co-occurrence of adjacent sparse local ternary patterns: a feature descriptor for texture and face image retrieval. Optik **157**, 877–889 (2018)
14. Nath, V.K., Hatibaruah, R., Hazarika, D.: An efficient multiscale wavelet local binary pattern for biomedical image retrieval. In: Mandal, J.K., Saha, G., Kandar, D., Maji, A.K. (eds.) Proceedings of the International Conference on Computing and Communication Systems. LNNS, vol. 24, pp. 489–497. Springer, Singapore (2018). https://doi.org/10.1007/978-981-10-6890-4_47
15. Ojala, T., Pietikäinen, M., Harwood, D.: A comparative study of texture measures with classification based on featured distributions. Pattern Recogn. **29**(1), 51–59 (1996)
16. Özuysal, M.: Ground texture classification with deep learning. In: 2018 26th Signal Processing and Communications Applications Conference (SIU), pp. 1–4. IEEE (2018)
17. Tan, X., Triggs, W.: Enhanced local texture feature sets for face recognition under difficult lighting conditions. IEEE Trans. Image Process. **19**(6), 1635–1650 (2010)
18. Topi, M., Timo, O., Matti, P., Maricor, S.: Robust texture classification by subsets of local binary patterns. In: Proceedings 15th International Conference on Pattern Recognition. ICPR-2000, vol. 3, pp. 935–938. IEEE (2000)
19. Verma, M., Raman, B.: Center symmetric local binary co-occurrence pattern for texture, face and bio-medical image retrieval. J. Vis. Commun. Image Represent. **32**, 224–236 (2015)
20. Verma, M., Raman, B.: Local tri-directional patterns: a new texture feature descriptor for image retrieval. Digit. Sig. Proc. **51**, 62–72 (2016)

An Approach of Transferring Pre-trained Deep Convolutional Neural Networks for Aerial Scene Classification

Nilakshi Devi(✉) and Bhogeswar Borah

Department of CSE, Tezpur University, Tezpur 784028, India
{nilakshid,bgb}@tezu.ernet.in

Abstract. Feature selection or feature extraction plays a vital role in image classification task. Since the advent of deep learning methods, significant efforts have been given by researchers to obtain an optimal feature set of images for improving classification performance. Though several deep architectures of Convolutional Neural Networks (CNNs) have been successfully designed but training such deep architectures with small datasets like aerial scenes often leads to overfitting hence affects the classification accuracy. To tackle this issue in past few works, pre-trained CNNs are adopted as feature extractor where features are directly transferred to train only the classification layer for classifying images on the target dataset. In this work, an approach of feature extraction is proposed where both "multi-layer" and "multi-model" features are extracted from pre-trained CNNs. "Multi-layer" features are concatenation of features from multiple layers within a same CNN and "Multi-model" are concatenation of features from different CNN models. The concatenated features are further reduced with some method to obtain an optimal feature set.

Keywords: Convolutional neural network · Feature extraction · Transfer learning

1 Introduction

Aerial scene classification has been receiving remarkable attention due to it's important role in a wide range of applications such as land mapping, natural hazards detection, vegetation mapping, environment monitoring and urban planning. Though CNNs have greatly improved the classification performance of aerial scenes compared to earlier methods with handcrafted features and conventional classifiers but there still exist some issues for CNN-based aerial scene classification. A good classification accuracy depends on the depth of the neural network which requires large number of images for proper training to well-optimize it's parameters. This creates an issue for aerial scene datasets which have not enough samples to train such deep architecture hence leads to overfitting problem for the network.

B. Deka et al. (Eds.): PReMI 2019, LNCS 11941, pp. 551–558, 2019.
https://doi.org/10.1007/978-3-030-34869-4_60

To alleviate the overfitting problem in aerial scene classification, lots of work [3,5,8,12] have been published in transfer learning where the features extracted from either convolutional layers or last fully connected layers of pre-trained CNN models are transferred for scene classification. In transfer learning, the pre-trained models that are trained with some large dataset such as Imagenet, are considered as feature extractor. The extracted features are then fed to either a traditional classifier or a neural network classification layer for classification. Some works have proved that the features extracted from convolutional layers are more generic than features from fully connected layers. In convolutional neural networks, lower layers learn features similar to gabor filter like edges, dots and corners which are not specific to a particular dataset [10] but these are global features applicable to many datasets. As proceeding towards the last layers (fully connected layers) of CNN, high level features which are the combination of low levels features like objects seems to be more specific. Hence each layer has different information that can be combined to obtain a high discriminative feature representation for an image for better performance in scene classification. Such multilayer features of pre-trained CNN are integrated by the proposed fusion strategy in paper [4], where the features extracted from convolutional layers are fused with the features of fully connected layers by a principal component analysis or spectral regression kernel discriminant analysis method. In this work, we have proposed an approach of concatenating both "multi-layer" features and "multi-model" features of pre-trained convolutional neural networks for transferring them to aerial scene classification task.

2 Brief Introduction of the Architecture of Convolutional Neural Networks

The architecture of a convolutional neural network is comprising of three components: convolutional layer, pooling layer and fully-connected layer as shown in Fig. 1. The basic advantage of CNN is that unlike other neural network, each neuron in convolutional layers receives input from only some of the neurons of it's previous layer which reduces the total trainable parameters of the network. The three types of layers in CNN are briefly described below:

Convolutional layer: This is the most significant layer for feature extraction which comprises of a set of filters, each of them extract a particular feature from that image. Each filter is independently convolved throughout the image which end up with the output of a set of feature maps equals to number of filters applied. The size of filter maps after a convolution operation is calculated using the formula: $(\mathrm{W}-\mathrm{F}+2\mathrm{P})/\mathrm{S}+1$; W is the input dimension, F is filter size, P is padding and S is stride. The stride is the number of pixels taken at a time to move the filter to fit into next block during convolution.

Pooling layer: This is the layer used for dimension reduction of high dimensional feature maps obtained after applying convolution operation.

Fully connected layer: This layer is mainly used for classification apart from feature extraction in some existing CNN architectures.

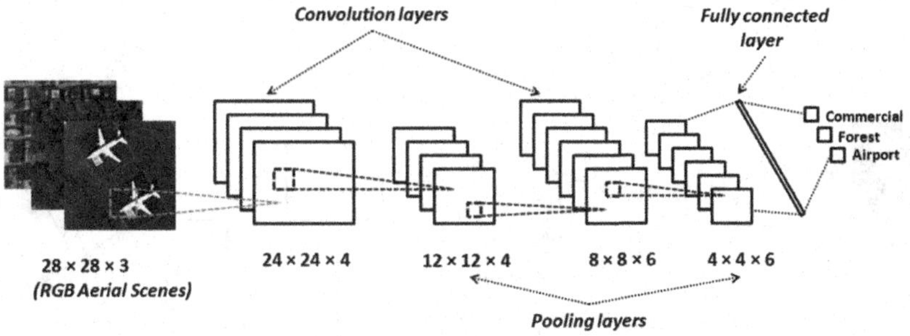

Fig. 1. Architecture of convolutional neural network

3 Methodology

Before taking step towards classification, one have to make sure whether the set of features extracted can well-represent that particular scene or object to be classified. Deep neural networks do feature extraction from the first layer to the last hidden layer at different levels of abstraction. The neurons of the last hidden layer are considered to extract the final features of an image and finally fed to the classification layer or output layer at the end of the neural network. Different types of classifiers can also be adopted instead of neural network classifier as done in many works. Hence the main concentration has to be given on feature extraction part of the deep neural network. In this paper, an approach of extracting features using pre-trained CNN models is proposed where two methods of feature extraction are defined namely, "multi-layer" feature extraction and "multi-model" feature extraction. In other words, we can say those extracted features are either "multi-layer" or "multi-model" features which are further reduced using one-dimensional (1D) convolution operation before fed to a classifier.

Mathematically, $F = \{f_i | i = 1, 2, ..., n\}$ be the feature set of either "multi-layer" or "multi-model" features of size n. After applying 1D convolution on the feature set F, x number of features are extracted from n number of features to reduce the size of final feature set $F' = \{f'_i | i = 1, 2, ..., x\}$ where $x < n$. The proposed framework consisting two architectures is shown in Fig. 2.

In case of "multi-layer" feature extraction, multiple layers of a CNN model are taken as feature extractors where features extracted from the last fully connected layer are concatenated with the convolutional/pooling features from middle or last convolutional/pooling layers. In case the extracted features are convolutional features then global average pooling is applied before proceeding to feature reduction step. Similarly, for "multi-model" feature extraction, features from the last convolutional or pooling layer of multiple CNN models are concatenated to construct a set of "multi-model" features. Finally, the feature set ("multi-layer" or "multi-model") is reduced using 1D convolution operation such that it does not produce a large number of trainable parameters in the single layer neural network for classification. The single layer neural network consists

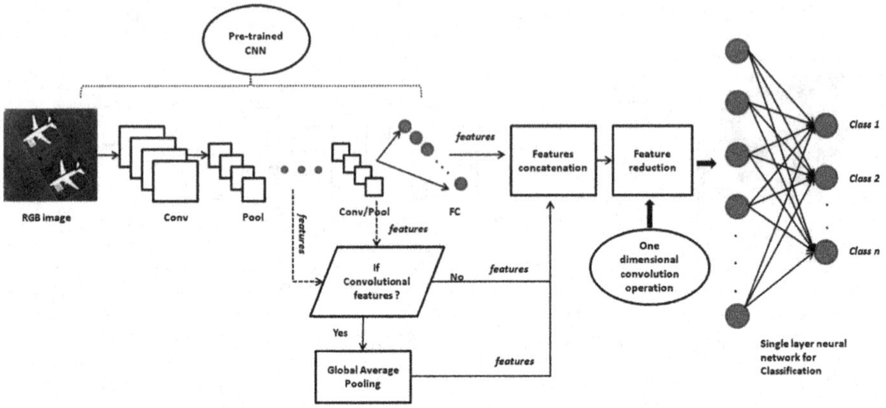

(a)

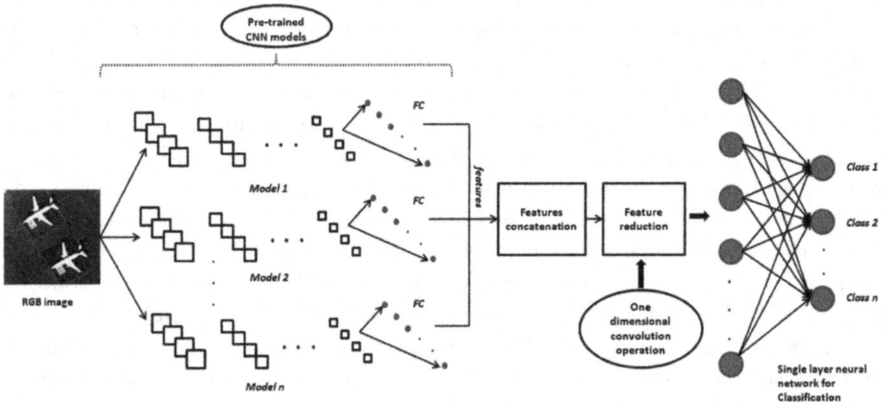

(b)

Fig. 2. Proposed framework (a) Architecture I (Multi-layer feature extraction) (b) Architecture II (Multi-model feature extraction)

of a fully connected layer and the total network parameters is calculated as number of features fed to it multiplied by number of scene classes (neurons in output layer) of respective datasets. In our proposed approach, instead of fine-tuning the entire CNN model, only the single layer neural network is trained using three target datasets. Hence before training as well as classification, the final set of "multi-layer" or "multi-model" features is build up by collecting the useful information from multiple layers or multiple models respectively which are considered as feature extractors, but without increasing it's size. Each of the two feature set is used to train the single layer neural network which acts as a classifier and compared with each other in terms of classification accuracy.

4 Experimental Results

4.1 Datasets Used

For evaluation of the proposed approach, three publicly available aerial scene datasets (containing RGB images) namely, RSSCN7 [13], UC Merced land-use [9] and WHU-RS [6] are used which consist of total 2800 images over 7 categories, 2100 images over 21 categories and 950 images with 19 aerial scene classes respectively. The aerial scene classes are like grassland, forest, farmland, river/lake, mountain, desert, industry, residential and so on.

4.2 Evaluation of Our Proposed Framework

The proposed approach is evaluated using five fold cross validation method where 80% of total images of each dataset are taken for training and the rest 20% for testing. Three existing CNN architectures are adopted to see the performance of our approach, namely: VGG19, Inception v3 and Inceptionresnet v2 which are pre-trained with a very large natural image dataset, Imagenet consisting 1000 image categories. The details of how features are extracted using the proposed framework are given below.

Multi-layer Feature Extraction

VGG19. The architecture of VGG19 can be viewed as five blocks: each of first two blocks have two convolution and one pooling operations and the last three blocks have four convolution followed by one pooling operation each. The features extracted from the last two blocks each having 512 features are concatenated to form a set of 1024 features. Then 1D convolution of 2×2 kernel size with stride 2 is applied to reduce the feature set to size of 512 features which is equal to the last feature extraction layer of the network.

Inception v3. The architecture of Inception v3 consists of five convolution and two pooling operations followed by 11 inception modules. Features extracted from eighth, ninth and eleventh inception module are concatenated to construct "multi-layer" feature set of total size 4096 features and here the feature set is also reduced to the size of last layer before the classification layer which is 2048 by applying 1D convolution with stride 2.

Inceptionresnet v2. Here, the features from layer named "mixed_7a" obtaining 2080 features are concatenated with the features from the last feature extraction layer named "conv_7b" obtaining 1536 features after applying global average pooling to each of the two layer to form the feature set of 3616 features which is reduced to size 1808.

Multi-model Feature Extraction. The features from the last fully connected layers of VGG19, Inception v3 and Inceptionresnet v2 are combined to form a feature set of 4096 features. Like, multi-layer features, this feature set is also reduced by applying 1D convolution operation with stride to 2048 features.

Table 1. Classification accuracies of recent state-of-the-art methods and the proposed methods on UC merced dataset

Method	Year	Classification accuracy (%)
GoogleNet + finetune [2]	2015	97.10
Multi-scale ADPM [5]	2016	94.86
LQPCANet [8]	2017	96.75
vgg19_EMR [7]	2017	97.90
Resnet512 (intermediate) + vgg19 [7]	2017	99.48
TEX-Nets [1]	2018	97.72
Inception-v3-CapsNet [12]	2019	99.05 ± 0.24
Architecture I (vgg19) (ours)	–	**98.52**
Architecture I (Inception v3) (ours)	–	**99.22**
Architecture I (Inceptionresnet v2) (ours)	–	**99.28**
Architecture II (ours)	–	**97.44**

The classification accuracies of our proposed framework with respect to each of three aerial scene datasets and their comparison with some recent state-of-the-art methods are shown in Tables 1, 2 and 3 respectively. The two proposed architectures beat most of the state-of-the-art works in terms of classification accuracy. One possible reason for which architecture I performs much better than architecture II is that in architecture II, there is a high chance that the

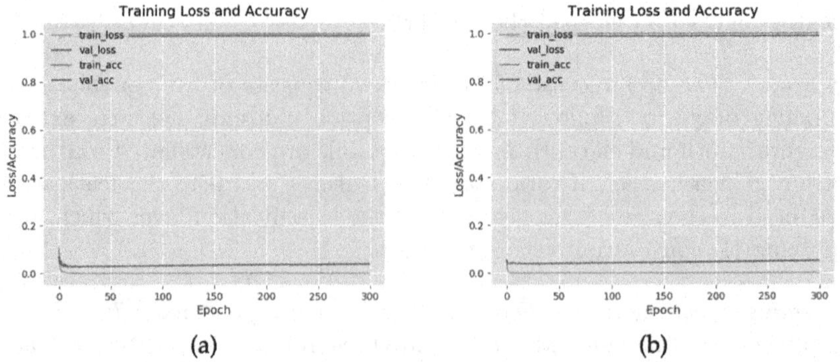

Fig. 3. Loss and accuracy curve of UC merced land use dataset (a) Architecture I (b) Architecture II

Table 2. Classification accuracies of recent state-of-the-art methods and the proposed methods on WHU_RS dataset

Method	Year	Classification accuracy (%)
Multi-scale ADPM [5]	2016	84.67
LQPCANet [8]	2017	96.22
TEX-Nets [1]	2018	98.20
Architecture I (vgg19) (ours)	–	**98.69**
Architecture I (Inception v3) (ours)	–	**98.97**
Architecture I (Inceptionresnet v2) (ours)	–	**98.99**
Architecture II (ours)	-	**98.82**

Table 3. Classification accuracies of recent state-of-the-art methods and the proposed methods on RSSCN7 dataset

Method	Year	Classification accuracy (%)
TEX-Nets [1]	2018	94.00
Global + Local [11]	2018	95.59 ± 0.49
Architecture I (vgg19) (ours)	–	**95.30**
Architecture I (Inception v3) (ours)	–	**95.77**
Architecture I (Inceptionresnet v2) (ours)	–	**95.74**
Architecture II (ours)	–	**93.74**

final feature set consists of redundant features. Features are collected from the last layer of different CNN models where each model may produce some common features. Due to this feature redundancy problem, the gap between the loss curves of training and validation (also called overfitting gap) in architecture II is slightly more than that of architecture I as shown in Fig. 3 that can also affect the classification results.

5 Conclusion

Each layer of a neural network extracts different information about an image so it is always a good idea of considering multiple layers of the network as feature extractors in transfer learning. Moreover, each state-of-the-art CNN model may not extract exactly same set of features of the image. So in the proposed framework, each of the architecture defines a way of feature extraction namely, "multi-layer" and "multi-model" feature extraction respectively using pre-trained CNNs. Each feature set are then reduced using convolution operation to lower the computation cost during classification. The implementation of the proposed approach is done in keras with tensorflow 1.7.0 as backend. The python version is 3.6.4 with cache memory 30720 KB and CPU speed 2.50 GHz.

References

1. Anwer, R.M., Khan, F.S., van de Weijer, J., Molinier, M., Laaksonen, J.: Binary patterns encoded convolutional neural networks for texture recognition and remote sensing scene classification. ISPRS J. Photogramm. Remote Sens. **138**, 74–85 (2018)
2. Castelluccio, M., Poggi, G., Sansone, C., Verdoliva, L.: Land use classification in remote sensing images by convolutional neural networks. arXiv preprint arXiv:1508.00092 (2015)
3. Hu, F., Xia, G.S., Hu, J., Zhang, L.: Transferring deep convolutional neural networks for the scene classification of high-resolution remote sensing imagery. Remote Sens. **7**(11), 14680–14707 (2015)
4. Li, E., Xia, J., Du, P., Lin, C., Samat, A.: Integrating multilayer features of convolutional neural networks for remote sensing scene classification. IEEE Trans. Geosci. Remote Sens. **55**(10), 5653–5665 (2017)
5. Liu, Q., Hang, R., Song, H., Zhu, F., Plaza, J., Plaza, A.: Adaptive deep pyramid matching for remote sensing scene classification. arXiv preprint arXiv:1611.03589 (2016)
6. Sheng, G., Yang, W., Xu, T., Sun, H.: High-resolution satellite scene classification using a sparse coding based multiple feature combination. Int. J. Remote Sens. **33**(8), 2395–2412 (2012)
7. Wang, G., Fan, B., Xiang, S., Pan, C.: Aggregating rich hierarchical features for scene classification in remote sensing imagery. IEEE J. Sel. Top. Appl. Earth Obs. Remote Sens. **10**(9), 4104–4115 (2017)
8. Wang, J., Luo, C., Huang, H., Zhao, H., Wang, S.: Transferring pre-trained deep cnns for remote scene classification with general features learned from linear pca network. Remote Sens. **9**(3), 225 (2017)
9. Yang, Y., Newsam, S.: Bag-of-visual-words and spatial extensions for land-use classification. In: Proceedings of the 18th SIGSPATIAL International Conference on Advances in Geographic Information Systems, pp. 270–279. ACM (2010)
10. Yosinski, J., Clune, J., Bengio, Y., Lipson, H.: How transferable are features in deep neural networks? In: Advances in Neural Information Processing Systems, pp. 3320–3328 (2014)
11. Zeng, D., Chen, S., Chen, B., Li, S.: Improving remote sensing scene classification by integrating global-context and local-object features. Remote Sens. **10**(5), 734 (2018)
12. Zhang, W., Tang, P., Zhao, L.: Remote sensing image scene classification using CNN-CapsNet. Remote Sens. **11**(5), 494 (2019)
13. Zou, Q., Ni, L., Zhang, T., Wang, Q.: Deep learning based feature selection for remote sensing scene classification. IEEE Geosci. Remote Sens. Lett. **12**(11), 2321–2325 (2015)

Encoding High-Order Statistics in VLAD for Scalable Image Retrieval

Alexy Bhowmick[1,2](✉), Sarat Saharia[2], and Shyamanta M. Hazarika[3]

[1] School of Technology, Assam Don Bosco University, Guwahati, India
alexy.bhowmick@gmail.com
[2] Department of Computer Science and Engineering, Tezpur University, Tezpur, India
[3] Department of Mechanical Engineering, IIT Guwahati, Guwahati, India

Abstract. We revisit the implicit design choices in the popular vector of locally aggregated descriptors (VLAD), which aggregates the *residuals* of local image descriptors. Since original VLAD ignores high-order statistics the resultant vector is not discriminative enough. We address this issue by exploiting high-order statistics for gaining complementary information. Our contributions are two-fold: First, we present a novel high-order VLAD (HO-VLAD) with increased discriminative power. Next, we propose a light-weight retrieval framework to demonstrate HO-VLAD's effectiveness for scalable image retrieval. Systematic experiments on two challenging public databases (INRIA Holidays, UKBench) exhibit a consistent improvement of performance with limited computational costs.

Keywords: VLAD · Residual · High-order · Kurtosis · Image retrieval

1 Introduction

Recent years have witnessed an exponential growth of multimedia data on the Web. We are interested in the compact encoding of local descriptors (*e.g.* SIFT [8]) of images/videos to design a super vector representation, and thereby address the challenge of efficient indexing and retrieval of similar images in large image databases. Given a query image, the goal is to retrieve candidate images depicting the same object (semantic concept) or scene from the database. The representation must be discriminative, sufficiently invariant to transformations (geometric, viewpoints, illuminations, and occlusion). Three important constraints have to be considered jointly: accuracy (quality), efficiency (speed), and memory usage (footprint) of the representation [5].

Recent works on Fisher Vector (FV) [11] and Vector of Locally Aggregated Descriptors (VLAD) [4,5] are considered significant contributions towards image retrieval and classification. VLAD is known to outperform Bag-of-Words (BoW) [12] and the more sophisticated FV in terms of computational cost and accuracy [5,7]. One of VLAD's issues is that it ignores high-order information of the distribution of descriptors [4,5]. As illustrated in Fig. 1a, two different descriptor

B. Deka et al. (Eds.): PReMI 2019, LNCS 11941, pp. 559–566, 2019.
https://doi.org/10.1007/978-3-030-34869-4_61

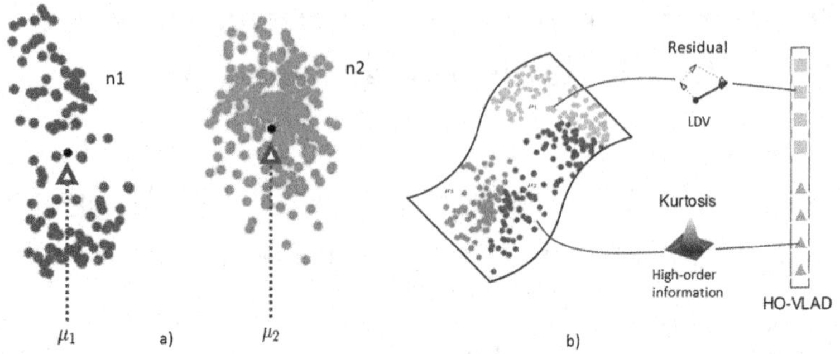

Fig. 1. (a) Two samples *n1* and *n2* (assigned to centroids μ_1 and μ_2) with dissimilar descriptor distributions but similar aggregated vectors (LDVs) by VLAD. (b) Discrimination is enhanced when distributional shape is encoded.

distributions may have similar aggregated vectors obtained by original VLAD; but the distribution of the sets of descriptors are dissimilar when observed by fourth-order statistic.

All descriptors do not contribute equally to the *residual* and an outlier can far outweigh the contribution of many inliers close to the centroid [3] (Fig. 1a). Lower-order statistics (*e.g.* residuals in VLAD) are not sufficient to capture the nature of the descriptor distribution. High-order statistics (2nd and 3rd order) have been used with VLAD to solve this problem in object categorization and action recognition [10]. We introduce HO-VLAD, a novel extension of VLAD which employs high-order information for scalable image retrieval. HO-VLAD captures the nature of typical non-Gaussian descriptor distributions and also takes into account the unequal contribution and effect of outliers.

Our contributions are two-fold:

- We present a novel *high-order* VLAD (HO-VLAD) that leverages on fourth-order statistics for increased discriminative power.
- We propose a light-weight framework for scalable image retrieval. The experiments and results demonstrate the proposed method's effectiveness.

The rest of the paper is as follows: Sect. 2 reviews the related works in brief. Section 3 introduces HO-VLAD encoding and the proposed framework. Section 4 reports the datasets, evaluation protocol, experiments and results. Lastly, Sect. 5 presents the conclusion and future works.

2 Related Works

This section reviews the existing state-of-the-art feature encoding methods namely, BoW, FV, and VLAD.

Bag-of-Words (BoW) [12] is the dominant model for image/video representation. It requires a pre-trained codebook $C = \{\mu_1, ..., \mu_k\}$ of k visual words.

Typically a k-means quantizer [12] maps high-dimensional local feature descriptors x, (e.g. SIFT [8]), to the nearest *centroid*:

$$NN(x) : x \mapsto q(x) = \arg\min_{\mu_i \in C} ||x - \mu_i|| \tag{1}$$

Here, *NN(x)* denotes the *nearest -neighborhood* mapping function. The BoW is the frequency histogram (*i.e. count*) of visual words, which is a *zeroth-order* statistic of the features.

Fisher Vector (FV) [11] extends the BoW by encoding *second-order* statistics (*i.e. mean, variances*) of the local descriptor distribution. Given a set of n local descriptors $\mathcal{X} = \{x_1, x_2, ...x_n\}$, $x_j \in \mathbb{R}^d$, the distribution is modeled as a Gaussian Mixture Model (GMM) (using Expectation-Maximization (EM)) $\theta = \{(\mu_k, \sum_k, \pi_k), i = 1, 2, ...k\}$ fitting the distribution of descriptors, where $(\mu_k, \sum_k, \pi_k)$ are the *mean*, *covariance*, and *prior* of the k-th component. The GMM 'soft' assigns each descriptor x_i to a mode k in the mixture with a posterior probability:

$$q_{ik} = \frac{\exp[\frac{-1}{2}(x_i - \mu_k)^T \sum_k^{-1}(x_i - \mu_k)]}{\sum_{t=1}^{k} \exp[\frac{-1}{2}(x_i - \mu_t)^T \sum_k^{-1}(x_i - \mu_t)]} \tag{2}$$

For each mode k, the *mean* and *covariance* deviation vectors are computed:

$$u_{jk} = \frac{1}{N\sqrt{\pi_k}} \sum_{i=1}^{N} q_{ik} \frac{x_{ji} - \mu_{jk}}{\sigma_{jk}}; v_{jk} = \frac{1}{N\sqrt{2\pi_k}} \sum_{i=1}^{N} q_{ik} \left[\left(\frac{x_{ji} - \mu_{jk}}{\sigma_{jk}} \right)^2 - 1 \right] \tag{3}$$

Finally, the u and v vectors are stacked together to construct the high-dimensional ($2dk$) FV representation.

$$v = [u_1^T, u_2^T, ...u_k^T, v_1^T, v_2^T, ...v_k^T]^T, \qquad u_i, v_i \in \mathbb{R}^{\mathrm{d}} \tag{4}$$

Vector of Locally Aggregated Descriptors (VLAD) [4,5], aggregates a set $\mathcal{X} = \{x_1, x_2, ...x_n\}$, $x_j \in \mathbb{R}^d$, of n local d-dimensional descriptors into a fixed-size, compact vector representation v. A codebook $\mathcal{C} = \{\mu_1, \mu_2, ..., \mu_k\}$, $\mu_i \in \mathbb{R}^d$ is obtained by applying *k-means* on the set of local descriptors of training samples. Each descriptor $x_j \in \mathbb{R}^d$ is mapped to its nearest centroid in the codebook:

$$NN(x) : x \mapsto q(x) = \arg\min_{\mu_i \in C} ||x_j - \mu_i||^2 \tag{5}$$

Typically $||.||^2$ (L_2 norm) is used to solve the minimization problem. VLAD encodes the *first-order* statistic, *i.e.* residual - the vector difference $(x_j - \mu_i)$, between a descriptor x_j and a centroid μ_i. The residuals are aggregated in a d-dimensional sub-vector v_i, called Local Difference Vector (LDV).

$$v_i = \sum_{x_j : NN(x_j) = \mu_i} (x_j - \mu_i) \tag{6}$$

The final VLAD encoding v for the query set $\mathcal{X}$ is obtained by concatenating all sub-vectors v_i, $i = 1, ..., k$ (*i.e.* k LDVs) forming a $D(= k \times d)$ dimensional image signature (*unnormalized VLAD*).

$$v = [v_1^T, v_2^T, ..., v_k^T]^T, \qquad v_i \in \mathbb{R}^{\mathrm{d}} \tag{7}$$

As final steps, VLAD vector v is first Power-normalized, then $L2$-normalized.

3 Encoding High-Order Statistics in VLAD

We introduce an effective method to augment original VLAD with high-order statistical information. As shown in Fig. 1, VLAD's discriminative power suffers because (a) it ignores high-order statistics, and (b) due to the effect of outliers. VLAD residuals describe the distribution but they are not sufficient to capture the nature of the distribution. This is because low-level descriptors (*e.g.* SIFT [8]) are not typically Gaussian in real life [10]. Consequently, we propose a high-order VLAD (HO-VLAD) which encodes fourth-order statistics *i.e kurtosis* of a distribution, to exploit complementary information.

Kurtosis represents the 'peakedness' or convexity of a probability distribution [2]. It is a measure of the outliers of a distribution. High kurtosis (*leptokurtic*, >3) indicates the data is heavy-tailed and there is profusion of outliers. Low kurtosis (*platykurtic*, <3) indicates lack of outliers. We design the *fourth-order* super vector as:

$$v_i^k = \frac{\frac{1}{N}\sum_{j=1}^{N}(x_j - \mu_i)^4}{\frac{1}{N}\sum_{j=1}^{N}(x_j - \mu_i)^4} \tag{8}$$

where, v_i^k indicates the kurtosis of the i-th cluster with centroid μ_i. After intra-normalization [1] separately, the residual v and kurtosis v^k vectors are concatenated to produce a (kd + k)-dimensional vector, *i.e.* the final HO-VLAD representation ($d = 128$ for SIFT). We consider intra-normalization due to its good performance [1,10]. Our method avoids soft weight computation and accommodates higher-order statistics in comparison with original VLAD.

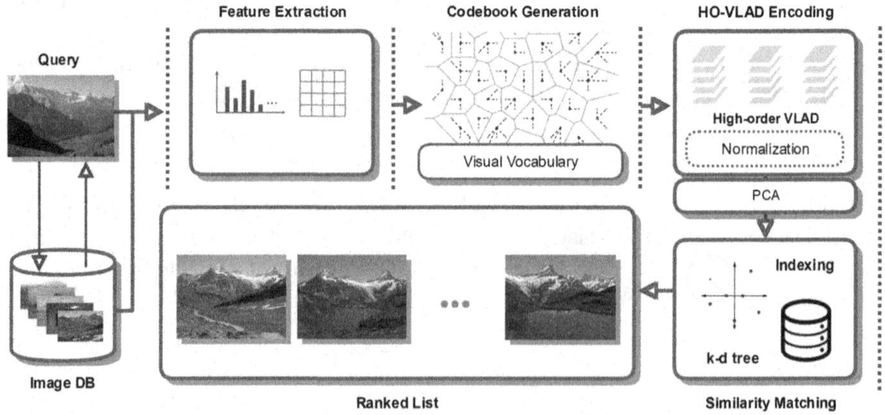

Fig. 2. The scalable image retrieval framework/pipeline for HO-VLAD computation and retrieval performance evaluation

HO-VLAD Algorithm. In keeping with the spirit of original VLAD, we incorporate higher-order information and formulate the HO-VLAD computation algorithm as stated below (Algorithm 1).

Retrieval Framework. The light-weight retrieval framework (Fig. 2) consists of four main components: (a) Local feature (SIFT [8]) extraction (b) Codebook generation (independent Flickr60K dataset[1]) (c) Proposed HO-VLAD encoding, normalization and dimensionality reduction (PCA) (d) Indexing and nearest neighbor search with k-d trees to produce ranked retrieval results.

Algorithm 1. HO-VLAD computation.

Input :
Local descriptors $\mathcal{X} = \{x_j\}_{j=1}^{n}$, $x_j \in \mathbb{R}^d$ extracted from a query image.
Codebook $\mathcal{C} = \{\mu_i\}_{i=1}^{k}$, $\mu_i \in \mathbb{R}^d$ learned by applying k-means on independent dataset.

Output:
$v = [v_1, v_2, ..., v_k, v_{K_1}, v_{K_2}, ..., v_{K_k}]$, the HO-VLAD representation of $\mathcal{X}$

for $i \leftarrow 1$ **to** k **do**
 $v_i = 0_d$

for $j \leftarrow 1$ **to** n **do**
 $i = \underset{i \in \{1,...,k\}}{\text{argmin}} \|x_j - \mu_i\|$
 $v_i = \sum_{x_j : NN(x_j) = \mu_i} (x_j - \mu_i)$

$v = [v_1^T, v_2^T, ... v_k^T]^T$

for $j \leftarrow 1$ **to** k **do**
 $i = \underset{i \in \{1,...,k\}}{\text{argmin}} \|x_j - \mu_i\|$
 $v_j^k = \dfrac{\frac{1}{N}\sum_{j=1}^{N}(x_j - \mu_i)^4}{\frac{1}{N}\sum_{j=1}^{N}(x_j - \mu_i)^4}$

for $j \leftarrow 1$ **to** D **do**
 $v_j = sign(v_j) \times |v_j|^\alpha$

$v = \left[\frac{v_1}{\|v_1\|_2}, \frac{v_2}{\|v_2\|_2}, ..., \frac{v_k}{\|v_k\|_2}\right]; v^k = \left[\frac{v_1}{\|v_1\|_2}, \frac{v_2}{\|v_2\|_2}, ..., \frac{v_k}{\|v_k\|_2}\right]$

$v = [v_1^T, v_2^T, ... v_k^T, v_1^{k^T}, v_2^{k^T}, ... v_k^{k^T}]^T$

---

[1] https://lear.inrialpes.fr/~jegou/data.php.

4 Experiments and Evaluations

We verify the effectiveness of our method based on experiments on benchmark datasets, and compare with state-of-the-art feature encoding methods.

4.1 Benchmark Datasets and Descriptors

For local feature extraction, we have employed the experimental setup similar to [5] and the feature extraction library[2]. More specifically, the regions of interest are extracted utilizing Hessian affine-invariant region detectors and described by the SIFT descriptor [8]. An independent dataset Flickr60k (67714 images, 140 M descriptors) was employed to train the codebook off-line for both Holidays and the UKB dataset. The evaluation is performed on two standard and publicly available image retrieval benchmarks:

INRIA Holidays: (See footnote 1) (1491 images, 4.456 M descriptors, 500 queries) [6] consists of personal holiday photos of 500 groups each representing a distinct scene. Mean Average Precision (mAP) is employed to evaluate retrieval accuracy, with the query removed (leave-one-out fashion) from the ranking list.

University of Kentucky Benchmark (UKB):[3] [9] is a collection of 10,200 images corresponding to 2,550 distinct classes and scenes of diverse categories. Every class is composed of 4 images; a query image and three groundtruth images. We use N-S score (ranging from 0–4) for retrieval accuracy.

4.2 Performance Evaluations and Analysis

Comparison with State-of-the-Art: Table 1 compares the results of our approach with the results from literature, in particular retrieval accuracies of BoW, FV, and VLAD, on the INRIA Holidays and UKB datasets. As can be seen from this table, an mAP of 0.611 is achieved on Holidays and a N-S score of 3.32 is obtained on UKBench with regular SIFT. The improvement provided by HO-VLAD over original VLAD is +4.6% on Holidays and +0.14 N-S score on UKB. For the sake of consistency, $k = 64$ is used in all experiments. The same SIFT descriptors as in [5] ensure a fair comparison. In our experiments, we found α = 0.5 remains a good choice for normalization parameter and gives optimal results. The scheme avoids multiple vocabularies and soft assignments.

Memory Footprint: Using floating point numbers for each element, each number requires 4 bytes of memory. For a VLAD vector describing an image I, the total memory usage with $k = 64$ and 128-D SIFT descriptors is 64 * 128 * 4 = 32,768 Bytes or 32 KB. HO-VLAD vector for the same parameters requires ((64 * 128) + 64) * 4 = 33,024 Bytes or 33 KB to describe the same image. We believe with this limited computation cost, a significant precision is obtained.

[2] http://www.vlfeat.org/.

[3] https://archive.org/details/ukbench.

Table 1. Comparison of proposed image representation HO-VLAD with state-of-the-art (mAP performance and N-S score).

Method	Codebook size	D(d*k)	*Holidays* (mAP(%))	*UKB* (score/4)
BoW [12]	$k = 20000$	20000	40.4	2.87
FV [11]	$k = 64$	16384	59.5	3.25
VLAD [4]	$k = 64$	8192	56.5	3.18
HO-VLAD*	$k = 64$	8256	**61.1**	**3.32**

*indicates our proposed algorithm.

Table 2. Comparative performance of FV, VLAD, and HO-VLAD on Holidays benchmark. Performance is given for full-dimensional D and PCA-reduced D′ descriptor.

Method ↓	|k|	D′ = D	D′(512)	D′(256)	D′(128)	D′(64)	D′(32)	D′(16)
FV	|64|	59.5	60.7	59.3	56.5	52.0	49.0	48.6
VLAD	|64|	56.5	56.1	53.6	53.5	47.0	44.0	44.2
HO-VLAD	|64|	**61.1**	**61.7**	**59.8**	**57.6**	**54.0**	**48.2**	**46.4**

Dimension Reduction: Table 2 shows the relative improvement of HO-VLAD is comparatively reduced when dimension reduction with PCA is applied. The gain shrinks significantly and one can observe that the dimension reduction reduces the gap between the different methods. For a finer vocabulary ($k = 256$), HO-VLAD attains 71.2% (Fig. 3a). It can be seen that even at low dimensions (D′ = 16), FV with $k = 256$ maintains a competitive accuracy (50.6%)(Fig. 3b).

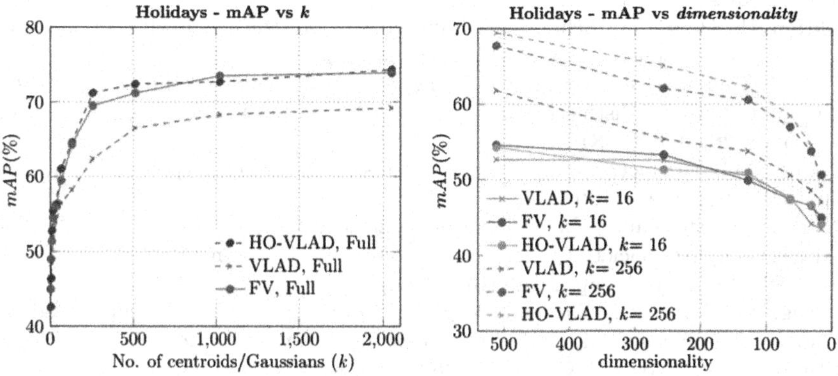

Fig. 3. Impact of (a) vocabulary size k (*left*) (b) dimensionality reduction (*right*) on accuracy (mAP on Holidays). Parameters: $\alpha = 0.5$.

Figure 3a shows the mAP score on Holidays as a function of k for full-sized VLAD descriptor. It is evident that with more number of centroids, the retrieval performance improves. For $k = 2048$, we achieve a mAP of 74.3%. However, for larger values of k, the cost of centroid assignment increases, hence we limit our analysis to $k = 2048$.

5 Conclusion

In this paper, we present HO-VLAD, a novel extension of the popular VLAD for scalable image retrieval. The proposed method encodes high-order statistics for greater discriminative encoding and considers effect of outliers. The tests on the image retrieval framework with a small-size codebook shows promising results on the benchmark INRIA Holidays and UKB datasets. Future work will examine comparisons with more recent encoding techniques, retrieval performance in presence of distractor images, and generating compact binary codes.

References

1. Arandjelovic, R., Zisserman, A.: All about VLAD. In: 2013 IEEE Conference on Computer Vision and Pattern Recognition, pp. 1578–1585 (2013)
2. Balanda, K.P., MacGillivray, H.L.: Kurtosis: a critical review. Am. Stat. **42**(2), 111–119 (1988)
3. Husain, S.S., Bober, M.: Improving large-scale image retrieval through robust aggregation of local descriptors. IEEE Trans. Pattern Anal. Mach. Intell. **39**(9), 1783–1796 (2016)
4. Jegou, H., Douze, M., Schmid, C., Prez, P.: Aggregating local descriptors into a compact image representation. In: 2010 IEEE Computer Society Conference on Computer Vision and Pattern Recognition, pp. 3304–3311, June 2010
5. Jegou, H., Perronnin, F., Douze, M., Snchez, J., Prez, P., Schmid, C.: Aggregating local image descriptors into compact codes. IEEE Trans. Pattern Anal. Mach. Intell. **34**(9), 1704–1716 (2012)
6. Jegou, H., Douze, M., Schmid, C.: Improving bag-of-features for large scale image search. Int. J. Comput. Vis. **87**(3), 316–336 (2010)
7. Liu, Z., Houqiang, L., Wengang, Z., Ting, R., Qi, T.: Making residual vector distribution uniform for distinctive image representation. IEEE Trans. Circ. Syst. Video Technol. **26**(2), 375–384 (2016)
8. Lowe, D.G.: Distinctive image features from scale-invariant keypoints. Int. J. Comput. Vis. **60**(2), 91–110 (2004)
9. Nister, D., Stewenius, H.: Scalable recognition with a vocabulary tree. In: 2006 IEEE Computer Society Conference on Computer Vision and Pattern Recognition (CVPR 2006), vol. 2, pp. 2161–2168 (2006)
10. Peng, X., Wang, L., Qiao, Y., Peng, Q.: Boosting VLAD with supervised dictionary learning and high-order statistics. In: Fleet, D., Pajdla, T., Schiele, B., Tuytelaars, T. (eds.) ECCV 2014. LNCS, vol. 8691, pp. 660–674. Springer, Cham (2014). https://doi.org/10.1007/978-3-319-10578-9_43
11. Perronnin, F., Dance, C.: Fisher kernels on visual vocabularies for image categorization. In: 2007 IEEE Conference on Computer Vision and Pattern Recognition, pp. 1–8, June 2007
12. Sivic, J., Zisserman, A.: Video Google: a text retrieval approach to object matching in videos. In: Proceedings Ninth IEEE International Conference on Computer Vision, vol. 2, pp. 1470–1477, October 2003

Image-Based Analysis of Patterns Formed in Drying Drops

Anusuya Pal[1], Amalesh Gope[2], and Germano S. Iannacchione[1](✉)

[1] Department of Physics, Worcester Polytechnic Institute, Worcester, USA
{apal,gsiannac}@wpi.edu
[2] Department of EFL, Tezpur University, Tezpur, Assam, India
amalesh@tezu.ernet.in

Abstract. Image processing and pattern recognition offer a useful and versatile method for optically characterizing drops of a colloidal solution during the drying process and in its final state. This paper exploits image processing techniques applied to cross-polarizing microscopy to probe birefringence and the bright-field microscopy to examine the morphological patterns. The bio-colloidal solution of interest is a mixture of water, liquid crystal (LC) and three different proteins [lysozyme (Lys), myoglobin (Myo), and bovine serum albumin (BSA)], all at a fixed relative concentration. During the drying process, the LC phase separates and becomes optically active detectable through its birefringence. Further, as the protein concentrates, it forms cracks under strain due to the evaporation of water. The mean intensity profile of the drying process is examined using an automated image processing technique that reveals three unique regimes- a steady upsurge, a speedy rise, and an eventual saturation. The high values of standard deviation show the complexity, the roughness, and inhomogeneity of the image surface. A semi-automated image processing technique is proposed to quantify the distance between the consecutive cracks by converting those into high contrast images. The outcome of the image analysis correlates with the initial state of the mixture, the nature of the proteins, and the mechanical response of the final patterns. The paper reveals new insights on the self-assembly of the macromolecules during the drying mechanism of any aqueous solution.

Keywords: Image processing · Drying drops · Liquid crystals

1 Introduction

Image processing and automated pattern recognition techniques are crucial elements in advance of (optical) imaging technology. As image acquisition has become more precise (in time and spatial resolution) and ubiquitous, the need to extract information from these images has grown exponentially. One of the many scientific areas impacted by such processing is in observing the drying drops of some complex fluids. Specifically, one such class of complex colloidal fluid relevant to forensics and medical sciences is biological fluid such as blood,

B. Deka et al. (Eds.): PReMI 2019, LNCS 11941, pp. 567–574, 2019.
https://doi.org/10.1007/978-3-030-34869-4_62

plasma serum, tear, urine, or simple protein solutions [1]. Recent research findings [2,3] have found that dried drops of serums can differentiate the state of health of a patient. Furthermore, the whole dried pattern appears to be connected to the initial state of the material present in these complex fluids [4,5]. For example, Deegan et al. have studied colloidal drops and found that the drops get pinned after depositing on the substrate. The particles start concentrating near the edge more than the central region, forming a ring called "coffee-ring effect" [6]. Due to this pinning and solvent loss, the drops are no longer able to shrink and ultimately, form cracks (patterns) under strain during the drying process.

This paper exploits image processing techniques applied to the cross-polarizing microscopy to probe birefringence and the bright-field microscopy to examine the crack patterns. The bio-colloidal solution of interest is a mixture of water, and three different proteins [lysozyme (Lys), myoglobin (Myo), and bovine serum albumin (BSA)], all at a fixed relative concentration with and without a nematic liquid crystal (LC). The patterns are characterized using three different criteria, (a) the optical outcome, (b) the textural dynamics, and, (c) statistical significance. This work uses both automated and semi-automated image processing techniques to quantify drying dynamics and dried morphology. Our analysis reveals that the topmost part of the BSA film is coated with LC droplets, whereas, the same LC droplets are randomly distributed underneath the Lys and Myo films. This paper demonstrates the versatility and usefulness of these image processing techniques and is adaptable to any similar imaging system.

2 Materials, Sample Preparation and Image Acquisition

Lysozyme (Lys), myoglobin (Myo) and bovine serum albumin (BSA) are globular proteins differing in mass, charge, and properties [4]. The presence of a heme group helps Myo to serve as an oxygen-binding protein. Lys and Myo have similar molecular masses of ∼14.3, and ∼17.0 KDa respectively, whereas BSA contains a higher molecular mass of ∼66.5 kDa [4]. The net charge of Lys, BSA, and Myo is positive, negative, and neutral respectively. The commercial powdered proteins were purchased from Sigma Aldrich, USA. Myo (Catalog no. M0630), Lys (L6876), and BSA (A2153) were used without any further purification. Distinct protein solutions containing each of Lys, Myo, and BSA (100 mg each) were prepared by dissolving those separately in 1 mL of de-ionized water. Following that, an optically active and polar LC, 5CB (328510) was heated above the transition temperature (35 °C), and 10 μL was added as a third component to the solution. A ∼1.3 μL of the solution was pipetted on the coverslip in the form of a circular drop, which was allowed to dry under ambient conditions [4].

A polarizing microscopy (Leitz Wetzlar, Germany) with a 5× objective lens at a fixed resolution of 2048 × 1536 pixels was used to capture the images [4,5]. An 8-bit digital camera (Amscope MU300) in color profile mode assigns intensity values from 0 to 255 in each color of red, green, and blue (RGB) pixels for image acquisition. After the deposition of the drop on the coverslip, the time-lapse

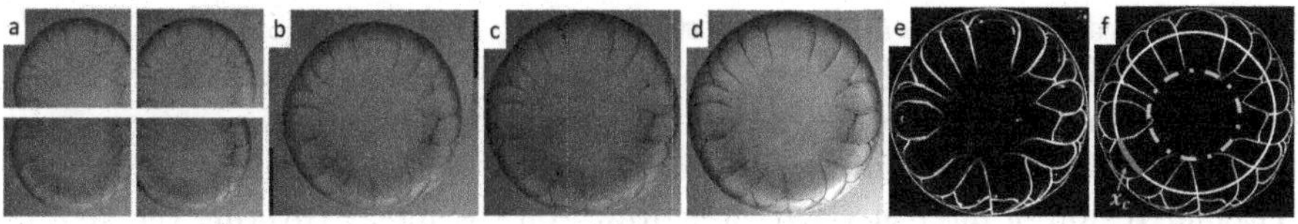

Fig. 1. A representative processing of images of the dried state of BSA drop to quantify the spacing between the cracks. (a) The raw microscopic images. (b) A stitched image. (c) An 8-bit gray image. (d) An adjusted brightness-contrast image. (e) A high contrast image after filtering through variance filter. (f) A final binary image depicting different regions, circular line-cut, and crack spacing (x_c). (Color figure online)

images in crossed polarizing configuration were recorded at every two seconds to monitor LC droplets' birefringence during the drying process. Each drop was deposited three times from the same sample to ensure the reproducibility. The final dried state was examined in bright-field configuration to understand different crack patterns in different protein drops with and without inclusion of LC droplets. Since it was challenging to view the entire ∼2 mm diameter drop using a 5× objective lens, different sections of the dried drops were imaged separately (Fig. 1a). Each section is stitched together (Fig. 1b) by using *Stitching* plugin [7] in Fiji [8]. We propose a semi-automated image-processing algorithm on the stitched images to determine the spacing between the consecutive cracks (x_c) in ImageJ [9]. We converted stitched bright-field images into gray (Fig. 1c). For these gray images, the range of monochromatic shades from black to white are displayed from 0 to 255 without partitioning into RGB sets of pixels. Then, the bright and contrast of the image were adjusted (Fig. 1d) and filtered with a variance filter (Fig. 1e), and processed into an 8-bit binary image (255 for pixels depicting the crack lines and 0 for pixels elsewhere else) (Fig. 1f) [4,5]. However, the filtering was not enough to process the images into exact binary images. The artifacts were, therefore manually removed by comparing (overlapping) the processed images on to the original images. However, future work will automate this procedure.

The morphology of the dried drop is such that it can be divided into two regions: rim and central. The rim is near the edge depicted by yellow dashed line in Fig. 1f. We converted all the images of the dried drops into a scaled stack. Three circular-cut lines (shown by white circular line in Fig. 1f) of different radii were made in each region by using *Oval Profile* plugin in ImageJ [4,5]. The intensity values were plotted along each circular line at every 0.1° as a function of arc-length along each circle. A script using '*Array.findMaxima*' was used to determine the positions of maximum intensity values. An estimate of the crack spacing, x_c (outlined by green color in Fig. 1f) is calculated by the consecutive maxima difference [4,5]. All the intensity values along the crack lines were cross-checked manually to see if there is an artifact. If found, were corrected to the consistent (and fixed) values to maintain uniformity along the crack lines. A threshold of ±0.005 mm were used as a standard range and x_c values outside this range was recorded and aggregated in each region to obtain an average ($\overline{x}_c$) [4,5]. However, there may be cracks that propagate along the drawn circle, leading to some uncertainty in the extracted crack distributions.

To quantify the birefringence of the LC droplets during the drying process, a circular region of interest (ROI) was drawn on the image covering the area in the drop, and another ROI was drawn on the image covering the background (coverslip) using the *Oval tool* in ImageJ [9]. Once the ROI was selected, the birefringence intensity of that ROI was measured. To ensure that the different size of ROI in the sample and the background do not affect the intensity measurement, *mean gray value* (the sum of the values of the pixels in the selection divided by the area of the pixels in the selection) in ImageJ [9] was chosen. A script was written for an automated image-processing algorithm which measures the *mean gray value* for each ROI in the images during the drying process. However, the differences in the *mean gray values* during image recording might affect both the background and the sample. To counter this, a correction (calibration) with the background i.e., a correction factor was determined. It could be done by subtracting the background gray values; however, it wouldn't have fixed an uneven background in the series of images. So, the ROI with the lowest background *mean gray value* was chosen from the whole set of images during the drying process as a reference. The lowest was also selected to avoid into running the risk of generating overexposed images (otherwise the correction factors would be smaller than 1). The corrected intensity of the sample (I_c) was then determined for each sample by dividing the mean gray values with the correction factor. The intensity was averaged over a range of 30 s and also averaged for three drops. Good reproducibility was found in the intensity profile, though the time was shifted (added or subtracted) to make the profile nearly overlap to each other. Note that the lamp intensity was kept fixed throughout the whole experiment done with each protein drop. And finally, the averaged corrected intensity values ($\overline{I}_c$) was plotted with time. In addition to $\overline{I}_c$, another textural parameter i.e., standard deviation (SD) was calculated and were averaged over 30 s to understand the emerging complexity during the drying process.

3 Results and Discussions

The upper panel of Fig. 2 reveals the variation of birefringence intensity ($\overline{I}_c$) of LC droplets during the drying process. The standard deviation (n = 3) is higher in all the transition compared to the other regimes due to the presence of different sized LC droplets in different drops. It demonstrates an archetypal response of LC droplets dissolved in protein solutions and displays three distinct regimes: (I) the initial process of getting a minimum contact angle configuration with a gradual increase of intensity, (II) the intensity gets amplified due to LC activities, and (III) the intensity is saturated in the final regime once the drop completes the evaporation process. It also suggests that the birefringence nature is independent of types of proteins and less efficient to capture the minute details of the LC activities during the drying process. The stage I completes in 11–14 min, and the

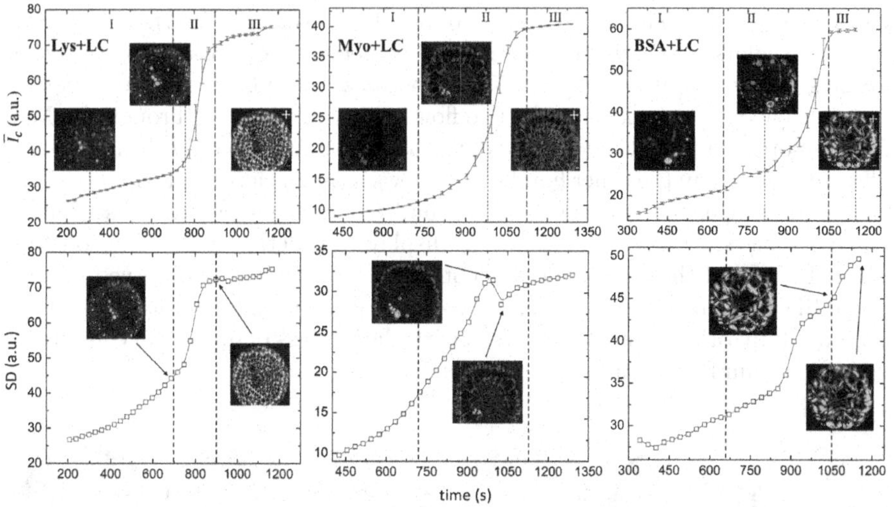

Fig. 2. Upper panel: Variation of birefringence intensity ($\overline{I}_c$) of LC droplets during the drying process. Three unique regimes are observed: (I) a steady upsurge, (II) a speedy rise, and, (III) an eventual saturation of intensity. Lower panel: variation of standard deviation (SD) of the images during different drying stages. (Color figure online)

rate of intensity in stage II is ~0.28 (for Lys), ~0.09 (for Myo), and ~0.08 (for BSA). A smooth transition regime in both Lys and Myo is observed, but not in BSA drops indicating a distinct final morphology in BSA. As mentioned earlier, BSA contains a higher molecular weight (~66.5 KDa) and the increase of mass in proteins decreases the rate of intensity change (one surprising finding is the drastic decrease from 0.28 to 0.09 in similar weighted proteins of Lys and Myo respectively). It is to note that LC has a strong dipole moment (polar) and the presence of a heme group without any disulfide bonds in Myo could interrupt the local interactions of Myo-LC differently than Lys-LC. Moreover, LC droplets wet the substrate (coverslip) differently for different proteins because of their difference in nature (mass, size, charges of proteins). Hence, it indicates that the mass is not only the contributing factor but involves many other interactions such as self-assembly among protein particles (protein-protein interaction), phase separation and self-assembly of LC droplets (protein-LC, LC-substrate), etc. [4].

The lower panel of Fig. 2 shows the SD (standard deviation of the image) of the same drops depicting a similar trend as a function of time. The $\overline{I}_c$ captures the whole process such as phase separation, formation of protein film, movement of fluid, water evaporation, etc., whereas, the SD is sensitive to changes in the image complexity [10]. The SD gives additional insights into the morphology in terms of changes in inhomogeneity and complexity. In Myo, it shows a difference in the evolution pattern from $\overline{I}_c$ by observing a dip and rise in the transition region. Though Myo protein film is faint red in color under cross-polarizing microscopy, it doesn't mean that it is optically active like the LC droplets (yellow in color). It means that Myo reflects a particular wavelength (red), and absorbs rest of the colors in the visible spectrum. This becomes advantageous over other proteins in our study i.e., Lys and BSA because the inhomogeneity is clearly distinguishable by calculating SD. The flow of the LC droplets through the Myo film promotes a dip, and the process of turning the film (into reddish) indicates a difference in SD. On the other hand, BSA shows a rise in the complexity of regime III. At this stage, the texture (SD) is found to be changing while distributing the LC droplets, whereas, the $\overline{I}_c$ fails to display such details and shows constant values. In spite of these differences, a common characteristic is observed in both $\overline{I}_c$ and SD by their increase through the drying process and saturation when the drops are fully dried. This is consistent with the process of droplet drying, phase separation, film formation, stain, and finally, film cracking.

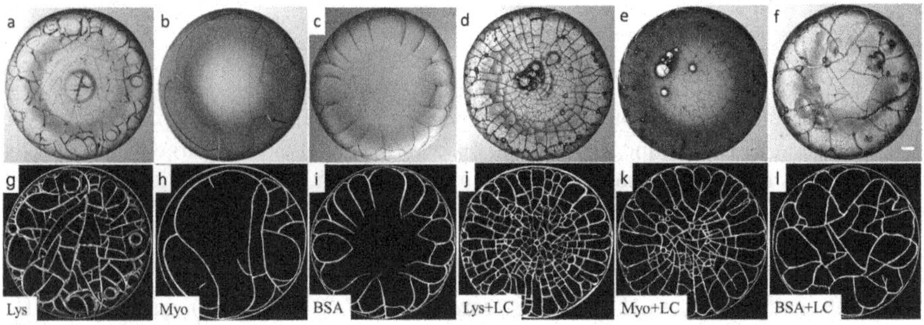

Fig. 3. Upper panel: After 24 hours, the images were captured in the bright-field configuration in the presence and absence of LC droplets (a–f). Lower panel: the processed binary images (g–l). The scale bar of 0.15 mm is shown in the right corner of [f].

Figure 3 shows the crack patterns and their processed binary images (in presence and absence of LC droplets). The common feature noticed in all the drops is the presence of "coffee-ring" [6] (upper panel of Fig. 3) confirming the general mechanism of the drying colloidal drops. The inclusion of the LC droplets affects the cracks' nature, and makes a rough protein film's surface. LC is believed to increase the stress in all the drops and the stress in turns produces more number of cracks. The central region of both Myo and BSA does not contain any cracks in the absence of the LC droplets (Fig. 3b, c). Contrary to that, the inclusion of the LC droplets in the protein drops leads to - (i) an ordered crack distribution and disturbs a 'mound'-like structure in the Lys drops (Fig. 3a, d), (ii) a formation of well-connected large and small crack domains in the Myo (Fig. 3b, e), and (iii) an overall increase of the cracks in the BSA and the formation of a few cracks in the central region (Fig. 3c, f).

To quantify the crack spacing (x_c) and to develop a physical mechanism of the underlying process of crack distributions, a non-parametric Mann-Whitney U and Kruskal Wallis tests were conducted. Any statistical difference in mean crack spacing ($\overline{x}_c$) is considered to be significant when $p < 0.05$. The significant

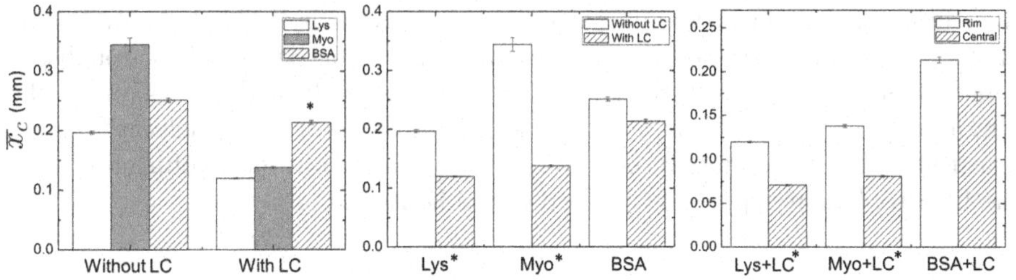

Fig. 4. The comparison of mean crack spacing ($\overline{x}_c$) in different conditions and environment. Left panel: comparison of $\overline{x}_c$ with and without LC droplets observed in the rim region. Middle panel: overall effect of the LC droplets in relation to $\overline{x}_c$ in all the protein drops. Right panel: comparison of $\overline{x}_c$ in the rim and the central region in different protein drops in presence of LC droplets. An asterisk [*] indicates significant pairs.

pairs are marked with an asterisk [*] in Fig. 4. All the experimental data are expressed as mean ± standard error (SE). The crack morphology discussed in Fig. 3 confirmed that the rim region of all the protein drops is specified for more number of cracks compared to the central region. We, therefore, conducted a series of statistical tests to examine the (significant) effects of LC droplets on all the proteins. The left panel of Fig. 4 shows the way LC droplets affect the x_c distributions in the rim region of all proteins. We did not observe any significant difference related to x_c distributions. The distribution is observed to be equally large in all the protein solutions without LC droplets. However, x_c is observed to be reduced (i.e., there is more number of cracks) when LC droplets are added. A pairwise comparison (using Dunn's procedure with a Bonferroni correction) of the proteins in the presence of LC droplets reveals that BSA still has significantly larger x_c compared to the other two protein drops (Lys and Myo). In other words, LC droplets do not have a significant effect in terms of x_c distributions on BSA even though the morphology shows a reduced x_c when LC is added (Fig. 3f, l). The middle panel of Fig. 4 further reveals that the presence of LC considerably reduces the x_c distributions (a large number of uniformed cracks) in all the proteins, however, the difference is found to be statistically significant in both Lys and Myo, but not in the BSA. Finally, we examined the way (inclusion of) LC droplets affect the x_c distributions in the rim and the central regions of all the protein drops (right panel of Fig. 4). The results confirm that the presence of LC considerably reduces the x_c in the central region of all the proteins. However, this reduction of x_c is observed to be significant in both Lys and Myo (but not in BSA). This outcome is evident as we have already seen that LC does not affect the BSA (in terms of x_c reduction). The quantification of x_c and the outcome of the statistical tests provide a convincing indication about the effect of LC on the proteins (under study) which otherwise is slightly misleading if the morphological observations are considered alone.

Following these findings discussed above, a physical mechanism is proposed. Our findings confirm that Lys and Myo follow a similar process of crack formation, unlike BSA. Each domain in these two proteins is buckled up, and most of the LC droplets are moved underneath the film from the cracks through the capillary action. The dark region at the center of each of these regions is the film attached to the coverslip. In contrast, BSA shows a radial defect (four-fan brushes) which is commonly observed at the LC-air interface confirming that the LC droplets are spread at the top of the BSA film [4].

4 Conclusions

Sessile drop drying is popularly used to explore the self-assembly and the phase separation of the particles in the colloidal drops. In this paper, we proposed an automated image-based mechanism to calculate birefringence of the optically active material and can be applied to any such colloidal drops. Further, the proposed semi-automated image processing mechanism of quantifying crack patterns can be applied to any drying system (with similar crack patterns) helping to understand the underlying physical mechanism. These methods have a broad range of applications quantifying image morphology in an efficient and economical way.

References

1. Brutin, D., Starov, V.: Recent advances in droplet wetting and evaporation. Chem. Soc. Rev. **47**(2), 558–585 (2018)
2. Chen, R., Zhang, L., Zang, D., Shen, W.: Blood drop patterns: formation and applications. Adv. Colloid Interface Sci. **231**, 1–14 (2016)
3. Bel'skaya, L.V., Sarf, E.A., Solonenko, A.P.: Morphology of dried drop patterns of saliva from a healthy individual depending on the dynamics of its surface tension. Surfaces **2**(2), 395–414 (2019)
4. Pal, A., Gope, A., Iannacchione, G.S.: A comparative study of the phase separation of a nematic liquid crystal in the self-assembling drying protein drops. MRS Adv. **4**(22), 1309–1314 (2019)
5. Pal, A., Gope, A., Kafle, R., Iannacchione, G.S.: Phase separation of a nematic liquid crystal in the self-assembly of lysozyme in a drying aqueous solution drop. MRS Commun. **9**(1), 150–158 (2019)
6. Deegan, R.D., Bakajin, O., Dupont, T.F., Huber, G., Nagel, S.R., Witten, T.A.: Capillary flow as the cause of ring stains from dried liquid drops. Nature **389**(6653), 827 (1997)
7. Preibisch, S., Saalfeld, S., Tomancak, P.: Globally optimal stitching of tiled 3D microscopic image acquisitions. Bioinformatics **25**(11), 1463–1465 (2009)
8. Schindelin, J., et al.: Fiji: an open-source platform for biological-image analysis. Nat. Methods **9**(7), 676–682 (2012)
9. Abràmoff, M.D., Magalhães, P.J., Ram, S.J.: Image processing with ImageJ. Biophotonics Int. **11**(7), 36–42 (2004)
10. Carreón, Y.J., Ríos-Ramírez, M., Moctezuma, R., González-Gutiérrez, J.: Texture analysis of protein deposits produced by droplet evaporation. Sci. Rep. **8**(1), 9580 (2018)

A Two Stage Framework for Detection and Segmentation of Writing Events in Air-Written Assamese Characters

Ananya Choudhury(✉) and Kandarpa Kumar Sarma

Department of Electronics and Communication Engineering, Gauhati University, Guwahati, India
ananya.apr@gmail.com, kandarpaks@gmail.com

Abstract. Gestural air-writing involves the process of writing continuous characters or words in free space using hand or finger motion. It differs from traditional pen-based writing from the fact that it does not contain delimiting points which helps in demarcation of valid writing segments. Thus, in gestural air-writing, detection of meaningful writing events from a continuous gestural sequence containing irrelevant writing movements is an intricate task which needs special attention. This paper presents an automatic method of gesture spotting and segmentation which identifies the meaningful air-written character segments confined within a continuous character pattern using a hybrid spatiotemporal and statistical feature set. A sliding window-based approach is employed for extracting the writing events from a continuous stream of hand-motion data, suppressing the superfluous idle data points. Consecutive writing events are then categorized into valid character segments and redundant ones. The relative performance of the proposed system is examined by taking various Assamese characters into consideration. Experimental results reveal that the proposed model achieves an overall segment error rate of 1.31%.

Keywords: Gestural air-writing · Gesture spotting · Gesture segmentation · Ligature · Human-machine interaction

1 Introduction

Human Computer Interaction (HCI) platforms have attained tremendous demand in today's world. In contrast to conventional human-machine interconnecting tools, gestural air-writing nowadays serves as an important substituent for natural and effortless HCI. It allows users to interact intuitively with computing devices by writing freely in an unrestricted and comfortable way [1].

However, gestural air-writing is different from conventional pen-based writing or 2D space handwriting in that the stroke flow is continuous and there are no intermediate pauses in between consecutive characters as well as different segments of a character [2]. For example, an in-air handwritten Assamese character is generally completed in a single continuous stroke, and some intermediate repositioning movements connect the adjacent strokes of the character. The connecting links which occurs in between adjacent characters as well as within individual characters are termed as

B. Deka et al. (Eds.): PReMI 2019, LNCS 11941, pp. 575–586, 2019.
https://doi.org/10.1007/978-3-030-34869-4_63

ligatures or movement epentheses. The presence of these ligatures makes the task of writing event detection and segmentation a challenging one. Moreover, these irrelevant connecting motions are diverse and vary widely depending upon users and their speed of articulation [1]. Therefore, the primary objective of this work is to implement a forward gestural character spotting and segmentation system which is user-independent and which shall form an important constituent for effective HCI.

Being a popular and intriguing research topic of current time, several in-air handwritten gesture spotting and segmentation techniques has been formulated by combining different algorithms. Chen et al. [3], Amma et al. [4] and Schick et al. [5] have utilized HMMs for modeling separate characters after which word recognition is performed by concatenating character HMMs with a repositioning HMM for describing the translocation that occurs between individual characters. Although HMM is a well-established method for modeling continuous characters, however with the increase in number of character patterns the model training becomes extensive as individual HMMs are constructed per label. Again, certain studies have considered alignment-based approaches for gesture spotting and recognition. For example, Frovola and Berman [6] and Jin et al. [7] have used most probable longest common sub-sequence (MPLCS) and dynamic time warping (DTW) approaches for segmenting out the potential character segments from gesture streams by measuring the similarity between extracted features of hand gesture template and a predefined template in the database. Alignment-based methods become computationally complex when the range of patterns increases, as these require warping of the temporal series with each and every template sequence in the database. Instead of HMMs or temporal alignment based methods for gestural sequence modeling, the proposed work achieves the same task by employing a distinctive feature set which determines the terminal points of character segments, extracts them and hence models the transience in their shapes, thus avoiding the need of extensive training or matching. Also, from survey it is seen that many systems have adopted preambles like manual gestures, physical buttons and other explicit signals for depicting the beginning and terminating points of character fragments in a temporal character stream. Murata and Shin [8] and Ayachi et al. [9] have utilized gestural commands, while Amma et al. [4] have used manual keys for temporal character segmentation. In contrast, our proposed method does not necessitate explicit preambles to specify the writing events in a continuous character sequence.

More specifically, the main contribution of this paper is to develop an efficient framework for gestural character spotting and segmentation by incorporating a two-stage approach. In the first stage, a window-based scheme is adopted to observe the fluctuations of a kinematic feature set and hence to spot the start and end points of different character segments existing inside a character pattern. In the second stage, spatiotemporal and statistical heuristics are applied to categorize the character segments obtained between terminal points as valid and ligature patterns.

The rest of the paper is organized as follows. In Sect. 2, the proposed methodology of air-written gesture spotting and segmentation is presented with elaborate description of all the individual processes involved. Section 3 discusses the results obtained through experimental evaluation on a large dataset with dynamic variations. Finally Sect. 4 concludes the paper and highlights some future scope of the present work.

2 Proposed System

The overall schematic block diagram of the proposed gestural air-writing detection and segmentation framework for spotting and identifying the valid strokes from a character sequence is shown in Fig. 1. The working of the first two modules of the proposed system i.e. hand segmentation and hand tracking are elaborated in [10].

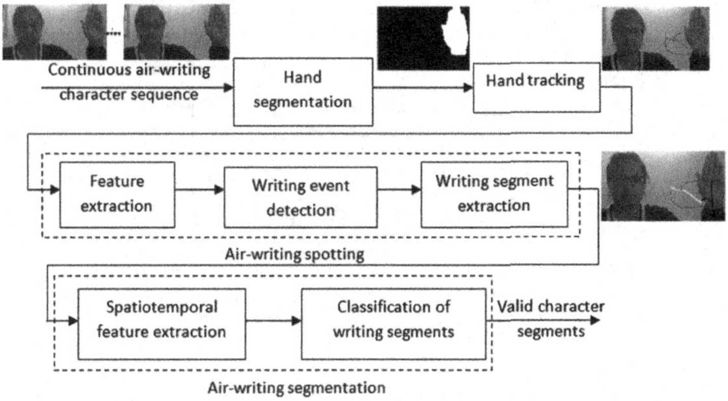

Fig. 1. Schematic block diagram of proposed air-writing spotting and segmentation system.

This work mainly concentrates on the air-writing spotting and segmentation tasks. The detailed working methodology of these processes is described as follows.

2.1 Gestural Air-Writing Spotting and Segmentation

The process of air-writing detection and segmentation requires recognizing the relevant stroke segments in a continuous character pattern. So for automatic spotting and segmentation of air-writing, we have amalgamated certain distinctive spatiotemporal features derived from hand tracking signals, and have employed a window-based approach for modeling the statistical variability in character patterns. The detailed functioning of the air-writing spotting and segmentation modules are described in the following sections.

Air-Writing Spotting. The workflow of the proposed air-writing detection module is shown in Fig. 2. Firstly, each sample of the hand trajectory is represented by positional coordinates $p_i = (x_i, y_i)$. Then, a set of motion features is derived and a sliding window is superimposed over this motion data to look for the presence of a writing event. The general characteristics of writing events include acute changes in writing direction and high average velocity in comparison to idling events. So, a window (of length L) is positioned at each designated frame by sliding it through a shift width (w), and the spotting algorithm observes the desired properties within these windows to determine the boundary points (frames) of all the writing segments inside a character pattern.

Here, we have empirically selected window length of 30 frames (2 s) and shift width of 2 frames (0.13 s).

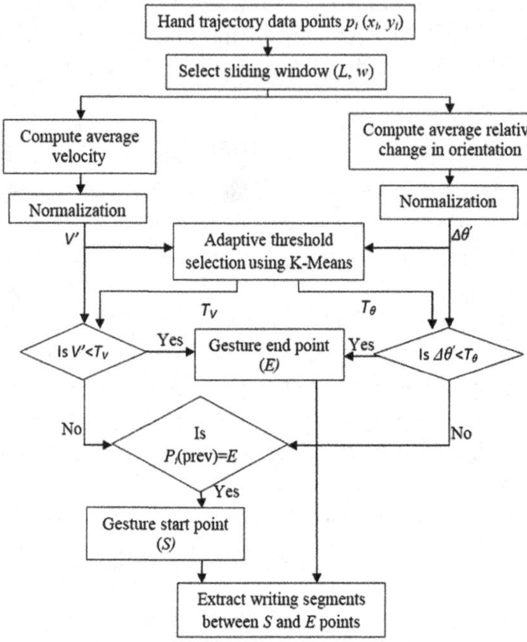

Fig. 2. Block diagram of proposed air-writing spotting module

On the selected sliding windows, the following motion features are estimated:

- Average velocity

$$\overline{v} = \frac{1}{L}\sum_{i=1}^{L} |v_i| \tag{1}$$

- Average angular change in orientation

$$\overline{\Delta\theta} = \sum_{i=1}^{L} \Delta\theta_i \tag{2}$$

where, L indicates the number of samples inside the window, v_i and $\Delta\theta_i$ denotes the velocity and angular change in orientation for a sample. These features are then normalized to reduce the effect of variations in writing style and speed. The normalized features are given by

$$V' = \frac{\overline{v}}{v_{\max}} \tag{3}$$

$$\Delta\theta' = \frac{\overline{\Delta\theta}}{\Delta\theta_{\max}} \tag{4}$$

where, v_{max} and $\Delta\theta_{\max}$ denotes the maximum velocity and angular change obtained within an observation window.

Practically, it is observed that while air-writing a character segment these feature values tend to increase, while towards the end of a writing segment the values of these parameters decreases and reaches a minimum value. However, since there are large number of character patterns having wide variations in writing speed and shape, so global threshold values for V' and $\Delta\theta'$ shall not be fruitful for demarcating the start and end points of all the character patterns in the database. So, the computed features (V' and $\Delta\theta'$) for a sample are compared with adaptively computed thresholds (T_V and T_θ respectively) determined using K-Means clustering algorithm. Hence, a trajectory sample is designated as start point (S) or end point (E) of a writing segment based on whether the kinematic parameters (V' and $\Delta\theta'$) are greater or less compared to their respective adaptive thresholds (T_V and T_θ). The resulting end points signify non-writing events and are thus eliminated from the character stream. Finally, the handwriting (HW) segments procured between each pair of boundary points are separated out for further processing.

Algorithm for determination of adaptive thresholds (T_V and T_θ) using K-Means clustering approach:

1. Choose a window W_{in} having length of 6 samples, and consider the feature values V'_i and $\Delta\theta'_i$ ($i \in 1$ to 6) inside this window.
2. Take two clusters CL_1 and CL_2 and initially assume their cluster centers (C_1 and C_2) as the 2nd and 4th feature points (i.e. V'_2, $\Delta\theta'_2$ and V'_4, $\Delta\theta'_4$).
3. Calculate the distance of all the feature points (V'_i and $\Delta\theta'_i$) within W_{in} from the cluster centers (C_1 and C_2) using Euclidean distance.
4. Assign each of the feature points to the cluster CL_1 or CL_2 with nearest cluster center.
5. Update the cluster centers (C_1 and C_2) by taking the mean of all the feature points within the respective clusters (CL_1 and CL_2).
6. Repeat steps 3, 4 and 5 until the cluster centers C_1 and C_2 do not change.
7. Compute adaptive thresholds (T_V and T_θ) by taking the mean of the converged cluster centers obtained in step 6.

Air-Writing Segmentation. After successful extraction of HW segments, the next task is to classify them into valid and ligature patterns so that the ligatures can be suppressed and the valid patterns can be passed on for recognition of the overall character pattern. The flow diagram of the gestural air-writing segmentation module is shown in Fig. 3.

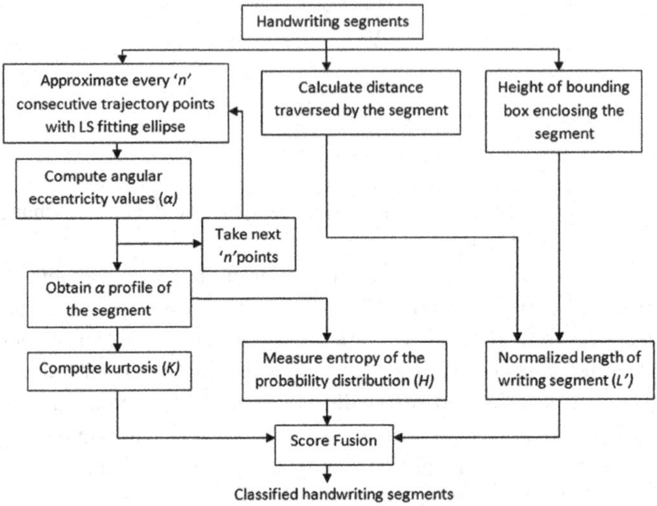

Fig. 3. Block diagram of proposed air-writing segmentation module

In our work, we are concerned with modeling the variations which occurs in the shape of Assamese characters. To consummate this task, a composite feature set is proposed comprising of kurtosis, entropy and normalized length of HW segment. These features are described as follows.

- Kurtosis (*K*): The kurtosis of a distribution is given by the ratio of fourth moment (m^4) and second moment (m^2) squared [11]

$$K(x) = \frac{m_4}{m_2^2} = \frac{m_4}{(\sigma^2)^2} = \frac{1}{N}\sum\left(\frac{x-\mu}{\sigma}\right)^4 = E(z^4) \quad (5)$$

where z (= $x - \mu/\sigma$) is standardized value, x represents the observation values, μ is the mean, σ is the standard deviation and E() is the expectation operator. The K value for a normal distribution is 3 and the K value for all other distributions reflects the variations of observations from a normal distribution. Equation (5) has a simpler interpretation and is given as [11]

$$K(x) = var(z^2) + 1 \quad (6)$$

where *var()* is the variance. Accordingly, kurtosis is inferred as the extent to which observations are dispersed away from the shoulders of a distribution (z^2 = 1, i.e. $z = \pm 1$).

In this study, we take scaled inverse kurtosis (K') as the first feature for character segmentation. It is given as

$$K'(x) = \frac{c}{K(x)} \tag{7}$$

where c is a scaling constant empirically taken as 6, and $K(x)$ is the kurtosis determined using Eq. (6). This implies that, if the observation values x of a character segment are more dispersed from the shoulders, then $K < 3$ and hence the K' value will be more compared to the segments exhibiting less dispersion which have $K \geq 3$.

- Entropy (H): Entropy gives a measure of uncertainty of probability distribution and is given by [12]

$$H = -\sum_i P_i \log P_i \tag{8}$$

where P_i denotes the probability of observation values.
- Normalized segment length (L'): The normalized length of HW segment is given by

$$L' = \frac{\|d\|}{h} \tag{9}$$

where d is the total distance traversed by the HW segment and h is the height of a minimum-area bounding rectangle [13] encompassing the segment.

So, in this procedure we firstly approximate every consecutive "n" points of the HW segment with a least square (LS) fitting ellipse [13] and compute its angular eccentricity values, which describe the variations in shape of the HW segments. The angular eccentricity of a point $P(x,y)$ on an ellipse with semi-major axis of length a and semi-minor axis of length b is given by

$$\alpha = \tan^{-1}\left(\frac{ay}{bx}\right) \tag{10}$$

From the α profile of a HW segment, the inverse kurtosis value ($K'(\alpha)$) and entropy of its probability distribution ($H(\alpha)$) is calculated using Eqs. (7) and (8). Simultaneously, the normalized length of the HW segment (L') is computed using Eq. (9). Now, in gestural air-writing of Assamese characters the valid character segments shall have more variations in α value than the ligatures. Ligatures are generally simple and have less variability. So, according to the concept of kurtosis, the valid HW segments will have K' value more compared to the ligature segments. Concurrently, the valid HW segments will have higher amount of uncertainty in α values i.e. more information content, and hence will have higher entropy value than ligature segments. Further, the ligature segments being simple in nature will inherently have shorter normalized length than valid segments. Here, we have considered combination of three features (K', H and L') for character segmentation, because in certain cases if a single feature fails to capture the distinguishing traits of valid and ligature patterns, then the score from the remaining features shall aid in correct character segmentation. Thus, the final score of a HW segment is given by the weighted sum of kurtosis, entropy and signal length.

$$score = \omega_1 K^{'} + \omega_2 H + \omega_3 L^{'} \tag{11}$$

Here, the weights ω_1, ω_2 and ω_3 are taken uniformly as 1.

Thus, the final score of all the HW segments inside a character pattern are computed using Eq. (11), and the decision threshold (T_D) is determined by taking the average of the first two smallest scores. Suppose, $C = \{C_1, C_2, C_3 ... C_n\}$ is a character pattern consisting of n HW segments, then a character segment is classified as valid ('1') or ligature pattern ('0') using the following decision equation

$$D(C_i) = \begin{cases} 1, & \text{score}(Ci) > T_D \\ 0, & \text{otherwise} \end{cases} \tag{12}$$

In case of characters consisting of only one HW segment, i.e. without any connecting links, only the kurtosis value (K) computed using Eq. (6) is taken as discriminating feature for classifying it into valid or ligature pattern. If the K value is less than 3, it implies that the character segment has wide variations and hence it will be designated as valid pattern, conversely it will be classified as ligature pattern.

3 Experimental Results and Discussions

We evaluate the performance of our proposed air-writing spotting and segmentation model on an Assamese character dataset. The experimental database consists of 46 Assamese characters, comprising of 11 vowels, 10 numerals and the first 25 consonants. Each character is written 20 times by the 2 subjects, thus forming an overall mixed dataset of 1840 patterns. Thus, the database encompasses dynamic variations in trajectory shape and duration of the character patterns. The video corpus of gestural characters is generated using a webcam with frame rate of 15 frames/s and resolution of 640 × 360. The following sections describe the results derived from our gestural air-writing detection and segmentation modules.

3.1 Air-Writing Spotting Results

In this section, we depict the profile distributions for the feature set (V' and $\Delta\theta'$) employed for gesture spotting and the corresponding writing event detection. Figure 4 (a) and (b) shows the distributions of normalized velocity (V') and relative angular change in orientation ($\Delta\theta'$) used for gesture spotting of a continuous character "ঠ". It is observed that during the commencement and eventual writing of a HW segment the feature values increase, while towards the end of the writing event these values tend to decline and attains a minimum point.

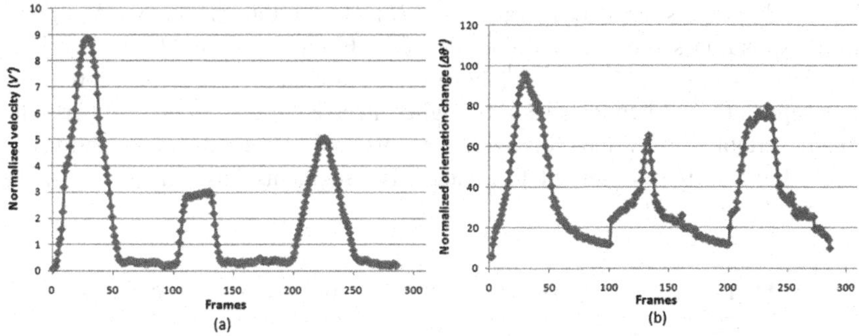

Fig. 4. Variation of (a) V' and (b) $\Delta\theta'$ profiles for continuous character "ঞ"

Figure 5 illustrates a few occurrences of the air-writing path of character pattern "ঞ" along with the start and end points (*S* and *E*) of writing segments obtained by employing the proposed feature set and window-based analysis. So, at the end of gesture spotting it is seen that we obtain three HW segments for the pattern "ঞ", which has to be classified as valid and ligature in the next stage.

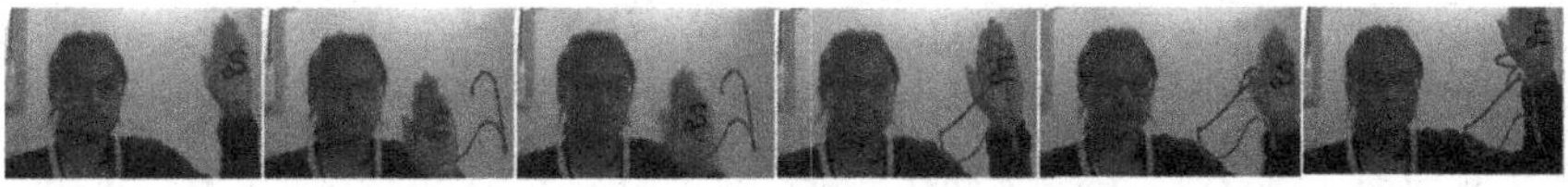

Fig. 5. Hand tracking output at few instances of continuous character pattern "ঞ" depicting the spotted start and end points (S and E)

3.2 Air-Writing Segmentation Results

Figure 6 shows the angular eccentricity (α) profile which outlines the variations in shape of all the three HW segments of the character pattern "ঞ".

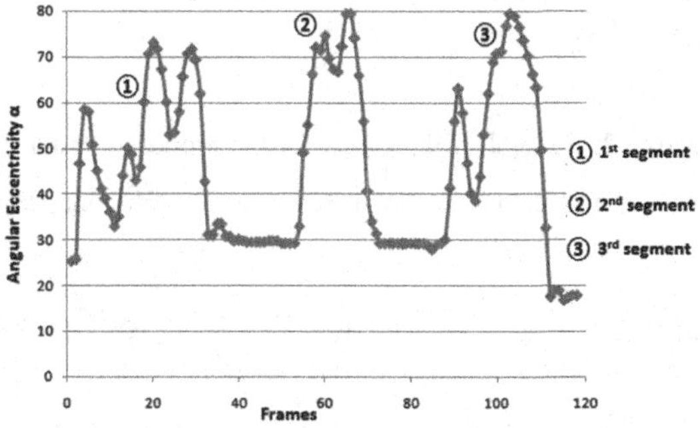

Fig. 6. Variation in angular eccentricity (α) values for continuous character "ঞ"

Table 1 highlights the features values (K', H and L') for all the three HW segments, their corresponding scores and the final decision. From Fig. 6 and Table 1, it is seen that the valid segments (1st and 3rd) have more variations in α values compared to ligature segments, and hence these have higher inverse kurtosis value. Similarly, there is more uncertainty in α values for valid segments, and hence it has more entropy than ligatures. Further, it is observed that valid HW segments have greater normalized length than ligature patterns.

Table 1. Character segment classification results for character pattern "ও"

HW segments	K'	H	L'	$Score$	T_D	Decision
1st segment	3.585	4.704	2.805	11.094	9.452	Valid
2nd segment	2.02	4.468	1.415	7.903		Ligature
3rd segment	3.00	5.143	2.859	11.002		Valid

3.3 Performance Evaluation

We evaluate the performance of our proposed system by calculating the overall Segment Error Rate (SER). For a character pattern C_i, the SER is given by

$$SER = \frac{E}{N_C} = \frac{S + D}{N_C} \tag{13}$$

where N_c is the total number of HW segments in the character pattern, E is the total number of segment errors in the character pattern. It is given by the sum of substitution (S) and deletion (D) errors [14]. Therefore, the overall SER is given by

$$\text{Overall SER} = \frac{\sum_{i=1}^{N} SER_i}{N} \times 100\% \tag{14}$$

where N indicates the total number of character patterns in the database.

The composite SER of our proposed system with and without score fusion is presented in Table 2.

Table 2. Overall SER results for the proposed system

Assamese character dataset	With score fusion	Without score fusion		
		K	H	L'
Overall SER	1.31%	3.5%	4.2%	8%

By comparing the results obtained using score fusion to the results acquired considering individual features, it is seen that there is a considerable improvement in performance while considering combined features. This is because, in some characters there are certain valid character segments which have very less variations, and in such cases the system will fail to resolve the ambiguities and falsely interpret it as ligature if only individual features are taken into consideration. The idea of aggregating three features into one platform helps in producing comparatively lesser SER than considering single features for classification.

4 Conclusion

In this paper, we have formulated a vision-based forward air-writing detection and segmentation mechanism for determining the principle character segments from continuous character patterns. For gesture spotting and character segmentation, we have implemented a sliding window-based approach and applied a mixed feature set by integrating spatiotemporal and statistical parameters. Experimental evaluation on continuous Assamese character patterns reveal that our proposed technique offers an overall segment error rate of around 1.3%, thereby demonstrating its effectiveness in extricating out legitimate portions from a character sequence. In this study, the efficiency of the spotting and segmentation paradigm has been validated for the vowels, numerals and a few consonants of Assamese vocabulary. Later, the remaining consonants and a few words shall be taken into consideration, and the database shall be upgraded by including more samples from different participants. In future work, a challenging task related to this field of study deals with the recognition of certain Assamese characters having similar trajectory pattern. In real world scenarios, the proposed system in conjunction with a recognition module can function as a complementary modality for applications such as communicative aid for hearing-impaired people, smart media controller, game controllers and yet more.

References

1. Choudhury, A., Sarma, K.K.: Visual gesture-based character recognition systems for design of assistive technologies for people with special necessities. In: Handmade Teaching Materials for Students with Disabilities, pp. 294–315. IGI Global, Hershey (2019)
2. Agarwal, C., Dogra, D.P., Saini, R., Roy, P.P.: Segmentation and recognition of text written in 3D using leap motion interface. In: 3rd Proceedings on IAPR Asian Conference on Pattern Recognition (ACPR), pp. 539–543. IEEE, Kuala Lumpur (2015)
3. Chen, M., AlRegib, G., Juang, B.-H.: Air-writing recognition—Part I: modeling and recognition of characters, words, and connecting motions. IEEE Trans. Hum.-Mach. Syst. **46** (3), 403–413 (2016)
4. Amma, C., Gehrig, D., Schultz, T.: Airwriting recognition using wearable motion sensors. In: Proceedings of Augmented Human International Conference. ACM (2010)
5. Schick, A., Morlock, D., Amma, C., Schultz, T., Stiefelhagen, R.: Vision-based handwriting recognition for unrestricted text input in mid-air. In: Proceedings of the 14th International Conference on Multimodal Interaction, pp. 217–220. ACM (2012)
6. Frolova, D., Stern, H., Berman, S.: Most probable longest common subsequence for recognition of gesture character input. IEEE Trans. Cybern. **43**(3), 871–880 (2013)

7. Jin, L., Yang, D., Zhen, L.-X., Huang, J.-C.: A novel vision-based finger-writing character recognition system. J. Circ. Syst. Comput. **16**(3), 421–436 (2007)
8. Murata, T., Shin, J.: Hand gesture and character recognition based on kinect sensor. Int. J. Distrib. Sens. Netw. **10**(7), 278460 (2014)
9. Ayachi, N., Kejriwal, P., Kane, L., Khanna, P.: Analysis of the hand motion trajectories for recognition of air-drawn symbols. In: Proceedings of 5th International Conference on Communication Systems and Network Technologies, pp. 505–510. IEEE, Gwalior (2015)
10. Choudhury, A., Sarma, K.K.: A novel approach for gesture spotting in an Assamese gesture-based character recognition system using a unique geometrical feature set. In: Proceedings of 5th International Conference on Signal Processing and Integrated Networks (SPIN), pp. 98–104. IEEE, Noida (2018)
11. Liang, Z., Wei, J., Zhao, J., Liu, H., Li, B., Shen, J., Zheng, C.: The statistical meaning of kurtosis and its new application to identification of persons based on seismic signals. Sensors **8**(8), 5106–5119 (2018)
12. Wang, Q.A.: Probability distribution and entropy as a measure of uncertainty. J. Phys. A: Math. Theor. **41**(6), 065004 (2008)
13. Bradski, G., Kaehler, A.: Learning OpenCV: Computer Vision with the OpenCV Library. O'Reilly Media Inc., Sebastopol (2008)
14. Elmezain, M., Al-Hamadi, A., Sadek, S., Michaelis, B.: Robust methods for hand gesture spotting and recognition using hidden markov models and conditional random fields. In: Proceedings of the 10th International Symposium on Signal Processing and Information Technology, pp. 131–136. IEEE, Luxor (2010)

Modified Parallel Tracking and Mapping for Augmented Reality as an Alcohol Deterrent

Prabhanshu Purwar, M. K. Bhuyan, Kangkana Bora(✉), and Debajit Sarma

Indian Institute of Technology Guwahati, Guwahati, India
{prabh174102058,mkb,kangkana.bora,s.debajit}@iitg.ac.in

Abstract. Drug addiction among the population is a major social menace. This paper proposes an innovative method to offer a promising deterrent against alcohol abuse using Augmented Reality (AR) as a tool. The proposed method aims at traumatizing an alcoholic person by augmenting multiple multimedia objects in the vicinity of a person. A modified Parallel Tracking and Mapping (PTAM) technique has been proposed to create an AR workspace. For this, we have proposed an improvement in the Bundle Adjustment. A novel concept of map independence is incorporated in the PTAM, providing freedom of augmenting multimedia in the desired position. The major contribution of this paper is that a versatile AR platform is developed for a monocular camera-based system for educating alcoholic persons for overcoming alcohol addiction.

Keywords: Alcohol deterrent · Graph SLAM · PTAM · Levenberg-Marquardt algorithm · Bundle adjustment

1 Introduction

In the name of modernization, alcohol has become a routine in our society [1]. Alcoholism generally leads to an increase in joblessness, assaults and mental health issues. There is also the larger impact on the community, schools, the workspace, the health care system and on society as a whole. To combat such pity conditions, we have introduced an Augmented Reality (AR) based platform which offers a promising deterrent against alcohol usage. The aim of this method is to address the problem of alcohol consumption via warnings, anti-alcohol slogan, and multimedia. All these warnings in a display are given with an objective to traumatize an alcoholic person. In this method, AR objects such as images, videos, animations, and 3D symbols (combinely termed as multimedia) are projected in a nearby workspace of the alcoholic person. The content of the multimedia can be guided upon by an anonymous person or system administrator/educator. Any kind of appropriate multimedia can be displayed to the alcoholic person via AR gadget. The administrator can set the relative positions of the multiple AR multimedia objects. He has full control over the placement

B. Deka et al. (Eds.): PReMI 2019, LNCS 11941, pp. 587–595, 2019.
https://doi.org/10.1007/978-3-030-34869-4_64

of multimedia objects/contents in a display window. Surrounding persons can also convey some anti-alcohol slogans or messages for the alcoholic person.

This work focuses on augmentation using a monocular camera, and so, only the closely related literatures in this particular domain are explored to propose our algorithm. Reid *et al.* [6] worked on mono Simultaneous Localization and mapping (SLAM) for real time localization and mapping of objects using freely moving camera. Thrun *et al.* [7] also used monocular camera to work on graph SLAM to acquire 3D maps of large scale urban workspace. To the best of our knowledge, our work is first-of-this-kind where monocular camera-based AR is applied as alcohol deterrent.

In achieving the objectives, an advance SLAM technique, *i.e.*, graph SLAM is employed along with Parallel Tracking and Mapping (PTAM) suiting the purpose of camera-based AR. The problem is quite challenging due to the limited field of view of the camera. The proposed method uses PTAM as its core [4], as the parallel nature of PTAM ensures real time implementation. We have modified the Levenberg-Marquardt Bundle Adjustment (BA) used in PTAM based on initial pose and convergence. Modification is inspired by the work of Transtrum *et al.* [8], which makes BA to perform faster, reflects positively in PTAM. Another major contribution of the work is the concept of map independence. Freedom is being assigned to the administrator to put multimedia anywhere and anytime irrespective to the instant of Map creation.

2 Proposed Methodologies

In order to make a versatile real time application, all the advantages of PTAM [4] are being incorporated in our proposed platform. The position of an AR object is being set up by the administrator. We used directions like "towards the north", "towards the northeast" and so, to represent position of an object in world coordinate system. AR object is being placed on the plane in the vicinity of the selected location.

Freedom is being assigned to the administrator to put multimedia anywhere and anytime irrespective of the instant of Map creation, this is termed as the 'map independence concept'. AR object is being placed on a plane in the vicinity of the selected location. A Gaussian with rotated cone-mirror like region is being used as the probability estimator. As soon as the region comes under the direct sight of view of the person (*i.e.,* first time in the map) the augmented object is placed there and the colour of location changes from white to green, as shown in Fig. 5.

Improvement in the BA is being made to make the overall system less computationally complex. There is a complementary relationship between the convergence speed and robustness to initial pose [8]. In situations where the stability of a pose is a primary concern (near AR objects), more priority is assigned to the poses rather than convergence speed. On the other hand, in the situations where speed is a primary concern (away from the AR objects), fast BA method should be employed.

Improvement is achieved using "geodesic acceleration" correction to the BA suggested by Transtrum *et al.* [8]. This has the tendency to improve success, fit quality and speed as being suggested. In addition, Bold acceptance and Broyden's update are included for a higher convergence speed.

Figure 1 shows the flow of our proposed scheme in terms of mapping threads. A map is created after initialization for the first time for PTAM [4].

Lock Pose in this flow diagram refers to the keyframes which get recorded when an observer is in direct sight of the AR object placed by the administrator.

"Improved global bundle adjustment" and "Improved local bundle adjustment" refer to bundle adjustment with geodesic acceleration while "Fast local bundle adjustment" refers to bundle adjustment with geodesic acceleration, along with Bold acceptance and Broyden's update [8].

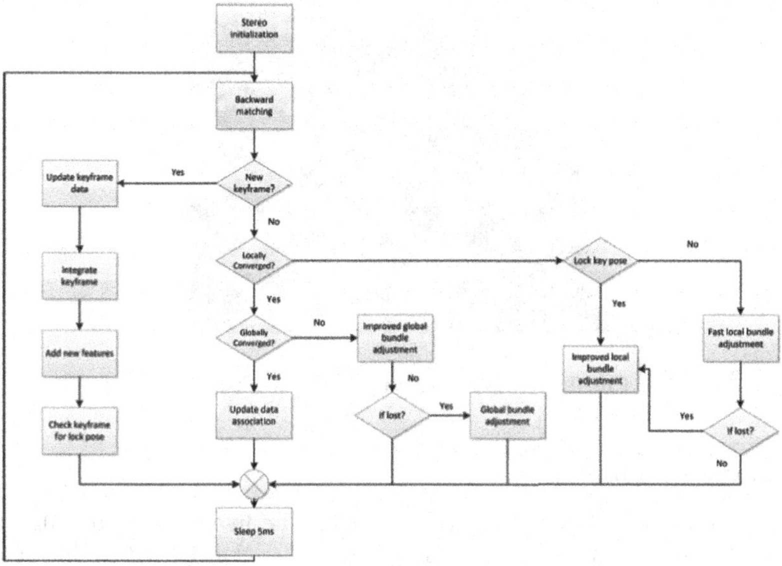

Fig. 1. Mapping thread flow diagram

Some other improvements are also performed for PTAM [4] to increase the speed and stability in the proposed work. For detection and decomposition of homography, the more accurate Zhang observation [9] is used in place of Fauger's observation [2,3]. To improve computational speed, Gram matrix accumulators are used rather than Least Square (LS) matrix. Gram matrix accumulator provides a computationally efficient way to perform matrix calculations such as multiplication which were more prominent than LS matrix used in original PTAM. Gram matrix with the sum for product (SoP) avoids large matrices. Rather than using straight-line in PTAM, pose of the camera is modeled using 3D Gaussian-like region. This method is employed to deal with pose uncertainty. PSD of a matrix is being checked using Cholesky Decomposition to avoid encountering any singular matrix, and further failure of the system. In order to get rid of the vanishing gradient effect, zero determinant check is made in our method. Additionally, to enhance applicability of the proposed scheme, multiple AR objects augmentation in the 3D map in different planes is implemented, along with real-time AR motion and animation. Experimental results are elaborately highlighted in the following section to show the efficacy of the proposed scheme.

3 Experimental Results

The experiment is mostly performed with a hand-held web camera in real-time. Figure 2 shows the set-up of the working workspace. All the experiments are performed in OpenCV platform, which uses TooN, libCVD, and GVars traditionally.

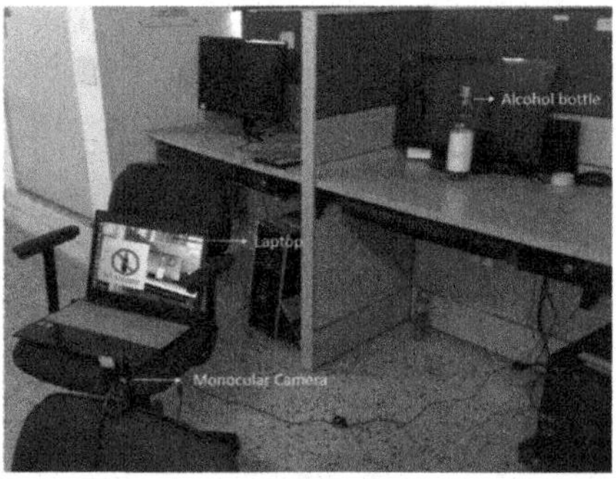

Fig. 2. Set up of the augmented workspace.

3.1 Initialization Step

Initialization steps are demonstrated in Fig. 3. The five-point stereo algorithm is employed for initialization, which requires 2D patches traced during smooth translation motion of the camera. Purple points represent the FAST features at a point, a transition of colour from yellow to red represents successful tracking of features in the preceding frames. The frame gets constructed, as soon as initialization completes successfully, upon which the AR object renders.

Fig. 3. Initialization steps. (Color figure online)

3.2 AR Object Augmentation

AR objects include 3D symbol and message, which are placed in the set of frames during initialization (see Fig. 4). Initially, the alcoholic person is not in the direct sight of view of the object placed by the administrator, and so, the estimated position of the object is depicted by a white circle in the 3D map (Fig. 4).

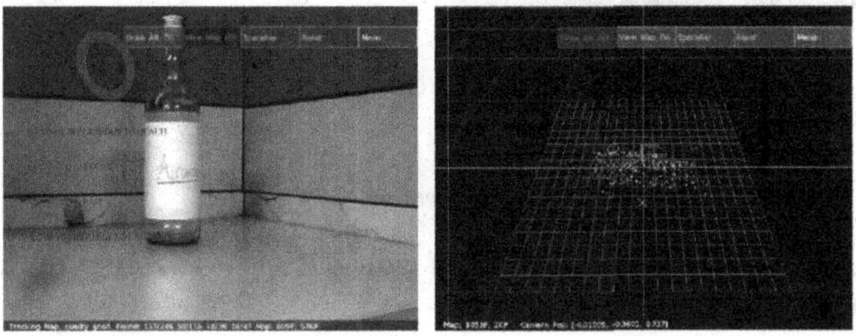

Fig. 4. AR object along with current position of the camera in 3D map.

As soon as, the alcoholic person is in direct sight of an AR object placed, colour of position changes from white to green in the 3D map and vice-versa (Fig. 5).

Fig. 5. Current position of the camera in 3D map when the person is in direct sight of view of the AR object set by the administrator.

3D map shows the exact trajectory and positions of the camera. Due to their simplicity and attractiveness, graphics are kept same as in PTAM (Fig. 6).

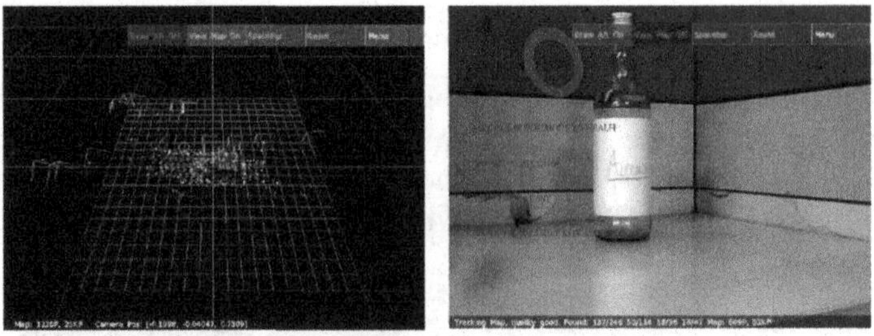

Fig. 6. Trajectory of the camera in the 3D map.

3.3 AR Object to Traumatize Person

Fig. 7 demonstrates the augmentation of multiple images in order to traumatize the person.

Fig. 7. Augmentation of multiple images in order to traumatize the person.

3.4 Moving AR 3D Slogan

Figures 8 and 9 demonstrate the augmentation of a message and 3D sign along with the motion in the 3D workspace. In our proposed method, full freedom is given for displaying of the moving objects and animation, where motion is independent of the camera movement (Fig. 8).

Rendering of the object is done in the 3D workspace which makes it independent of the camera position as shown in Fig. 9, where the position of the AR object is stagnant with respect to the camera movement.

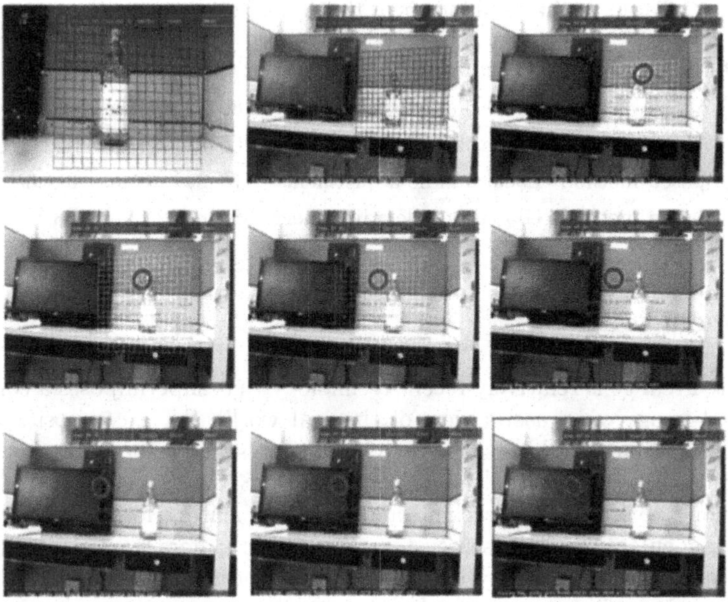

Fig. 8. Augmentation of a message and 3D sign along with the motion in the 3D workspace.

Fig. 9. Stagnant position of an AR object with the camera motion.

3.5 Bundle Adjustment Timing Comparison

It is very difficult to achieve most computationally efficient individual steps in PTAM [4] due to difference in scene structure, map size and asynchronous nature of method. The proposed method employs bundle adjustment in PTAM, and so, the overall computational time is reduced in our proposed method. Table 1 shows bundle adjustment time for both PTAM and our proposed scheme. In this framework, we have concentrated upon increasing stability as well as the speed of the algorithm in multiple AR workspace. The reason for the degradation of performance in 50–99 frames is that during these frames our algorithm gives precedence to the stability of the next AR object being introduced rather than speed. In contrary, the original PTAM was designed for a single workspace and the other workspaces remains unstable.

Table 1. Bundle adjustment timing for various map sizes

Method applied	Key-frames	2–49	50–99	100–149
PTAM	Local bundle adjustment	189 ms	280 ms	460 ms
PTAM	Global bundle adjustment	400 ms	1.9 s	7.3 s
Proposed method	Local bundle adjustment	176 ms	290 ms	430 ms
Proposed method	Global bundle adjustment	360 ms	2.1 s	6.8 s

4 Conclusion

Augmented reality is relatively recent technology that superimposes a computer -generated image on a user's view of the real world. So, a composite view is obtained. This paper addresses a very important social issue with the help of augmented reality. We made use of the advance SLAM technique, *i.e.*, graph SLAM along with parallel tracking and mapping suiting for the implementation in monocular camera-based system. As PTAM suffers from the problem of the computational complexity of BA, it is efficiently addressed in this paper. The proposed platform is successfully tested by augmenting images and objects; generating animations; and creating multiple AR objects in a 3D workspace. In terms of social importance, the proposed AR platform provides an efficient way to traumatize an alcoholic person. The proposed platform can be effectively employed as an assistance for prevention of alcoholism.

However, there is still a scope to make the proposed AR system more user friendly for most common people. Also the motion blur along with the repeated structure produce a large number of outliers, and these issues mainly constitute future research in this direction [4].

References

1. Ambekar, A., et al.: Magnitude of substance use in India. Technical report Ministry of Social Justice, Empowerment, Government of India. A National Survey on Extent, and Pattern of Substance Use in India (2019)
2. Devernay, F., Faugeras, O.: Machine Vision and Applications. Morgan Kaufman, Elsevier, Burlington (2001)
3. Hartley, R.I., Zisserman, A.: Multiple View Geometry in Computer Vision, 2nd edn. Cambridge University Press, Cambridge (2004)
4. Klein, G., Murray, D.: Parallel tracking and mapping for small AR workspaces. In: Proceedings of Sixth IEEE and ACM International Symposium on Mixed and Augmented Reality (ISMAR 2007). ACM (2007)
5. Noar, S.M., et al.: Pictorial cigarette pack warnings: a meta-analysis of experimental studies. J. Tob. Control **25**, 341–354 (2016)
6. Reid, I.D., et al.: MonoSLAM: real-time single camera SLAM. IEEE Trans. Pattern Anal. Mach. Intell. **6**, 1052–1067 (2007)

7. Thrun, S., Montemerlo, M.: The graph SLAM algorithm with applications to large-scale mapping of urban structures. Int. J. Robot. Res. **25**, 403–429 (2006)
8. Transtrum, M.K., Sethna, J.P.: arXiv preprint arXiv:1201.5885 (2012)
9. Zhang, F.: Matrix Theory: Basic Results and Techniques. Springer, Heidelberg (2001)

Detection of Handwritten Document Forgery by Analyzing Writers' Handwritings

Priyanka Roy(✉) and Soumen Bag

Department of Computer Science and Engineering, Indian Institute of Technology, (Indian School of Mine), Dhanbad 826004, Jharkhand, India
priyankarooy@gmail.com, soumen@iitism.ac.in

Abstract. Since digitization is yet to be adopted globally, handwritten documents are still in use in many places. Handwritten documents are prone to get forged thanks to acts like the versatility of tampering which are very frequent among skilful fraudsters. Our research work focuses on one of the major problems to detect whether a document is treated as false or not based on an analysis of the handwriting of the content writers. Mostly, legal documents are scripted authentically by a single person. If the content is a combination of more than one person, then it will be treated like a forged document. The proposed work is formulated as a binary classification problem. Various contour related sliding window based features are extracted from word images of the corresponding handwritten document. The same writer with different handwriting styles are also considered here as well. Bagging meta-classifier is trained for classification of the extracted features. The accuracy of this proposed work is 89.64% on IAM dataset is quite sound. We have also tested our method on IDRBT check image dataset. However, since there is a lack of direct implementation on this particular problem we could not make a comparative analysis of the proposed method.

Keywords: Bagging · Contour based feature · Forged document · Handwritten document

1 Introduction

Document falsification is one of the most wanted juridical crimes in countries. Criminals somehow escape from laws due to the existence of insufficient rules in the law of enforcement. Therefore, the examination techniques of forensic documents cannot meet the desired results. Exact implementation unit of a particular problem is not always available since problems being discovered may differ from the existing ones. The article aims to seek about finding forgery activity, especially in legitimate documents. Practically, human beings are usually in writing and written documents. Legal contracts like wills, bills, etc. are handwritten in

B. Deka et al. (Eds.): PReMI 2019, LNCS 11941, pp. 596–605, 2019.
https://doi.org/10.1007/978-3-030-34869-4_65

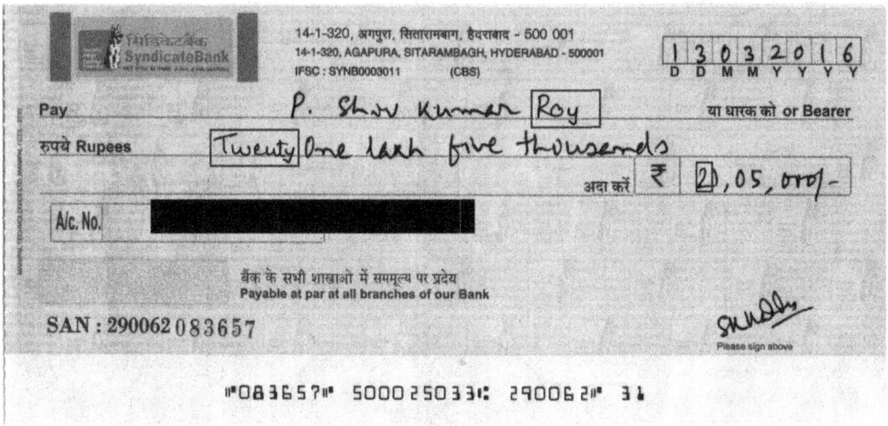

Fig. 1. Deceptive handwritten bank cheque. The deception is indicated by rectangular boxes. Reference is from IDRBT database [1].

most cases. A fraudster always tries to deceive and affect other persons or organizations either by means of his/her own interest or for the interest of others. A pictorial demonstration is shown in the Fig. 1. A deceptive bank cheque, where deception is marked by boxes. If the mark was not presented to the particular word no one can conclude that the cheque is forged. The cheque is drafted by two different writers with two different pens of similar ink. This type of forgery is perceptually intractable. The forgery identification problem is also solved with the help of ink analysis methods [7,12]. However, the issue associated with ink based method is that it classifies wrongly when the same writer writes on the same document using two different pen of similar ink. As a result, a non-forged document happens to be detected as forged. It has been surveyed that written strokes contain much information about a writer. Hence, writing style analysis for forgery detection is also important besides ink analysis. The primary concern here is to identify this kind of fraudulation by means of analyzing the handwriting of the writers. Along with this to establish the objective, we also have introduced two new features, namely histogram of correlation coefficients and radius of curvature histogram.

The present work is not similar to the writer identification problem or writer retrieval problem. Typically, writer identification is the determination of the author from a set of several word samples of different authors [8]. It has been seen that two types of writer identification problems exist mainly. Informally, it is the assignment of a writer for a query image. One is dependent on the text of writers to be written. Another one is text-independent writer identification, where prior knowledge of the text of writer to be identified is not required. The proposal is based on text-independent writer identification. Here we attempt to construct a forensic implementation to identify forgery in handwritten doc-

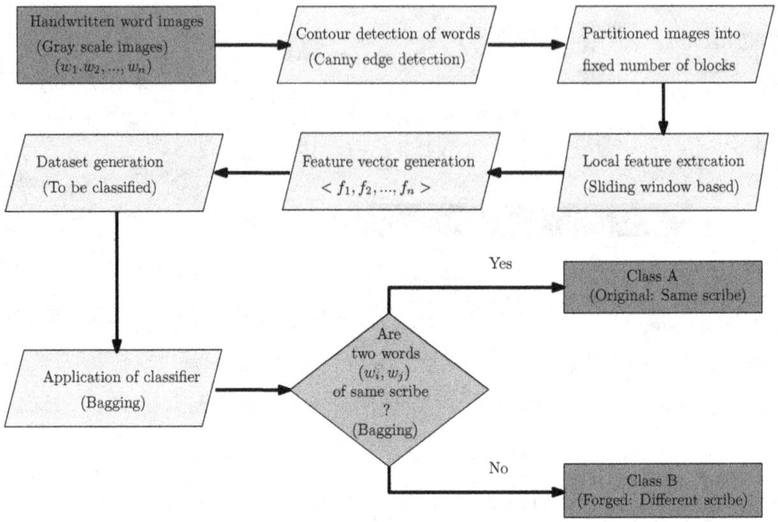

Fig. 2. Proposed conceptual model for fraudulation detection.

uments particularly in financial or authentic handwritten documents owing to writer identification verification.

Over the first two decades, researchers are investing their time in finding writers of historical documents and contemporal documents. Bensefia and Paquet [3] have proposed a writer identification approach by extracting grapheme clustering based features and cosine similarity measurement for classification. Sameh [2] suggests a method of writer identification of Arabic manuscript. Features like Gradient distribution features, window-based contour chain code distribution are extracted. Nearest neighbor (NN) classifier is employed for classification. The accuracy of their proposal depends on a number of writers; the effect increases with the increasing number of writers. Sheng and his colleague have proposed a curvature free joint features: run-lengths of the local binary pattern (LBPruns) and a cloud of line distribution (COLD) features for writer identification [9]. Many more approaches on writer identification are exist based on textural, curvature, grapheme, and modern CNN application-based feature analysis [4–6, 10]. Hence, here only we focus on our motive of forgery identification. There is no exact research available on financial documents. Thereafter, the present work has been designed to solve the problem of forgery identification on handwritten legal documents only by examining the writer's handwriting with the application of image processing and machine learning techniques.

The remaining portions of this paper are documented as follows: Sect. 2 narrates the proposed methodology for the detection of forgery in a handwritten document by analyzing the handwriting of the writer. The corresponding outcomes of the experiment and the conclusion of this study are illustrated in Sect. 3 and 4 respectively.

2 Proposed Methodology

The focus of this work lies on the computation of various local features from the corresponding word images with the computation of two new features. On the basis of these features, classification has been planned to perform the detection of forged documents. Features have been extracted solely by designing sliding window based technique moving from top-left to top-right and end up to bottom-right like raster scanner movement. The proposed work has been tested on IAM dataset containing 10,000 word images of 50 different persons. Figure 2 is the conceptual model of the proposed methodology. Each rectangle box outlines the phases of the method. The steps are illustrated in the following subsections.

a b

Fig. 3. (a) Original image and (b) Background eliminated image.

2.1 Handwritten Word Extraction and Gray Scaling

The first and foremost step is to extract words of the handwritten documents which is considered as suspected archive. In laboratory, due to the aid of various handwritten segmentation techniques the process of word extraction has become easier. Here, horizontal and vertical projections are employed for the segmentation. A huge variation of writing style of each individual writer has been taken. Segmented words go through the step of pre-processing of noise removal and gray scale conversion. The entire treatment has been performed on gray scale images exclusively. A word is the main object, a linear separation of object from background is necessary. Thus Otsu thresholding [11] is applied to extract foreground object from the back ground surface. Figure 3 shows sample image for background elimination. Whichever ink color is used for writing purpose, the draft is only tested on gray plane. The aim of this proposal completely depends on writing styles. Hence, the ink does not matter that much.

2.2 Feature Extraction

Survey of few associated research papers helps us to established the work using diverse set of contour related features. Canny edge detection algorithm helps in contouring the object shape. Single feature does not hold enough potential for classification problem. Being as an binary classification, there is an equal need of adaptation of multiple features. In aggregate, 540 features have been quantified initially. Primarily, image has been divided into fixed number of blocks. Features are calculated from those blocks which have single connected component. Sliding

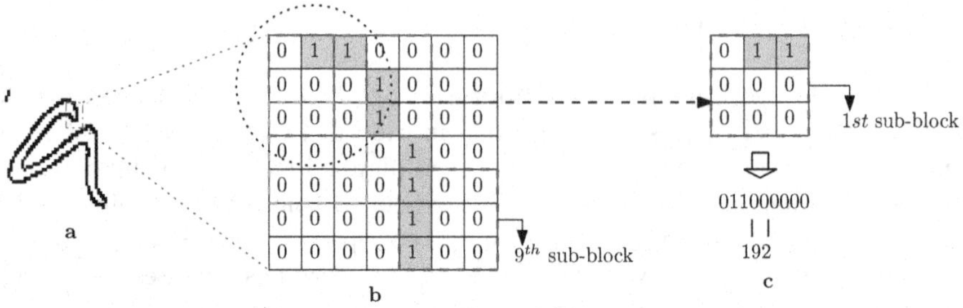

Fig. 4. Shows a LCP construction process. (a) Contour form of the handwritten letter image, (b) 7×7 block with 9 overlapping sub-blocks describing the 2^9 several situations (binary patterns), (c) Binary pattern of the selected (marked by dotted circle) sub-blocks.

window of size 3×3 is used to capture the appearances of the object positions. Concise description of the computed features are demonstrated below.

Window Based Local Contour Pattern Histogram: The process of extracting local contour pattern is adapted from the work of Tang *et al.* [14]. The study of literatures conveys that local contour pattern actually carries information about writer. Thus, points on contour of the every arc corresponding to the considered block have dominating characteristics to help in separating the writers. A block is named as local contour pattern (LCP). The LCP is further divided into 3 overlapping windows. Two adjacent blocks share a common pixel-width line horizontally as well as vertically at two ends in some points. In this way, a block of size $n \times n$ contains $m = (\frac{n-1}{2})^2$ number of overlapping sub-blocks. The elements of blocks are either 1 or 0 as the image is in binary image and a sub-block has $2^9 = 512$ different situations. The pattern of each sub-block is concatenated to form a binary string of length 9 and its corresponding decimal equivalent is taken for deciding the situation of the sub-block. On that wise, each and every sub-block must be in one of the 512 situations. Figure 4 illustrates the LCP construction process for a single block containing 9 overlapping sub-blocks. Thus, the histogram has been designed for counting the occurrences of the situations. A histogram of a particular block contains at most 512 different situations for m different sub-blocks. These steps are repeated for all blocks and then the frequencies at whole are combined to build the final histogram of LCP. Finally, the values of each bin are normalized by total number of non-zero blocks. Cumulative distribution of the resultant histogram has been discerned for constructing feature vector of length 512.

Window Based Contour Differential Chain Code Histogram: Differential chain code deals with the differences of successive directions of relative positions. Differential chain codes are more informative and useful than normal chain code and is computed by taking the modulo 8 of the difference of current element to its previous element of the chain code, where 8 is the connectivity.

Histogram of Correlation Coefficients: In statistics, correlation is a measure of linear dependency of two quantities. Pearson's correlation coefficient is widely used in pattern recognition and image processing. Basically it measures how two data are more apposite to each other or diverse from them. As a single correlation coefficient is not enough to measure the relationship between two objects, we calculate three different correlation coefficient values, namely horizontal, vertical, and diagonal. Row-wise application of 2-D correlation coefficient which processes on two objects, gives a separate degree of relation of the pixel points of the image named horizontal values. The vertical correlation coefficient has been calculated by applying the correlation method on the columns (exclusively on vertical pixel positions). The diagonal correlation values are computed by operating correlation of diagonal positional pixels. Eight bin values from each of the correlation coefficient histograms have been extracted as features of 24 lengths.

Radius of Curvature Histogram: A set of points spatially distributed on a plane are approximated by fitting a curve through them. The curve which should has a radius and thus it also has a centre. In a spatial domain an image (in the current research handwritten words or strokes) is nothing but a set of different concatenated arcs. Therefore, radius of curvature approximation has been implemented on the strokes [13]. The approximation procedure is applied only in the blocks which have connected arc.

A histogram has been designed to count the frequency of similar type of curves occurs. Four higher rated features are extracted from the histogram for generating feature vector.

Features of different lengths are accommodated to form the final feature vector of length 540 from a particular image. Those steps are repeated for all images of the dataset.

3 Experimental Result and Discussion

3.1 Dataset Generation and Classification

Dataset Generation: The formulation of the problem is binary classification problem. One class is designated for indicating no changes of scribe of written words, i.e., a document is fully written by a single person. Other side, another class is assigned for indicating the words of the corresponding document are written by different persons. The extracted word-features are grouped by measuring absolute distances between words in pairs. For clear understanding, two separated group of datasets are created as describe below.

1. **Within Same Person Distance:** The absolute distances between word-features of same person are treated as a feature vector of a genuine person. The corresponding feature vectors accompanied with a class value 0 indicating that this is a positive class. Positive means no interception of other imposters is there in the corresponding document, i.e., the document is authentic.

2. **Between Different Persons Distance:** This case includes absolute distances between features of different words written by different persons and a class value 1 is assigned with the feature vector. In this fashion, data instances have been prepared for each category for further classification.

In general sense, the feature set of a word w written by person i and person j is represented as $F_w{}^i = \{f_1, f_2, f_3, ..., f_n\}$ and $F_w{}^j = \{f_1, f_2, f_3, ..., f_n\}$, respectively, where n indicates the total number of features extracted. Mathematically, a data instance is actually a vector of length $n+1$. The vector form can be represented as $<|f_1{}^i - f_1{}^j|, |f_2{}^i - f_2{}^j|, |f_3{}^i - f_3{}^j|, ..., |f_n{}^i - f_n{}^j|, Class\ value(0/1)>$. The notation indicates that elements of this vector are the absolute differences between the extracted features.

Classification Strategy and Classifier Adapted: The state of the art is to detect two different writing styles of more than one person in a copy, if found, the copy has to be considered as forged document otherwise, not. All the data instances are grouped into two cases as illustrated in previous point. K-left out strategy has been involved to separate these data sets again into smaller data sets. 10-person left out strategy is adopted. Based on this strategy, the construction of training sets and test sets have been realized. The proposed method is tested on IAM handwritten dataset of 50 distinguishable writers. There are 10 different words corresponding to individual writer. In response to the dataset, these 10 different words are also of different writing styles of the corresponding writers. Purposefully, 5 training sets have been generated by keeping data of 10 fixed numbers of writers from the set as test dataset. So, a training set contains data instances of 40 other writers and remaining instances of 10 writers are kept as test data exclusively. Along this line, overall 5 disjoint training and 5 corresponding test sets have been made.

Selection of appropriate classifier is another challenging task, thus trial-and error method helps researchers to choose a suitable classifier. The performance of the process evaluated on the basis of bagging accompanied with REPTree classifier with the selection of 5-fold cross validation. During branch break, no restriction has been possessed on the tree pruning. Bagging performs as a meta-estimator that randomizes the training sets into different sets and build networks. It provides the ultimate prediction of classification either by averaging the results of every set or by means of voting. Bagging is a necessary classifier for data with diverse variations. The judgment of classification has been recorded on the basis of 500 iterations considering minimum error rate. 500 iterations are corresponding to 500 REPTree formations. Each and every internal neural network is tested on the applied test sets for the documentation of final outcomes.

3.2 Performance of Bagging

Table 1 shows the overall classification accuracy in response to 5-fold cross validation technique. Besides, the performances also checked against different sizes of feature vectors. The accomplition of reducing the feature length has been

Table 1. Performance of the classification for correctly identification of different handwritten word of two persons (IAM dataset).

Writers left out	Training data	Classification rate (%) Attribute = 540	Classification rate (%) Selected attribute = 39
P_1 to P_{10}	Data ($P_{11}..P_{50}$)	87.33	91.08
P_{11} to P_{20}	Data ($P_1..P_{10}, P_{21}..P_{50}$)	86.67	91.88
P_{21} to P_{30}	Data ($P_1..P_{20}, P_{31}..P_{50}$)	76.83	86.20
P_{31} to P_{40}	Data ($P_1..P_{30}, P_{41}..P_{50}$)	86.30	90.86
P_{41} to P_{50}	Data ($P_1..P_{40}$)	79.71	88.18
Average		83.77	89.64

Table 2. Performance of the classification for correctly identification of different handwritten word of two persons. Accuracy of IDRBT Dataset.

Writers left out	Training data	Classification rate (%) Attribute = 540	Classification rate (%) Selected attribute = 25
P_1	Data ($P_2..P_9$)	56.85	56.85
P_2	Data ($P_1, P_3..P_9$)	62.08	62.35
P_3	Data ($P_1, P_2, P_4..P_9$)	58.43	59.58
P_4	Data ($P_1..P_3, P_5..P_9$)	59.83	60.15
P_5	Data ($P_1..P_4, P_6, P_9$)	57.21	58.09
P_6	Data ($P_1..P_5, P_7..P_9$)	62.54	62.79
P_7	Data ($P_1..P_6, P_8, P_9$)	62.25	63.88
P_8	Data ($P_1..P_7, P_9$)	58.11	59.47
P_9	Data ($P_1..P_8$)	61.94	60.77
Average		59.91	60.44

performed by attribute selection method and the consequences are recorded thereafter. It is observed that the result increases from 83.77% to 89.64% with reducing the number of attributes from 540 to 39 respectively. Huge reduction of attributes is occurred due to most of the features from LCP features are zeros but few of them have good discriminating values. So, indeed that this set of features has been agreed to fulfill the purpose.

As our main objective is to detect forgery on legal handwritten documents, the proposed technique is also tested on the check image of IDRBT dataset [1]. Consequences is documented in Table 2. For 9 writers, single writer left out strategy and 10-fold cross validation has been applied to compute the results. Accuracy decreases in this case though the data set is fully synthetic. Because of dearthness of original dataset the authors are unable to test the methodology. Moreover, we compare our performances with a most recent work [15] with their best outcome is tabulated in Table 3. They have shown accuracy of 83.94% for IAM dataset of 50 writers by computing 644 geometrical and structural features. The result of our work is relatively superior. A graphical representation is shown in Fig. 5 for both datasets. Comparatively, attribute selection techniques help to speed up the classification process with least number of iterations.

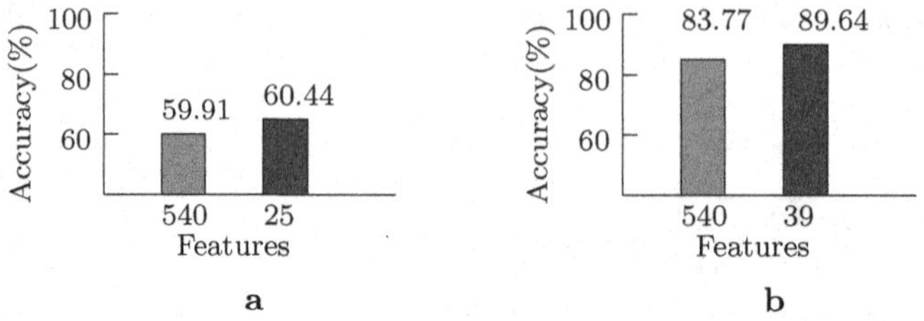

Fig. 5. Average classification accuracy plot. (a) IDRBT dataset and (b) IAM dataset.

Table 3. Relative comparison of the proposed method with a existing method.

Method	Classification accuracy (%)
Wadhwa *et al.* [15]	83.94
Proposed method	**89.64**

4 Conclusion

We have solved the problem of handwritten fraudulation detection by analyzing the writing styles of different persons. Writing style capturing features are extracted exclusively as well. The major portion of this study involves image processing and machine learning techniques in a non-destructive manner, where the document remains undamaged. We have introduced two new features– correlation based histogram and radius of curvature based histogram. Apart from the accuracy, information preservation and nonparticipation of expensive apparatus make this process more attractive.

The intention of the work lies on the document fraudulation identification based on handwriting analysis which we aimed to solve and executed on IAM dataset of 50 writers. The proposed method is also applied on IDRBT cheque image dataset. Notably, the performance rate is quite high when applied on standard IAM dataset, but it is reduced in case of IDRBT dataset. Hence, it is necessary to implement the work on practical datasets to a large extent to safeguard the betterment of the proposal in future works of research. The work can be extended with a novel approach of combining the ink based features and the writing style based features in order to fetch a higher rate of accuracy.

Acknowledgement. The work is sponsored by the project "Design and Implementation of Multiple Strategies to Identify Handwritten Forgery Activities in Legal Documents" (No. ECR/2016/001251, Dt.16.03.2017), SERB, Govt. of India.

References

1. IDRBT Cheque Image Dataset. http://www.idrbt.ac.in/icid.html
2. Awaida, M.S.: Text independent writer identification of Arabic manuscripts and the effects of writers increase. In: Proceedings of the International Conference on Computer Vision and Image Analysis Applications (2015)
3. Bensefia, A., Paquet, T.: Writer verification based on a single handwriting word samples. EURASIP J. Image Video Process. **34**(1) (2016)
4. Brink, A.A., Smit, J., Bulacu, M.L., Schomaker, L.R.B.: Writer identification using directional ink-trace width measurements. Pattern Recogn. **45**, 162–171 (2012)
5. Chen, Z., Yu, H., X., Wu, A., Zheng. W.S.: Letter-level writer identification. In: Proceedings of the International Conference on Automatic Face and Gesture Recognition, pp. 381–388 (2018)
6. Christlein, V., Gropp, M., Fiel, S., Maier, A.: Unsupervised feature learning for writer identification and writer retrieval. In: Proceedings of the International Conference on Document Analysis and Recognition, pp. 991–997 (2017)
7. Dansena, P., Bag, S., Pal, R.: Differentiating pen inks in handwritten bank cheques using multi-layer perceptron. In: Shankar, B.U., Ghosh, K., Mandal, D.P., Ray, S.S., Zhang, D., Pal, S.K. (eds.) PReMI 2017. LNCS, vol. 10597, pp. 655–663. Springer, Cham (2017). https://doi.org/10.1007/978-3-319-69900-4_83
8. Fiel, S., Sablatnig, R.: Writer retrieval and writer identification using local features. In: Proceedings of the IAPR International Workshop on Document Analysis Systems, pp. 145–149 (2012)
9. He, S., Schomaker, L.: Writer identification using curvature-free features. Pattern Recogn. **63**, 451–464 (2017)
10. Hosoe, M., Yamada, T., Kato, K., Yamamoto, K.: Offline text-independent writer identification based on writer-independent model using conditional autoencoder. In: Proceedings of the International Conference on Frontiers in Handwriting Recognition, pp. 441–446 (2018)
11. Otsu, N.: A threshold selection method from gray-level histograms. IEEE Trans. Syst. Man Cybern. **9**(1), 62–66 (1979)
12. Roy, P., Bag, S.: Identification of fraudulent alteration by similar pen ink in handwritten bank cheque. In: Chaudhuri, B.B., Nakagawa, M., Khanna, P., Kumar, S. (eds.) Proceedings of 3rd International Conference on Computer Vision and Image Processing. AISC, vol. 1024, pp. 183–195. Springer, Singapore (2019). https://doi.org/10.1007/978-981-32-9291-8_16
13. Rutter, J., W.: Geometry of crves (2000)
14. Tang, T., Wu, X., Bu, W.: Offline text-independent writer identification using stroke fragment and contour based features. In: Proceedings of the International Conference on Biometrics, pp. 1–6 (2012)
15. Wadhwa, A., Meheshwari, M., Dansena, P., Bag, S.: Geometrical and structural features for forensics in handwritten bank cheques. In: Proceedings of the IEEE India Council International Conference (2018)

Exploring Convolutional Neural Network for Multi-spectral Face Recognition

Anish K. Prabhu, Narayan Vetrekar, and R. S. Gad(✉)

Department of Electronics, Goa University, Taleigao-Plateau, Goa, India
{elect.anishprabhu,elect.ntvetrekar,rsgad}@unigoa.ac.in
https://www.unigoa.ac.in

Abstract. Multi-spectral face recognition has procured noteworthy consideration over a most recent times because of its potential capacity to acquire spatial and spectral information over the electromagnetic range, which cannot be obtained using traditional visible imaging techniques. With the advances in deep learning, Convolutional Neural Network (CNN) based approach has become an essential method in the field of face recognition. In this work, we present two face recognition techniques using face image at nine unique spectra ranging from Visible (VIS) to Near-Infra-Red (NIR) range of the electromagnetic spectrum. This paper is based on the application of using CNN as feature extractor along with Support Vector Machine (SVM) and k-Nearest Neighbor (k-NN) as a classifier on the images of nine different spectra ranging from 530 nm–1000 nm. The obtained performance evaluation results show highest $Rank - 1$ recognition rate of 84.52% using CNN-KNN, demonstrating the significance of using CNN extracted features for improved accuracy.

Keywords: Face recognition · Multi-spectral imaging · Convolutional Neural Network · Support Vector Machine

1 Introduction

Biometrics is one of the widely developed area of bio-engineering, as it is the computerized technique of recognizing individuals depending on a physiological or behavioral traits. The biometric system based on physiological or behavioral traits consists of fingerprint, voice, gate, signature, iris, retina, face, etc. Among these available biometric systems, face recognition has gained noteworthy attention in the field of biometric, due to which it has found its applicability in various sectors such as surveillance, authentication, secure access control, law enforcement, border security, etc. Face recognition is receiving significant attention mainly due to its non-invasive method of face image acquisition that allow to capture face image at different stand-off distance [8]. In spite of all these advantages of face biometric over other biometric traits, the performance of face recognition system is hampered by the variations in illuminations, expressions,

B. Deka et al. (Eds.): PReMI 2019, LNCS 11941, pp. 606–613, 2019.
https://doi.org/10.1007/978-3-030-34869-4_66

poses, background, occlusion, etc. [3]. The effect of variations in illumination conditions is well addressed by employing multi-spectral face imaging, thereby makes use of more than one electromagnetic spectrum (For instant Thermal and visible broad spectrum). The principle behind utilizing multi-spectral imaging is to extract complementary discriminant information over various spectrum bands (reflectance and/or emittance) for improved performance [5]. One of the main characteristic features that make multi-spectral imaging more appropriate approach to obtain robust face recognition is its ability extract distinctive features of the individuals.

The majority of work available in the literature have employed various tools and techniques to perform face recognition across multi-spectral range. Li et al. [7] have employed Local Binary Patterns (LBP) to extract the textural features from Near-Infra-Red (NIR) face images, which is based on rotation invariance and robustness image blur but this approach poses serious limitation when there is change in the face location in an image due to head movement. However, the limitation of variation in the pose was addressed in the method proposed by Zhang et al. [13] by using the potential of Gabor-Directional Binary Code (GDBC) for multi-spectral face recognition. In another work, Farokhi et al. [4] introduced a methodology based on Zernike moments and Hermite kernels (ZMHK) to address the limitation of face detection problem in NIR face images, thereby obtaining an effective and superior face detection. Although, the above methods have presented the promising results, but they are highly dependent on manual feature selection process. In the recent times, various deep learning techniques have been utilized in the face recognition domain working across visible and Near-Infra-Red (NIR) spectrum. Zhang et al. [14] proposed Convolutional Neural Network (CNN) architecture for NIR images, with improved face recognition rate as compared to GDBC and ZMHK methods mentioned above. In another work, CNN based model mainly identified as NIRFaceNet was proposed by Peng et al. [10] worked on NIR face images of Chinese Academy of Sciences' Institute of Automation (CASIA) NIR database. In the similar line with the previous work, a CNN based model also known as (a.k.a.) DeepFace [11] was trained on four million facial images, proposed by Facebook's Artificial Intelligence (AI) group to perform face recognition in visible wavelength range.

The previous works have shown limited attention towards the use of deep learning technique for multi-spectral face recognition domain, with few works centered towards NIR face recognition. In this work, we present a deep learning based approach to automatically extract features from Visible (VIS) as well as Near-Infra-Red (NIR) face images using CNN and then processing independently the extracted feature sets with Support Vector Machine (SVM) and K-Nearest Neighbor (KNN) classifiers for multi-spectral face recognition. The experimental evaluation results in the form of recognition rate at $Rank-1$ is presented on multi-spectral face database of 6048 images collected across nine narrow spectral band images corresponding to 530 nm, 590 nm, 650 nm, 710 nm, 770 nm, 830 nm, 890 nm, 950 nm, 1000 nm spanning from 530 nm to 1000 nm spectrum range. In the due course of time the major contributions can be summarized as follows:

- Explore the potential of Convolutional Neural Network (CNN) to extract the individual spectral band features for robust performance on the multi-spectral face database i.e. *SpecVis* having face images of 168 subjects across nine spectral bands.
- Use of pre-trained network *Alexnet* [6] as a feature extractor of face images across *nine* spectral bands ranging from 530 nm–1000 nm and Support Vector Machine (SVM), k-Nearest Neighbor (k-NN) as classifiers.
- Experimental results in the form of recognition rate at $Rank-1$ across individual spectral band for improved performance accuracy.

In the reminder of the paper, Sect. 2 describes the multi-spectral face database employed and the preprocessing techniques used, Sect. 3 presents the proposed method of feature extraction based on Convolutional Neural Network (CNN) processing independently with two different methods such as Support Vector Machine (SVM) and k-Nearest Neighbor (k-NN) using multi-spectral face database of 168 subjects, Sect. 4 presents the detailed experimental evaluation protocol and related experimental results along with major observations, and final conclusive remarks along with future work is summarized in Sect. 5.

2 Database

This section of paper explains the details related to the publicly available database *SpecVis* [12] utilized in our work for performance analysis. The database comprises of 168 subjects collected using in-house custom-build multi-spectral imaging sensor across nine narrow spectrum bands including 530 nm, 590 nm, 650 nm, 710 nm, 770 nm, 830 nm, 890 nm, 950 nm, 1000 nm spanning from Visible to Near-Infra-Red spectrum range. The 168 subjects consists of 96 male and 72 female collected in two sessions with two samples each. Between each sessions a time difference of about 1 to 3 weeks is maintained during the data collection. In total, 6048 sample multi-spectral images are collected which corresponds to 168 Subjects × 2 Sessions × 2 Samples × 9 Bands = 6048 Samples. Table 1 summarizes the total number of samples collected.

Table 1. *SpecVis* database describing total number of sample images

Subjects	Sessions	Samples	Bands	Total images
168	2	2	9	6048

2.1 Pre-processing of Images

Images acquired using multi-spectral sensor have a broader field of view, which in turn includes the background scene apart from the facial data. Hence, we pre-process the multi-spectral facial images that includes face normalization and

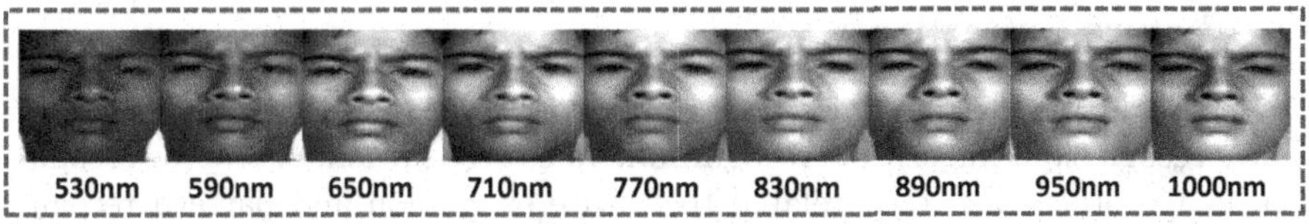

Fig. 1. Multi-spectral face database across nine narrow spectral bands

enhancement. In case of normalization, we used eye-coordinate based approach to detect the facial region and then perform translation and rotation correction to perform geometric alignment. Following the normalization technique, we employ contrast enhancement to improve the facial features. Further, the cropped facial image is resized to a common 120 × 120 uniform dimension before performing feature extraction and classification. Figure 1 illustrates the multi-spectral facial database after performing pre-processing technique.

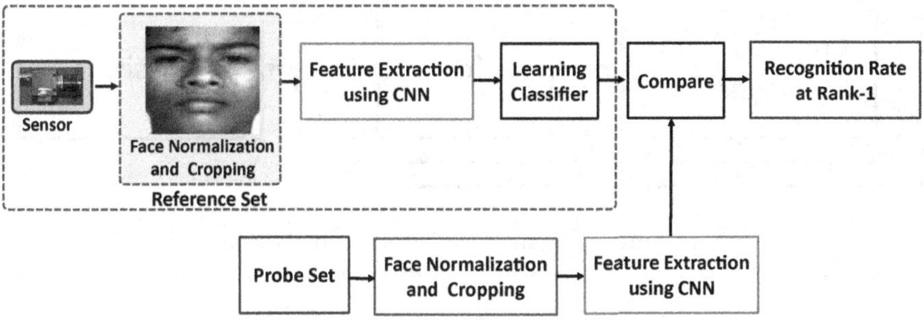

Fig. 2. Proposed method to extract CNN based features corresponding to individual spectral bands and classification

3 Methodology

This section presents the details about the feature extraction method and classification approach employed in this work for multi-spectral face recognition. The methodology consists of reference set and probe set. For reference set, we first perform the individual spectral band face normalization and cropping (Refer Sect. 2.1), followed by CNN based feature extraction, which further processed to learn the classifier. During the probe set, the respective individual spectral bands processed to compare it against the learned reference set to obtained recognition rate at $Rank-1$. Figure 2 presents the illustration of feature extraction and classification approach employed in this work. Further, the details related to the proposed method is explained using following sub-sections: Feature extraction and classification.

3.1 Feature Extraction

CNN has had pivotal outcomes over the previous decade in several fields such as face recognition, fingerprint recognition, voice recognition, character recognition, image processing, etc., related to feature extraction and pattern recognition for robust performance [2,9]. This accomplishment of CNN has provoked researcher community to build larger models to tackle complex problems which were impractical with classical Artificial Neural Networks (ANN). One of the most critical aspects of CNN is that it can also be used as a feature extractor as it acquires unique and complex features when the input image propagates into the deeper layers of the network.

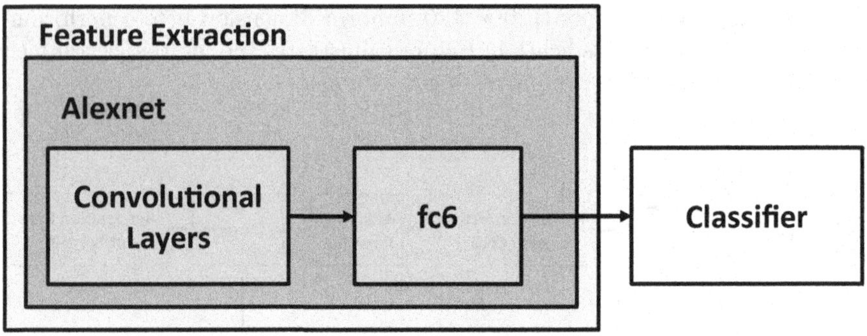

Fig. 3. Feature extraction using CNN and classification

Figure 3 depicts the implementation of pre-trained CNN model Alexnet as a feature extractor. On the pre-processed multi-spectral images of face region in each band, we apply an existing pre-trained CNN model i.e. Alexnet proposed by Krizhevsky et al. [6]. Alexnet comprises of 5 convolutional layers and 3 fully-connected layers. We extract the feature set from the first fully-connected layer (feature layer 'fc6') of the pre-trained Alexnet model. The same procedure is followed for remaining of the nine spectral bands to compute the feature vector. For each of the nine bands, we calculate the feature vector for the training set by the use of deep learning technique.

3.2 Classification

The next step of the proposed method is to efficiently perform the classification based on the feature vector generated by pre-trained Alexnet model (Discussed in The previous Sect. 3.1). In this paper, we have employ independently the multi-class Support Vector Machine (SVM) [1] and K-Nearest Neighbor (k-NN) classifier separately across individual spectral bands to compute improved recognition rate at $Rank - 1$. SVM and k-NN classifiers are broadly utilized in the past for robust classification accuracy.

4 Experimental Results

In this segment of the paper, we present experimental evaluation protocol and quantitative results using multi-spectral facial database of 168 subjects across nine spectral bands. Specifically, we first extract features using pre-trained Convolutional Neural Network (CNN) and then learn the classifier independently using Support Vector Machine (SVM) and k-Nearest Neighbor (k-NN) classifier. The $Rank-1$ recognition rate is obtained independently across individual spectral band using proposed methods and compared against the state-of-the-art multi-spectral face recognition methods.

Table 2. Rank-1 recognition rates (%) across individual nine spectral bands using state-of-the-art methods and proposed methods.

Algorithm	Individual spectral band								
	530 nm	590 nm	650 nm	710 nm	770 nm	830 nm	890 nm	950 nm	1000 nm
LBP-SVM	17.26	25.3	59.52	59.82	42.86	39.29	40.48	35.12	42.86
LBP-KNN	20.24	27.68	61.01	59.23	44.05	41.07	40.18	35.71	49.11
CNN	56.55	63.1	65.48	61.9	59.23	50.6	67.56	63.69	66.07
CNN-SVM	60.71	72.62	76.49	77.08	72.02	73.81	76.19	74.7	79.46
CNN-KNN	66.37	79.46	80.65	84.52	76.79	78.87	80.95	80.65	82.14

4.1 Evaluation Protocol

In this section, we explain the experimental evaluation protocol followed for multi-spectral face recognition in this work. We split the *SpecVis* database into two sets: reference set and probe set. Reference set consists of 168 subjects along with their samples corresponding to session-1 data and probe set consists of respective 168 subjects along with their samples corresponding to session-2. Since the performance is obtained across individual spectral bands, the protocol remains same while computing the recognition rate at $Rank-1$ across individual spectral bands.

4.2 Observations

Based on the evaluation protocol, we present our results across state-of-the-art methods and proposed method. Table 2 summarizes the $Rank-1$ recognition rate obtained across individual spectral bands. From the obtained results, we make our major observations as follow:

- The overall $Rank-1$ recognition rate across individual spectral bands shows reasonable performance using the proposed method.
- The highest recognition rate of 84.52% is obtained for CNN-KNN approach using 710 nm spectrum band and the lowest performance of 17.26% recognition rate is obtained with LBP-SVM using 530 nm spectrum band.

- Of the classification method used in proposed approach, CNN-KNN outperforms CNN-SVM across all the individual spectral bands.

To summarize, the proposed method based on extracting CNN features for multi-spectral facial database improves the performance accuracy across individual spectral band compared to the state-of-the-art methods, demonstrating the significance of employing CNN based feature extraction.

5 Conclusion

Face recognition is always a challenging task under varying illumination conditions. In recent times, due to the advancement in sensing technologies, multi-spectral imaging have shown greater potential to under varying illumination conditions to perform better. Although, multi-spectral imaging across individual spectral bands obtains the discriminative feature information, the performance of individual bands are not consistent using state-of-the-art texture descriptor methods. Hence, in this paper, we present the CNN based feature extraction technique along with two independent classifiers such as Support Vector Machine (SVM) and k-Nearest Neighbor (k-NN) to obtain an improved performance across individual spectral bands. The results in the form of recognition rate at $Rank - 1$ is obtained based on the publicly available $SpecVis$. The observed results demonstrated the consistent improvement in the recognition rate across individual spectral band. Proposed method of CNN-KNN has appeared as the algorithm with highest rank-1 recognition rates across all nine bands. Also, the recognition rate of 84.5% for the 710 nm band is the highest rank-1 recognition rate obtained using CNN-KNN approach.

Acknowledgement. This work is supported by University Grant Commission, India (Grant No. 40-664/2012 (SR)).

References

1. Ahuja, Y., Yadav, S.K.: Multiclass classification and support vector machine. Glob. J. Comput. Sci. Technol. Interdisc. **12**, 14–20 (2012)
2. Albawi, S., Mohammed, T.A., Al-Zawi, S.: Understanding of a convolutional neural network. In: 2017 International Conference on Engineering and Technology (ICET), pp. 1–6, August 2017
3. Bebis, G., Gyaourova, A., Singh, S., Pavlidis, I.: Face recognition by fusing thermal infrared and visible imagery. Image Vis. Comput. **24**, 727–742 (2006)
4. Farokhi, S., Sheikh, U.U., Flusser, J., Yang, B.: Near infrared face recognition using Zernike moments and hermite kernels. Inform. Sci. **316**, 234–245 (2015)
5. Ghiass, R.S., Arandjelović, O., Bendada, A., Maldague, X.: Infrared face recognition: a comprehensive review of methodologies and databases. Pattern Recogn. **47**(9), 2807–2824 (2014)
6. Krizhevsky, A., Sutskever, I., Hinton, G.E.: ImageNet classification with deep convolutional neural networks. Neural Inform. Process. Syst. **25** (2012)

7. Li, S.Z., Chu, R., Liao, S., Zhang, L.: Illumination invariant face recognition using near-infrared images. IEEE Trans. Pattern Anal. Mach. Intell. **29**(4), 627–639 (2007)
8. Nicolo, F., Schmid, N.: Long range cross-spectral face recognition: matching SWIR against visible light images. IEEE Trans. Inform. Forensics Secur. **7**, 1717–1726 (2012)
9. Patel, R., Yagnik, S.B.: A literature survey on face recognition techniques. Int. J. Comput. Trends Technol. (IJCTT) **5**, 189–194 (2013)
10. Peng, M., Wang, C., Chen, T., Liu, G.: NIRFaceNet: a convolutional neural network for near-infrared face identification. Information **7**, 61 (2016)
11. Taigman, Y., Yang, M., Ranzato, M., Wolf, L.: DeepFace: closing the gap to human-level performance in face verification. In: 2014 IEEE Conference on Computer Vision and Pattern Recognition, pp. 1701–1708, June 2014
12. Vetrekar, N.T., Raghavendra, R., Raja, K.B., Gad, R.S., Busch, C.: Extended spectral to visible comparison based on spectral band selection method for robust face recognition. In: 2017 12th IEEE International Conference on Automatic Face Gesture Recognition (FG 2017), pp. 924–930, May 2017
13. Zhang, B., Zhang, L., Zhang, D., Shen, L.: Directional binary code with application to polyu near-infrared face database. Pattern Recogn. Lett. **31**(14), 2337–2344 (2010)
14. Zhang, X., Peng, M., Chen, T.: Face recognition from near-infrared images with convolutional neural network. In: 2016 8th International Conference on Wireless Communications Signal Processing (WCSP), pp. 1–5, October 2016

Superpixel Correspondence for Non-parametric Scene Parsing of Natural Images

Veronica Naosekpam[1], Alexy Bhowmick[1(✉)], and Shyamanta M. Hazarika[2]

[1] School of Technology, Assam Don Bosco University, Guwahati, Assam, India
alexy.bhowmick@gmail.com
[2] Indian Institute of Technology, Guwahati, Assam, India

Abstract. Scene parsing refers to the task of labeling every pixel in an image with the class label it belongs to. In this paper, we propose a novel scalable non-parametric scene parsing system based on superpixels correspondence. The non-parametric approach requires almost no training and can scale up to datasets with thousands of labels. This involves retrieving a set of images similar to the query image, followed by superpixel matching of the query image with the retrieval set. Finally, our system warps the annotation results of superpixel matching, and integrates multiple cues in a Markov Random Field (MRF) to obtain an accurate segmentation of the query image. Our non-parametric scene parsing achieves promising results on the LabelMe Outdoor dataset. The system has limited parameters, and captures contextual information naturally in the retrieval and alignment procedure.

Keywords: Scene · Scene parsing · Non-parametric · Label transfer · Partial correspondence

1 Introduction

In the recent decade, scene parsing has become an active area of research. The idea behind scene parsing is to label each pixel in an image to the category or the class it belongs to. It can be said that scene parsing is one step ahead of the traditional image segmentation. Semantic labels can be *stuffs* such as - grass or sky, as well as *things*, such as - person or building [8]. Traditionally there two approaches of parsing an image: (1) Parametric and (2) Non-Parametric [13].

The traditional parametric approach consists of model training phase using the training set which is then tested using the test data [5]. Though this approach is effective and gives highly accurate result, it does not account for the dynamic nature of the world where the size of the data keep increasing dynamically. Parametric approach work with "closed universe" datasets where size of the data is fixed. Once the data size varies, the model has to be trained again which is an overhead. The non-parametric approach [2,8,13] provides several advantages

B. Deka et al. (Eds.): PReMI 2019, LNCS 11941, pp. 614–622, 2019.
https://doi.org/10.1007/978-3-030-34869-4_67

over the parametric approach. Instead of training the model, semantic labels are directly transfered to the query image from the retrieval set. It works with "Open universe" datasets whose size can vary at all time. Since it is a data-driven approach, it requires almost no training. In non-parametric scene parsing, the semantic label transfer can be done at pixel level (dense correspondence) [8] or the superpixel level (partial correspondence) [2,13].

A typical pipeline of non-parametric scene parsing is shown in Fig. 1. The image database comprises of training and test images and their corresponding annotations or the ground-truth labels. There are three major steps involved in non-parametric scene parsing: (1) *Scene Retrieval*: Given a query image, a set of similar images are retrieved from the training images; (2) *Scene Correspondence*: It is the alignment of the query image with the retrieved images (results from the scene retrieval); for correspondence of points to be used for labeling. And (3) *Markov Random Field* (MRF) *Inference*: It performs the task of spatial smoothing or aggregation of labels to obtain better segmentation results.

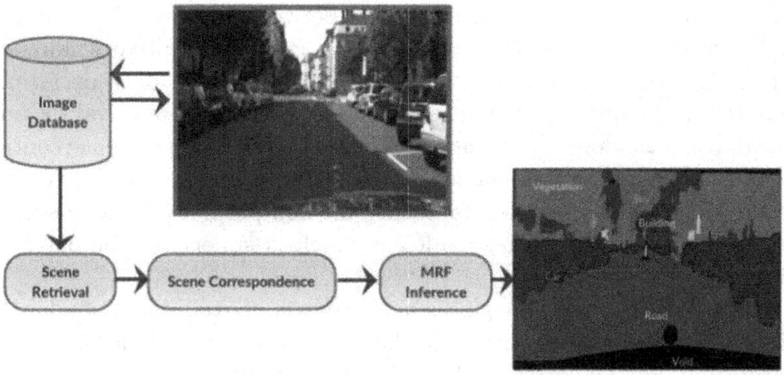

Fig. 1. A typical non-parametric scene parsing pipeline

In this work, our aim is to propose an improved context based non-parametric framework for the scene parsing problem. Our entire scene parsing pipeline heavily relies on the correct retrieval of the similar image set to achieve a correct labels of the query image. We aim to get a fairly similar set of images as a retrieval set of the query image in order to perform the parsing. Our scene retrieval module has been built by taking into account the spatial features of the query image. The label transfer is performed based on partial correspondence using super-pixels. An improved super-pixels features extraction techniques are used for correspondence in the partial scene correspondence module. Regarding the MRF inference module, off-the-shelf graph based MRF inferencing technique has been used. We demonstrate our system on SIFT Flow (LabelMe Outdoor) dataset. Experimental evaluation is done based on per-pixel and per-class accuracy.

The remainder of the paper is as follows: Sect. 2 starts with the summarization of related works. It is followed by a presentation of the improved framework and experimental evaluation in Sects. 3 and 4. Finally, conclusions are drawn and outlook on future work is presented in Sect. 5.

2 Related Works

Various literature on non-parametric scene parsing have been studied [2,8,13]. For label transfer from the retrieved images to the query images, the existing state-of-the-art approaches differ in the methods of retrieval of similar images, the nature of correspondence algorithms, and the choice of the features used.

The earliest works on non-parametric scene parsing are based on dense correspondence [8]. Dense correspondence is a pixel-by-pixel alignment of the query image with the retrieval set using dense local features such as SIFT. The per-pixel local information are transferred from an image in the retrieval set to the query image. Liu et al. [8] introduced the concept of "label transfer" based on local dense SIFT Flow matching algorithm [9]. ANN (Approximate Nearest Neighbour) bilateral matching transfers label at pixel level by integrating prior knowledge based on local partial similarity between images [17]. Collageparsing [15] is an approach that extracts mid-level content adaptive windows from the retrieved images and the query image. The label transfer is performed by matching the query image's content adaptive windows with retrieval set images' content-adaptive windows. As research on dense scene correspondence continued, it is found that dense correspondence is complex and computationally expensive and hence, the recent research approaches on non-parametric scene segmentation are based on partial correspondence which is super-pixel or patch based approach.

In the non-parametric scene parsing based on partial correspondence, the query image is aligned to the images in the retrieval set by considering a group of pixels at a time. Partial correspondence labels cohesive groups of pixels together without loss of geometrical information of the image and in turn significantly reduces the number of elements to process. Label transfer based on partial correspondence was first introduced by Tighe and Lazebnik [13]. The superpixels of the query image are labelled using an MRF model, based on similar superpixels in the query's nearest neighbor. It also performs simultaneous labeling of geometric and semantic classes of the query image. Their work is extended by Eigen et al. [2] where per-descriptor weights for each superpixels' segment are learnt in order to minimize classification error which in turn helps to nullify the biasing of the semantic classes towards common classes.

In order to detect interesting objects in a scene, Tighe and Lazebnik [14] proposed to augment their SuperParsing work with pre-trained exemplar SVMs [10]. Although the overall per-pixel accuracy improved, it failed to detect rare classes. Yang et al. [16] added rare classes in the retrieval set for a more balanced super-pixels classification in a feedback manner in order to refine the matching at the super-pixel level. The super-pixels in the rare classes are populated using exemplars. Various classifiers (ensemble approach) are combined to perform superpixels based scene correspondence in [4]. The likelihood scores of various classifiers are combined in order to label each superpixel of the query image with a balanced score. Label transfer via efficient filtering [11] performed parsing by first sampling of superpixels based on some similarity score and transfering labels via filtering using a Gaussian kernel which encodes how similar two superpixels are in terms of their feature vector.

3 Non-parametric Scene Parsing

3.1 Scene Retrieval

The scene retrieval component aims to find a subset of the training set similar to the query image. It is a critical step in the pipeline as the overall scene parsing performance is highly dependent on correct scene retrieval. There is no chance of recovery in later stages in case of an incorrect retrieval. We used a combination of local and global features for image retrieval. The global feature used for obtaining the retrieval set is Gist. The local features used to obtain the retrieval set are VLAD (Vector of Locally Aggregated Descriptors) [6] and SPM (Spatial Pyramid Matching) [7]. Gist is low dimensional global feature which gives the summary of an image. VLAD is an extension of Bag of Visual Words which gives a more discriminative representation. SPM incorporates the spatial information of an image. The approximate nearest neighbor images for the retrieval set are obtained using KD-tree indexing. Our improved scene retrieval module performs better retrieval of the similar images. For each of the three mentioned features, the first three most similar images are selected as the content of the retrieval set. Hence, the total number of images in the retrieval set is $k = 9$ for each query image.

3.2 Partial Scene Correspondence

The approach followed for scene alignment is the partial scene correspondence using superpixels. It labels large, cohesive groups of pixels at once, decreases the number of elements to process and at the same time, preserves the geometric information. Superpixels are segmented from the query image and the retrieval set using fast graph based segmentation [3]. The advantage of using this approach is that the it segments a scene or an image into a combination of coarse and fine regions. The scale of segmentation for the query as well as for the retrieval set is set to 200.

Five types of features are used for representing each super-pixel, (1) SIFT histogram (1024D), RGB color histogram (128D), HSV color histogram (128D), Location histogram (36D) and PHOG histogram of the superpixel's bounding box (168D). For SIFT histogram, SIFT features are extracted and encoded by 5 words from a vocabulary of size 1024 using LLC algorithm. 128-D color histogram are obtained by quantizing color features from a vocabulary of 128 words. 36-D location histogram is obtained by quantizing (x, y) location into a 6×6 grid. In order to include contextual information, the superpixels are masked by

20 pixels and the same type of features are used for representing the dilated superpixels. So, the total number of features for each superpixel is $(1024+128+128+36+168) \times 2 = 2968$. Then, similar to [16], the classification cost of each input superpixel $s_i \in Q$ with reference to the K-Nearest Neighbour ($N_k(i)$) in Retrieval Set R (s_j, x_j, y_j) using Kernel $K(x_i, x_j)$ is given by:

$$P(y_i = c|s_i) = 1 - \frac{\sum_{j \in N_k(i), y_j} K(x_i, x_j)}{\sum_{j \in N_k(i)} K(x_i, x_j)} \tag{1}$$

Kernel functions are real valued functions that quantifies the similarity between two feature spaces. They map data into higher dimensional space. $K(x_i, x_j) > 0$ is the similarity of x_i and $x_j \in X$. To avoid operating in the high dimensional space, a feature space is chosen in which the dot product can be computed directly using a nonlinear function in the input space $K(x_i, x_j) = <\theta(x_i), \theta(x_j)>$ (called the kernel trick).

The kernel function is the Chi-square kernel. It is a kernel based on Chi-square distribution.

$$K(x_i, x_j) = 1 - \sum_{i,j=1}^{N} \frac{(x_i - x_j)}{\frac{1}{2}(x_i + x_j)} \tag{2}$$

3.3 MRF Inference

In order to fuse global semantic constraints, a random field model is used. For this, a four connected MRF (Markov Random Field) is chosen whose energy function is given by the equation:

$$E(Y) = \sum_{p} E_{data}(y_p) + \lambda \sum_{pq} E_{smooth}(y_p, y_q) \tag{3}$$

where p and q are pixels, λ is the pairwise energy weight. The data term of one pixel is given by the result of Eq. 1 and the smoothness cost is given by label variant cost as $E_{smooth}(y_p, y_q) = d(p,q).\mu(c,c')$. d(p,q) is the color dissimilarity between two neighbouring pixels and $\mu(c,c')$ is the penalty of assigning c and c' to adjacent pixels. The objective is to find the labeling that optimizes the energy function.

For semantic labeling, inferencing on E(Y) is performed via Alpha-Beta Swap [1]. It is a graph-cut algorithm that minimizes or optimizes the E(Y) by exchanging labels between an arbitrary set of pixels labeled α and another arbitrary set of pixels labeled β.

4 Experiments

Here we evaluate and compare our method with state-of-the-art approaches on SIFT Flow (LabelMe Outdoor) dataset. It consists of 2688 outdoor scene images with 2488 images as training set and 200 images as test set as in [8]. The images are 256 × 256 pixels with 33 labels. The metric used for evaluation are: (1) *Per-pixel accuracy*: The percent of correctly labeled pixels in total and (2) *Per-class accuracy*: The average percent of correctly labeled pixels in each class.

The average per-pixel recognition rate r_p is calculated as

$$r_p = \frac{1}{\Sigma_i m_i} \Sigma_i \Sigma_{p \in \Lambda_i} 1(o(p) = a(p), a(p) > 0), \tag{4}$$

where, for a pixel p in image i, the ground-truth label is a(p) and system output is o(p); for unlabeled pixels, a(p) = 0. The symbol Λ_i represents the image lattice for test image i and m = $\Sigma_{p \in \Lambda_i} 1(a(p) > 0)$ is the number of labeled pixels for image i. It is to be noted that some pixels may remain unlabeled.

The average per-class recognition rate r_c is

$$r_c = \frac{\Sigma_i \Sigma_{p \in \Lambda_i} 1(o(p)=a(p), a(p)=l)}{\Sigma_i \Sigma_{p \in \Lambda_i} 1(a(p)=l)}, l = 1, 2, 3, ...L \tag{5}$$

For each query, the number of retrieval set is set to 9. Parameters for superpixels segmentation are set to: k = 200, sigma = 0.8 and minimum-area = 25. The nearest neighbour (K) superpixels for label transfer is set to 100. The MRF parameters are set to: $\lambda = 6$ and $\alpha = 0.7$. It has been identified that 5 of the 33 labels are common classes while the rest of the labels are rare classes. Our result is compared with recent works in Table 1.

Table 1. Comparison with the state-of-the-art systems on LMO dataset

	Per-pixel	Per-class
Liu et al. [8]	74.75%	N/A
Tighe et al. [13]	73.2%	29.1%
Gould et al. [5]	65.7%	14.2%
Razzaghi et al. [12]	75.84%	31.3%
Eigen et al. [2]	75.3%	39.2%
Our approach	73.5%	21.719%

The class distribution graph of the labels is shown in Fig. 2. From the plot, it is been shown that building, sky, road, mountain, sea and tree are the most common class distribution. It also shows that the system failed to detect rare classes such as desert, bus, balcony, bird, cow, pole, moon and sun. Some of our labeling results along with the ground truth is shown in Fig. 3.

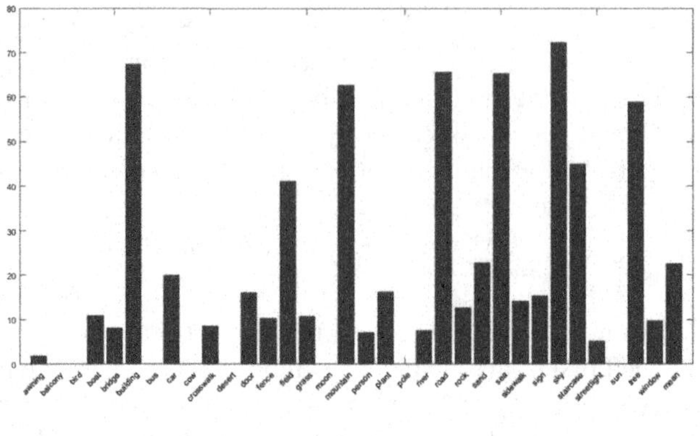

Fig. 2. Classes of objects in the dataset and their frequency.

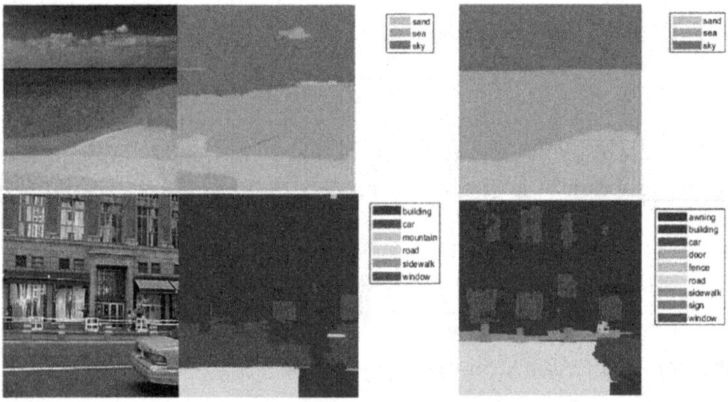

Fig. 3. Some results on SIFT flow dataset. On the left are query images, and results of scene parsing. On the right are ground truth images for comparison.

5 Conclusion

In this work, we present a novel non-parametric scene parsing framework whose overall per-pixel labeling accuracy is improved by accurate retrieval and partial correspondence. By combining various features for similar image retrieval, we have boosted the strength of the partial correspondence matching. Evaluation results have shown that our system obtains compatible performance on SIFT Flow dataset. Dividing the query image and retrieval set into non-uniform grids based on salient region detection and performing partial correspondence on corresponding grids will be a good research direction.

References

1. Boykov, Y., Veksler, O., Zabih, R.: Fast approximate energy minimization via graph cuts. In: Proceedings of the Seventh IEEE International Conference on Computer Vision, vol. 1, pp. 377–384. IEEE (1999)
2. Eigen, D., Fergus, R.: Nonparametric image parsing using adaptive neighbor sets. In: IEEE Conference on Computer Vision and Pattern Recognition, pp. 2799–2806. IEEE (2012)
3. Felzenszwalb, P.F., Huttenlocher, D.P.: Efficient graph-based image segmentation. Int. J. Comput. Vis. **59**(2), 167–181 (2004)
4. George, M.: Image parsing with a wide range of classes and scene-level context. In: Proceedings of the IEEE Conference on Computer Vision and Pattern Recognition, pp. 3622–3630 (2015)
5. Gould, S., Zhang, Y.: PatchMatchGraph: building a graph of dense patch correspondences for label transfer. In: Fitzgibbon, A., Lazebnik, S., Perona, P., Sato, Y., Schmid, C. (eds.) ECCV 2012. LNCS, vol. 7576, pp. 439–452. Springer, Heidelberg (2012). https://doi.org/10.1007/978-3-642-33715-4_32
6. Jégou, H., Douze, M., Schmid, C., Pérez, P.: Aggregating local descriptors into a compact image representation. In: CVPR 2010–23rd IEEE Conference on Computer Vision and Pattern Recognition, pp. 3304–3311. IEEE Computer Society (2010)
7. Lazebnik, S., Schmid, C., Ponce, J.: Beyond bags of features: spatial pyramid matching for recognizing natural scene categories. In: 2006 IEEE Computer Society Conference on Computer Vision and Pattern Recognition (CVPR 2006), vol. 2, pp. 2169–2178. IEEE (2006)
8. Liu, C., Yuen, J., Torralba, A.: Nonparametric scene parsing via label transfer. IEEE Trans. Pattern Anal. Mach. Intell. **33**(12), 2368–2382 (2011)
9. Liu, C., Yuen, J., Torralba, A., Sivic, J., Freeman, W.T.: SIFT flow: dense correspondence across different scenes. In: Forsyth, D., Torr, P., Zisserman, A. (eds.) ECCV 2008. LNCS, vol. 5304, pp. 28–42. Springer, Heidelberg (2008). https://doi.org/10.1007/978-3-540-88690-7_3
10. Malisiewicz, T., Gupta, A., Efros, A.A., et al.: Ensemble of exemplar-svms for object detection and beyond. In: ICCV, vol. 1, p. 6. Citeseer (2011)
11. Najafi, M., Taghavi Namin, S., Salzmann, M., Petersson, L.: Sample and filter: nonparametric scene parsing via efficient filtering. In: Proceedings of the IEEE Conference on Computer Vision and Pattern Recognition, pp. 607–615 (2016)
12. Razzaghi, P., Samavi, S.: A new fast approach to nonparametric scene parsing. Pattern Recogn. Lett. **42**, 56–64 (2014)
13. Tighe, J., Lazebnik, S.: SuperParsing: scalable nonparametric image parsing with superpixels. In: Daniilidis, K., Maragos, P., Paragios, N. (eds.) ECCV 2010. LNCS, vol. 6315, pp. 352–365. Springer, Heidelberg (2010). https://doi.org/10.1007/978-3-642-15555-0_26
14. Tighe, J., Lazebnik, S.: Finding things: Image parsing with regions and per-exemplar detectors. In: Proceedings of the IEEE Conference on Computer Vision and Pattern Recognition, pp. 3001–3008 (2013)
15. Tung, F., Little, J.J.: CollageParsing: nonparametric scene parsing by adaptive overlapping windows. In: Fleet, D., Pajdla, T., Schiele, B., Tuytelaars, T. (eds.) ECCV 2014. LNCS, vol. 8694, pp. 511–525. Springer, Cham (2014). https://doi.org/10.1007/978-3-319-10599-4_33

16. Yang, J., Price, B., Cohen, S., Yang, M.H.: Context driven scene parsing with attention to rare classes. In: Proceedings of the IEEE Conference on Computer Vision and Pattern Recognition, pp. 3294–3301 (2014)
17. Zhang, H., Fang, T., Chen, X., Zhao, Q., Quan, L.: Partial similarity based non-parametric scene parsing in certain environment. In: CVPR 2011, pp. 2241–2248. IEEE (2011)

Power Line Segmentation in Aerial Images Using Convolutional Neural Networks

Sumeet Saurav, Prashant Gidde(✉), Sanjay Singh, and Ravi Saini

CSIR - Central Electronics Engineering Research Institute (CSIR-CEERI), Pilani, India
sumeet@ceeri.res.in, prashantpsgl@gmail.com

Abstract. Visual inspection of transmission and distribution networks is often carried out by various electricity companies on a regular basis to maintain the reliability, availability, and sustainability of electricity supply. Till date the widely used technique for carrying out an inspection is done manually either using foot patrol and/or helicopter operated manually. However, recently due to the widespread use of quadcopters/UAVs powered by deep learning algorithms, there have been requirements to automate the visual inspection of the power lines. With this objective in mind, this paper presents an approach towards automatic autonomous vision-based power line segmentation in optical images captured by Unmanned Aerial Vehicle (UAV) using deep learning backbone for data analysis. Power line segmentation is often considered as a first step required for power line inspection. Different state-of-the-art semantic segmentation techniques available in the literature have been used and a comparative analysis has been done in terms of the Jaccard index on two different power line databases. This paper also presents a new power line database captured using UAV along with the baseline results. Experimental results show that out of the four deep learning-based segmentation architectures used in our experiments the Nested U-Net architecture out-performed others in terms of line segmentation accuracy in various background scenarios.

Keywords: U-Net · Nested U-Net · Transfer learning · Unmanned Aerial Vehicle (UAV) · Semantic segmentation · Power line inspection

1 Introduction

Visual inspection of the transmission and distribution networks is often carried out by the electricity companies regularly to maintain the reliability, availability, and sustainability of electricity supply. Traditional methods used for inspection of power networks which are being followed from decades makes use of field surveys and airborne surveys [1]. Moreover, during emergency situations or on regular basis the inspection is usually carried out by a team of inspectors traveling either on foot or by helicopters to visually inspect the power lines with the help of binoculars and sometimes with Infrared (IR) and corona detection cameras [2]. The major limitation involved in using the above-mentioned methodology is that the method is quite slow, it is expensive and involves danger and is also limited by the visual observation skill of

B. Deka et al. (Eds.): PReMI 2019, LNCS 11941, pp. 623–632, 2019.
https://doi.org/10.1007/978-3-030-34869-4_68

the inspectors [3]. Therefore, in order to overcome these limitations of the traditional methods of power line inspection, recently a number of studies have been conducted to automate the visual inspection by using automated helicopters, flying robots, and/or climbing robots [3]. In this paper, we also propose a methodology for automatic inspection of the power lines in images captured by UAVs using Deep Learning (DL) as a backbone for carrying out the analysis.

The rest of the paper is structured as follows: Sect. 2 details different existing relevant literature reviews followed by a description of the proposed methodology in Sect. 3. Section 3 also highlights different blocks used in our proposed methodology which includes discussion about different pre-processing and post-processing techniques and deep learning architectures used in our experiments. Next, in Sect. 4, we highlight the experimental setup which includes a description of the databases, details about different experiments carried out on the databases and discussion related to the results. Finally, we conclude the paper with a brief conclusion mentioned in Sect. 5.

2 Related Works

Although a vision-based approach powered by deep learning algorithms seems to be the most impeccable approach for power line inspection, but there are only a few works available in the literature dealing with power line inspection. This is mainly attributed to the lack of power line database available in the literature for performing experiments. From the best of our knowledge, the first work related to the use of computer vision in power line inspection has been reported in [4]. In this work, the authors have done a survey study related to the use of computer vision in the detection of power lines, an inspection of power lines, detection and inspection of insulators, power line corridor maintenance, and pylon detection. In another work reported in [5], the authors have devised a method called CBS (Circle Based Search) for power line detection. This method was validated using several tests on real and synthetic images, obtaining satisfactory results in both cases. In [6] author proposed method, named PLineD for power line detection and inspection using UAV captured visual camera images. The database consists of 82 images from various background scenarios of power lines captured using hexacopter UAV. In [7], proposed a technique for a multi-class classification for power infrastructure detection and classification using deep learning approaches. The database consists of 150 pictures taken from UAV. The proposed method achieved 75% F-score for multi-class classification and an 88% F-score achieved for pylon detection. For power line recognition, an F-score of 70% obtained on 11 unseen images. In [8], the authors have proposed secure autonomous navigation approaches for transmission line inspection. The tower detection performed using a faster region-based convolution neural network and power line segmentation achieved using a fully convolutional neural network. Finally constructed UAV platform was evaluated in a practical environment.

3 Proposed Methodology

The block diagram representation of the proposed methodology used in this work has been shown in Fig. 1. During training, we provide the CNN architecture the training and corresponding ground-truth mask images. Once the training gets completed the trained network should be able to predict the binary mask corresponding to any unseen test image containing power lines. The predicted mask obtained from the trained model is overlying on the original input image in order to visualize the segmented power line image as shown in the figure below.

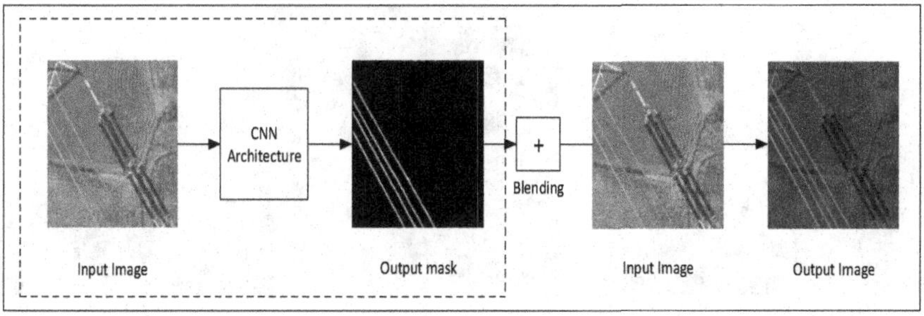

Fig. 1. The methodology used for power line segmentation

3.1 Datasets and Ground-Truth Preparation

In the proposed work, different experiments have been carried out on two power line databases. The first database (referred to in this work as SR-RGB database) is generated with the cooperation of Turkish Electricity Transmission Company (TEIAS) and has been obtained from [9]. In this database, videos were captured from actual aircraft flown over Turkey at 21 different locations during different days of the season. The captured images available in the database have different background scenarios, and weather conditions along with different lighting conditions. Due to varying conditions, the database contains several difficult scenes where low contrast causes invisibility for the power line. At present, the database contains 4000 Infrared (IR) and 4000 Visible light (VL) images having a resolution size of 128 × 128. Out of 4000 images from each category, 2000 images contain power lines and the rest of the images do not contain power lines. We have only used the RGB images available in the database in our experiments. Moreover, since the size of the images was too small, therefore the images were first super-resolved to a size of 512 × 512 using the technique presented in [10]. Sample images from the database and there corresponding super-resolved version has been shown in Fig. 2. As can be seen from the figure, the quality of the super-resolved image is comparable to that of the original image and hence we used the super-resolution technique presented in [10].

The second database used in experiments is an in-house database that consists of 530 power line images captured by a UAV and we refer to this database as the NAL-RGB database. The dataset consists of various background scenarios such as agrarian and rural areas. The resolution of the images in the database is 5472 × 3078 pixels. Therefore, due to the limitation of the GPU memory, we first cropped the images to a size of 512 × 512 pixels, which are non-overlapping image patches from the original input aerial image as shown in Fig. 3. Using this operation we obtained a total of 40227 images. From the total images, the images which do not contain any power lines were removed and the final database consists of 3568 images containing power lines.

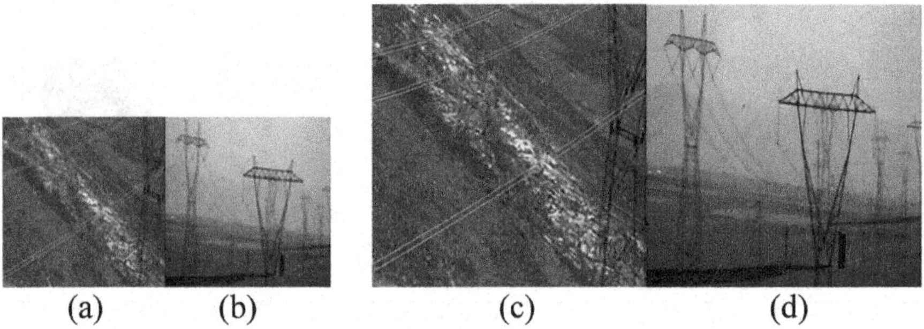

Fig. 2. (a) and (b) are the original images; (c) and (d) are 4x super-resolved images

Once the database gets prepared, the next step involved in the generation of ground-truth required for training the deep learning architectures. To generate the ground–truth mask from the images available in both the datasets we used the VGG image annotator (VIA) [11] which is publically available free of cost. The sample image along with its annotation has been shown in Fig. 4 and the sample annotated image with its binary mask is shown in Fig. 5.

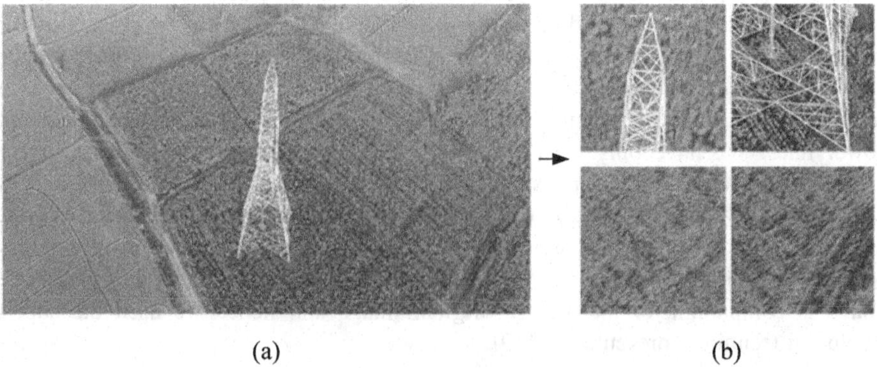

Fig. 3. (a) Original image with size 5472 × 3078 (b) Cropped image with size 512 × 512

3.2 Deep Learning Architectures

Four different state-of-the-art deep learning inspired architectures used for semantic segmentation have been used in this work. The first architecture called U-Net is proposed for biomedical image segmentation [12]. The architecture consists of an Encoder and a Decoder as shown in Fig. 6. As shown in the figure, the encoder consists of four blocks, wherein each block contains two 3 $\times$ 3 convolutional layers with ReLU activation function followed by a 2 $\times$ 2 max pool layer with stride 2 used for the down-sampling operation. The number of feature channels gets doubled after each successive down-sampling operation, starting with 64 feature maps for the first block, 128 for the second, and so on. The purpose of this contracting path is to capture the context of the input image in order to be able to do segmentation. The decoder also consists of four blocks, wherein each block contains the up-sampling (de-convolution) layer and the concatenation layer followed by two 3 $\times$ 3 convolution layer. After each up-sampling operation, it halves the number of feature channels and concatenates the higher resolution features from the encoder and up-sampled feature from decoder for better localization. The final layer consists of a 1 $\times$ 1 convolution layer to map the feature vector to the desired class. The output of this model is a pixel-by-pixel mask that shows the class of each pixel.

Fig. 4. Images with their annotations

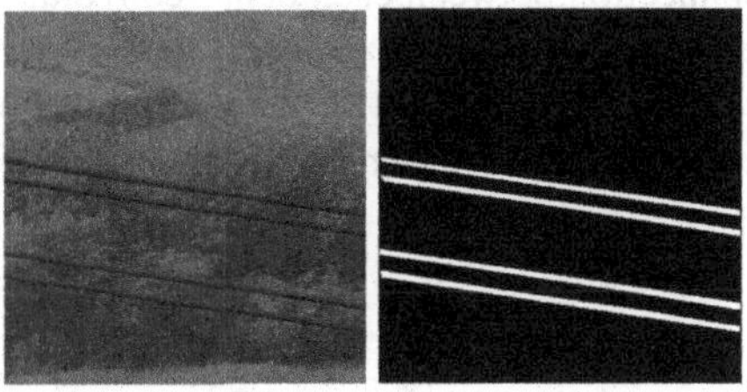

Fig. 5. Images with their binary masks

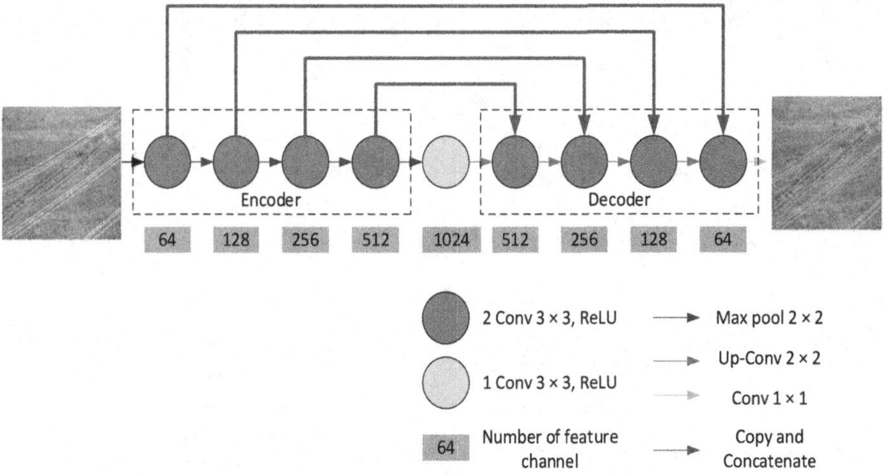

Fig. 6. U-Net architecture.

The second architecture called UNet-11 is an improved version of existing U-Net architecture. It consists of the VGG11 network as an encoder and further details of the architecture can be found in [13]. VGG11 network contains 7 convolutional layers along with 5 max pool layers. Each successive convolution layer is followed by the ReLU activation function. All convolutional layer uses 3 × 3 kernels wherein max pool layer is used to reduce the size of the feature map by 2. The third architecture called UNet-16 is also an improved version of U-Net and is reported in [13]. It consists of the VGG16 network as an encoder in U-Net architecture. VGG16 network contains 13 convolutional layers, with each successive layer followed by the ReLU activation function. All convolutional layer uses 3 × 3 kernels wherein max pool layer is used to reduce the size of the feature map by 2. The final architecture called Nested U-Net is an improved and modified version of U-Net architecture [14]. Nested U-Net as the name implies makes use of nested and dense skip connection between encoder and decoder apart from the typical skip connection used in U-Net. Dense skip connection is used to improve the flow of gradient. Nested U-Net consists of dense convolution block helpful in collecting the semantic level of feature map from the encoder part. The block-level representation of the Nested U-Net architecture has been shown in Fig. 7. In the figure, the green box indicates the dense convolution block which follows the consecutive dense convolution layer. The red line and orange line indicate the skip and nested connection between each dense convolution layer. Deep supervision is performed after each convolution block visualized using the blue line. In deep supervision, dense convolution layer 0_1, 0_2, 0_3 and 0_4 are added which is finally used for pixel-wise segmentation. Further details about the Nested U-Net could be found in [14].

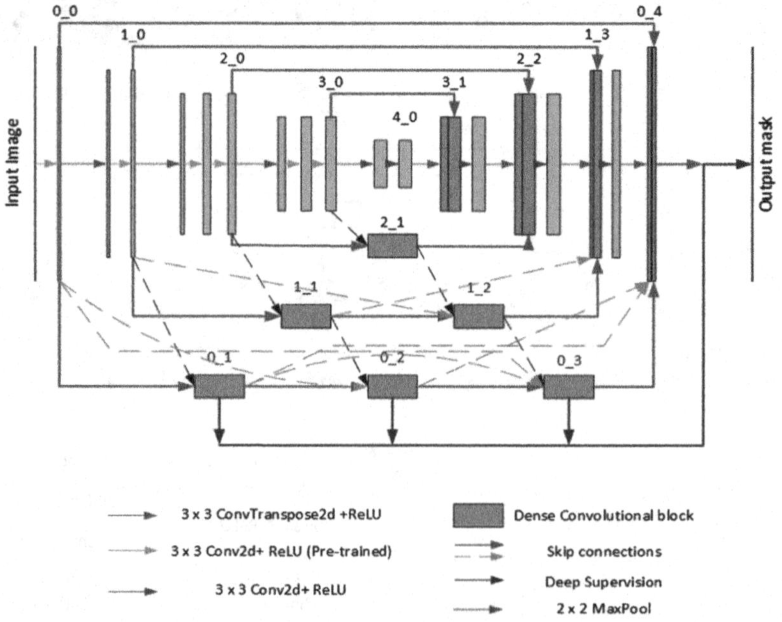

Fig. 7. Nested U-Net architecture (Color figure online)

4 Experimental Results and Discussion

In this section, we discuss different experiments performed on two power line database. All this experiment is performed in the PyTorch environment on GTX GeForce 1080Ti GPU. For U-Net and Nested U-Net, we used the open-source implementation available at github.com/Nested-UNet, whereas for UNet-11 and UNet-16 we implemented the architecture on our own. The evaluation metric used is called the Jaccard index (Intersection over Union) which is defined as a similarity measure between a finite number of sets.

4.1 Results on SR-RGB Database

As discussed in Sect. 3.1 the SR-RGB dataset consists of 2000 images out of which 499 images were blurred and were found difficult to annotate so we removed those images. After removing blurred images, the SR-RGB dataset consists of 1501 power line images. We selected 1200 images for training and 301 images were used for validation purposes.

The training and validation plot of different deep learning architectures such as U-Net, UNet-11, UNet-16, and Nested U-Net has been shown in Fig. 8. These plots have been obtained by training these networks using a batch size of 4, learning rate value of 1e-4 with Adam optimizer. The values of these hyperparameters have been obtained by performing a number of experiments using different values of these hyperparameters.

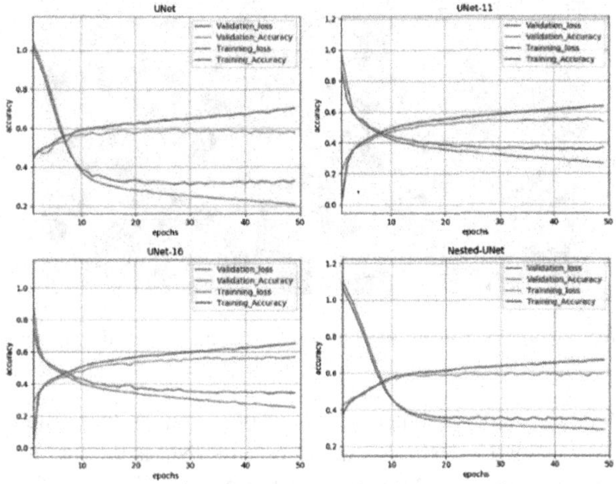

Fig. 8. Training and validation plot of different deep learning architectures on SR-RGB database.

The Jaccard index value obtained on validation images from the database corresponding to U-Net, UNet-11, UNet-16, and Nested U-Net is 0.59, 0.59, 0.60, and 0.59 respectively. From here, we can find out that all the architectures performed equally well on this dataset. The segmented output image obtained after performing the blending operation on the predicted output mask corresponding to the input test image using the Nested U-Net architecture has been shown in Fig. 9. From the out segmented result, it is clear that the trained model is perfectly capable of segmenting the power lines contained in the input test image.

4.2 Results on NAL-RGB Database

The NAL-RGB database consists of 3568 RGB images having a resolution of 512 × 512 pixels. From 3568 images, we have used 2850 images for training and the remaining 718 images have been used for validation/testing purposes.

The training and validation plot of different deep learning architectures such as U-Net, UNet-11, UNet-16, and Nested U-Net has been shown in Fig. 10. These plots have been obtained by training these networks using a batch size of 4, learning rate value of 1e-4 with Adam optimizer. The values of these hyperparameters have been obtained by performing a number of experiments using different values of these hyperparameters.

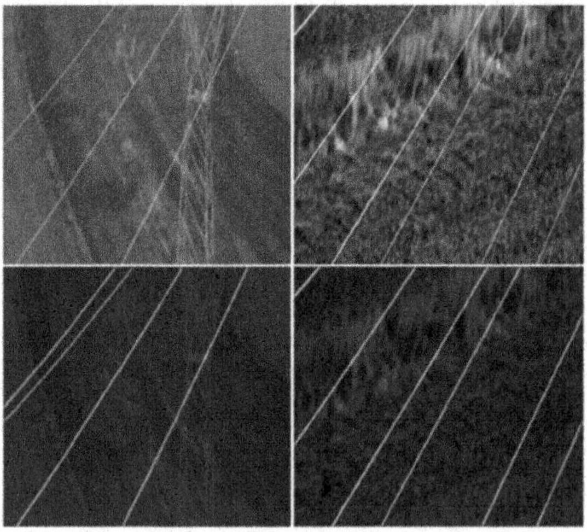

Fig. 9. Visual results obtained using the Nested U-Net trained model on SR-RGB database.

The Jaccard index value obtained on validation images from the database corresponding to U-Net, UNet-11, UNet-16, and Nested U-Net is 0.64, 0.66, 0.67, and 0.70 respectively. From here, we can find out that the Nested U-Net performed well compared to the other three architectures. This is mainly attributed to the deep supervision used in Nested U-Net architecture. The segmented output image obtained after performing the blending operation on the predicted output mask corresponding to the input test image using the Nested U-Net architecture has been shown in Fig. 11. From the segmented output result, it is clear that the trained model is perfectly capable of segmenting the power lines contained in the input test image.

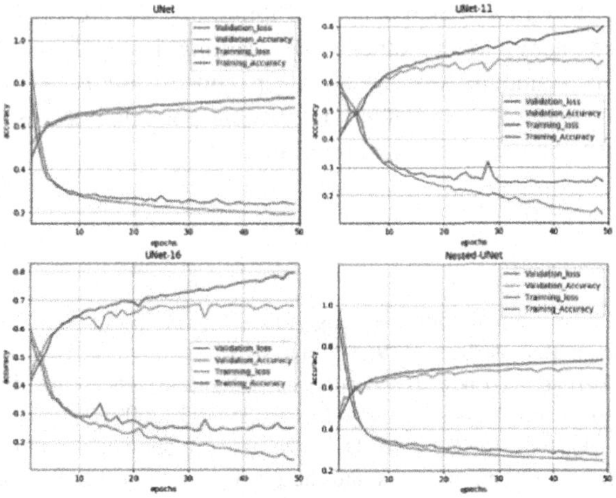

Fig. 10. Training and validation plot of different deep learning architectures on the NAL-RGB database.

Fig. 11. Visual results obtained using the Nested U-Net trained model on NAL-RGB database.

5 Conclusion

In this paper, we have presented a methodology for the automatic segmentation of the power line in UAV image using deep learning backbone for data analysis. Power line segmentation is often considered as the first step required for power line inspection. We have also introduced a new database captured using UAV and presented baseline results using different deep learning architectures available in the literature for semantic segmentation. Different deep learning architectures were trained and validated on two power line database and a comparative analysis was done using the Jaccard index as the evaluation metric. From the experiments, we found that the Nested U-Net performed relatively well compared to the other deep learning inspired image segmentation architectures. This is mainly due to the deep supervision used in Nested U-Net architecture. Thus, the proposed methodology could potentially be used for automatic inspection of power lines in UAVs captured images.

References

1. Matikainen, L., et al.: Remote sensing methods for power line corridor surveys. ISPRS J. Photogr. Remote Sens. **119**, 10–31 (2016)
2. Katrasnik, J., Pernus, F., Likar, B.: A survey of mobile robots for distribution power line inspection. IEEE Trans. Power Delivery **25**(1), 485–493 (2009)
3. Nguyen, V.N., Jenssen, R., Roverso, D.: Automatic autonomous vision-based power line inspection: a review of current status and the potential role of deep learning. Int. J. Electr. Power Energy Syst. **99**, 107–120 (2018)
4. Mirallès, F., Pouliot, N., Montambault, S.: State-of-the-art review of computer vision for the management of power transmission lines. In: Proceedings of the 2014 3rd International Conference on Applied Robotics for the Power Industry, pp. 1–6. IEEE, October 2014
5. Cerón, A., Prieto, F.: Power line detection using a circle based search with UAV images. In: 2014 International Conference on Unmanned Aircraft Systems (ICUAS), pp. 632–639. IEEE, May 2014
6. Santos, T., et al.: PLineD: vision-based power lines detection for unmanned aerial vehicles. In: 2017 IEEE International Conference on Autonomous Robot Systems and Competitions (ICARSC), pp. 253–259. IEEE, April 2017
7. Varghese, A., Gubbi, J., Sharma, H., Balamuralidhar, P.: Power infrastructure monitoring and damage detection using drone captured images. In: 2017 International Joint Conference on Neural Networks (IJCNN), pp. 1681–1687. IEEE, May 2017
8. Hui, X., Bian, J., Zhao, X., Tan, M.: Vision-based autonomous navigation approach for unmanned aerial vehicle transmission-line inspection. Int. J. Adv. Rob. Syst. **15**(1), 1729881417752821 (2018)
9. Yetgin, Ö.E., Gerek, Ö.N.: Powerline image dataset (infrared-IR and visible light-VL). Mendeley Data, v7 (2017). http://dx.doi.org/10.17632/n6wrv4ry6v.7
10. Yamanaka, J., Kuwashima, S., Kurita, T.: Fast and accurate image super resolution by deep CNN with skip connection and network in network. In: Liu, D., Xie, S., Li, Y., Zhao, D., El-Alfy, E.S. (eds.) Neural Information Processing. LNCS, vol. 10635, pp. 217–225. Springer, Cham (2017). https://doi.org/10.1007/978-3-319-70096-0_23
11. Dutta, A., Zisserman, A.: The VGG image annotator (VIA). arXiv preprint arXiv:1904.10699 (2019)
12. Ronneberger, O., Fischer, P., Brox, T.: U-net: convolutional networks for biomedical image segmentation. In: Navab, N., Hornegger, J., Wells, W.M., Frangi, A.F. (eds.) MICCAI 2015. LNCS, vol. 9351, pp. 234–241. Springer, Cham (2015). https://doi.org/10.1007/978-3-319-24574-4_28
13. Iglovikov, V., Shvets, A.: Ternausnet: U-net with VGG11 encoder pre-trained on imagenet for image segmentation. arXiv preprint arXiv:1801.05746 (2018)
14. Zhou, Z., Rahman Siddiquee, M.M., Tajbakhsh, N., Liang, J.: UNet++: a nested U-net architecture for medical image segmentation. In: Stoyanov, D., et al. (eds.) DLMIA/ML-CDS 2018. LNCS, vol. 11045, pp. 3–11. Springer, Cham (2018). https://doi.org/10.1007/978-3-030-00889-5_1

Author Index

Printed in the United States
By Bookmasters